Studies in Systems, Decision and Control

Volume 614

The series "Studies in Systems, Decision and Control" (SSDC) covers both new developments and advances, as well as the state of the art, in the various areas of broadly perceived systems, decision making and control–quickly, up to date and with a high quality. The intent is to cover the theory, applications, and perspectives on the state of the art and future developments relevant to systems, decision making, control, complex processes and related areas, as embedded in the fields of engineering, computer science, physics, economics, social and life sciences, as well as the paradigms and methodologies behind them. The series contains monographs, textbooks, lecture notes and edited volumes in systems, decision making and control spanning the areas of Cyber-Physical Systems, Autonomous Systems, Sensor Networks, Control Systems, Energy Systems, Automotive Systems, Biological Systems, Vehicular Networking and Connected Vehicles, Aerospace Systems, Automation, Manufacturing, Smart Grids, Nonlinear Systems, Power Systems, Robotics, Social Systems, Economic Systems and other. Of particular value to both the contributors and the readership are the short publication timeframe and the worldwide distribution and exposure which enable both a wide and rapid dissemination of research output.

Indexed by SCOPUS, DBLP, WTI Frankfurt eG, zbMATH, SCImago.

All books published in the series are submitted for consideration in Web of Science.

Sophie C. F. Hendrikse · Jan Treur ·
Sander L. Koole
Editors

New Analysis and Modeling Directions in Social Interaction Science

Emergent Multimodal Interpersonal Synchrony, Affiliation, and Bonding

Springer

Editors
Sophie C. F. Hendrikse
Department of Methodology and Statistics
School of Social and Behavioral Sciences
Tilburg University
Tilburg, The Netherlands

Sander L. Koole
Department of Clinical Psychology
Vrije Universiteit Amsterdam
Amsterdam
Noord-Holland, The Netherlands

Jan Treur
Social AI Group, Department of Computer Science
Vrije Universiteit Amsterdam,
Amsterdam, Noord-Holland, The Netherlands

ISSN 2198-4182 ISSN 2198-4190 (electronic)
Studies in Systems, Decision and Control
ISBN 978-3-031-99967-3 ISBN 978-3-031-99968-0 (eBook)
https://doi.org/10.1007/978-3-031-99968-0

This Springer imprint is published by the registered company Springer Nature Switzerland AG
The registered company address is: Gewerbestrasse 11, 6330 Cham, Switzerland

Preface

This book is about the magic of how people connect during social interaction and how connectedness affects social interaction. A situation of connectness can have various forms and labels such as affiliation, friendship, relationship, relatedness, or bonding. The magical process of connecting can also happen in different ways and through different types of influences. Two often studied ways are connecting induced by homophily and connecting induced by interpersonal synchrony. Homophily concerns two persons having the same opinions or preferences for music, for example. Interpersonal synchrony refers to showing similar or compatible patterns of behavior such as movements, emotions, or verbal actions, meaning there is mutual coordination of behavior. Although usually addressed separately, we bring these two often studied influences (homophiliy and interpersonal synchrony) on connecting together in one book.

From an abstract perspective, both concern some form of similarity: of being the same or acting the same. In both cases, it has been recognized that there is an underlying circular dynamic pattern over time, formulated in an abstract way by: from connectedness via social contagion to strengthened similarity, and from similarity via mental and behavioral adaptation to strengthened connectedness. To analyze such dynamic and adaptive patterns over time, we have used an adaptive dynamical system perspective, modeled as agent-based self-modeling temporal-causal networks.

Usually, notions such as homophily and synchrony are addressed from an objective external observer perspective, i.e., the researcher analyzes gathered time-series from human and/or virtual participants. However, within one of the two partners in social interaction, there is not such an external observer viewpoint. Instead, a person can only perceive subjectively if homophily or synchrony occurs. Using our agent-based perspective, we introduce here the concepts of subjectively perceived homophily and subjectively perceived synchrony.

Moreover, to get some more clarity on the connecting process, we considered it as adaptivity of mental processes and behavior within social interaction, so: how do mental processes and behavior change during connecting? Looking from that perspective, various forms of adaptivity were found and distinguished within social interaction: different forms on the same timescale and different forms on different timescales, different forms for perceiving, responding and acting, different forms for different modalities, and different forms for both first-order and second-order adaptivity.

For our perspective of computational analysis and formalization, the network-oriented modeling approach by self-modeling temporal-causal networks used supported integration of such different forms of adaptivity well. Based on all the above, a unified computational architecture is contributed and in line with this architecture applications are discussed for analysis and support for socially induced adaptivity for elderly, and for human social interaction with bonding bots that support interpersonal emotion regulation.

Thus, summarizing the above, the main contributions in the different parts of this volume are:

- Bringing together in one book both homophily-induced and synchrony-induced connectedness approaches.
- Introducing subjectively perceived homophily and subjectively perceived synchrony.
- Introducing computational analysis by higher-order adaptive dynamical systems represented as self-modeling networks.
- Introducing analysis of connectedness by distinguishing different forms of mental and behavioral adaptation.
- Introducing a multi-timescale perspective on different forms of mental and behavioral adaptation.
- Unification of homophily-induced connectedness and synchrony-induced connectedness.
- Application to computational analysis of joint activities of elderly and how their mental capabilities can benefit from that in an adaptive manner.
- Application to obtain bonding bots that support interpersonal emotion regulation.

Note that different chapters may have some overlap as each of them is meant to be self-contained so that it can be downloaded and read independently.

We as editors are very grateful to NWO for their funding of part of the work presented in this volume by NWO Open Competition project 'The Rhythm of Relating: How Emotional Sharing Emerges From Interpersonal Synchrony in Movements, Physiology, and Neural Activations' (Project number 406.18.GO.024) and to all authors of chapters for their contributions.

Tilburg, The Netherlands — Sophie C. F. Hendrikse
Amsterdam, The Netherlands — Jan Treur
Amsterdam, The Netherlands — Sander L. Koole
September 2024

Contents

Introduction

In this part, first in "Introduction to the Role of Adaptivity in Social Interaction Science", an introduction is presented of the topic of this volume and the different parts and chapters of it. In "Modeling Coevolution of Dynamics, Adaptivity and Control of Adaptivity by Self-Modeling Networks", the network-oriented modeling approach used is introduced and the self-modeling principle that enables modeling of multilevel adaptive networks in a network-oriented manner.

Introduction to the Role of Adaptivity in Social Interaction Science

Sophie C. F. Hendrikse, Jan Treur, and Sander L. Koole

Abstract In this chapter an overview is given of the field of social interaction through the lens of an adaptive dynamical system perspective. It briefly reviews relevant concepts and causal effects such as social contagion, homophily-induced bonding, synchrony-induced affiliation and bonding, multimodality, and multi-order adaptivity.

Keywords Attunement · Connectedness · Multimodal · Social interaction · Adaptive · Dynamical system

1 Introduction

Virtually everyone is familiar with the feeling of being on the same wavelength with someone else. For example, this feeling may relate to finding out that you both appreciate similar types of music or movies, but it can also concern that you notice to be in sync in terms of showing mutually attuned behavioral patterns. Such a feeling appears to reflect an intuitive mutual understanding that allows people to coordinate their behavior spontaneously and seemingly without effort. The processes where people mutually coordinate their behavior have been studied under various labels, including homophily [74], alignment [102], mimicry [9], interpersonal coordination

S. C. F. Hendrikse
Department of Methodology and Statistics, Tilburg School of Social and Behavioral Sciences, Tilburg University, Tilburg, The Netherlands
e-mail: S.C.F.Hendrikse@tilburguniversity.edu

J. Treur (✉)
Social AI Group, Department of Computer Science, Vrije Universiteit Amsterdam, Amsterdam, The Netherlands
e-mail: j.treur@vu.nl

S. L. Koole
Department of Clinical Psychology, Vrije Universiteit Amsterdam, Amsterdam, The Netherlands
e-mail: s.l.koole@vu.nl

S. C. F. Hendrikse et al. (eds.), *New Analysis and Modeling Directions in Social Interaction Science*, Studies in Systems, Decision and Control 614,
https://doi.org/10.1007/978-3-031-99968-0_1

(dynamics), coregulation [23], concordance [12], coupling [36], matching [11], and mirroring [37]. We use the term 'interpersonal synchrony' or simply 'synchrony', to refer to mutually coordinated behavioral patterns over time displayed by two or more agents [6, 8, 20, 24, 32, 68, 84, 100, 102]. The term 'homophily' is used here if it concerns states or characteristics of two persons that are (approximately) the same at a given point in time, such as an emotion, opinion or a preference [74].

2 Coevolution of Similarity and Induced Connectedness

2.1 Homophily-Induced Connectedness Adaptivity

Already in early research in the area of social networks an important focus was on the relationship between connectedness between persons and the extent of similarity of them. For example, in McPherson et al. [74] it is put forward that the relationship between connectedness and similarity is mentioned by Aristotle by writing that people 'love those who are like themselves' ([5], p. 1371) and Plato by writing 'similarity begets friendship' ([86], p. 837). Systematic observations of group formation and network ties by social scientists began in the 1920s and 1930s. For example, [16, 59, 131] reported that school children formed friendships and play groups at higher rates if they were similar on demographic characteristics [74]. An important contribution to the sociological literature is [71] about forming friendships. They use the term 'homophily' and to describe the process they also quoted the metaphoric expression.

The phrase 'Birds of a feather flock together' [71] has often been used in the literature since then, e.g., [74]. For a further historical review, see [34]. Although in such literature often the emphasis may seem to be on the causal effect that the extent of homophily has on (strengthened) connectedness such as friendship, usually there is still a causal effect in the opposite direction as well. By social contagion, persons that have strong connectedness can strengthen their extent of homophily. Unfortunately, the above metaphor is not applicable to describe this opposite causal effect: it is very unlikely that feathers of birds can change due to flocking together. Nevertheless, for many cases a bidirectional causal mechanism between connecting and homophily is considered realistic and sometimes described by the term coevolution, which means that over time

- connectedness evolves via mental and behavioral adaptation under influence of the extent of homophily
- homophily evolves via social contagion under influence of the extent of connectedness.

Especially when homophily is considered for personal characteristics or states that can change over time, such as preferences for types of movies or music, it is assumed that these can become attuned to each other due to social contagion. In turn, social contagion depends on strength of connectedness, so in this way indeed the above circular feedback loop occurs.

In recent years this circular coevolution process in relation to homophily has been modeled from an adaptive dynamical system [87] perspective, often in a network-oriented manner, see, for example, [13, 58, 64, 72, 73, 104, 106, 108, 117, 128]. Simulations based on such adaptive computational network models have demonstrated how such a circular coevolution process can explain the emergence of clusters or communities, or segregation in society, e.g., [40, 62].

Perhaps because homophily-induced connectedness adaptivity often has been addressed from a social science or sociology perspective, in computational models for coevolution as indicated above usually no internal mental states are incorporated. In particular, the underlying criterion of homophily is usually based on objective states in the social world as considered from an external observer perspective. However, in reality, for one of the persons in a social interaction to notice a form of homophily, a process of perception and interpretation plays an important role. Therefore, in some recent work on homophily-induced connectedness adaptivity this subjective perspective is addressed. Here, the connectedness is induced by subjective homophily, not by objective homophily [79, 80, 121], see also this volume, Part III, Chaps. 7, 8 and 10: [81, 82, 123]. In such a case also manipulation of homophily can be modelled, for example, by faking that you like the same type of movies as your friend does [14], see also this volume, Part VI, Chap. 9 [15].

2.2 *Synchrony-Induced Connectedness Adaptivity*

In the analysis in the above Sect. 2.1, 'homophily' can also be replaced by 'synchrony'. Then, the concept 'homophily-induced connectedness' is replaced by 'synchrony-induced connectedness', a concept which has been explored extensively in literature such as [57, 109, 126, 127, 133]. Like for homophily as discussed in the above Sect. 2.1, also for synchrony an overall circular pattern of coevolution can be found over time:

- connectedness evolves via mental and behavioral adaptation under influence of the extent of synchrony
- synchrony evolves via social contagion under influence of the extent of connectedness.

Recently, this type of coevolution processes involving subjective perception of synchrony have been analysed computationally in Hendrikse et al. [43, 44, 50–52]; see also this volume Parts IV and V, Chaps. 11–16. To introduce this concept of subjectivity for synchrony, based on the temporal factorisation principle from [116], the notion of subjective synchrony detection was postulated to indicate how a person can actually notice being in synchrony with someone else.

Adaptive self-modeling network models were very useful to address the complexity of this type of adaptive dynamical systems. The developed adaptive dynamical system modeling approach to synchrony-based connectedness has been successfully applied to model complex social phenomena such as adaptive interpersonal emotion regulation and adaptive collective phenomena involving elderly, see also this volume, Part VI, Chaps. 17, 18 and 19.

Within the wider area focusing on synchrony, research has documented how people spontaneously synchronize their activities in a variety of ways, including their movements [10, 20, 83], vocal pitch [99, 113, 134], linguistic expressions [27, 38, 103], their physiological responses [84] and neural responses [102]. These interpersonal dynamics play a role in many relationships over the lifetime. Already in utero, infants synchronize their heart beats with their mother's heart beats [129]. After birth, children develop autonomous emotion regulation from physiological synchrony that they exhibit with their caretaker [31]. Moreover, when adults form close relationships, they are prone to synchronize their physiological responses [84]. Neural synchrony can also elevate within parent–child relationships [31, 32, 92], romantic relationships [65] and among more distant relationships like students [26]. Strangers may even synchronize their physiological responses by merely watching the same event [69]. Interpersonal synchrony is thus a pervasive pattern in social life.

Concerning the important psychological consequences for adaptation of the social interaction behaviour, it has been found, for instance, that people bond more with each other after they were experimentally induced to synchronize their movements [109]. Furthermore, leading people to synchronize by singing or laughing together has been shown to stimulate more cooperative behavior [133]. Perhaps most impressively, experimentally inducing synchrony in brain-to-brain activity has been found to improve joint performance on cognitive tasks [85]. These and related studies demonstrate that, indeed, interpersonal synchrony is not merely an epiphenomenon, but that it actually shapes interpersonal dynamics in an adaptive manner. Consequently, it is important to learn more about the nature of interpersonal synchrony and how via coevolution processes it leads to mental and behavioural adaptivity. This volume [45] offers much material for this, thereby incorporating a computational dimension as well.

2.3 Toward a Unified Perspective on Homophily-Induced Connectedness Adaptivity and Synchrony-Induced Connectedness Adaptivity

In the analysis covered in this volume [45], in the first place contributions were collected both addressing homophily-induced connectedness adaptivity (mainly by the five chapters in Part III) and synchrony-induced connectedness adaptivity (mainly by the six chapters in Parts IV and V). In this way, for each of the perspectives foundational computational analyses are obtained. In Part VII, similarities between the two perspectives are analysed and a more unified perspective is described.

3 The Multidisciplinary Field of Synchrony

Especially in recent years, interpersonal synchrony has become a major focus of research across multiple scientific disciplines, including communication studies [39], developmental neuroscience [31], movement science [21], psychology [124], psychophysiology [84], psychiatry [75], and social-cognitive neuroscience [23]. This variety of disciplines attests to the integrative potential of the notion of interpersonal synchrony. However, it also comes with important challenges.

When people interact, their behavior tends to become dynamically and reciprocally adapted over time. In this temporal coordination process, interpersonal synchrony is a core phenomenon. People who are dancing together [66], participating in lively, smooth conversations [70] or just walking side by side in the same rhythm [130] are all examples of the role interpersonal synchrony plays in everyday life. Interpersonal synchrony is fundamental in the regulation of close relationships between humans [31, 68, 101] and promotes enriching, fulfilling social exchanges. When people try to achieve positive relationships with each other, they are more eager to synchronize [76].

Research indicates that such interpersonal synchrony causes mutual beneficial outcomes. Regarding movement synchrony, for instance, social-psychological experiments have demonstrated that enhancing movement synchrony between people results in more sensitivity in reactions towards each other, cooperation, helping, compassion and liking [57, 126, 127, 133]. Thus, interpersonal synchrony spontaneously emerges in a variety of social settings and facilitates relationship building.

Different scientific disciplines each tend to have their own research paradigms, including specific ways of measuring and manipulating certain variables, conventions in data analysis, and preferred kinds of theorizing. This diversity in research paradigms makes it difficult to relate the findings on interpersonal synchrony in one discipline to the findings on interpersonal synchrony in another discipline. This poses a great challenging in developing an integrated scientific understanding of interpersonal synchrony.

A case in point is the different modalities within which interpersonal synchrony may emerge. Interpersonal synchrony has been documented within multiple response modalities, such as movements, language utterances, physiological activations, and neural responses [67]. These response modalities have so far been mostly examined separately. The latter is understandable, given that each response modality requires a different type of measurement technique. For example, movement synchrony is usually assessed for people who are engaged in conversation, often in naturalistic settings like psychotherapy or collaborative tasks [88]. In movement synchrony studies, people are typically seated, but otherwise can move freely. By contrast, brain-to-brain synchrony is typically assessed in controlled laboratory environments, where participants can be mounted with equipment such as EEG caps [7, 26, 33, 78]. In EEG studies, participants' freedom of movement is often restricted so as to minimize artifacts in the EEG recordings.

Little research so far has examined the relation between interpersonal synchrony in different modalities. Nevertheless, there are likely to be meaningful connections between different modalities of interpersonal synchrony [68, 102]. For instance, the In-Sync model [68] proposes that moment-to-moment changes in movement synchrony may pave the way for the engagement of higher-level cognitive processes, which operate on a slower timescale and presumably are reflected in brain-to-brain synchrony. Likewise, the multilevel model [102] suggests that motor synchrony, cognitive synchrony, and emotional synchrony are jointly regulated by a core alignment mechanism. These and related models highlight the need for further conceptual and methodological integration in the scientific study of interpersonal synchrony.

4 Methods for Detecting Interpersonal Synchrony

Human interaction dynamics are complex. Human interactions can be captured as multiple signals that evolve over time (i.e., time-series), in which events follow on each other quickly. This cascading of events means that activations of certain modalities at a certain time point are dependent on the activations at previous time points, i.e., the data of each person is dependent over time. Moreover, it is challenging to separate signals from noise. Based on chance alone, one would expect synchronous patterns to emerge every now and then, a phenomenon known as pseudo synchrony [77]. Many statistical methods have been proposed to detect synchrony, each of them with different assumptions and capturing different characteristics of the data.

4.1 Methods from the Inter-Brain Synchrony Domain

In the domain of inter-brain synchrony, Czeszumski et al. [23] distinguished the following statistical methods for synchrony detection: coupling/connectivity measures (e.g., phase-locking value, phase lag index, phase coherence, wavelet transform coherence), correlation and dependency analysis, graph theory measures (e.g., modularity, smallworldness) and information flow measures (e.g., Granger causality, partial directed coherence). Correlation methods are often distance-based, and graph measures work on networks that are created from the interpersonal synchrony scores. Information flow measures are directed interpersonal synchrony detection methods, meaning they can capture causality in driver-response interactions between time series.

4.2 Methods from the Movement Synchrony Domain

In the domain of movement synchrony, methods based on recurrence analysis, correlations or spectral analysis have typically been used [25]. Recurrence analysis captures non-linear patterns in the data, whereas correlations can detect linear connectedness. Methods within the group of correlation methods concern for example the time-lagged cross-correlation and peak picking algorithms. Schoenherr et al. [97, 98] distinguished the following (linear) correlation-based methods for movement synchrony: cross-lagged correlation, cross-lagged regression, windowed cross-correlation, windowed cross-lagged correlation (versions based on strength and frequency) and windowed crosslagged regression. These methods with a time lag compare the two signals at different time lags to detect synchrony.

4.3 Methods from the Physiological Synchrony Domain

Finally, in the domain of physiological synchrony, studies have relied on a yet another set of statistical methods, which partly overlaps with the previous domains. For instance, a review by Palumbo et al. [84] distinguished differential equation models, signal matching, cross-recurrence quantification analysis, correlations, wavelet coherence, multilevel models and generalized partial directed coherence.

4.4 How to Choose a Method

As we can see, both between response modalities and within response modalities, diverse analytic methods have been proposed for assessing interpersonal synchrony. The available methods differ in the data patterns and their underlying assumptions that they can capture. For instance, when people are in a conversation, they are more synchronized when one of them talks and the other one listens, meaning there is an antiphase. This antiphase means that opposite actions (listening and talking) actually characterize more interpersonal language synchrony. In contrast, during a dance performance, interpersonal synchrony is meaningfully the highest when dancers their movements are as much the same as possible. The same actions are, in this example, a sign of more interpersonal movement synchrony. When a specific analysis/detection method is selected, one should be aware of the data characteristics such an analysis method captures and whether this is in line with the desired nature of synchrony that needs to be measured.

Depending on the definition of interpersonal synchrony in a certain context, an analysis method needs to be selected. To date, there exist no conclusive guidelines about when to select which analysis method in which context. Some simulation studies have compared a couple of interpersonal detection methods (e.g., [60, 61,

67]). However, these simulation studies compared only a subset of the available methods at once and have often relied on a specific data generation method that cannot be extrapolated to the whole field of interpersonal synchrony detection methods. Moreover, how interpersonal synchrony detection methods can function and perform within an adaptive dynamical system has, to our knowledge, never been evaluated.

To summarize, interpersonal synchrony is complex and can be analyzed using a diversity of interpersonal synchrony detection methods. On the one hand, this offers great potential because many interpersonal synchrony detection methods have been proposed. On the other hand, this makes it more difficult for researchers how to decide which methods to use. One first step toward methodological integration would be to bundle the interpersonal synchrony detection methods into user-friendly software to facilitate other researchers to more systematically compare these interpersonal synchrony detection methods. Such a comparison can be done from an empirical perspective using empirically obtained time series. However, when agent-based computational models for social interaction are studied, the use of such synchrony detection methods within these agent models can be analyzed in a comparative manner.

5 Modeling Adaptive Social Interaction Using AI

The detection of interpersonal synchrony in time series from empirical data sets takes only into account the values of the specific time series themselves. However, such synchrony patterns arise in a complex social and psychological context, and this context information is lost when only the synchrony patterns themselves are quantified. The adaptation of interaction behaviors, for example, results from such synchrony patterns embedded in the broader context. This narrowing down of the scope of analysis has advantages as then the mathematical, statistical and algorithmic techniques used for synchrony detection get full attention. However, the analysis of how the emergence and causal effects of synchrony relate to factors in the context can also be valuable. To address this empirically is challenging, but as an alternative it can be addressed computationally, in a computer simulation. Computer simulations allow researchers to widen the focus to model the mental processes underlying social interaction, in particular, the role of interpersonal synchrony and adaptivity of the interaction behavior.

5.1 Dynamics and Adaptivity

The considered social interaction processes are dynamical because various circular patterns and emergent phenomena can arise like mirroring and contagion of each others emotions and actions. Moreover, it also leads to adaptivity because, in relation to these emergent interaction phenomena, behavior of agents will change over time, such as their affiliation and cooperation, e.g., [57, 58, 109]. Therefore, it concerns a complex agent-based adaptive dynamical system, which is in general challenging to model but still feasible. One element of such a complex adaptive dynamical system, is detection of interpersonal synchrony from a subjective agent perspective. For this detection, the interpersonal synchrony detection methods discussed in the previous Sect. 4 provide good input. Overall, to explore and analyse this computationally, a modeling approach is needed that can capture the adaptive causal mechanisms and pathways involved. One modeling approach that meets these requirements is the adaptive network-oriented modeling approach by Treur [118, 119], see also this volume Part I, Chap. 2: [122].

5.2 Network Concepts

In this network-oriented modeling approach there are a few characteristics that define a so-called temporal-causal network. There are states (also called nodes or vertices) that represent a certain concept, such as an action, sensory representation, belief or feeling. These states are connected with each other through causal connections or edges to represent the causal relationship between them within their dynamical process: each state has an activation level that can dynamically change over time under influence of other connected states. Such a temporal-causal network model can be seen as a directed graph with certain characteristics. These characteristics specify the structure of such a temporal-causal network. Its connections have weights (*connectivity characteristics*). Each state has a combination function that defines the aggregation that is applied on its incoming influences from other states (*aggregation characteristics*). Furthermore, each state has an own speed factor that defines how quick it changes for its given aggregated influence (*timing characteristics*).

Connectivity characteristics can be used to distinguish stronger and weaker causal relations and positive and negative ones. Aggregation characteristics can be used to modulate how a state responds to certain (multiple) incoming impacts. For example, the characteristic for an excitability threshold can be higher or lower to indicate that the state is harder or easier to activate. Timing characteristics can be used to differentiate between speeds of change for different states, see also Sect. 5.4.

5.3 Network Adaptivity of Different Orders

In reality, often not only the states but also the characteristics of a temporal-causal network can evolve over time, which means that there is adaptivity. To achieve adaptivity, a self-modeling network, also known as a reified network, can be designed [118, 119] which was inspired by the area of reflection and metalevel architectures within AI; e.g., [6, 15, 16, 23]. This approach involves including new states in the network, referred to as self-model states, which account for the adaptive network characteristics. While the original temporal-causal network stays at the base level, these additional self-model states are depicted at a higher level called the self-model level or reification level. In this way, the network model can be extended with first-order, second-order, third-order, and so on, self-model levels. All network characteristics like connection weights, combination functions, their parameters, and speed factors can become adaptive by adding a self-model state for them. For instance, when a logistic function is chosen as combination function, its excitability threshold can be made adaptive through a self-model state at the first-order self-model level. Likewise, connection weights and speed factors can be made adaptive through including self-model states for them at the first-order self-model level. These self-model states have their own network characteristics.

This network-oriented modeling approach has turned out to be very useful to provide the required environment to handle not only causality and related dynamics of the states but also adaptivity of different orders, as highlighted above. Moreover, this approach is generic in the sense that it is able to model any smooth adaptive dynamical system [43, 120], see also this volume [45], Part IV, Chap. 12 .

5.4 Modeling Different Timescales

Timing characteristics based on speed factors per state can be used to incorporate different time scales for different states, for example, fast for direct responding, slow for short term adaptation and still slower for long-term adaptation. More specifically, different types of adaptation on different time scales (e.g., short-term adaptation vs long-term adaptation) can be modeled by using different speed factors for the different self-model states as shown in Hendrikse et al. [43], see also this volume [45], Part IV, Chap. 12. Here, induced by detected emerging synchrony, different time scales for short-term affiliation and long-term bonding are used where repeated episodes of the short-term adaptation can boost the long-term adaptation. In such a way, by using proper speed factors any combination of any number of time scales can be integrated in a network model.

As a further example, a second-order self-model may include a second-order self-model state that represents the speed factor for the dynamics of the first-order self-model state that, for example, represents the adaptation of the connection weight at the base level. Also such second-order self-model states have their own network characteristics such as, for example, speed factors which can be used to give them their own time scale.

6 Overview of This Volume

An important research aim for the area covered in this volume [45] is to computationally model social model social interaction dynamics by adaptive agent models. A number of topics within this area are distinguished and addressed by the following parts:

Part I	Introduction
Part II	Agent-Based Analysis of the Emergence of Multimodal Interpersonal Synchrony During Social Interaction
Part III	Adaptive Agent-Based Simulation of Social Interaction and the Emergence of Homophily and Related Bonding
Part IV	Exploiting Subjective Synchrony Analysis for Computational Analysis of Social Interaction Adaptivity
Part V	Computational Analysis of Social Interaction Adaptivity Induced by Subjective Interpersonal Synchrony
Part VI	Computational Analysis of Real-World Scenarios for Synchrony-Induced Social Interaction Adaptivity
Part VII	Finalising

In the remainder of this chapter, each of these parts is briefly discussed in Sects. 6.1–6.7.

6.1 Part I: Introduction

This part provides an introduction to the area of social interaction science addressed in this volume [45] and to the computational modeling approach used. Thus, in addition to the general introduction to the area in current chapter, Chap. 2 [122] is included with an introduction to the adaptive dynamical system modeling approach based on self-modeling network models used throughout the volume.

6.2 Part II: Agent-Based Analysis of the Emergence of Multimodal Interpersonal Synchrony During Social Interaction

Part II of this volume [45] consists of three chapters. Chapter 3 addresses the design of an agent model in which by social contagion interpersonal synchrony can arise at multiple response modalities. Chapter 4 presents an agent model of how coregulation of emotions between two agents can occur and how this may be applied in human-bot interaction. Chapter 5 describes how agents can learn to synchronize their actions and emotions and visualizes this adaptive agent model through avatars.

Interpersonal synchrony emerges in several modalities, including bodily movements, facial expressions, vocal pitch, language, and physiological responses [68, 102]. One of the aims was to explore how to simulate multimodal interpersonal synchrony in interacting agents. Specifically, in this volume Chap. 3 [53] the underlying mechanisms were modeled that contribute to the emergence of interpersonal synchrony among two agents who possess internal cognitive and affective states. This shows how social contagion and interpersonal synchrony for multiple response modalities can be captured in agent models through general mechanisms from psychology and neuroscience. To examine how these multiple interpersonal synchronies can emerge during social interaction, agent models were designed and simulation experiments conducted. In an example simulation one agent was exposed to a repeated stimulus, leading the other agent to progressively synchronize its movements, affective responses, conscious emotions, and verbal actions with the exposed agent.

The study contributed in this volume Chap. 4 [28] analyses the effects of emotional coregulation in the social interaction between a distressed human and a bot. Through the coregulation premises, an emotion contagion process impacts the emotional system, which initially increases the distress level of the bot due to the influence of the human's emotions. Configuring the bot with optimal emotion regulation decreases its distress levels as it can self-regulate its emotions, which affects the human's emotions, regulating them by the emotion contagion process. The simulation results show how this coregulation process occurs.

After the development of agents that can synchronize multiple response modalities, another aim has been to explore how agents can learn to synchronize. Potential applications of adaptive computational models for such processes can be found in areas like online coaching and therapy, where considering adaptivity of the social interaction behaviour is an interesting challenge. Therefore, in this volume Chap. 5 [41] an adaptive computational modeling approach is contributed capturing the synchrony emerging in social interaction between two individuals acting as agents. This synchronization occurs within multiple modalities such as movements, speech and emotions, modeled from a joint decision making perspective.

6.3 Part III: Adaptive Agent-Based Simulation of Social Interaction and the Emergence of Homophily and Related Bonding

In this part of this volume [45], first in Chap. 6 a first analysis is made of the interaction between attunement modeled as homophily, bonding based on homophily, and adaptive mirroring based on Hebbian learning during joint decision making. In Chap. 7 it is shown how from an agent-based perspective subjectively perceived and interpreted homophily can be used as a basis for the bonding, instead of objective forms of homophily. Next, in Chap. 8 it is shown how synchronous joint action can go hand in hand with behavioural adaptivity based on subjective homophily leading to more coordination. In Chap. 9, it is shown how subjective homophily detection can be used to explain how bonding based on faked homophily can take place. Finally, in Chap. 10 adaptive bonding based on homophily is analysed where the criterion for subjective homophily is adaptive and depends on the connections someone already has.

More specifically, in Chap. 6 [112] the interplay of social contagion, Hebbian learning of mirroring and bonding is analysed in the context of joint decision making. A second-order adaptive network model is presented and scenarios are explored using this computational model. The results show that Hebbian learning of mirroring connections plays an important role in decision making, and that also bonding can be an important element in making a joint decision, but also that conversely joint decisions strengthen the bonding and the mirroring connections.

In Chap. 7 [123], the role of subjective elements and control in social interaction adaptation is analyzed computationally. In particular, it is analyzed: (1) how the coevolution of social contagion and bonding by homophily may be controlled by the persons involved, and (2) how subjective representation states (e.g., what they know or believe) can play a role in this coevolution and its control. To address this, a second-order adaptive social network model is presented in which persons do have a form of control over the coevolution process, and in relation to this, their bonding depends on their subjective representation states about themselves and about each other underlying subjective homophily, and social contagion depends on their subjective representation states about their connections.

Chapter 8 [81] presents an adaptive network model in the context of joint action and social bonding. Exploration of mechanisms for mental and social network models is presented, specifically focusing on adaptation by bonding based on subjective homophily and Hebbian learning during joint rhythmic action. The chapter provides a comprehensive explanation of these concepts and their role in controlled adaptation within illustrative scenarios.

In Chap. 9 [15], a second-order adaptive network model based on subjective homophily is introduced for presenting a user with content they like on a social media platform. The platform's method is using a fake state to trigger faked homophily for the user. Simulation results are presented for different scenarios for different starting values of the user and the platform. In all scenarios the platform's method can move towards the state of the user in order to achieve a form of bonding through faked homophily. After the user indeed connects to the platform, it becomes more easy for the platform to move the user's preferences in a desired direction so that the platform can offer more adequate content more efficiently.

Chapter 10 [82] presents a multi-adaptive network model integrating multiple adaptation mechanisms, specifically focusing on five types of such adaptation mechanisms. Two of them address first-order adaptation by learning of responding on others and first-order adaptation by bonding with others based on homophily. Three other adaptation mechanisms addressed are second-order adaptation of the speed of both Hebbian learning and bonding by homophily, and second-order adaptation of the homophily tipping point. This adaptive tipping point makes that the bonding by (subjective) homophily is context-sensitive depending on control of adaptation based on which connections someone already has. The chapter provides a comprehensive explanation of these concepts and their role in controlled adaptation within diverse contextual scenarios.

6.4 Part IV: Exploiting Subjective Synchrony Detection for Computational Analysis of Social Interaction Adaptivity

Part IV of this volume [45] considers how subjective interpersonal synchrony detection and adaptivity of interaction can be captured in adaptive agent models (Chaps. 11, 12 and 13). Chapter 11 reports on the development of an adaptive agent model in which agents can detect themselves in a subjective manner how well they synchronize with each other at multiple response modalities and how this interpersonal synchrony detection adapts their experienced affiliation. Chapter 12 is a refinement of this adaptive agent model for different time scales, in which the interplay of interpersonal synchrony, short-term affiliation and long-term bonding in agent models is modeled and evaluated. Chapter 13 addresses transference and differentiation in relationship-specific and relationship-independent social interaction adaptivity. It is described how both relationship-specific and relationship-independent adaptivity develops over time in an experimental setting with four synchronizing agents.

As discussed in Sects. 2 and 3, during social interaction people respond to temporal patterns of synchrony by changed interaction behaviour. Based on this observation, the temporal factorisation principle [116] which is foundational for dynamical systems, suggests that humans possess the ability to perceive and detect such patterns of synchrony, and due to them adapt their social interaction behaviour over longer periods of time. An important aim was to explore adaptive agent models that are able to indeed detect subjective interpersonal synchrony with other agents and consequently can adapt the interaction behavior based on this form of interpersonal synchrony detection. It was examined how adaptive agents are capable of recognizing temporal patterns of synchrony, and that this detection can contribute to the development of stronger connections or affiliations.

In the study described in Chapter 11 [54] a modeling approach was developed that introduces detector states for both intrapersonal and interpersonal synchrony for multiple modalities. These detector states influence interpersonal affiliation in an adaptive manner. Through multiple simulation experiments, in which stochastic stimuli were manipulated, the modeling approach was evaluated.

In Chapter 12 [47] it is explored how different timescales can be used in particular for differentiation between short-term adaptation (like affiliation) and long-term adaptation (like bonding) and how these can be facilitated through interpersonal synchrony. Although in separate research lines, studies have linked interpersonal synchrony to both short-term adaptation [1, 57, 109, 111, 133] and long-term adaptation [31, 89] in interaction behavior. A logical theoretical and modeling next step was to integrate the dynamics of interpersonal synchrony, short-term affiliation and long-term bonding in a single complex model. A novel way of computational modeling was introduced, using both first-order and second-order adaptation, which captures the dynamics of short-term and long-term adaptivity induced by synchronization at different time scales. This modelling approach encompasses movement, affect, and verbal modalities, addressing both intrapersonal synchrony (within an individual) and interpersonal synchrony (between individuals). The behavior of the proposed approach has been evaluated through simulations involving various stimuli and communication conditions. This can provide valuable insights into the underlying mechanisms of social interaction dynamics.

Patterns of interpersonal behavior can transfer between different relationships and differentiate into relationship-specific and relationship-independent adaptations. This is addressed in Chap. 13. Such transference of social interaction behavior fits within attachment theory. Attachment theory states that adaptations gained in a specific relationship can also affect other relationships [17, 93–96]. Furthermore, transference is also a key component in classical psychoanalysis, and has been a focus in modern social-cognitive empirical research [2–4, 22, 63]. Such relationship-independent adaptations (transference) offer more general application areas in social interaction dynamics, like virtual agents that can learn over interactions with multiple users. Such an adaptive agent modelling perspective can be beneficial because when a virtual agent then learns adaptations in a certain social interaction, this learned behavior can then be applied in new social interactions. To explore how transference can take place over relationships and how relationship-specific and relationship-independent adaptation in social interactions can be distinguished, in Chap. 13 four agents were designed, each equipped with relationship-specific and relationship-independent learning or adaptation capabilities. Their performances over multiple simulations in which the agents were allowed to randomly interact (for random lengths of time and randomly selected modalities) with one another, were evaluated.

6.5 *Part V: Further Computational Analysis of Social Interaction Adaptivity Induced by Subjective Interpersonal Synchrony Detection*

In this part of this volume [45], computational analysis is discussed for some further specific topics, refining the methods and models developed in Part IV, Chaps. 11, 12, and 13: time lags for synchrony detection, transitions in synchrony, and group bonding in relation to group synchrony. Chapter 14 describes simulation experiments in which time lags in interpersonal synchrony detection were manipulated and how this affects the outcomes of interpersonal synchrony scores and the behavioral adaptivity of agents. Moreover, Chap. 15 addresses on the role that transitions in interpersonal synchrony play on top of interpersonal synchrony itself in the development of affiliation between agents. Chapter 16 addresses how group synchrony and group bonding mutually affect each other.

In Chap. 14 , time lags in the used interpersonal synchrony detection methods within adaptive agent models were introduced and analysed. In general, people can sense their own processes faster than the processes of interaction partners [91, 105, 110]. This means that, although internal and external sensing are processes that occur in parallel, there is a time lag in the sensing of these processes. Specifically, to model and analyze these human processes, an adaptive agent model was developed that explores the role of time lags in various forms of subjective synchrony detection and its influence on behavioral adaptivity. Through multiple simulation experiments, it was evaluated whether patterns of subjective time-lagged synchrony detection and its impact on related behavioral adaptivity are in line with theoretical expectations. Such an adaptive agent model contributes to our understanding of how persons perceive and respond to synchrony with temporal delays.

Because human behavior is complex, it is not a one-to-one relation from more detected interpersonal synchrony to more social interaction adaptation within the dynamics. Maybe counterintuitively, disruptions in interpersonal synchrony may positively affect behavioral adaptivity in social interaction as well [35]. Furthermore, interpersonal liking is highest when interpersonal synchrony and unpredictability are balanced during a mirror game [90]. These findings indicate that not only the absolute amount of detected interpersonal synchrony matters, although this is how most research investigates the role of interpersonal synchrony. It seems to be also of importance when interpersonal synchrony arises, highlighting the rhythmic nature in which interpersonal synchrony occurs to reach its effects on behavioral outcomes. When the timing of interpersonal synchrony is right, it might suffice instead of having high levels of interpersonal synchrony all the time. In other words, this work suggests that regarding adaptivity of social interaction behaviour, transitions in interpersonal synchrony detection can play an important role as well in addition to interpersonal synchrony detection. To analyse this computationally, in Chap. 15 an agent modelling approach was developed that addresses (subjective) detection of transitions in interpersonal synchrony. Specifically, it was studied whether the development of affiliation between two agents was favored when these agents were able to detect interpersonal synchrony transition on top of interpersonal synchrony instead of solely interpersonal synchrony.

In Chap. 16, the relationship between group synchrony and group bonding were analysed through a novel adaptive dynamical system model. By simulating multimodal interactions within a group of four agents, the study uncovers patterns in group cohesion in the sense of emerging multimodal group synchrony and related group bonding. Findings include different patterns for emerging group synchrony and group bonding. The obtained insights offer an understanding of group interaction dynamics.

6.6 Part VI Computational Analysis of Real-World Scenarios for Synchrony-Induced Social Interaction Adaptivity

In this part of this volume [45], the methods and models developed in Part IV, Chaps. 11, 12, and 13 are applied to analyse three real-world cases. The first two concern events with elderly where synchrony-induced adaptivity occurs: Chaps. 17–18. With the aging population increasing, there is a crucial need to focus on the health and social interaction of older adults. These two studies offer a valuable contribution to understanding and improving social interactions and memory recall among the elderly. The third case applies the contributions from Part IV, Chaps. 11 and 12 for human-bot interaction and interpersonal emotion regulation.

In Chap. 17, an adaptive network model for computationally recreating, simulating and subsequently inspecting a real-life scenario with elderly is contributed. The obtained first-order adaptive network model makes use of subjective detection states of interpersonal synchrony for multiple modalities.

Chapter 18, explores second-order adaptive network modeling for joint action and memory recall among elderly through detecting interpersonal synchrony. Based on research of the significance of social interaction and memory use for the elderly, as well as the role of interpersonal synchrony in joint action, this chapter analyses computationally how to enhance positive effects of social interactions among older individuals. The research examines the concept of interpersonal synchrony and its impact on joint action, memory, and emotional well-being in elderly populations. Through simulation experiments and analysis, the study demonstrates the potential benefits for music in memory recall for older adults with cognitive decline, highlighting the importance of social interaction and emotional resonance.

Chapter 19 [30] applies such synchrony detector states for different modalities in an adaptive agent model to analyse human-bot interaction. This was done to help guide designers in assessing to what extent mapping human-like mental processes and behaviors on their companion bots has valuable effects. The analysis results found here provide a promising perspective for interpersonal emotion regulation using such human-like bots. The bot's multimodal interactions helped to regulate the human's emotions by effects of emerging synchrony during the interaction even under less favorable situations such as poor individual emotion regulation capabilities.

6.7 Part VII Finalising

In this part of this volume [45], in Chaps. 20 and 21 some unification and reflections are contributed on what has been achieved and what still may be achieved in future work.

In Chap. 42 [36] a computational architecture is described that unifies and subsumes the various case studies presented in other chapters in this volume. In particular, it offers a unified perspective on both homophily-induced connectedness adaptivity (Sect. 2.1 and Part III of this volume) and synchrony-induced connectedness adaptivity (Sect. 2.2 and Parts IV–VI of this volume).

Chapter 21 discusses in a comparative manner over all chapters of this volume the computational modeling of homophily and interpersonal synchrony and their related adaptivity of social interaction behaviour and the way to model these processes by self-modeling network models. With further prospects and research themes.

References

1. Accetto, M., Treur, J., Villa, V.: An adaptive cognitive-social model for mirroring and social bonding during synchronous joint action. Proc. Comput. Sci. **145**, 3–12 (2018)
2. Andersen, S. M., Berk, M. S.: The social-cognitive model of transference: experiencing past relationships in the present. Curr. Dir. Psychol. Sci. **7**(4), 109–115 (1998a)
3. Andersen, S. M., Berk, M. S.: Transference in everyday experience: implications of experimental research for relevant clinical phenomena. Rev. Gen. Psychol. **2**(1), 81–120 (1998b)
4. Andersen, S. M., Przybylinski, E.: Experiments on transference in interpersonal relations: implications for treatment. Psychotherapy **49**(3), 370 (2012)
5. Aristotle: Rhetoric. Nichomachean ethics. In Aristotle in 23 volumes. Rackman transl. Cambridge: Harvard Univ. Press (1934).
6. Atzil-Slonim, D., Soma, C.S., Zhang, X., Paz, A., Imel, Z.E.: Facilitating dyadic synchrony in psychotherapy sessions: systematic review and meta-analysis. Psychother. Res. **33**(7), 898–917 (2023)
7. Babiloni, F., Astolfi, L.: Social neuroscience and hyperscanning techniques: past, present and future. Neurosci. Biobehav. Rev. **44**, 76–93 (2014)
8. Basso, J.C., Satyal, M.K., Rugh, R.: Dance on the brain: enhancing intra-and interbrain synchrony. Front. Hum. Neurosci. **14**, 584312 (2021)
9. Bourgeois, P., Hess, U.: The impact of social context on mimicry. Biol. Psychol. **77**(3), 343–352 (2008)
10. Bernieri, F.J.: Coordinated movement and rapport in teacher-student interactions. J. Nonverbal Behav. **12**, 120–138 (1988)
11. Bernieri, F.J., Rosenthal, R.: Interpersonal coordination: behavior matching and interactional synchrony. In: Feldman, R.S., Rimé, B. (eds.) Studies in Emotion & Social Interaction. Fundamentals of Nonverbal Behavior, pp. 401–432. Cambridge University Press (1991)
12. Bilodeau-Houle, A., Morand-Beaulieu, S., Bouchard, V., Marin, M.F.: Parent–child physiological concordance predicts stronger observational fear learning in children with a less secure relationship with their parent. J. Exp. Child Psychol. **226**, 105553 (2023)
13. Blankendaal, R., Parinussa, S., Treur, J.: A temporal-causal modelling approach to integrated contagion and network change in social networks. In: Proceedings of the Twenty-Second European Conference on AI, ECAI'16, pp. 1388–1396. IOS Press (2016)
14. Bössenecker, G., Mreijen, P., Treur, J.: A second-order adaptive network model for connecting social media platforms and users through bonding by faked homophily. Procedia Comput. Sci. **213**, 32–40 (2022). https://doi.org/10.1016/j.procs.2022.11.035
15. Bössenecker, G., Mreijen, P., Treur, J: Connecting social media platforms and users through bonding by faked homophily: a second-order adaptive network model. In: Hendrikse, S.C.F., Treur, J., Koole, S.L. (eds.) New Analysis and Modeling Directions in Social Interaction

Science: Emergent Multimodal Interpersonal Synchrony, Affiliation, and Bonding, Ch. 9. Springer Nature (2025)
16. Bott, H.: Observation of play activities in a nursery school. Genet. Psychol. Monogr. **4**, 44–88 (1928)
17. Bowlby, J.: Attachment. Basic books (2008)
18. Brazier, F.M.T., Treur, J.: Compositional modelling of reflective agents. Int. J. Hum. Comput. Stud **50**(5), 407–431 (1999)
19. Butler, E.A., Randall, A.K.: Emotional coregulation in close relationships. Emot. Rev, **5**(2), 202–210 (2013)
20. Cirelli, L.K.: How interpersonal synchrony facilitates early prosocial behavior. Curr. Opin. Psychol. **20**, 35–39 (2018)
21. Chauvigné, L.A., Walton, A., Richardson, M.J., Brown, S.: Multi-person and multisensory synchronization during group dancing. Hum. Mov. Sci. **63**, 199–208 (2019)
22. Chen, S., Andersen, S. M.: Relationships from the past in the present: significant-other representations and transference in interpersonal life. In: Advances in Experimental Social Psychology, vol. 31, pp. 123–190. Academic Press (1999)
23. Czeszumski, A., Eustergerling, S., Lang, A., Menrath, D., Gerstenberger, M., Schuberth, S. et al.: Hyperscanning: a valid method to study neural inter-brain underpinnings of social interaction. Front. Hum. Neurosci. 14, 39 (2020)
24. Czeszumski, A., Hsin-Yi Liang, S., Dikker, S., König, P., Lee, C.-P., Koole, S. L., Kelsen, B.: Cooperative behavior evokes interbrain synchrony in the prefrontal and temporoparietal cortex: a systematic review and meta-analysis of fNIRS Hyperscanning Studies. eNeuro **9**(2) (2022). https://doi.org/10.1523/ENEURO.0268-21.2022
25. Delacherche, E., Chetouani, M., Mahdhaoui, A., Saint-Georges, C., Viaux, S., Cohen, D.: Interpersonal synchrony: a survey of evaluation methods across disciplines. IEEE Trans. Affect. Comput. **3**(3), 349–365 (2012). https://doi.org/10.1109/T-AFFC.2012.12
26. Dikker, S., Wan, L., Davidesco, I., Kaggen, L., Oostrik, M., McClintock, J., et al.: Brain-to-brain synchrony tracks real-world dynamic group interactions in the classroom. Curr. Biol. **27**(9), 1375–1380 (2017)
27. Doré, B.P., Morris, R.R.: Linguistic synchrony predicts the immediate and lasting impact of text-based emotional support. Psychol. Sci. **29**(10), 1716–1723 (2018)
28. Eler, E., Dimopoulos, F., Timmers, M., Treur, J., Koole, S.L.: Interpersonal emotion regulation through attunement of emotions in human-bot interaction: an agent-based network model. In: Hendrikse, S.C.F., Treur, J., Koole, S.L. (eds.) New Analysis and Modeling Directions in Social Interaction Science: Emergent Multimodal Interpersonal Synchrony, Affiliation, and Bonding, Ch. 4. Springer Nature (2025)
29. Eler, E., Treur, J., Hendrikse, S.C.F., Donker, T., Koole, S.L.: An adaptive network model for interpersonal emotion regulation in multimodal human-bot interaction. In: Nguyen, N.T., et al. (eds.) Computational Collective Intelligence. Proc. ICCCI 2024. Lecture Notes in Computer Science, vol. 14811, pp. 67–79. Springer Nature, Cham (2024)
30. Eler, E., Treur, J., Hendrikse, S.C.F., Donker, T., Koole, S.L.: Interpersonal emotion regulation via multimodal interaction with a bonding bot: an adaptive network model. In: Hendrikse, S.C.F., Treur, J., Koole, S.L. (eds.) New Analysis and Modeling Directions in Social Interaction Science: Emergent Multimodal Interpersonal Synchrony, Affiliation, and Bonding, Ch. 19. Springer Nature (2025)
31. Feldman, R.: Parent–infant synchrony: biological foundations and developmental outcomes. Curr. Dir. Psychol. Sci. **16**, 340–345 (2007)
32. Feldman, R.: Parent-infant synchrony: a biobehavioral model of mutual influences in the formation of affiliative bonds. Monogr. Soc. Res. Child Dev. **77**(2), 42–51 (2012)
33. Ferrari, M., Quaresima, V.: A brief review on the history of human functional near-infrared spectroscopy (fNIRS) development and fields of application. Neuroimage **63**(2), 921–935 (2012)
34. Freeman, L.: Some antecedents of social network analysis. Connections **19**, 39–42 (1996)

35. García, E., Di Paolo, E.A.: Embodied coordination and psychotherapeutic outcome: beyond direct mappings. Front. Psychol. **9**, 1257 (2018)
36. Goldstein, P., Weissman-Fogel, I., Shamay-Tsoory, S.G.: The role of touch in regulating inter-partner physiological coupling during empathy for pain. Sci. Rep. **7**(1), 3252 (2017)
37. Gorman, J.C., Amazeen, P.G., Crites, M.J., Gipson, C.L.: Deviations from mirroring in interpersonal multifrequency coordination when visual information is occluded. Exp. Brain Res. **235**, 1209–1221 (2017)
38. Goode, J., Robinson, J.D.: Linguistic synchrony in parasocial interaction. Commun. Stud. **64**(4), 453–466 (2013)
39. Hadley, L. V., Ward, J.A.: Synchrony as a measure of conversation difficulty: Movement coherence increases with background noise level and complexity in dyads and triads. PLoS One, **16**(10), e0258247 (2021)
40. Heijmans, P., van Stijn, J., Treur, J.: Modeling cultural segregation of the queer community through an adaptive social network model. In: Fourth International Congress on Information and Communication Technology: ICICT 2019, London, vol. 2, pp. 233–248. Springer Singapore (2020)
41. Hendrikse, S.C.F., Kluiver, S., Treur, J., Wilderjans, T.F., Dikker, S., Koole, S.L.: Learning to synchronize: adaptive joint decision-making by virtual agents. In: Hendrikse, S.C.F., Treur, J., Koole, S.L. (eds.) New Analysis and Modeling Directions in Social Interaction Science: Emergent Multimodal Interpersonal Synchrony, Affiliation, and Bonding, Ch. 5. Springer Nature (2025)
42. Hendrikse, S.C.F., Treur, J.: From a multilevel multi-timescale computational architecture for synchrony-induced and homophily-induced attunement to designing bonding bots. In: Hendrikse, S.C.F., Treur, J., Koole, S.L. (eds.) New Analysis and Modeling Directions in Social Interaction Science: Emergent Multimodal Interpersonal Synchrony, Affiliation, and Bonding, Ch. 20. Springer Nature (2025)
43. Hendrikse, S.C.F., Treur, J., Koole, S.L.: Modeling emerging interpersonal synchrony and its related adaptive short-term affiliation and long-term bonding: a second-order multi-adaptive neural agent model. Inter. J. Neural Syst. **33**(7), 2350038 (2023)
44. Hendrikse, S.C.F., Treur, J., Koole, S.L.: Relationship-specific and relationship-independent behavioural adaptivity in affiliation and bonding: a multi-adaptive dynamical systems approach. Cogn. Syst. Res. **84**, 101182 (2024)
45. Hendrikse, S.C.F., Treur, J., Koole, S.L. (eds.): New Analysis and Modeling Directions in Social Interaction Science: Emergent Multimodal Interpersonal Synchrony, Affiliation, and Bonding. Springer Nature (2025a)
46. Hendrikse, S.C.F., Treur, J., Koole, S.L.: Introduction to the role of adaptivity in social interaction science. In: Hendrikse, S.C.F., Treur, J., Koole, S.L. (eds.) New Analysis and Modeling Directions in Social Interaction Science: Emergent Multimodal Interpersonal Synchrony, Affiliation, and Bonding, Ch. 1. Springer Nature (2025b)
47. Hendrikse, S.C.F., Treur, J., Koole, S.L.: The adaptive interplay of interpersonal synchrony, short-term affiliation and long-term bonding: a multi-timescale multi-adaptive agent model. In: Hendrikse, S.C.F., Treur, J., Koole, S.L. (eds.) New Analysis and Modeling Directions in Social Interaction Science: Emergent Multimodal Interpersonal Synchrony, Affiliation, and Bonding, Ch. 12. Springer Nature (2025c)
48. Hendrikse, S.C.F., Treur, J., Koole, S.L.: A multi-timescale multi-adaptive dynamical system approach for relationship-specific and relationship-independent adaptivity in affiliation and bonding. In: Hendrikse, S.C.F., Treur, J., Koole, S.L. (eds.) New Analysis and Modeling Directions in Social Interaction Science: Emergent Multimodal Interpersonal Synchrony, Affiliation, and Bonding, Ch. 13. Springer Nature (2025d)
49. Hendrikse, S.C.F., Treur, J., Wilderjans, T.F., Dikker, S., Koole, S.L.: On the same wavelengths: emergence of multiple synchronies among multiple agents. In: Proceedings of the 22nd International Workshop on Multi-Agent-Based Simulation, MABS'21. Lecture Notes in AI, vol. 13128, pp. 57–71. Springer Nature (2022a)

50. Hendrikse, S.C.F., Treur, J., Wilderjans, T.F., Dikker, S., Koole, S.L.: Switching in and out of sync: a controlled adaptive network model of transition dynamics in the effects of interpersonal synchrony on affiliation. In: Proceedings of the 11th International Conference on Complex Networks and their Applications. Studies in Computational Intelligence, vol. 1078, pp. 81–95. Springer Nature Switzerland, Cham (2022b)
51. Hendrikse, S.C.F., Treur, J., Wilderjans, T.F., Dikker, S., Koole, S.L.: Becoming attuned to each other over time: a computational neural agent model for the role of time lags in subjective synchrony detection and related behavioral adaptivity. Proceedings of the 15th International Conference on Brain Informatics, BI'22. Lecture Notes in AI, vol. 13406, pp. 369–383. Springer Nature (2022c)
52. Hendrikse, S.C.F., Treur, J., Wilderjans, T.F., Dikker, S., Koole, S.L.: On becoming in sync with yourself and others: an adaptive agent model for how persons connect by detecting intrapersonal and interpersonal synchrony. Hum.-Centric Intell. Syst. **3**(2), 123–146 (2023)
53. Hendrikse, S.C.F., Treur, J., Wilderjans, T.F., Dikker, S., Koole, S.L.: The emergence of multimodal interpersonal synchrony during social interaction: a network model. In: Hendrikse, S.C.F., Treur, J., Koole, S.L. (eds.) New Analysis and Modeling Directions in Social Interaction Science: Emergent Multimodal Interpersonal Synchrony, Affiliation, and Bonding, Ch. 3. Springer Nature (2025a)
54. Hendrikse, S.C.F., Treur, J., Wilderjans, T.F., Dikker, S., Koole, S.L.: Growing connectedness through synchrony: the role of subjective analysis of intrapersonal synchrony and interpersonal synchrony within an adaptive agent model. In: Hendrikse, S.C.F., Treur, J., Koole, S.L. (eds.) New Analysis and Modeling Directions in Social Interaction Science: Emergent Multimodal Interpersonal Synchrony, Affiliation, and Bonding, Ch. 11. Springer Nature (2025b)
55. Hendrikse, S.C.F., Treur, J., Wilderjans, T.F., Dikker, S., Koole, S.L.: The role of timing in synchrony detection and social interaction adaptivity: including time lags in a second-order adaptive agent model. In: Hendrikse, S.C.F., Treur, J., Koole, S.L. (eds.) New Analysis and Modeling Directions in Social Interaction Science: Emergent Multimodal Interpersonal Synchrony, Affiliation, and Bonding, Ch. 14. Springer Nature (2025c)
56. Hendrikse, S.C.F., Treur, J., Wilderjans, T.F., Dikker, S., Koole, S.L.: The effect of transition dynamics in synchrony on social interaction adaptivity: a multi-adaptive network model. In: Hendrikse, S.C.F., Treur, J., Koole, S.L. (eds.) New Analysis and Modeling Directions in Social Interaction Science: Emergent Multimodal Interpersonal Synchrony, Affiliation, and Bonding, Ch. 15. Springer Nature (2025d)
57. Hove, M.J., Risen, J.L.: It's all in the timing: interpersonal synchrony increases affiliation. Soc. Cogn. **27**(6), 949–960 (2009)
58. Hsieh, C.S., König, M.D., Liu, X.: A structural model for the coevolution of networks and behavior. Rev. Econ. Stat. **104**(2), 355–367 (2022)
59. Hubbard R.M.: A method of studying spontaneous group formation. In: Thomas, D.S. (ed.) Some New Techniques for Studying Social Behavior, pp. 76–85. Child Dev. Monogr (1929)
60. Hülsemann, M.J., Naumann, E., Rasch, B.: Quantification of phase-amplitude coupling in neuronal oscillations: comparison of phase-locking value, mean vector length, modulation index, and generalized-linear-modeling-cross-frequency coupling. Front. Neurosci. **13**, 573 (2019)
61. Jirsa, V., Müller, V.: Cross-frequency coupling in real and virtual brain networks. Front. Comput. Neurosci. **7**, 78 (2013)
62. Kappert, C., Rus, R., Treur, J.: On the emergence of segregation in society: network-oriented analysis of the effect of evolving friendships. In: Computational Collective Intelligence: 10th International Conference, ICCCI 2018, Bristol, UK, September 5–7, 2018, Proceedings, Part I 10, pp. 178–191. Springer International Publishing (2018)
63. Kernberg, O. F., Yeomans, F. E., Clarkin, J. F., Levy, K. N.: Transference focused psychotherapy: overview and update. Int. J. Psychoanal. **89**(3), 601–620 (2008)
64. King, A., Smith, D., Webb, B.: Minimal specialization: coevolution of network structure and dynamics. Phys. A **651**, 130000 (2024)

65. Kinreich, S., Djalovski, A., Kraus, L., Louzoun, Y., Feldman, R.: Brainto-brain synchrony during naturalistic social interactions. Sci. Rep, **7**(1), 17060 (2017)
66. Koch, S.C., Fischman, D.: Embodied enactive dance/movement therapy. Am. J. Dance Ther. **33**(1), 57 (2011)
67. Koole, S.L., Atzil-Slonim, D., Butler, E., Dikker, S., Tschacher, W., Wilderjans, T.: In sync with your shrink: grounding psychotherapy in interpersonal synchrony. In: Applications of Social Psychology, pp. 161–184. Routledge (2020)
68. Koole, S.L., Tschacher, W.: Synchrony in psychotherapy: a review and an integrative framework for the therapeutic alliance. Front. Psychol. **7**, 862 (2016)
69. Konvalinka, I., Xygalatas, D., Bulbulia, J., Schjødt, U., Jegindø, E. M., Wallot, S., Roepstorff, A.: Synchronized arousal between performers and related spectators in a fire-walking ritual. Proc. Natl.Acad. Sci. U.S.A, **108**(20), pp. 8514–8519 (2011)
70. Koudenburg, N., Postmes, T., Gordijn, E.H.: Beyond content of conversation: the role of conversational form in the emergence and regulation of social structure. Pers. Soc. Psychol. Rev. **21**(1), 50–71 (2017)
71. Lazarsfeld, P.F., Merton, R.K..: Friendship as a social process: a substantive and methodological analysis. In: Berger, M. (ed.) Freedom and Control in Modern Society, pp. 18–66. Van Nostrand, New York (1954)
72. Liu, J., Huang, S., Aden, N.M., Johnson, N.F., Song, C.: Emergence of polarization in coevolving networks. Phys. Rev. Lett. **130**(3), 037401 (2023)
73. Luo, S., Du, Y., Liu, P., Xuan, Z., Wang, Y.: A study on coevolutionary dynamics of knowledge diffusion and social network structure. Expert Syst. Appl. **42**(7), 3619–3633 (2015)
74. McPherson, M., Smith-Lovin, L., Cook, J.M.: Birds of a feather: homophily in social networks. Annu. Rev. Sociol. **27**, 415–444 (2001)
75. Mende, M.A., Schmidt, H.: Psychotherapy in the framework of embodied cognition—does interpersonal synchrony influence therapy success?. Frontiers in Psychiatry, **12**, 562490 (2021)
76. Miles, L.K., Lumsden, J., Richardson, M.J., Macrae, C.N.: Do birds of a feather move together? Group membership and behavioral synchrony. Exp. Brain Res. **211**(3–4), 495–503 (2011)
77. Moulder, R.G., Boker, S.M., Ramseyer, F., Tschacher, W.: Determining synchrony between behavioral time series: an application of surrogate data generation for establishing falsifiable null-hypotheses. Psychol. Methods **23**(4), 757–773 (2018). https://doi.org/10.1037/met0000172
78. Mu, Y., Cerritos, C., Khan, F.: Neural mechanisms underlying interpersonal coordination: a review of hyperscanning research. Soc. Pers. Psychol. Compass **12**(11), e12421 (2018)
79. Mukeriia, Y., Treur, J., Hendrikse, S.C.F.: An adaptive network model for learning and bonding during a varying in rhythm synchronous joint action. In: Proceedings of the 12th International Conference on Complex Networks and Their Applications. Studies in Computational Intelligence, vol. 1144, pp. 41–52. Springer Nature Switzerland, Cham (2023)
80. Mukeriia, Y., Treur, J., Hendrikse, S.C.F.: A multi-adaptive network model for human Hebbian learning, synchronization and social bonding based on adaptive homophily. Cogn. Syst. Res. **84**, 101187 (2024)
81. Mukeriia, Y., Treur, J., Hendrikse, S.C.F.: The interplay of bonding based on subjective homophily and learning during a varied rhythmic action: an adaptive network model. In: Hendrikse, S.C.F., Treur, J., Koole, S.L. (eds.) New Analysis and Modeling Directions in Social Interaction Science: Emergent Multimodal Interpersonal Synchrony, Affiliation, and Bonding, Ch. 8. Springer Nature (2025a)
82. Mukeriia, Y., Treur, J., Hendrikse, S.C.F.: Human learning, attunement and social bonding based on adaptive homophily: a multi-adaptive network model. In: Hendrikse, S.C.F., Treur, J., Koole, S.L. (eds.) New Analysis and Modeling Directions in Social Interaction Science: Emergent Multimodal Interpersonal Synchrony, Affiliation, and Bonding, Ch. 10. Springer Nature (2025b)
83. Oullier, O., De Guzman, G.C., Jantzen, K.J., Lagarde, J., Kelso, J.A.S.: Social coordination dynamics: measuring human bonding. Soc. Neurosci. **3**(2), 178192 (2008)

84. Palumbo, R.V., Marraccini, M.E., Weyandt, L.L., Wilder-Smith, O., McGee, H.A., Liu, S., Goodwin, M.S.: Interpersonal autonomic physiology: a systematic review of the literature. Pers. Soc. Psychol. Rev. **21**(2), 99–141 (2017)
85. Pan, Y., Novembre, G., Song, B., Zhu, Y., Hu, Y.: Dual brain stimulation enhances interpersonal learning through spontaneous movement synchrony. Soc. Cogn. Affect. Neurosci. **16**(1–2), 210–221 (2021)
86. Plato: Laws. Plato in Twelve Volumes, vol. 11. Bury translator. Harvard University Press, Cambridge (1968)
87. Port, R.F., Van Gelder, T.: Mind as Motion: Explorations in the Dynamics of Cognition. MIT Press, Cambridge, MA (1995)
88. Ramseyer, F.T.: Exploring the evolution of nonverbal synchrony in psychotherapy: The idiographic perspective provides a different picture. Psychother. Res. **30**(5), 622–634 (2020)
89. Ramseyer, F., Tschacher, W.:. Nonverbal synchrony in psychotherapy: coordinated body movement reflects relationship quality and outcome. J. Consult. Clin. Psychol. **79**(3), 284 (2011)
90. Ravreby, I., Shilat, Y., Yeshurun, Y.: Liking as a balance between synchronization, complexity and novelty. Sci. Rep. **12**(1), 1–12 (2022). https://doi.org/10.1038/s41598-022-06610-z
91. Rayner, K., Smith, T.J., Malcolm, G.L., Henderson, J.M.: Eye movements and visual encoding during scene perception. Psychol. Sci. **20**(1), 6–10 (2009)
92. Reindl, V., Gerloff, C., Scharke, W., Konrad, K.: Brain-to-brain synchrony in parent-child dyads and the relationship with emotion regulation revealed by fNIRS-based hyperscanning. NeuroImage **178**, 493–502 (2018)
93. Salter Ainsworth, M. D., Bowlby, J.: Child Care and the Growth of Love. Penguin Books, London (1965)
94. Salter Ainsworth, M. D.: Infancy in Uganda. Johns Hopkins, Baltimore (1967)
95. Salter Ainsworth, M. D., Blehar, M., Waters, E., Wall, S.: Patterns of Attachment. Erlbaum, Hillsdale, NJ (1978)
96. Salter Ainsworth, M. D., Bowlby, J.: An ethological approach to personality development. Am. Psychol. **46**(4), 333–341 (1991)
97. Schoenherr, D., Paulick, J., Strauss, B.M., Deisenhofer, A.K., Schwartz, B., Rubel, J.A., et al.: Identification of movement synchrony: validation of windowed cross-lagged correlation and-regression with peak-picking algorithm. PloS ONE **14**(2), e0211494 (2019a)
98. Schoenherr, D., Paulick, J., Worrack, S., Strauss, B.M., Rubel, J.A., Schwartz, B., et al.: Quantification of nonverbal synchrony using linear time series analysis methods: lack of convergent validity and evidence for facets of synchrony. Behav. Res. Methods 1–23 (2019b)
99. Schoenherr, D., Strauss, B., Stangier, U., Altmann, U.: The influence of vocal synchrony on outcome and attachment anxiety/avoidance in treatments of social anxiety disorder. Psychotherapy **58**(4), 510 (2021)
100. Semin, G.R.: Grounding communication: synchrony. In: Kruglanski, A.W., Higgins, E.T. (eds.) Social Psychology: Handbook of Basic Principles, pp. 630–649. The Guilford Press (2007)
101. Semin, G.R., Cacioppo, J.T.: Grounding Social Cognition: Synchronization, Coordination, and Co-regulation. Cambridge University Press, New York (2008)
102. Shamay-Tsoory, S.G., Saporta, N., Marton-Alper, I.Z., Gvirts, H.Z.: Herding brains: a core neural mechanism for social alignment. Trends Cogn. Sci. **23**(3), 174–186 (2019)
103. Shapira, N., Atzil-Slonim, D., Tuval-Mashiach, R., Shapira, O.: Measuring linguistic synchrony in psychotherapy. In: Proceedings of the Eighth Workshop on Computational Linguistics and Clinical Psychology, pp. 158–176 (2022)
104. Sharpanskykh, A., Treur, J.: Modelling and analysis of social contagion in dynamic networks. Neurocomputing **146**, 140–150 (2014)
105. Shelton, J., Kumar, G.: Comparison between auditory and visual simple reaction times. Neurosci. Med. **1**(1), 30–32 (2010)
106. Snijders, T., Steglich, C., Schweinberger, M.: Modeling the coevolution of networks and behavior. In: van Montfort, K., Oud, J., Satorra, A. (eds.) Longitudinal Models in the Behavioral and Related Sciences, pp. 41–71. Taylor and Francis (2017)

107. Sommariva, S., Sorrentino, A., Piana, M., Pizzella, V., Marzetti, L.: A comparative study of the robustness of frequency-domain connectivity measures to finite data length. Brain Topogr. **32**, 675–695 (2019)
108. Su, J., Liu, B., Li, Q., Ma, H.: Coevolution of opinions and directed adaptive networks in a social group. J. Artif. Soc. Soc. Simul. **17**(2) (2014)
109. Tarr, B., Launay, J., Dunbar, R.I.M.: Silent disco: dancing in synchrony leads to elevated pain thresholds and social closeness. Evol. Hum. Behav. **37**(5), 343–349 (2016)
110. Thompson, J.J., Blair, M.R., Henrey, A.J.: Over the hill at 24: persistent age-related cognitive-motor decline in reaction times in an ecologically valid video game task begins in early adulthood. PLoS ONE **9**(4), e94215 (2014)
111. Tichelaar, C., Treur, J.: Network-oriented modeling of the interaction of adaptive joint decision making, bonding and mirroring. In: International Conference on Theory and Practice of Natural Computing pp. 328–343. Cham: Springer International Publishing (2018)
112. Tichelaar, C., Treur, J.: Bonding by homophily and adaptive mirroring in joint decision making: a second-order adaptive network model. In: Hendrikse, S.C.F., Treur, J., Koole, S.L. (eds.) New Analysis and Modeling Directions in Social Interaction Science: Emergent Multimodal Interpersonal Synchrony, Affiliation, and Bonding, Ch. 6. Springer Nature (2025)
113. Tomprou, M., Kim, Y.J., Chikersal, P., Woolley, A.W., Dabbish, L.A.: Speaking out of turn: how video conferencing reduces vocal synchrony and collective intelligence. PLoS ONE **16**(3), e0247655 (2021)
114. Treur, J.: On the use of reflection principles in modelling complex reasoning.Int. J. Intell. Syst. **6**(3), 277–294 (1991)
115. Treur, J.: Temporal semantics of meta-level architectures for dynamic control of reasoning. In: Fribourg, L., Turini, F. (eds) Logic Program Synthesis and Transformation—Meta-Programming in Logic. Proceedings META LOPSTR 1994. Lecture Notes in Computer Science, vol. 883, pp. 353–376. Springer, Berlin, Heidelberg (1994)
116. Treur, J.: Temporal factorisation: a unifying principle for dynamics of the world and of mental states. Cogn. Syst. Res. **8**(2), 57–74 (2007)
117. Treur, J.: Mathematical analysis of the emergence of communities based on coevolution of social contagion and bonding by homophily. Appl. Netw. Sci. **4**(1), **39**(2019)
118. Treur, J.: Modeling higher-order adaptivity of a network by multilevel network reification. Netw. Sci. **8**, S110–S144 (2020)
119. Treur, J.: Network-Oriented Modeling for Adaptive Networks: Designing Higher-order Adaptive Biological, Mental and Social Network Models. Springer Nature Cham, Cham, Switzerland (2020b)
120. Treur, J.: On the dynamics and adaptivity of mental processes: relating adaptive dynamical systems and self-modeling network models by mathematical analysis. Cogn. Syst. Res. **70**, 93–100 (2021)
121. Treur, J.: Controlled social network adaptation: subjective elements in an objective social world. In: Yang, X.S., Sherratt, S., Dey, N., Joshi, A. (eds) Proceedings of Sixth International Congress on Information and Communication Technology, ICICT'21. Lecture Notes in Networks and Systems, vol. 235, pp. 263–274. Springer Nature (2022)
122. Treur, J.: Modeling coevolution of dynamics, adaptivity and control of adaptivity by self-modeling networks. In: Hendrikse, S.C.F., Treur, J., Koole, S.L. (eds.) New Analysis and Modeling Directions in Social Interaction Science: Emergent Multimodal Interpersonal Synchrony, Affiliation, and Bonding, Ch. 2. Springer Nature (2025a)
123. Treur, J.: A second-order adaptive network model for bonding based on subjectively perceived homophily. In: Hendrikse, S.C.F., Treur, J., Koole, S.L. (eds.) New Analysis and Modeling Directions in Social Interaction Science: Emergent Multimodal Interpersonal Synchrony, Affiliation, and Bonding, Ch. 7. Springer Nature (2025b)
124. Tschacher, W., Ramseyer, F., Koole, S.L.: Sharing the now in the social present: duration of nonverbal synchrony is linked with personality. J. Pers. **86**(2), 129–138 (2018)
125. Vacharkulksemsuk, T., Fredrickson, B.L.: Strangers in sync: achieving embodied rapport through shared movements. J. Exp. Soc. Psychol. **48**(1), 399–402 (2012)

126. Valdesolo, P., DeSteno, D.: Synchrony and the social tuning of compassion. Emotion **11**(2), 262–266 (2011)
127. Valdesolo, P., Ouyang, J., DeSteno, D.: The rhythm of joint action: synchrony promotes cooperative ability. J. Exp. Soc. Psychol. **46**(4), 693–695 (2010)
128. van den Beukel, S., Goos, S.H., Treur, J.: An adaptive temporal-causal network model for social networks based on the homophily and more-becomes-more principle. Neurocomputing **338**, 361–371 (2019)
129. Van Leeuwen, P., Geue, D., Thiel, M., Cysarz, D., Lange, S., Romano, M. C., Grönemeyer, D.H.: Influence of paced maternal breathing on fetal–maternal heart rate coordination. Proceedings of the National Academy of Sciences, **106**(33), 13661–13666 (2009)
130. Van Ulzen, N.R., Lamoth, C.J.C., Daffertshofer, A., Semin, G.R., Beek, P.J.: Characteristics of instructed and uninstructed interpersonal coordination while walking side-by-side. Neurosci. Lett. **432**(2), 88–93 (2008)
131. Wellman, B.: The school child's choice of companions. J. Educ. Res. **14**, 126–132 (1929)
132. Weyhrauch, R.W.: Prolegomena to a theory of mechanized formal reasoning. Artif. Intell. **13**(1–2), 133–170 (1980)
133. Wiltermuth, S.S., Heath, C.: Synchrony and cooperation. Psychol. Sci. **20**(1), 1–5 (2009)
134. Yokozuka, T., Miyamoto, H., Kasai, M., Miyake, Y., Nozawa, T.: The relationship between turn-taking, vocal pitch synchrony, and rapport in creative problem-solving communication. Speech Commun. **129**, 33–40 (2021)

Modeling Coevolution of Dynamics, Adaptivity and Control of Adaptivity by Self-modeling Networks

Jan Treur

Abstract Networks provide an intuitive, declarative way of modeling which has turned out to be suitable for many types of applications that involve complex dynamics. In many cases also adaptivity plays a role. Using algorithmic or procedural descriptions for the adaptation processes as is often the approach followed, easily leads to less declarative and less transparent forms of modeling. This chapter exploits the notion of self-modeling network that has been developed recently to avoid this. According to this approach, adaptivity is obtained by adding a self-model to a given base network, with network states that represent part of the base network's structure. This results in a two-level network. The self-modeling construction can easily be iterated so that multiple orders of adaptation can be covered as well. In particular, a three-level self-modeling network can be used to integrate dynamics, adaptivity and control of adaptivity in a unified manner in one network. In this chapter, it is shown how this can provide useful building blocks to design network models for social interaction dynamics, adaptivity, and control of adaptivity.

Keywords Network-oriented modeling · Self-modeling network · Network reification · Adaptive network model · Controlled adaptation

1 Introduction

A network modeling approach based on self-modeling can be used to model not only the dynamics of nodes or states of a network but also the adaptation of its network structure and the control of that adaptation. In this chapter, it will be discussed how such self-modeling networks [104, 105] can be designed.

J. Treur (✉)
Social AI Group, Department of Computer Science, Vrije Universiteit Amsterdam, Amsterdam, Netherlands
e-mail: j.treur@vu.nl

S. C. F. Hendrikse et al. (eds.), *New Analysis and Modeling Directions in Social Interaction Science*, Studies in Systems, Decision and Control 614,
https://doi.org/10.1007/978-3-031-99968-0_2

Self-modeling networks are networks that include a self-model for part of their own network structure based on nodes that represent certain network structure characteristics, for example, connection weights or excitability thresholds. Such a self-modeling process for a network is also called network reification. This relates to a long-standing tradition in other areas of AI, namely that of meta-programming and metalevel architectures; e.g., [18, 30, 82, 81, 118]. By using self-models within a network, adaptation of the network structure can be modeled by the dynamics of the self-model states representing the network structure. The latter can be specified by declarative means in a network-oriented manner in the form of mathematical relations and functions; thus, adaptivity of the network structure can be specified in a declarative manner. To support the modeler, a dedicated software environment for self-modeling networks is available; see Treur [104, Ch. 9] and Treur [106].

In this chapter, first in Sect. 2 some background knowledge is discussed concerning dynamics and adaptivity in social and mental processes. Next, in Sect. 3 the network-oriented modeling approach based on self-modeling networks will be briefly introduced. In Sect. 4 it is discussed how various adaptation principles can be modeled. In Sect. 5, it is briefly pointed out how this approach can be applied to a number of concepts concerning learning within and of an organization. Finally, Sect. 6 is a discussion.

2 Background on Social and Brain Networks and Their Dynamics

As the focus here is on dynamics and adaptivity related to social interaction and its underlying mental processes, in this section some historical background of this is discussed.

2.1 Social Networks and Their Dynamics and Adaptivity

First, the area of social networks is discussed. Already in early research in that area an important focus was on the causal relationships between connectedness between persons and the extent of similarity of them, for example, in the following quote from [68]:

> A pattern as powerful and pervasive as the relationship between association and similarity did not go unnoticed in classical Western thought. In Aristotle's Rhetoric and Nichomachean Ethics, he noted that people 'love those who are like themselves' [5, p. 1371]. Plato observed in Phaedrus that 'similarity begets friendship'. [73, p. 837]. [68, p. 416]

For a further historical review, see [33]. Systematic observations of group formation and network ties by social scientists began in the 1920s and 1930s. For example, [17, 58, 117] reported that school children formed friendships and play groups at higher rates if they were similar on demographic characteristics [68]. An important contribution to the sociological literature is [63] about forming friendships. They use the term 'homophily' and to describe bonding by homophily they also introduced the metaphor

> Birds of a feather flock together [63]

which has often been used in the literature since then [68].

Although in such literature and the above quotes, often the emphasis may seem on the causal effect that the extent of similarity has on (strengthening of) connectedness such as friendship, there often is still a causal effect in the opposite direction as well. As network nodes that are considered for similarity can have their own dynamics too, persons that have strong connectedness can strengthen their extent of similarity. Unfortunately, the above metaphor is not applicable to describe this opposite causal effect: it is very unlikely that feathers of birds can change due to their flocking together. This is because in such cases similarity is considered for characteristics that are assumed static, like cultural background, but in many other cases similarity is considered for characteristics that can change over time such as preferences for types of music, movies or sport activities. In the latter cases, often social contagion plays an important role in such changes. Therefore, for many cases a circular causal mechanism is considered realistic, sometimes described by the term coevolution, which means that over time

- connectedness evolves via mental and behavioral adaptation under influence of the extent of similarity
- similarity evolves via social contagion under influence of the extent of connectedness.

In recent years often such circular coevolution processes have been modeled from an adaptive dynamical system perspective in a network-like manner, as is shown, for example, in [2, 14, 57, 62, 65, 66, 70, 77, 80, 83, 85, 86, 91, 115]. See also this volume [43], Chapter Bonding by Homophily and Adaptive Mirroring in Joint Decision Making: A Second-Order Adaptive Network Model [87]. Simulations based on such adaptive computational network models have demonstrated how such a circular coevolution process can explain the emergence of clusters or communities, or segregation in society, e.g., [39, 60].

Moreover, in line with the temporal factorisation principle from [89], also a mental state for subjective homophily detection has been postulated and analysed in Treur [109] and used in Bössenecker et al. [21] to analyse computationally how also faked homophily can lead to (real) bonding; see also this volume [43], Chapters A Second-Order Adaptive Network Model for Bonding Based on Subjectively Perceived Homophily, The Interplay of Bonding Based on Subjective Homophily and Learning During a Varied Rhythmic Action: An Adaptive Network Model, Connecting Social Media Platforms and Users Through Bonding by Faked Homophily: A Second-Order Adaptive Network Model and Human Learning, Attunement and Social Bonding Based on Adaptive Homophily: A Multi-Adaptive Network Model: [22, 111].

If in the above analysis, synchrony of two persons is considered instead of similarity, the concept 'bonding by homophily' can be replaced by 'bonding by synchrony', a concept which has been explored in literature such as [56, 84, 114 – 116, 121], see also this volume [43], Chapter Introduction to the Role of Adaptivity in Social Interaction Science [44]. Like for bonding by homophily as discussed above, also for bonding by synchrony such an overall circular pattern of coevolution can be found over time:

- connectedness evolves via mental and behavioral adaptation under influence of the extent of synchrony
- synchrony evolves via social contagion under influence of the extent of connectedness.

Recently, this type of coevolution processes have been analysed computationally in Hendrikse et al. [40, 41, 42, 52]; see also this volume [43], Chapters Growing Connectedness through Synchrony: The Role of Subjective Analysis of Intrapersonal Synchrony and Interpersonal Synchrony within an Adaptive Agent Model, The Emergence of Group Synchrony and Adaptivity for Group Bonding: A Multi-Adaptive Network Model: [45, 46, 55]. To achieve this, based on the temporal factorisation principle from [89], the notion of subjective synchrony detection was postulated to indicate how a person can notice being in synchrony with someone else. Here, adaptive self-modeling network models were very useful to address the complexity of this type of adaptive dynamical systems. This developed adaptive dynamical system approach to bonding by synchrony has been successfully applied to model complex social phenomena such as adaptive interpersonal emotion regulation and adaptive collective phenomena involving elderly in Vervoordeldonk et al. [116], Xu et al. [123], and for human-bot interaction; see also this volume [43], Chapters Reliving Memory and Religious Experience Through Multimodal Synchrony and Adaptation: An Adaptive Network Model, Joint Action and Memory Recall for Elderly by Detecting Interpersonal Synchrony: A Higher-Order Adaptive Network Model and Interpersonal Emotion Regulation via Multimodal Interaction with a Bonding Bot: An Adaptive Network Model.

2.2 Networks in the Brain and Their Adaptivity

Mental processes can have their own forms of adaptivity, often related to forms of plasticity of the underlying brain network structures. Some early work can be found, for example, in contributions by Beale [9] and Bain [8] who as a first step have put forward the cyclic structure of networks ('circuits') in the brain:

> I venture to conclude that the typical anatomical arrangement of a nervous mechanism is not a cord with two ends—a point of origin and terminal extremity, but a cord without an end—a continuous circuit. [9, p. 392]

Bain [8] commented on this and points at the role of connection ('cell-crossing') points:

> He couples with this appearance the doctrine (maintained by him, although disputed by others) that the nerves terminate in loops, and consequently form an unbroken nervous circuit. He then suggests that the cell-crossing is the place where the inner bendings of a great many independent circuits come into close neighbourhood, and affect one another by a process of the nature of electrical induction. Any one of the circuits being active, or excited, would impart excitement to all that came near it in the same cell. [8, pp. 118–119]

Adaptation of connectivity in the brain: synaptic plasticity

As one step further, also plasticity of brain networks was discussed in Bain [8]:

> Now to each impression, each sensation or thought, there corresponds physically a group or series of nerve-currents; when two impressions concur, or closely succeed one another, the nerve currents find some bridge or place of continuity, better or worse, according to the abundance of nerve matter available for the transition. In the cells or corpuscles where the currents meet and join, there is, in consequence of the meeting, a *strengthened connexion* or *diminished obstruction*—a preference track for that line over lines where no continuity has been established. [8, p. 117, original italics]

He continues in more detail in:

> This is merely a hypothetical rendering of the facts: yet it is a very probable rendering. In the nature and number of the nerve elements, and their mode of connexion, there is nothing hypothetical; and there is no departure from fact or strong probability, in assigning special and distinct tracks for the currents connected with each separate sensation, idea, emotion, or other conscious state. As to the precise mode of the plastic growth that unites separate impressions into trains and aggregates in the memory,—we know that the corpuscles or crossings are the points that must be operated upon; that a flow of healthy blood must co-operate to the effect; and that the process takes time. It is evidently a species of growth: but the precise molecular change effected in the lines of strengthened communication, or diminished obstruction, we can describe only as increasing the conducting tendency in those lines, as compared with the collateral openings where no such operation has taken place. [8, pp. 117–118]
>
> I can suppose that, at first, each one of the circuits would affect all others indiscriminately; but that, in consequence of two of them being independently made active at the same moment (which is the fact in acquisition), a strengthened connexion or diminished obstruction would arise between these two, by a change wrought in the intervening cell-substance; and that, afterwards, the induction from one of these circuits would not be indiscriminate, but select; being comparatively strong towards one, and weaker towards the rest. [8, p. 119]

Much later, this perspective on connectivity adaptation has been presented as well by Hebb [38]. For example, in this form:

> When an axon of cell A is near enough to excite B and repeatedly or persistently takes part in firing it, some growth process or metabolic change takes place in one or both cells such that A's efficiency, as one of the cells firing B, is increased. [38, p. 62]

This type of connectivity adaptation is often referred as Hebbian learning, but considering the above, might as well or even better be called Bainian learning. For more on this issue, see also Wilkes and Wade [119] and Cardno [25]. To indicate this type of adaptation, often the following popular phrasing is used as well:

> What fires together, wires together [61, 78]

Adaptation of intrinsic neuron characteristics in the brain: nonsynaptic plasticity

The brain network does not only show connectivity adaptation (synaptic plasticity). More recently, in the neuroscientific literature, a distinction was made between synaptic and nonsynaptic (or intrinsic) plasticity. Synaptic plasticity concerns plasticity of connectivity in the brain. This notion is described above, following [8, p. 117, 38, 78]. The intrinsic nonsynaptic adaptation of (neural) states has been addressed in more detail more recently in literature such as [26, 29, 125]. For example:

> Learning-related cellular changes can be divided into two general groups: modifications that occur at synapses and modifications in the intrinsic properties of the neurons. While it is commonly agreed that changes in strength of connections between neurons in the relevant networks underlie memory storage, ample evidence suggests that modifications in intrinsic neuronal properties may also account for learning related behavioral changes. [26, p. 30]

The following quote indicates that long-lasting modifications in excitability of neurons can evolve over time:

> Long-lasting modifications in intrinsic excitability are manifested in changes in the neuron's response to a given extrinsic current (generated by synaptic activity or applied via the recording electrode). [26, p. 30]

In this chapter for these two types of adaptivity, two examples of first-order adaptation principles are considered: Hebbian learning for connection weights and excitability modulation for the excitability threshold of states. The latter form of adaptation has been related, for example, to homeostatic regulation [120] and also to how deviant dopamine levels during sleep make that dreams can make use of more associations due to easier excitable neurons; e.g., [16, 20]. Moreover, both (synaptic and nonsynaptic) forms of adaptation can work together, e.g., [4, 20, 36, 64], see for example:

> So, circuit plasticity could be due to the interplay of synaptic and intrinsic mechanisms and this balance could be different for short and long-term plasticity. In this view, ensembles could be quickly formed by fast intrinsic mechanisms but then be modified by slower synaptic plasticity, perhaps in relation with learning. [4, p. 18/25]
>
> To explain how neuronal ensembles are generated, we propose an "iceberg" model, by which the increased neuronal excitability that results from repeated input stimulation makes subthreshold connections become suprathreshold, enhancing the postsynaptic effect of already existing synapses, and generating a neuronal ensemble. [36, p. 4/10]

Based on this perspective, the model proposed in Budding et al. [20] explains how creativity can be based on dopamine-induced (DA) enhanced associativity during dreaming that is 'exported' to the awake life. More specifically, the model uses dopamine-induced enhanced intrinsic excitability by lowering socalled **T**-states for adaptive excitability thresholds (see also Sect. 4.1) as short-term adaptation during dreaming (only short-lived) which in turn boosts longer-term adaptation based on Hebbian learning that can cross the border from dreaming to being awake:

> The longer-term adaptation of connection weights by Hebbian learning creates the plasticity that is needed, as it has the potential to cross the border from sleeping to being awake (depending also on dream recall). This is in contrast to the DA-dependent **T**-state suppression which is a short-term adaptation that will not cross this border. [20, p. 42]

In a different context, a similar perspective on short-term adaptation boosting long-term adaptation is used in Hendrikse et al. [40], see also this volume [43, Ch. 12]. Here, induced by detected emerging synchrony, also different time scales for short-term affiliation and long-term bonding are used where repeated episodes of the short-term adaptation can boost the long-term adaptation.

For more literature on this form of (nonsynaptic) learning or adaptation of enhanced excitability and its interplay with Hebbian (synaptic) learning, see, for example, [3, 28, 29, 35, 64, 88, 124].

2.3 Context-Sensitive Control of Adaptation Processes in the Brain: Metaplasticity

The two first-order adaptation principles discussed above refer to what in neuroscientific literature is called *plasticity*. However, this is usually not a constant feature. For an organism, in some circumstances it is better to learn (and change) fast, but in other circumstances, it is better to stay stable and let what has been learnt in the past persist: the Plasticity Versus Stability Conundrum [79, p. 773]. Under which circumstances and to which extent such plasticity actually takes place is controlled by a form of socalled *metaplasticity*; e.g., [1, 34, 67, 75, 79]. Such control can address 'The Plasticity Versus Stability Conundrum' by only making plasticity happen in contextual circumstances when it is important for the person to change and otherwise stabilize it. In literature as mentioned, various studies show how adaptation (as described, for example, by Hebbian learning), is modulated by accelerating the adaptation process or decelerating or even blocking it. Among the reported factors affecting plasticity in such a way are stimulus exposure, activation, previous experiences, and stress. For example, in Robinson et al. [75] the following compact quote can be found summarizing that increasing stimulus exposure makes that the adaptation speed increases:

> Adaptation accelerates with increasing stimulus exposure [75, p. 2]

Also, for learning of mental models by individuals or for learning within or of organisations such context-sensitive control of adaptation is a crucial concept, e.g., [23, 24]. For example, depending on the context, a manager can decide to organise organisational processes where individuals learn from other individuals. In Mukeriia et al. [70]; Treur [104, Ch. 6] different variants of second-order adaptation to control bonding by adaptive homophily are addressed; see also this volume [43], Chapter Human Learning, Attunement and Social Bonding Based on Adaptive Homophily: A Multi-Adaptive Network Model.

3 Network-Oriented Modeling by Temporal-Causal Networks

In this section, the network-oriented modeling approach using self-modeling networks to model adaptive networks is briefly introduced in two steps. In this approach *nodes* Y in a network have activation values $Y(t)$ that are dynamic over time t; therefore, they serve as state variables and will usually be called *states*. To specify these dynamics, the states are considered to affect each other by the connections within the network. Following [90], a basic *network structure* is characterised by:

- **connectivity characteristics**

  Connections from a state X to a state Y and their *weights* $\boldsymbol{\omega}_{X,Y}$
- **aggregation characteristics**

  For any node Y, some *combination function* $\mathbf{c}_Y(\ldots)$ defines aggregation that is applied to the *single impacts* $\boldsymbol{\omega}_{X_i,Y}X_i(t)$ on Y from its incoming connections from states $X_1, \ldots, X_k$.
- **timing characteristics**

  Each state Y has a *speed factor* $\boldsymbol{\eta}_Y$ defining how fast it changes upon given impact

Here, the states X_i and Y have activation levels $X_i(t)$ and $Y(t)$ that vary (often within the [0, 1] interval) over time, described by real numbers t. Note that speed factors can be used to design network models where different states may change on different time scales.

The dynamics of such networks are described by the following difference (or differential) equations that incorporate in a canonical manner the network characteristics $\boldsymbol{\omega}_{X,Y}$, $\mathbf{c}_Y(\ldots)$, $\boldsymbol{\eta}_Y$:

$$Y(t+\Delta t) = Y(t) + \boldsymbol{\eta}_Y\big[\mathbf{c}_Y\big(\boldsymbol{\omega}_{X_1,Y}X_1(t), \ldots, \boldsymbol{\omega}_{X_k,Y}X_k(t)\big) - Y(t)\big]\Delta t \tag{1}$$

for any state Y and where $X_1, \ldots, X_k$ are the states from which Y gets its incoming connections. The Eq. (1) are useful for simulation purposes and also for analysis of properties of the emerging behaviour of such network models. The overall combination function $\mathbf{c}_Y(\ldots)$ for state Y is taken as the weighted average of some of the available basic combination functions $\mathbf{c}_j(\ldots)$ by specified weights $\boldsymbol{\gamma}_{j,Y}$, and parameters $\boldsymbol{\pi}_{1,j,Y}$, $\boldsymbol{\pi}_{2,j,Y}$ of $\mathbf{c}_j(\ldots)$, for Y:

$$\mathbf{c}_Y(V_1, \ldots, V_k) = \frac{\boldsymbol{\gamma}_{1,Y}\, \mathrm{c}_1(V_1, \ldots, V_k) + \ldots + \boldsymbol{\gamma}_{m,Y}\, \mathrm{c}_m(V_1, \ldots, V_k)}{\boldsymbol{\gamma}_{1,Y} + \ldots + \boldsymbol{\gamma}_{m,Y}} \tag{2}$$

Such Eqs. (1) and (2) are hidden in the dedicated software environment that can be used for simulation and analysis; see Treur [104, Ch. 9]. This software environment is freely downloadable from URL https://www.researchgate.net/publication/368775720

Combination functions are similar to the functions used in a static manner in the deterministic Structural Causal Model perspective described, for example, in [69, 71, 122]. However, in the Network-Oriented Modelling approach described here they are used in a dynamic manner. For example, [71, p. 203], denotes nodes by V_i and combination functions by f_i (although he uses a different term for these functions). In the following quote he points at the issue of under-specification concerning aggregation of multiple connections, as in the often used graph representations the specification of combination functions f_i for nodes V_i, is lacking:

> Every causal model M can be associated with a directed graph (…) This graph merely identifies the endogenous and background variables that have a direct influence on each V_i; it does not specify the functional form of f_i. [71, p. 203]

Therefore, to obtain a full specification of a network model in addition to graph representations for connectivity, at least aggregation in terms of combination functions has to be addressed. Indeed, this is is done for the way network models are specified here, to avoid this problem of under-specification. Therefore, aggregation in terms of combination functions is part of the definition of the network structure, in addition to connectivity in terms of connections and their weights and timing in terms of speed factors.

As part of the software environment, a large number >70 of useful basic combination functions are included in a Combination Function Library, and also a facility to easily indicate any function composition of any available basic combination functions in the library. For a few examples of basic combination functions, see Table 1. Here $V_1, \ldots, V_k$ are variables for the single impacts $\boldsymbol{\omega}_{X_i,Y}X_i(t)$. For mathematical analysis of such functions, see Treur [102, 93] or Treur [104, Ch. 11 and 12].

For synchrony detection states in addition combination functions have been applied that work on two time series for a certain sliding time window as input and determine some measure for them, such as a form of correlation for these time series; see, for example, this volume [43, Ch. 14 and 15].

Table 1 Examples of basic combination functions from the library

Name	Notation	Formula	Parameters
Identity	$\mathbf{id}(V)$	V	–
Scaled sum	$\mathbf{ssum}_{\lambda}(V_1, \ldots, V_k)$	$\frac{V_1+\ldots+V_k}{\lambda}$	Scaling factor $\boldsymbol{\lambda}$
Euclidean	$\mathbf{eucl}_{n,\lambda}(V_1, \ldots, V_k)$	$\sqrt[n]{\frac{V_1^n+\ldots+V_k^n}{\lambda}}$	Order $\boldsymbol{n}$ Scaling factor $\boldsymbol{\lambda}$
Advanced logistic	$\mathbf{alogistic}_{\sigma,\tau}(V_1, \ldots, V_k)$	$\left[\frac{1}{1+e^{-\sigma(V_1+\ldots+V_k-\tau)}} - \frac{1}{1+e^{\sigma\tau}}\right](1+e^{-\sigma\tau})$	Steepness $\boldsymbol{\sigma} > 0$ Threshold $\boldsymbol{\tau}$
Stepmod	$\mathbf{stepmod}_{\rho,\delta}(V_1, \ldots, V_k)$	0 if t mod $\boldsymbol{\rho} < \boldsymbol{\delta}$, else 1	Repetition $\boldsymbol{\rho}$ Duration $\boldsymbol{\delta}$

The above characteristics $\boldsymbol{\omega}_{X,Y}$, $\boldsymbol{\gamma}_{j,Y}$, $\boldsymbol{\pi}_{i,j,Y}$, $\boldsymbol{\eta}_Y$ enable to design network models and their dynamics in a declarative manner, based on mathematically defined functions and relations for them. Note that for each state Y, all characteristics $\boldsymbol{\omega}_{X,Y}$, $\boldsymbol{\gamma}_{j,Y}$, $\boldsymbol{\pi}_{i,j,Y}$, $\boldsymbol{\eta}_Y$ mentioned above affect the activation level of Y, as also can be seen from Eqs. (1) and (2). Each of these characteristics has that influence in its own way from a specific role, either for connectivity, for aggregation or for timing.

4 Modeling Adaptivity by Self-Modeling Networks

Realistic network models are usually adaptive: often some of their network characteristics $\boldsymbol{\omega}_{X,Y}$, $\boldsymbol{\gamma}_{j,Y}$, $\boldsymbol{\pi}_{i,j,Y}$, $\boldsymbol{\eta}_Y$ change over time. For example, as discussed in Sect. 2 for mental networks often the connections are assumed to change by Bainian or Hebbian learning [8, 38] and for social networks, it is often assumed that connections between persons change, for example through a bonding by homophily principle [68, 72, 77].

In many cases, adaptive networks are modeled in a hybrid manner by considering two different types of separate models that interact with each other: a network model for the base network and the dynamics of its states, and a separate numerical model for the adaptivity of (some of) the network structure characteristics of the base network. The latter dynamic model is usually specified in a format outside the context of network modeling: in the form of some adaptation-specific procedural or algorithmic programming specification used to run the difference or differential equations underlying the network adaptation process. In contrast, by including *self-models*, introduced in Treur [99], a network-oriented conceptualisation like what was described above, is applied to cover adaptive networks as well using declarative descriptions of adaptation principles based on mathematically defined functions and relations.

4.1 Self-Modeling and Its Magic

The approach using self-models was inspired in a metaphorical sense by the more general idea of self-referencing or 'Mise en abyme', sometimes also called 'the Droste-effect' after the famous Dutch chocolate brand which uses this effect in packaging and advertising of their products since 1904. For some examples, see Fig. 1. For more explanation, see for example

https://en.wikipedia.org/wiki/Mise_en_abyme

https://en.wikipedia.org/wiki/Droste_effect

This effect occurs in art when within an artwork a small copy of the same artwork is included. This can be applied graphically in paintings or photographs, or in sculptures. In these cases, such self-models are static, they cannot change over time. But how about cases in which self-models can change over time? And even more, about

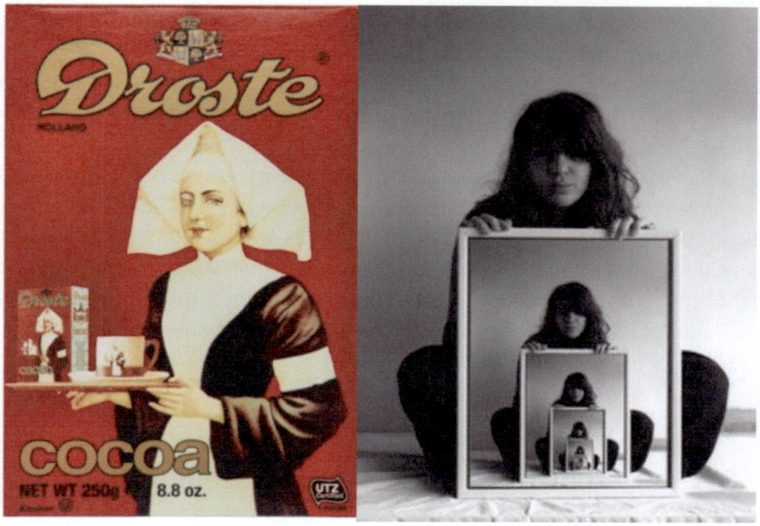

Fig. 1 Two examples of the Mise en abyme or Droste-effect for pictures. http://michel.parpere.pagesperso-orange.fr/pedago/voc/mise%20en%20abyme.htm; https://www.instagram.com/p/CCYmVLMpGPo/

cases in which changes of such self-models also magically change the real situation that is modeled by the self-model?

There are actually cases in art where a self-model can change. More specifically, the idea of self-modeling is sometimes used within literature (story-within-the-story), theater (theater-within-the-theater), or movies (movie-within-the-movie). In each of these cases the self-model extends over time and can change. But can such a change also magically change the real situation modeled by the self-model? As an example, imagine that a writer is writing a novel with as main character (to a certain extent autobiographically) a writer writing a novel. Then some development of this main character while writing a novel may take place because of new insights obtained by the deepened reflection and analysis performed for this novel. In fact the real writer may use the main character as some alter ego that explores possibilities for the real writer in a kind of fantasy or internal simulation format that would be one bridge too far to try them out in real life. In that case the main character in its development may come to new insights and changes that in turn provide new insights for the real writer too, so that he or she can also change accordingly.

This idea of self-models that change over time and where these changes magically lead to similar changes of the real situation has actually been applied to develop the notion of self-modeling network model. Inspiration for this also came from the area of reflection and metalevel architectures from AI; e.g. [19, 97, 98, 118]. Self-modeling networks contain self-models as subnetworks with states with activations that can change over time, and changes of these state activations make the network structure change accordingly.

This concept was introduced in Treur [99] to model adaptation for a network by adding *self-models* to it. It was extended by introducting higher-order self-models in Treur [101, 92] and further extended by illustrations by case studies and mathematical analysis in Treur [104, 105], resulting in multilevel *self-modeling networks*, also called *reified* networks.

The idea works through the addition of new states to the network (called *self-model states*) which represent network characteristics by network states, e.g., connection weights or excitability thresholds. Then the impacts of these characteristics on a state Y as mentioned above can be modelled as impacts from such self-model states. This brings the impacts from these characteristics on a state Y in the standard form of a network model where via connections node activations affect activations of other nodes.

4.2 Different Types of Self-Models and Embedding Them in the Network

More specifically, adding a self-model for a base network can be done in the way that for some of the network structure characteristics $\boldsymbol{\omega}_{X,Y}$, $\boldsymbol{\gamma}_{j,Y}$, $\boldsymbol{\pi}_{i,j,Y}$, $\boldsymbol{\eta}_Y$ for connectivity, aggregation and timing, additional network states $\mathbf{W}_{X,Y}$, $\mathbf{C}_{j,Y}$, $\mathbf{P}_{i,j,Y}$, $\mathbf{H}_Y$ (*self-model states* or *reification states*) are introduced and connected to the other states:

(a) **Connectivity self-model**

- Self-model states $\mathbf{W}_{X,Y}$ are added representing connectivity characteristics, in particular connection weights $\boldsymbol{\omega}_{X,Y}$

(b) **Aggregation self-model**

- Self-model states $\mathbf{C}_{j,Y}$ are added representing aggregation characteristics, in particular combination function weights $\boldsymbol{\gamma}_{j,Y}$
- Self-model states $\mathbf{P}_{i,j,Y}$ are added representing aggregation characteristics, in particular combination function parameters $\boldsymbol{\pi}_{i,j,Y}$

(c) **Timing self-model**

- Self-model states $\mathbf{H}_Y$ are added representing timing characteristics, in particular speed factors $\boldsymbol{\eta}_Y$.

Note that the names using the letters **W**, **C**, **P** and **H** can also be chosen in a different manner. For example, for combination function parameter self-model states often names are used that refer to the specific parameter, for example, **T** for excitability threshold parameter $\boldsymbol{\tau}$, and **M** for persistence parameter $\boldsymbol{\mu}$. The process of adding a self-model to a base network is also called *network reification* and the resulting self-modeling network is sometimes called a *reified network*. If such self-model states are dynamic, they describe adaptive network characteristics. In a graphical 3D-format, such self-model states are depicted at a next level (also called *self-model level* or *reification level*), where the original network is at a *base level.*

Embedding self-model states in the network

Having self-model states to model a specific form of adaptation in a network-oriented manner is only a first step. To fully specify a certain adaptation principle by a self-modeling network, the dynamics of each self-model state itself and its effect on a corresponding target state Y have to be specified in a network-oriented manner by the three general standard types of network characteristics (a) *connectivity*, (b) *aggregation*, and (c) *timing*:

(a) **Specifying connectivity for the self-model states in a self-modeling network**

For the self-model states, their *connectivity* in terms of their incoming and outgoing connections has two different functions:

- **Effectuating its special effect from its specific role**

  The *outgoing downward connections* from the self-model states $\mathbf{W}_{X,Y}$, $\mathbf{C}_{j,Y}$, $\mathbf{P}_{i,j,Y}$, $\mathbf{H}_Y$ to state Y represent the specific impact (their special effect from their specific role) each of these self-model states has on Y. These downward impacts are standard per role, and make that the dynamic values $\mathbf{W}_{X,Y}(t)$, $\mathbf{C}_{j,Y}(t)$, $\mathbf{P}_{i,j,Y}(t)$, $\mathbf{H}_Y(t)$ at t are actually used for the adaptive characteristics of the base network in Eqs. (1) and (2).
- **Indicating the input for the adaptation principle as specified in (b)**

  The *incoming upward or leveled connections* to a self-model state are used to specify the *input* needed for the particular adaptation principle that is addressed.

(b) **Specifying aggregation for the self-model states in a self-modeling network**

For the self-model states, specification of their aggregation characteristics has one main aim:

- **Expressing the adaptation principle by a mathematical function**

  For the *aggregation* of the incoming impacts for a self-model state, provided as indicated in (a), a specific combination function is chosen *to express the adaptation principle* in a declarative mathematical manner.

(c) **Specifying Timing for the Self-Model States in a Self-Modeling Network**

For the self-model states, specification of their timing characteristics has one main aim:

- **Expressing the adaptation speed for the adaptation principle by a value**

  Finally, like any other state, self-model states have their own *timing* in terms of speed factors. These speed factors are used as the means to express the adaptation speed. In most cases adaptation processes take place on slower time scales than the time scale of direct responding at the base level. Also different types of adaptation may take place on different time scales. For example, the model introduced in Hendrikse et al. [40] uses two different adaptation timescales, for short-term adaptation (modeled by **T**-states) and for long-term adaptation (modeled by **W**-states), which both are slower than the timescale for responding at the base level, see also this volume [43], Chapter The Adaptive Interplay of Interpersonal Synchrony, Short-Term Affiliation and Long-Term Bonding: A Multi-Timescale Multi-Adaptive Agent Model [45] and Sect. 1.5 in Chapter Introduction to the Role of Adaptivity in Social Interaction Science [44].

As a base network extended by including a self-model is also a temporal-causal network model itself, as has been shown in Treur [104, Ch. 10], this self-modeling construction can easily be applied iteratively to include self-models of multiple self-modeling (or reification) levels. For example, a second-order self-model state $\mathbf{H}_{\mathbf{W}_{X,Y}}$ can represent the speed factor of first-order self-model state $\mathbf{W}_{X,Y}$ so that the adaptation speed of $\mathbf{W}_{X,Y}$ becomes adaptive. This can provide higher-order adaptive network models and has turned out quite useful to model in a unified form by a second-order self-model metaplasticity: context-sensitive control of adaptation, as discussed in more detail in Sect. 4. Such $\mathbf{H}_{\mathbf{W}}$-states can use yet another time scale.

5 Modeling Adaptation Principles by Self-Model States

In this section, it will be pointed out how the modeling approach for self-modeling network models described in Sects. 3 and 4 can be applied to model adaptation principles discussed in Sect. 2. By making self-model states change over time in a proper manner, a useful method to model any adaptation principle is obtained. This does not only apply to first-order adaptive networks, but also to second-order or higher-order adaptive networks, for example to model control by using second-order self-models.

5.1 Modeling First-Order Adaptation Principles by First-Order Self-Models

As discussed in Sect. 2.2, within neuroscience a distinction is made between synaptic and nonsynaptic (intrinsic) plasticity. Synaptic plasticity concerns plasticity of connectivity in the brain: [8, pp. 117–121, 38, 78]. The nonsynaptic adaptation of intrinsic excitability of (neural states has been addressed in more detail more recently, e.g., [26, 29, 125]. In this chapter for these two types of adaptivity, two examples of first-order adaptation principles are considered: hebbian learning and bonding by homophily for connection weights and excitability modulation for the excitability threshold of states.

Connectivity adaptation

A well-known adaptation principle of the first type (addressing adaptive connectivity) is often explained by, see also Sect. 2.2:

> What fires together, wires together [78, 61]

Within a self-modeling network, this can be modeled by using a *connectivity self-model* based on self-model states $\mathbf{W}_{X,Y}$ representing connection weights $\boldsymbol{\omega}_{X,Y}$. These self-model states need incoming and outgoing connections to let them function within the network. To incorporate the 'firing together' part, for the self-model state's connectivity, incoming connections from the connected states X and Y to $\mathbf{W}_{X,Y}$ are used; see Fig. 2 (upward arrows in blue). These upward connections have weight 1 here. Also a connection from $\mathbf{W}_{X,Y}$ to itself with weight 1 is used to model persistence of the learnt effect; in pictures they are usually left out. In addition, an outgoing connection from $\mathbf{W}_{X,Y}$ to state Y is used to indicate where this self-model state $\mathbf{W}_{X,Y}$ has its causal effect; see in Fig. 2 the (pink) downward arrow. The downward connection indicates that the at the base level the value of $\mathbf{W}_{X,Y}$ is actually used for the connection weight of the connection from X to Y. For the aggregation characteristics of the self-model state $\mathbf{W}_{X,Y}$, one of the options for a learning principle is defined by combination function $\mathbf{hebb}_{\boldsymbol{\mu}}(V_1, V_2, W)$ from Table 2. This adaptation principle and the combination function **hebb** have been used very often, for example, in this volume [43, Ch. 5, 6, 7, 8]. But several other options for combination functions for Hebbian learning are explored and analysed mathematically in Treur [104, Ch. 14], see also Treur [100].

The connectivity adaptation principle for bonding by homophily in social networks can be modeled by the same self-model state $\mathbf{W}_{X,Y}$ and the same network picture as in Fig. 2. But what differs is the combination function for aggregation. An often used example for such a combination function is $\mathbf{slhomo}_{\boldsymbol{\alpha},\boldsymbol{\tau}}(V_1, V_2, W)$ in the second row of Table 2; for example, see this volume [43, Ch. 6–10]. Several alternative options for combination functions describing bonding by homophily are explored and analysed mathematically in Treur [104, Ch. 13], see also Treur [103].

Also, the adaptation principle for synchrony-induced affilation or bonding makes use of such self-model states $\mathbf{W}_{X,Y}$, but this time with incoming connections from

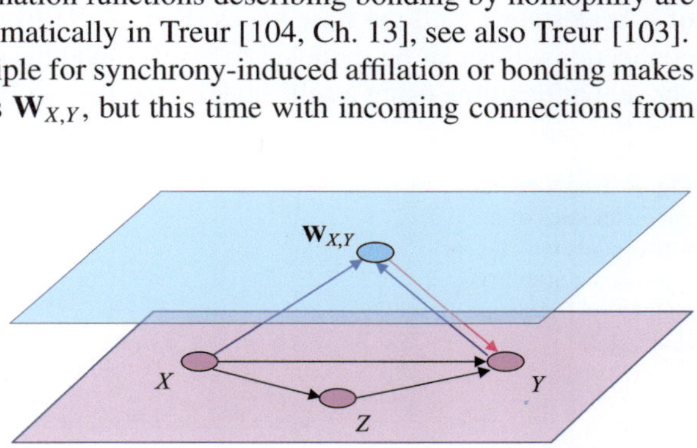

Fig. 2 Connectivity of the self-model state $\mathbf{W}_{X,Y}$ for the Hebbian learning adaptation principle for a connection $X \rightarrow Y$ (see Sect. 2.2) or the bonding by homophily adaptation principle (see Sect. 2.1)

Table 2 Combination functions for connectivity self-model states $\mathbf{W}_{X,Y}$ used to model first-order adaptation and timing self-model states $\mathbf{H}_{\mathbf{W}_{X,Y}}$ used to model second-order adaptation

Self-model states	Combination functions	Parameters and variables
Hebbian learning $\mathbf{W}_{X,Y}$	$\mathbf{hebb}_{\boldsymbol{\mu}}(V_1, V_2, W) = V_1V_2(1 - W) + \mu W$	$\boldsymbol{\mu}$ persistence factor V_1, V_2 activation levels of the connected states; W activation level of state $\mathbf{W}_{X,Y}$
Bonding by homophily $\mathbf{W}_{X,Y}$	$\mathbf{slhomo}_{\boldsymbol{\alpha},\boldsymbol{\tau}}(V_1, V_2, W) = \alpha(\tau - \lvert V_1 - V_2\rvert)W(1 - W) + W$	$\boldsymbol{\alpha}$ modulation factor, $\boldsymbol{\tau}$ tipping point V_1, V_2 activation levels of the connected states; W activation level of $\mathbf{W}_{X,Y}$
Bonding by synchrony $\mathbf{W}_{X,Y}$	$\mathbf{alogistic}_{\boldsymbol{\sigma},\boldsymbol{\tau}}(V_1, \ldots, V_k, W) = \left[\frac{1}{1+e^{-\sigma(V_1+\ldots+V_k+W-\tau)}} - \frac{1}{1+e^{\sigma\tau}}\right](1 + e^{-\sigma\tau})$	$\boldsymbol{\sigma}$ steepness, $\boldsymbol{\tau}$ excitability threshold $V_1, \ldots, V_k$ impacts from states for synchrony detection; W activation level of $\mathbf{W}_{X,Y}$
Control of adaptation $\mathbf{H}_{\mathbf{W}_{X,Y}}$	$\mathbf{alogistic}_{\boldsymbol{\sigma},\boldsymbol{\tau}}(V_1, \ldots, V_k) = \left[\frac{1}{1+e^{-\sigma(V_1+\ldots+V_k-\tau)}} - \frac{1}{1+e^{\sigma\tau}}\right](1 + e^{-\sigma\tau})$	$\boldsymbol{\sigma}$ steepness, $\boldsymbol{\tau}$ excitability threshold $V_1, \ldots, V_k$ impacts from base states representing stimulus exposure context

subjective synchrony detection states and an **alogistic** combination function to aggregate the adaptive effects; see the third row of Table 2. For all this, and further applications, see this volume [43, Ch. 11–19].

Intrinsic excitability adaptation

Although connectivity adaptation is most often addressed in the literature, it more recently has been pointed out that also other characteristics can be made adaptive, such as excitability; see Sect. 2.2. As here the adaptation depends on activation of a base state Y and the base states (here X, Z) from which it gets its incoming connections, this can be modeled in a self-modeling network in a similar form as above, but this time using a self-model state $\mathbf{T}_Y$; see Fig. 3.

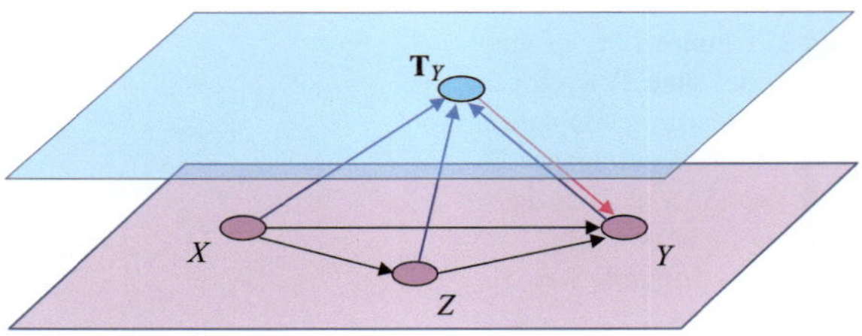

Fig. 3 Connectivity characteristics of a self-model state $\mathbf{T}_Y$ for intrinsic excitability adaptation for a state Y, see (Sect. 2.2)

In this case, based on literature as referred in Sect. 2.2 it is assumed that exposure enhances excitability, which means that it decreases the excitability threshold. To achieve this, for the self-model state $\mathbf{T}_Y$ a monotonically increasing combination function can be used, while the connection weights from X, Y, Z to $\mathbf{T}_Y$ are negative; examples of monotonically increasing combination functions are the logistic sum functions and the Euclidean function (with odd order $\boldsymbol{n}$) from Table 1. In this case, the (pink) downward connection from $\mathbf{T}_Y$ to Y indicates that the value of $\mathbf{T}_Y$ is used for the threshold value of the logistic sum function of base state Y.

Similarly to adaptive connectivity self-model states $\mathbf{W}_{X,Y}$ addressed above, to model bonding by synchrony aggregation self-model states $\mathbf{T}_Y$ with incoming connections from synchrony detection states have been used for adaptive excitability thresholds. For all this, and further applications, see this volume [43, Ch. 12–19].

5.2 *Modeling Second-Order Adaptation Principles for Control of Adaptation by Second-Order Self-Model States*

The two first-order adaptation principles discussed in Sect. 4.1 refer to *plasticity*. It was shown how they can be described by a first-order self-model for connectivity or aggregation characteristics of the base network, in this case in particular for the connection weights (connectivity) or the excitability thresholds (aggregation). As discussed in Sect. 2.3, adaptation (as described, for example, by Hebbian learning), is modulated by metaplasticity accelerating the adaptation process or decelerating or even blocking it. Here we consider one specific second-order adaptation principle for such control of first-order adaptation:

> Adaptation accelerates with increasing stimulus exposure [75, p. 2]

This describes a form of metaplasticity that controls the speed of adaptation (learning rate). This principle can be modeled by a (dynamic) second-order self-model for timing characteristics (speed factors) of a first-order self-model for the first-order adaptation. Such a second-order is based on self-model states $\mathbf{H}_{\mathbf{W}_{X,Y}}$ or $\mathbf{H}_{\mathbf{T}_{X,Y}}$ for adaptive learning speed of any of the two types of (synaptic or intrinsic) learning discussed in Sect. 2.2, or $\mathbf{H}_{\mathbf{WT}_{X,Y}}$ for both types combined. The principle formulated above indicates that the activation level of these second-order self-model states should depend in a monotonically increasing manner on the activation levels of the base states involved: these base states are Y itself and the states X, Z from which Y gets incoming connections. This makes that the connectivity of this timing self-model (for both forms of learning) is as shown in Fig. 4: the (positive, blue) upward connections from the base states X, Y and Z to the self-model state $\mathbf{H}_{\mathbf{WT}_{X,Y}}$ are used to express the part of the principle in (6) referring to ‘stimulus exposure’. For the aggregation, for $\mathbf{H}_{\mathbf{WT}_{X,Y}}$, a Euclidean combination function (with odd order $\boldsymbol{n}$) or a logistic sum combination function can be used to get the monotonic effect as needed;

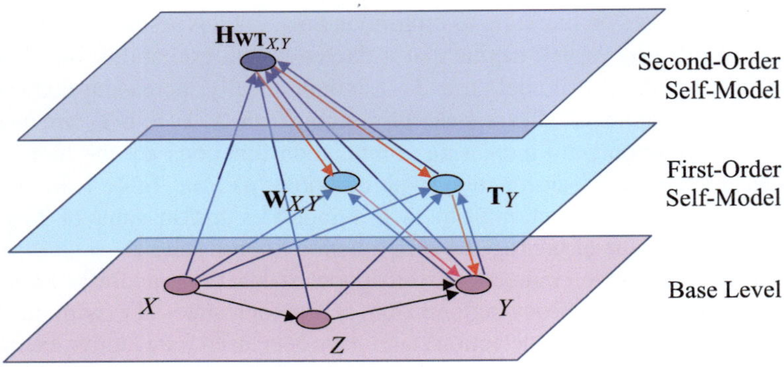

Fig. 4 Connectivity of a second-order self-model state $\mathbf{H_{WT}}_{X,Y}$ for the second-order adaptation principle for control of first-order self-model states $\mathbf{W}_{X,Y}$ and $\mathbf{T}_Y$ for Hebbian learning and Excitability adaptation, see Sect. 2.3

see also Table 2, fourth row. The (blue) upward connections from $\mathbf{W}_{X,Y}$ and $\mathbf{T}_Y$ (with negative and positive weight, respectively) to the self-model state $\mathbf{H_{WT}}_{X,Y}$ indicate a counterbalancing effect that makes that the learning speed is limited depending on a high learnt level as represented by a high value of $\mathbf{W}_{X,Y}$ and a low value of $\mathbf{T}_Y$. The downward (pink) connections from $\mathbf{H_{WT}}_{X,Y}$ to $\mathbf{W}_{X,Y}$ and $\mathbf{T}_Y$ indicate that the value of $\mathbf{H_{WT}}_{X,Y}$ is actually used as speed factor for $\mathbf{W}_{X,Y}$ and $\mathbf{T}_Y$.

This type of second-order timing adaptivity for first-order adaptation based on self-model states $\mathbf{H_W}_{X,Y}$ or $\mathbf{HT}_Y$ has been applied in this volume [43, Ch. 6–8 and 10–19].

6 Discussion

Most material for this chapter comes from Treur [104, 105, 110]. For many domains network models provide an intuitive, declarative way of modeling supported by graphical representations. Connections between nodes in a network can be used as a format to model different types of relations occurring in real-world situations. Once network models are represented within a computing device, these relations can be used for some types of computational processes to generate *within-network dynamics*. However, as relations in real-world domains often change over time themselves too, network models for realistic situations also need facilities to change their structure: *network adaptation*. Self-modeling networks enable to model such changes in network structure relatively easily. They include nodes that represent specific network characteristics of the network itself and in this way form a self-model of part of the network's own structure. By using a self-model, the adaptation of the network structure can be modeled through within-network dynamics of this self-model.

To analyse the scope of applicability of this network-oriented modeling approach based on self-modeling networks following [6], in this volume [43, Ch. 12] it has been shown that any smooth (state-determined) dynamical system as defined in Ashby [6] and also used in Port and Van Gelder [74] can be (re)modeled by a network model. These methods can also be applied to adaptive processes: any description of an adaptation process by a dynamical system or by first-order differential equations can be rewritten as a self-model in a self-modeling network. This is used in Hendrikse et al. [40] and this volume [43, Ch. 12] to show that any adaptive dynamical system can be modeled as a self-modeling network.

More background on the use of an architecture with multiple adaptation or control levels from biological and philosophical perspectives can be found in Bechtel and Bich [10], Bevilacqua et al. [11], Bich and Bechtel [12, 13], Treur [95, 96]. Although in this volume [43] only two adaptation or control levels are used, in other research where the role of environment-induced epigenetic changes for the development of mental disorders such as Schizophrenia, ASPD, Binge eating, Insomnia, Psychopathy, SUD is analysed, a fifth-order control architecture turned out useful; e.g., [7, 15, 27, 37, 59, 76].

Finally, analysis of stationary points and equilibria for self-modeling network models has been addressed in Treur [90, Ch. 12, 104, Ch. 11–14, 94, 107]. Validation and parameter tuning for such network models has been addressed in Treur [90, Ch. 14, 108].

References

1. Abraham, W.C., Bear, M.F.: Metaplasticity: the plasticity of synaptic plasticity. Trends Neurosci. **19**(4), 126–130 (1996)
2. Accetto, M., Treur, J., Villa, V.: An adaptive cognitive-social model for mirroring and social bonding during synchronous joint action. Proc. BICA 2018. Procedia Comput. Sci. **145**, 3–12 (2018)
3. Aizenman, C.D., Linden, D.J.: Rapid, synaptically driven increases in the intrinsic excitability of cerebellar deep nuclear neurons. Nat. Neurosci. **3**, 109–111 (2000)
4. Alejandre-García, T., Kim, S., Pérez-Ortega, J., Yuste, R.: Intrinsic excitability mechanisms of neuronal ensemble formation. eLife **11**, e77470 (2022)
5. Aristotle: Rhetoric. Nichomachean ethics. In: Aristotle in 23 volumes. Rackman (transl.). Harvard University Press, Cambridge (1934)
6. Ashby, W.R.: Design for a Brain, 2nd extended edn. Chapman and Hall, London (1960)
7. Augustinus, F., Hendrikse, S.C.F., Treur, J.: Psychopathic genes: a multi-level adaptive dynamic modelling approach. In: Proceedings of the 8th Computational Methods in Systems and Software Conference, CoMeSySo'24. Lecture Notes in Networks and Systems, vol. 1471, pp. 477–498. Springer Nature (2024)
8. Bain, A.: Mind and Body: The Theories of Their Relation. HS King & Company. Farnborough Eng. Gregg International (1873)
9. Beale, L.S.: Indications of the paths taken by the nerve-currents as they traverse the caudate nerve-cells of the spinal cord and encephalon. In: Proceedings of the Royal Society of London, vol. 13, pp. 386–392 (1864). https://doi.org/10.1098/rspl.1863.0084
10. Bechtel, W., Bich, L.: Organisms need mechanisms; mechanisms need organisms. In: Cordovil, J.L., Santos, G., Vecchi, D. (eds.) New Mechanism: Explanation, Emergence and

Reduction. History, Philosophy and Theory of the Life Sciences, vol. 35, pp. 85–108. Springer Nature (2024)
11. Bevilacqua, A., Wilkinson, S.J., Dimelow, R., Murabito, E., Rehman, S., Nardelli, M., et al.: Vertical systems biology: from DNA to flux and back. In: Hetherington, A., Grierson, C. (eds.) Practical Systems Biology, pp. 79–106. Taylor & Francis (2008)
12. Bich, L., Bechtel, W.: Control mechanisms: explaining the integration and versatility of biological organisms. Adapt. Behav. **30**(5), 389–407 (2022)
13. Bich, L., Bechtel, W.: Organization needs organization: understanding integrated control in living organisms. Stud. Hist. Philos. Sci. **93**, 96–106 (2022)
14. Blankendaal, R., Parinussa, S., Treur, J.: A temporal-causal modelling approach to integrated contagion and network change in social networks. In: Proceedings of the Twenty-second European Conference on AI, ECAI'16, pp. 1388–1396. IOS Press (2016)
15. Blei, Z.A., Hendrikse, S.C.F., Treur, J.: Modeling epigenetic modification due to childhood abuse and its relation to ASPD: a fifth-order adaptive network model. In: Proceedings of the 15th Biologically Inspired Cognitive Architectures for AI Conference, BICA*AI'24. Studies in Computational Intelligence, vol. 477, pp. 59–70. Springer Nature (2024)
16. Boot, N., Baas, M., Van Gaal, S., Cools, R., De Dreu, C.K.W.: Creative cognition and dopaminergic modulation of fronto-striatal networks: integrative review and research agenda. Neurosci. Biobehav. Rev. **78**, 13–23 (2017)
17. Bott, H.: Observation of play activities in a nursery school. Genet. Psychol. Monogr. **4**, 44–88 (1928)
18. Bowen, K.A., Kowalski, R.: Amalgamating language and meta-language in logic programming. In: Clark, K., Tarnlund, S. (eds.) Logic Programming, pp. 153–172. Academic Press, New York (1982)
19. Brazier, F.M.T., Treur, J.: Compositional modelling of reflective agents. Int. J. Hum.-Comput. Stud. **50**(5), 407–431 (1999)
20. Budding, D., Doornkamp, S., Treur, J.: A second-order adaptive mental network model relating dreaming to creativity. Cogn. Syst. Res. **80**, 37–49 (2023)
21. Bössenecker, G., Mreijen, P., Treur, J.: A second-order adaptive network model for connecting social media platforms and users through bonding by faked homophily. Procedia Comput. Sci. **213**, 32–40 (2022)
22. Bössenecker, G., Mreijen, P., Treur, J.: Connecting social media platforms and users through bonding by faked homophily: a second-order adaptive network model. In: Hen-drikse, S.C.F., Treur, J., Koole, S.L. (eds.) New Analysis and Modeling Directions in Social Interaction Science: Emergent Multimodal Interpersonal Synchrony, Affiliation, and Bonding, Ch. 9. Springer Nature (2025)
23. Canbaloğlu, G., Treur, J., Wiewiora, A.: Computational modeling of the role of leadership style for its context-sensitive control over multilevel organisational learning. In: Proceedings ICICT'22. Lecture Notes in Networks and Systems, vol. 447, pp. 223–239. Springer Nature (2023a)
24. Canbaloğlu, G., Treur, J., Wiewiora, A. (eds.): Computational Modeling of Multilevel Organisational Learning and its Control Using Self-Modeling Network Models, this volume. Springer Nature (2023b)
25. Cardno, J.A.: Bain and physiological psychology. Aust. J. Psychol. **7**(2), 108–120 (1955). https://doi.org/10.1080/00049535508256103
26. Chandra, N., Barkai, E.: A non-synaptic mechanism of complex learning: modulation of intrinsic neuronal excitability. Neurobiol. Learn. Mem. **154**, 30–36 (2018)
27. Călin, R.S., Hendrikse, S.C.F., Treur, J.: Environmental and Epigenetic Influences in Binge Eating: An Adaptive Network Modeling Approach. To appear (2024)
28. Daoudal, G., Debanne, D.: Long-term plasticity of intrinsic excitability: learning rules and mechanisms. Learn. Mem. **10**, 456–465 (2003)
29. Debanne, D., Inglebert, Y., Russier, M.: Plasticity of intrinsic neuronal excitability. Curr. Opin. Neurobiol. **54**, 73–82 (2019)

30. Demers, F.N., Malenfant, J.: Reflection in logic, functional and object oriented programming: a short comparative study. In: IJCAI'95 Workshop on Reflection and Meta-Level Architecture and their Application in AI, pp. 29–38 (1995)
31. Eler, E., Treur, J., Hendrikse, S.C.F., Donker, T., Koole, S. L.: An adaptive network model for interpersonal emotion regulation in multimodal human-bot interaction. In: Proceedings of ICCCI'24. Lecture Notes in AI, vol. 14811, pp. 67–79. Springer Nature (2024)
32. Eler, E., Treur, J., Hendrikse, S.C.F., Donker, T., Koole, S.L.: Interpersonal emotion regulation via multimodal interaction with a bonding bot: an adaptive network model. In: Hendrikse, S.C.F., Treur, J., Koole, S.L. (eds.) New Analysis and Modeling Directions in Social Interaction Science: Emergent Multimodal Interpersonal Synchrony, Affiliation, and Bonding, Ch. 19. Springer Nature (2025)
33. Freeman, L.: Some antecedents of social network analysis. Connections **19**, 39–42 (1996)
34. Garcia, R.: Stress, metaplasticity, and antidepressants. Curr. Mol. Med. **2**, 629–638 (2002)
35. Hansel, C.: Contiguity in perception: origins in cellular associative computations. Trends Neurosci. **47**(3), 170–180 (2024)
36. Hansel, C., Yuste, R.: Neural ensembles: role of intrinsic excitability and its plasticity. Front. Cell. Neurosci. **18**, 1440588 (2024)
37. Hassouna, S., Hendrikse, S.C.F., Treur, J.: The Dynamics of Epigenetic Influence in Insomnia: A Higher-Order Adaptive Modeling Perspective. In: Itthipuripat, S., Ascoli, G.A., Li, A., Pat, N., Kuai, H. (eds) Brain Informatics. Proceedings BI 2024. Lecture Notes in Computer Science, vol. 15541, pp. 312–325. Springer Nature (2025)
38. Hebb, D.O.: The Organisation of Behavior: A Neuropsychological Theory. Wiley (1949)
39. Heijmans, P., van Stijn, J., Treur, J.: Modeling cultural segregation of the queer community through an adaptive social network model. In: Proceedings of the Fourth International Congress on Information and Communication Technology, ICICT'19, vol. 2. Advances in Intelligent Systems and Computing, vol. 1027, pp. 233–248. Springer Singapore (2019)
40. Hendrikse, S.C.F., Treur, J., Koole, S.L.: Modeling emerging interpersonal synchrony and its related adaptive short-term affliation and long-term bonding: a second-order multi-adaptive neural agent model. Inter. J. Neural Syst. **33**(7), 2350038 (2023)
41. Hendrikse, S.C.F., Treur, J., Koole, S.L.: Relationship-specific and relationship-independent behavioural adaptivity in affiliation and bonding: a multi-adaptive dynamical systems approach. Cogn. Syst. Res. **84**, 101182 (2024)
42. Hendrikse, S.C.F., Treur, J., Wilderjans, T.F., Dikker, S., Koole, S.L.: On becoming in sync with yourself and others: an adaptive agent model for how persons connect by detecting intrapersonal and interpersonal synchrony. Hum.-Centric Intell. Syst. **3**(2), 123–146 (2023)
43. Hendrikse, S.C.F., Treur, J., Koole, S.L. (eds.): New Analysis and Modeling Directions in Social Interaction Science: Emergent Multimodal Interpersonal Synchrony, Affiliation, and Bonding. Springer Nature (2025a)
44. Hendrikse, S.C.F., Treur, J., Koole, S.L.: Introduction to the role of adaptivity in social interaction science. In: Hendrikse, S.C.F., Treur, J., Koole, S.L. (eds.) New Analysis and Modeling Directions in Social Interaction Science: Emergent Multimodal Interpersonal Synchrony, Affiliation, and Bonding, Ch. 1. Springer Nature (2025b)
45. Hendrikse, S.C.F., Treur, J., Koole, S.L.: The adaptive interplay of interpersonal synchrony, short-term affiliation and long-term bonding: a multi-timescale multi-adaptive agent model. In: Hendrikse, S.C.F., Treur, J., Koole, S.L. (eds.) New Analysis and Modeling Directions in Social Interaction Science: Emergent Multimodal Interpersonal Synchrony, Affiliation, and Bonding, Ch. 12. Springer Nature (2025c)
46. Hendrikse, S.C.F., Treur, J., Koole, S.L.: A multi-timescale multi-adaptive dynamical system approach for relationship-specific and relationship-independent adaptivity in affiliation and bonding. In: Hendrikse, S.C.F., Treur, J., Koole, S.L. (eds.) New Analysis and Modeling Directions in Social Interaction Science: Emergent Multimodal Interpersonal Synchrony, Affiliation, and Bonding, Ch. 13. Springer Nature (2025d)

47. Hendrikse, S.C.F., Treur, J., Koole, S.L.: Discussion and concluding remarks. In: Hendrikse, S.C.F., Treur, J., Koole, S.L. (eds.) New Analysis and Modeling Directions in Social Interaction Science: Emergent Multimodal Interpersonal Synchrony, Affiliation, and Bonding, Ch. 21. Springer Nature (2025e)
48. Hendrikse, S.C.F., Treur, J., Wilderjans, T.F., Dikker, S., Koole, S.L.: On the same wavelengths: emergence of multiple synchronies among multiple agents. In: Proceedings of the 22nd International Workshop on Multi-Agent-Based Simulation, MABS'21. Lecture Notes in AI, vol. 13128, pp. 57–71. Springer Nature (2022a)
49. Hendrikse, S.C.F., Treur, J., Wilderjans, T.F., Dikker, S., Koole, S.L.: Becoming attuned to each other over time: a computational neural agent model for the role of time lags in subjective synchrony detection and related behavioral adaptivity. In: Proceeedings of the 15th International Conference on Brain Informatics, BI'22. Lecture Notes in AI, vol. 13406, pp. 369–383. Springer Nature (2022b)
50. Hendrikse, S.C.F., Treur, J., Wilderjans, T.F., Dikker, S., Koole, S.L.: In sync with yourself and with others: detection of intra- and interpersonal synchrony within an adaptive agent model. In: Face2face: Advancing the Science of Social Interaction. Royal Society, London (2022c). https://www.researchgate.net/publication/358964043
51. Hendrikse, S.C.F., Treur, J., Wilderjans, T.F., Dikker, S., Koole, S.L.: Switching in and out of sync: a controlled adaptive network model of transition dynamics in the effects of interpersonal synchrony on affiliation. In: Proceedings of the 11th International Conference on Complex Networks and their Applications. Studies in Computational Intelligence, vol. 1078, pp. 81–95. Springer Nature Switzerland, Cham (2022d)
52. Hendrikse, S.C.F., Treur, J., Wilderjans, T.F., Dikker, S., Koole, S.L.: The emergence of multimodal interpersonal synchrony during social interaction: a network model. In: Hendrikse, S.C.F., Treur, J., Koole, S.L. (eds.) New Analysis and Modeling Directions in Social Interaction Science: Emergent Multimodal Interpersonal Synchrony, Affiliation, and Bonding, Ch. 3. Springer Nature (2025a)
53. Hendrikse, S.C.F., Treur, J., Wilderjans, T.F., Dikker, S., Koole, S.L.: Growing connectedness through synchrony: the role of subjective analysis of intrapersonal synchrony and interpersonal synchrony within an adaptive agent model. In: Hendrikse, S.C.F., Treur, J., Koole, S.L. (eds.) New Analysis and Modeling Directions in Social Interaction Science: Emergent Multimodal Interpersonal Synchrony, Affiliation, and Bonding, Ch. 11. Springer Nature (2025b)
54. Hendrikse, S.C.F., Treur, J., Wilderjans, T.F., Dikker, S., Koole, S.L.: The role of timing in synchrony detection and social interaction adaptivity: including time lags in a second-order adaptive agent model. In: Hendrikse, S.C.F., Treur, J., Koole, S.L. (eds.) New Analysis and Modeling Directions in Social Interaction Science: Emergent Multimodal Interpersonal Synchrony, Affiliation, and Bonding, Ch. 13. Springer Nature (2025c)
55. Hendrikse, S.C.F., Treur, J., Wilderjans, T.F., Dikker, S., Koole, S.L.: The effect of transition dynamics in synchrony on social interaction adaptivity: a multi-adaptive network model. In: Hendrikse, S.C.F., Treur, J., Koole, S.L. (eds.) New Analysis and Modeling Directions in Social Interaction Science: Emergent Multimodal Interpersonal Synchrony, Affiliation, and Bonding, Ch. 14. Springer Nature (2025d)
56. Hove, M.J., Risen, J.L.: It's all in the timing: interpersonal synchrony increases affiliation. Soc. Cogn. **27**(6), 949–960 (2009)
57. Hsieh, C.S., König, M.D., Liu, X.: A structural model for the coevolution of networks and behavior. Rev. Econ. Stat. **104**(2), 355–367 (2022)
58. Hubbard R.M.: A method of studying spontaneous group formation. In: Thomas, D.S. (ed.) Some New Techniques for Studying Social Behavior, pp. 76–85. Child Dev. Monogr (1929)
59. Jaganjac, I., Hendrikse, S.C.F., Treur, J.: An adaptive dynamical system model for development of schizophrenia: epigenetics and false memories. Cogn. Syst. Res. 88, 101288 (2024)
60. Kappert, C., Rus, R., Treur, J.: On the emergence of segregation in society: network-oriented analysis of the effect of evolving friendships. In: Proceedings of the 10th International

Conference on Computational Collective Intelligence, ICCCI'18. Lecture Notes in Computer Science, vol. 11055, pp. 178–191. Springer International Publishing (2018)
61. Keysers, C., Gazzola, V.: Hebbian learning and predictive mirror neurons for actions, sensations and emotions. Philos. Trans. R. Soc. B: Biol. Sci. **369**(1644), 20130175 (2014)
62. King, A., Smith, D., Webb, B.: Minimal specialization: coevolution of network structure and dynamics. Phys. A **651**, 130000 (2024)
63. Lazarsfeld, P.F., Merton, R.K.: Friendship as a social process: a substantive and methodological analysis. In: Berger, M. (ed.) Freedom and Control in Modern Society, pp. 18–66. Van Nostrand, New York (1954)
64. Lisman, J., Cooper, K., Sehgal, M., Silva, A.J.: Memory formation depends on both synapse-specific modifications of synaptic strength and cell-specific increases in excitability. Nat. Neurosci. **21**, 309–314 (2018)
65. Liu, J., Huang, S., Aden, N.M., Johnson, N.F., Song, C.: Emergence of polarization in coevolving networks. Phys. Rev. Lett. **130**(3), 037401 (2023)
66. Luo, S., Du, Y., Liu, P., Xuan, Z., Wang, Y.: A study on coevolutionary dynamics of knowledge diffusion and social network structure. Expert Syst. Appl. **42**(7), 3619–3633 (2015)
67. Magerl, W., Hansen, N., Treede, R.D., Klein, T.: The human pain system exhibits higher-order plasticity (metaplasticity). Neurobiol. Learn. Mem. **154**, 112–120 (2018)
68. McPherson, M., Smith-Lovin, L., Cook, J.M.: Birds of a feather: homophily in social networks. Annu. Rev. Sociol. **27**, 415–444 (2001)
69. Mooij, J.M., Janzing, D., Schölkopf, B.: From Differential equations to structural causal models: the deterministic case. In: Nicholson, A., Smyth, P. (eds.) Proceedings of the 29th Annual Conference on Uncertainty in Artificial Intelligence (UAI-13), pp. 440–448. AUAI Press (2013)
70. Mukeriia, Y., Treur, J., Hendrikse, S.C.F.: A multi-adaptive network model for human hebbian learning, synchronization and social bonding based on adaptive homophily. Cogn. Syst. Res. **84**, 101187 (2024)
71. Pearl, J.: Causality. Cambridge University Press (2000)
72. Pearson, M., Steglich, C., Snijders, T.: Homophily and assimilation among sport-active adolescent substance users. Connections **27**(1), 47–63 (2006)
73. Plato: Laws. Plato in Twelve Volumes, vol. 11. Bury translator. Harvard University Press, Cambridge (1968)
74. Port, R.F., Van Gelder, T.: Mind as Motion: Explorations in the Dynamics of Cognition. MIT Press, Cambridge, MA (1995)
75. Robinson, B.L., Harper, N.S., McAlpine, D.: Meta-adaptation in the auditory midbrain under cortical influence. Nat. Commun. **7**, 13442 (2016)
76. Schlöder, D., Hendrikse, S.C.F., Treur, J.: Higher-order adaptive dynamical system modeling of the role of epigenetics in substance use disorder. In: Proceedings of the 15th Biologically Inspired Cognitive Architectures for AI Conference, BICA*AI'24. Studies in Computational Intelligence, vol. 477, pp. 358–368. Springer Nature (2024)
77. Sharpanskykh, A., Treur, J.: Modelling and analysis of social contagion in dynamic networks. Neurocomputing **146**, 140–150 (2014)
78. Shatz, C.J.: The developing brain. Sci. Am. **267**, 60–67 (1992). https://doi.org/10.1038/scientificamerican0992-60
79. Sjöström, P.J., Rancz, E.A., Roth, A., Hausser, M.: Dendritic excitability and synaptic plasticity. Physiol. Rev. **88**, 769–840 (2008)
80. Snijders, T., Steglich, C., Schweinberger, M.: Modeling the coevolution of networks and behavior. In: van Montfort, K., Oud, J., Satorra, A. (eds.) Longitudinal Models in the Behavioral and Related Sciences, pp. 41–71. Taylor and Francis (2017)
81. Sterling, L., Beer, R.: Metainterpreters for expert system construction. J. Log. Program. **6**, 163–178 (1989)
82. Sterling, L., Shapiro, E.: The Art of Prolog, Ch. 17, pp. 319–356. MIT Press (1996)
83. Su, J., Liu, B., Li, Q., Ma, H.: Coevolution of opinions and directed adaptive networks in a social group. J. Artif. Soc. Soc. Simul. **17**(2) (2014)

84. Tarr, B., Launay, J., Dunbar, R.I.M.: Silent disco: dancing in synchrony leads to elevated pain thresholds and social closeness. Evol. Hum. Behav. **37**(5), 343–349 (2016)
85. Tichelaar, C., Treur, J.: Network-oriented modeling of the interaction of adaptive joint decision making, bonding and mirroring. In: Fagan, D., Martín-Vide, C., O'Neill, M., Vega-Rodríguez, M.A. (eds) Proceedings of the 7th International Conference on Theory and Practice of Natural Computing, TPNC'18. Lecture Notes in Computer Science, vol. 11324, pp. 328–343. Springer Publishers (2018)
86. Tichelaar, C., Treur, J.: On the adaptive interplay of mirroring and bonding by homophily in joint decision making: a second-order adaptive network model. In: Proceedings of the 8th International Conference on Computer-Human Interaction Research and Applications, CHIRA'24. Communications in Computer and Information Science, vol 2371, pp. 280–298. Springer Nature (2025)
87. Tichelaar, C., Treur, J.: Bonding by homophily and adaptive mirroring in joint decision making: a second-order adaptive network model. In: Hendrikse, S.C.F., Treur, J., Koole, S.L. (eds.) New Analysis and Modeling Directions in Social Interaction Science: Emergent Multimodal Interpersonal Synchrony, Affiliation, and Bonding, Ch. 6. Springer Nature (2025)
88. Titley, H.K., Brunel, N., Hansel, C.: Toward a neurocentric view of learning. Neuron **95**, 19–32 (2017)
89. Treur, J.: Temporal factorisation: a unifying principle for dynamics of the world and of mental states. Cogn. Syst. Res. **8**(2), 57–74 (2007)
90. Treur, J.: Network-Oriented Modeling: Addressing Complexity of Cognitive, Affective and Social Interactions. Springer Publishers (2016)
91. Treur, J.: Mathematical analysis of the emergence of communities based on coevolution of social contagion and bonding by homophily. Appl. Netw. Sci. **4**(1), 39 (2019)
92. Treur, J.: Modeling higher-order adaptivity of a network by multilevel network reification. Netw. Sci. **8**, S110–S144 (2020)
93. Treur, J.: Analysis of a network's asymptotic behavior via its structure involving its strongly connected components. Netw. Sci. **8**(S1), S82–S109 (2020)
94. Treur, J.: On the dynamics and adaptivity of mental processes: relating adaptive dynamical systems and self-modeling network models by mathematical analysis. Cogn. Syst. Res. **70**, 93–100 (2021)
95. Treur, J.: On structure, dynamics, and adaptivity for biological and mental processes: a higher-order adaptive dynamical system modeling perspective. In: Proceedings of the Annual Meeting of the Cognitive Science Society, vol. 46 (2024a)
96. Treur, J.: Using multilevel temporal factorisation to analyse structure and dynamics for higher-order adaptive and evolutionary processes. In: International Conference on Computational Collective Intelligence. Lecture Notes in AI, vol 14811, pp. 378–392. Springer Nature Switzerland, Cham (2024b)
97. Treur, J.: On the use of reflection principles in modelling complex reasoning. Int. J. Intell. Syst. **6**(3), 277–294 (1991)
98. Treur, J.: Temporal semantics of meta-level architectures for dynamic control of reasoning. In: Fribourg, L., Turini, F. (eds.) Logic Program Synthesis and Transformation—Meta-Programming in Logic. Proceedings META LOPSTR 1994. Lecture Notes in Computer Science, vol. 883, pp. 353–376. Springer, Berlin, Heidelberg (1994)
99. Treur, J.: Network reification as a unified approach to represent network adaptation principles within a network. In: Theory and Practice of Natural Computing: Proceeedings of the 7th International Conference TPNC'18. Lecture Notes in Computer Science, vol. 11324, pp. 344–358. Springer International Publishing (2018a)
100. Treur, J.: Relating an adaptive network's structure to its emerging behaviour for Hebbian learning. In: Martín-Vide, C., Vega-Rodríguez, M.A., Fagan, D., O'Neill, M. (eds.) Theory and Practice of Natural Computing: 7th International Conference, TPNC 2018, Proceedings. Lecture Notes in Computer Science, vol. 11324. pp. 359–373. Springer Publishers (2018b)
101. Treur, J.: Multilevel network reification: representing higher order adaptivity in a network. In: Aiello, L., Cherifi, C., Cherifi, H., Lambiotte, R., Lió, P., Rocha, L. (eds) Complex

Networks and Their Applications VII: Volume 1 Proceedings The 7th International Conference on Complex Networks and Their Applications, COMPLEX NETWORKS'18. Studies in Computational Intelligence, vol. 812, pp. 635–651. Springer International Publishing (2019a)

102. Treur, J.: Relating emerging network behaviour to network structure. In: Complex Networks and Their Applications VII: Volume 1 Proceedings of the 7th International Conference on Complex Networks and Their Applications COMPLEX NETWORKS'18. Studies in Computational Intelligence, vol. 812, pp. 619–634. Springer International Publishing (2019c)
103. Treur, J.: Relating an adaptive social network's structure to its emerging behaviour based on homophily. In: Complex Networks and Their Applications VII: Volume 2 Proceedings of the 7th International Conference on Complex Networks and Their Applications COMPLEX NETWORKS 2018. Studies in Computational Intelligence, vol. 813, pp. 341–356. Springer International Publishing (2019d)
104. Treur, J.: Network-Oriented Modeling for Adaptive Networks: Designing Higher-order Adaptive Biological, Mental and Social Network Models. Springer Nature Cham, Cham, Switzerland (2020b)
105. Treur, J.: Modeling multi-order adaptive processes by self-modeling networks. In: Tallon-Ballesteros, A.J., Chen, C.-H. (eds.) Machine Learning and Intelligent Systems: Proceedings of MLIS 2020. Frontiers in Artificial Intelligence and Applications, vol. 332, pp. 206–217. IOS Press (2020c)
106. Treur, J.: With a little help: a modeling environment for self-modeling network models. In: Treur, J., van Ments, L. (eds.) Mental Models and their Dynamics, Adaptation and Control: A Self-Modeling Network Modeling Approach, Ch. 17. Springer Nature Switzerland (2022a)
107. Treur, J.: Where is this leading me: stationary point and equilibrium analysis of self-modeling network models. In: Treur, J., van Ments, L. (eds.) Mental Models and their Dynamics, Adaptation and Control: A Self-Modeling Network Modeling Approach, Ch. 18. Springer Nature Switzerland (2022b)
108. Treur, J.: Does this suit me: validation of self-modeling network models by parameter tuning. In: Treur, J., van Ments, L. (eds.) Mental Models and their Dynamics, Adaptation and Control: A Self-Modeling Network Modeling Approach, Ch. 19. Springer Nature Switzerland (2022c)
109. Treur, J.: Controlled social network adaptation: subjective elements in an objective social world. In: Yang, X.S., Sherratt, S., Dey, N., Joshi, A. (eds.) Proceedings of Sixth International Congress on Information and Communication Technology, ICICT'21. Lecture Notes in Networks and Systems, vol. 235, pp. 263–274. Springer Nature (2022e)
110. Treur, J.: Modeling dynamics, adaptivity and control by self-modeling networks. In: Canbaloğlu, G., Treur, J., Wiewiora, A. (eds.) Computational Modeling of Multilevel Organisational Learning and Its Control Using Self-modeling Network Models, Ch. 3, pp. 33–50. Springer Nature (2023)
111. Treur, J.: A second-order adaptive network model for bonding based on subjectively perceived homophily. In: Hendrikse, S.C.F., Treur, J., Koole, S.L. (eds.) New Analysis and Modeling Directions in Social Interaction Science: Emergent Multimodal Interpersonal Synchrony, Affiliation, and Bonding, Ch. 7. Springer Nature (2025b)
112. Vacharkulksemsuk, T., Fredrickson, B.L.: Strangers in sync: achieving embodied rapport through shared movements. J. Exp. Soc. Psychol. **48**(1), 399–402 (2012)
113. Valdesolo, P., DeSteno, D.: Synchrony and the social tuning of compassion. Emotion **11**(2), 262–266 (2011)
114. Valdesolo, P., Ouyang, J., DeSteno, D.: The rhythm of joint action: synchrony promotes cooperative ability. J. Exp. Soc. Psychol. **46**(4), 693–695 (2010)
115. van den Beukel, S., Goos, S.H., Treur, J.: An adaptive temporal-causal network model for social networks based on the homophily and more-becomes-more principle. Neurocomputing **338**, 361–371 (2019)
116. Vervoordeldonk, D., Treur, J., Hendrikse, S.C.F., Roelofsma, P.H.M.P.: An adaptive network model of the interplay of synchrony and adaptation for reliving memory and religious experience through multimodal interaction. In: Silhavy, R., Silhavy, P. (eds) Artificial Intelligence and System Engineering. Proceedings CoMeSySo 2024. Lecture Notes in Networks and Systems, vol. 1490, pp. 59–93. Springer Nature (2024)

117. Wellman, B.: The school child's choice of companions. J. Educ. Res. **14**, 126–132 (1929)
118. Weyhrauch, R.W.: Prolegomena to a theory of mechanized formal reasoning. Artif. Intell. **13**(1–2), 133–170 (1980)
119. Wilkes, A.L., Wade, N.J.: Bain on neural networks. Brain Cogn. **33**(3), 295–305 (1997). https://doi.org/10.1006/brcg.1997.0869
120. Williams, A.H., O'Leary, T., Marder, E.: Homeostatic regulation of neuronal excitability. Scholarpedia **8**, 1656 (2013)
121. Wiltermuth, S.S., Heath, C.: Synchrony and cooperation. Psychol. Sci. **20**(1), 1–5 (2009)
122. Wright, S.: Correlation and causation. J. Agric. Res. **20**, 557–585 (1921)
123. Xu, Y., Hendrikse, S.C.F., Treur, J., Roelofsma, P.H.M.P.: Adaptive network modeling for joint action and memory recall for elderly by detecting interpersonal synchrony. Cogn. Syst. Res. **88**, 101280 (2024)
124. Zhang, W., Linden, D.J.: The other side of the engram: experience-driven changes in neuronal intrinsic excitability. Nat. Rev. Neurosci. **4**, 885–900 (2003)
125. Zhang, A., Li, X., Gao, Y., Niu, Y.: Event-driven intrinsic plasticity for spiking convolutional neural networks. IEEE Trans. Neural Netw. Learn. Syst. (2021). https://doi.org/10.1109/tnnls.2021.3084955

Agent-Based Analysis of Social Interaction and the Emergence of Interpersonal Synchrony

In this part, first in "The Emergence of Multimodal Interpersonal Synchrony During Social Interaction: A Network Model", it is shown by agent-based modeling and simulation how multimodal interaction between agents can lead to emerging multimodal synchrony of their movements, affects and verbal actions. The behavior displayed by the simulations was consistent with theory and empirical evidence from the psychological and neuroscience literature. Next, in "Interpersonal Emotion Regulation through Attunement of Emotions in Human-Bot Interaction: an Agent-Based Network Model", it is shown how attuning emotions to each other during social interaction can lead to a form of interpersonal emotion regulation. This study analyses the effects of coregulation of emotions in the interaction between a distressed human and a bot. An emotion contagion process impacts the emotional system, which initially increases the distress level of the bot due to the influence of the human's emotions. Configuring the bot with optimal emotion regulation capabilities decreases its distress levels as it can self-regulate its emotions, which in turn affects the human's emotions, regulating them by the emotion contagion process. Finally, in "Learning to Synchronize: Adaptive Joint Decision-Making by Virtual Agents" it is shown how emerging interpersonal synchrony can go hand in hand with adaptivity leading to more interpersonal coordination. The aim of this chapter was (1) to develop an adaptive computational model for the type of synchrony that can occur in social interaction for two persons modeled as agents, and (2) to visualize the two persons by avatars as virtual agents during their social interaction. Programming in the Unity 3D engine was done to virtualize the social interaction process between virtual agents during a therapeutic session as example scenario.

The Emergence of Multimodal Interpersonal Synchrony During Social Interaction: A Network Model

Sophie C. F. Hendrikse, Jan Treur, Tom F. Wilderjans, Suzanne Dikker, and Sander L. Koole

Abstract People spontaneously synchronize their mental states and behavioral actions when they interact. This chapter models general mechanisms that can lead to the emergence of interpersonal synchrony by multiple agents with internal cognitive and affective states. In our simulations, one agent was exposed to a repeated stimulus and the other agent started to synchronize consecutively its movements, affects, conscious emotions and verbal actions with the exposed agent. The behavior displayed by the agents was consistent with theory and empirical evidence from the psychological and neuroscience literature. These results shed new light on the

---

S. C. F. Hendrikse
Department of Methodology and Statistics, Tilburg School of Social and Behavioral Sciences, Tilburg University, Tilburg, The Netherlands
e-mail: S.C.F.Hendrikse@tilburguniversity.edu

J. Treur (✉)
Social AI Group, Department of Computer Science, Vrije Universiteit Amsterdam, Amsterdam, The Netherlands
e-mail: j.treur@vu.nl

T. F. Wilderjans
Methodology and Statistics Research Unit, Institute of Psychology, Faculty of Social and Behavioural Sciences, Leiden University, Leiden, The Netherlands

Research Group of Quantitative Psychology and Individual Differences, Faculty of Psychology and Educational Sciences, Katholieke Universiteit (KU) Leuven, Leuven, Belgium

T. F. Wilderjans
e-mail: t.f.wilderjans@leidenuniv.nl

S. Dikker
NYU—Max Planck Center for Language, Music and Emotion, New York University, New York, USA

S. L. Koole
Amsterdam Emotion Regulation Lab, Department of Clinical Psychology, Vrije Universiteit Amsterdam, Amsterdam, The Netherlands
e-mail: s.l.koole@vu.nl

S. C. F. Hendrikse et al. (eds.), *New Analysis and Modeling Directions in Social Interaction Science*, Studies in Systems, Decision and Control 614,
https://doi.org/10.1007/978-3-031-99968-0_3

emergence of interpersonal synchrony in a wide variety of settings, from close relationships to psychotherapy. Moreover, the present work could provide a basis for future development of socially responsive virtual agents.

Keywords Social agent model · Emergent synchrony patterns · Social simulations · In-Sync model

1 Introduction

People spontaneously synchronize their movements, affective responses, and verbal actions when they interact with each other. Such interpersonal synchrony has been related to a variety of positive outcomes in social settings. For instance, Miles et al. [31] found that synchrony was the most pronounced for minimal groups of people who were most divergent in terms of their artistic taste, suggesting that synchrony might serve as a tool to overcome/bridge social distance and intergroup differences. Elevated levels of movement synchrony have further been shown to foster social affiliation [10], cooperation [57] and compassion [51]. Moreover, interpersonal synchrony has been suggested to be a key component of a good therapeutic alliance, or working relationship, between patients and therapists in psychotherapy [29].

Crucially, interpersonal synchrony has mostly been examined separately in the fields of movement science, psychophysiology and (cognitive) linguistics, respectively, without relating them to each other, e.g. [27, 37, 40]. Nevertheless, it is plausible that associations between these different types of interpersonal synchrony exist [4, 28, 32, 35, 44]. To date, only a few theories have attempted to integrate the different modalities of interpersonal synchrony into one model. One of these theories is the Interpersonal Synchrony (In-Sync) model [28]. The In-Sync model seeks to explain how two actors mutually synchronize their behaviors and experiences (for example, patient and therapist during psychotherapy). At its core, the model supposes that higher-level synchrony processes of language and emotion regulation are affected by more elementary synchrony processes of movement and physiology.

The present chapter presents computational simulation experiments in which two agents synchronize with each other by an emergent process. The aim is to examine how agent-based simulations created through general mechanisms derived and operationalized from theories and findings in psychology (see Sect. 2.1) can achieve the emergence of interpersonal synchrony. These interpersonal synchrony patterns will be evaluated against the principles of the In-Sync model and other theories of interpersonal coordination, as discussed in more detail in Sect. 2.2. The simulations are based on agent models where the internal agent processes are, in addition to the interactions between the agents, also modeled in some detail. To achieve this, we use a network-oriented agent modeling approach, as previously presented in [48, 49]. In this approach, the internal agent processes are modeled by a dynamic interplay of mental states based on general psychological (and neural) mechanisms. Section 3 presents the multi-agent model in terms of its scientific background and architecture. Section 4 describes the simulation methods, including parameter specifications. Section 5 presents the results of the performed social simulations, followed by concluding remarks in Sect. 6.

2 Psychological Background

2.1 General Psychological Mechanisms Used to Design the Agent Models

When individuals prepare to execute a certain action, like a movement, they assess what the effect of this action will be as part of their decision-making process. According to [2, 3], through a prediction loop such internal simulations generate an internal sensory representation of the likely outcome of an action. In other words, mental predictions of actions are done before these actions are actually executed, these predictions are triggered by an action preparation which can be activated via a stimulus–response effect or based on a similar action observed in other individuals [6, 7, 21]. In the latter case, mirror neurons are a relevant general mechanism that have received extensive empirical support from patient, brain stimulation and brain imaging studies [22]. Mirror neurons are neurons that would fire both when individuals prepare for their own action or body change to perform and when individuals observe a corresponding action by somebody else [23, 42]. Incoming connections (from sensory representation states) to mirror neurons (modeled as preparation states) are called mirroring links in the current chapter. The preparation of the action by mirror neurons serves as a starting point for the internal simulation of the prediction loop. The twin concepts of internal simulations and mirror neurons provide a neurobiological explanation for the attunement of actions and emotions [23]. The precise role of mirror neurons in human behavior is still being investigated. Nevertheless, the notions of internal simulations and mirroring serve as usable and operationalizable constructs for a simulation model on interpersonal synchrony.

2.2 Emerging Synchrony Among Individuals

Of the three interpersonal synchrony—movement, physiological/affective and language synchrony—types that we consider here, movement synchrony is probably the most well-documented. Controlled experiments have shown that people naturally synchronize their movements, such as in finger-tapping paradigms [40]. More generally, people display a consistent tendency to synchronize their movements with familiar and unfamiliar others, in both structured and unstructured environments [5, 41, 54].

Physical experiences like movements have been argued to serve as grounding or scaffolding for higher-order mental processes [1, 24, 33, 55]. For example, emotional language comprehension has been shown to emerge faster when people's (facial) movements are congruent with the emotions from the text comprehension task [9]. Synchrony in language (also known as 'linguistic alignment', or 'accommodation') has also been reported to occur at several levels of representation, ranging from low-level speech properties to syntactic structure [25, 36, 38], and accommodation in speech is also well-documented.

Finally, a third type of interpersonal synchrony occurs for physiological responses. Physiological synchrony is shown to be important throughout development: Human infants in the uterus already adapt their physiological responses to their mother's [26, 53], and physiological synchrony between children and their caretaker during childhood prepares children to individually regulate their emotions [5]. Additionally, physiological synchrony emerges in close relationships between adults [34] and when unacquainted adults are involved in a collective ritual [27].

An influential account of affect is proposed by [43] and states that physiological changes form the basis for core affect. Core affect is defined as 'a neurophysiological state consciously accessible as the simplest raw (nonreflective) feelings evident in moods and emotions'. Core affect is situated on two dimensions (valence and arousal) and fluctuates immediately after an event occurs to prepare people to act. Conscious emotions are enabled by core affect [43] and they arise when there is a conflict among lower-order processes (such as motor expressions and physiological responses). These conscious emotions at a higher-order level help to solve the insufficiency at the lower-order levels [8]. Verbal actions rely on language that can be generated once emotions are experienced in conscious awareness [8]. Once people put their feelings into words, this in turn can influence their emotional states [46]. Based on previous findings, we expect that the following patterns will be obtained in the agent simulations:

- Synchrony between people can be found in the form of comparable patterns over time.
- At the (intra-)individual level, movement will emerge first followed by conscious emotions (that are enabled by affect) to end with verbal actions.
- At the inter-individual level, movement synchrony will be followed by affective synchrony and language synchrony will emerge in the end.
- The different types of synchronies will be interdependent. Concretely, we expect that when the movements are disabled, the emergence of other types of synchrony will be complicated, as stated by the embodied cognition theories about synchrony.

3 The Two-Agent Model

3.1 General Approach of Agent Modeling

To model the emergence of interpersonal synchrony, two agents were designed with internal mental processes modeled as a dynamic interplay of mental states based on the network-oriented modeling approach from [47], see also this volume [14], Chap. 2: [50]. The structure of the two modeled agents is displayed in Fig. 1. The structure of the model is based on the general internal simulation and mirror neuron mechanisms outlined in Sect. 2.1 in order to test an emerging interplay of the different synchronies outlined in Sect. 2.2. These mechanisms are modeled by causal relations between mental states (e.g., the sensory representation of a stimulus, emotions, preparations). The mental states are represented by nodes with values that change over time (also simply called states) and the causal relations by connections between them, enabling interactive dynamics. In this way the mental states create an emergent mental process by which their activations dynamically change over time. These dynamical changes are affected by the input states and result in the output states, which in turn affect the input states of the agent itself and/or the other agent. Here the agent's input (sensing) states concern, for example, hearing the relevant verbal actions or seeing the relevant movements of the other agent. The output (execution) states concern the agents' actions and body states that are visible to the external world, including, for instance, the execution of a movement, the (facial) expression of affect and looking at the other agent. The interplay between internal (cognitive and affective) mental states involves, for example, the representations of the other agent's movement, the preparation of the agent's own movement, affect, and conscious emotion.

These causal networks are conceptualized by a labeled graph based on the following labels:

- A *connection weight* $\boldsymbol{\omega}_{X,Y}$ is associated with each connection from a state X to state Y; this denotes the strength of that connection.
- A *combination function* $\mathbf{c}_Y$ for each state Y; this defines the aggregation of the impact from all incoming connections on that state.
- A *speed factor* $\boldsymbol{\eta}_Y$ for each state Y to time the effect of the impact in a state-specific manner.

Each of these labels contain specific parameter values that need to be tuned. As can be seen, each agent has the same 15 states, from which 7 internal (invisible for the outside world; colored white), 3 input and 5 external states. A brief explanation of all the states of agent A is provided in Table 1 and all states of two agents A and B are presented in Table 4 from the Appendix section. Furthermore, the role matrices used in our MATLAB software environment as described in [48, 49] for the different experiments are also available in this Appendix (part 7).

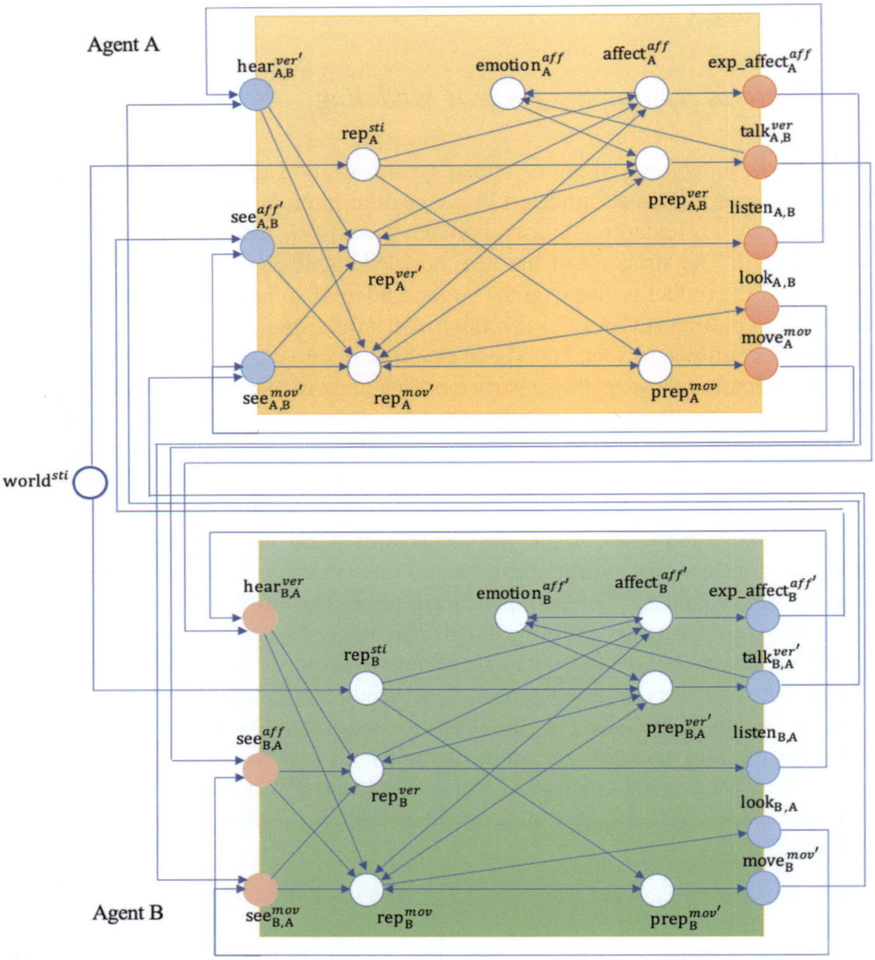

Fig. 1 The model for the two agents and the stimulus from the world

3.2 *Conceptual Representations for the Agent Model*

First, agents A and B can receive input from a stimulus $world^{sti}$ for some predefined time periods, meaning the activation of $world^{sti}$ alternates between 0 (*sti* not present) and 1 (*sti* present), with *sti* being an instance of a stimulus. The model structure of all agents is the same. Therefore, for the sake of simplicity, we focus in the current chapter on the model of an agent A with respect to a single other agent B. Within the presented scenario, this external stimulus only influences A by a causal relation to the sensory representation rep_A^{sti} of this stimulus which in turn directly triggers the agent's three internal preparation states: $prep_A^{mov}$, $affect_A^{aff}$ and $prep_{A,B}^{ver}$, with *mov*, *aff*

Table 1 The description of all states of agent A

State	Explanation
world^{sti}	World state for stimulus *sti*
$\text{see}_{A,B}^{aff'}$	Agent A receives the visual cues of the affective expression *aff'* of agent B
$\text{see}_{A,B}^{mov'}$	Agent A receives the visual cues of the agent B movement *mov'*
$\text{hear}_{A,B}^{ver'}$	Agent A receives the verbal cues *ver'* of agent B by hearing them
rep_{A}^{sti}	Sensory representation state for stimulus *sti* in agent A
$\text{rep}_{A}^{ver'}$	Sensory representation state of agent B verbal action *ver'* in agent A
$\text{rep}_{A}^{mov'}$	Sensory representation state of agent B movement *mov'* in agent A
prep_{A}^{mov}	Preparation state for movement *mov* in agent A
affect_{A}^{aff}	Preparation state for affective response *aff* in agent A
$\text{prep}_{A,B}^{ver}$	Preparation state for verbal action *ver* of agent A to agent B
move_{A}^{mov}	Execution state for movement *mov* in agent A
emotion_{A}^{aff}	The conscious emotional state for *aff* in agent A
$\text{exp\_affect}_{A}^{aff}$	The expression of the affective response *aff* in agent A
$\text{talk}_{A,B}^{ver}$	Verbal action *ver* of agent A to agent B
$\text{look}_{A,B}$	Agent A looks at agent B
$\text{listen}_{A,B}$	Agent A listens to agent B

and *ver* being an instance of movement, affective response and a verbal action of agent A, respectively. Note that $\text{prep}_{A,B}^{ver}$ has an A, B subscript as this state denotes the preparation of only those verbal actions from agent A to B. The agent model also includes a sensory representation state for the verbal actions *ver'* ($\text{rep}_{A}^{ver'}$) and the movements *mov'* of any other agent ($\text{rep}_{A}^{mov'}$), with the prime symbol ′ indicating that the behavior comes from the other agent B (without prime means that it comes from agent A). Each of these two sensory representation states is affected by the agent's three input states: hearing the verbal cues *ver'* ($\text{hear}_{A,B}^{ver'}$), seeing the affective expressions *aff'* ($\text{see}_{A,B}^{aff'}$) and seeing the movements *mov'* ($\text{see}_{A,B}^{mov'}$) of any other agent.

As *mov'* is assumed to be a movement similar to *mov*, $\text{rep}_{A}^{mov'}$ is directly connected with prep_{A}^{mov}, affect_{A}^{aff} and $\text{prep}_{A,B}^{ver}$ and these (reciprocal) causal links show how the sensory representation of lower movement processes serve as a base for (higher) internal action preparations and the connections from each preparation state to $\text{rep}_{A}^{mov'}$ in turn reflect the feedback from the higher-order to the lower-order processes. Furthermore, $\text{rep}_{A}^{mov'}$ directly influences the gaze direction of the agent ($\text{look}_{A,B}$), because the sensory representation of the other agent's movement *mov'* can enhance the agent's visual focus. State $\text{rep}_{A}^{ver'}$ causally affects affect_{A}^{aff} and $\text{prep}_{A,B}^{ver}$, but not the preparation state for movement because the latter is a lower-order process. The execution of $\text{listen}_{A,B}$ is also directly triggered by $\text{rep}_{A}^{ver'}$ because the sensory representation of the other agent's verbal action can alert the agent to listen more focused.

The mental processes, starting from the representation states, ultimately lead to both conscious emotions (internal state $\text{emotion}_{\text{A}}^{aff}$) and executed actions in the physical world ($\text{look}_{\text{A,B}}$, $\text{listen}_{\text{A,B}}$, $\text{move}_{\text{A}}^{mov}$, $\text{exp\_affect}_{\text{A}}^{aff}$, $\text{talk}_{\text{A,B}}^{ver}$).

Regarding the preparation states, $\text{prep}_{\text{A}}^{mov}$ influences both the actual execution of the movement $\text{move}_{\text{A}}^{mov}$ and its representation $\text{rep}_{\text{A}}^{mov'}$, thereby highlighting the dynamic interplay of one's representation of *mov'* and one's own execution *mov*. The state $\text{affect}_{\text{A}}^{aff}$ directly influences $\text{rep}_{\text{A}}^{mov'}$, the expression of the affective response *aff* ($\text{exp\_affect}_{\text{A}}^{aff}$) and the conscious $\text{emotion}_{\text{A}}^{aff}$ of *aff*. The fact that the sensory representation of *ver'* does not get feedback from $\text{affect}_{\text{A}}^{aff}$ reflects the higher-order process of language compared to the lower-level affective changes. State $\text{prep}_{\text{A,B}}^{ver}$ has an immediate effect on $\text{rep}_{\text{A}}^{mov'}$, $\text{rep}_{\text{A}}^{ver'}$ and the execution of the verbal action $\text{talk}_{\text{A,B}}^{ver}$. State $\text{talk}_{\text{A,B}}^{ver}$ in turn triggers $\text{emotion}_{\text{A}}^{aff}$ which in turn directly affects $\text{prep}_{\text{A,B}}^{ver}$. This feedback loop reflects the need of conscious emotions to initiate verbal actions *ver* and at the meantime how these verbal actions *ver* themselves further shape the conscious emotions of the affective response *aff*.

The Coupling Between the Sensing and Execution States

In a situation with multiple agents, we can connect the execution states of any agent A to the sensory states of any agent B. Specifically, each of agent A's sensing states receives two input connections, one from their own execution state and one from an execution state of agent B. Specifically, $\text{hear}_{\text{A,B}}^{ver'}$ gets input from $\text{listen}_{\text{A,B}}$ because listening is required to receive auditory cues and from the execution $\text{talk}_{\text{B,A}}^{ver'}$ of the verbal action *ver'* of agent B. Both $\text{see}_{\text{A,B}}^{aff'}$ and $\text{see}_{\text{A,B}}^{mov'}$ are activated by $\text{look}_{\text{A,B}}$ to capture the need of looking at agent B to get visual cues from this agent. Regarding the input connections from Agent B, the expression $\text{exp\_affect}_{\text{B}}^{aff'}$ of the affective response for *aff'* is directly linked to state $\text{see}_{\text{A,B}}^{aff'}$ for seeing *aff'* and similarly the execution state $\text{move}_{\text{B}}^{mov'}$ of the movement *mov'* is directly linked to $\text{see}_{\text{A,B}}^{mov'}$.

General Mechanisms

In line with the general psychological mechanisms outlined in Sect. 2.1, in addition to stimulus–response links, each agent contains prediction loops and mirroring links. Both direct and indirect effects are predicted through the prediction loops. The prediction loops from $\text{prep}_{\text{A}}^{mov}$ to $\text{rep}_{\text{A}}^{mov'}$ and from $\text{prep}_{\text{A,B}}^{ver}$ to $\text{rep}_{\text{A}}^{ver'}$ consist of the prediction of direct effects (i.e., what the effect of the execution of a certain action will be on the sensory representation of this action). The indirect effects (i.e., the effect of a preparation of a certain action on the representation of another action) regards the prediction loops from $\text{prep}_{\text{A,B}}^{ver}$ to $\text{rep}_{\text{A}}^{mov'}$ and from $\text{affect}_{\text{A}}^{aff}$ to $\text{rep}_{\text{A}}^{mov'}$. Concerning the mirroring links, the connection from $\text{rep}_{\text{A}}^{mov'}$ to $\text{prep}_{\text{A}}^{mov}$ mirrors the action movement and the connection from $\text{rep}_{\text{A}}^{ver'}$ to $\text{prep}_{\text{A,B}}^{ver}$ mirrors the verbal action of another agent. In principle the connection from $\text{see}_{\text{A,B}}^{aff'}$ to $\text{affect}_{\text{A}}^{aff}$ through both $\text{rep}_{\text{A}}^{mov'}$ and $\text{rep}_{\text{A}}^{ver'}$ can also be interpreted as a mirroring path, because every sensory representation is a) influenced by $\text{see}_{\text{A,B}}^{aff'}$ and b) internally influences $\text{affect}_{\text{A}}^{aff}$ through the representation states of both the verbal action ($\text{rep}_{\text{A}}^{ver'}$) and the movement ($\text{rep}_{\text{A}}^{mov'}$). It is argueable that the model could also in addition contain a $\text{rep}_{\text{A}}^{aff'}$, however, we have decided to not include this additional state for reasons of simplicity. There were already multiple pathways from $\text{see}_{\text{A,B}}^{aff'}$ to $\text{affect}_{\text{A}}^{aff}$ itself.

Connection Weights

All the connection weight $\boldsymbol{\omega}_{X,Y}$ values were fixed on the value 1, except the $\boldsymbol{\omega}_{X,Y}$ of all incoming connections to $\text{affect}_{\text{A}}^{aff}$ and $\text{affect}_{\text{B}}^{aff'}$. The incoming connections to these affect states from respectively $\text{rep}_{\text{A}}^{mov'}$ and $\text{rep}_{\text{B}}^{mov}$ were set to 2 and all the other connections were set to 0.5. In this way, the less complex sensorimotor processes serve as a foundation or 'grounding' [1] for the affective responses. Furthermore, the connection weight $\boldsymbol{\omega}_{\text{world}^{sti},\text{rep}_{\text{B}}^{sti}}$ is set to 0 because agent B did not receive a *sti*.

3.3 Numerical Representations for the Agent Model

We used the software environment described in [48, 49]. In this software environment, the conceptual representations of the multi-agent model are mapped onto their associated numerical representations as follows:

- $Y(t)$ denotes the activation value for state Y of an agent at time point t; this is a real number, usually in the range [0, 1].
- The single causal impact $\textbf{impact}_{X,Y}(t) = \boldsymbol{\omega}_{X,Y}X(t)$ defines at each time point t the single impact from state X connected to state Y on state Y, where $\boldsymbol{\omega}_{X,Y}$ is the weight of the connection from X to Y.
- Aggregating of multiple single causal impacts through combination function $\mathbf{c}_Y(\ldots)$ is defined by

$$\begin{aligned}\textbf{aggimpact}_Y(t) &= \mathbf{c}_Y\big(\textbf{impact}_{X_1,Y}(t), \ldots, \textbf{impact}_{X_k,Y}(t)\big)\\ &= \mathbf{c}_Y\big(\boldsymbol{\omega}_{X_1,Y}X_1(t), \ldots, \boldsymbol{\omega}_{X_k,Y}X_k(t)\big)\end{aligned}$$

  for the states $X_1, \ldots, X_k$ from which Y has incoming connections.
- *The speed factor* $\boldsymbol{\eta}_Y$ *determines how the effect of* $\textbf{aggimpact}_Y(t)$ *on state* Y *is exerted gradually over time:*

$$Y(t + \Delta t) = Y(t) + \boldsymbol{\eta}_Y\big[\textbf{aggimpact}_Y(t) - Y(t)\big]\Delta t$$

- *This leads to the following difference or differential equation for* Y*:*

$$Y(t + \Delta t) = Y(t) + \boldsymbol{\eta}_Y\big[\mathbf{c}_Y\big(\boldsymbol{\omega}_{X_1,Y}X_1(t), \ldots, \boldsymbol{\omega}_{X_k,Y}X_k(t)\big) - Y(t)\big]\Delta t, \text{ or}$$

$$\frac{dY(t)}{dt} = \boldsymbol{\eta}_Y\big[\mathbf{c}_Y\big(\boldsymbol{\omega}_{X_1,Y}X_1(t), \ldots, \boldsymbol{\omega}_{X_k,Y}X_k(t)\big) - Y(t)\big]$$

Table 2 The two combination functions used

Name	Formula	Parameters
Advanced logistic sum combination function **alogistic**$_{\sigma,\tau}(V_1, \ldots, V_k)$	$\left[\frac{1}{1+e^{-\sigma(V_1+\cdots+V_k-\tau)}} - \frac{1}{1+e^{\sigma\tau}}\right](1 + e^{-\sigma\tau})$	Steepness σ Excitability threshold τ
Step-mod function **stepmod**$_{\rho,\delta}(V_1, \ldots, V_k)$	for time t if $mod(t, \rho) < \delta$ then $x = 0$, else 1	Repeated time interval ρ Duration of value 0 δ

All agent states use the advanced logistic sum combination function $\mathbf{c}_Y(\ldots)$, whereas the world state for stimulus *sti* uses the step-modulo combination function $\mathbf{c}_Y(\ldots)$; see Table 2.

4 Simulation Method for the Agents

We will present three scenarios. All initial state values were set to 0 in all simulations. Regarding the first/main simulation, as specified with the input parameters of the step-modulo combination function (see Table 3), agent A is exposed to an external stimulus *sti* for 150 time units every time after 150 time units without stimulation. This process (a total of 300 time units) is repeated until the end of the simulations. We deliberately do not further specify stimulus *sti* because there are numerous situations that can provoke interpersonal synchrony. Such repeated stimuli may concretely regard, for instance, daily life events that one person shares with another person, a series of therapy sessions or a dance choreography, meaning these agent simulations might have a wide variety of applications. The activation level of *sti* can vary over specific applications, however, for the sake of simplicity, we have decided to use activation level 1 over all simulations. The (length of the) stimulus *sti* intervals for this main social simulation were selected such that, as can be seen in the Appendix section Fig. 2, 6, 7, 8 and 9, (most of) the mental states ended in their equilibrium phase for both the stimulus *sti* present and absent periods. In Fig. 2 and the Appendix Fig. 6, 7, 8, 9, 10, 11, 12 and 13 is it shown that emerging limit cycle behavior occurs right from the start.

The characteristics of the two agents have been set according to the homeostatic regulation of neuronal excitability principle, which refers to the adaptation of neurons' internal properties to control a desired activation level; e.g., [56]. More specifically, as agent B (the agent not directly receiving the stimulus) gets less incoming activation than agent A, we have mimicked this principle by putting some of the excitability threshold and steepness values from the advanced logistic sum combination function of agent B lower than for agent A. The speed factors of almost all states equaled 0.5. The speed factor of the stimulus was set to 2 to ensure the fast appearance and disappearance of the stimulus. In contrast, the speed factors of the looking direction and the focus of listening were set to 0.2 (possible scale range: 0

Table 3 The values for the main characteristics of the model: speed factors $\boldsymbol{\eta}$ and combination function parameters $\boldsymbol{\sigma}$ and $\boldsymbol{\tau}$ for each agent state and combination function parameters $\boldsymbol{\rho}$ and $\boldsymbol{\delta}$ for the stimulus *sti*

State	η	σ	τ	State	η	σ	τ
$see_{A,B}^{aff'}$	0.5	1	0.4	$see_{B,A}^{aff}$	0.5	1	0.4
$see_{A,B}^{mov'}$	0.5	1	0.4	$see_{B,A}^{mov}$	0.5	1	0.4
$hear_{A,B}^{ver'}$	0.5	1	0.4	$hear_{B,A}^{ver}$	0.5	1	0.4
rep_{A}^{sti}	0.5	20	0.6	rep_{B}^{sti}	0.5	20	0.6
$rep_{A}^{ver'}$	0.5	1	0.4	rep_{B}^{ver}	0.5	1	0.4
$rep_{A}^{mov'}$	0.5	1	3	rep_{B}^{mov}	0.5	1	2
$prep_{A}^{mov}$	0.5	4	0.8	$prep_{B}^{mov'}$	0.5	2	0.3
$affect_{A}^{aff}$	0.5	5	0.8	$affect_{B}^{aff'}$	0.5	2	0.6
$prep_{A,B}^{ver}$	0.5	1	3.5	$prep_{B,A}^{ver'}$	0.5	1	2.5
$move_{A}^{mov}$	0.5	4	0.3	$move_{B}^{mov'}$	0.5	4	0.1
$emotion_{A}^{aff}$	0.5	2	0.3	$emotion_{B}^{aff'}$	0.5	2	0.1
$exp\_affect_{A}^{aff}$	0.5	2	0.4	$exp\_affect_{B}^{aff'}$	0.5	2	0.4
$talk_{A,B}^{ver}$	0.5	4	0.3	$talk_{B,A}^{ver'}$	0.5	4	0.3
$look_{A,B}$	0.2	1	0.3	$look_{B,A}$	0.2	1	0.5
$listen_{A,B}$	0.2	2	0.3	$listen_{B,A}$	0.2	1	0.4
		ρ	**δ**				
$world^{sti}$	2	300	150				

to 1, with higher speed factors indicating that the specific state will change quicker) for both agents because in the real world these actions often do not rapidly change.

The second simulation consists of shorter stimulus and non-stimulus intervals (each of them lasting 10 time units instead of 150), and this is the only difference from the main simulation. This is a representative example for the cases where no equilibria are reached within the stimulus and non-stimulus periods. The third simulation is exactly the same as the main simulation except that the states related to movement were disabled in both agents (see Appendix, part 8). The aim of this third simulation was to test whether synchrony can emerge without movement.

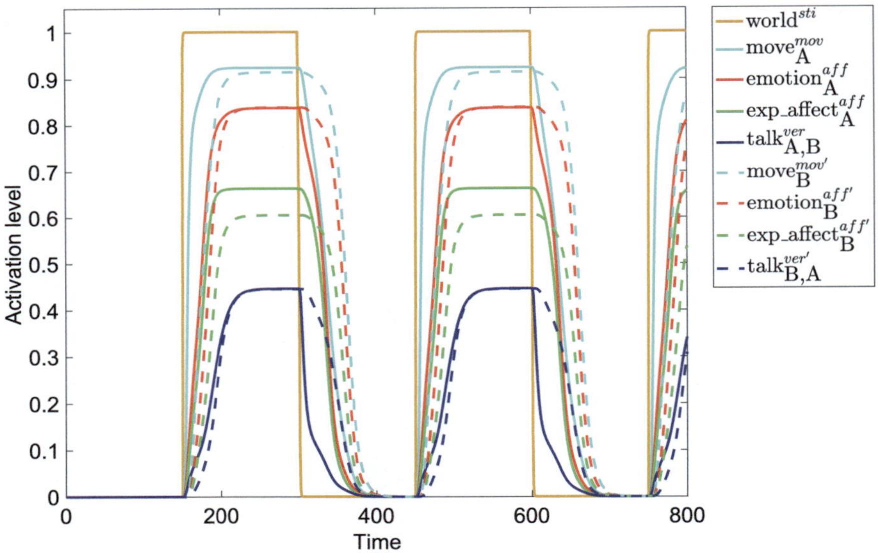

Fig. 2 Activation levels over 800 time units (with 150 non-stimulus time units alternated with 150 stimulus time units) for the execution states of the two agents. Corresponding states of agent A (solid lines) and B (dashed lines) are indicated in the same color, whereas the stimulus *sti* is indicated in yellow

5 Analysis of the Two-Agent Model

5.1 Main Simulation

To validate the two-agent model, we derived some testable predictions from the literature (as discussed in Sect. 2). First, the occurrence of a limit cycle. Synchrony is quantified as the same states of agent A and B that exhibit comparable patterns over time. Therefore, we focus for each agent on corresponding actions. Equivalent states of agent A (solid lines) and B (dashed lines) are colored the same in all figures. As shown in Figs. 2, 6, 7, 8, 9, 10, 11, 12 and 13 from the Appendix section regarding the main simulation, all the states of both agent A and B are activated after the stimulus *sti* is presented to agent A and deactivated when *sti* is no longer present, resulting in a limit cycle for the model.

This finding demonstrates that emergent synchrony patterns between agents can originate through communication/interaction when only one agent actually receives a stimulus *sti*. Note that such synchrony patterns would not be able to emerge when agent B cannot receive sensing information from agent A as an input (i.e., see or hear agent A). These emergent synchrony patterns remain consistent over time, indicated by the repetitive (equally high) peaks of each of two agents' states across the stimulus *sti* episodes. Furthermore, as can be seen in Figs. 2, 6, 7, 8 and 9 from the Appendix, each of the state activations of agent A precedes activation of the equivalent state of agent B, except the input and representation states. The reasoning behind the latter is that the input states, and thereby the representation states that are directly dependent on the input states, of agent A require input from the output states of agent B and thus cannot precede these states of agent B in terms of activation. The analysis of the simulation with shorter stimulus and non-stimulus intervals, a more extensive evaluation of the sequence of the different types of synchrony and the simulation without movement are studied in Sects. 5.2, 5.3 and 5.4 respectively.

5.2 Analysis of Equilibria

In the simulation with the shorter stimulus intervals, the same sequence of state activations is maintained, see Figs. 3, 14, 15, 16 and 17 from the Appendix section. However, the maximal state activations are not already obtained after the first episode but only a number of episodes later, with emerging limit cycle behavior still occurring after that time. No equilibria are obtained in the presence and absence of *sti*. This simulation with the shorter stimulus intervals is a typical example in which synchrony strengthens over time. In the future, variable *sti* intervals and/or levels might be incorporated to explore how the emerging synchrony and common ground evolve.

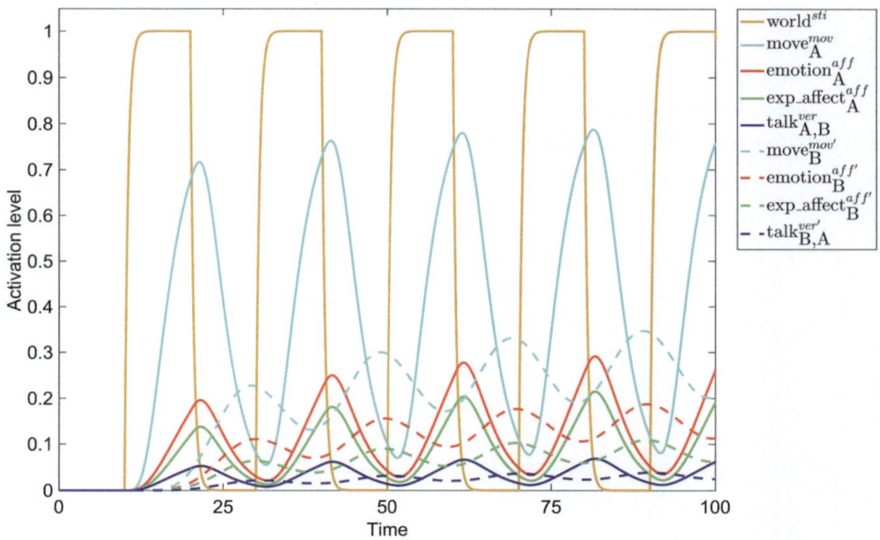

Fig. 3 Activation levels over 100 time units (with 10 non-stimulus time units alternating with 10 stimulus time units) for the execution states of two agents. Corresponding states of agent A (solid lines) and B (dashed lines) are indicated in the same color, whereas the stimulus *sti* is indicated in yellow

5.3 Analysis of the Sequence of Emergent Synchrony

A second theoretical prediction deals with the sequence in which states get activated. In order to study emerging synchrony, we examined in more detail the sequence of the (starting of the) activation levels of the preparation and the execution states of both agents of the main simulation. In the left panel of Fig. 4, one can see that the preparation states of agent A get activated in the following order: $\mathrm{prep}_{\mathrm{A}}^{mov}$, $\mathrm{affect}_{\mathrm{A}}^{aff}$, and $\mathrm{prep}_{\mathrm{A,B}}^{ver}$. This sequence fits well with the empirically based theoretical foundation described in Sect. 2.2 of the main chapter, stating that lower-order processes such as movement precede higher-order language processes, e.g. [1, 23, 31, 53]. At a few time points later, the preparation states of agent B start to elevate in the same sequence displaying a comparable pattern as those preparation states of agent A did. The maximum reached state values of $\mathrm{prep}_{\mathrm{A,B}}^{ver}$ and $\mathrm{prep}_{\mathrm{B,A}}^{ver'}$ are the same, whereas $\mathrm{prep}_{\mathrm{A}}^{mov}$ and $\mathrm{affect}_{\mathrm{A}}^{aff}$ have a slightly larger peak value than $\mathrm{prep}_{\mathrm{B}}^{mov'}$ and $\mathrm{affect}_{\mathrm{B}}^{aff'}$ respectively (see Appendix Fig. 8).

In the right panel of Fig. 4 it is demonstrated that similar emergent synchrony patterns as obtained for the preparation states are observed for the execution states. All execution states of agent B are activated in the same sequence as these states of agent A but a few time points later and corresponding states follow comparable patterns over time. First, the movements $\mathrm{move}_{\mathrm{A}}^{mov}$ and $\mathrm{move}_{\mathrm{B}}^{mov'}$ synchronize, followed by the conscious emotions $\mathrm{emotion}_{\mathrm{A}}^{aff}$ and $\mathrm{emotion}_{\mathrm{B}}^{aff'}$. Thereafter, synchrony emerges for the expressions of the affective responses $\mathrm{exp\_affect}_{\mathrm{A}}^{aff}$ and $\mathrm{exp\_affect}_{\mathrm{B}}^{aff'}$, reflecting the slowness of robust changes in physical appearances compared to more fluctuating conscious emotions. Finally, the verbal actions $\mathrm{talk}_{\mathrm{A,B}}^{ver}$ and $\mathrm{talk}_{\mathrm{B,A}}^{ver'}$ start to synchronize. The peak values of almost all corresponding states of agent A and B are (approximately) the same (Fig. 2 in the main chapter). Only $\mathrm{exp\_affect}_{\mathrm{A}}^{aff}$ reached higher maximum states values than the corresponding $\mathrm{exp\_affect}_{\mathrm{B}}^{aff'}$, possibly because agent

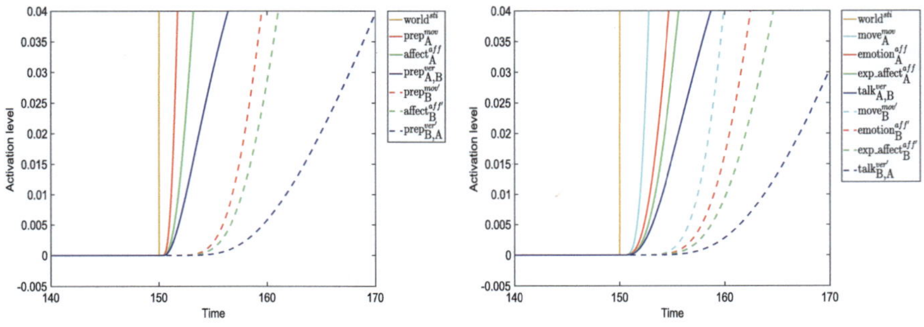

Fig. 4 Activation levels of the preparation (left panel) and execution (right panel) states of agent A (solid lines) and B (dashed lines) over time; time is zoomed in at the period 140–170 time units, with 150 being the time unit at which the stimulus *sti* for the first time is present (yellow solid line)

A needs to transfer the stimulus *sti* information to agent B through stronger expressions of affect. Moreover, $\text{affect}_{\text{A}}^{aff}$ was higher than $\text{affect}_{\text{B}}^{aff'}$, and the affective response is the only state directly connected to the expression of these affect states.

5.4 Analysis of the Agents When Lower-Order Movement is Lacking

As an additional validation of the model, we evaluated the behavior of the two agents when their representation and preparation states of movement were disabled. In such a simulation, based on the theory and empirical research results presented in Sect. 2.2, we predict that no synchrony, or smaller degrees of synchrony, will emerge as the lower-order movement processes are assumed to be necessary for emerging synchrony at higher levels. To this end, the speed factors of $\text{rep}_{\text{A}}^{mov'}$, $\text{rep}_{\text{B}}^{mov}$, $\text{prep}_{\text{A}}^{mov}$ and $\text{prep}_{\text{B}}^{mov'}$ were set at 0. In this way, these states and the execution of the movements $\text{move}_{\text{A}}^{mov}$ and $\text{move}_{\text{B}}^{mov\prime}$ of both agents A and B cannot arise, because this execution state is completely dependent on its preparation state.

It appears from Figs. 5, 20, 21, 22 and 23 from the Appendix that all the states have a maximum activation level of 0.2 at most, which is much lower than when the representation and preparations states of both agents were enabled (see Figs. 5, 6, 7, 8 and 9 from the Appendix). Agent A has still some—although lower—activations in all other states, except $\text{see}_{\text{A,B}}^{aff'}$, $\text{see}_{\text{A,B}}^{mov'}$ and $\text{look}_{\text{A,B}}$. In contrast, agent B only achieves some small activations in the states $\text{see}_{\text{B,A}}^{aff}$, $\text{hear}_{\text{B,A}}^{ver}$ and $\text{rep}_{\text{B}}^{ver}$. All other states of agent B, including the affective and verbal preparation and execution states (Fig. 5 and Appendix Fig. 22), are (almost) not activated anymore. This means that the same states of the two agents do not show comparable trajectories over time anymore, so synchrony does not longer emerge. We can conclude that the simulation results are in line with the predictions derived from Sect. 2.2. As such, these results underscore the need for lower-order movement processes to enable other types of emergent synchrony between two agents at higher levels, such as expressions of affective responses and verbal actions.

6 Discussion

Our agent models demonstrate the emergence of movement, expression of affective response, and language synchrony in two agents. Moreover, conscious emotional synchrony occurred as well. The social simulations based on the described general psychological mechanisms succeeded in reproducing emerging synchrony patterns that are widely reported in the literature, in the same order as predicted by several theories about synchrony. Obviously, certain parameter settings for these general mechanisms represent certain types of persons. The example simulation settings

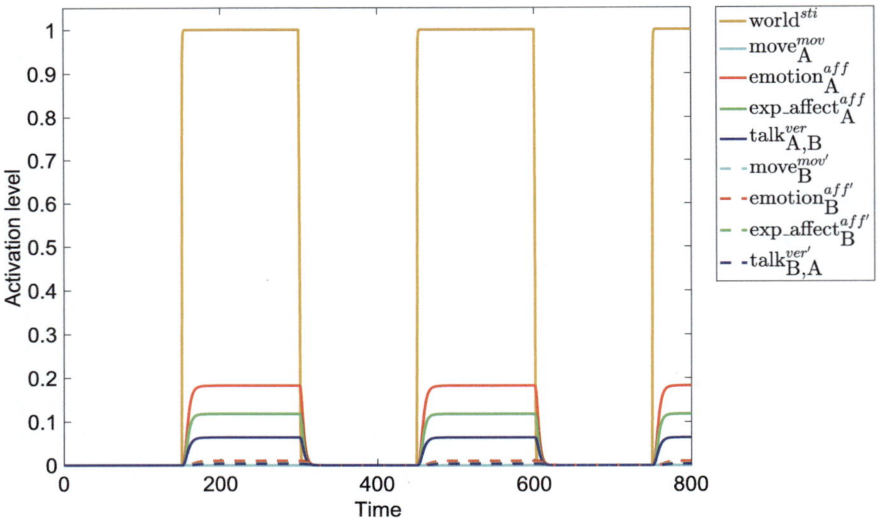

Fig. 5 Simulation of the 2 agents without movement over 800 time units (with 150 non-stimulus time units alternated with 150 stimulus time units) for the execution states. Corresponding states of agent A (solid lines) and B (dashed lines) are indicated in the same color, whereas the stimulus *sti* is indicated in yellow

describe a specific dyad. When the parameter settings are tuned differently, other types of dyads can be represented as well. Thus, by considering realistic input, internal and output states in the agent models, we were able to capture a complex process of mental representations of the physical world and perceived similarities that do occur during interpersonal synchrony under specific circumstances. In particular, the importance of embodiment through movements for cognition is demonstrated [1]. This chapter is based on material from [17].

Interpersonal synchrony belongs to a broader class of synchrony patterns that is observed in the natural life and behavioral sciences [39, 45]. This means that our simulations can potentially be extended to other domains as well. Based on our agent modeling and the conceptual In-Sync model, therapeutic sessions between therapists and patients can be an interesting application field. Agent-based computational models like our model can be the basis for the development of virtual agents that might be used in settings to interact with humans.

There are also some limitations that could be explored in future work. First, we modeled only two agents. How synchrony patterns evolve in, for example, triads and with different stimulus episodes across agents are potential future simulations. Indeed, at a later stage, synchrony in groups of agents was studied [30], see also this volume [14], Chap. 16.

Second, future work should verify whether the same synchrony patterns in the agents hold on empirical data. Third, the current agents are non-adaptive: for example, the excitability thresholds are fixed over time. Adaptive agents might be able to automatically tune their synchrony behavior in varying situations, thereby maintaining their equilibria states in even more unpredictable environments.

The work presented in this chapter was indeed a point of departure for several explorations for emerging synchrony in adaptive agent models, see, for example, [11–13, 18] or various chapters in this volume [14], Chaps. 11–15.

Fourth, the internal states of our agent models could always be extended, for example by including more specific states. A typical example would be to include a representation state of the affective expression of other agents or to separate the representation states for one's own movements and the movements of other agents. Fifth, we did not include some anticipation theories from psychology in our agent models. Based on some finger tapping experiments [52], it would also be possible that the follower (in our case agent B) anticipates on the leader (agent A) and thereby becomes effectively the leader in synchrony. Sixth, in the future, variable stimulus intervals and/or levels might be incorporated to explore how the emerging synchrony and common ground evolve. Finally, future work is needed in which more extensive analysis on the interplay between limit cycles, equilibria and synchronization is conducted.

Appendix 1

See Table 4.

Appendix 2

See Figs. 6, 7, 8 and 9.

Appendix 3

See Figs. 10, 11, 12 and 13.

Appendix 4

See Figs. 14, 15, 16 and 17.

Table 4 The description of all the states

State	Explanation
world^{sti}	world state for stimulus *sti*
$\text{see}_{A,B}^{aff'}$	Agent A receives the visual cues of the affective expression *aff'* of agent B
$\text{see}_{A,B}^{mov'}$	Agent A receives the visual cues of the agent B movement *mov'*
$\text{hear}_{A,B}^{ver'}$	Agent A receives the verbal cues *ver'* of agent B by hearing them
rep_{A}^{sti}	sensory representation state for stimulus *sti* in agent A
$\text{rep}_{A}^{ver'}$	sensory representation state of agent B verbal action *ver'* in agent A
$\text{rep}_{A}^{mov'}$	sensory representation state of agent B movement *mov'* in agent A
prep_{A}^{mov}	preparation state for movement *mov* in agent A
affect_{A}^{aff}	Preparation state for affective response *aff* in agent A
$\text{prep}_{A,B}^{ver}$	Preparation state for verbal action *ver* of agent A to agent B
move_{A}^{mov}	execution state for movement *mov* in agent A
emotion_{A}^{aff}	The conscious emotional state for *aff* in agent A
$\text{exp\_affect}_{A}^{aff}$	The expression of the affective response *aff* in agent A
$\text{talk}_{A,B}^{ver}$	Verbal action *ver* of agent A to agent B
$\text{look}_{A,B}$	Agent A looks at agent B
$\text{listen}_{A,B}$	Agent A listens to agent B
$\text{see}_{B,A}^{aff}$	Agent B receives the visual cues of the affective expression *aff* of agent A
$\text{see}_{B,A}^{mov}$	Agent B receives the visual cues of the agent A movement *mov*
$\text{hear}_{B,A}^{ver}$	Agent B receives the verbal cues *ver* of agent A by hearing them
rep_{B}^{sti}	sensory representation state for stimulus *sti* in agent B
rep_{B}^{ver}	sensory representation state of agent A verbal action *ver* in agent B
rep_{B}^{mov}	sensory representation state of agent A movement *mov* in agent B
$\text{prep}_{B}^{mov'}$	preparation state for movement *mov'* in agent B
$\text{affect}_{B}^{aff'}$	Preparation state for affective response *aff'* in agent B
$\text{prep}_{B,A}^{ver'}$	Preparation state for verbal action *ver'* of agent B to agent A
$\text{move}_{B}^{mov'}$	execution state for movement *mov'* in agent B
$\text{emotion}_{B}^{aff'}$	The conscious emotional state for *aff'* in agent B
$\text{exp\_affect}_{B}^{aff'}$	The expression of the affective response *aff'* in agent B
$\text{talk}_{B,A}^{ver'}$	Verbal action *ver'* of agent B to agent A
$\text{look}_{B,A}$	Agent B looks at agent A
$\text{listen}_{B,A}$	Agent B listens to agent A

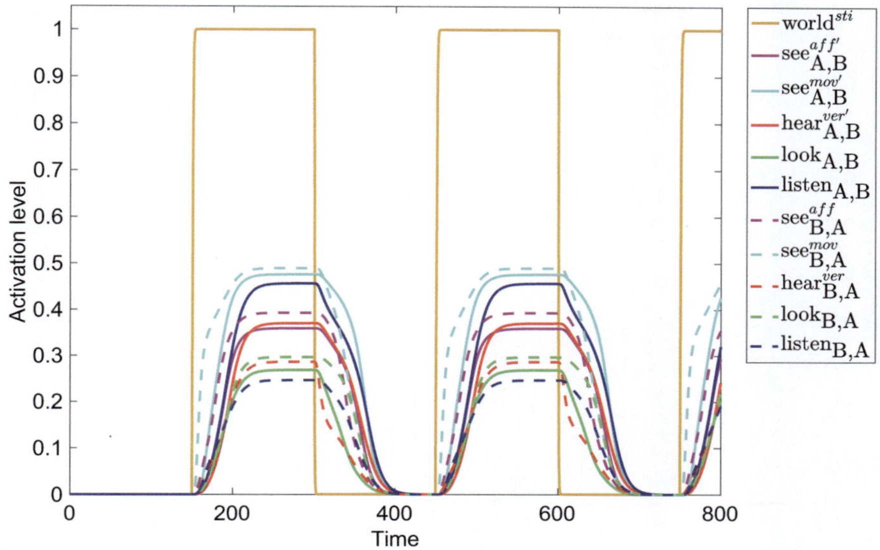

Fig. 6 Activation levels over 800 time units (with 150 non-stimulus time units alternated with 150 stimulus time units) for several states of the two agents: the input, look and listen. Corresponding states of agent A (solid lines) and B (dashed lines) are indicated in the same color, whereas the stimulus *sti* is indicated in yellow

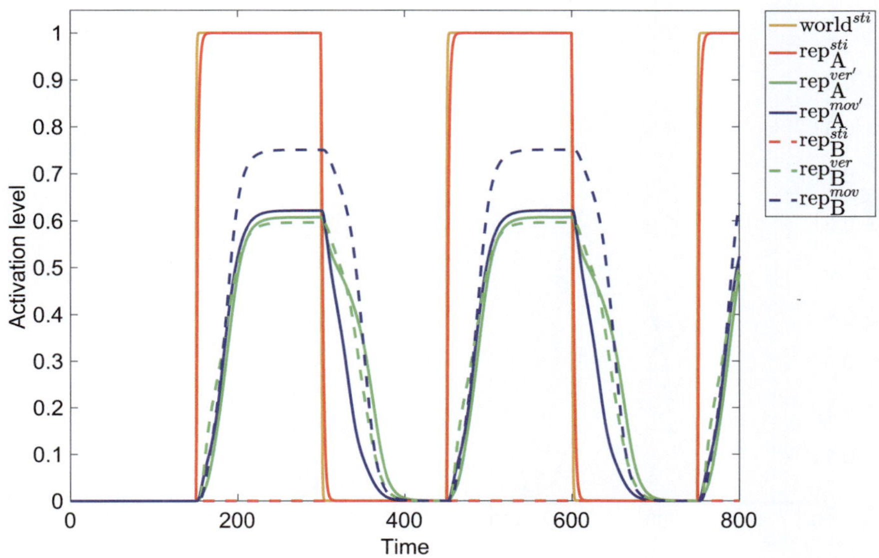

Fig. 7 Activation levels over 800 time units (with 150 non-stimulus time units alternated with 150 stimulus time units) for the representation states of the two agents. Corresponding states of agent A (solid lines) and B (dashed lines) are indicated in the same color, whereas the stimulus *sti* is indicated in yellow

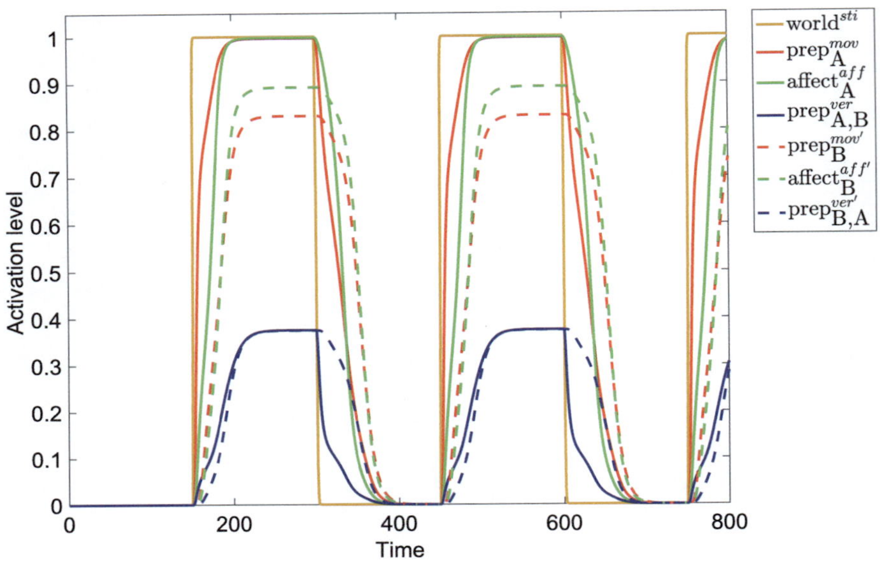

Fig. 8 Activation levels over 800 time units (with 150 non-stimulus time units alternated with 150 stimulus time units) for the preparation states of the two agents. Corresponding states of agent A (solid lines) and B (dashed lines) are indicated in the same color, whereas the stimulus *sti* is indicated in yellow

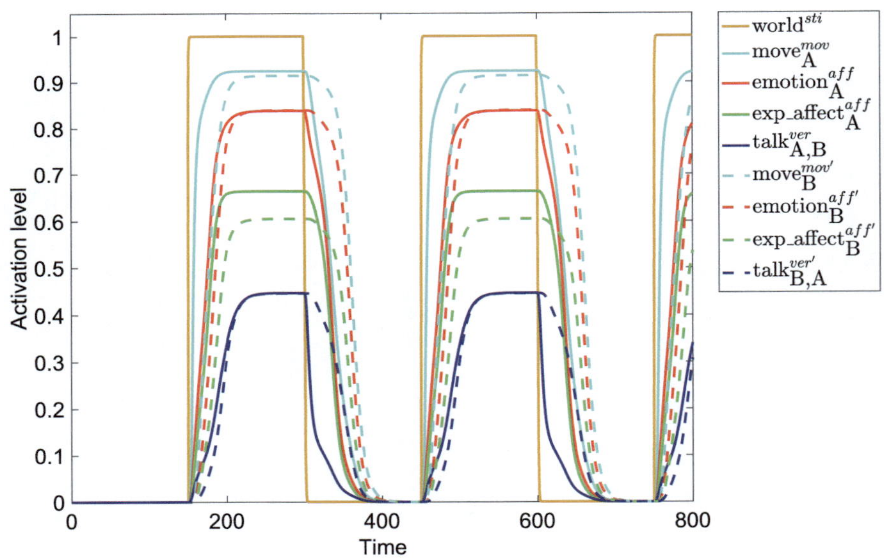

Fig. 9 Activation levels over 800 time units (with 150 non-stimulus time units alternated with 150 stimulus time units) for the execution states of the two agents. Corresponding states of agent A (solid lines) and B (dashed lines) are indicated in the same color, whereas the stimulus *sti* is indicated in yellow

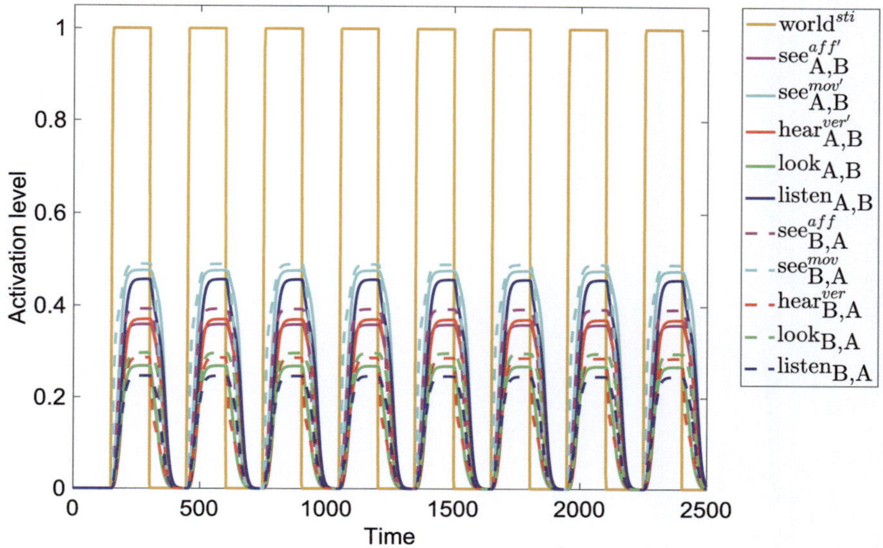

Fig. 10 Activation levels over 2500 time units (with 150 non-stimulus time units alternated with 150 stimulus time units) for several states of the two agents: the input, look and listen. Corresponding states of agent A (solid lines) and B (dashed lines) are indicated in the same color, whereas the stimulus *sti* is indicated in yellow

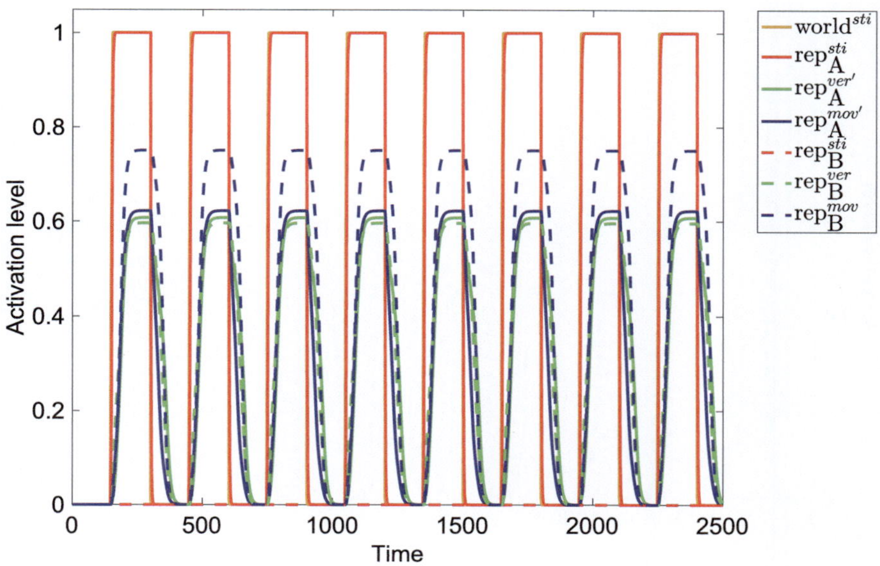

Fig. 11 Activation levels over 2500 time units (with 150 non-stimulus time units alternated with 150 stimulus time units) for the representation states of the two agents. Corresponding states of agent A (solid lines) and B (dashed lines) are indicated in the same color, whereas the stimulus *sti* is indicated in yellow

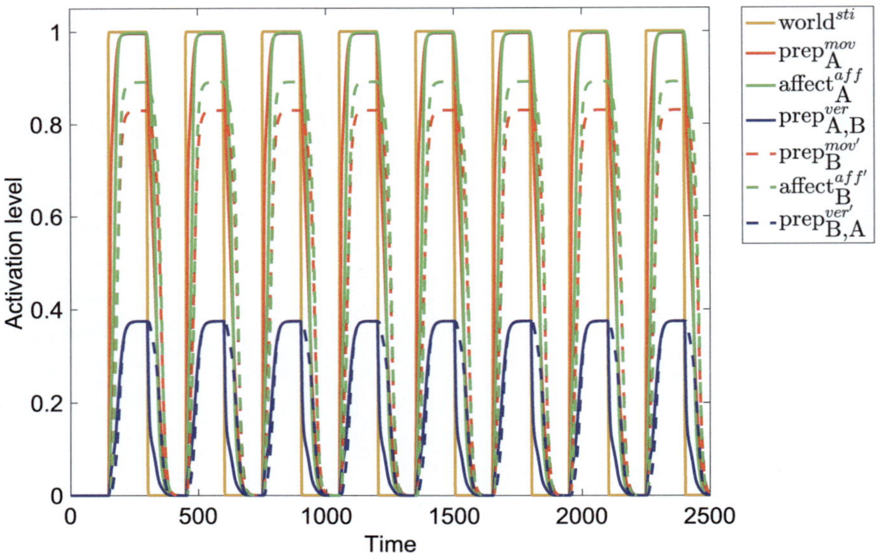

Fig. 12 Activation levels over 2500 time units (with 150 non-stimulus time units alternated with 150 stimulus time units) for the preparation states of the two agents. Corresponding states of agent A (solid lines) and B (dashed lines) are indicated in the same color, whereas the stimulus *sti* is indicated in yellow

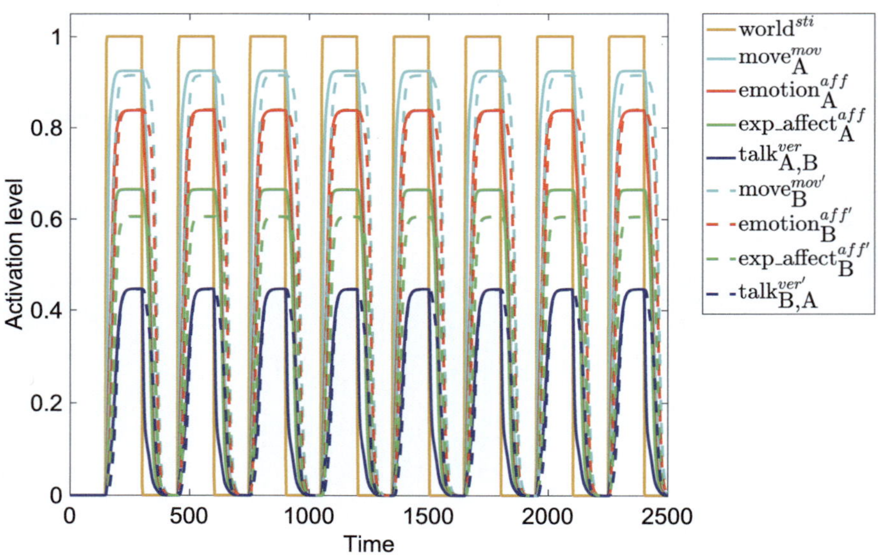

Fig. 13 Activation levels over 2500 time units (with 150 non-stimulus time units alternated with 150 stimulus time units) for the execution states of the two agents. Corresponding states of agent A (solid lines) and B (dashed lines) are indicated in the same color, whereas the stimulus *sti* is indicated in yellow

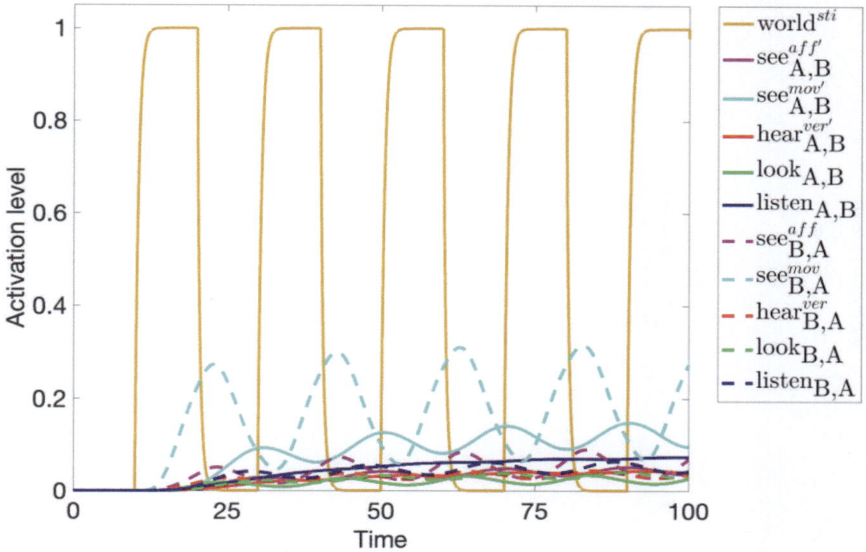

Fig. 14 Activation levels over 100 time units (with 10 non-stimulus time units alternating with 10 stimulus time units) for several states of two agents: the input, look and listen. Corresponding states of agent A (solid lines) and B (dashed lines) are indicated in the same color, whereas the stimulus *sti* is indicated in yellow

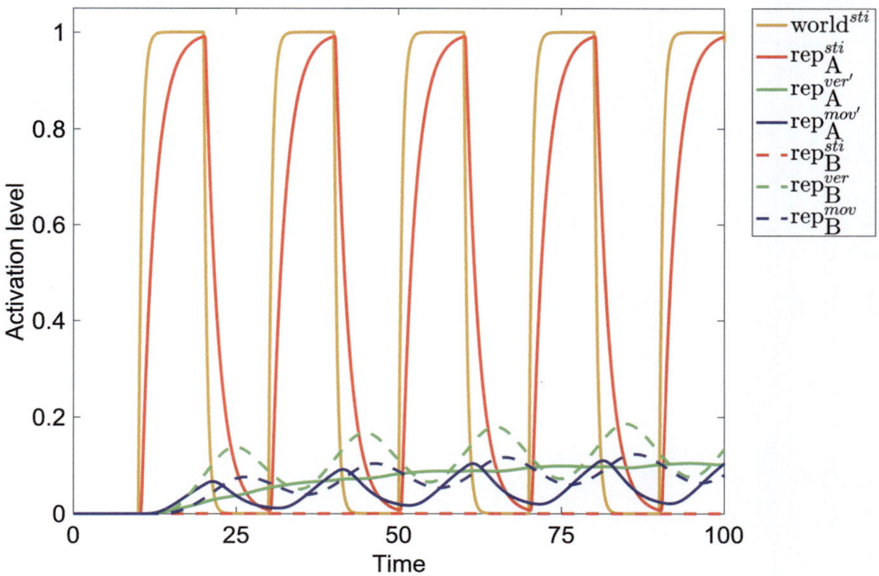

Fig. 15 Activation levels over 100 time units (with 10 non-stimulus time units alternating with 10 stimulus time units) for the representation states of two agents. Corresponding states of agent A (solid lines) and B (dashed lines) are indicated in the same color, whereas the stimulus *sti* is indicated in yellow

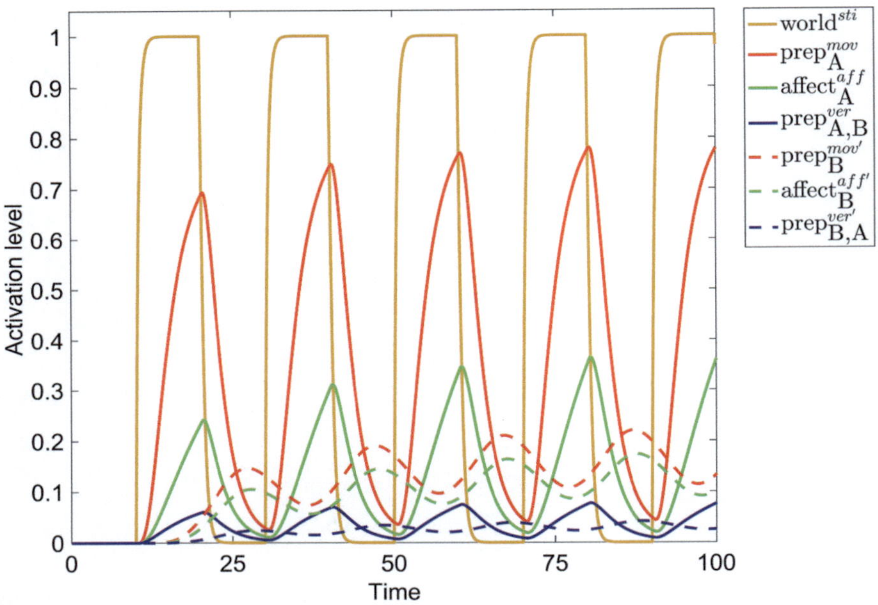

Fig. 16 Activation levels over 100 time units (with 10 non-stimulus time units alternating with 10 stimulus time units) for the preparation states of two agents. Corresponding states of agent A (solid lines) and B (dashed lines) are indicated in the same color, whereas the stimulus *sti* is indicated in yellow

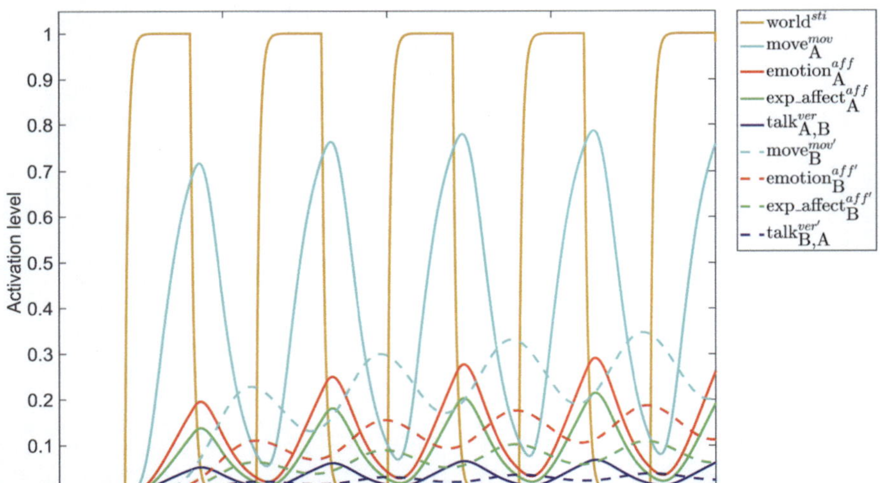

Fig. 17 Activation levels over 100 time units (with 10 non-stimulus time units alternating with 10 stimulus time units) for the execution states of two agents. Corresponding states of agent A (solid lines) and B (dashed lines) are indicated in the same color, whereas the stimulus *sti* is indicated in yellow

Appendix 5

See Figs. 18 and 19.

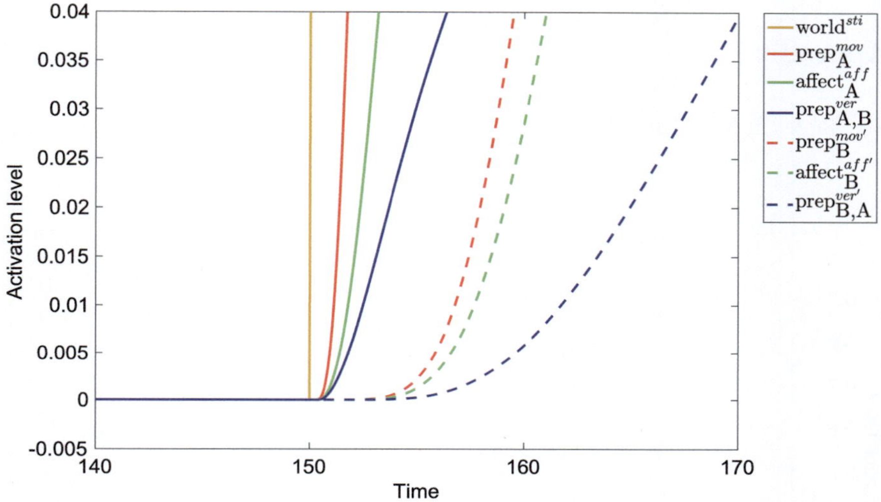

Fig. 18 Activation levels of the preparation states of agent A (solid lines) and B (dashed lines) over time; time is zoomed in at the period 140–170 time units, with 150 being the time unit at which the stimulus *sti* for the first time is present (yellow solid line)

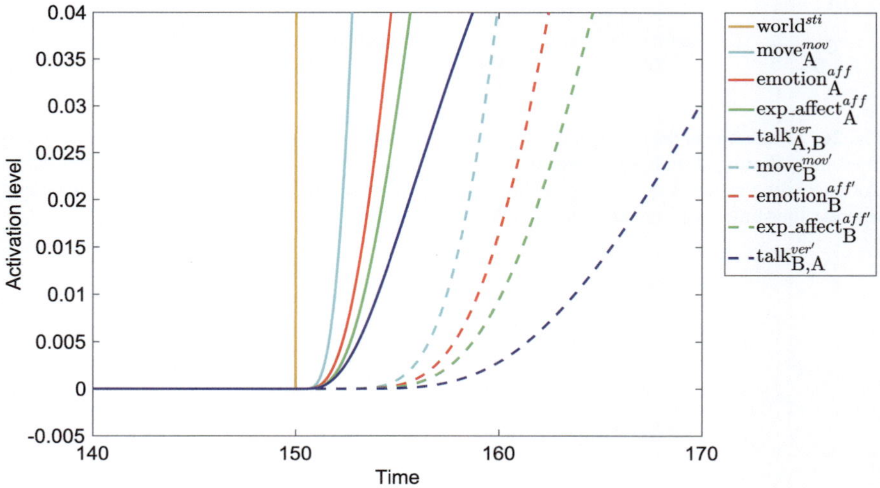

Fig. 19 Activation levels of the execution states of agent A (solid lines) and B (dashed lines) over time; time is zoomed in at the period 140–170 time units, with 150 being the time unit at which the stimulus *sti* for the first time is present (yellow solid line)

Appendix 6

See Figs. 20, 21, 22 and 23.

Appendix 7

See Tables 5, 6, 7, 8 and 9.

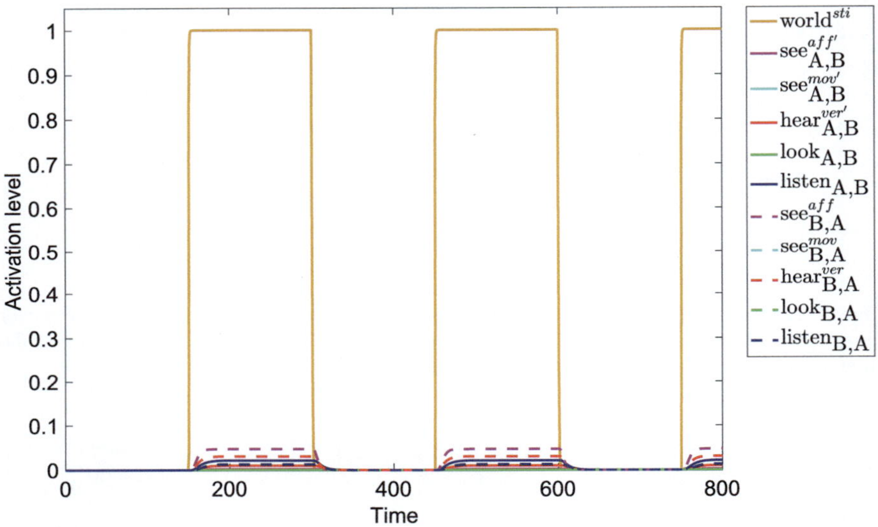

Fig. 20 Simulation of the two agents without movement over 800 time units (with 150 non-stimulus time units alternated with 150 stimulus time units) for several states: the input, look and listen. Corresponding states of agent A (solid lines) and B (dashed lines) are indicated in the same color, whereas the stimulus *sti* is indicated in yellow

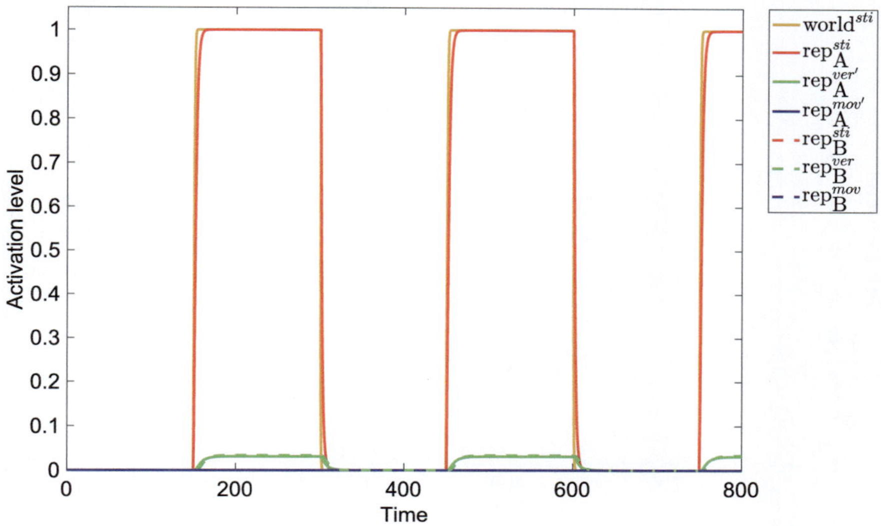

Fig. 21 Simulation of the two agents without movement over 800 time units (with 150 non-stimulus time units alternated with 150 stimulus time units) for the representation states. Corresponding states of agent A (solid lines) and B (dashed lines) are indicated in the same color, whereas the stimulus *sti* is indicated in yellow

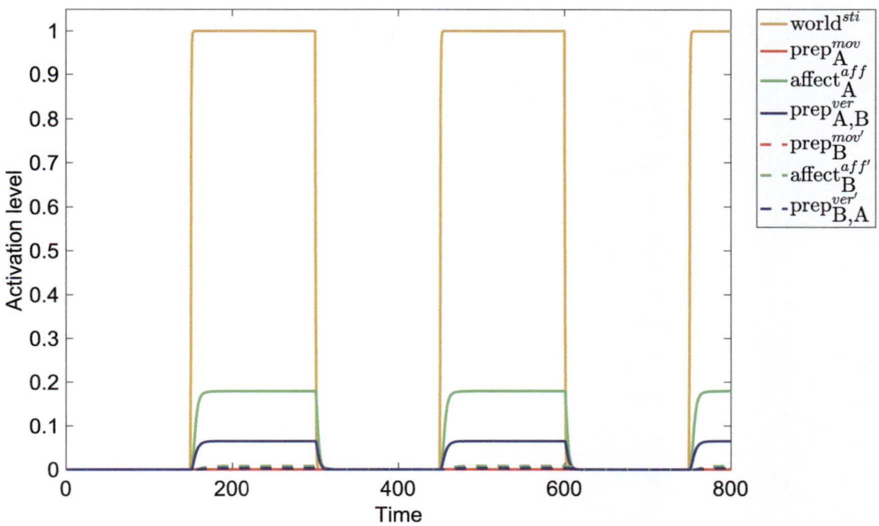

Fig. 22 Simulation of the two agents without movement over 800 time units (with 150 non-stimulus time units alternated with 150 stimulus time units) for the preparation states. Corresponding states of agent A (solid lines) and B (dashed lines) are indicated in the same color, whereas the stimulus *sti* is indicated in yellow

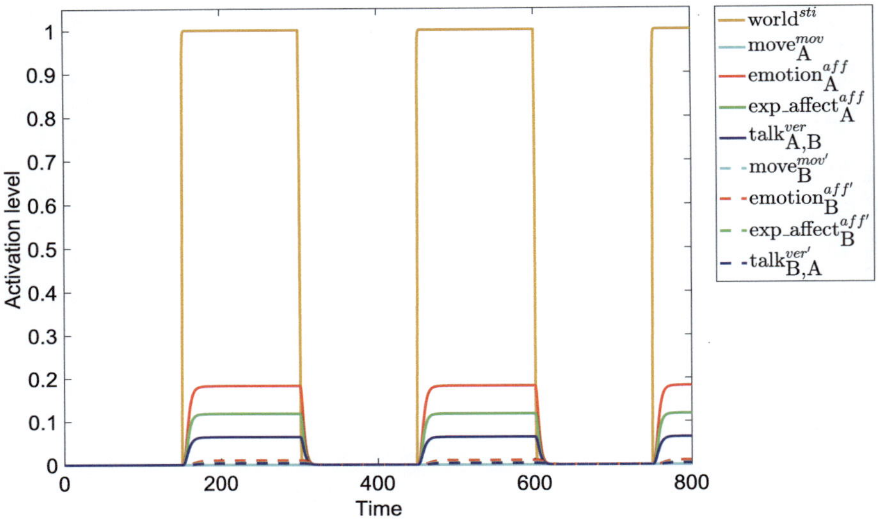

Fig. 23 Simulation of the two agents without movement over 800 time units (with 150 non-stimulus time units alternated with 150 stimulus time units) for the execution states. Corresponding states of agent A (solid lines) and B (dashed lines) are indicated in the same color, whereas the stimulus *sti* is indicated in yellow

Table 5 Connectivity role matrices **mb** (for base connections) of the main simulation

mb base connectivity	State	1	2	3	4	5	6
X_1	world^{sti}	1					
X_2	see$^{aff'}_{A,B}$	15	28				
X_3	see$^{mov'}_{A,B}$	15	26				
X_4	hear$^{ver'}_{A,B}$	16	29				
X_5	rep$^{sti}_{A}$	1	2	3	4		
X_6	rep$^{ver'}_{A}$	2	3	4	10		
X_7	rep$^{mov'}_{A}$	2	3	4	8	9	10
X_8	prep$^{mov}_{A}$	5	7	6			
X_9	affect$^{aff}_{A}$	5	7	6			
X_{10}	prep$^{ver}_{A,B}$	5	6	7	12		
X_{11}	move$^{mov}_{A}$	8					
X_{12}	emotion$^{aff}_{A}$	9	10				

(continued)

Table 5 (continued)

mb base connectivity	State	1	2	3	4	5	6
X_{13}	$\text{exp\_affect}_A^{aff}$	9					
X_{14}	$\text{talk}_{A,B}^{ver}$	10					
X_{15}	$\text{look}_{A,B}$	7					
X_{16}	$\text{listen}_{A,B}$	6					
X_{17}	$\text{see}_{B,A}^{aff}$	30	13				
X_{18}	$\text{see}_{B,A}^{mov}$	30	11				
X_{19}	$\text{hear}_{B,A}^{ver}$	31	14				
X_{20}	rep_B^{sti}	1	17	18	19		
X_{21}	rep_B^{ver}	17	18	19	25		
X_{22}	rep_B^{mov}	17	18	19	23	24	25
X_{23}	$\text{prep}_B^{mov'}$	20	22	21			
X_{24}	$\text{affect}_B^{aff'}$	20	22	21			
X_{25}	$\text{prep}_{B,A}^{ver'}$	20	21	22	27		
X_{26}	$\text{move}_B^{mov'}$	23					
X_{27}	$\text{emotion}_B^{aff'}$	24	25				
X_{28}	$\text{exp\_affect}_B^{aff'}$	24					
X_{29}	$\text{talk}_{B,A}^{ver'}$	25					
X_{30}	$\text{look}_{B,A}$	22					
X_{31}	$\text{listen}_{B,A}$	21					

Table 6 Connectivity role matrices **mcw** (for connection weights) of the main simulation

mcw connection weights	State	1	2	3	4	5	6
X_1	world^{sti}	1					
X_2	$\text{see}_{A,B}^{aff'}$	1	1				
X_3	$\text{see}_{A,B}^{mov'}$	1	1				
X_4	$\text{hear}_{A,B}^{ver'}$	1	1				
X_5	rep_A^{sti}	1	0	0	0		
X_6	$\text{rep}_A^{ver'}$	1	1	1	1		
X_7	$\text{rep}_A^{mov'}$	1	1	1	1	1	1

(continued)

Table 6 (continued)

mcw connection weights	State	1	2	3	4	5	6
X_8	$\text{prep}_{\text{A}}^{mov}$	1	1	1			
X_9	$\text{affect}_{\text{A}}^{aff}$	0.5	2	0.5			
X_{10}	$\text{prep}_{\text{A,B}}^{ver}$	1	1	1	1		
X_{11}	$\text{move}_{\text{A}}^{mov}$	1					
X_{12}	$\text{emotion}_{\text{A}}^{aff}$	1	1				
X_{13}	$\text{exp\_affect}_{\text{A}}^{aff}$	1					
X_{14}	$\text{talk}_{\text{A,B}}^{ver}$	1					
X_{15}	$\text{look}_{\text{A,B}}$	1					
X_{16}	$\text{listen}_{\text{A,B}}$	1					
X_{17}	$\text{see}_{\text{B,A}}^{aff}$	1	1				
X_{18}	$\text{see}_{\text{B,A}}^{mov}$	1	1				
X_{19}	$\text{hear}_{\text{B,A}}^{ver}$	1	1				
X_{20}	$\text{rep}_{\text{B}}^{sti}$	0	0	0	0		
X_{21}	$\text{rep}_{\text{B}}^{ver}$	1	1	1	1		
X_{22}	$\text{rep}_{\text{B}}^{mov}$	1	1	1	1	1	1
X_{23}	$\text{prep}_{\text{B}}^{mov'}$	1	1	1			
X_{24}	$\text{affect}_{\text{B}}^{aff'}$	0.5	2	0.5			
X_{25}	$\text{prep}_{\text{B,A}}^{ver'}$	1	1	1	1		
X_{26}	$\text{move}_{\text{B}}^{mov'}$	1					
X_{27}	$\text{emotion}_{\text{B}}^{aff'}$	1	1				
X_{28}	$\text{exp\_affect}_{\text{B}}^{aff'}$	1					
X_{29}	$\text{talk}_{\text{B,A}}^{ver'}$	1					
X_{30}	$\text{look}_{\text{B,A}}$	1					
X_{31}	$\text{listen}_{\text{B,A}}$	1					

Table 7 Timing role matrix **ms** (for speed factors) and initial values **iv** of the main simulation

	State	ms speed factors	iv initial values
X_1	$world^{sti}$	2	0
X_2	$see_{A,B}^{aff'}$	0.5	0
X_3	$see_{A,B}^{mov'}$	0.5	0
X_4	$hear_{A,B}^{ver'}$	0.5	0
X_5	rep_{A}^{sti}	0.5	0
X_6	$rep_{A}^{ver'}$	0.5	0
X_7	$rep_{A}^{mov'}$	0.5	0
X_8	$prep_{A}^{mov}$	0.5	0
X_9	$affect_{A}^{aff}$	0.5	0
X_{10}	$prep_{A,B}^{ver}$	0.5	0
X_{11}	$move_{A}^{mov}$	0.5	0
X_{12}	$emotion_{A}^{aff}$	0.5	0
X_{13}	$exp\_affect_{A}^{aff}$	0.5	0
X_{14}	$talk_{A,B}^{ver}$	0.5	0
X_{15}	$look_{A,B}$	0.2	0
X_{16}	$listen_{A,B}$	0.2	0
X_{17}	$see_{B,A}^{aff}$	0.5	0
X_{18}	$see_{B,A}^{mov}$	0.5	0
X_{19}	$hear_{B,A}^{ver}$	0.5	0
X_{20}	rep_{B}^{sti}	0.5	0
X_{21}	rep_{B}^{ver}	0.5	0
X_{22}	rep_{B}^{mov}	0.5	0
X_{23}	$prep_{B}^{mov'}$	0.5	0
X_{24}	$affect_{B}^{aff'}$	0.5	0
X_{25}	$prep_{B,A}^{ver'}$	0.5	0
X_{26}	$move_{B}^{mov'}$	0.5	0
X_{27}	$emotion_{B}^{aff'}$	0.5	0
X_{28}	$exp\_affect_{B}^{aff'}$	0.5	0
X_{29}	$talk_{B,A}^{ver'}$	0.5	0
X_{30}	$look_{B,A}$	0.2	0
X_{31}	$listen_{B,A}$	0.2	0

Table 8 Aggregation role matrices **mcfw** (for combination function weights) of the main simulation

mcfw combination function weights	State	1 alogistic	2 step mod
X_1	world^{sti}		1
X_2	$\text{see}_{A,B}^{aff'}$	1	
X_3	$\text{see}_{A,B}^{mov'}$	1	
X_4	$\text{hear}_{A,B}^{ver'}$	1	
X_5	rep_{A}^{sti}	1	
X_6	$\text{rep}_{A}^{ver'}$	1	
X_7	$\text{rep}_{A}^{mov'}$	1	
X_8	prep_{A}^{mov}	1	
X_9	affect_{A}^{aff}	1	
X_{10}	$\text{prep}_{A,B}^{ver}$	1	
X_{11}	move_{A}^{mov}	1	
X_{12}	emotion_{A}^{aff}	1	
X_{13}	$\text{exp\_affect}_{A}^{aff}$	1	
X_{14}	$\text{talk}_{A,B}^{ver}$	1	
X_{15}	$\text{look}_{A,B}$	1	
X_{16}	$\text{listen}_{A,B}$	1	
X_{17}	$\text{see}_{B,A}^{aff}$	1	
X_{18}	$\text{see}_{B,A}^{mov}$	1	
X_{19}	$\text{hear}_{B,A}^{ver}$	1	
X_{20}	rep_{B}^{sti}	1	
X_{21}	rep_{B}^{ver}	1	
X_{22}	rep_{B}^{mov}	1	
X_{23}	$\text{prep}_{B}^{mov'}$	1	
X_{24}	$\text{affect}_{B}^{aff'}$	1	
X_{25}	$\text{prep}_{B,A}^{ver'}$	1	
X_{26}	$\text{move}_{B}^{mov'}$	1	
X_{27}	$\text{emotion}_{B}^{aff'}$	1	
X_{28}	$\text{exp\_affect}_{B}^{aff'}$	1	
X_{29}	$\text{talk}_{B,A}^{ver'}$	1	
X_{30}	$\text{look}_{B,A}$	1	
X_{31}	$\text{listen}_{B,A}$	1	

Table 9 Aggregation role matrices **mcfp** (for combination function parameters) of the main simulation

mcfp combination function parameters	State	1 alogistic		2 step mod	
		σ	τ	ρ	δ
X_1	$world^{sti}$			300	150
X_2	$see^{aff'}_{A,B}$	1	0.4		
X_3	$see^{mov'}_{A,B}$	1	0.4		
X_4	$hear^{ver'}_{A,B}$	1	0.4		
X_5	rep^{sti}_{A}	20	0.6		
X_6	$rep^{ver'}_{A}$	1	0.4		
X_7	$rep^{mov'}_{A}$	1	3		
X_8	$prep^{mov}_{A}$	4	0.8		
X_9	$affect^{aff}_{A}$	5	0.8		
X_{10}	$prep^{ver}_{A,B}$	1	3.5		
X_{11}	$move^{mov}_{A}$	4	0.3		
X_{12}	$emotion^{aff}_{A}$	2	0.3		
X_{13}	$exp\_affect^{aff}_{A}$	2	0.4		
X_{14}	$talk^{ver}_{A,B}$	4	0.3		
X_{15}	$look_{A,B}$	1	0.3		
X_{16}	$listen_{A,B}$	2	0.3		
X_{17}	$see^{aff}_{B,A}$	1	0.4		
X_{18}	$see^{mov}_{B,A}$	1	0.4		
X_{19}	$hear^{ver}_{B,A}$	1	0.4		
X_{20}	rep^{sti}_{B}	20	0.6		
X_{21}	rep^{ver}_{B}	1	0.4		
X_{22}	rep^{mov}_{B}	1	2		
X_{23}	$prep^{mov'}_{B}$	2	0.3		
X_{24}	$affect^{aff'}_{B}$	2	0.6		
X_{25}	$prep^{ver'}_{B,A}$	1	2.5		
X_{26}	$move^{mov'}_{B}$	4	0.1		
X_{27}	$emotion^{aff'}_{B}$	2	0.1		
X_{28}	$exp\_affect^{aff'}_{B}$	2	0.4		
X_{29}	$talk^{ver'}_{B,A}$	4	0.3		
X_{30}	$look_{B,A}$	1	0.5		
X_{31}	$listen_{B,A}$	1	0.4		

References

1. Barsalou, L.W.: Grounded cognition. Annu. Rev. Psychol. **59**, 617–645 (2008)
2. Damasio, A.R.: Descartes' error: emotion, reason, and the human brain. Penguin, London (1994)
3. Damasio, A.R.: The feeling of what happens: body and emotion in the making of consciousness, Harvest Harcourt, San Diego, CA (1999)
4. Dumas, G., Nadel, J., Soussignan, R., Martinerie, J., Garnero, L.: Inter-brain synchronization during social interaction. PLoS ONE **5**(8), e12166 (2010)
5. Feldman, R.: Parent–infant synchrony: Biological foundations and developmental outcomes. Curr. Dir. Psychol. Sci. **16**, 340–345 (2007)
6. Gallese, V., Goldman, A.: Mirror neurons and the simulation theory of mindreading. Trends Cogn. Sci. **2**, 493–501 (1998)
7. Goldman, A.I.: Simulating minds: the philosophy, psychology, and neuroscience of mindreading. Oxford Univ. Press, New York (2006)
8. Grandjean, D., Sander, D., Scherer, K.R.: Conscious emotional experience emerges as a function of multilevel, appraisal-driven response synchronization. Conscious. Cogn. **17**(2), 484–495 (2008)
9. Havas, D.A., Glenberg, A.M., Rinck, M.: Emotion simulation during language comprehension. Psychon. Bull. Rev. **14**(3), 436–441 (2007)
10. Hove, M.J., Risen, J.L.: It's all in the timing: Interpersonal synchrony increases affiliation. Soc. Cogn. **27**(6), 949–960 (2009)
11. Hendrikse, S.C.F., Kluiver, S., Treur, J., Wilderjans, T.F., Dikker, S., Koole, S.L.: How virtual agents can learn to synchronize: an adaptive joint decision-making model of psychotherapy. Cogn. Syst. Res. **79**, 138–155 (2022)
12. Hendrikse, S.C.F., Treur, J., Koole, S.L.: Modeling emerging interpersonal synchrony and its related adaptive short-term affliation and long-term bonding: a second-order multi-adaptive neural agent model. Inter. J. Neural. Syst. **33**(7), 2350038 (2023)
13. Hendrikse, S.C.F., Treur, J., Koole, S.L.: Relationship-specific and relationship-independent behavioural adaptivity in affiliation and bonding: a multi-adaptive dynamical systems approach. Cogn. Syst. Res. **84**, 101182 (2024)
14. Hendrikse, S.C.F., Treur, J., Koole, S.L. (eds.): New analysis and modeling directions in social interaction science: emergent multimodal interpersonal synchrony, affiliation, and bonding. Springer (2025a)
15. Hendrikse, S.C.F., Treur, J., Koole, S.L.: The adaptive interplay of interpersonal synchrony, short-term affiliation and long-term bonding: a multi-timescale multi-adaptive agent model. In: Hendrikse, S.C.F., Treur, J., Koole, S.L. (eds.) New analysis and modeling directions in social interaction science: emergent multimodal interpersonal synchrony, affiliation, and bonding, Chap. 12. Springer (2025b)
16. Hendrikse, S.C.F., Treur, J., Koole, S.L.: A multi-timescale multi-adaptive dynamical system approach for relationship-specific and relationship-independent adaptivity in affiliation and bonding. In: Hendrikse, S.C.F., Treur, J., Koole, S.L. (eds.) New analysis and modeling directions in social interaction science: emergent multimodal interpersonal synchrony, affiliation, and bonding, Chap. 13. Springer (2025c)
17. Hendrikse, S. C. F., Treur, J., Wilderjans, T. F., Dikker, S., Koole, S.L.: On the same wavelengths: emergence of multiple synchronies among multiple agents. In: Proceedings of the 22nd International Workshop on Multi-Agent Systems and Agent-Based Simulation, MABS'21 Lecture Notes in Computer Scienc, vol 13128, pp. 57–71. Springer, Cham (2021)
18. Hendrikse, S.C.F., Treur, J., Wilderjans, T.F., Dikker, S., Koole, S.L.: On becoming in sync with yourself and others: an adaptive agent model for how persons connect by detecting intrapersonal and interpersonal synchrony. Hum.-Centric Intell. Syst. **3**(2), 123–146 (2023)
19. Hendrikse, S.C.F., Treur, J., Wilderjans, T.F., Dikker, S., Koole, S.L.: Growing connectedness through synchrony: the role of subjective analysis of intrapersonal synchrony and interpersonal synchrony within an adaptive agent model. In: Hendrikse, S.C.F., Treur, J., Koole, S.L.

(eds.) New analysis and modeling directions in social interaction science: emergent multimodal interpersonal synchrony, affiliation, and bonding, Chap. 11. Springer (2025d)
20. Hendrikse, S.C.F., Treur, J., Wilderjans, T.F., Dikker, S., Koole, S.L.: The role of timing in synchrony detection and social interaction adaptivity: including time lags in a second-order adaptive agent model. In: Hendrikse, S.C.F., Treur, J., Koole, S.L. (eds.) New analysis and modeling directions in social interaction science: emergent multimodal interpersonal synchrony, affiliation, and bonding, Chap. 13. Springer (2025e)
21. Hesslow, G.: Conscious thought as simulation of behaviour and perception. Trends Cogn. Sci. **6**, 242–247 (2002)
22. Heyes, C., Catmur, C.: What happened to mirror neurons? Perspectives on psychological science e1745691621990638 (2021)
23. Iacoboni, M.: Mirroring people: the new science of how we connect with others. Farrar, Straus & Giroux, New York (2008)
24. IJzerman, H., Koole, S.L.: From perceptual rags to metaphoric riches: bodily, social, and cultural constraints on sociocognitive metaphors. Psychol. Bull. **137**, 355–361 (2011). https://doi.org/10.1037/a0022373
25. Ireland, M.E., Pennebaker, J.W.: Language style matching in writing: synchrony in essays, correspondence, and poetry. J. Pers. Soc. Psychol. **99**, 549–571 (2010). https://doi.org/10.1037/a0020386
26. Ivanov, P.C., Ma, Q.D.Y., Bartsch, R.P.: Maternal–fetal heartbeat phase synchronization. Proc. Natl. Acad. Sci. **106**(33), 13641–13642 (2009)
27. Konvalinka, I., Xygalatas, D., Bulbulia, J., Schjødt, U., Jegindø, E. M., Wallot, S, Roepstorff, A.: Synchronized arousal between performers and related spectators in a fire-walking ritual. Proc. Natl. Acad. Sci. **108**(20), 8514–8519 (2011)
28. Koole, S.L., Tschacher, W.: Synchrony in psychotherapy: a review and an integrative framework for the therapeutic alliance. Front. Psychol. **7**, e862 (2016)
29. Koole, S.L., Tschacher, W., Butler, E., Dikker, S., Wilderjans, T.F.: In sync with your shrink. In: Forgas, J.P., Crano, W.D., Fiedler, K. (eds.) Applications of Social Psychology, pp. 161–184. Taylor and Francis, Milton Park (2020)
30. Mattera, F., Hendrikse, S.C.F., Treur, J.: An adaptive network model for the emergence of group synchrony and behavioral adaptivity for group bonding. In Proceeding of the 12th International Conference on Complex Networks and Their Applications. Studies in Computational Intelligence, vol. 1144, pp. 53–66. Springer, Cham (2024)
31. Miles, L.K., Lumsden, J., Richardson, M.J., Macrae, C.N.: Do birds of a feather move together? Group membership and behavioral synchrony. Exper. Brain Res. **211**(3–4), 495–503 (2011)
32. Nguyen, T., Abney, D. H., Salamander, D., Bertenthal, B., Hoehl, S.: Social touch is associated with neural but not physiological synchrony in naturalistic mother-infant interactions. BioRxiv (2021)
33. Niedenthal, P.M.: Embodying emotion. Science **316**, 1002–1005 (2007)
34. Palumbo, R.V., Marraccini, M.E., Weyandt, L.L., Wilder-Smith, O., McGee, H.A., Liu, S., Goodwin, M.S.: Interpersonal autonomic physiology: a systematic review of the literature. Pers. Soc. Psychol. Rev. **21**(2), 99–141 (2017)
35. Pan, Y., Novembre, G., Song, B., Zhu, Y., Hu, Y.: Dual brain stimulation enhances interpersonal learning through spontaneous movement synchrony. Soc. Cogn. Affect. Neurosci. **16**(1–2), 210–221 (2021)
36. Pickering, M.J., Garrod, S.: Toward a mechanistic psychology of dialogue. Behav. Brain Sci. **27**, 169–190 (2004). https://doi.org/10.1017/S0140525X04000056
37. Pickering, M.J., Garrod, S.: Alignment as the basis for successful communication. Res. Lang. Comput. **4**(2), 203–228 (2006)
38. Pickering, M.J., Garrod, S.: Understanding dialogue: language use and social interaction. Cambridge University Press (2021)
39. Pikovsky, A., Rosenblum, M., Kurths, J.: Synchronization: a universal concept in nonlinear sciences. Cambridge University Press, New York, NY (2003)

40. Repp, B.H., Su, Y.-H.: Sensorimotor synchronization: a review of recent research (2006–2012). Psychon. Bull. Rev. **20**(3), 403–452 (2013)
41. Richardson, M.J., Marsh, K.L., Isenhower, R.W., Goodman, J.R.L., Schmidt, R.C.: Rocking together: dynamics of intentional and unintentional interpersonal coordination. Hum. Mov. Sci. **26**(6), 867–891 (2007)
42. Rizzolatti, G., Sinigaglia, C.: Mirrors in the brain: how our minds share actions and emotions. Oxford University Press (2008)
43. Russell, J.A.: Core affect and the psychological construction of emotion. Psychol. Rev. **110**(1), 145 (2003)
44. Shamay-Tsoory, S.G., Saporta, N., Marton-Alper, I.Z., Gvirts, H.Z.: Herding brains: a core neural mechanism for social alignment. Trends Cogn. Sci. **23**, 174–186 (2019)
45. Strogatz, S.: Sync: the emerging science of spontaneous order. Hyperion Books, New York, NY (2003)
46. Torre, J.B., Lieberman, M.D.: Putting feelings into words: affect labeling as implicit emotion regulation. Emot. Rev. **10**(2), 116–124 (2018)
47. Treur, J.: Network-oriented modeling: addressing complexity of cognitive. Springer Publishers, Affective and Social Interactions (2016)
48. Treur, J.: A modeling environment for reified temporal-causal networks: modeling plasticity and metaplasticity in cognitive agent models. In: Proc. of the 22nd International Conference on Principles and Practice of Multi-Agent Systems, PRIMA'19. Lecture Notes in Computer Science, vol 11873, pp. 487–495. Springer, Cham (2019)
49. Treur, J.: Network-oriented modeling for adaptive networks: designing higher-order adaptive biological, mental and social network models. Springer Nature, Cham (2020)
50. Treur, J.: Modeling coevolution of dynamics, adaptivity and control of adaptivity by self-modeling networks. In: Hendrikse, S.C.F., Treur, J., Koole, S.L. (eds.), New Analysis and modeling directions in social interaction science: emergent multimodal interpersonal synchrony, affiliation, and bonding, Chap. 2. Springer (2025)
51. Valdesolo, P., DeSteno, D.: Synchrony and the social tuning of compassion. Emotion **11**(2), 262–266 (2011)
52. Van Der Steen, M.C., Keller, P.E.: The adaptation and anticipation model (ADAM) of sensorimotor synchronization. Front. Hum. Neurosci. **7**, e253 (2013)
53. Van Leeuwen, P., Geue, D., Thiel, M., Cysarz, D., Lange, S., Romano, M.C., Grönemeyer, D.H.: Influence of paced maternal breathing on fetal–maternal heart rate coordination. Proc. Nat. Acad. Sci. **106**(33), 13661–13666 (2009)
54. van Ulzen, N.R., Lamoth, C.J.C., Daffertshofer, A., Semin, G.R., Beek, P.J.: Characteristics of instructed and uninstructed interpersonal coordination while walking side-by-side. Neurosci. Lett. **432**(2), 88–93 (2008)
55. Williams, L.E., Huang, J.Y., Bargh, J.A.: The scaffolded mind: higher mental processes are grounded in early experience of the physical world. Eur. J. Soc. Psychol. **39**(7), 1257–1267 (2009)
56. Williams, A. H., O'Leary T., Marder, E.: Homeostatic regulation of neuronal excitability. Scholarpedia **8**, 1656 (2013)
57. Wiltermuth, S.S., Heath, C.: Synchrony and cooperation. Psych. Sci. **20**(1), 1–5 (2009)

Interpersonal Emotion Regulation Through Attunement of Emotions in Human-Bot Interaction: An Agent-Based Network Model

Edgar Eler, Filippos Dimopoulos, Marieke Timmers, Jan Treur, and Sander L. Koole

Abstract The study reported in this chapter analyses the effects of emotion coregulation in the interaction between a distressed human and a bot. Through the coregulation premises, an emotion contagion process impacts the emotion system, which initially increases the distress level of the bot due to the influence of the human's emotions. Configuring the bot with optimal emotion regulation decreases its distress levels as it can self-regulate its emotions, which affects the human's emotions, regulating them by the emotion contagion process. The results show how this coregulation process occurs.

1 Introduction

Emotions are commonly strong and unpredictable in the early stages of human development. Caregivers must interact with their children to control those emotions, avoiding potentially harmful bodily responses [1]. Over the years, through cognitive development, most humans improve their internal emotion control mechanisms and become able to self-regulate their emotions [2]. Nonetheless, due to the dynamics of human life, several situations can trigger deeper emotion states that affect

E. Eler · F. Dimopoulos · M. Timmers · J. Treur (✉)
Social AI Group, Department of Computer Science, Vrije Universiteit Amsterdam, Amsterdam, The Netherlands
e-mail: j.treur@vu.nl

E. Eler
e-mail: e.eler@vu.nl

F. Dimopoulos
e-mail: f.dimopoulos@student.vu.nl

E. Eler · S. L. Koole
Department of Clinical Psychology, Vrije Universiteit Amsterdam, Amsterdam, The Netherlands
e-mail: s.l.koole@vu.nl

S. C. F. Hendrikse et al. (eds.), *New Analysis and Modeling Directions in Social Interaction Science*, Studies in Systems, Decision and Control 614,
https://doi.org/10.1007/978-3-031-99968-0_4

self-regulation mechanisms. Higher levels of distress can decrease an individual's capacity to process and control emotions. This scenario expands to a societal level when the increase in mental health problems is considered [3].

One potential coping mechanism for distressed humans is social interaction, by which they engage in interpersonal emotion regulation [4]. The mere presence of others can be seen as a potential solution to help regulate emotions. By affecting each other's emotions, a pattern of mutual emotion regulation is developed, which can result in stability over time. This can be defined as coregulation [5].

Although interpersonal emotion regulation can be beneficial, social interaction can be challenging in some situations. Aspects of personal life, socioeconomic conditions, global events, and changes in the surrounding environment can negatively impact an individual's ability to interact with others [6]. Advances in artificial intelligence and virtualization enable the development of synthetic agents that can provide personal companionship to humans dealing with those situations [7].

This chapter presents an agent-based network model that simulates the social interactions between a distressed human and a bot that supports interpersonal emotion regulation, based on attunement of emotions by the emotion contagion process. Following this section, the theoretical background is provided. The third section describes the computational network model used. The fourth section presents the results of the simulations for different settings. The fifth and sixth sections include the conclusion of our research and a discussion around future work and ambitions based on this project.

2 Background

Emotion theorists have put forward different definitions of what emotion is. According to the American Psychological Association (APA), emotion is defined as "a complex reaction pattern, involving experiential, behavioural and physiological elements." In the Dictionary of Cognitive Psychology [8], although not formally defined, it is described as a mental state. Subsequently, according to the James-Lange theory [9] and [10], emotional stimuli first induce peripheral physiological variations, which occur without consciousness of affect. These bodily responses are further interpreted by the brain to produce the feeling state of an emotion [11].

Over the last decades, emotion regulation has become a more and more interesting research topic for diverse fields, such as psychology and neuroscience, where a point of interest is the mechanism that governs the emotions biology-wise, and, more recently, artificial intelligence, where a recreation of the emotion regulation that takes place by an interaction between prefrontal cortex and amygdala [12, 13] is addressed computationally. Most of this literature focuses on individual emotion regulation, but recently the interest in interpersonal emotion regulation has grown as well.

In the remainder of this section, the agent modeling approach by (self-modeling) temporal-causal networks used is briefly introduced [14–15], see also this volume

Chapter Modeling Coevolution of Dynamics, Adaptivity and Control of Adaptivity by Self-Modeling Networks [16]. In this approach *nodes* Y in a network have activation values $Y(t)$ that are dynamic over time t; therefore, they serve as state variables and will usually be called *states*. To specify these dynamics, the states are considered to affect each other by the connections within the network. Following [14], a basic *network structure* is characterised by:

- **connectivity characteristics**

  Connections from a state X to a state Y and their *weights* $\boldsymbol{\omega}_{X,Y}$.
- **aggregation characteristics**

  For any node Y, some *combination function* $\mathbf{c}_Y(\ldots)$ defines aggregation that is applied to the *single impacts* $\boldsymbol{\omega}_{X_i,Y}X_i(t)$ on Y from its incoming connections from states $X_1, \ldots, X_k$.
- **timing characteristics**

  Each state Y has a *speed factor* $\boldsymbol{\eta}_Y$ defining how fast it changes upon given impact.

Here, the states X_i and Y have activation levels $X_i(t)$ and $Y(t)$ that vary (often within the [0, 1] interval) over time, described by real numbers t. The dynamics of such networks are described by the following difference (or differential) equations that incorporate in a canonical manner the network characteristics $\boldsymbol{\omega}_{X,Y}$, $\mathbf{c}_Y(\ldots)$, $\boldsymbol{\eta}_Y$:

$$Y(t + \Delta t) = Y(t) + \boldsymbol{\eta}_Y\left[\mathbf{c}_Y\left(\boldsymbol{\omega}_{X_1,Y}X_1(t), \ldots, \boldsymbol{\omega}_{X_k,Y}X_k(t)\right) - Y(t)\right]\Delta t \qquad (1)$$

for any state Y and where $X_1, \ldots, X_k$ are the states from which Y gets its incoming connections. The Eqs. (1) are useful for simulation purposes and also for analysis of properties of the emerging behaviour of such network models. The overall combination function $\mathbf{c}_Y(\ldots)$ for state Y is taken as the weighted average of some of the available basic combination functions $\mathbf{c}_j(\ldots)$ by specified weights $\boldsymbol{\gamma}_{j,Y}$, and parameters $\boldsymbol{\pi}_{1,j,Y}$, $\boldsymbol{\pi}_{2,j,Y}$ of $\mathbf{c}_j(\ldots)$, for Y:

$$\mathbf{c}_Y(V_1, \ldots, V_k) = \frac{\boldsymbol{\gamma}_{1,Y}\mathbf{c}_1(V_1, \ldots, V_k) + \ldots + \boldsymbol{\gamma}_{m,Y}\mathbf{c}_m(V_1, \ldots, V_k)}{\boldsymbol{\gamma}_{1,Y} + \ldots + \boldsymbol{\gamma}_{m,Y}} \qquad (2)$$

Such Eqs. (1) and (2) are hidden in the dedicated software environment that can be used for simulation and analysis; see [15, 17]

https://www.researchgate.net/publication/368775720

For the combination functions used here, see Table 1.

Table 1 Combination functions from the library used here

Name	Notation	Formula	Parameters
Euclidean	$\mathbf{eucl}_{n,\lambda}(V_1, \ldots, V_k)$	$\sqrt[n]{\frac{V_1^n+\ldots+V_k^n}{\lambda}}$	Order $\boldsymbol{n}$ Scaling factor $\boldsymbol{\lambda}$
Advanced logistic	$\mathbf{alogistic}_{\sigma,\tau}(V_1, \ldots, V_k)$	$\left[\frac{1}{1+e^{-\sigma(V_1+\ldots+V_k-\tau)}} - \frac{1}{1+e^{\sigma\tau}}\right](1+e^{-\sigma\tau})$	Steepness $\boldsymbol{\sigma} > 0$ Threshold $\boldsymbol{\tau}$

3 The Introduced Social Agent Model

The social multi-agent model introduced in this chapter was designed by considering the mechanisms involved in the internal emotion regulation processes of two agents *a* and *b*, a human and a bot, and the dynamics of their interaction for emotion contagion. The cyclic mechanisms for individual emotion regulation are based on the conceptual causal model from Treur [14, Ch. 3]. The conceptual representation of the social agent model is illustrated in Fig. 1, in which the circles represent the mental states, the arrows represent the connections, and the shaded areas in the background represent the individuals.

The individual emotion regulation process is modelled in the same way for both agents *a* and *b*. For each agent:

- the stimulus from the world/environment is represented by the *ws* states
- the *ps* state represents the preparation for the emotion response
- the *es* state represents the expressed emotion
- the *fs* state represents the felt emotion
- the *cs* state represents the internal emotion regulation control.

The connections that involve the emotion regulation control states *cs* are displayed as dotted arrows to illustrate the emotion control processes of monitoring the stimulus (incoming connections) and suppressing the emotion response *ps* and felt emotion *fs* (outgoing connections).

The interpersonal emotion regulation process conceptualized in Fig. 1 has the premise that the expressed emotion *es* of both individuals will be the interface by

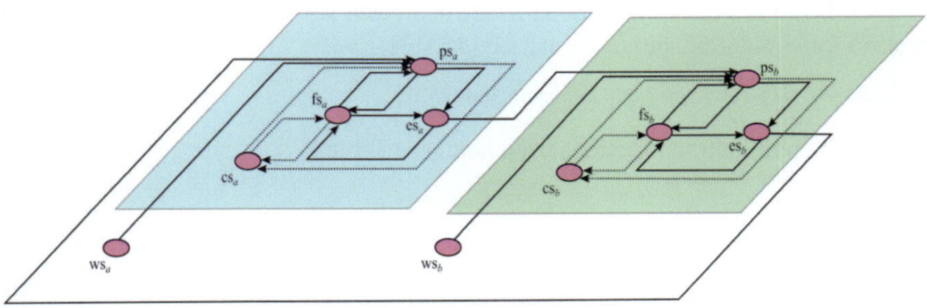

Fig. 1 Social (multi-)agent model for the human-bot interaction

which the emotions of one individual will impact the emotions of the other individual and vice-versa. Skipping for the sake of simplicity the process of sensing and representing incoming stimuli, there are direct connections from the *es* state of one agent to the emotion response preparation state *ps* of the other. This preparation state is used as the primary state that receives the external stimuli (from *ws* and *ps*), the suppression effect from the emotion regulation control state *cs* and the impact of the emotion feeling state *fs*. The latter feeling state receives impacts from the emotion preparation state *ps*, the expressed emotion state es, and from the regulation control state *cs* that suppresses it. The expressed emotion state *es* receives impact from the emotion preparation and feeling states *ps* and *fs*. As the expressed emotion and the emotion felt mutually affect each other, which also impacts the emotion preparation state in a bi-directional manner [14], some loops were included in the model to simulate the cyclic form in which those states affect each other.

The base idea of this scenario is a distressed human (agent *a*) with a poorly functioning individual emotion regulation system that gets in contact with a bot (agent *b*), and due to the interaction with the bot, the human's emotion is regulated. The tables in Fig. 2 are the role matrices of the designed social (multi-)agent model. They show the values of the network characteristics (serving as the model parameters) defined for the main simulation. In this scenario, the human had its ability to regulate emotions individually reduced due to its distress situation, so the table *mcwv* illustrated in Fig. 2, shows that the two incoming connections for the emotion regulation control state of the human (cs_a) have weight 0.1 while the bot has its internal emotion regulation fully operational, with the two incoming connections to the bot's emotion regulation control state (cs_b) have the maximum weight 1.0.

4 Simulation Results

The expected outcome of the interaction between the distressed human and the bot is an initial increase in the bot's distress levels due to the emotion contagion from the high level of distress of the human. Once the well-functioning internal emotion regulation control of the bot gets activated, its distress level drops significantly. Due to that, by emotion contagion also the human's emotion gets regulated. This will support the idea of interpersonal emotion regulation through coregulation as described by [18]. Figure 3 illustrates the expected results, with the curves representing the feeling states fs_a of the human (agent *a*) and fs_b of the bot (agent *b*).

Simulation 1—Main Scenario

For the main scenario, the simulation was performed following all network characteristics illustrated in Fig. 2. The human is considered to have a high distress level, while having very poor individual emotion control, and the bot with a perfectly operational individual emotion control. The results for Simulation 1 are displayed in Fig. 4. The results start with a steep and rapidly increasing activation level in all non-control states. After this, once the control state cs_b of the bot gets activated, the

mb	base connectivity	1	2	3	4
X_1	wsa	X_1			
X_2	fsa	X_3	X_4	X_5	
X_3	psa	X_1	X_2	X_4	X_{10}
X_4	csa	X_2	X_3		
X_5	esa	X_2	X_3		
X_6	wsb	X_6			
X_7	fsb	X_8	X_9	X_{10}	
X_8	psb	X_5	X_6	X_7	X_9
X_9	csb	X_7	X_8		
X_{10}	esb	X_7	X_8		

mcwv	connection weights	1	2	3	4
X_1	wsa	1			
X_2	fsa	1	-1	0.2	
X_3	psa	0.5	0.5	-1	1
X_4	csa	0.1	0.1		
X_5	esa	1	1		
X_6	wsb	1			
X_7	fsb	1	-1	0.2	
X_8	psb	1	1	1	-1
X_9	csb	1	1		
X_{10}	esb	1	1		

msv	speed factors	1
X_1	wsa	0
X_2	fsa	1
X_3	psa	1
X_4	csa	0.005
X_5	esa	0.5
X_6	wsb	0
X_7	fsb	1
X_8	psb	1
X_9	csb	0.005
X_{10}	esb	0.5

mcfwv	combination function weights	1 eucl	2 alogistic
X_1	wsa	1	
X_2	fsa		1
X_3	psa		1
X_4	csa		1
X_5	esa		1
X_6	wsb	1	
X_7	fsb		1
X_8	psb		1
X_9	csb		1
X_{10}	esb		1

mcfpv	combination function parameters	1		2	
	function	eucl		alogistic	
	parameter	n	λ	σ	τ
X_1	wsa	1	1		
X_2	fsa			8	0.95
X_3	psa			5	1.1
X_4	csa			5	0.5
X_5	esa			5	0.5
X_6	wsb	1	1		
X_7	fsb			8	0.95
X_8	psb			5	1.1
X_9	csb			5	0.5
X_{10}	esb			5	0.5

iv	initial values	1
X_1	wsa	1
X_2	fsa	0
X_3	psa	0
X_4	csa	0
X_5	esa	0
X_6	wsb	1
X_7	fsb	0
X_8	psb	0
X_9	csb	0
X_{10}	esb	0

Fig. 2 Network characteristics representing the multi-agent model

feeling state fs_b of the bot immediately starts to decrease, which by emotion contagion also impacts the human feeling state fs_a. A tipping point occurs at time point 441.5, significantly dropping all states' values.

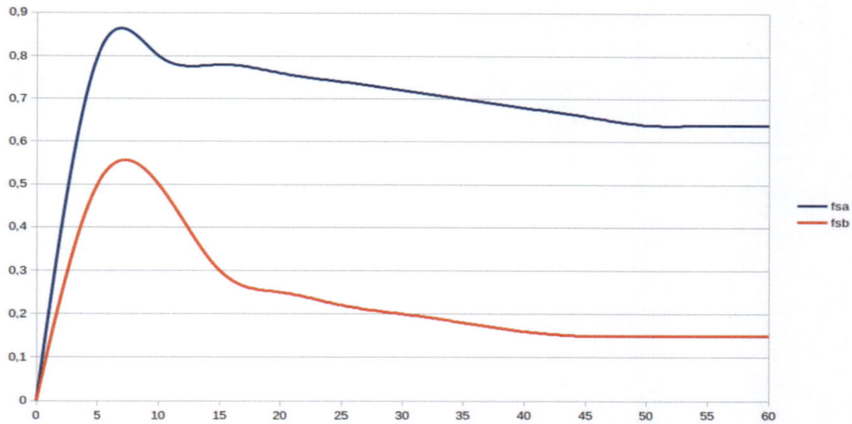

Fig. 3 Expected results for the main simulation

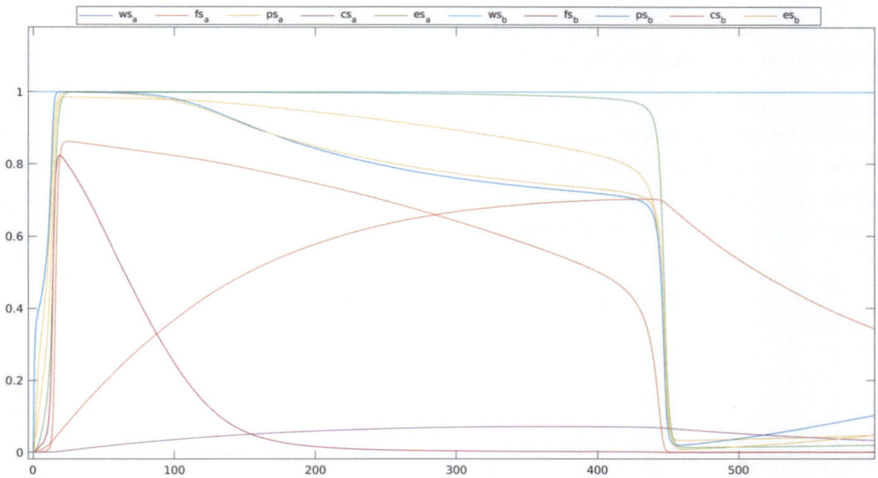

Fig. 4 Main simulation result

Simulation 2—Human with Better Emotion Regulation Control

In Simulation 2, we increased the emotion regulation control of the human by increasing the weight of the two incoming connections of cs_a from 0.1 to 0.5, which resulted in a similar outcome to Simulation 1, but with the emotion being regulated almost four times faster, which is noticeable by the tipping point moving from the time point 441 to around 120. Figure 5 illustrates those results.

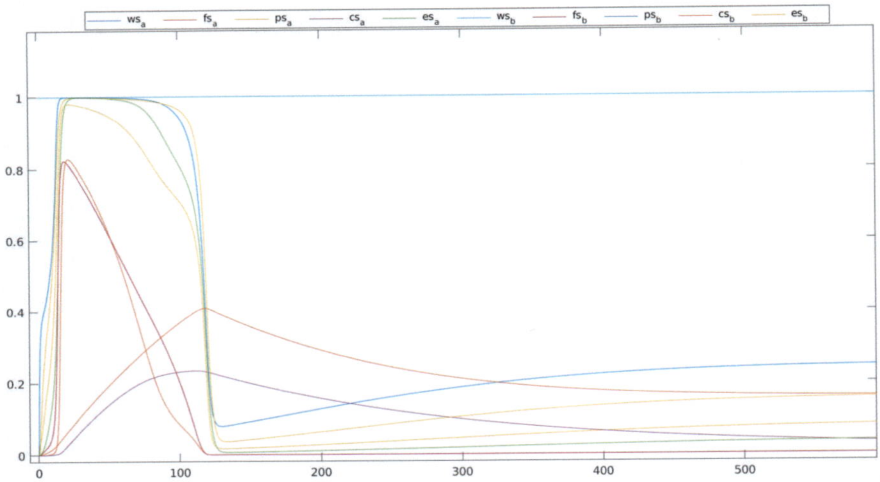

Fig. 5 Simulation results with the human configured with an increased emotion control

Simulation 3—Bot with Poorer Emotion Regulation Control

A third simulation was performed for further validation of the model, but now decreasing the emotion regulation control of the bot. In this simulation, the effects of the changes in an individual's emotions have less impact on the other's emotions, which makes this a more realistic scenario when compared to the expected results. Figure 6 illustrates this simulation.

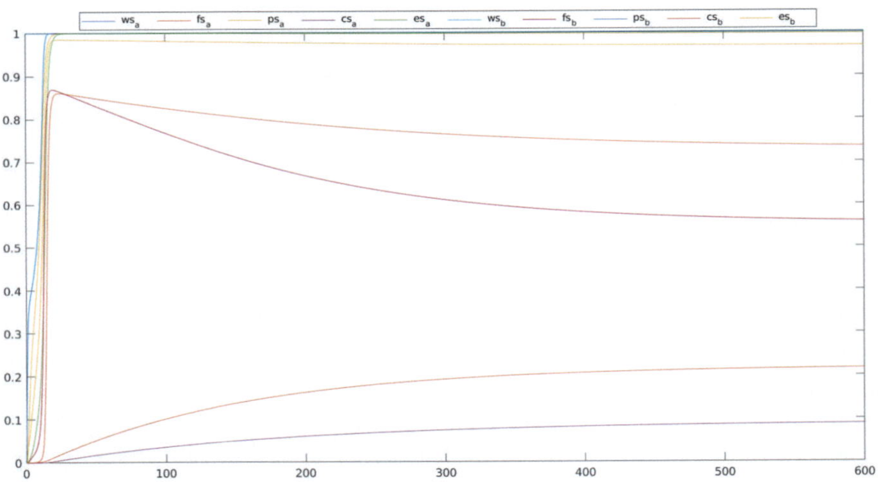

Fig. 6 Simulation results with the bot configured with a poorer emotion regulation control

5 Verification of the Model by Mathematical Analysis

Verification by mathematical analysis can be done to check correctness of an AI model; e.g., [19]. To verify the results obtained from the simulations against a network model design, analysis of stationary points can be done [20]. For any state Y, a time point in which it is stationary can be selected, a time t where approximately $dY(t)/dt = 0$, or approximately $Y(t + \Delta t) = Y(t)$. Given Eq. (1) from Sect. 2 and a speed factor $\boldsymbol{\eta}_Y$ that is nonzero, this is equivalent to approximately

$$\mathbf{c}_Y\big(\boldsymbol{\omega}_{X_1,Y}X_1(t), \ldots, \boldsymbol{\omega}_{X_k,Y}X_k(t)\big) - Y(t) = 0$$

For this t, from the simulation values and network characteristics (using the values and weights of the connected states in the combination function), the aggregated impact $\mathbf{c}_Y\big(\boldsymbol{\omega}_{X_1,Y}X_1(t), \ldots, \boldsymbol{\omega}_{X_k,Y}X_k(t)\big)$. Then by comparing that to the value $Y(t)$ from the simulation, the deviation can be found:

$$\text{Deviation} = \mathbf{c}_Y\big(\boldsymbol{\omega}_{X_1,Y}X_1(t), \ldots, \boldsymbol{\omega}_{X_k,Y}X_k(t)\big) - Y(t)$$

Tables 2, 3, and 4 display the absolute values of these deviations.

In Table 2, most of the results performed by the simulation match the results obtained from the verification formula, except for the state cs_b, which is highly decreasing its value in the time point 27, so it is not stationary. This is also applicable to the results obtained in Table 3. Simulation 3 results in the best matching among all simulations, as described in Table 4. The deviation values are more consistent because, in time points selected from Simulation 3, the states are more stationary than in the other simulations.

6 Validation by Parameter Tuning for the Expected Results

To validate the model according to Treur [21], data was extracted from the expected results described in Sect. 4 and illustrated in Fig. 3. Six distinct time points of the *fs* and *cs* states of both individuals were selected to represent expected data to be compared with the results obtained by the model. Table 5 displays the values of the *fs* and *cs* states for the selected time points.

To compare the expected results to the results obtained by the model, a computational iterative parameter tuning process was performed, running Simulation 1 and obtaining the deviations between the simulated data and the expected data from Table 5. The speed factor parameters $\boldsymbol{\eta}_{cs_a}$, $\boldsymbol{\eta}_{es_a}$, $\boldsymbol{\eta}_{cs_b}$, $\boldsymbol{\eta}_{es_b}$ of the *cs* and *es* states of both individuals were selected to be tuned by Simulated Annealing [22] in order to reach simulation values close to the expected data. Those parameters started with a random value of 0.5 and could get any value between 0 and 1 over the iterations.

Table 2 Verification of the agent model for simulation 1

	ws_a	fs_a	ps_a	cs_a	es_a	ws_b	fs_b	ps_b	cs_b	es_b
Maxima										
Time point	27	27	27	27	27	27	27	27	27	27
State value	1	0.860604	0.983853	0.000632922	0.998291	1	0.782143	0.999657	0.0939761	0.998228
Aggimpact	1	0.86558205331	0.999599612034	0.001684590169	0.999219780921	1	0.775906251462	0.999999072862	0.998712672848	0.998712672848
Deviation	0	0.00497805331	0.015746612034	0.001051668169	0.000928780921	0	0.006236748538	0.000342072862	0.904736572848	0.000484672848
Minima										
Time point	1168	1168	1168	1168	1168	1168	1168	1168	1168	1168
State value	1	0.000590792	0.093897	0.00512246	0.0437967	1	0.00082221	0.25034	0.160733	0.159971
Aggimpact	1	0.000590920134	0.085907780713	0.00003927452	0.000564674225	1	0.000822337324	0.370455615714	0.00321993606	0.00321993606
Deviation	0	0.000000128134	0.007989219287	0.00508318548	0.043232025775	0	0.000000127324	0.120115615714	0.15751306394	0.15675106394

Table 3 Results of the verification of the agent model for simulation 2

	ws_a	fs_a	ps_a	cs_a	es_a	ws_b	fs_b	ps_b	cs_b	es_b
Maxima										
Time point	22	22	22	22	22	22	22	22	22	22
State value	1	0.827915	0.979885	0.0332754	0.990933	1	0.811244	0.999712	0.0710132	0.996856
Aggimpact	1	0.826032257842	0.999401092419	0.4085351643	0.998954171125	1	0.805955223948	0.999999351634	0.998980219085	0.998980219085
Deviation	0	0.001882742158	0.019516092419	0.3752597643	0.008021171125	0	0.005288776052	0.000287351634	0.927967019085	0.002124219085
Minima										
Time point	598	598	598	598	598	598	598	598	598	598
State value	1	0.000272811	0.0802717	0.0329552	0.0362027	1	0.000782667	0.245267	0.158441	0.155139
Aggimpact	1	0.000230428065	0.067304828865	0.000190105539	0.000452357302	1	0.000783319838	0.297179171286	0.003071373016	0.003071373016
Deviation	0	0.000042382935	0.012966871135	0.032765094461	0.035750342698	0	0.000000652838	0.051912171286	0.155369626984	0.152067626984

Table 4 Results of the verification of the agent model for simulation 3

	ws_a	fs_a	ps_a	cs_a	es_a	ws_b	fs_b	ps_b	cs_b	es_b
Maxima										
Time point	23	23	23	23	23	23	23	23	23	23
State value	1	0.861055	0.983893	0.00472622	0.996088	1	0.866407	0.999837	0.0170026	0.998299
Aggimpact	1	0.861343555915	0.999587269006	0.001685446746	0.999222837234	1	0.865224124617	0.999999740222	0.009319172361	0.999344496597
Deviation	0	0.000288555915	0.015694269006	0.003040773254	0.003134837234	0	0.001182875383	0.000162740222	0.007683427639	0.001045496597
Minima										
Time point	598	598	598	598	598	598	598	598	598	598
State value	1	0.73439	0.966465	0.0887453	0.997337	1	0.559823	0.997971	0.216742	0.994569
Aggimpact	1	0.644741941855	0.998619114642	0.001447913094	0.99754296147	1	0.5597749555	0.9998789599	0.005517176954	0.992323213233
Deviation	0	0.089648058145	0.032154114642	0.087297386906	0.00020596147	0	0.0000480445	0.0019079599	0.211224823046	0.002245786767

Table 5 Expected data for states at given time points

	10	20	50	100	300	500
fs_a	0.05	0.3	0.9	0.75	0.7	0.65
cs_a	0.00	0.01	0.02	0.05	0.08	0.09
fs_b	0.01	0.28	0.6	0.45	0.4	0.35
cs_b	0.02	0.06	0.16	0.3	0.4	0.5

After tuned by this Simulated Annealing algorithm, the parameters obtained the value as follows: $\boldsymbol{\eta}_{cs_a} = 0.9797$, $\boldsymbol{\eta}_{es_a} = 0.5357$, $\boldsymbol{\eta}_{cs_b} = 0.0013$, $\boldsymbol{\eta}_{es_b} = 0.1469$. The lowest deviation found by the Simulated Annealing is Root Mean Square Error RMSE = 0.1298, illustrated in Fig. 7.

Figure 8 illustrates the results of Simulation 1 when applied the tuned parameters, including the reference points (with *ref_* prefix) of the expected data from Table 5.

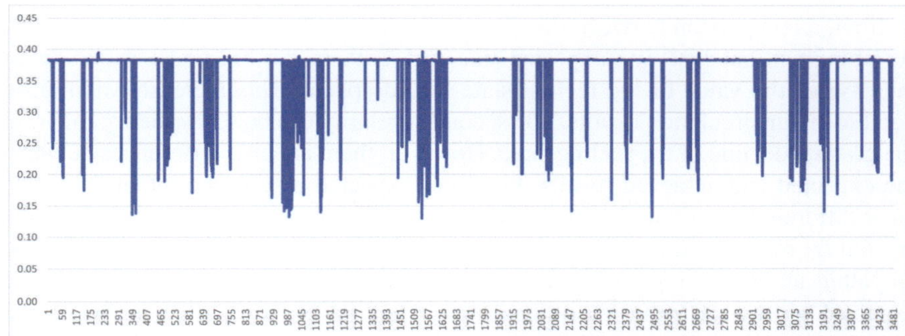

Fig. 7 RMSE errors over the Simulated Annealing process

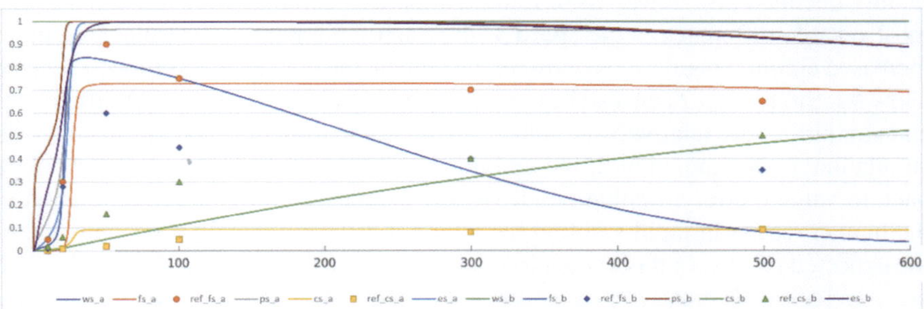

Fig. 8 Simulation 1 after applying tuned parameters

7 Discussion

The material for this chapter is mainly from Dimopoulos et al. [23]. In this section, first, the focus will be on the comparison between the expected results and the achieved results. This comparison allows for further elaboration on the next steps in making the model more realistic. Lastly, possibilities for further research will be discussed as this will bring more specific information about how to make the emotion-regulation model more realistic. The simulation results consisted of three different simulations:

1. Human with reduced individual emotion regulation and bot with perfect individual emotion regulation
2. Human with better individual emotion regulation and bot with perfect individual emotion regulation
3. Human with reduced individual emotion regulation and bot with reduced individual emotion regulation.

First, a comparison between the expected results and simulation 1 will be made. As aforementioned, it was expected that first both in the bot and the human an increase of the value of the feeling state would appear. This increase was indeed notable in our simulation results. The second similarity can be seen when comparing the gradual decline of the feeling state. However, there are also differences between the expected and achieved results. First, our expected results show that there is a large difference between the value of fs_a and fs_b. However, in our simulation results fs_a and fs_b reach an almost similar value at their peaks, which may support a good empathic understanding from the bot of the human's emotion state. Second, the expected results show that the value of fs_b would have a steep decline, but would stabilise at a value above 0. The final results show a steep decline, but the value of fs_b rapidly stabilises at 0. Similarly, it was expected that fs_a would decrease its value and stabilise at a value above zero. Nonetheless, at a tipping point the value of fs_a rapidly declined all the way to 0. If simulation 2 is compared to simulation 1, the most notable difference is the speed with which the emotion is regulated. Due to increased emotion regulation control in the human, the bot needs less time to achieve a low fs_a. Simulation 2 can also be compared to the expected results. The differences and similarities are relatively the same as when comparing the expected results with simulation 1. In simulation 2, the emotion regulation still appears to make fs_a go back to value 0 after a tipping point. Hence, the increase of emotion regulation in the human results in rapid emotion regulation. Lastly, simulation 3 is compared to the expected results. In this simulation the emotion regulation control of the bot had been decreased. Therefore, the emotion regulation occurs steadily and as a result fs_a stabilises at a value higher than 0. Unlike the other two simulations, simulation 3 thus does not have a tipping point. Consequently, this simulation shows many similarities with the expected results.

A next step for this model can be to extend the model by adding additional ways of coregulation which can make it more realistic. Currently, the model represents emotion regulation by the bot based on social contagion. State es_b does not represent any specific action such as a hug or a pat on the shoulder, that would affect the fs_a in the human but just the expressed emotion. The consequence of adding such other actions as well could be that the fs_a in the human could be affected in different ways by the different actions. This may contribute to optimising the bots response in its attempt to regulate the human emotion fs_a. Furthermore, the model can be made adaptive so that by learning the emotion regulation is improved over time. In this chapter, it was assumed that internal emotion regulation in the human is very poor and internal emotion regulation in the bot is excellent. In short, this means that in the current model the human does not have the quality to regulate its own emotions well, while the bot has the power to both regulate its own emotions and those of the human. However, in reality the human most likely has more capacities to regulate its own emotions. In order to explore this, additional states in this model may be specified. In the current model, this was not done because for this first step we wanted to keep the model simple.

Moreover, note that the agent models in the current chapter are nonadaptive and do not have multimodal interaction capabilities. In the meantime also a more refined agent-based network model for human-bot interaction has been designed, taking into account both multimodal interaction and adaptivity [24]. In that refined network model:

- the two agents interact via different modalities and can synchronize for these modalities, like in this volume, Chapter The Emergence of Multimodal Interpersonal Synchrony During Social Interaction: A Network Model [25]
- the agents apply synchrony-induced adaptivity to adapt their social interaction behaviour over time, as described in this volume, Chapters Introduction to the Role of Adaptivity in Social Interaction Science and The Adaptive Interplay of Interpersonal Synchrony, Short-Term Affiliation and Long-Term Bonding: A Multi-Timescale Multi-Adaptive Agent Model [26, 27].

For more details of this refined network model, see also this volume Chapter Interpersonal Emotion Regulation via Multimodal Interaction with a Bonding Bot: An Adaptive Network Model [28].

References

1. Paley, B., Hajal, N.J.: Conceptualizing emotion regulation and coregulation as family-level phenomena. Clin. Child. Fam. Psychol. Rev. **25**(1), 19–43 (2022)
2. Sivers, J.A.: Adolescence as a pivotal period for emotion regulation development. Curr. Opin. Psychol. **44**, 258–263 (2022)
3. Hetherton, T.F., Wagner, D.D.: Cognitive neuroscience of self-regulation failure. Trends Cogn. Sci. **15**(3), 132–139 (2011)
4. Williams, W.C., Morelli, S.A., Ong, D.C., Zaki, J.: Interpersonal emotion regulation: implications for affiliation, perceived support, relationships, and well-being. J. Pers. Soc. Psychol. **115**(2), 224 (2018)

5. Butler, E.A., Randall, A. K.: Emotional coregulation in close relationships. Emot Rev, 5(2), 202–210 (2013)
6. Gren, K.H., Van De Groep, S., Sweijen, S.W., Becht, A.I., Buijzen, M., De Leeuw, R.N., et al.: Mood and emotional reactivity of adolescents during the COVID-19 pandemic: short-term and long-term effects and the impact of social and socioeconomic stressors. Sci. Rep. **11**(1), 11563 (2021)
7. Berdasco, A., López, G., Diaz, I., Quesada, L., Guerrero, L.A.: User experience comparison of intelligent personal assistants: Alexa, Google Assistant, Siri and Cortana. Proc. MDPI **31**(1), 51 (2019)
8. Oatley, K.: Emotion: Theories (2000)
9. James, W.: Principles of Psychology (1890)
10. Lange, C.G.: The mechanism of the emotions. The classical psychologists, 672–684. (1885)
11. Critchley, H.D.: Psychophysiology of neural, cognitive and affective integration: fMRI and autonomic indicants. Int. J. Psychophysiol. **73**(2), 88–94 (2009)
12. Gross, J.J.: Emotion regulation: current status and future prospects. Psychol. Inq. **26**(1), 1–26 (2015). https://doi.org/10.1080/1047840X.2014.940781
13. Phelps, E.A., Delgado, M.R., Nearing, K.I., LeDoux, J.E.: Extinction learning in humans: role of the Amygdala and vmPFC. Neuron **43**(6), 897–905 (2004). https://doi.org/10.1016/j.neuron.2004.08.042
14. Treur, J.: Network-Oriented Modeling: Addressing Complexity of Cognitive, Affective and Social Interaction. Springer, Cham (2016)
15. Treur, J.: Network-Oriented Modeling for Adaptive Networks: Designing Higher-order Adaptive Biological, Mental and Social Network Models. Springer Nature Cham, Cham, Switzerland (2020b)
16. Treur, J.: Modeling coevolution of dynamics, adaptivity and control of adaptivity by self-modeling networks. In: Hendrikse, S.C.F., Treur, J., Koole, S.L. (eds.) New Analysis and Modeling Directions in Social Interaction Science: Emergent Multimodal Interpersonal Synchrony, Affiliation, and Bonding, this volume, Ch. 2. Springer Nature (2025)
17. Treur, J.: A modeling environment for reified temporal-causal networks: modeling plasticity and metaplasticity in cognitive agent models. In: Baldoni, M., Dastani, M., Liao, B., Sakurai, Y., Zalila Wenkstern, R. (eds.) PRIMA 2019: Principles and Practice of Multi-Agent Systems. PRIMA 2019. Lecture Notes in Computer Science, vol. 11873, pp. 487–495. Springer Nature, Cham (2019)
18. Butler, E.A., Randall, A.K.: Emotional coregulation in close relationships. Emotion Review **5**(2), 202–210 (2013)
19. Cornelissen, F., Jonker, C.M., Treur, J.: Compositional verification of knowledge-based systems: a case study for diagnostic reasoning. In: Plaza, E., Benjamins, R. (eds) Knowledge Acquisition, Modeling and Management. Proceedings EKAW 1997. Lecture Notes in Computer Science, vol 1319, pp. 65–80. Springer, Berlin, Heidelberg. (1997)
20. Treur, J.: Where is this leading me: stationary point and equilibrium analysis of self-modeling network models. In: Treur, J., van Ments, L. (eds.) Mental Models and their Dynamics, Adaptation and Control: A Self-Modeling Network Modeling Approach, Ch. 18. Springer Nature Switzerland (2022a)
21. Treur, J.: Does this suit me: validation of self-modeling network models by parameter tuning. In: Treur, J., van Ments, L. (eds.) Mental Models and their Dynamics, Adaptation and Control: A Self-Modeling Network Modeling Approach, Ch. 19. Springer Nature Switzerland (2022b)
22. Kirkpatrick, S., Gelatt, C.D., Vecchi, M.P.: Optimization by simulated annealing. Science, New Series **220**, 671–680 (1983)
23. Dimopoulos, F., Eler, E., Timmers, M., Treur, J., Koole, S.L.: An agent-based network model for interpersonal emotion regulation in a human-bot interaction. In: Nguyen, N.T., et al. (eds.) Advances in Computational Collective Intelligence. Proceedings of the 15th International Conference on Computational Collective Intelligence, ICCCI'23. Communications in Computer and Information Science, vol. 1864, pp. 68–79. Springer Nature (2023)

24. Eler, E., Treur, J., Hendrikse, S.C.F., Donker, T., Koole, S. L.: An adaptive network model for interpersonal emotion regulation in multimodal human-bot interaction. In: Proceedings of the 16th International Conference on Computational Collective Intelligence, ICCCI'24. Lecture Notes in AI, vol. 14811, pp. 67–79. Springer Nature (2024)
25. Hendrikse, S.C.F., Treur, J., Wilderjans, T.F., Dikker, S., Koole, S.L.: The emergence of multimodal interpersonal synchrony during social interaction: a network model. In: Hendrikse, S.C.F., Treur, J., Koole, S.L. (eds.) New Analysis and Modeling Directions in Social Interaction Science: Emergent Multimodal Interpersonal Synchrony, Affiliation, and Bonding, this volume, Ch. 3. Springer Nature (2025)
26. Hendrikse, S.C.F., Treur, J., Koole, S.L.: Introduction to the role of adaptivity in social interaction science. In: Hendrikse, S.C.F., Treur, J., Koole, S.L. (eds.) New Analysis and Modeling Directions in Social Interaction Science: Emergent Multimodal Interpersonal Synchrony, Affiliation, and Bonding, this volume, Ch. 1. Springer Nature (2025b)
27. Hendrikse, S.C.F., Treur, J., Koole, S.L.: The adaptive interplay of interpersonal synchrony, short-term affiliation and long-term bonding: a multi-timescale multi-adaptive agent model. In: Hendrikse, S.C.F., Treur, J., Koole, S.L. (eds.) New Analysis and Modeling Directions in Social Interaction Science: Emergent Multimodal Interpersonal Synchrony, Affiliation, and Bonding, this volume, Ch. 12. Springer Nature (2025c)
28. Eler, E., Treur, J., Hendrikse, S.C.F., Donker, T., Koole, S. L.: An adaptive network model for interpersonal emotion regulation in multimodal human-bot interaction. In: Hendrikse, S.C.F., Treur, J., Koole, S.L. (eds.) New Analysis and Modeling Directions in Social Interaction Science: Emergent Multimodal Interpersonal Synchrony, Affiliation, and Bonding, this volume, Ch. 19. Springer Nature (2025)

Learning to Synchronize: Adaptive Joint Decision-Making by Virtual Agents

Sophie C. F. Hendrikse, Sem Kluiver, Jan Treur, Tom F. Wilderjans, Suzanne Dikker, and Sander L. Koole

Abstract Joint decision-making can be seen as the synchronization of actions and emotions, usually via nonverbal interaction between people while they show empathy. The aim of the current chapter was (1) to develop an adaptive computational model for the type of synchrony that can occur in joint decision-making for two persons modeled as agents, and (2) to visualize the two persons by avatars as virtual agents during their decision-making. How to model joint decision-making computationally while taking into account adaptivity is rarely addressed, although such models based on psychological literature have a lot of future applications like online coaching and

---

S. C. F. Hendrikse
Department of Methodology and Statistics, Tilburg School of Social and Behavioral Sciences, Tilburg University, Tilburg, The Netherlands
e-mail: S.C.F.Hendrikse@tilburguniversity.edu

S. Kluiver · J. Treur (✉)
Social AI Group, Department of Computer Science, Vrije Universiteit Amsterdam, Amsterdam, The Netherlands
e-mail: j.treur@vu.nl

S. Kluiver
e-mail: semkluiver@xs4all.nl

T. F. Wilderjans
Faculty of Social and Behavioural Sciences, Methodology and Statistics Research Unit, Institute of Psychology, Leiden University, Leiden, the Netherlands

Research Group of Quantitative Psychology and Individual Differences, Faculty of Psychology and Educational Sciences, Katholieke Universiteit (KU) Leuven, Leuven, Belgium

T. F. Wilderjans
e-mail: t.f.wilderjans@fsw.leidenuniv.nl

S. Dikker
NYU – Max Planck Center for Language, Music and Emotion, New York University, New York, USA

S. L. Koole
Amsterdam Emotion Regulation Lab, Department of Clinical Psychology, Vrije Universiteit Amsterdam, Amsterdam, The Netherlands
e-mail: s.l.koole@vu.nl

S. C. F. Hendrikse et al. (eds.), *New Analysis and Modeling Directions in Social Interaction Science*, Studies in Systems, Decision and Control 614,
https://doi.org/10.1007/978-3-031-99968-0_5

therapeutics. We used an adaptive network-oriented modelling approach to build an adaptive joint decision-making model in an agent-based manner and simulated multiple scenarios of such joint decision-making processes using a dedicated software environment that was implemented in MATLAB. It does not only address first-order adaptation but also second-order adaptation to cover the metaplasticity principle 'Adaptation accelerates with increasing stimulus exposure'. Programming in the Unity 3D engine was done to virtualize this process as nonverbal interaction between virtual agents, their internal and external states, and the scenario. Although our adaptive joint decision model has general application areas, we have selected a therapeutic session as example scenario to visualize and interpret the example simulations.

1 Introduction

Whenever people come into contact with each other, they tend to spontaneously synchronize or align their nonverbal behavior, physiology and brain signals. The importance of such interpersonal synchrony has been established in multiple social settings. As an example, higher levels of nonverbal synchrony promote cooperation [59] and social affiliation [60]. Furthermore, interpersonal synchrony may foster a good working relationship between clients and their therapist during psychotherapy [61]. A concept closely related to synchrony is facial mimicry. Facial mimicry refers to the matching of individuals' facial expressions with their emotional experiences [15]. Indeed, facial mimicry is a central component within all social interactions [19, 32]. More broadly, mimicry is labeled as the matching or imitation of nonverbal behavior, and it can range from facial expressions including pupil dilation [41] to body postures [9]. Although such movement and facial mimicry (including emotional expressions) are linked to each other [42], they are not the same. Facial mimicry in itself carries information about the expresser's appraisal of the event [25, 57], that directly impacts the mimicry. In contrast, body movements themselves do not directly contain such appraisal information, although the receiver can infer an emotional state from such signals based on their own interpretation [19]. Facial mimicry plays a role for both emotional and cognitive empathy [15].

In accordance with the above, for joint decision-making processes the outcome consists of (1) a joint action, (2) a common positive feeling about this action and (3) an empathic understanding of this action and feeling [53]. In other words, a successful joint decision-making process can be seen as the synchronization or alignment of both actions and positive feelings together with a mutual empathic understanding. Throughout this process, nonverbal interactions play an important role. Moreover, these alignment processes themselves can become attuned by a form of learning or adaptation. Computational modeling of these complex and dynamic processes by means of (virtualized) agents is challenging and has been addressed only partially.

In earlier work [16, 53], non-adaptive joint decision-making based on non-verbal interactions has been modeled within and between agents. In a previous chapter of Part II of this volume [29], synchronization at different action modalities was modeled in a dyad of agents, which can be seen as the modeling of joint actions and positive emotion states of agents, see this volume, Chapter The Emergence of Multimodal Interpersonal Synchrony During Social Interaction: A Network Model [31]. However, these agent models are not adaptive, and the simulations of these models have not been visualized by avatars. Other agent models that address emotion regulation processes (but not joint decision-making) have been successfully developed together with their accompanied avatars [13]. By virtualization of the agents, the human-likeness of the agent model can be better demonstrated and viewers can relate more strongly to the generated interaction patterns. Therefore, our aim here is to (1) extend the non-adaptive nonverbal joint decision-making model from [53] to an adaptive model and (2) visualize these adaptive joint decision-making processes by means of avatars. To verify the developed computational model, we conducted multiple simulation experiments and our main simulation was visualized by virtual agents displayed as avatars. As an illustrative visualized scenario, the focus was on the adaptive development of nonverbal closeness of contact between client and therapist over multiple therapy sessions as a central joint decision.

In this chapter, Sect. 2 provides an overview of the background knowledge used to design the introduced model. Section 3 briefly summarizes the modeling approach used, after which the second-order adaptive network model is introduced in Sect. 4. In Sect. 5, the main example simulation is discussed in more detail, including its visualization. Section 6 elaborates on some other simulation results with alternative parameter settings for different personal characteristics. In Sect. 7 a final discussion is provided and the complete specification of the model is shown in the Appendix.

2 Background Knowledge

For the design of the introduced computational model, different types of neural or psychological mechanisms found in the literature have been used as building blocks. A number of these mechanisms are of a general nature, whereas other ones are more specifically related to joint decision-making. An overview of them is given in this section.

2.1 General Psychological Mechanisms as Building Blocks

Several general mechanisms from psychology and neuroscience were used as building blocks to create the internal structure of the virtual agents in order to achieve human-likeness.

2.1.1 Mirroring

The principle of mirroring describes that a person's preparation state for a certain action is also activated when the corresponding action of another person is observed. It can be explained by mirror neurons and mirroring links: the connection from the sensory representation state of the action of the other person to the preparation state of a person's own action [34]. People rely on such mirroring in their nonverbal communication and therefore, mirroring should play an important role in the design of a joint decision process of virtual agents.

2.1.2 Emotion Integration

In addition to mirroring, a virtual agent needs to decide to conduct or not conduct a certain action. Humans use a prediction loop to internally simulate the predicted effect of an action on their emotional response and feeling states [11, 12]. The latter states play an important role as 'somatic markers' for making decisions. First, internal simulation of the considered action takes place by a prediction link that activates a sensory representation of the predicted effect of the action. Next, an emotional response and feelings are generated by using links for emotion association to this predicted effect. For the latter, two types of loops can be used: *body loops* (involving expression of the emotional response and sensing this expression) and *as-if body loops* (internal simulation of the emotional response, without actually expressing it); e.g., [6, 11, 12, 46]. In turn, such an associated feeling affects the preparation for the considered action, which makes it a cyclic causal pathway.

2.1.3 Self-Other Distinction

Furthermore, we have relied on control and self-other distinction [7, 10, 34, pp. 196–203, 35, 37, 47]. These involve neurons which are suggested to have a function in control (allowing or suppressing) action execution after preparation has taken place. In single-cell recording experiments with epileptic patients, cells were found that are active when the person prepares an own action that is executed, but shut down when the action is only observed. This finding leads to the hypothesis that these cells may be involved in the functional distinction between a preparation state generated in order to actually perform the action, and a preparation state generated to interpret an observed action (or both, in case of imitation). More specifically, this has been shown in work reported in research by Mukamel and colleagues [44], Fried et al. [20]; see also [34–36, 39], see also [34, pp. 201–203]. Some of the main findings are that neurons with mirror neuron properties were found in all sites in the mesial frontal cortex where recording took place (approximately 12% of all recorded neurons); half of them related to hand-grasping, and the other half to emotional face expressions. A subset of neurons was found that shows behavior that relates to execution of the action: they have excitatory responses during action execution and inhibitory responses during action observation [35, p. 30]. In Iacoboni [34–36], such types of neurons have been termed super mirror neurons, to indicate the control function they may have with respect to the execution of an action. Some of such cells are sensitive to a specific person, so that an observed action can also be attributed to the person that was observed (self-other distinction) [34, pp. 201–202]. It is suggested that the types of social interaction seen in persons with an autism spectrum disorder can be related to reduced self-other distinction and control of imitation [7, 22].

2.1.4 Plasticity and Metaplasticity

Within the cognitive neuroscience literature, two types of (first-order) adaptation or *plasticity* are often considered, one for connection weights and one for intrinsic neuronal properties such as excitability thresholds; for example, see [8]. In this chapter, one example of a first-order adaptation principle is considered: Hebbian learning for connection weights. This is a well-known adaptation principle addressing adaptive connectivity going back to [5, pp. 117–121] that is explained in [26, p. 62] by: 'When an axon of cell A is near enough to excite B and repeatedly or persistently takes part in firing it, some growth process or metabolic change takes place in one or both cells such that A's efficiency, as one of the cells firing B, is increased'. This is sometimes simplified (neglecting the phrase 'as one of the cells firing B') to: 'What fires together, wires together' [38, 49]. For more background, see also [28] which is included as Chapter From a Multilevel Multi-timescale Computational Architecture for Synchrony-Induced and Homophily-Induced Attunement to Designing Bonding Bots in this volume [29].

The 'plasticity versus stability conundrum' describes how an organism adjusts its plasticity over time in a context-sensitive manner [51, p. 773]. Under which circumstances and to which extent plasticity actually takes place is controlled by socalled *metaplasticity*; e.g., [3, 21, 48, 51]. In the aforementioned literature, various studies have shown how adaptation (as described, for example, by Hebbian learning) is modulated by accelerating the adaptation process or decelerating or even blocking it. Among the reported factors affecting plasticity in such a way are stimulus exposure, activation, previous experiences, and stress. For example, Robinson and colleagues noted that 'Adaptation accelerates with increasing stimulus exposure' [48, p. 2]. This indeed describes a form of metaplasticity that controls the speed of adaptation (learning rate). This principle for metaplasticity also has been applied in the present computational model.

2.2 *Building Blocks for the Joint Decision-Making Context*

As mentioned in Sect. 1, a component of interpersonal synchrony regards (facial/emotional and movement) mimicry. Emotional mimicry is defined as the matching or imitation of each other's (facial) emotions. The Contextual Model of Emotional

Mimicry relies on two assumptions, namely (1) a shared mind is the foundation of emotional mimicry [45] and, (2) there is no mimicry of the facial movements themselves but of the meaning of these movements [19]. Concretely, the shared mind assumption (1) means that both the expresser and mimicker have a minimal affiliation and perspective in common that provoke the mimicry. Assumption (2) implies that although the mimicry itself goes automatically, emotional mimicry is goal-instead of stimulus-driven. In other words, emotional mimicry regards a top-down process that originates from prior experiences in social interactions regarding the interpretation of other's facial movement [19]. These theory and findings also mean that when individuals do not want to become emotionally closer, they will (unconsciously) not mimic the emotions of others.

Several studies have demonstrated a bidirectional association between facial mimicry and emotional empathy. Emotional empathy is described as the emotional reaction on and inference of another's emotional state, meaning that the emotional states of both persons are congruent, but also the ability to differentiate between the self and other [17]. People who score higher on empathic traits mimic more consistently with each other [62, 63]. Conversely, a shared emotional state (contributing to emotional empathy) is promoted by the mimicking of each other's facial expressions [52]. In sum, empathy and facial mimicry appear to be involved together during tasks.

A system closely related to the emotion system is approach behavior and its motivation. Approach motivation refers to the strong desire to move forward [23, 40] and, vice versa, avoidance motivation is the desire to move away. Both approach and avoidance behavior and their motivation can vary in their intensity [23]. Traits, moods and external stimuli can all elicit approach behavior [23]. Although it is often thought that approach motivation is linked to positive states (e.g., happiness) and, in contrast, avoidance motivation to negative states (e.g., sadness) [64], approach motivation can also be related to negative states like anger [65]. This implies that the approach motivation and the emotion system are two separate systems, they are connected to each other, but not the same.

There is also experimental evidence that the body posture of leaning forwards enhances approach motivation [66]. In one experiment, approach motivation was indirectly measured through the relative left frontal cortical activity and three body posture conditions (leaning, upright and reclining) were created. It turned out that participants who leaned forward with their arms extended (a body posture that is normally used to grasp a desired object) displayed higher activation in the relative left frontal cortical area than those who were leaning backwards. In other research, leaning backwards inhibited approach-motivated anger [24, 40]. In sum, approach behaviour in the sense of a body posture of leaning forwards characterizes (and thereby visualizes) the state of elevated approach motivation. Approach motivation in itself reflects the desire to move towards a goal and is therefore always involved during joint decision-making. In such joint decision-making, the (connected) separate emotion system and the linked empathy play a role as well.

3 Method: Self-Modeling Network Modeling

The designed virtual agents and their simulations are based on models of their internal mental processes. To achieve this, the adaptive network-oriented modeling approach presented in [54, 55] has been used to create a dynamic and adaptive interplay of mental states, see also this volume [29], Chapter Modeling Coevolution of Dynamics, Adaptivity and Control of Adaptivity by Self-Modeling Networks [56].

Following [55], a network model is characterized by (here X and Y denote nodes of the network, also called states; they have time-dependent activation values $X(t)$ and $Y(t)$):

- *Connectivity characteristics*

  Connections from a state X to a state Y and their weights $\boldsymbol{\omega}_{X,Y}$

- *Aggregation characteristics*

  For any state Y, some combination function $\mathbf{c_Y(...)}$ defines the aggregation that is applied to the impacts $\boldsymbol{\omega}_{X,Y}X(t)$ on Y from its incoming connections from states X

- *Timing characteristics*

  Each state Y has a speed factor $\boldsymbol{\eta}_Y$ defining how fast it changes for a given (aggregated) causal impact.

The following generic difference (or related differential) equations that are used for simulation purposes and also for analysis of such temporal-causal networks incorporate these network characteristics $\boldsymbol{\omega}_{X,Y}$, $\mathbf{c}_Y(\ldots)$, $\boldsymbol{\eta}_Y$ in a standard numerical format:

$$Y(t + \Delta t) = Y(t) + \boldsymbol{\eta}_Y\left[\mathbf{c}_Y\left(\boldsymbol{\omega}_{X_1,Y}X_1(t), \ldots, \boldsymbol{\omega}_{X_k,Y}X_k(t)\right) - Y(t)\right]\Delta t \quad (1)$$

for any state Y and where X_1 to X_k are the states from which Y gets its incoming connections. Within the dedicated software environment described in Treur [55], Ch. 9, a large number of currently around 50 useful basic combination functions are included in a combination function library. The above concepts enable to design network models and their dynamics in a declarative manner, based on mathematically defined functions and relations. The examples of basic combination functions that are applied in the model introduced here can be found in Table 1.

Realistic network models are usually adaptive: often not only their states but also some of their network characteristics change over time. By using a *self-modeling network* (also called a *reified* network), a similar network-oriented conceptualization can also be applied to *adaptive* networks to obtain a declarative description using mathematically defined functions and relations for them as well; see [55]. This works through the addition of new states to the network (called *self-model states*) which represent (adaptive) network characteristics. In the graphical 3D-format as shown in Sect. 4, such additional states are depicted at a next level (called *self-model level* or *reification level*), where the original network is at the *base level*.

Table 1 The combination functions used in the introduced network model

	Notation	Formula	Parameters
Advanced logistic sum	$\mathbf{alogistic}_{\sigma,\tau}(V_1, \ldots, V_k)$	$\left[\frac{1}{1+e^{-\sigma(V_1+\ldots+V_k-\tau)}} - \frac{1}{1+e^{\sigma\tau}}\right](1+e^{-\sigma\tau})$	Steepness $\boldsymbol{\sigma} > 0$ Excitability threshold $\boldsymbol{\tau}$
Stepmod	$\mathbf{stepmod}_{\rho,\delta}(\ldots)$	1 if time t mod $\boldsymbol{\rho} > \boldsymbol{\delta}$, else 0	Repetition time $\boldsymbol{\rho}$ Step time $\boldsymbol{\delta}$
Hebbian learning	$\mathbf{hebb}_{\mu}(V_1, V_2, V_3)$	$V_1 * V_2(1 - V_3) + V_3$	V_1, V_2 activation levels of the connected states; V_3 activation level of the self-model state for the connection weight Persistence factor $\boldsymbol{\mu}$

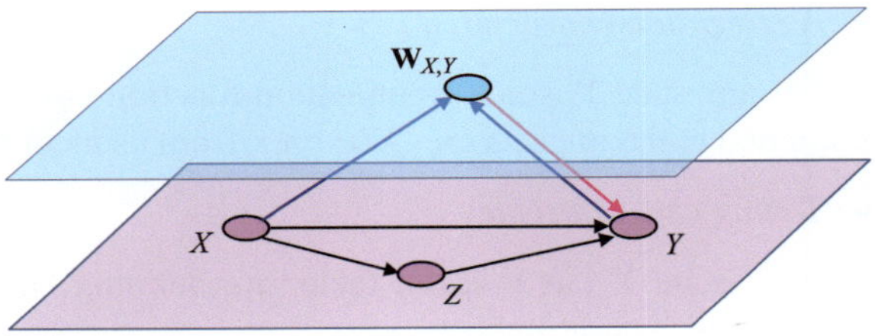

Fig. 1 Connectivity of a self-model for Hebbian learning

As an example, the weight $\boldsymbol{\omega}_{X,Y}$ of a connection from state X to state Y can be represented (at a next self-model level) by a self-model state named $\mathbf{W}_{X,Y}$. Such states are generally called **W**-states. For Hebbian learning [26] this self-model state is connected as shown in Fig. 1. By using the function $\mathbf{hebb}_{\mu}(\ldots)$ from Table 1 as combination function, based on generic difference Eq. (1), this self-model state $\mathbf{W}_{X,Y}$ models Hebbian learning as its behavior.

Similarly, all other network characteristics from $\boldsymbol{\omega}_{X,Y}$, $\mathbf{c}_Y(\ldots)$ and $\boldsymbol{\eta}_Y$ can be made adaptive by including self-model states for them. For example, an adaptive speed factor $\boldsymbol{\eta}_Y$ can be represented by a self-model state named $\mathbf{H}_Y$. Such states are generally called **H**-states.

As the outcome of such a process of network reification is also a network model itself, as has been shown in detail in Treur [55], Ch. 10, this self-modeling network construction can easily be applied iteratively to obtain multiple orders of self-models at multiple (first-order, second-order, etc.) self-model levels, for example, to model metaplasticity as discussed in Sect. 2; e.g., [3]. For instance, a second-order self-model may include a second-order self-model state $\mathbf{H}_{\mathbf{W}_{X,Y}}$ representing the speed factor $\boldsymbol{\eta}_{\mathbf{W}_{X,Y}}$ for the (learning) dynamics of first-order self-model state $\mathbf{W}_{X,Y}$ which in turn represents the adaptation of connection weight $\boldsymbol{\omega}_{X,Y}$; see Fig. 2 for the connectivity. Such states are generally called $\mathbf{H}_{\mathbf{W}}$-states; they can be considered learning rates of the learning modeled by the concerning states $\mathbf{W}_{X,Y}$. This can be used to model the second-order adaptation principle 'Adaptation accelerates with increasing

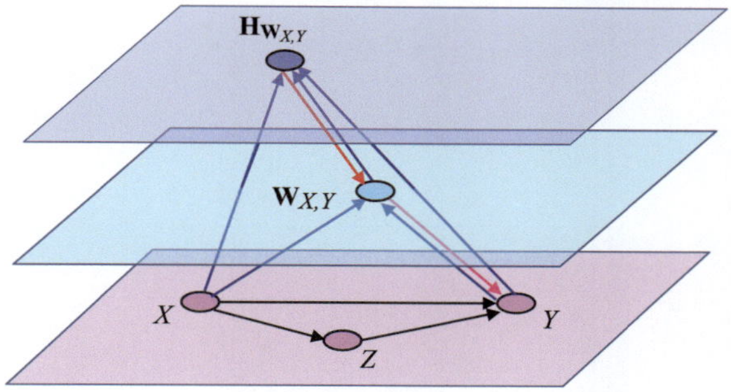

Fig. 2 Connectivity of a second-order self-model for the second-order adaptation principle 'Adaptation accelerates with increasing stimulus exposure' with a first-order self-model for Hebbian learning

stimulus exposure' [48] discussed in Sect. 2. As combination function for such $\mathbf{H}_{\mathbf{W}}$-states the function $\mathbf{alogistic}_{\sigma,\tau}(\ldots)$ from Table 1 can be used. This second-order adaptation can be interpreted as a context-sensitive form of control over the first-order adaptation: the plasticity only occurs in specific (relevant) contexts.

The two cases of self-model states depicted in Figs. 1 and 2 have been used in the model discussed in Sect. 4 in order to obtain a second-order adaptive network model for joint decision-making applying context-sensitive control of Hebbian learning.

4 The Adaptive Network Model for Joint Decision-Making

Recall from Sect. 1 that our research aims are to (1) extend the non-adaptive nonverbal joint decision-making model from [53] to an adaptive model and (2) visualize these adaptive joint-decision-making models by avatars. The current section addresses (1) whereas (2) is addressed in the next section.

The adaptive joint decision model introduced here consists of three levels: a base level, a first-order self-model level to model plasticity and a second-order self-model level to model metaplasticity to control the plasticity. For the base level it takes the nonadaptive model from [53] as a point of departure, see also [58]. The first-order self-modeling level and second-order self-model level are added here. The overall network connectivity for the base level of a single agent is displayed in the blue (base) plane in Fig. 3. A novelty introduced here is that we made the joint decision model second-order adaptive, meaning that the aspect of learning and the control over it is incorporated within the model. To achieve this form of adaptivity, we relied on Hebbian learning modeled by self-model **W**-states at the middle level (first-order self-modeling level) above the base level (the green plane in Fig. 3).

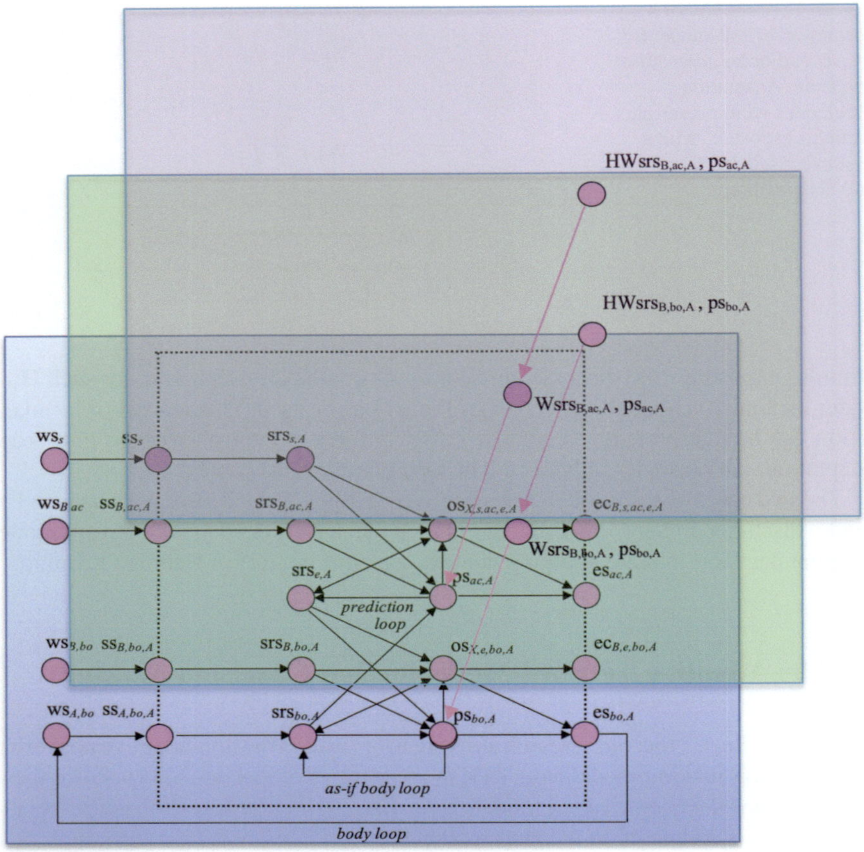

Fig. 3 Connectivity for the network architecture of a single agent A

On top of that, we added $\mathbf{H_W}$-states, which are in turn controlling the learning rates for the Hebbian learning. These $\mathbf{H_W}$-states are at the third level of the model (second-order self-modeling level) displayed by the pink plane in Fig. 3. In Fig. 4, the interaction connections between the two agents are displayed.

The literature briefly discussed in Sect. 2 has been used as a basis for the neurologically inspired adaptive network model presented here:

- Decision-making is based on *emotions associated* to *predicted effects* of action options
- Both the tendency to go for an action and the associated emotion are transferred between agents via *mirroring processes* using *internal simulation*
- These mirroring processes at the same time induce a gradual process of mutually *tuning* the considered actions and their associated emotions
- The outcome of such a joint decision process in principle involves three elements:

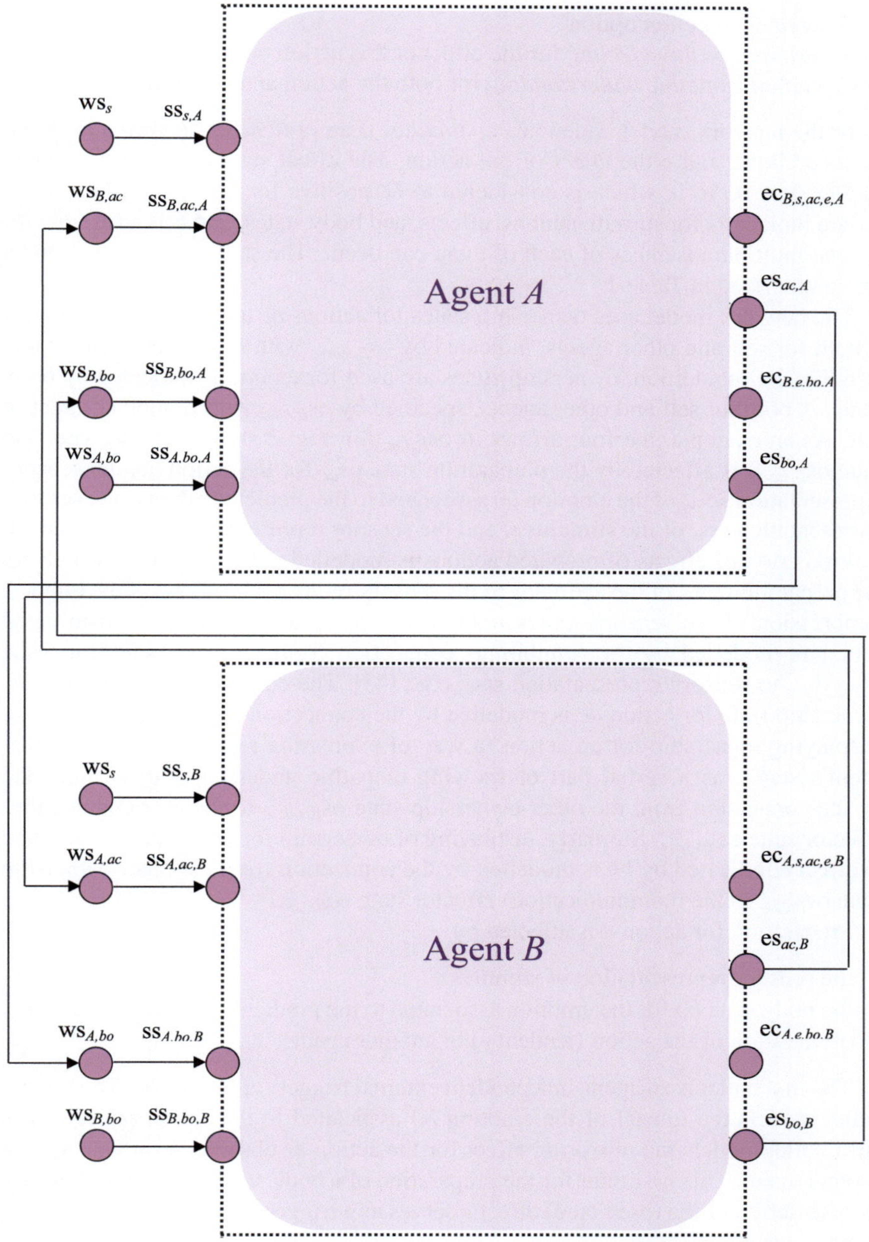

Fig. 4 Network architecture of the interaction between the two virtual agents, see also [58]

- a *common action* option
- a *shared positive feeling* for the effect of this action option
- *mutual empathic understanding* for both the action and the feeling.

In the network model, *s* denotes a *stimulus*, *a* an *option* for an *action ac* to be decided about, and *e* the *effect* of the action. The effect state *e* has an associated *feeling* state *bo* to it, which is considered to be positive for the agent. So, *s*, *ac*, *e*, *bo*, are indicators for stimuli, actions, effects, and body states, and *B* is a variable for agents; multiple instances of each of them can occur. The states used in the model are summarized in Table 1.

The network model uses ownership states for actions *ac* and their related effects *e*, both for self and other agents, indicated by $os_{B,s,ac,e}$ with *B* another agent or self (see Fig. 3). In addition, ownership states are used for emotions indicated by body state *bo*, both for self and other agents, specified by $os_{B,e,bo}$ with *B* another agent or self. As an example, the four arrows to $os_{B,s,ac,e}$ in Fig. 3 show that an ownership state $os_{B,s,ac,e}$ is affected by the preparation state ps_{ac} for the action *ac*, the sensory representation srs_{bo} of the emotion *bo* associated to the predicted effect *e*, the sensory representation srs_s of the stimulus *s*, and the sensory representation srs_B of agent *B*.

Prediction of effects of prepared actions is modelled using the connection from the preparation ps_{ac} of the action *ac* to the sensory representation srs_e of the effect *e*. Suppression of the sensory representation of a predicted effect of a self-initiated action is modelled by the (inhibiting) connection from the self-ownership state $os_{Self,s,ac,e}$ to sensory representation srs_e; e.g., [43]. The control exerted by the self-ownership state for action *ac* is modelled by the connection from $os_{Self,s,ac,e}$ to es_{ac}. Displaying ownership for an action (a way of expressing recognition of the other agent's states, as a verbal part of showing empathic understanding) is modelled by the connection from the other-ownership state $os_{B,s,ac,e}$ to the (communication) effector state $ec_{B,s,ac,e}$. Similarly, displaying of ownership for an emotion associated to effect *e* indicated by *bo* is modelled by the connection from the other-ownership state $os_{B,e,bo}$ to the (communication) effector state $ec_{B,e,bo}$.

Preparation for action *a* is affected by

- the sensory representation of stimulus *s*
- the body state *bo* for the emotion associated to the predicted effect *e* of the action
- observation of the action (tendency) in another agent.

The first bullet is an agent-independent external trigger for the action. The second bullet models the impact of the emotion *bo* associated to the action effect *e*. The third bullet models the mirroring effect for the action as observed as a tendency in another agent. This is similar for the preparation of a body state *bo*; here the sensory representation of the (predicted) effect *e* serves as a trigger, and the emotion state of another agent is mirrored.

Ownership states for an action *ac* or body state *bo* keep track of an agent *B*'s context with respect to the action or body state. This context concerns both the agent self and the other agents; it is a basis for attribution of an action or emotion to an agent and thus covers self-other distinction. Moreover, a self-ownership is used to control execution of prepared actions or body states. For example, in case the agent *B* is self, the ownership state for action *ac* strengthens the initiative to perform action *ac* as a self-generated action: executing a prepared action depends on whether a certain activation level of the ownership state for the agent self is available for this action. This is how control over the execution of the action (like a go/no-go decision) is exerted and can, for example, be used to veto the action in a stage of preparation. Expression of ownership of the other agent to the other agent represents acknowledgement of an agent that it has noticed the state of the other agent: a verbal part of an empathic response. These communications depend on the other-ownership states.

Plasticity was modeled using Hebbian learning [26] for two mirroring connections for each agent:

- for the mirroring connection from an agent *A*'s sensory representation state $\text{srs}_{B,ac,A}$ of a agent *B*'s action *ac* (tendency) to preparation state $\text{ps}_{ac,A}$ of *A* for *ac*
- for the mirroring connection from an agent *A*'s sensory representation state $\text{srs}_{B,bo,A}$ of *B*'s emotional body state *bo* to preparation state $\text{ps}_{bo,A}$ of *A* for the same emotion *bo*.

This form of learning was modeled by the (first-order) self-model states $\mathbf{W}_{\text{srs}_{B,ac,A},\text{ps}_{ac,A}}$ and $\mathbf{W}_{\text{srs}_{B,bo,A},\text{ps}_{bo,A}}$ in the middle plane; they use the combination function $\mathbf{hebb}_{\boldsymbol{\mu}}$ (see Table 1). Via the (upward) connections from $\text{srs}_{B,ac,A}$ and $\text{ps}_{ac,A}$ to $\mathbf{W}_{\text{srs}_{B,ac,A},\text{ps}_{ac,A}}$ it is monitored whether these base states are 'firing together' (see Sect. 2). Accordingly, the value of self-model state $\mathbf{W}_{\text{srs}_{B,ac,A},\text{ps}_{ac,A}}$ is updated using the **W**-state's combination function $\mathbf{hebb}_{\boldsymbol{\mu}}(\ldots)$, thus obtaining 'wiring together'. The resulting value of $\mathbf{W}_{\text{srs}_{B,ac,A},\text{ps}_{ac,A}}$ is used by $\text{ps}_{ac,A}$ via the (downward) connection of $\mathbf{W}_{\text{srs}_{B,ac,A},\text{ps}_{ac,A}}$ to $\text{ps}_{ac,A}$. Similarly, the Hebbian learning mechanism for the other adaptive connection concerning mirroring of emotions was modeled. By making the weights of these mirroring connections adaptive based on Hebbian learning, over time the agents get more responsive to each other and due to that it will become easier for them to reach a joint decision than for the nonadaptive case described in [16, 53] (Table 2).

However, we did not assume that plasticity always occurs no matter what. Instead, we assumed that the extent of plasticity is context-sensitive, which is a much more realistic assumption; e.g., [3, 48, 51], see also Sect. 2. To model this, second-order self-model states $\mathbf{H}_{\mathbf{W}_{\text{srs}_{B,ac,A},\text{ps}_{ac,A}}}$ and $\mathbf{H}_{\mathbf{W}_{\text{srs}_{B,bo,A},\text{ps}_{bo,A}}}$ (and similarly for the other agent's **W**-states) were added that represent the adaptation speed (learning rate) of the **W**-states. They affect the adaptive dynamics of the **W**-states through the downward (pink) connections to them. No plasticity occurs when these $\mathbf{H}_{\mathbf{W}}$-states have value 0, and the higher their values, the higher the adaptation speed. In this way, the second-order adaptation principle 'Adaptation accelerates with increasing stimulus exposure' [48] was modeled (see also Sect. 2). More specifically, the (upward) connections from

Table 2 States and their explanation

State nr and name	explanation
X_1 ws_s	Stimulus *s* in the world
X_2 $\mathrm{ws}_{A,ac}$	*A* tending to do action *ac*
X_3 $\mathrm{ws}_{B,ac}$	*B* tending to do action *ac*
X_4 $\mathrm{ws}_{A,bo}$	Body state *bo* of *A*
X_5 $\mathrm{ws}_{B,bo}$	Body state *bo* of *B*
X_6 $\mathrm{ss}_{s,A}$	Sensing stimulus *s* by agent *A*
X_7 $\mathrm{ss}_{s,B}$	Sensing stimulus *s* by agent *B*
X_8 $\mathrm{ss}_{B,ac,A}$	Sensing *B* tending to do action *ac* as experienced by agent *A*
X_9 $\mathrm{ss}_{A,ac,B}$	Sensing *A* tending to do action *ac* as experienced by agent *B*
X_{10} $\mathrm{ss}_{B,bo,A}$	Sensing body state *bo* of *B* as experienced by agent *A*
X_{11} $\mathrm{ss}_{A,bo,B}$	Sensing body state *bo* of *A* as experienced by agent *B*
X_{12} $\mathrm{ss}_{A,bo,A}$	Sensing own body state *bo* of *A* as experienced by agent *A*
X_{13} $\mathrm{ss}_{B,bo,B}$	Sensing own body state *bo* of *B* as experienced by agent *B*
X_{14} $\mathrm{srs}_{s,A}$	Sensory representation state for stimulus *s* by agent *A*
X_{15} $\mathrm{srs}_{s,B}$	Sensory representation state for stimulus *s* by agent *B*
X_{16} $\mathrm{srs}_{e,A}$	Sensory representation state for action effect *e* by agent *A*
X_{17} $\mathrm{srs}_{e,B}$	Sensory representation state for action effect *e* by agent *B*
X_{18} $\mathrm{srs}_{B,ac,A}$	Sensory representation state for *B* tending to do action *ac* perceived by agent *A*
X_{19} $\mathrm{srs}_{A,ac,B}$	Sensory representation state for *A* tending to do action *ac* perceived by agent *B*
X_{20} $\mathrm{srs}_{B,bo,A}$	Sensory representation state for body state *bo* of *B* by agent *A*
X_{21} $\mathrm{srs}_{A,bo,B}$	Sensory representation state for body state *bo* of *A* by agent *B*
X_{22} $\mathrm{srs}_{A,bo,A}$	Sensory representation state for own body state *bo* of *A* by agent *A*
X_{23} $\mathrm{srs}_{B,bo,B}$	Sensory representation state for own body state *bo* of *B* by agent *B*
X_{24} $\mathrm{ps}_{ac,A}$	Preparation state for action *ac* by agent *A*
X_{25} $\mathrm{ps}_{ac,B}$	Preparation state for action *ac* by agent *B*
X_{26} $\mathrm{ps}_{bo,A}$	Preparation state for emotional response *bo* by agent *A*
X_{27} $\mathrm{ps}_{bo,B}$	Preparation state for emotional response *bo* by agent *B*
X_{28} $\mathrm{os}_{B,s,ac,e,A}$	Other-Ownership state for doing action *ac* in the context of *B*, *s* and *e* by agent *A*
X_{29} $\mathrm{os}_{A,s,ac,e,B}$	Other-Ownership state for doing action *ac* in the context of *A*, *s* and *e* by agent *B*
X_{30} $\mathrm{os}_{B,e,bo,A}$	Other-Ownership state for emotion *bo* in the context of *B* and *e* by agent *A*
X_{31} $\mathrm{os}_{A,e,bo,B}$	Other-Ownership state for emotion *bo* in the context of *A* and *e* by agent *B*
X_{32} $\mathrm{os}_{A,s,ac,e,A}$	Self-Ownership state for doing action *ac* in the context of *A*, *s* and *e* by agent *A*
X_{33} $\mathrm{os}_{B,s,ac,e,B}$	Self-Ownership state for doing action *ac* in the context of *B*, *s* and *e* by agent *B*
X_{34} $\mathrm{os}_{A,e,bo,A}$	Self-Ownership state for emotion *bo* in the context of *A* and *e* by agent *A*
X_{35} $\mathrm{os}_{B,e,bo,B}$	Self-Ownership state for emotion *bo* in the context of *B* and *e* by agent *B*
X_{36} $\mathrm{ec}_{B,s,ac,e,A}$	Communication of action *ac* in the context of *B*, *s* and *e* by agent *A*
X_{37} $\mathrm{ec}_{A,s,ac,e,B}$	Communication of action *ac* in the context of *A*, *s* and *e* by agent *B*
X_{38} $\mathrm{ec}_{B,e,bo,A}$	Communication of emotion *bo* in the context of *B* and *e* by agent *A*
X_{39} $\mathrm{ec}_{A,e,bo,B}$	Communication of emotion *bo* in the context of *A* and *e* by agent *B*
X_{40} $\mathrm{es}_{ac,A}$	Execution state of action *ac* by agent *A*
X_{41} $\mathrm{es}_{ac,B}$	Execution state of action *ac* by agent *B*
X_{42} $\mathrm{es}_{bo,A}$	Execution state of body state *bo* by agent *A*
X_{43} $\mathrm{es}_{bo,B}$	Execution state of body state *bo* by agent *B*
X_{44} $\mathbf{W}_{\mathrm{srs}_{B,ac,A}\text{-}\mathrm{ps}_{ac,A}}$	First-order connectivity self-model state representing the weight ω of $\mathrm{srs}_{B,ac,A} \rightarrow \mathrm{ps}_{ac,A}$
X_{45} $\mathbf{W}_{\mathrm{srs}_{A,ac,B}\text{-}\mathrm{ps}_{ac,B}}$	First-order connectivity self-model state representing the weight ω of $\mathrm{srs}_{A,ac,B} \rightarrow \mathrm{ps}_{ac,B}$
X_{46} $\mathbf{W}_{\mathrm{srs}_{B,bo,A}\text{-}\mathrm{ps}_{bo,A}}$	First-order connectivity self-model state representing the weight ω of $\mathrm{srs}_{B,bo,A} \rightarrow \mathrm{ps}_{bo,A}$
X_{47} $\mathbf{W}_{\mathrm{srs}_{A,bo,B}\text{-}\mathrm{ps}_{bo,B}}$	First-order connectivity self-model state representing the weight ω of $\mathrm{srs}_{A,bo,B} \rightarrow \mathrm{ps}_{bo,B}$
X_{48} $\mathbf{H}_{\mathbf{W}_{\mathrm{srs}_{B,ac,A}\text{-}\mathrm{ps}_{ac,A}}}$	Second-order timing self-model state representing the speed factor (learning rate) η of the adaptive weight ω of the connection from $\mathrm{srs}_{B,ac,A}$ to $\mathrm{ps}_{ac,A}$
X_{49} $\mathbf{H}_{\mathbf{W}_{\mathrm{srs}_{A,ac,B}\text{-}\mathrm{ps}_{ac,B}}}$	Second-order timing self-model state representing the speed factor (learning rate) η of the adaptive weight ω of the connection from $\mathrm{srs}_{A,ac,B}$ to $\mathrm{ps}_{ac,B}$
X_{50} $\mathbf{H}_{\mathbf{W}_{\mathrm{srs}_{B,bo,A}\text{-}\mathrm{ps}_{bo,A}}}$	Second-order timing self-model state representing the speed factor (learning rate) η of the adaptive weight ω of the connection from $\mathrm{srs}_{B,bo,A}$ to $\mathrm{ps}_{bo,A}$
X_{51} $\mathbf{H}_{\mathbf{W}_{\mathrm{srs}_{A,bo,B}\text{-}\mathrm{ps}_{bo,B}}}$	Second-order timing self-model state representing the speed factor (learning rate) η of the adaptive weight ω of the connection from $\mathrm{srs}_{A,bo,B}$ to $\mathrm{ps}_{bo,B}$

base states $\mathrm{srs}_{B,ac,A}$ and $\mathrm{ps}_{ac,A}$ to the related $\mathbf{H}_{\mathbf{W}}$-state monitor the exposure at the base level and adapt the level of the $\mathbf{H}_{\mathbf{W}}$-states accordingly. To this end, the $\mathbf{H}_{\mathbf{W}}$-states use a common logistic combination function, which is monotonically increasing. The full specification of the model in terms of role matrices that can be directly executed (thus supporting reproducibility) can be found in the Appendix Section at the end of the chapter.

5 Simulation Results of the Main Scenario Including Their Visualizations

In this section the results of our main simulated example scenario and its visualization by avatars are discussed.

5.1 Example Scenarios

A wide variety of individual and situational differences can be observed in the real world. Accordingly, there are many possible outcomes for joint decision-making [16]. From a modeling perspective, all these differences can be captured by different settings for the network characteristics $\boldsymbol{\omega}_{X,Y}$, $\mathbf{c}_Y(\ldots)$ and $\boldsymbol{\eta}_Y$ defining the model. For example, if an agent shows poor mirroring, this may be due to weak internal mirroring links (agent characteristic) or just because it is almost dark so that visibility of the other agent is poor (situational characteristic). In the former case, this can be modeled by giving the internal mirroring connection (see Fig. 3) a low weight, whereas in the latter case the inter-agent connection (see Fig. 4) can be given a low weight. In such a way the variety of different combinations of values for the network characteristics $\boldsymbol{\omega}_{X,Y}$, $\mathbf{c}_Y(\ldots)$ and $\boldsymbol{\eta}_Y$ can reflect or match the variety of individual and situational differences in the real world.

In the first scenario, outlined in the current section, we have chosen consecutive therapeutic sessions as the repetitive stimulus s representing the therapy context. Agent A is visualized as a male client and agent B as a female therapist. The central joint decision during these therapeutic sessions regards approach behavior, represented by leaning forwards and backwards. In our example visualizations, a key difference between the therapist and client regards that the therapist (agent B) is relatively eager to conduct the specific action closeness of contact and the client initially does not want this closeness in the contact. This action approach motivation serves as a visualization for the $\mathrm{es}_{\mathrm{ac}}$ state of each agent. Moreover, the facial expression that goes from neutral (most negative affect) to smiling (most positive affect) regards the visualization of the $\mathrm{es}_{\mathrm{bo}}$ state of each agent. This body state $\mathrm{es}_{\mathrm{bo}}$ shows how both therapist and client feel about the decision over time.

The three scenarios considered in this chapter in Sects. 5 and 6 vary on the responsiveness upon the external stimulus s for the closeness of the contact between therapist and client; ws_s represents the therapy sessions, it has value 1 during a session and 0 when no session occurs at that moment; see Figs. 3 and 4. The differences in this stimulus responsiveness (or eagerness) have been modeled by differences in weights of the connection from stimulus representation to response preparation: the weights $\boldsymbol{\omega}_{\mathrm{srs}s,\mathrm{A},\mathrm{ps}ac,\mathrm{A}}$ and $\boldsymbol{\omega}_{\mathrm{srs}s,\mathrm{B},\mathrm{ps}ac,\mathrm{B}}$ of the connections from stimulus representations $\mathrm{srs}_{s,\mathrm{A}}$ and $\mathrm{srs}_{s,\mathrm{B}}$ to the respective response preparations $\mathrm{ps}_{ac,\mathrm{A}}$ and $\mathrm{ps}_{ac,\mathrm{B}}$ of the client A and therapist B. In particular, variations in this responsiveness strength from each side for this closeness were made as shown in Table 3. In this table, the numbers refer to the weights of these connections from the stimulus representation srs_s to the closeness action preparation ps_{ac}. Here, in Scenarios 1 and 3 it is assumed that the therapist is experienced and has a high initial responsiveness as part of her professional repertoire. For example, in Scenario 1 addressed in the current section, while this responsiveness of therapist is high (weight 1), the weight of the corresponding connection of the client is low (weight 0.1). This is based on the assumption that over the years the therapist has become experienced in responding to the type of stimuli

Table 3 Variations of the stimulus responsiveness strengths for the three scenarios considered

	Responsiveness strength for therapy context *s*	
	Therapist	Client
Scenario 1	1	0.1
Scenario 2	0.1	0.1
Scenario 3	1	0.7

during therapy sessions. In contrast, therapy sessions are assumed to be new for the client.

Note that the above only concerns the responsiveness upon the general therapy context stimulus *s*. In addition, during the sessions also responsiveness upon (dynamic) signals that are exchanged between therapist and client plays an important role, which in particular takes place in mirroring. While the aforementioned connections from stimulus *s* to response preparations are assumed nonadaptive, the mirroring connections were modeled as adaptive and therefore can and preferably will strengthen within and over sessions.

The overall views on the simulations for the three scenarios indicated in Table 3 are depicted in Fig. 5. As this is not easy to read, in subsequent pictures in Sects. 3 and 4, parts of them will be shown according to specific views in order to illustrate different phenomena that occur.

Nevertheless, in Fig. 5 it roughly can be seen that:

- in Scenario 1 (upper graph) with a therapist that is highly responsive for the therapy context there is some breakthrough in the fourth session,
- in Scenario 2 with a weakly responsive therapist (middle graph) there is no breakthrough at all,
- in Scenario 3 (lower graph) where the therapist and client are both responsive there is already a breakthrough in the second session.

What exactly such a breakthrough is, will be explained in Sect. 3 for Scenario 1 and in Sect. 6 for the other two scenarios.

5.2 Visualization Method

The scenarios are visualized in Unity using the free assets [1, 2] for the male and for the female agents, respectively. In each scenario the female agent represents the therapist whilst the male agent represents the client. The assets from the room in the background were provided by the Unity asset [14]. The room's design was slightly changed from the provided example to give it more of a therapy room look. Animations were either included with the character assets, or were taken from Adobe [4]. Code that controls the flow of the sessions and controls the characters was written

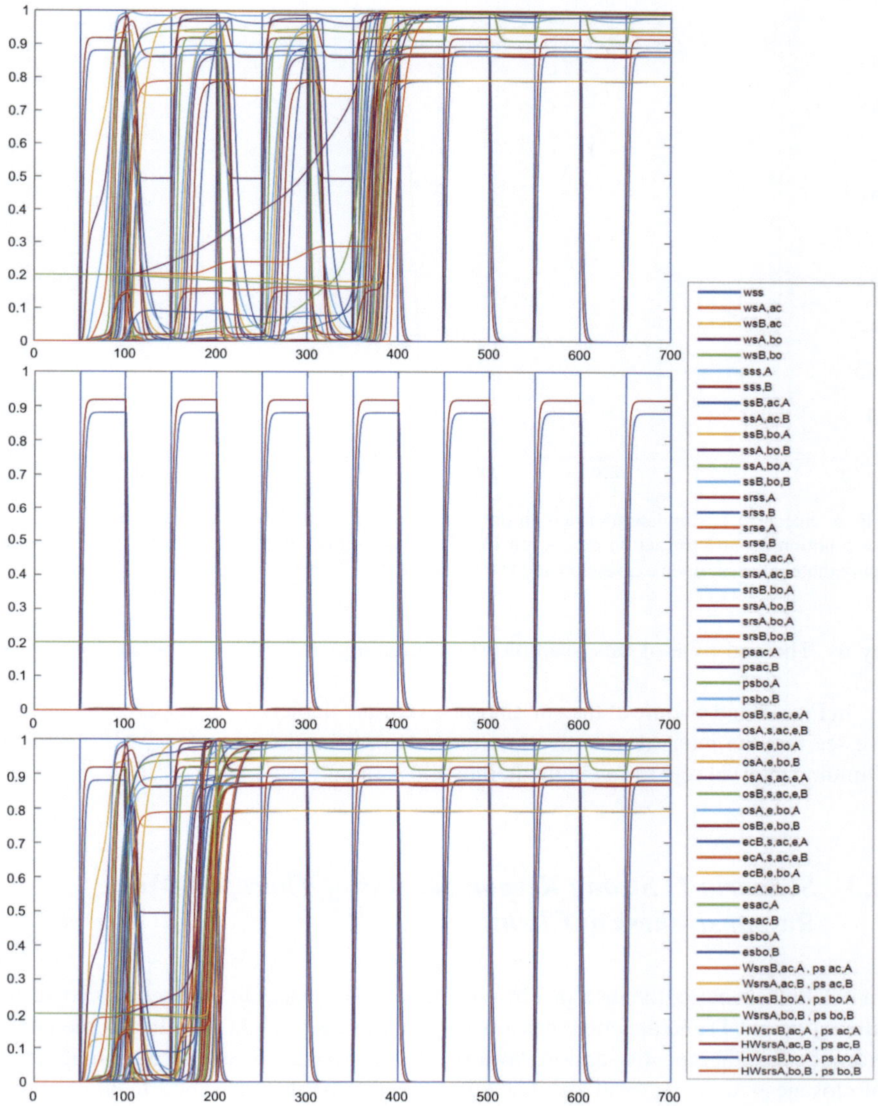

Fig. 5 Overall view on the three simulation scenarios. Upper graph: scenario 1 with breakthrough in the fourth session. Middle graph: scenario 2 without any breakthrough. Lower graph: scenario 3 with breakthrough in the second session. A breakthrough can be seen as the es_{ac} and es_{bo} of the client that go upwards

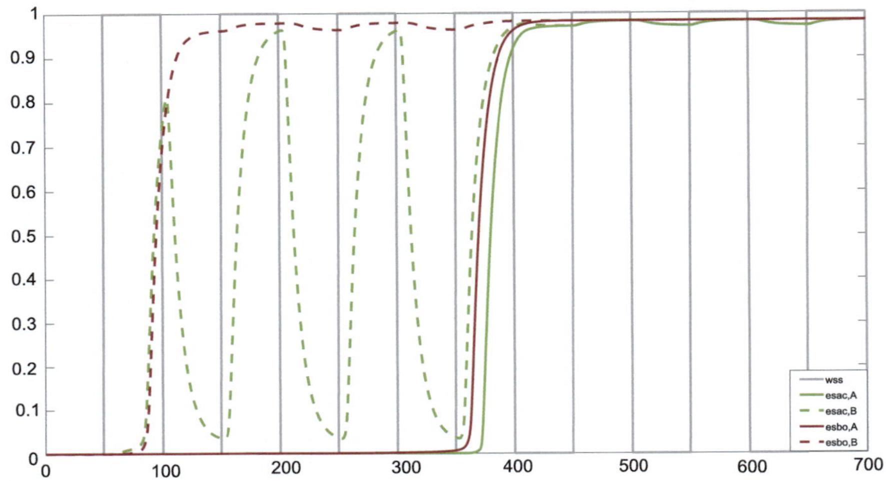

Fig. 6 Scenario 1: simulation case with only the expression and execution states (es_{bo} and es_{ac}) for emotion *bo* and for action *ac* chosen for the virtualization. It can be seen that the therapist immediately has an open expression and stance, while the client takes a few sessions to open up

by us. The end time of this example simulation equaled 700 and the step size was 0.5.

In Figs. 5 and 6 some of the simulation results are depicted. As displayed in Fig. 5, we see that the interval of both stimulus and non-stimulus periods equals 50. Each stimulus interval regards a single therapeutic session.

5.3 Scenario 1: Strong Responsiveness of Therapist, Weak Responsiveness of Client

In this first scenario, the therapist has strong responsiveness to therapy context stimulus *s* (weight 1) and the client has low responsiveness (weight 0.1). In this scenario, from the first therapeutic session onwards, the therapist starts to execute a high level of closeness in contact and also has a high feeling body state (good feeling) about this closeness of contact; see the upper graph in Fig. 5 for the overall simulation and Figs. 6, 7 and 8 for specific views. The dashed lines refer to states of agent B (the therapist), the solid lines to agent A (the client). Corresponding states between the two agents have the same color from Fig. 6 onwards.

In contrast, the client does not display any closeness of contact or happy feeling at all during the first three therapeutic sessions. However, in the fourth therapeutic session, there is a breakthrough, in which the client starts feeling better about the closeness in contact with the therapist and almost immediately afterwards starts to execute this closeness in contact too. At this point the adaptivity based on

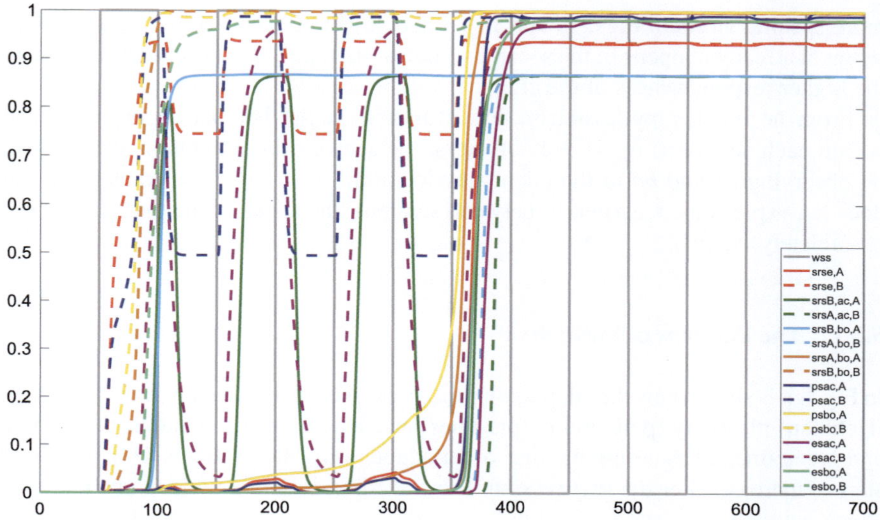

Fig. 7 Scenario 1: the representation and preparation states indicated by srs and ps together with the expression and execution states

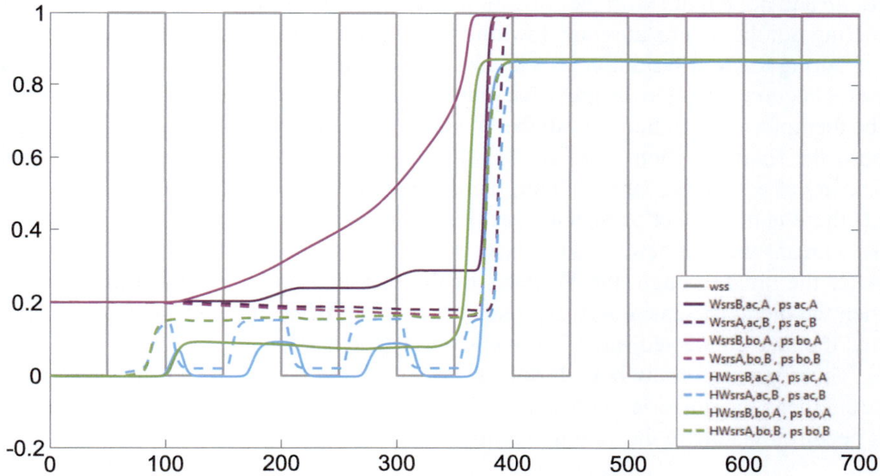

Fig. 8 Scenario 1: the adaptation **H**-states and $\mathbf{H_W}$-states from the first- and second-order self-model level

Hebbian learning has made the mirroring connections strong enough to achieve a joint decision, which was not possible with the initial settings.

In the next therapeutic sessions, both therapist and client are close in their contacts and feel good about this closeness in the therapeutic relationship (joint decision). One of the alternative simulations is Scenario 3 (see the lower graph in Fig. 5 and also more specific views in Figs. 13 and 14), where it can be seen that the breakthrough moment already happens in the second stimulus (therapeutic session) episode due to the higher responsiveness of the client.

From the first therapy session on, almost all states of the therapist become activated within each session (Fig. 7). Nevertheless, the sensory representation states based on observing *ac* and *bo* of the client stay low until the fourth session as the client does not express them earlier. In between sessions, the states of the therapist do not (completely) vanish.

5.3.1 The Patterns of Plasticity

In Fig. 8 it is shown how the adaptation processes take place based on the **W**-states for (Hebbian) plasticity (pink and purple lines) and $\mathbf{H_W}$-states for metaplasticity (blue and green lines). Regarding the first-order adaptability, the **W**-states from the client start to increase after the first session. From the first session onwards, the therapist shows activation of expression of both *ac* and bo, which are sensed by the client (see Fig. 8). As can be seen, the mirroring links strengthen by Hebbian learning during the sessions after session 1: for action *ac* slow (purple line) and for feeling *bo* a bit faster (pink line). This goes hand in hand with an increase in the preparation states for *ac* and bo, but not with the execution states to express or execute the action *ac* or feeling bo; the latter states stay low during the period of three sessions (see Fig. 8).

During the fourth session, as a breakthrough both the **W**-states and the execution states become high; see the green (ac) and bordeaux red (bo) lines in Fig. 8. Regarding the therapist (dashed lines), both the **W**-states for *ac* and *bo* tend to decrease slightly until the fourth session (see Fig. 8). This has two reasons: (1) due to the lack of sensing of activations of the client, there are no activations of the connected states, (2) there is no perfect persistence, as the persistence factors $\boldsymbol{\mu}$ are 0.995 and not 1. This means that extinction takes place: per time unit 0.5% of the learnt value is lost. After the breakthrough, the **W**-states of the therapist do increase sharply because then the therapist senses very high execution values of *ac* and *bo* from the client, so that the own connected states become strongly activated.

These patterns show how during successive therapy sessions a learning process strengthens the mirroring within the client, which has to reach a certain level before it becomes visible in the action execution and emotion expression.

5.3.2 The Patterns of Metaplasticity

Learning itself manifests differently depending on circumstances. In this case the learning depends on the exposure to activation, according to the following metaplasticity principle discussed earlier.

Fig. 9 In this screenshot, the therapist (left) and client (right) walk to their seats in order to start a new session

'Adaptation accelerates with increasing stimulus exposure' [48]. This metaplasticity principle is modeled by the $\mathbf{H}_\mathbf{W}$-states. For the client, these $\mathbf{H}_\mathbf{W}$-states are depicted by the solid green and light blue lines (for *bo* and *ac*, respectively) in Fig. 8. As can be seen, the $\mathbf{H}_\mathbf{W}$-state for *ac* follows the therapy sessions as during these episodes the therapist's actions are sensed. The $\mathbf{H}_\mathbf{W}$-state for *bo* follows a different pattern as the emotion-related states (srs, ps, es states) have a less variable tendency over time. Note that the breakthrough in the fourth session goes together with a steep increase of $\mathbf{H}_\mathbf{W}$-states. The $\mathbf{H}_\mathbf{W}$-states of the therapist follow a similar pattern, although there are some slight differences.

5.3.3 The Virtualizations

For the virtualizations, see Figs. 9 and 10. The feeling is visualized through the facial expressions and the closeness in contact as leaning backwards and forwards.

5.4 Simulation Results for the Alternative Scenarios 2 and 3

5.4.1 Scenario 2: Weak Responsiveness of Both Therapist and Client

In this scenario (see Figs. 11 and 12) both client and therapist have a low responsiveness upon the general therapy context *s*: both have weight 0.1 for the stimulus–response connection for closeness of contact. This models, for example, a therapist who has not much experience yet, or for other reasons is not able to be responsive upon the therapy context.

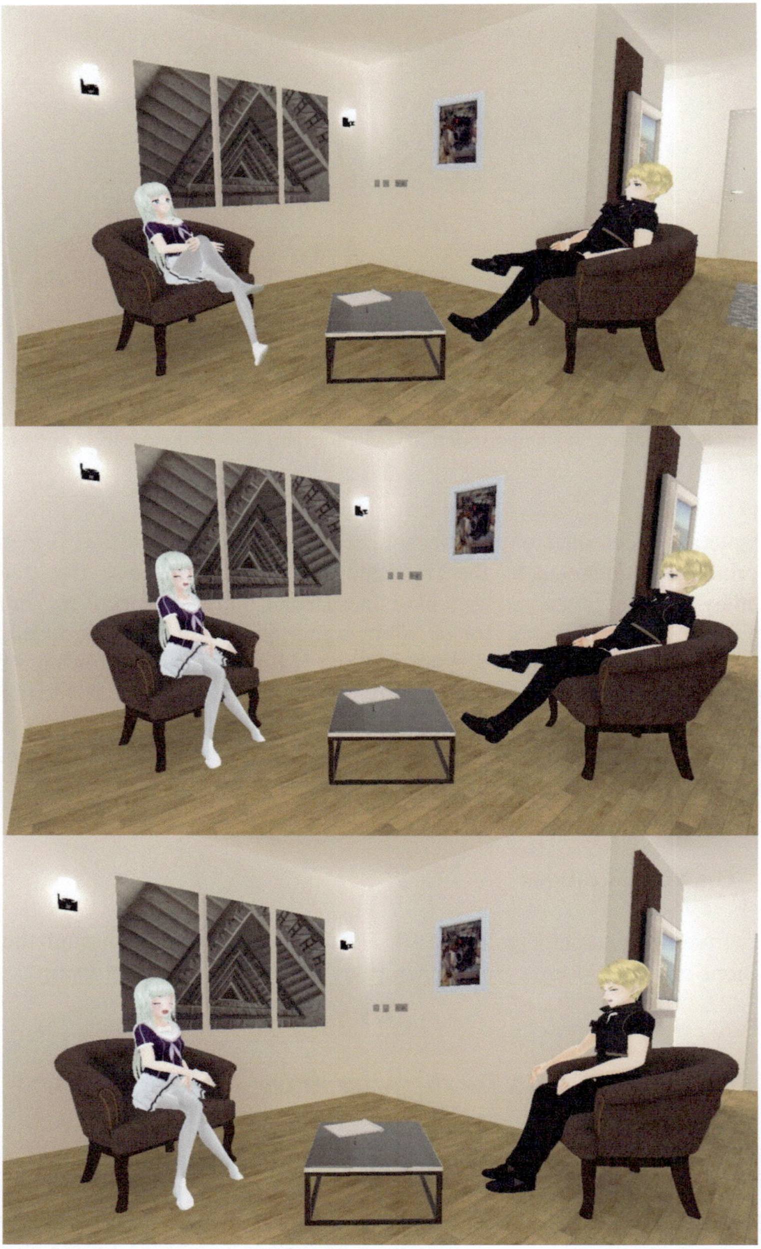

Fig. 10 Screenshots taken of the visualization of scenario 1 for three sessions (from upper to lower) with the therapist (left) and client (right). The therapist has an active posture and happy expression soon after the beginning, in contrast to the client who has to develop that over a number of sessions

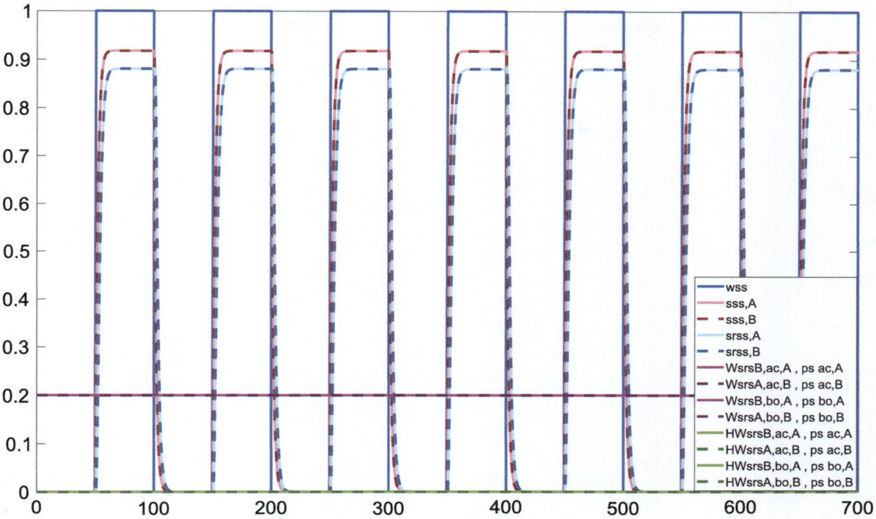

Fig. 11 Scenario 2: all states that do not stay close to 0. Client and therapist both have low responsiveness to the therapy context: the connections from srs_s states to ps_{ac} states have weight 0.1. The states for therapy context stimulus *s* and the **W**- and $\mathbf{H_W}$-states modeling plasticity and metaplasticity. The **W**-states are constant, so no learning takes place; the $\mathbf{H_W}$-states stay very close to 0

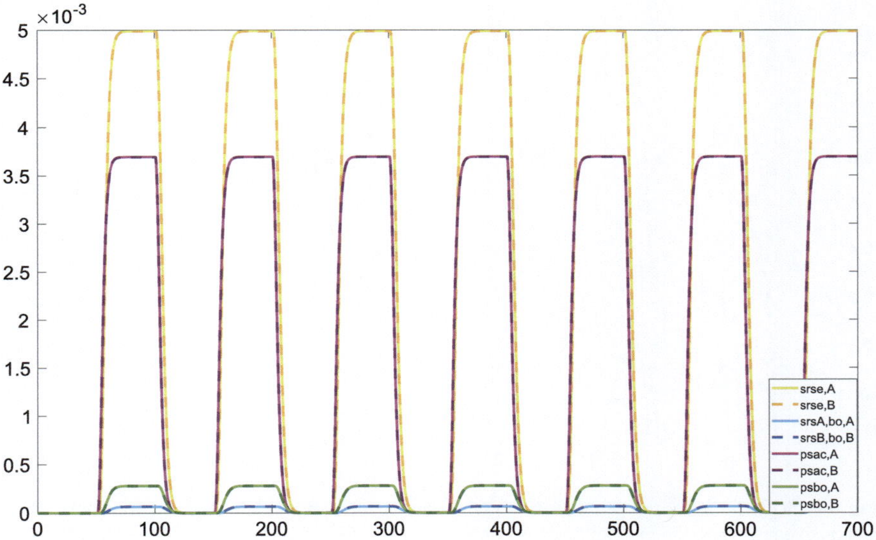

Fig. 12 Scenario 2: the representation and preparation states. Note that the values of all states depicted here are very low: below $5 * 10^{-3}$

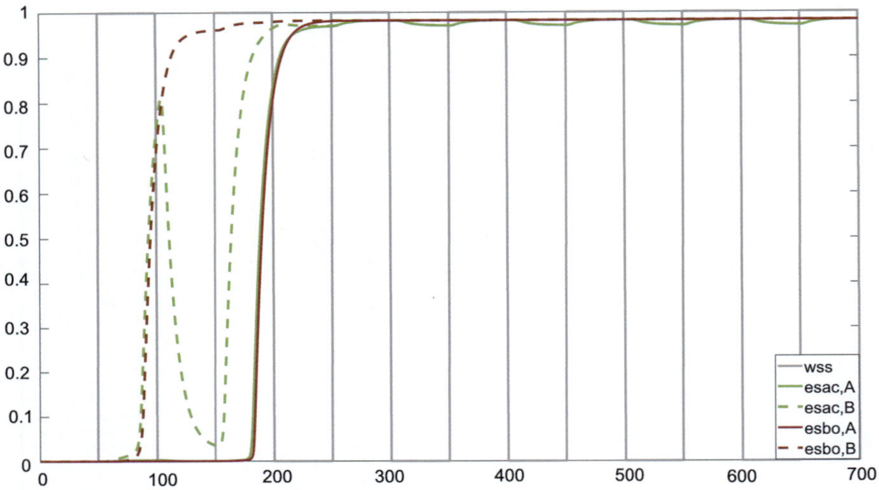

Fig. 13 Scenario 3: general therapy context stimulus *s* and the actions and expressions of both persons as used in the visualizations. Client (0.7) and therapist (1.0) connections from srs_s to ps_{ac} for responsiveness have high weights. Roughly a similar pattern as for scenario 1 but much faster: breakthrough already in session 2

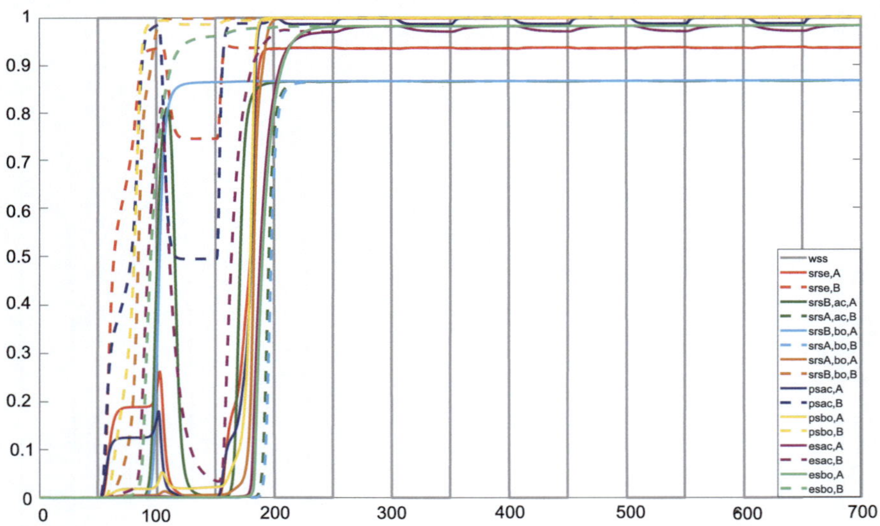

Fig. 14 Scenario 3: the representation and preparation states. Both client and therapist have a high responsiveness upon the general therapy context: client weight 0.7 and therapist weight 1.0 for the stimulus–response connection from srs_s to ps_{ac}

Some of the states reach values around 0.9, but there are very few of them and they all directly relate to the stimulus *s*: only the sensing state $ss_{s,A}$ and $ss_{s,B}$ for both agents, and the sensory representation states $srs_{s,A}$ and $srs_{s,B}$; see Fig. 11. Regarding the plasticity, as can be seen, no learning takes place: all **W**-states remain constant at 0.2 all the time, also within the therapy sessions. This happens because the adaptation speed represented by the $\mathbf{H_W}$-states equals 0. The dominant impact on a $\mathbf{H_W}$-state is via its incoming connection with negative weight −0.5 from the related **W**-state, which overrules the positive impact from the other incoming connections. All other states stay below an activation value of $5 * 10^{-3}$; see Fig. 12, where solely the preparation and predicted effect sensory representation states for *bo* and *ac* are displayed. All other states still follow a similar pattern being slightly activated during the therapy sessions but with still lower activation levels $<5 * 10^{-5}$.

5.4.2 Scenario 3: Strong Responsiveness of Therapist, Moderate Responsiveness of Client

In this scenario the client has a higher responsiveness to the therapy context: value 0.7 for the weight of the connection from the srs_s states to ps_{ac} states; see Figs. 13, 14 and 15. Compared to Scenario 1, it can be seen that now the adaptation within the client goes much faster so that already in the second session a breakthrough is achieved. For the rest, the pattern is similar to that of Scenario 1.

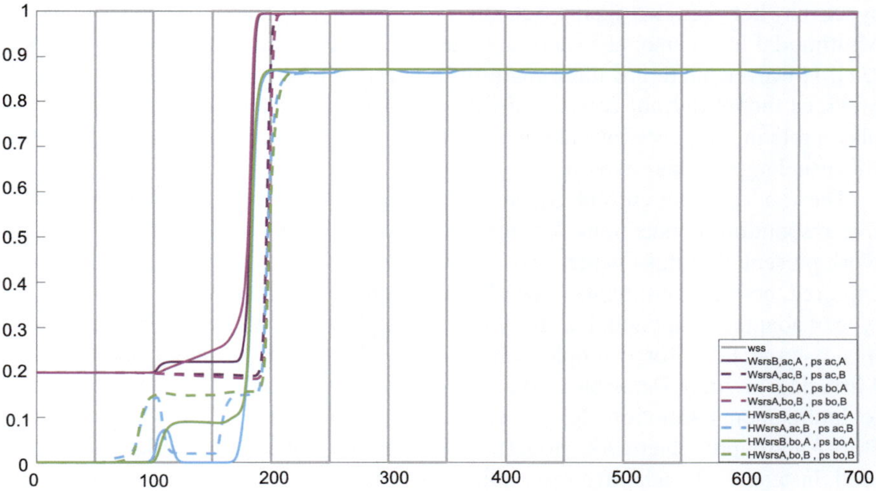

Fig. 15 Scenario 3: the self-model states modeling plasticity and metaplasticity. Both client and therapist have a high responsiveness upon the general therapy context: client weight 0.7 and therapist weight 1.0 for the connection from srs_s to ps_{ac}

6 Discussion and Conclusions

This chapter is based on material from [27]. Our aim was to simulate an adaptive joint decision-making process as a specific form of synchrony between two persons modeled as virtual agents and visualize both the execution of the action and body states (feeling) of each virtual agent. To illustrate the approach, we have used a therapeutic setting and the joint decision regarding the closeness of contact between therapist and client. From our simulations, it turned out that, initially, it was not possible to reach a joint decision—namely close contact—but after a number of sessions it was possible, mainly due to learning on the side of the client. Our model was made adaptive using a Hebbian learning principle applied to the four mirroring connections (for both actions and emotions) for the two virtual agents [26]. Due to this, the agents get more responsive to each other over time and so that it will become easier for them to reach a joint decision than for the nonadaptive case described in [16, 53].

To make it more realistic, the learning speed itself was not assumed to be constant but was made adaptive in a context-sensitive manner to model metaplasticity [3]. The modeling approach used is based on states with time-varying activation levels that were connected to each other through temporal-causal relationships. There are sensing, internal and execution states. Using the self-modeling network modeling option, the first- and second-order adaptivity was modeled based on the same temporal-causal network modeling principles [54, 55].

In our previous work [30] we developed a computational model that addressed synchrony between two agents. See also this volume [29] Chapter The Emergence of Multimodal Interpersonal Synchrony During Social Interaction: A Network Model [31]. However, there are three main differences with the current chapter: (1) this previous model did not consider the more complex internal mental processes that play a role in joint decision-making, (2) it was not adaptive, and (3) no visualization by virtual agents was developed.

The model in the current chapter addresses only one type of adaptivity: adaptive responding connections for mirroring, based on Hebbian learning. After the work presented in this chapter was finished, also several other variations have been explored, both concerning the type of connection that is adaptive and concerning the type of adaptation applied. For many of these explorations also chapters are included this volume [29], for example, Chapters Bonding by Homophily and Adaptive Mirroring in Joint Decision Making: A Second-Order Adaptive Network Model to Interpersonal Emotion Regulation via Multimodal Interaction with a Bonding Bot: An Adaptive Network Model), in Part III–VI. Among the variations considered, instead of the adaptive responding connections, are representing connections, executing connections and connections between agents, and also excitability thresholds for representing states and executing states. Moreover, variations made for the type of adaptation cover (subjective) homophily-induced adaptivity of connectedness, (subjective) synchrony-induced adaptivity, synchrony-transitions-induced adaptivity, and excitability modulation.

On the basis of three illustrative simulations, we conclude that we succeeded to model an adaptive joint decision process with an application in a human-like situation, namely a therapeutic setting. These findings might serve as a foundation for the development of virtual support for therapies in the future. This research opens a number of directions for further research. First, our agent models relied solely on nonverbal communication, but also language plays an important role in humans' communication. Therefore, future agent models could be extended to verbal communication, for which it can be explored whether Large Language Models (LLM) might be integrated, see, for example [18, 33, 50]. Second, we only visualized one body state, facial expressions that ranged from neutral to positive. The emotional response system is much more differentiated in humans and the distinction between, for example, different approach-motivated states like attraction and anger was beyond the scope of the current research. Future agent models could incorporate more emotions like anger, disgust and sadness. Third, it is also possible to let human participants verify the realism of the expressed actions and emotions (such as closeness and happiness) in the visualizations.

Appendix: Full Specification

In this Appendix a full specification of the computational model by role matrices is provided. The network characteristics are set on the values used in the main simulation (Scenario 1) presented in this chapter. See Figs. 16, 17, 18, 19, 20 and 21.

mb base connectivity	State	1	2	3	4
X_1	ws_s	X_1			
X_2	$ws_{A,ac}$	X_{40}			
X_3	$ws_{B,ac}$	X_{41}			
X_4	$ws_{A,bo}$	X_{42}			
X_5	$ws_{B,bo}$	X_{43}			
X_6	$ss_{s,A}$	X_1			
X_7	$ss_{s,B}$	X_1			
X_8	$ss_{B,ac,A}$	X_3			
X_9	$ss_{A,ac,B}$	X_2			
X_{10}	$ss_{B,bo,A}$	X_5			
X_{11}	$ss_{A,bo,B}$	X_4			
X_{12}	$ss_{A,bo,A}$	X_4			
X_{13}	$ss_{B,bo,B}$	X_5			
X_{14}	$srs_{s,A}$	X_6			
X_{15}	$srs_{s,B}$	X_7			
X_{16}	$srs_{e,A}$	X_{24}	X_{32}		
X_{17}	$srs_{e,B}$	X_{25}	X_{33}		
X_{18}	$srs_{B,ac,A}$	X_8			
X_{19}	$srs_{A,ac,B}$	X_9			
X_{20}	$srs_{B,bo,A}$	X_{10}			
X_{21}	$srs_{A,bo,B}$	X_{11}			
X_{22}	$srs_{A,bo,A}$	X_{12}	X_{26}		
X_{23}	$srs_{B,bo,B}$	X_{13}	X_{27}		
X_{24}	$ps_{ac,A}$	X_{14}	X_{18}	X_{22}	
X_{25}	$ps_{ac,B}$	X_{15}	X_{19}	X_{23}	
X_{26}	$ps_{bo,A}$	X_{16}	X_{20}	X_{22}	
X_{27}	$ps_{bo,B}$	X_{17}	X_{21}	X_{23}	
X_{28}	$os_{B,s,ac,e,A}$	X_{18}			
X_{29}	$os_{A,s,ac,e,B}$	X_{19}			
X_{30}	$os_{B,e,bo,A}$	X_{20}			
X_{31}	$os_{A,e,bo,B}$	X_{21}			
X_{32}	$os_{A,s,ac,e,A}$	X_{14}	X_{18}	X_{16}	X_{24}
X_{33}	$os_{B,s,ac,e,B}$	X_{15}	X_{19}	X_{17}	X_{25}
X_{34}	$os_{A,e,bo,A}$	X_{16}	X_{20}	X_{22}	X_{26}
X_{35}	$os_{B,e,bo,B}$	X_{17}	X_{21}	X_{23}	X_{27}
X_{36}	$ec_{B,s,ac,e,A}$	X_{28}			
X_{37}	$ec_{A,s,ac,e,B}$	X_{29}			
X_{38}	$ec_{B,e,bo,A}$	X_{30}			
X_{39}	$ec_{A,e,bo,B}$	X_{31}			
X_{40}	$es_{ac,A}$	X_{32}	X_{24}		
X_{41}	$es_{ac,B}$	X_{33}	X_{25}		
X_{42}	$es_{bo,A}$	X_{34}	X_{26}		
X_{43}	$es_{bo,B}$	X_{35}	X_{27}		
X_{44}	$\mathbf{W}_{srsB,ac,A\text{ , }ps\ ac,A}$	X_{18}	X_{24}	X_{44}	
X_{45}	$\mathbf{W}_{srsA,ac,B\text{ , }ps\ ac,B}$	X_{19}	X_{25}	X_{45}	
X_{46}	$\mathbf{W}_{srsB,bo,A\text{ , }ps\ bo,A}$	X_{20}	X_{26}	X_{46}	
X_{47}	$\mathbf{W}_{srsA,bo,B\text{ , }ps\ bo,B}$	X_{21}	X_{27}	X_{47}	
X_{48}	$\mathbf{H}w_{srsB,ac,A\text{ , }ps\ ac,A}$	X_{18}	X_{24}	X_{44}	X_{48}
X_{49}	$\mathbf{H}w_{srsA,ac,B\text{ , }ps\ ac,B}$	X_{19}	X_{25}	X_{45}	X_{49}
X_{50}	$\mathbf{H}w_{srsB,bo,A\text{ , }ps\ bo,A}$	X_{20}	X_{26}	X_{46}	X_{50}
X_{51}	$\mathbf{H}w_{srsA,bo,B\text{ , }ps\ bo,B}$	X_{21}	X_{27}	X_{47}	X_{51}

Fig. 16 Connectivity role matrix **mb** (for base connections) of the main simulation

mcw connection weights	State	1	2	3	4
X_1	ws_s	1			
X_2	$ws_{A,ac}$	1			
X_3	$ws_{B,ac}$	1			
X_4	$ws_{A,bo}$	1			
X_5	$ws_{B,bo}$	1			
X_6	$ss_{s,A}$	1			
X_7	$ss_{s,B}$	1			
X_8	$ss_{B,ac,A}$	1			
X_9	$ss_{A,ac,B}$	1			
X_{10}	$ss_{B,bo,A}$	1			
X_{11}	$ss_{A,bo,B}$	1			
X_{12}	$ss_{A,bo,A}$	1			
X_{13}	$ss_{B,bo,B}$	1			
X_{14}	$srs_{s,A}$	1			
X_{15}	$srs_{s,B}$	1			
X_{16}	$srs_{e,A}$	1	-0.2		
X_{17}	$srs_{e,B}$	1	-0.2		
X_{18}	$srs_{B,ac,A}$	1			
X_{19}	$srs_{A,ac,B}$	1			
X_{20}	$srs_{B,bo,A}$	1			
X_{21}	$srs_{A,bo,B}$	1			
X_{22}	$srs_{A,bo,A}$	1	1		
X_{23}	$srs_{B,bo,B}$	1	1		
X_{24}	$ps_{ac,A}$	0.1	X_{44}	1	
X_{25}	$ps_{ac,B}$	1	X_{45}	1	
X_{26}	$ps_{bo,A}$	1	X_{46}	1	
X_{27}	$ps_{bo,B}$	1	X_{47}	1	
X_{28}	$os_{B,s,ac,e,A}$	1			
X_{29}	$os_{A,s,ac,e,B}$	1			
X_{30}	$os_{B,e,bo,A}$	1			
X_{31}	$os_{A,e,bo,B}$	1			
X_{32}	$os_{A,s,ac,e,A}$	1	1	1	1
X_{33}	$os_{B,s,ac,e,B}$	1	1	1	1
X_{34}	$os_{A,e,bo,A}$	1	1	1	1
X_{35}	$os_{B,e,bo,B}$	1	1	1	1
X_{36}	$ec_{B,s,ac,e,A}$	1			
X_{37}	$ec_{A,s,ac,e,B}$	1			
X_{38}	$ec_{B,e,bo,A}$	1			
X_{39}	$ec_{A,e,bo,B}$	1			
X_{40}	$es_{ac,A}$	1	1		
X_{41}	$es_{ac,B}$	1	1		
X_{42}	$es_{bo,A}$	1	1		
X_{43}	$es_{bo,B}$	1	1		
X_{44}	$\mathbf{W}_{srsB,ac,A\ ,\ ps\ ac,A}$	1	1	1	
X_{45}	$\mathbf{W}_{srsA,ac,B\ ,\ ps\ ac,B}$	1	1	1	
X_{46}	$\mathbf{W}_{srsB,bo,A\ ,\ ps\ bo,A}$	1	1	1	
X_{47}	$\mathbf{W}_{srsA,bo,B\ ,\ ps\ bo,B}$	1	1	1	
X_{48}	$\mathbf{Hw}_{srsB,ac,A\ ,\ ps\ ac,A}$	1	1	-0.5	1
X_{49}	$\mathbf{Hw}_{srsA,ac,B\ ,\ ps\ ac,B}$	1	1	-0.5	1
X_{50}	$\mathbf{Hw}_{srsB,bo,A\ ,\ ps\ bo,A}$	1	1	-0.5	1
X_{51}	$\mathbf{Hw}_{srsA,bo,B\ ,\ ps\ bo,B}$	1	1	-0.5	1

Fig. 17 Connectivity role matrix **mcw** (for connection weights) of the main simulation

Fig. 18 Aggregation role matrix **mcfw** (for combination function weights) of the main simulation

mcfw combination function weights	**State**	**1 alogistic**	**2 step mod**	**3 hebb**
X_1	ws_s		1	
X_2	$ws_{A,ac}$	1		
X_3	$ws_{B,ac}$	1		
X_4	$ws_{A,bo}$	1		
X_5	$ws_{B,bo}$	1		
X_6	$ss_{s,A}$	1		
X_7	$ss_{s,B}$	1		
X_8	$ss_{B,ac,A}$	1		
X_9	$ss_{A,ac,B}$	1		
X_{10}	$ss_{B,bo,A}$	1		
X_{11}	$ss_{A,bo,B}$	1		
X_{12}	$ss_{A,bo,A}$	1		
X_{13}	$ss_{B,bo,B}$	1		
X_{14}	$srs_{s,A}$	1		
X_{15}	$srs_{s,B}$	1		
X_{16}	$srs_{e,A}$	1		
X_{17}	$srs_{e,B}$	1		
X_{18}	$srs_{B,ac,A}$	1		
X_{19}	$srs_{A,ac,B}$	1		
X_{20}	$srs_{B,bo,A}$	1		
X_{21}	$srs_{A,bo,B}$	1		
X_{22}	$srs_{A,bo,A}$	1		
X_{23}	$srs_{B,bo,B}$	1		
X_{24}	$ps_{ac,A}$	1		
X_{25}	$ps_{ac,B}$	1		
X_{26}	$ps_{bo,A}$	1		
X_{27}	$ps_{bo,B}$	1		
X_{28}	$os_{B,s,ac,e,A}$	1		
X_{29}	$os_{A,s,ac,e,B}$	1		
X_{30}	$os_{B,e,bo,A}$	1		
X_{31}	$os_{A,e,bo,B}$	1		
X_{32}	$os_{A,s,ac,e,A}$	1		
X_{33}	$os_{B,s,ac,e,B}$	1		
X_{34}	$os_{A,e,bo,A}$	1		
X_{35}	$os_{B,e,bo,B}$	1		
X_{36}	$ec_{B,s,ac,e,A}$	1		
X_{37}	$ec_{A,s,ac,e,B}$	1		
X_{38}	$ec_{B,e,bo,A}$	1		
X_{39}	$ec_{A,e,bo,B}$	1		
X_{40}	$es_{ac,A}$	1		
X_{41}	$es_{ac,B}$	1		
X_{42}	$es_{bo,A}$	1		
X_{43}	$es_{bo,B}$	1		
X_{44}	$\mathbf{W}_{srsB,ac,A\ ,\ ps\ ac,A}$			1
X_{45}	$\mathbf{W}_{srsA,ac,B\ ,\ ps\ ac,B}$			1
X_{46}	$\mathbf{W}_{srsB,bo,A\ ,\ ps\ bo,A}$			1
X_{47}	$\mathbf{W}_{srsA,bo,B\ ,\ ps\ bo,B}$			1
X_{48}	$\mathbf{H}w_{srsB,ac,A\ ,\ ps\ ac,A}$	1		
X_{49}	$\mathbf{H}w_{srsA,ac,B\ ,\ ps\ ac,B}$	1		
X_{50}	$\mathbf{H}w_{srsB,bo,A\ ,\ ps\ bo,A}$	1		
X_{51}	$\mathbf{H}w_{srsA,bo,B\ ,\ ps\ bo,B}$	1		

mcfw combination function weights	State	1 alogistic		2 step mod		3 hebb
X_1	ws_s			100	50	
X_2	$ws_{A,ac}$	6	0.5			
X_3	$ws_{B,ac}$	6	0.5			
X_4	$ws_{A,bo}$	6	0.5			
X_5	$ws_{B,bo}$	6	0.5			
X_6	$ss_{s,A}$	5	0.5			
X_7	$ss_{s,B}$	5	0.5			
X_8	$ss_{B,ac,A}$	5	0.5			
X_9	$ss_{A,ac,B}$	5	0.5			
X_{10}	$ss_{B,bo,A}$	5	0.5			
X_{11}	$ss_{A,bo,B}$	5	0.5			
X_{12}	$ss_{A,bo,A}$	5	0.5			
X_{13}	$ss_{B,bo,B}$	5	0.5			
X_{14}	$srs_{s,A}$	5	0.5			
X_{15}	$srs_{s,B}$	5	0.5			
X_{16}	$srs_{e,A}$	5	0.2			
X_{17}	$srs_{e,B}$	5	0.2			
X_{18}	$srs_{B,ac,A}$	5	0.5			
X_{19}	$srs_{A,ac,B}$	5	0.5			
X_{20}	$srs_{B,bo,A}$	5	0.5			
X_{21}	$srs_{A,bo,B}$	5	0.5			
X_{22}	$srs_{A,bo,A}$	5	0.6			
X_{23}	$srs_{B,bo,B}$	5	0.6			
X_{24}	$ps_{ac,A}$	5	1			
X_{25}	$ps_{ac,B}$	5	1			
X_{26}	$ps_{bo,A}$	5	0.9			
X_{27}	$ps_{bo,B}$	5	0.9			
X_{28}	$os_{B,s,ac,e,A}$	8	0.7			
X_{29}	$os_{A,s,ac,e,B}$	8	0.7			
X_{30}	$os_{B,e,bo,A}$	8	0.7			
X_{31}	$os_{A,e,bo,B}$	8	0.7			
X_{32}	$os_{A,s,ac,e,A}$	8	2.5			
X_{33}	$os_{B,s,ac,e,B}$	8	2.5			
X_{34}	$os_{A,e,bo,A}$	8	2.5			
X_{35}	$os_{B,e,bo,B}$	8	2.5			
X_{36}	$ec_{B,s,ac,e,A}$	20	0.5			
X_{37}	$ec_{A,s,ac,e,B}$	20	0.5			
X_{38}	$ec_{B,e,bo,A}$	20	0.5			
X_{39}	$ec_{A,e,bo,B}$	20	0.5			
X_{40}	$es_{ac,A}$	5	1.2			
X_{41}	$es_{ac,B}$	5	1.2			
X_{42}	$es_{bo,A}$	5	1.2			
X_{43}	$es_{bo,B}$	5	1.2			
X_{44}	$\mathbf{W}_{srsB,ac,A\ ,\ ps\ ac,A}$					0.995
X_{45}	$\mathbf{W}_{srsA,ac,B\ ,\ ps\ ac,B}$					0.995
X_{46}	$\mathbf{W}_{srsB,bo,A\ ,\ ps\ bo,A}$					0.995
X_{47}	$\mathbf{W}_{srsA,bo,B\ ,\ ps\ bo,B}$					0.995
X_{48}	$\mathbf{H}\mathbf{w}_{srsB,ac,A\ ,\ ps\ ac,A}$	3	1.6			
X_{49}	$\mathbf{H}\mathbf{w}_{srsA,ac,B\ ,\ ps\ ac,B}$	3	1.6			
X_{50}	$\mathbf{H}\mathbf{w}_{srsB,bo,A\ ,\ ps\ bo,A}$	3	1.6			
X_{51}	$\mathbf{H}\mathbf{w}_{srsA,bo,B\ ,\ ps\ bo,B}$	3	1.6			

Fig. 19 Aggregation role matrix **mcfp** (for combination function parameters) of the main simulation

Fig. 20 Timing role matrix **ms** (for speed factors) of the main simulation

ms speed factors	State	1
X_1	ws_s	2
X_2	$ws_{A,ac}$	0.5
X_3	$ws_{B,ac}$	0.5
X_4	$ws_{A,bo}$	0.5
X_5	$ws_{B,bo}$	0.5
X_6	$ss_{s,A}$	0.5
X_7	$ss_{s,B}$	0.5
X_8	$ss_{B,ac,A}$	0.5
X_9	$ss_{A,ac,B}$	0.5
X_{10}	$ss_{B,bo,A}$	0.5
X_{11}	$ss_{A,bo,B}$	0.5
X_{12}	$ss_{A,bo,A}$	0.5
X_{13}	$ss_{B,bo,B}$	0.5
X_{14}	$srs_{s,A}$	0.5
X_{15}	$srs_{s,B}$	0.5
X_{16}	$srs_{e,A}$	0.5
X_{17}	$srs_{e,B}$	0.5
X_{18}	$srs_{B,ac,A}$	0.5
X_{19}	$srs_{A,ac,B}$	0.5
X_{20}	$srs_{B,bo,A}$	0.5
X_{21}	$srs_{A,bo,B}$	0.5
X_{22}	$srs_{A,bo,A}$	0.5
X_{23}	$srs_{B,bo,B}$	0.5
X_{24}	$ps_{ac,A}$	0.5
X_{25}	$ps_{ac,B}$	0.5
X_{26}	$ps_{bo,A}$	0.5
X_{27}	$ps_{bo,B}$	0.5
X_{28}	$os_{B,s,ac,e,A}$	0.5
X_{29}	$os_{A,s,ac,e,B}$	0.5
X_{30}	$os_{B,e,bo,A}$	0.5
X_{31}	$os_{A,e,bo,B}$	0.5
X_{32}	$os_{A,s,ac,e,A}$	0.5
X_{33}	$os_{B,s,ac,e,B}$	0.5
X_{34}	$os_{A,e,bo,A}$	0.5
X_{35}	$os_{B,e,bo,B}$	0.5
X_{36}	$ec_{B,s,ac,e,A}$	0.1
X_{37}	$ec_{A,s,ac,e,B}$	0.1
X_{38}	$ec_{B,e,bo,A}$	0.1
X_{39}	$ec_{A,e,bo,B}$	0.1
X_{40}	$es_{ac,A}$	0.1
X_{41}	$es_{ac,B}$	0.1
X_{42}	$es_{bo,A}$	0.1
X_{43}	$es_{bo,B}$	0.1
X_{44}	$\mathbf{W}_{\text{srsB,ac,A , ps ac,A}}$	X_{48}
X_{45}	$\mathbf{W}_{\text{srsA,ac,B , ps ac,B}}$	X_{49}
X_{46}	$\mathbf{W}_{\text{srsB,bo,A , ps bo,A}}$	X_{50}
X_{47}	$\mathbf{W}_{\text{srsA,bo,B , ps bo,B}}$	X_{51}
X_{48}	$\mathbf{H}w_{\text{srsB,ac,A , ps ac,A}}$	0.5
X_{49}	$\mathbf{H}w_{\text{srsA,ac,B , ps ac,B}}$	0.5
X_{50}	$\mathbf{H}w_{\text{srsB,bo,A , ps bo,A}}$	0.5
X_{51}	$\mathbf{H}w_{\text{srsA,bo,B , ps bo,B}}$	0.5

Fig. 21 Initial values **iv** of the main simulation

iv initial values	State	1
X_1	ws_s	0
X_2	$ws_{A,ac}$	0
X_3	$ws_{B,ac}$	0
X_4	$ws_{A,bo}$	0
X_5	$ws_{B,bo}$	0
X_6	$ss_{s,A}$	0
X_7	$ss_{s,B}$	0
X_8	$ss_{B,ac,A}$	0
X_9	$ss_{A,ac,B}$	0
X_{10}	$ss_{B,bo,A}$	0
X_{11}	$ss_{A,bo,B}$	0
X_{12}	$ss_{A,bo,A}$	0
X_{13}	$ss_{B,bo,B}$	0
X_{14}	$srs_{s,A}$	0
X_{15}	$srs_{s,B}$	0
X_{16}	$srs_{e,A}$	0
X_{17}	$srs_{e,B}$	0
X_{18}	$srs_{B,ac,A}$	0
X_{19}	$srs_{A,ac,B}$	0
X_{20}	$srs_{B,bo,A}$	0
X_{21}	$srs_{A,bo,B}$	0
X_{22}	$srs_{A,bo,A}$	0
X_{23}	$srs_{B,bo,B}$	0
X_{24}	$ps_{ac,A}$	0
X_{25}	$ps_{ac,B}$	0
X_{26}	$ps_{bo,A}$	0
X_{27}	$ps_{bo,B}$	0
X_{28}	$os_{B,s,ac,e,A}$	0
X_{29}	$os_{A,s,ac,e,B}$	0
X_{30}	$os_{B,e,bo,A}$	0
X_{31}	$os_{A,e,bo,B}$	0
X_{32}	$os_{A,s,ac,e,A}$	0
X_{33}	$os_{B,s,ac,e,B}$	0
X_{34}	$os_{A,e,bo,A}$	0
X_{35}	$os_{B,e,bo,B}$	0
X_{36}	$ec_{B,s,ac,e,A}$	0
X_{37}	$ec_{A,s,ac,e,B}$	0
X_{38}	$ec_{B,e,bo,A}$	0
X_{39}	$ec_{A,e,bo,B}$	0
X_{40}	$es_{ac,A}$	0
X_{41}	$es_{ac,B}$	0
X_{42}	$es_{bo,A}$	0
X_{43}	$es_{bo,B}$	0
X_{44}	$\mathbf{W}_{srsB,ac,A\ ,\ ps\ ac,A}$	0.2
X_{45}	$\mathbf{W}_{srsA,ac,B\ ,\ ps\ ac,B}$	0.2
X_{46}	$\mathbf{W}_{srsB,bo,A\ ,\ ps\ bo,A}$	0.2
X_{47}	$\mathbf{W}_{srsA,bo,B\ ,\ ps\ bo,B}$	0.2
X_{48}	$\mathbf{H}w_{srsB,ac,A\ ,\ ps\ ac,A}$	0
X_{49}	$\mathbf{H}w_{srsA,ac,B\ ,\ ps\ ac,B}$	0
X_{50}	$\mathbf{H}w_{srsB,bo,A\ ,\ ps\ bo,A}$	0
X_{51}	$\mathbf{H}w_{srsA,bo,B\ ,\ ps\ bo,B}$	0

References

1. A2 Games: Satomi: Anime-Style Character for Games and VR-Chat (2020). https://assetstore.unity.com/packages/3d/characters/humanoids/satomi-anime-style-character-for-games-and-vrchat-175748
2. A2 Games: Rin New: Anime-Style Character for Games and VR-Chat (2021). https://assetstore.unity.com/packages/3d/characters/humanoids/rin-new-anime-style-character-for-games-and-vrchat-174995
3. Abraham, W.C., Bear, M.F.: Metaplasticity: the plasticity of synaptic plasticity. Trends Neurosci. **19**(4), 126–130 (1996)
4. Adobe: Adobe Mixamo (2008). https://www.mixamo.com/
5. Bain, A.: Mind and Body: The Theories of Their Relation. HS King & Company. Farnborough Eng. Gregg International (1873)
6. Boukouvala, A: Imitation of affects and mirror neurons: exploring empathy in Spinoza's theory and contemporary neuroscience. Philosophia **45**, 1007–1017 (2017)
7. Brass, M., Spengler, S.: The inhibition of imitative behaviour and attribution of mental states. In: Striano, T., Reid, V. (eds.) Social Cognition: Development, Neuroscience, and Autism, pp. 52–66. Wiley-Blackwell (2009)
8. Chandra, N., Barkai, E.: A non-synaptic mechanism of complex learning: modulation of intrinsic neuronal excitability. Neurobiol. Learn. Mem. **154**, 30–36 (2018)
9. Chartrand, T.L., Bargh, J.A.: The chameleon effect: the perception-behavior link and social interaction. J. Pers. Soc. Psychol. **76**, 893–910 (1999)
10. Courtney, A.L, Meyer, M.L.: Self-other representation in the social brain reflects social connection. J. Neurosci. **40**(29), 5616–5627 (2020). https://doi.org/10.1523/JNEUROSCI.2826-19.2020
11. Damasio, A.R.: Descartes' Error: Emotion, Reason, and the Human Brain. Vintage Publishing (1994)
12. Damasio, A.R.: The Feeling of What Happens: Body and Emotion in the Making of Consciousness. Houghton Mifflin Harcourt (1999)
13. De Jong, F.C., Eler, E., Rass, L., Treur, R.M., Treur, J., Koole, S.L.: From mental network models to virtualisation by avatars: a first software implementation. In: Proceedings of the 12th Annual Conference on Biologically Inspired Cognitive Architectures, BICA'21. Studies in Computational Intelligence, vol. 1032, pp. 75–88. Springer Nature (2022)
14. DevDen: Devden archviz, vol. 1, Scotland (2020). https://assetstore.unity.com/packages/3d/environments/urban/devden-archviz-vol-1-scotland-158539
15. Drimalla, H., Landwehrd, N., Hessa, U., Dziobek, I.: From face to face: the contribution of facial mimicry to cognitive and emotional empathy. Cogn. Emot. **33**(8), 1672–1686 (2019)
16. Duell, R., Treur, J.: A computational analysis of joint decision-making processes. In: Proceedings of the International Conference on Social Informatics, SocInfo'12, Lecture Notes in Computer Science, **7710**, 292–308. Springer (2012)
17. Eisenberg, N., Fabes, R.A.: Empathy: conceptualization, measurement, and relation to prosocial behavior. Motiv. Emot. **14**(2), 131–149 (1990)
18. Filienko, D., Wang, Y., Jazmi, C.E., Xie, S., Cohen, T., De Cock, M., Yuwen, W.: Toward large language models as a therapeutic tool: comparing prompting techniques to improve GPT-delivered problem-solving therapy. arXiv preprint arXiv:2409.00112 (2024)
19. Fischer, A., Hess, U.: Mimicking emotions. Curr. Opin. Psychol. **17**, 151–155 (2017)
20. Fried, I., Mukamel, R., Kreiman, G.: Internally generated preactivation of single neurons in human medial frontal cortex predicts volition. Neuron **69**, 548–562 (2011)
21. Garcia, R.: Stress, metaplasticity, and antidepressants. Curr. Mol. Med. **2**, 629–638 (2002)
22. Hamilton, A.F.d.C., Brindley, R.M., Frith, U.: Imitation and action understanding in autistic spectrum disorders: how valid is the hypothesis of a deficit in the mirror neuron system? Neuropsychologia **45**, 1859–1868 (2007)

23. Harmon-Jones, E., Gable, P.A., Price, T.F.: Does negative affect always narrow and positive affect always broaden the mind? Considering the influence of motivational intensity on cognitive scope. Curr. Dir. Psychol. Sci. **22**(4), 301–307 (2013)
24. Harmon-Jones, E., Peterson, C.K.: Supine body position reduces neural response to anger evocation. Psychol. Sci. **20**, 1209–1210 (2009)
25. Hareli, S., Hess, U.: The social signal value of emotion. Cognit. Emot. **3**, 385–389 (2012)
26. Hebb, D.O.: The Organization of Behavior: A Neuropsychological Theory. John Wiley and Sons, New York (1949)
27. Hendrikse, S.C.F., Kluiver, S., Treur, J., Wilderjans, T.F., Dikker, S., Koole, S.L.: How virtual agents can learn to synchronize: an adaptive joint decision-making model of psychotherapy. Cogn. Syst. Res. **79**, 138–155 (2023)
28. Hendrikse, S.C.F., Treur, J.: From a multilevel multi-timescale computational architecture for synchrony- and homophily-induced attunement to designing bonding bots. In: Hendrikse, S.C.F., Treur, J., Koole, S.L. (eds.) New Analysis and Modeling Directions in Social Interaction Science: Emergent Multimodal Interpersonal Synchrony, Affiliation, and Bonding, Ch. 20. Springer Nature (2025)
29. Hendrikse, S.C.F., Treur, J., Koole, S.L. (eds.): New Analysis and Modeling Directions in Social Interaction Science: Emergent Multimodal Interpersonal Synchrony, Affiliation, and Bonding. Springer Nature (2025)
30. Hendrikse, S.C.F., Treur, J., Wilderjans, T.F., Dikker, S., Koole, S.L.: On the same wavelengths: emergence of multiple synchronies among multiple agents. In: Proceedings of the 22nd International Workshop on Multi-Agent Systems and Agent-Based Simulation, MABS'21 Lecture Notes in Computer Scienc, vol. 13128, pp. 57–71. Springer International Publishing, Cham (2022)
31. Hendrikse, S.C.F., Treur, J., Wilderjans, T.F., Dikker, S., Koole, S.L.: The emergence of multimodal interpersonal synchrony during social interaction: a network model. In: Hendrikse, S.C.F., Treur, J., Koole, S.L. (eds.) New Analysis and Modeling Directions in Social Interaction Science: Emergent Multimodal Interpersonal Synchrony, Affiliation, and Bonding, Ch. 3. Springer Nature (2025)
32. Hess, U., Fischer, A.: Emotional mimicry as social regulation. Pers. Soc. Psychol. Rev. **17**(2), 142–157 (2013)
33. Hodson, N., Williamson, S.: Can large language models replace therapists? Evaluating performance at simple cognitive behavioral therapy tasks. JMIR AI **3**(1), e52500 (2024)
34. Iacoboni, M.: Mirroring people: the new science of how we connect with others. Farrar, Straus and Giroux (2008a)
35. Iacoboni, M.: Mesial frontal cortex and super mirror neurons. Behav. Brain Sci. **31**, 30 (2008b)
36. Iacoboni, M., Dapretto, M.: The mirror neuron system and the consequences of its dysfunction. Nat. Rev. Neurosci. **7**, 942–951 (2006)
37. Kahl, S., Kopp, S.: A predictive processing model of perception and action for self-other distinction. Front. Psychol. **9**(2421) (2018). https://doi.org/10.3389/fpsyg.2018.02421
38. Keysers, C., Gazzola, V.: Hebbian learning and predictive mirror neurons for actions, sensations and emotions. Philos. Trans. R. Soc. Lond. B Biol. Sci. **369**, 20130175 (2014)
39. Keysers, C., Gazzola, V.: Social neuroscience: mirror neurons recorded in humans. Curr. Biol. **20**(2010), 253–254 (2010)
40. Koole, S.L., Veenstra, L., Domachowska, I., Dillon, K.P., Schneider, I.K.: Embodied anger management: approach-oriented postures moderate whether trait anger becomes translated into state anger and aggression. Motiv. Sci. **8**(2), 174 (2022)
41. Kret, M.E., Fischer, A.H., De Dreu, C.K.W.: Pupil mimicry correlates with trust in in-group partners with dilating pupils. Psychol. Sci. **26**, 1401–1410 (2015). https://doi.org/10.1177/0956797615588306
42. Moody, E.J., McIntosh, D.N.: Mimicry of dynamic emotional and motor-only stimuli. Soc. Psychol. Pers. Sci. **2**, 679–686 (2011)
43. Moore, J., Haggard, P.: Awareness of action: inference and prediction. Conscious. Cogn. **17**, 136–144 (2008)

44. Mukamel, R., Ekstrom, A.D., Kaplan, J., Iacoboni, M., Fried, I.: Single-neuron responses in humans during execution and observation of actions. Curr. Biol. **20**, 750–756 (2010)
45. Oatley, K.: On the sharing of mind. In: Hess, U., Fischer, A.H. (eds) Emotional Mimicry in Social Context, pp. 7–26. Cambridge University Press (2015)
46. Poppa, T., Bechara, A.: The somatic marker hypothesis: revisiting the role of the 'body-loop' in decision-making. Curr. Opin. Behav. Sci. **19**, 61–66 (2018)
47. Quesque, F., Brass, M.: The role of the temporoparietal junction in self-other distinction. Brain Topogr. **32**, 943–955 (2019). https://doi.org/10.1007/s10548-019-00737-5
48. Robinson, B.L., Harper, N.S., McAlpine, D.: Meta-adaptation in the auditory midbrain under cortical influence. Nat. Commun. **7**, article 13442 (2016)
49. Shatz, C.J.: The developing brain. Sci. Am. **267**, 60–67 (1992). https://doi.org/10.1038/scientificamerican0992-60
50. Shen, H., Li, Z., Yang, M., Ni, M., Tao, Y., Yu, Z., et al.: Are Large Language Models Possible to Conduct Cognitive Behavioral Therapy? arXiv preprint arXiv:2407.17730 (2024)
51. Sjöström, P.J., Rancz, E.A., Roth, A., Hausser, M.: Dendritic excitability and synaptic plasticity. Physiol. Rev. **88**, 769–840 (2008)
52. Stel, M., van Baaren, R.B., Vonk, R.: Effects of mimicking: acting prosocially by being emotionally moved. Eur. J. Soc. Psychol. **38**(6), 965–976 (2008)
53. Treur, J.: Modelling joint decision making processes involving emotion-related valuing and empathic understanding. In: Kinny, D., Hsu, J.Y., Governatori, G., Ghose, A.K. (eds.) In: Kinny, D., Hsu, J.Yj., Governatori, G., Ghose, A.K. (eds) Agents in Principle, Agents in Practice. Proceedings of the 14th International Conference on Principles and Practice of Multi-Agent Systems, PRIMA'2011. Lecture Notes in Computer Science, vol. 7047, pp. 410–423. Springer, Berlin, Heidelberg (2011)
54. Treur, J.: A modeling environment for reified temporal-causal networks: modeling plasticity and metaplasticity in cognitive agent models. In: Proceedings of the 22nd International Conference on Principles and Practice of Multi-Agent Systems, PRIMA'19. Lecture Notes in Computer Science, vol. 11873, pp. 487–495. Springer, Cham (2019)
55. Treur, J.: Network-oriented modeling for adaptive networks: designing higher-order adaptive biological, mental and social network models. Springer Nature (2020)
56. Treur, J.: Modeling coevolution of dynamics, adaptivity and control of adaptivity by self-modeling networks. In: Hendrikse, S.C.F., Treur, J., Koole, S.L. (eds.) New Analysis and Modeling Directions in Social Interaction Science: Emergent Multimodal Interpersonal Synchrony, Affiliation, and Bonding, Ch. 2. Springer Nature (2025)
57. Van Kleef, G.A.: How emotions regulate social life: the emotions as social information (EASI) model. Curr. Direct. Psychol. Sci. **2009**(18), 184–188 (2009)
58. Van Ments, L., Treur, J.: Modeling adaptive cooperative and competitive metaphors as mental models for joint decision-making. Cogn. Syst. Res. **69**(4), 67–82 (2021)
59. Wiltermuth, S.S., & Heath, C.: Synchrony and cooperation. Psychol. Sci. **20**(1), 1–5 (2009)
60. Hove, M.J., & Risen, J.L.: It's all in the timing: Interpersonal synchrony increases affiliation. Soc. Cog. **27**(6), 949–960 (2009)
61. Koole, S.L., Atzil-Slonim, D., Butler, E., Dikker, S., Tschacher, W., & Wilderjans, T.: In sync with your shrink: Grounding psychotherapy in interpersonal synchrony. App. Soc. Psychol. (pp. 161–184). Routledge. A concept (2020)
62. Dimberg, U., Andréasson, P., & Thunberg, M.: Emotional empathy and facial reactions to facial expressions. J. Psychophysiol. (2011)
63. Sonnby–Borgström, M.: Automatic mimicry reactions as related to differences in emotional empathy. Scandinavian J Psychol. **43**(5), 433–443 (2002)
64. Watson, D.: Mood and temperament. Guilford Press (2000)
65. Carver, C.S., & Harmon-Jones, E.: Anger is an approach-related affect: evidence and implications. Psychol. Bulletin. **135**(2), 183 (2009)
66. Price, T.F., & Harmon-Jones, E.: Approach motivational body postures lean toward left frontal brain activity. Psychophysiol. **48**(5), 718–722 (2011)

Modeling Social Interaction Adaptivity Induced by Homophily and by Subjectively Perceived Homophily

In this part, the coevolution of social contagion and bonding by homophily during social interaction is analyzed. First, in "Bonding by Homophily and Adaptive Mirroring in Joint Decision Making: a Second-Order Adaptive Network Model", in the context of emotionally grounded joint decision making, a combination of two types of adaptivity is addressed: bonding by homophily for between-agent connections and Hebbian learning for within-agent connections that play a role in the mirroring process underlying social contagion. In the subsequent four chapters, the concept of subjective homophily detection is addressed, where the criterion for similarity is not based on objective actions of the two agents in a dyad, but on subjective perceptions and interpretations each agent makes of these actions. Here "A Second-Order Adaptive Network Model for Bonding Based on Subjectively Perceived Homophily" is an introduction to this subjective perspective. A second-order adaptive social network model is introduced in which during social interaction their bonding depends on their subjective representation states about themselves and about each other, and in relation to this the two persons have a form of control over the coevolution process. In "The Interplay of Bonding based on Subjective Homophily and Learning During a Varied Rhythmic Action: An Adaptive Network Model", this subjective perspective is applied for a dyad of agents in a rhythmic joint action. Moreover, in "Connecting Social Media Platforms and Users through Bonding by Faked Homophily: A Second-Order Adaptive Network Model" a similar idea is applied in the context of the adaptive connections between a social media platform and its users, where by manipulation of the perceived states a form of faked homophily is created. Finally, in "Human Learning, Attunement and Social Bonding based on Adaptive Homophily: A Multi-Adaptive Network Model" the (subjective) criterion of similarity for bonding is itself made (second-order) adaptive: the tipping point for homophily depends on the connections someone already has. This provides different, adaptive dynamics depending on the context of the already existing bonds.

Bonding by Homophily and Adaptive Mirroring in Joint Decision Making: A Second-Order Adaptive Network Model

Caroline F. Tichelaar and Jan Treur

Abstract In this chapter, the adaptive interplay between bonding by homophily and adaptive mirroring by Hebbian learning within the context of joint decision-making is explored by second-order adaptive network modeling. Second-order adaptive mechanisms were applied to capture the adaptive dynamic interplay between these processes. Several simulation experiments were conducted across different scenarios to investigate the influence of various characteristics on formation and maintenance of social bonds, learning, and decision-making. The obtained model may be applied for computational analysis of collaborative human-bot interactions in order to jointly make decisions while bonding and adapting.

1 Introduction

Social interactions are a fundamental aspect of society, often leading to collective decision-making where individuals must decide on actions together. The complexity of this process depends on various factors, including interpersonal dynamics—such as shared interests, mutual understanding, and agreement on viewpoints—and intrapersonal traits like one's ability to learn from others and sensitivity to internal and external influences. Human-like and social characteristics of joint decision making, also including bonding and adaptivity, can be useful to be analysed for collaborative human-(ro)bot interaction.

To computationally model, analyse and simulate such collective processes, the network-oriented modeling approach for adaptive networks introduced in [43] and extended to multiple levels of adaptivity in [42, 43] is employed. This method constructs a network of states and connections that emulates dynamics and adaptivity of human interactions and behavior in a biologically inspired manner. The resulting adaptive computational network model encompasses collaborative decision-making,

C. F. Tichelaar · J. Treur (✉)
Department of Computer Science, Social AI Group, Vrije Universiteit Amsterdam, Amsterdam, The Netherlands
e-mail: j.treur@vu.nl

S. C. F. Hendrikse et al. (eds.), *New Analysis and Modeling Directions in Social Interaction Science*, Studies in Systems, Decision and Control 614,
https://doi.org/10.1007/978-3-031-99968-0_6

learning of connectivity [2], p 117, Hebb [13] of mirroring connections [22], and bonding based on shared interests [23]. By integrating these elements, the model sheds light on the interplay between decision-making, learning from others, and the formation of bonds based on similarities.

Our model integrates joint decision-making, bonding, and mirroring through different types of biologically inspired adaptive mechanisms. The obtained model operates on first-order adaptation, allowing plasticity of connections between agents and within agents to evolve. The two first-order adaptation mechanisms applied are bonding by homophily, sometimes phrased as 'Birds of a feather flock together' [23] and Hebbian learning, sometimes phrased as 'Neurons that fire together, wire together' [37]. In addition, our approach also addresses second-order adaptivity to obtain metaplasticity [1], where the speed of the first-order adaptation itself becomes dynamic in a context-sensitive manner. This reflects the influence of prior experiences and contextual factors on relationship dynamics. The metaplacticity principle used here is 'Adaptation accelerates with increasing stimulus exposure' as formulated in [34]. By incorporating the three first- and second-order adaptivity principles, our model aims to provide deeper insights into the emergent behaviors in social interaction, with a particular focus on bonding, joint decision-making, and Hebbian learning.

2 Background Literature

The subject of this chapter delves into a multifaceted and adaptive interplay between processes rooted in different fields, such as adaptive bonding, social contagion, mirroring, Hebbian learning, and decision-making. These elements have been previously discussed in various parts of the literature, which we will explore here.

In the realm of Social Science, the principle of bonding based on homophily posits that social connections tend to form more frequently among individuals who share similarities than among those with differences [23]. In simpler terms, when two people are alike, they tend to strengthen their bond. This principle generally holds across various types of relationships and often intertwines with the concept of social contagion, which, in some respects, works in the opposite direction: strong connections tend to make people more similar. Together, these two principles create a circular relationship between 'being connected' and 'being similar'.

From a perspective rooted in Social Neuroscience, the social contagion principle draws on the existence of mirror neurons. These specialized neurons, identified by Rizzolatti et al. [30, 32, 33] in macaques, respond to actions both performed by oneself and observed in others. The human brain houses a similar mirror neuron system in the inferior frontal gyrus (IFG) and the ventral premotor cortex. This system not only connects environmental perception to internal sensorimotor representations but also plays a role in understanding the emotional aspect of actions. Reciprocal connections with regions like the medial prefrontal cortex and the limbic system allow mirror neurons to represent underlying intentions and emotions. The IFG's activation differs when observing social interactions compared to solitary activities, as demonstrated by Iacoboni et al. [21].

This mirror neuron system likely operates based on Hebbian learning, a learning principle proposed by Donald Hebb [13], following the principle of learning of connectivity based on activations described in [2], pp. 117. Essentially, Hebbian learning refers to the strengthening of synaptic connections between cells when one cell consistently contributes to the firing of another [13, 37]. In the context of mirroring, observing someone's actions activates cells in sensory cortices, followed by activation of mirror neurons in the relevant region, creating a causal link in cell firing [22]. Additionally, the 'ideomotor model', discussed by [20], suggests that sensory representations of actions are linked to motor plans through Hebbian learning. According to this perspective, external stimuli typically initiate actions, emphasizing the importance of stimuli-motor response relationships in modeling.

Concerning decision-making, Damasio's Somatic Marker Hypothesis (SMH) provides insights into the role of the prefrontal cortex in decision-making processes. Damasio [9, 10] proposed that the ventromedial prefrontal cortex integrates emotionally-based body states with external stimulus information, referred to as somatic markers. These markers can lead to bodily changes and experiences when activated by external stimuli, forming a 'body-loop'. Alternatively, in the 'as-if body-loop', the bodily changes are skipped even though somatic marking occurs. Parts of the basal ganglia are also implicated in the SMH network, manipulating responses originating from ventromedial cortices. According to Damasio, somatic markers play an important role in decision-making. Another theory, value-based decision-making, emphasizes the role of the amygdala. The amygdala, with its extensive connections to the prefrontal cortex and sensory regions, assigns values to stimuli, shaping decisions and behavior [24].

In order to reach collective decisions and coordinate actions, there must exist a mechanism whereby one person's actions exert an influence on the actions of others. Sebanz et al. [36] conducted experiments to investigate whether individuals' actions are impacted when they are aware of the actions of someone else. Their findings confirmed that being aware of another person's actions indeed does have an effect on one's own actions. According to [11], this suggests that having knowledge of potential actions by others makes stimuli related to one's own actions more significant and amplifies the necessity to monitor the actions of others. Rangel et al. [30] put forward a model for decision-making processes, which can be deconstructed into five fundamental steps. These steps encompass creating a representation of the decision problem, assessing the worth of different actions, selecting an action based on this assessment, gauging the desirability of outcomes following the implementation of the decision, and using the assessment of outcomes to enhance future decision-making.

Furthermore, this chapter considers that the Hebbian learning and adaptive mirroring processes could themselves be subject to adaptation in a context-sensitive manner, a concept known as metaplasticity [1]. By [34] the second-order adaptation principle 'Adaptation accelerates with increasing stimulus exposure' is described, which shows how adaptation can be context-sensitive depending on perceived circumstances. Moreover, as discussed by [29], metaplasticity can refer to the modulation of synaptic plasticity affected by the history of synaptic activity, allowing the adaptation process to evolve dynamically over time in a more broad context-sensitive manner. This multiorder adaptation mechanism plays a critical role in the complexity of social interactions, enabling the model to reflect how past experiences influence the ongoing capacity for learning and connection, ultimately providing a richer and more nuanced simulation of social behavior.

3 The Adaptive Network Modeling Approach Used

In this section the modeling approach based on self-modeling networks to address higher-order adaptivity used is briefly explained. As introduced in [41] and extended and further explained in [42–45], the structure of a temporal-causal network model is described by nodes X and Y, also called states, with real number activation values $X(t)$ and $Y(t)$ over time t and different types of network characteristics (Table 1, first row):

- **Connectivity characteristics** Connections $X \rightarrow Y$ from a state X to a state Y and weights $\boldsymbol{\omega}_{X,Y}$ of them

Table 1 Network structure, dynamics and adaptivity: concepts and their formalisation

	Concepts	Formalisation
Network structure	States and connections Network characteristics · connectivity · Aggregation · Timing	X and Y and connections $X \rightarrow Y$ between them $\boldsymbol{\omega}_{X,Y}$, $\mathbf{c}_Y(..)$, $\boldsymbol{\eta}_Y$ Real numbers for connection weights: $\boldsymbol{\omega}_{X,Y}$ Mathematical combination functions: $\mathbf{c}_Y$ Real numbers for speed factors: $\boldsymbol{\eta}_Y$
Network dynamics	Canonical difference or differential equation for state values over time	$Y(t+\Delta t) = Y(t) + \boldsymbol{\eta}_Y[\mathbf{c}_Y(\boldsymbol{\omega}_{X_1,Y}X_1(t), \ldots, \boldsymbol{\omega}_{X_k,Y}X_k(t)) - Y(t)]\Delta t$ $\frac{dY(t)}{dt} = \boldsymbol{\eta}_Y[\mathbf{c}_Y(\boldsymbol{\omega}_{X_1,Y}X_1(t), \ldots, \boldsymbol{\omega}_{X_k,Y}X_k(t)) - Y(t)]$
Network adaptivity	Canonical difference or differential equation using adaptive network connection weights	$Y(t+\Delta t) = Y(t) + \boldsymbol{\eta}_Y[\mathbf{c}_Y(\mathbf{W}_{X_1,Y}(t)X_1(t), \ldots, \mathbf{W}_{X_k,Y}(t)X_k(t)) - Y(t)]\Delta t$ $\frac{dY(t)}{dt} = \boldsymbol{\eta}_Y[\mathbf{c}_Y(\mathbf{W}_{X_1,Y}(t)X_1(t), \ldots, \mathbf{W}_{X_k,Y}(t)X_k(t)) - Y(t)]$

- **Aggregation characteristics** For each state Y, the impacts $\boldsymbol{\omega}_{X_i,Y}X_i\,(t)$ on Y from its incoming connections from states X_i are aggregated by a combination function $\mathbf{c}_Y$, thus obtaining $\mathbf{aggimpact}_Y(t) = \mathbf{c}_Y(\boldsymbol{\omega}_{X_1,Y}X_1(t), \ldots, \boldsymbol{\omega}_{X_k,Y}X_k(t))$.
- **Timing characteristics** Each state Y has a speed factor $\boldsymbol{\eta}_Y$ defining how fast it changes for given aggregated impact.

Based on the network structure defined by the above characteristics, the following standard difference equation defines the network's dynamics incorporating these network characteristics $\boldsymbol{\omega}_{X,Y}$, $\mathbf{c}_Y(..)$ and $\boldsymbol{\eta}_Y$:

$$Y(t + \Delta t) = Y(t) + \boldsymbol{\eta}_Y[\mathbf{aggimpact}_Y(t) - Y(t)]\Delta t \quad (1)$$

for any state Y and where X_1 to X_k are the states from which Y gets its incoming connections (Table 1, second row).

This makes that the dynamics of a network model with n states defines a dynamical system model that can be described by n coupled difference or differential equations; for the network model introduced in Sect. 4, $n = 22$. The software environment described in [45], Chapter 9 currently includes around 70 useful basic combination functions in a library.

To model adaptive processes, adaptive network models can be used: for them both a network's states but also (some of) its network characteristics can change over time. A *self-modelling network* (also called a *reified* network), enables to apply a network-oriented conceptualization to adaptive networks as well; see [42, 45]. This works by adding new states to the network (called *self-model* or *reification* states) which represent one of the (adaptive) network characteristics. In the graphical 3D-format (Fig. 1) such additional states are depicted at a next level (called *self-model* level), where the original network is at the *base level*.

As an example, the weight $\boldsymbol{\omega}_{X,Y}$ of a connection from state X to state Y can be represented (at a next self-model level) by a self-model state named $\mathbf{W}_{X,Y}$. In such a case, in equations, for the connectivity characteristics $\boldsymbol{\omega}_{X_i,Y}$ for a base level state Y the values $\mathbf{W}_{X_i,Y}(t)$ are used as shown in the following equation:

$$Y(t + \Delta t) = Y(t) + \boldsymbol{\eta}_Y[\mathbf{c}_Y(\mathbf{W}_{X_1,Y}(t)X_1(t), \ldots, \mathbf{W}_{X_k,Y}(t)X_k(t)) - Y(t)]\Delta t \quad (2)$$

Moreover, in [44], Chap. 10 it was shown that the outcome of a process of network self-modeling results also in a temporal-causal network model itself. Therefore, the self-modelling network construction can easily be applied iteratively to obtain multiple orders of self-models at multiple (first-order, second-order, …) self-model levels. For example, a second-order self-model state $\mathbf{H}_{\mathbf{W}_{X,Y}}$ can be added representing the adaptive speed factor $\boldsymbol{\eta}_{\mathbf{W}_{X,Y}}$ for the first-order self-model state $\mathbf{W}_{X,Y}$ at the level below, which in turn represents the adaptive connection weight $\boldsymbol{\omega}_{X,Y}$. Such higher-order self-model levels can be used to model (context-sensitive) control over adaptation. In Eq. (1) for self-model state $Y = \mathbf{W}_{X,Z}$ this can be incorporated by substituting $\mathbf{H}_{\mathbf{W}_{X,Y}}(t)$ for $\boldsymbol{\eta}_{\mathbf{W}_{X,Y}}$.

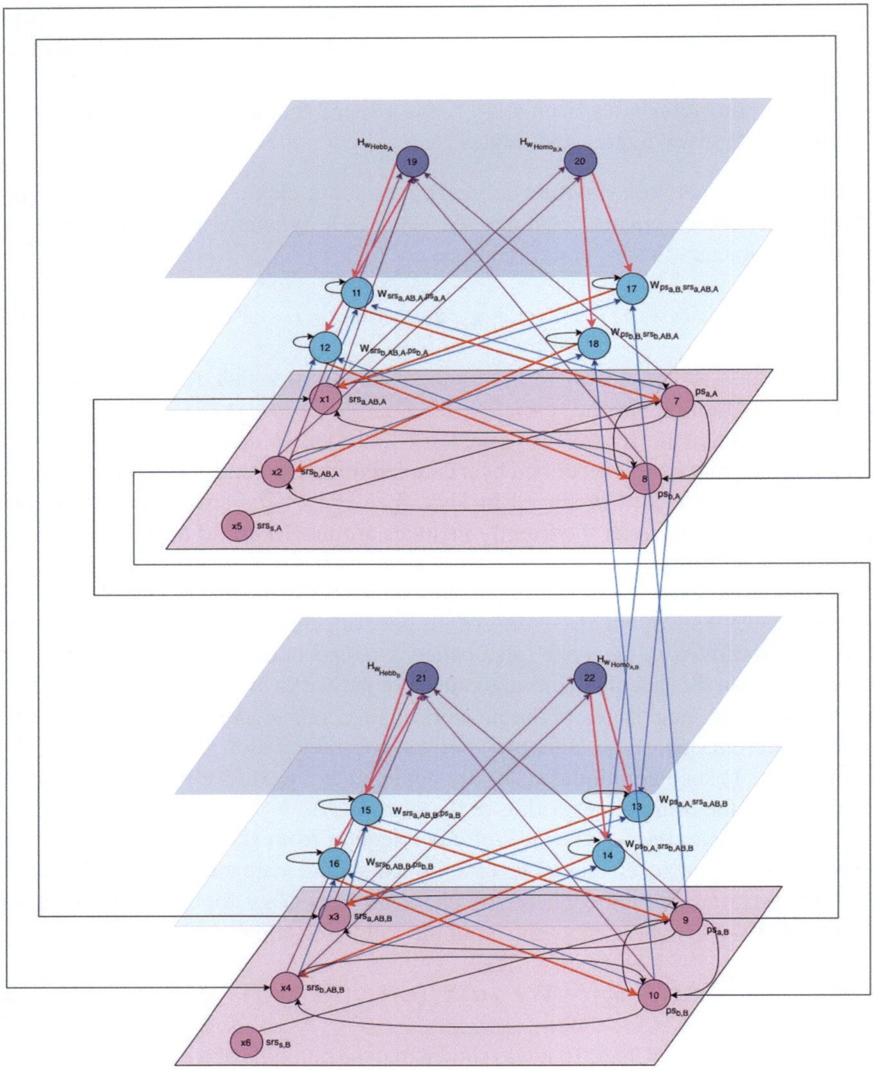

Fig. 1 Conceptual representation of the second-order adaptive network model. The circles are representing states and the arrows represent the connections between the states

In this way adaptive dynamical systems can be modeled in a network-oriented manner. As a modeling format, this does not introduce any fundamental limitation, as in [15] it has been shown in detail by mathematical analysis how any (smooth) adaptive dynamical system can be represented as a temporal-causal self-modeling network model. For more details about this approach, see also this volume [16], Chaps. 2 and 12 [17, 49].

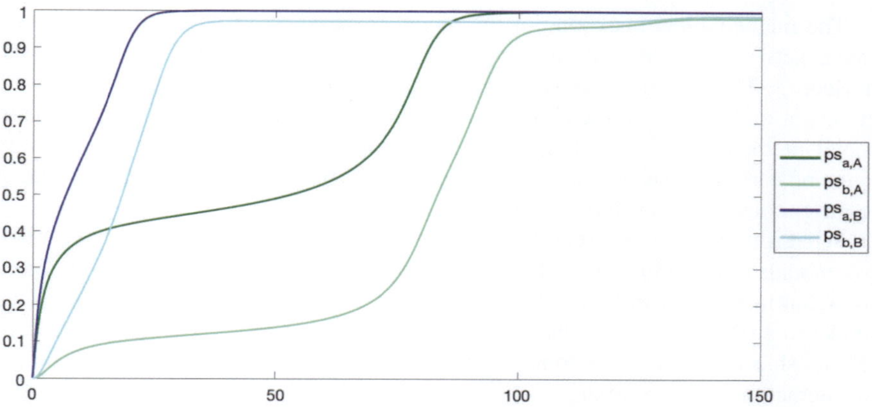

Fig. 2 Preparation state values over time for scenario 1

4 The Proposed Second-Order Adaptive Network Model

The model presented in this work applies a second-order adaptive framework to simulate the adative dynamics of social interactions. So, not only first-order adaptation but also second-order adaptation is addressed, where not only the interaction between agents is modeled, and the adaptation of the interaction, but also the adaptation speed of these adaptations becomes adaptive itself. The core concept revolves around how social connections, influenced by homophily, evolve over time and how the rate of this evolution adapts based on the agents' experiences. Accordingly, in this model each agent is represented by a three-level structure:

- **Base Level**: Represents the fundamental states and processes of the agent, such as their preferences or initial tendencies and the mechanisms for responding, mirroring and interacting.
- **First-Order Self-model Level**: The states of this level are **W**-states $\mathbf{W}_{X,Y}$ that capture the weights of connections between agents or between mental states within an agent, reflecting how these relationships strengthen or weaken over time based on the interactions.
- **Second-Order Self-model Level**: The states of this level are $\mathbf{H_W}$-states $\mathbf{H}_{\mathbf{W}_{X,Y}}$ that control the adaptation speed of the corresponding first-order **W**-states, allowing the model to dynamically adjust how quickly connections or relationships evolve.

The adaptive nature of the model is primarily governed by the two first-order adaptation principles for Hebbian learning and the bonding by homophily, as discussed in Sect. 2. Hebbian learning allows the model to simulate how repeated positive interactions between mental states within an agent lead to stronger connections, while bonding by homophily ensures that agents with similar characteristics are more likely to form and maintain connections.

The mathematical backbone of the model is based on self-modeling temporal-causal networks where each state and connection can change over time, as discussed in Sect. 3. These changes are driven by combination functions that aggregate the effects of multiple incoming connections, thus allowing for a realistic simulation of social interaction dynamics. For instance, in a scenario where two agents consistently agree on various issues, the weights of connections between them increases, leading to faster consensus in future interactions.

What sets this second-order adaptive network model apart from first-order adaptive models is the inclusion of adaptive learning rates at the second-order level. These rates can increase or decrease based on the overall state of the agent, allowing the model to simulate complex phenomena such as the acceleration or deceleration of relationship dynamics in a context-sensitive manner based on the current context of interactions. For example, if an agent experiences that a certain (social) context occurs, the model can increase the speed at which it adapts, reflecting a kind of social learning. This enhanced model is particularly useful for simulating social systems where the dynamics are not static but instead evolve as agents learn and adapt to their environment, but do that in a context-sensitive manner in dependence on that environment: the Plasticity Versus Stability Conundrum [38].

Base Network Structure

The network consists of nodes representing states (e.g., sensory representation states, action preparation states, emotion preparation states) and directed edges indicating causal connections. A conceptual representation of the model can be found in Fig. 1. An overview of the states can be found in Table 2. The model includes at the base level for both agents (the pink planes):

- **Mirroring Links**: These represent Hebbian learning for connections from sensory representations to preparation states, enabling one agent to influence the other's actions and emotions adaptively.
- **Homophily Links**: These connections reflect the similarity-based bonding between agents, where agents with similar states strengthen their mutual connections.
- **Stimulus Representation**: Sensory representations of external stimuli influence agents' action preparation states, modulated by individual preferences and characteristics.

First- and second-order self-model levels network structure

In the blue-grey and purple planes in Fig. 1 for both agents the first- and second-order self-model levels are depicted for the two adaptation levels, respectively.

At the first-order self-model level (middle planes), the **W**-states are included, where each $\mathbf{W}_{X,Y}$ for base states X and Y has two upward incoming connections from both X and Y (for monitoring them) and one downward connection to indicate its causal effect exerted to base state Y. In addition each of the **W**-states has a connection to itself for persistency.

Table 2 Description of the states in the model

State Nr	Name	Explanation
X1	$srs_{a,AB,A}$	Sensory representation of action *a* performed by B and action *a* performed by A, both as represented by A
X2	$srs_{b,AB,A}$	Sensory representation of emotion *b* of B and of emotion *b* of A, both as presented by A
X3	$srs_{a,AB,B}$	Sensory representation of action *a* performed by A and of action *a* performed by B, both as represented by B
X4	$srs_{b,AB,B}$	Sensory representation of emotion *b* of A and emotion *b* of B, both as represented by B
X5	$srs_{s,A}$	Sensory representation of stimulus *s* as represented by A
X6	$srs_{s,B}$	Sensory representation of stimulus *s* as represented by B
X7	$ps_{a,A}$	Preparation of action *a* by A
X8	$ps_{b,A}$	Preparation of emotion *b* by A
X9	$ps_{a,B}$	Preparation of action *a* by B
X10	$ps_{b,B}$	Preparation of emotion *b* by B
X11	$\mathbf{W}_{srs_{a,AB,A},ps_{a,A}}$	Self-model state representing the weight of the connection from X1 to X7 (Hebbian learning)
X12	$\mathbf{W}_{srs_{b,AB,A},ps_{b,A}}$	Self-model state representing the weight of the connection from X2 to X8 (Hebbian learning)
X13	$\mathbf{W}_{ps_{a,A},srs_{a,AB,B}}$	Self-model state representing the weight of the connection from X7 to X3 (bonding by homophily)
X14	$\mathbf{W}_{ps_{b,A},srs_{b,AB,B}}$	Self-model state representing the weight of the connection from X8 to X4 (bonding by homophily)
X15	$\mathbf{W}_{srs_{a,AB,B},ps_{a,B}}$	Self-model state representing the weight of the connection from X3 to X9 (Hebbian learning)
X16	$\mathbf{W}_{srs_{b,AB,B},ps_{b,B}}$	Self-model state representing the weight of the connection from X4 to X10 (Hebbian learning)
X17	$\mathbf{W}_{ps_{a,B},srs_{a,AB,A}}$	Self-model state representing the weight of the connection from X1 to X9 (bonding by homophily)
X18	$\mathbf{W}_{ps_{b,B},srs_{b,AB,A}}$	Self-model state representing the weight of the connection from X2 to X10 (bonding by homophily)
X19	$\mathbf{H}\mathbf{w}_{hebbA}$	Self-model state representing the speed factor of the Hebbian Learning connections in A
X20	$\mathbf{H}\mathbf{w}_{homoA,B}$	Self-model state representing the speed factor of the Homophily connections from A to B
X21	$\mathbf{H}\mathbf{w}_{hebbB}$	Self-model state representing the speed factor of the Hebbian Learning connections in B
X22	$\mathbf{H}\mathbf{w}_{homoB,A}$	Self-model state representing the speed factor of the Homophily connections from A to B

At the second-order self-model level, the $\mathbf{H}_{\mathbf{W}}$-states are included, distinguished in an abstracted manner for hebbian learning and bonding by homophily. Each of them has four upward incoming connections from the relating base states (for monitoring them), this contextualizes activation of the $\mathbf{H}_{\mathbf{W}}$-states according to the 'Adaptation accelerates with increasing stimulus exposure' principle for metaplasticity. Each $\mathbf{H}_{\mathbf{W}}$-state has two downward connections to indicate its causal effect exerted to the two related **W**-states.

Mathematical Formalisation

The second-order adaptive network model addressing social interactions is built upon a network of states connected through adaptive and non-adaptive links. The dynamics of these states are governed by combination functions and differential equations, which model the evolution of base states, causal relationships between them and their adaptation speeds in a context-sensitive manner.

As also briefly indicated in Sect. 3 and Table 1, for state Y by $Y(t)$ the activation value of Y at time t is denoted Each state is influenced by incoming connections from other states $X_1, X_2,\ldots, X_k$, with connection weights $\boldsymbol{\omega}_{X_i,Y}\ (t)$ that can be static or adaptive. The impact on state Y from states $X_1, X_2,\ldots, X_k$ is aggregated using a combination function $\mathbf{c}_Y\ (\cdot)$. The general form of this aggregation is (see also Table 1):

$$\mathbf{aggimpact}_Y(t) = \mathbf{c}_Y\big(\boldsymbol{\omega}_{X_1,Y}(t)X_1(t), \ldots, \boldsymbol{\omega}_{X_k,Y}(t)X_k(t)\big)$$

A state Y evolves over time according to the following differential equation:

$$\mathrm{d}Y(t)/\mathrm{d}t = \boldsymbol{\eta}_Y(t)\,[\mathbf{aggimpact}_Y(t) - Y(t)]$$

where $\boldsymbol{\eta}_Y$ is the speed factor that determines how quickly the state Y responds to changes in the aggregated impact, which can either be adaptive or nonadaptive as well. Note that this standard differential equation is used for all 22 states at all levels, so also for the first- and second-order self-model states that model first- and second-order adaptation. Thus, a dynamical system for a system of 22 coupled differential equation is simulated, where each of these equations is an instantiation (by substituting the proper network characteristics) of the general differential equation above. Moreover, note that:

- For the base level states, 8 of the 20 connections between them are adaptive; the speed factors of all 10 base states are nonadaptive
- For the first-order self-model level, for all 8 states, their speed factors are adaptive; all connections to them are nonadaptive
- For the second-order self-model level, all connections and speed factors are nonadaptive

Furthermore, for the model introduced here, at the different levels the following combination functions are used:

- For all base level states the advanced logistic sum combination function $\mathbf{alogistic}_{\sigma,\tau}$ is used, see Table 3, first row.
- At the first-order self-model level, all **W**-states for the Hebbian learning principle use the combination function $\mathbf{hebb}_{\mu}$, see Table 3, second row. Moreover, at the first-order self-model level, all **W**-states for the bonding by homophily principle use the combination function $\mathbf{slhomo}_{\alpha,\tau}$, see Table 3, third row.

Table 3 The combination function used in the proposed network model

	Notation	Formula	Parameters
Advanced logistic sum	$\mathbf{alogistic}_{\sigma,\tau}(V_1, \ldots, V_k)$	$[\frac{1}{1+e^{-\sigma(V_1+\cdots+V_k-\tau)}} - \frac{1}{1+e^{\sigma\tau}})](1+e^{-\sigma\tau})$	Steepness σ Excitability threshold τ
Hebbian learning	$\mathbf{hebb}_{\mu}(V_1, V_2, W)$	$V_1V_2\,(1-W) + \mu W$	Persistence μ
Bonding by homophily	$\mathbf{slhomo}_{\alpha,\tau}(V_1, V_2, W)$	$\alpha(\tau-\lvert V_1-V_2\rvert)W(1-W) + W$	Modulation factor α Tipping point τ

- At the second-order self-model level, all $\mathbf{H_W}$-states for the metaplasticity principle 'Adaptation accelerates with increasing stimulus exposure' [34] use combination function $\mathbf{alogistic}_{\sigma,\tau}$, which is monotonically increasing and self-normalising for the [0, 1] interval.

4.1 Simulation Results

To evaluate the second-order adaptive model, four distinct scenarios were simulated, each with specific variations. The scenarios were designed to analyze the interplay between bonding and Hebbian learning on joint decision-making. In the Appendix in Sect. 6 all values for the network characteristics used for Scenario 1 are listed. The observations from the simulations are interpreted using the following criteria.

For the purpose of this model, a positive decision by agent *A* is identified when the preparation state $ps_{a,A}$ reaches a high value, while a low value indicates a negative decision. The same logic applies to agent *B*. A joint decision—whether positive or negative—occurs when the preparation states $ps_{a,A}$ and $ps_{a,B}$ of both agents converge to similar values. Additionally, an emotionbased joint decision is recognized when the emotion preparation states $ps_{b,A}$ and $ps_{b,B}$ are also closely aligned, indicating that the agents not only reach a consensus on the action but also share similar emotional motivations.

In terms of bonding, this process is considered successful when the homophily connections shows high weights. Hebbian learning is deemed effective when the Hebbian learning connections demonstrate significant increases in weight, reflecting robust learning from repeated interactions.

Scenario 1: Normal Situation

In the normal scenario, the simulation shows that both agents reach a joint decision, supported by high levels of bonding and Hebbian learning, which can be seen in Fig. 3. The alignment in their preparation states and emotions indicates that the agents not only agree on actions but also share a common emotional grounding, showed by Fig. 2. This scenario confirms the model's expected behavior, where balanced parameters result in effective decision-making and strong social bonds. The course of the adaptation speed of the adaptive links as modeled as the $\mathbf{H_W}$-states can be found in Fig. 4.

Scenario 2: Low Hebbian Learning Persistence

In the scenario with reduced Hebbian learning ($\mu = 0.2$), the simulations reveal that the agents do not reach a joint decision, as the values for the preparations states differ amongst the two agents, as seen in Fig. 5. In Fig. 6, we can see that only agent B develops higher values for the learning, albeit less high than in our baseline scenario 1. In both agents we do not see high development in the bonding weights. In Fig. 7, we can see the development of the $\mathbf{H_W}$-states over time.

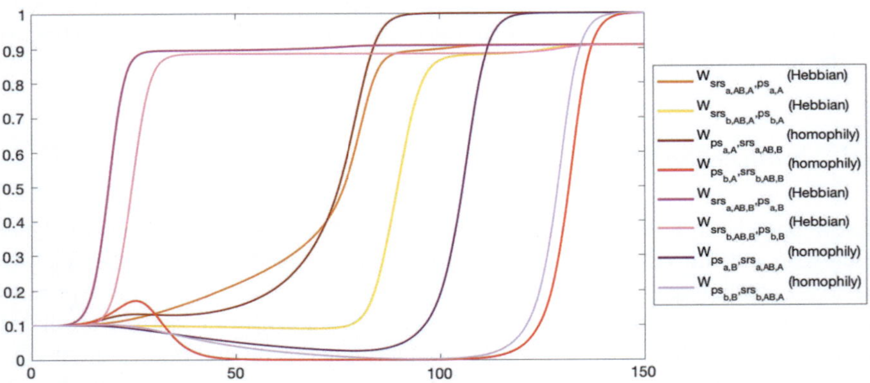

Fig. 3 **W**-state values over time for Scenario 1

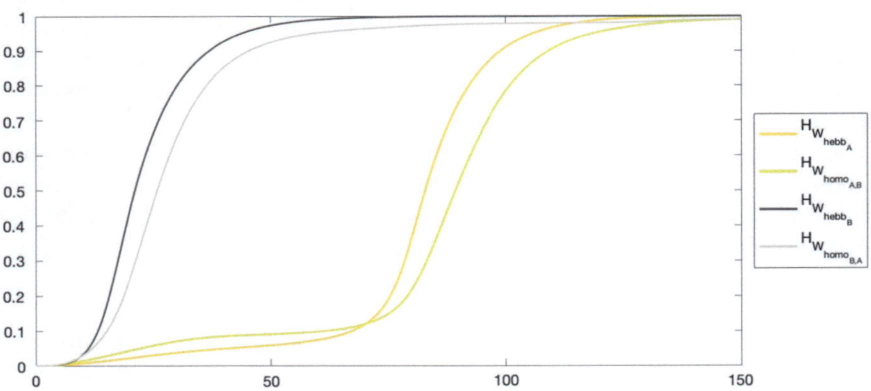

Fig. 4 $\mathbf{H_W}$-state values over time for Scenario 1

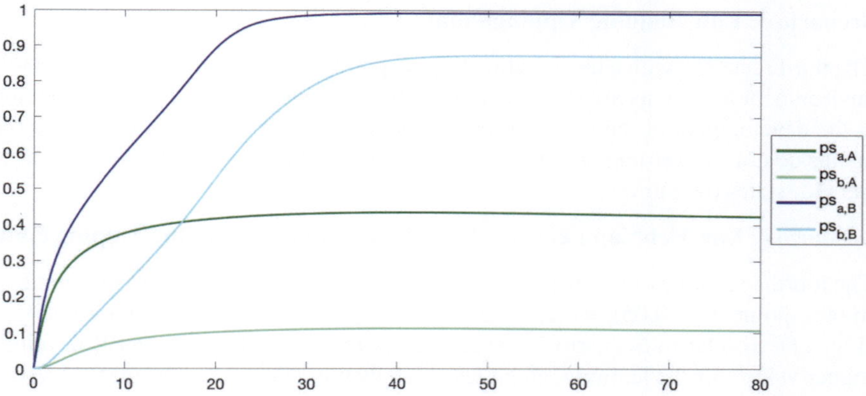

Fig. 5 Preparation state values over time for Scenario 2

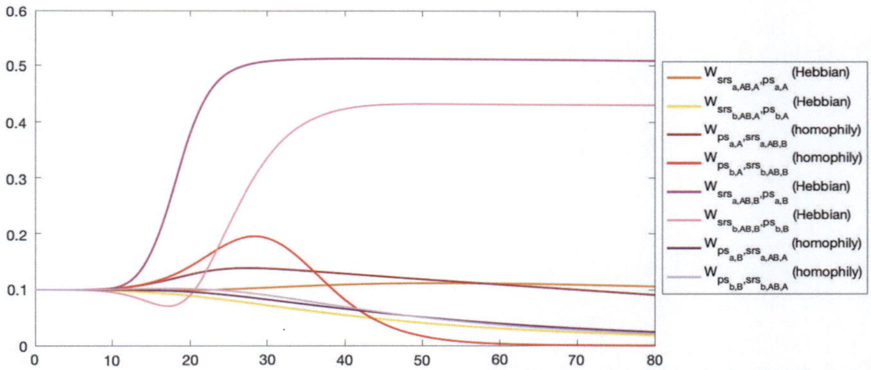

Fig. 6 **W**-state values over time for Scenario 2

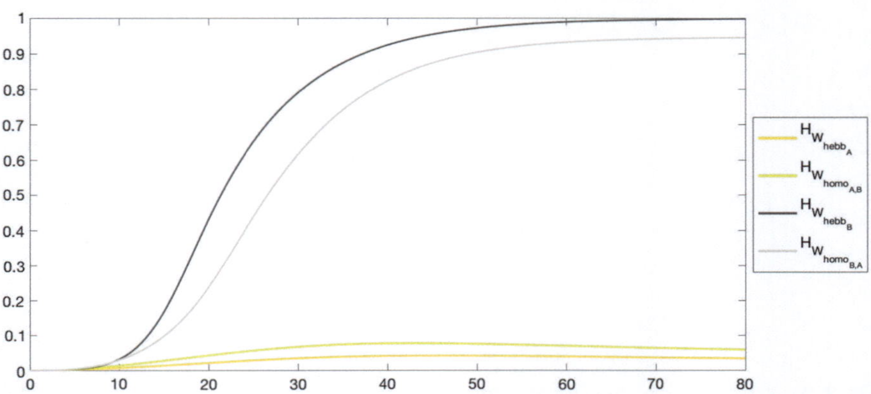

Fig. 7 $\mathbf{H_W}$-state values over time for Scenario 2

Scenario 3: Low Bonding Tipping Point

The third scenario, with a low bonding tipping point ($\boldsymbol{\tau} = 0.05$), illustrates that in this environment the agents are still able to form a joint decision (Fig. 8). When looking at the development of the values of our adaptive links in Fig. 9, we can see that the agents do not form strong bonding values. In Fig. 10, we can see the development of the $\mathbf{H_W}$-states over time.

Scenario 4: Low Hebbian Learning Persistence and Low Bonding Tipping Point

The fourth scenario, with with reduced Hebbian learning ($\boldsymbol{\mu} = 0.2$) and a low bonding tipping point ($\boldsymbol{\tau} = 0.05$), we can see that our agents do not reach a joint decision (Fig. 11). Similar to Scenario 2, we can see in Fig. 12 that only agent B develops higher values for the learning, albeit less high than in our baseline scenario 1. Also,

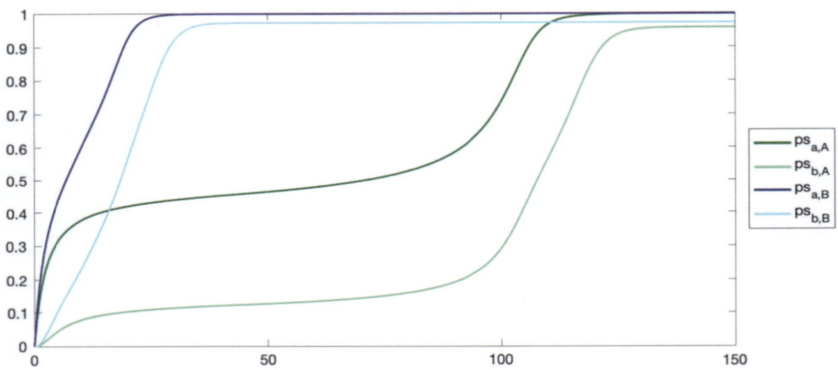

Fig. 8 Preparation state values over time for Scenario 3

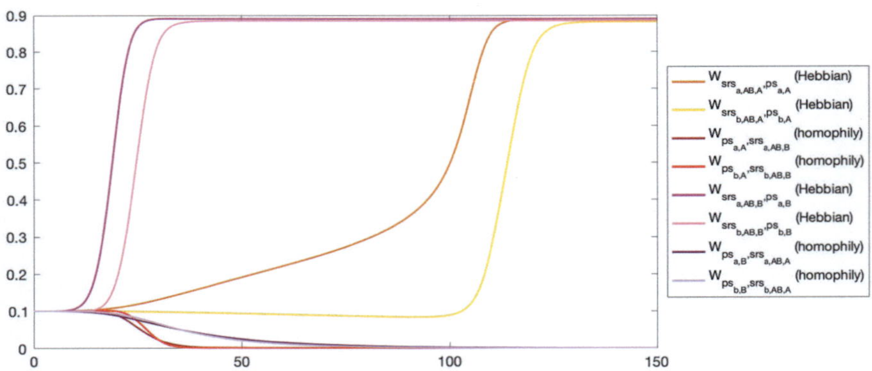

Fig. 9 **W**-state values over time for Scenario 3

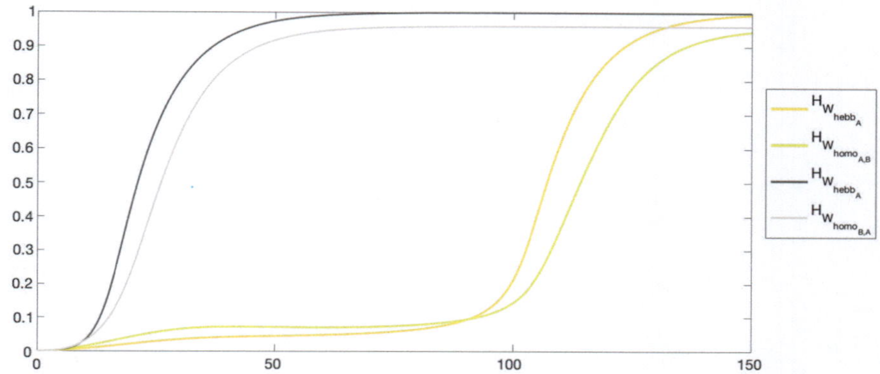

Fig. 10 $\mathbf{H_W}$-state values over time for Scenario 3

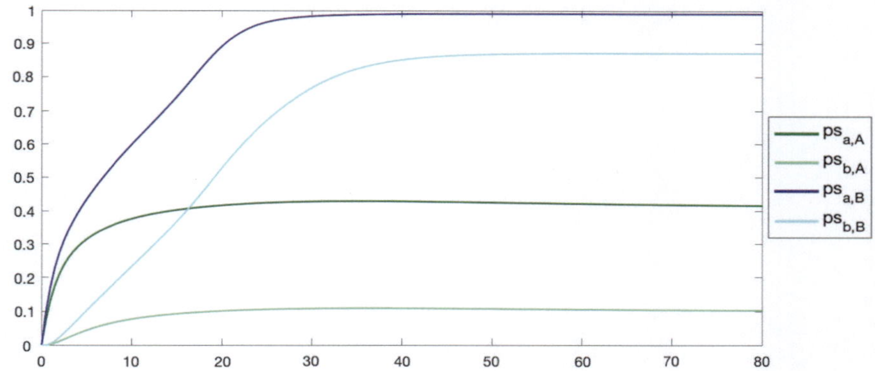

Fig. 11 Preparation state values over time for Scenario 4

in both agents we do not see high development in the bonding weights. In Fig. 13, we can see the development of the $\mathbf{H_W}$-states over time.

Overall Observations

The comparative analysis across the four scenarios underscores the robustness and adaptability of the model under different parameter settings. This is evident from the varying outcomes observed in each scenario regarding joint decision making, bonding, and Hebbian learning adaptations.

In the normal scenario, the simulations align with the theoretical expectations, showcasing effective decision-making and strong social bonds, as illustrated by high levels of bonding and synchronized Hebbian learning.

In the scenario with low Hebbian learning, the results are in line with expectations based on the previous model as the agents fail to achieve a joint decision, highlighted by the disparate preparation states and suboptimal development of bonding

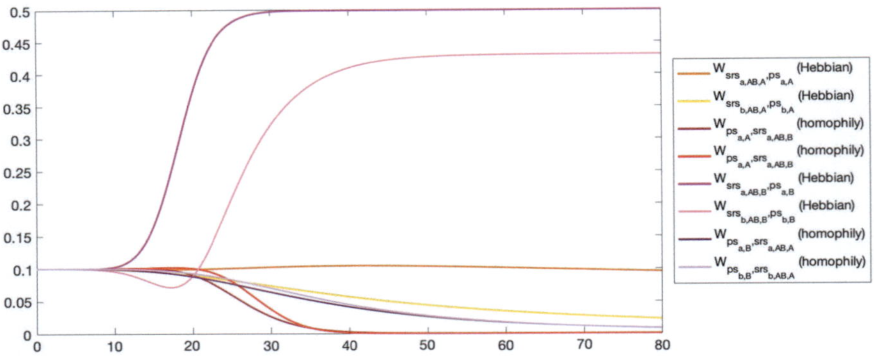

Fig. 12 **W**-state values over time for Scenario 4

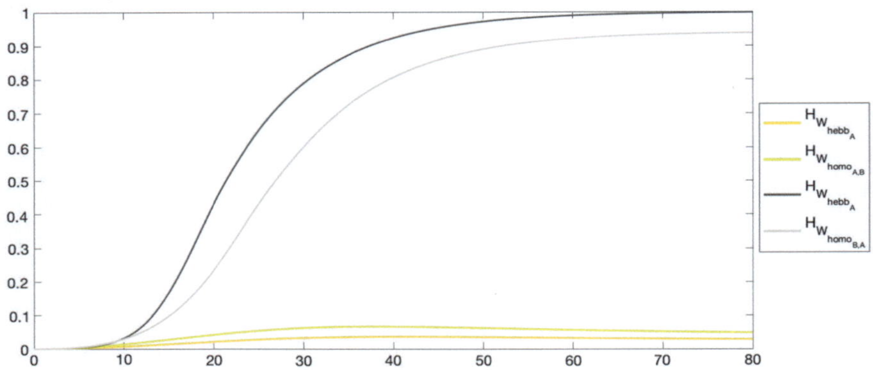

Fig. 13 $\mathbf{H_W}$-state values over time for Scenario 4

and learning values. This outcome suggests that significantly reduced learning rates hinder the agents' ability to synchronize their decisions effectively, impacting the model's efficiency under these modified conditions.

The scenario with a low bonding tipping point presents a mixed outcome where, despite a more selective bonding process, agents manage to reach a joint decision. This indicates that even under stringent bonding conditions, the model can still facilitate effective decision-making, although the bonding values themselves do not develop strongly.

Finally, the combined scenario of low Hebbian learning and a low bonding tipping point presents the most challenging environment, resulting in the agents' inability to form a joint decision. This underscores the compounded negative impact of both reduced learning rates and stringent bonding criteria on the model's functionality.

5 Discussion and Conclusion

The material for this chapter came from [40]. In this study, we have extended the exploration of intricate dynamics between bonding, Hebbian learning of mirroring, and joint decision-making in social interactions by designing a second-order adaptive network model. By introducing a second-order adaptive mechanism, our model allows the speed of first-order adaptations to dynamically adjust based on prior experiences and contextual factors. Through a series of simulation experiments across four distinct scenarios, we investigated how varying parameters influence the formation and maintenance of social bonds, learning processes, and decision-making outcomes.

The first scenario, conducted under normal conditions, confirmed the model's ability to foster coordinated action and emotional states among agents. The simulations showed that, with balanced parameters, both Hebbian learning and bonding developed strongly, leading to effective joint decision-making. These results align with theoretical expectations, demonstrating the model's capacity to simulate the mutual reinforcement between bonding and decision-making processes effectively.

In the second scenario, where the Hebbian learning persistence was reduced, the agents failed to achieve a joint decision. This result contrasts sharply with the first scenario, highlighting the critical role that Hebbian learning plays in synchronizing decisions among agents. Despite the model's second-order adaptivity providing some resilience, the significant reduction in learning rates hindered the development of both bonding and learning values, ultimately impairing the agents' ability to reach a consensus.

The third scenario examined the effects of a lower bonding tipping point, where agents were required to exhibit a greater degree of similarity to form bonds. Despite this increased selectivity, agents were still capable of making joint decisions, although the bonds that formed were weaker. This scenario suggests that while lower bonding tipping points may reduce the frequency of bond formation, the connections that do develop are stable enough to support cooperative outcomes. The model's ability to simulate environments with high selectivity underscores its versatility in representing diverse social contexts.

Finally, the fourth scenario, which combined low Hebbian learning with a low bonding tipping point, presented the most challenging conditions for the model. Under these settings, the agents were unable to reach a joint decision, reflecting the compounded negative effects of both reduced learning and stringent bonding criteria. This scenario highlights the importance of maintaining a balance between learning and bonding for the model's effectiveness, as extreme alterations can significantly impair its functionality.

Overall, this research demonstrates the utility of a second-order adaptive model in capturing the complex interplay between bonding, Hebbian learning, and joint decision-making. The findings across the four scenarios underscore the model's robustness under balanced conditions while also revealing its limitations when faced with extreme parameter settings. While second-order adaptivity adds a layer of resilience, the balance between learning and bonding parameters remains crucial for optimal performance.

The obtained network model differs from earlier work in [39] in a number of repects. First of all, the network design was different, now based on a self-modeling network structure, the dedicated specification format by role matrices introduced in [48], and the 3D visualisation of the overall model shown in Fig. 1. Next, also the dedicated software environment for self-modeling networks introduced in [44] was used to run the simulations, see also [45]. Therefore no programming was needed. While the original model only operates on first-order adaptation—allowing connections between agents to evolve—our approach here introduces a second-order adaptation mechanism as well. This advancement incorporates meta-level adaptation, where the speed of the first-order adaptation itself becomes dynamic, reflecting the influence of prior experiences and contextual factors on relationship dynamics. Thus, our model aims to provide deeper insights into the emergent behaviors in social networks, with a particular focus on bonding, joint decision-making, and Hebbian learning.

More in particular, the following distinguishes the current model from the earlier approach:

- For the first-order adaptation, the adaptation principle was applied here in a declarative manner, specified by a mathematical function available in the library and executed by the general difference equation in the software environment. In contrast, for [39] the dynamics for the connection weights had to be programmed separately and had no network structure nor visualisation like Fig. 1 at all.
- For the second-order adaptation principle, also this was incorporated here in a declarative manner. This was also executed by the same standard difference equation without any need for programming. In contrast, in [39] no attempt was made to include second-order adaptation since that would have meant still more programming, for adaptation of the code for the first-order adaptation as the first-order adaptation was already programmed and not a network. This created a bareer to even begin with addressing that next level of adaptation. It is clearly shown here that the self-modeling network modeling format does not come with such bareers to address higher-order forms of adaptivity and therefore the step to higher-order adaptivity is made much easier.

The network model presented here has the use of Hebbian learning for adaptive mirroring responding links in common with the previous Chap. 5 [14] in this volume [16]. However, in contrast to Chap. 5, in the current chapter this is combined with another type of adaptivity, namely homophily-induced connectedness adaptivity. Note also that from these two types of adaptivity, the Hebbian learning takes place within each agent, for connections from a representation state to a preparation state of the same agent. However, this does not hold for the homophily-induced adaptivity as that works on connections from a preparation state of one agent to a repesentation state of the other agent. The first type of adaptivity can be considered as a within-agent or subjective form of adaptivity with that agent as 'subject'. But, the second type of adaptivity is not subjective: it is a form of between-agent or inter-agent adaptivity from the perspective of an 'objective' social world, as if 'the world' creates connectness between persons. This is how practically all literature on homophily-induced connectedness adaptivity addresses it, usually from a social science or sociology perspective. A interesting question here is, in how far it matters in which way and in how far a person perceives homophily. For example, you may have the impression that a friend likes the same type of music as you do and due that you get more connected. But maybe your friend only gave you that impression to please you and does not really like the same type of music. From a social psychology perspective, a subjective approach can also be considered so that these subjective phenomena can be addressed as well. Such a subjective perspective on homophily-induced connectedness adaptivity was introduced in [46] and is developed further in the remainder of Part III of this volume [16], Chaps. 7–10 [8, 26, 27, 50].

Future work could focus on applying this model to more diverse social contexts and incorporating additional adaptive mechanisms to further enhance its realism and applicability. Moreover, exploring the impact of environmental factors on the dynamics of bonding and decision-making could offer deeper insights into the resilience and adaptability of social networks. Furthermore, further work already has been done where the homophily tipping point is made (second-order) adaptive too: [25], see also this volume [16], Chap. 10 [27].

For the modeling approach used and the way it addresses network structure, dynamics, and adaptivity, in some further analysis has been made, thereby addressing on the one hand a conceptual perspective such as the relation between dynamics, structure and adaptivity in dynamic and adaptive processes. On the other hand, the conceptual analysis is in [47, 48] related to the self-modeling network modeling approach to model adaptive dynamical systems, illustrated in particular for a self-modeling network architecture for multilevel control.

Also to analyse the role of epigenetics for the effect of environmental circumstances on the development of mental disorders [28] such a multilevel (in this case fifth-order) control architecture has turned out very useful, see, for example, [7, 18, 19, 35]. This is in line with the importance of multiple levels of control as put forward in philosophy [3, 5, 6] and biology, see for example [4].

6 Appendix Role Matrices Specifying the Network Model

In each role matrix each row indicates for a given state what the causal impacts are on that state from that role. These are the role matrices for Simulation 1.

Role matrix 1. Specifies the base connectivity.

mb	base connectivity	1	2	3	4	level
X_1	$srs_{a,AB,A}$	X_9	X_7			Base level
X_2	$srs_{b,AB,A}$	X_{10}	X_8			
X_3	$srs_{a,AB,B}$	X_7	X_9			
X_4	$srs_{b,AB,B}$	X_8	X_{10}			
X_5	$srs_{s,A}$	X_5				
X_6	$srs_{s,B}$	X_6				
X_7	$ps_{a,A}$	X_1	X_5	X_8		
X_8	$ps_{b,A}$	X_2	X_7			
X_9	$ps_{a,B}$	X_3	X_6	X_{10}		
X_{10}	$ps_{b,B}$	X_4	X_9			
X_{11}	$\mathbf{W}_{srs_{a,AB,A},ps_{a,A}}$	X_1	X_7	X_{11}		First-order
X_{12}	$\mathbf{W}_{srs_{b,AB,A},ps_{b,A}}$	X_2	X_8	X_{12}		
X_{13}	$\mathbf{W}_{ps_{a,A},srs_{a,AB,B}}$	X_7	X_3	X_{13}		
X_{14}	$\mathbf{W}_{ps_{b,A},srs_{b,AB,B}}$	X_8	X_4	X_{14}		
X_{15}	$\mathbf{W}_{srs_{a,AB,B},ps_{a,B}}$	X_3	X_9	X_{15}		
X_{16}	$\mathbf{W}_{srs_{b,AB,B},ps_{b,B}}$	X_4	X_{10}	X_{16}		
X_{17}	$\mathbf{W}_{ps_{a,B},srs_{a,AB,A}}$	X_9	X_1	X_{17}		
X_{18}	$\mathbf{W}_{ps_{b,B},srs_{b,AB,A}}$	X_{10}	X_2	X_{18}		
X_{19}	$\mathbf{H}_{\mathbf{W}_{hebbA}}$	X_1	X_2	X_7	X_8	Second-order
X_{20}	$\mathbf{H}_{\mathbf{W}_{homoA,B}}$	X_1	X_2			
X_{21}	$\mathbf{H}_{\mathbf{W}_{hebbB}}$	X_3	X_4	X_9	X_{10}	
X_{22}	$\mathbf{H}_{\mathbf{W}_{homoB,A}}$	X_3	X_4			

Role matrix 2. Specifies the weights of the connections between the states including the connections to the **W**-states. Note that in the first column it is indicated that for base states X_1 to X_4 the states X_{17}, X_{18}, X_{13}, and X_{14} are self-model states for the weights of these connections as indicated in **mb**. This makes these weights adaptive. Similarly for base states X_7 to X_{10} and self-model states X_{11}, X_{12}, X_{15}, and X_{16}.

mcw connection weights		1	2	3	4	level
X_1	$srs_{a,AB,A}$	X_{17}	1			Base level
X_2	$srs_{b,AB,A}$	X_{18}	1			
X_3	$srs_{a,AB,B}$	X_{13}	1			
X_4	$srs_{b,AB,B}$	X_{14}	1			
X_5	$srs_{s,A}$	1				
X_6	$srs_{s,B}$	1				
X_7	$ps_{a,A}$	X_{11}	1	1		
X_8	$ps_{b,A}$	X_{12}	1			
X_9	$ps_{a,B}$	X_{15}	1	1		
X_{10}	$ps_{b,B}$	X_{16}	1			
X_{11}	$\mathbf{W}_{srs_{a,AB,A},ps_{a,A}}$	1	1	1		First-order
X_{12}	$\mathbf{W}_{srs_{b,AB,A},ps_{b,A}}$	1	1	1		
X_{13}	$\mathbf{W}_{ps_{a,A},srs_{a,AB,B}}$	1	1	1		
X_{14}	$\mathbf{W}_{ps_{b,A},srs_{b,AB,B}}$	1	1	1		
X_{15}	$\mathbf{W}_{srs_{a,AB,B},ps_{a,B}}$	1	1	1		
X_{16}	$\mathbf{W}_{srs_{b,AB,B},ps_{b,B}}$	1	1	1		
X_{17}	$\mathbf{W}_{ps_{a,B},srs_{a,AB,A}}$	1	1	1		
X_{18}	$\mathbf{W}_{ps_{b,B},srs_{b,AB,A}}$	1	1	1		
X_{19}	$\mathbf{H}_{\mathbf{W}_{hebbA}}$	1	1	1	1	Second-order
X_{20}	$\mathbf{H}_{\mathbf{W}_{homoA,B}}$	1	1			
X_{21}	$\mathbf{H}_{\mathbf{W}_{hebbB}}$	1	1	1	1	
X_{22}	$\mathbf{H}_{\mathbf{W}_{homoB,A}}$	1	1			

Role matrix 3. This shows the speed factors and initial values for all states. Note that for the **W**-states no values are indicated but $\mathbf{H}_\mathbf{W}$-states X_{19} to X_{22} as the speeds of the **W**-states are adaptive, represented by the corresponding $\mathbf{H}_\mathbf{W}$-states.

msiv speed factors **ms** and initial values **iv**		**ms**		**iv**	**level**
X_1	$srs_{a,AB,A}$	0.5		0	Base level
X_2	$srs_{b,AB,A}$	0.5		0	
X_3	$srs_{a,AB,B}$	0.5		0	
X_4	$srs_{b,AB,B}$	0.5		0	
X_5	$srs_{s,A}$	0		1	
X_6	$srs_{s,B}$	0		1	
X_7	$ps_{a,A}$	0.5		0	
X_8	$ps_{b,A}$	0.5		0	
X_9	$ps_{a,B}$	0.5		0	
X_{10}	$ps_{b,B}$	0.5		0	
X_{11}	$\mathbf{W}_{srs_{a,AB,A},ps_{a,A}}$	X_{19}		0.1	First-order
X_{12}	$\mathbf{W}_{srs_{b,AB,A},ps_{b,A}}$	X_{19}		0.1	
X_{13}	$\mathbf{W}_{ps_{a,A},srs_{a,AB,B}}$	X_{22}		0.1	
X_{14}	$\mathbf{W}_{ps_{b,A},srs_{b,AB,B}}$	X_{22}		0.1	
X_{15}	$\mathbf{W}_{srs_{a,AB,B},ps_{a,B}}$	X_{21}		0.1	
X_{16}	$\mathbf{W}_{srs_{b,AB,B},ps_{b,B}}$	X_{21}		0.1	
X_{17}	$\mathbf{W}_{ps_{a,B},srs_{a,AB,A}}$	X_{20}		0.1	
X_{18}	$\mathbf{W}_{ps_{b,B},srs_{b,AB,A}}$	X_{20}		0.1	
X_{19}	$\mathbf{H}_{\mathbf{W}_{hebbA}}$	0.1		0	Second-order
X_{20}	$\mathbf{H}_{\mathbf{W}_{homoA,B}}$	0.1		0	
X_{21}	$\mathbf{H}_{\mathbf{W}_{hebbB}}$	0.1		0	
X_{22}	$\mathbf{H}_{\mathbf{W}_{homoB,A}}$	0.1		0	

Role matrix 4. Specifies the combination functions for all states.

mcfw combination function weights		**alogistic**	**hebb**	**slhomo**	**level**
X_1	$srs_{a,AB,A}$	1			Base level
X_2	$srs_{b,AB,A}$	1			
X_3	$srs_{a,AB,B}$	1			
X_4	$srs_{b,AB,B}$	1			
X_5	$srs_{s,A}$	1			
X_6	$srs_{s,B}$	1			
X_7	$ps_{a,A}$	1			
X_8	$ps_{b,A}$	1			
X_9	$ps_{a,B}$	1			
X_{10}	$ps_{b,B}$	1			
X_{11}	$\mathbf{W}_{srs_{a,AB,A},ps_{a,A}}$		1		First-order
X_{12}	$\mathbf{W}_{srs_{b,AB,A},ps_{b,A}}$		1		
X_{13}	$\mathbf{W}_{ps_{a,A},srs_{a,AB,B}}$			1	
X_{14}	$\mathbf{W}_{ps_{b,A},srs_{b,AB,B}}$			1	
X_{15}	$\mathbf{W}_{srs_{a,AB,B},ps_{a,B}}$		1		
X_{16}	$\mathbf{W}_{srs_{b,AB,B},ps_{b,B}}$		1		
X_{17}	$\mathbf{W}_{ps_{a,B},srs_{a,AB,A}}$			1	
X_{18}	$\mathbf{W}_{ps_{b,B},srs_{b,AB,A}}$			1	
X_{19}	$\mathbf{H}_{\mathbf{W}_{hebbA}}$	1			Second-order
X_{20}	$\mathbf{H}_{\mathbf{W}_{homoA,B}}$	1			
X_{21}	$\mathbf{H}_{\mathbf{W}_{hebbB}}$	1			
X_{22}	$\mathbf{H}_{\mathbf{W}_{homoB,A}}$	1			

Role matrix 5. Specifies the parameter values for the combination functions.

mcfp combination function parameters		**alogistic** steepness σ	**alogistic** threshold τ	**hebb** persistence factor μ		**slhomo** modulation factor α	**slhomo** tipping point τ	**level**
X_1	$srs_{a,AB,A}$	5	0.7					Base level
X_2	$srs_{b,AB,A}$	5	0.7					
X_3	$srs_{a,AB,B}$	5	0.7					
X_4	$srs_{b,AB,B}$	5	0.7					
X_5	$srs_{s,A}$	5	0.3					
X_6	$srs_{s,B}$	5	0.3					
X_7	$ps_{a,A}$	4	1.2					
X_8	$ps_{b,A}$	4	0.9					
X_9	$ps_{a,B}$	4	1.1					
X_{10}	$ps_{b,B}$	4	0.8					
X_{11}	$\mathbf{W}_{srs_{a,AB,A},ps_{a,A}}$			0.9				First-order
X_{12}	$\mathbf{W}_{srs_{b,AB,A},ps_{b,A}}$			0.9				
X_{13}	$\mathbf{W}_{ps_{a,A},srs_{a,AB,B}}$					1	0.4	
X_{14}	$\mathbf{W}_{ps_{b,A},srs_{b,AB,B}}$					1	0.4	
X_{15}	$\mathbf{W}_{srs_{a,AB,B},ps_{a,B}}$			0.9				
X_{16}	$\mathbf{W}_{srs_{b,AB,B},ps_{b,B}}$			0.9				
X_{17}	$\mathbf{W}_{ps_{a,B},srs_{a,AB,A}}$					1	0.4	
X_{18}	$\mathbf{W}_{ps_{b,B},srs_{b,AB,A}}$					1	0.4	
X_{19}	$\mathbf{H}_{\mathbf{W}_{hebbA}}$	4	1.6					Second-order
X_{20}	$\mathbf{H}_{\mathbf{W}_{homoA,B}}$	4	0.8					
X_{21}	$\mathbf{H}_{\mathbf{W}_{hebbB}}$	4	1.6					
X_{22}	$\mathbf{H}_{\mathbf{W}_{homoB,A}}$	4	0.8					

References

1. Abraham, W.C., Bear, M.F.L.: Metaplasticity: the plasticity of synaptic plasticity. Trends Neurosci. **19**(4), 126–130. https://doi.org/10.1016/s0166-2236(96)80018-x. PMID:8658594 (1996)
2. Bain, A.: Mind and body: the theories of their relation. HS King & Company 1873 Farnborough Eng. Gregg International (1873)
3. Bechtel, W., Bich, L.: Organisms need mechanisms; mechanisms need organisms. In: Cordovil, J.L., Santos, G., Vecchi, D. (eds.), New Mechanism: Explanation, Emergence and Reduction. History, Philosophy and Theory of the Life Sciences, vol. 35, pp. 85–108. Springer Nature (2024)
4. Bevilacqua, A., Wilkinson, S. J., Dimelow, R., Murabito, E., Rehman, S., Nardelli, M., van Eunen, K., Rossell, S., Bruggemen, F.J., Bluthgen, N., De Vos, D., Westerhoff, H.V.: Vertical systems biology: from DNA to flux and back. In: Hetherington, A.M., Grierson, C.S. (eds.) (2008). Practical Systems Biology, pp. 79–106. Taylor & Francis (2008)
5. Bich, L., Bechtel, W.: Control mechanisms: explaining the integration and versatility of biological organisms. Adapt. Behav. **30**(5), 389–407 (2022)
6. Bich, L., Bechtel, W.: Organization needs organization: understanding integrated control in living organisms. Stud. Hist. Philos. Sci. **93**, 96–106 (2022)
7. Blei, Z.A., Hendrikse, S.C.F., Treur, J.: Modeling epigenetic modification due to childhood abuse and its relation to ASPD: a fifth-order adaptive network model. In: Proceedings of 15th Biologically Inspired Cognitive Architectures for AI Conference, BICA*AI'24. Studies in Computational Intelligence, vol. 477, pp. 59–70. Springer Nature (2024)
8. Bössenecker, G., Mreijen, P., Treur, J: connecting social media platforms and users through bonding by faked homophily: a second-order adaptive network model. In: Hendrikse, S.C.F., Treur, J., Koole, S.L. (eds.), New Analysis and Modeling Directions in Social Interaction Science: Emergent Multimodal Interpersonal Synchrony, Affiliation, and Bonding, Ch 9. Springer Nature (2025)
9. Damasio, A.R.: Descartes' error and the future of human life. Sci. Am. **271**(4), 144–144 (1994)
10. Damasio, A.R.: The somatic marker hypothesis and the possible functions of the prefrontal cortex. Philos. Trans. R. Soc. London. Ser. B: Biol. Sci. **351**(1346), 1413–1420 (1996)
11. Frith, C.D., Singer, T.: The role of social cognition in decision making. Philos. Trans. R. Soc. London. Ser. B: Biol. Sci. **363**(1511), 3875–3886 (2008)
12. Hebb, D.O.: The organization of behavior: a neuropsychological theory. Psychology press (2005)
13. Hebb, D.O.: The Organization of Behavior: A Neuropsychological Theory. Psychology Press (1949)
14. Hendrikse, S.C.F., Kluiver, S., Treur, J., Wilderjans, T.F., Dikker, S., Koole, S.L.: Learning to synchronize: adaptive joint decision-making by virtual agents. In: Hendrikse, S.C.F., Treur, J., Koole, S.L. (eds.), New Analysis and Modeling Directions in Social Interaction Science: Emergent Multimodal Interpersonal Synchrony, Affiliation, and Bond-ing, Ch 5. Springer Nature (2025)
15. Hendrikse, S.C.F., Treur, J., Koole. S.L.: Modeling emerging interpersonal synchrony and its related adaptive short-term affiliation and long-term bonding: a second-order multi-adaptive neural agent model. Int. J. Neural Syst. **33**(7), 2350038, 1–41 (2023). https://doi.org/10.1142/S0129065723500387
16. Hendrikse, S.C.F., Treur, J., Koole, S.L. (eds.), New Analysis and Modeling Directions in Social Interaction Science: Emergent Multimodal Interpersonal Synchrony, Affiliation, and Bonding. Springer Nature (2025)
17. Hendrikse, S.C.F., Treur, J., Koole, S.L.: The adaptive interplay of interpersonal synchrony, short-term affiliation and long-term bonding: a multi-timescale multi-adaptive agent model. In: Hendrikse, S.C.F., Treur, J., Koole, S.L. (eds.), New Analysis and Modeling Directions in Social Interaction Science: Emergent Multimodal Interpersonal Synchrony, Affiliation, and Bonding, Ch 12. Springer Nature (2025)

18. Hoffmans, C., Hendrikse, S.C.F., Treur, J.: Modeling the adaptive interplay of epigenetics and insecurity in social media in the development of Anorexia Nervosa. In: Proceedings of 15th Biologically Inspired Cognitive Architectures for AI Conference, BICA*AI'24. Studies in Computational Intelligence, vol. 477, pp. 134–145. Springer Nature (2024)
19. Huisman, L., Ong, C., Van de Werken, M., Treur, J.: The role of epigenetics in OCD: a multi-order adaptive network model for DNA-methylation pathways and the development of OCD. In: Proceedings of the 20th International Conference on AI Applications and Innovations, AIAI'24. Advances in Information and Communication Technology, vol. 711, pp. 226–240. Springer Nature Switzerland (2024)
20. Iacoboni, M.: Imitation, empathy, and mirror neurons. Annu. Rev. Psychol. **60**, 653–670 (2009)
21. Iacoboni, M., Lieberman, M.D., Knowlton, B.J., Molnar-Szakacs, I., Moritz, M., Throop, C.J., Fiske, A.P.: Watching social interactions produces dorsomedial prefrontal and medial parietal bold FMRI signal increases compared to a resting baseline. Neuroimage **21**(3), 1167–1173 (2004)
22. Keysers, C., Perrett, D.I.: Demystifying social cognition: a Hebbian perspective. Trends Cogn. Sci. **8**(11), 501–507 (2004)
23. McPherson, M., Smith-Lovin, L., Cook, J.M.: Birds of a feather: homophily in social networks. Ann. Rev. Sociol. **27**(1), 415–444 (2001)
24. Morrison, S.E., Salzman, C.D.: Re-valuing the amygdala. Curr. Opin. Neurobiol. **20**(2), 221–230 (2010)
25. Mukeriia, Y., Treur, J., Hendrikse, S.C.F.: A multi-adaptive network model for human Hebbian learning, synchronization and social bonding based on adaptive homophily. Cogn. Syst. Res. **84**, 101187 (2024)
26. Mukeriia, Y., Treur, J., Hendrikse, S.C.F.: The interplay of bonding based on subjective homophily and learning during a varied rhythmic action: an adaptive network model. In: Hendrikse, S.C.F., Treur, J., Koole, S.L. (eds.), New Analysis and Modeling Directions in Social Interaction Science: Emergent Multimodal Interpersonal Synchrony, Affiliation, and Bonding, Ch 8. Springer Nature (2025)
27. Mukeriia, Y., Treur, J., Hendrikse, S.C.F.: human learning, attunement and social bonding based on adaptive homophily: a multi-adaptive network model. In: Hendrikse, S.C.F., Treur, J., Koole, S.L. (eds.), New Analysis and Modeling Directions in Social Interaction Science: Emergent Multimodal Interpersonal Synchrony, Affiliation, and Bonding, Ch 10. Springer Nature (2025)
28. Nigg, J.T.: Considerations toward an epigenetic and common pathways theory of mental disorder. J. Psychopathol. Clin. Sci. **132**(3), 297–313 (2023)
29. Parsons, R.G.: Behavioral and neural mechanisms by which prior experience impacts subsequent learning. Neurobiol. Learn. Mem. **154**, 22–29 (2018)
30. Rangel, A., Camerer, C., Montague, P.R.: A framework for studying the neurobiology of value-based decision making. Nat. Rev. Neurosci. **9**(7), 545–556 (2008)
31. Rizzolatti, G., Craighero, L.: The mirror-neuron system. Annu. Rev. Neurosci. **27**, 169–192 (2004)
32. Rizzolatti, G., Fadiga, L., Gallese, V., Fogassi, L.: Premotor cortex and the recognition of motor actions. Cogn. Brain Res. **3**(2), 131–141 (1996)
33. Rizzolatti, G., Fadiga, L., Matelli, M., Bettinardi, V., Paulesu, E., Perani, D., Fazio, F.: Localization of grasp representations in humans by pet: 1. observation versus execution. Exp. Brain Res. **111**, 246–252 (1996)
34. Robinson, B.L., Harper, N.S., McAlpine, D.: Meta-adaptation in the auditory midbrain under cortical influence. Nat. Commun. **7**, e13442 (2016)
35. Schlöder, D., Hendrikse, S.C.F., Treur, J.: Higher-order adaptive dynamical system modeling of the role of epigenetics in substance use disorder. In: Proceedings of 15th Biologically Inspired Cognitive Architectures for AI Conference, BICA*AI'24. Studies in Computational Intelligence, vol. 477, pp. 358–368. Springer Nature (2024)
36. Sebanz, N., Knoblich, G., Prinz, W.: Representing others' actions: just like one's own? Cognition **88**(3), B11–B21 (2003)
37. Shatz, C.J.: The developing brain. Sci. Am. **267**(3), 60–67 (1992)

38. Sjöström, P.J., Rancz, E.A., Roth, A., Hausser, M.: Dendritic excitability and synaptic plasticity. Physiol. Rev. **88**, 769–840 (2008)
39. Tichelaar, C., Treur, J.: Network-oriented modeling of the interaction of adaptive joint decision making, bonding and mirroring. In: Fagan, D., Martín-Vide, C., O'Neill, M., Vega-Rodríguez, M.A. (eds) Theory and Practice of Natural Computing. Proceedings TPNC 2018. Lecture Notes in Computer Science, vol. 11324, pp. 328–343. Springer, Cham (2018)
40. Tichelaar, C., Treur, J.: On the adaptive interplay of mirroring and bonding by homophily in joint decision making: a second-order adaptive network model. In: Proceedings of the 8th International Conference on Computer-Human Interaction Research and Applications, CHIRA'24. Communications in Computer and Information Science, vol. 2371, pp. 280–298. Springer Nature (2025)
41. Treur, J.: Network-Oriented Modeling: Addressing Complexity of Cognitive. Springer Publishers, Affective and Social Interactions (2016)
42. Treur, J.: Modeling higher-order adaptivity of a network by multilevel network reification. Netw. Sci. **8**, 110–144 (2020)
43. Treur, J.: Network reification as a unified approach to represent network adaptation principles within a network. In: Proceedings of the 7th International Conference on Theory and Practice of Natural Computing, TPNC'18. Lecture Notes in Computer Science, vol. 11324, pp. 344–358. Springer Publishers (2018)
44. Treur, J.: A modeling environment for reified temporal-causal networks: modeling plasticity and metaplasticity in cognitive agent models. In: PRIMA 2019: Principles and Practice of Multi-Agent Systems: 22nd International Conference 2019, Proceedings. Lecture Notes in Computer Science, vol. 11873, pp. 487–495. Springer International Publishing (2019)
45. Treur, J.: Network-oriented modeling for adaptive networks: designing higher-order adaptive biological, mental and social network models. Springer Nature (2020)
46. Treur, J.: Controlled social network adaptation: subjective elements in an objective social world. In: Yang, X.S., Sherratt, S., Dey, N., Joshi, A. (eds.) Proceedings of Sixth International Congress on Information and Communication Technology, ICICT'21. Lecture Notes in Networks and Systems, vol. 235, pp 263–274. Springer Nature (2022). https://doi.org/10.1007/978-981-16-2377-6_26
47. Treur, J.: On structure, dynamics, and adaptivity for biological and mental processes: a higher-order adaptive dynamical system modeling perspective. In: Proceedings of the 46th Annual Meeting of the Cognitive Science Society: Dynamics of Cognition, CogSci'24 (2024)
48. Treur, J.: Using multilevel temporal factorisation to analyse structure and dynamics for higher-order adaptive and evolutionary processes. In: Proceedings of the 16th International Conference on Computational Collective Intelligence, ICCCI'24. Lecture Notes in AI, Springer Nature (2024)
49. Treur, J.: Modeling coevolution of dynamics, adaptivity and control of adaptivity by self-modeling networks. In: Hendrikse, S.C.F., Treur, J., Koole, S.L. (eds.), New Analysis and Modeling Directions in Social Interaction Science: Emergent Multimodal Interpersonal Synchrony, Affiliation, and Bonding, Ch 2. Springer Nature (2025)
50. Treur, J.: A second-order adaptive network model for bonding based on subjectively perceived homophily. In: Hendrikse, S.C.F., Treur, J., Koole, S.L. (eds.), New Analysis and Modeling Directions in Social Interaction Science: Emergent Multimodal Interpersonal Synchrony, Affiliation, and Bonding, Ch 7. Springer Nature (2025)

A Second-Order Adaptive Network Model for Bonding Based on Subjectively Perceived Homophily

Jan Treur

Abstract In this chapter, the role of subjective elements and control in social network adaptation is analysed computationally. In particular, it is analysed: (1) how the coevolution of social contagion and bonding by homophily may be based on subjective representation states (e.g., what they know or believe) of the persons involved, and (2) how by changing or manipulating such subjective representation states, bonding can be controlled by the persons involved. To address this, a second-order adaptive social network model is presented in which persons do use subjective representation states about themselves and about each other and their connection, and have a form of control over the coevolution process, as both their bonding and social contagion depend on their subjective representation states.

Keywords Controlled social network adaptation · Bonding by homophily

1 Introduction

Social networks often do not only show dynamics *within* the nodes or states of the network but also dynamics *of* the network, where the latter is usually called network adaptation. These combined dynamics are sometimes referred to as coevolution of the network states and the network connections. An often studied case for social networks is the coevolution of similarity or being alike via social contagion (for the dynamics of the network nodes or states) and bonding by homophily (for the dynamics of the weights of the network connections). The bonding by homophily adaptation principle expresses how 'being alike' strengthens the connection between two persons, also explained as 'birds of a feather flock together'; e.g., [17, 20]. On the other hand, social contagion makes that network states affect each other through their connections, which implies that the stronger two persons are connected, the more

J. Treur (✉)
Social AI Group, Department of Computer Science, Vrije Universiteit Amsterdam, Amsterdam, The Netherlands
e-mail: j.treur@vu.nl

S. C. F. Hendrikse et al. (eds.), *New Analysis and Modeling Directions in Social Interaction Science*, Studies in Systems, Decision and Control 614,
https://doi.org/10.1007/978-3-031-99968-0_7

they will become alike [15]. This makes circular, reciprocal causal relations between the two processes. It has been found in simulations that, as in the real world, the emerging behaviour of adaptive network models based on coevolution of these two processes, often shows a form of clustering, seggregation or community formation; e.g., [2, 3, 11, 21, 25, 26, 39, 40].

Usually, in literature as mentioned, these social processes are considered without taking into account subjective elements for the persons involved. For example, do the persons themselves actually know in how far they are alike? Do they *have* to know that to let the bonding work properly? Do they know their connections? Are persons able to have some control over their bonding? Or are they just willless victims of objective social laws independent of what they know or what they want? Such subjective aspects are lacking in (computational) research on bonding by homophily as mentioned, as usually these coevolution processes are addressed exclusively from the perspective of an external observer of an objective social world. Note that in other Social Science literature such as [6, 13, 41] from a wider perspective also the role of cognitive and cultural interpretation in social dynamics is emphasized.

In the current chapter, it is assumed that such subjective elements indeed do matter and it is analysed computationally how some of them can play their role in the coevolution process. More specifically, it is analysed:

(1) how the coevolution of similarity via social contagion and connectedness via 'bonding by homophily' may be controlled by the persons involved, and
(2) how subjective states representing what they know about themselves, about others and about their connections play a role in this coevolution and its control.

To this end a second-order adaptive social network model has been developed in which persons have control over the coevolution process, and their bonding and social contagion depend on subjective representations of the involved persons about themselves and each other, and about their connections. Based on this chapter, the next three chapters will also make use of this subjective perspective on homophily-induced connectedness adaptivity: this volume [10], Part III Chaps. 8, 9, and 10 (The Interplay of Bonding based on Subjective Homophily and Learning During a Varied Rhythmic Action: An Adaptive Network Model, Connecting Social Media Platforms and Users through Bonding by Faked Homophily: A Second-Order Adaptive Network Model, Human Learning, Attunement and Social Bonding based on Adaptive Homophily: A Multi-Adaptive Network Model).

In this chapter, in Sect. 2 the higher-order adaptive network-oriented modeling approach from [28] used is briefly explained. In Sect. 3 the designed second-order adaptive social network model is presented. Section 4 addresses simulation results for a case study on adaptation in tetradic relationships. Finally, Sect. 5 is a discussion section, where, among others, it is discussed how the model can predict that faking your properties can be an effective way to achieve a desired bonding.

2 Higher-Order Adaptive Network Models

In this section, the network-oriented modeling approach used is briefly introduced. Following [24, 28], a temporal-causal network model is characterised by:

- **Connectivity characteristics** Connections from a state X to a state Y and their weights $\boldsymbol{\omega}_{X,Y}$
- **Aggregation characteristics** For any node Y, some combination function $\mathbf{c}_Y(..)$ defines aggregation that is applied to the impacts $\boldsymbol{\omega}_{X_i,Y}X_i(t)$ on Y from its incoming connections from states $X_1, \ldots, X_k$
- **Timing characteristics** Each state Y has a speed factor $\boldsymbol{\eta}_Y$ defining how fast it changes for given causal impact.

The following difference (or differential) equations that are useful for simulation purposes and also for analysis of temporal-causal networks incorporate these network characteristics $\boldsymbol{\omega}_{X,Y}$, $\mathbf{c}_Y(..)$, $\boldsymbol{\eta}_Y$:

$$Y(t+\Delta t) = Y(t) + \boldsymbol{\eta}_Y\left[\mathbf{c}_Y(\boldsymbol{\omega}_{X_1,Y}X_1(t), \ldots, \boldsymbol{\omega}_{X_k,Y}X_k(t)) - Y(t)\right]\Delta t \quad (1)$$

for any state Y and where $X_1, \ldots, X_k$ are the states from which it gets its incoming connections. Within the software environment described in [27], a large number > 70 of useful combination functions are included in a combination function library. The three combination functions from this library used for states Y in the introduced network model are:

- the *Euclidean combination function* $\mathbf{eucl}_{n,\lambda}(V_1, \ldots, V_k)$ defined by

$$\mathbf{eucl}_{n,\lambda}(V_1, \ldots, V_k) = \sqrt[n]{\frac{V_k^n + \ldots + V_k^n}{\lambda}} \quad (2)$$

  where $\boldsymbol{n}$ is the order and $\boldsymbol{\lambda}$ a scaling factor and $V_1, \ldots, V_k$ are the impacts from the states from which the considered state Y gets incoming connections.

- the *advanced logistic sum combination function* $\mathbf{alogistic}_{\sigma,\tau}(V_1, \ldots, V_k)$ defined by:

$$\mathbf{alogistic}_{\sigma,\tau}(V_1, \ldots, V_k) = \left[\frac{1}{1+\mathrm{e}^{-\sigma(V_1+\ldots+V_k-\tau)}} - \frac{1}{1+\mathrm{e}^{\sigma\tau})}\right] \left(1+e^{-\sigma\tau}\right) \quad (3)$$

where $\boldsymbol{\sigma}$ is a steepness parameter and $\boldsymbol{\tau}_{\mathbf{log}}$ a threshold parameter and $V_1, \ldots, V_k$ are the impacts from the states from which the considered state Y gets incoming connections

- the *simple linear homophily combination function* $\mathbf{slhomo}_{\alpha,\tau_{\mathbf{homo}}}(V_1, V_2, W)$ defined by

$$\mathbf{slhomo}_{\alpha,\tau_{\mathbf{homo}}}(V_1, V_2, W) = W + \alpha W(1 - W)(\tau_{\mathbf{homo}} - |V_1 - V_2|) \tag{4}$$

where α is an amplification parameter and $\tau_{\mathbf{homo}}$ the tipping point parameter and V_1, V_2 are a person's representations of the two persons' states involved and W represents the weight of their connection.

In Sect. 3, the combination function $\mathbf{eucl}_{n,\lambda}(\ldots)$ will be used to model social contagion and formation of internal state representations, $\mathbf{slhomo}_{\alpha,\tau_{\mathbf{homo}}}(V_1, V_2, W)$ to model homophily-induced bonding by internal connection weight representations, and **alogistic**(. . .) to model control of the bonding. Note that the homophily tipping point $\tau_{\mathbf{homo}}$ is the point where the difference between the states of the two individuals (represented by $|V_1—V_2|$) turns an increase of bonding (outcome $>W$) into a decrease (outcome $<W$), and conversely.

The above concepts enable to design network models and their dynamics in a declarative manner, based on mathematically defined functions and relations. Realistic network models are usually adaptive: often some of their network characteristics change over time. By using *self-modeling networks* (or *network reification*), a similar network-oriented conceptualisation can also be applied to *adaptive* networks to obtain a declarative description using mathematically defined functions and relations for them as well; see [28]. This works through the addition of new states to the network (called *reification states* or *self-model states*) which represent network characteristics by network states. If such self-model states are dynamic, they describe adaptive network characteristics. In a graphical 3D-format, such self-model states are depicted at a next level (*reification level*), where the original network is at a *base level*. As an example, the weight $\boldsymbol{\omega}_{X,Y}$ of a connection from state X to state Y can be represented (at a next reification level) by a self-model state named $\mathbf{W}_{X,Y}$ (objective representation) or $\mathbf{RW}_{X,Y}$ (subjective representation). Similarly, all other network characteristics from $\boldsymbol{\omega}_{X,Y}$, $\mathbf{c}_Y(..)$, $\boldsymbol{\eta}_Y$ can be made adaptive by including self-model states for them.

A self-modeling network model is also a temporal-causal network model itself, as has been proven in [28] Ch 10, this self-modeling construction can easily be applied iteratively to obtain multiple self-model levels. This can provide higher-order adaptive network models, and has turned out quite useful to model, for example, plasticity and metaplasticity in the form of a second-order adaptive mental network with three levels, one base level and a first- and a second-order self-model level; e.g., [1, 16] and [30]. For more details of this modeling approach, see also this volume [8], Chap. 2 (Modeling Coevolution of Dynamics, Adaptivity and Control of Adaptivity by Self-Modeling Networks) [34].

As shown in [7, 29] any smooth adaptive dynamical system can be represented in the self-modeling network format used here, see also this volume [8], Chap. (The Adaptive Interplay of Interpersonal Synchrony, Short-Term Affiliation and Long-Term Bonding: A Multi-Timescale Multi-Adaptive Agent Model) [9]. This indicates that the modeling format used does not introduce any fundamental limitation.

In the current chapter, multi-level network self-modeling will be applied for higher-order adaptive social network models in particular, inspired among others by [35, 36].

3 A Network Model for Controlled Social Network Adaptation

This section presents the introduced network model for controlled social network adaptation by using subjective representations. This network model integrates three types of interacting processes:

- The social network's within-network dynamics based on social contagion
- First-order social network adaptation based on bonding by homophily
- Second-order social network adaptation to control the network adaptation.

The above three types of processes have been modeled by a second-order adaptive network architecture based on multi-level self-modeling as described in Sect. 2, with connectivity as depicted in Fig. 1. In this 3D picture, each of the three planes models one of the three types of processes mentioned above.

The types of states and connections used at and between the three levels within this network model are shown in Tables 1 and 2. Here *A* and *B* are variables over persons and *Y* is a type of state of a person, for example, how much the person likes to watch Netflix series. At the base level, social contagion is modelled by connections $\mathbf{S}_{A,Y} \rightarrow \mathbf{S}_{B,Y}$. Each person has subjective internal representation states of other persons' states *Y* (and the state of her or himself) and of his or her connections to others. This is modeled by the first-order self-model. A person *B*'s internal representation state for person *A* having state *Y* is modeled by state representation $\mathbf{RS}_{A,B,Y}$. A person *A*'s subjective representation of his or her connection to *B* is modeled by connection weight representation $\mathbf{RW}_{A,B,Y}$. There are two pathways that contribute to formation of state representations $\mathbf{RS}_{A,B,Y}$. First, these representations can be communicated between persons. For example, if *A* communicates his or her subjective representation $\mathbf{RS}_{A,A,Y}$ of the own state $\mathbf{S}_{A,Y}$ to *B*, this is modeled by a connection $\mathbf{RS}_{A,A,Y} \rightarrow \mathbf{RS}_{A,B,Y}$. A second pathway for a person *B* to get information on person *A*'s state is through observation of $\mathbf{S}_{A,Y}$ by *B*. This is modeled by a connection $\mathbf{S}_{A,Y} \rightarrow \mathbf{RS}_{A,B,Y}$.

As indicated, person *A*'s representation of her or his connection to person *B* is modeled by $\mathbf{RW}_{A,B,Y}$. It is assumed that the adaptive change of the represented connections depends on the internal representation states $\mathbf{RS}_{A,B,Y}$. As the changes considered here are based on a homophily principle for state *Y*, this adaptation is supported by connections $\mathbf{RS}_{A,A,Y} \rightarrow \mathbf{RW}_{A,B,Y}$ and $\mathbf{RS}_{B,A,Y} \rightarrow \mathbf{RW}_{A,B,Y}$. The connection representations $\mathbf{RW}_{A,B,Y}$ in turn affect the social contagion within the social network, which is modeled by downward connections $\mathbf{RW}_{A,B,Y} \rightarrow \mathbf{S}_{B,Y}$.

To control the social network adaptation processes, two types of control actions are considered in particular:

- controlling the communication of state *Y* from person *A* to person *B*, modeled by control states $\mathbf{CC}_{A,B,Y}$

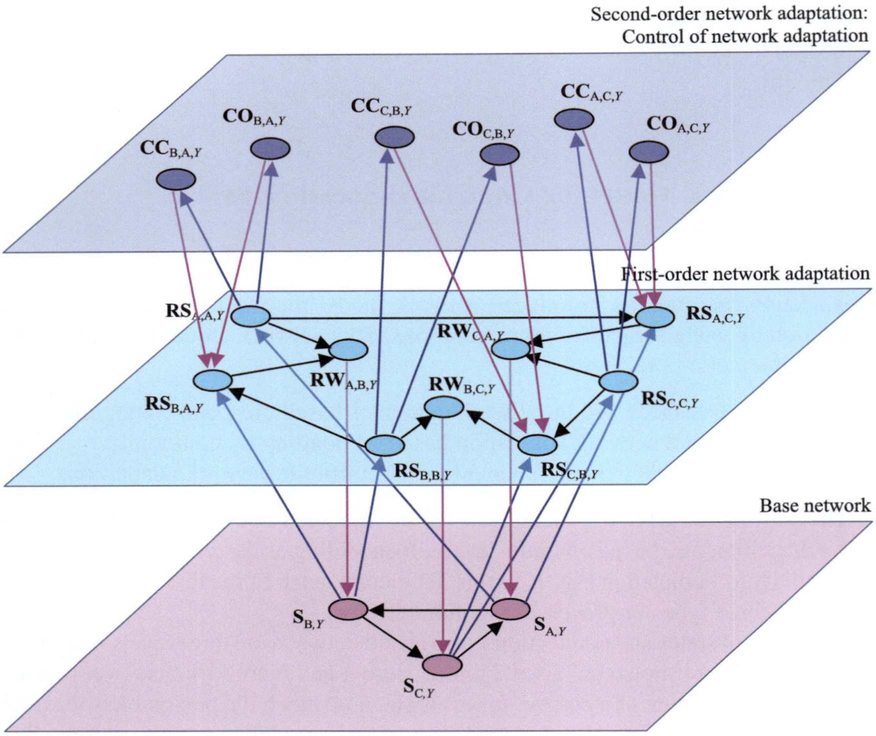

Fig. 1 Overview of the connectivity of the second-order adaptive social network model for three example persons A, B and C

- controlling the observation of state Y from person A by person B, modeled by control states $\mathbf{CO}_{A,B,Y}$

Activation of a communication control state $\mathbf{CC}_{A,B,Y}$ makes that the connection $\mathbf{RS}_{A,A,Y} \rightarrow \mathbf{RS}_{A,B,Y}$ from A's state $\mathbf{RS}_{A,A,Y}$ to B's state $\mathbf{RS}_{A,B,Y}$ gets a high value (1 or close to 1) so that the transfer of information by communication happens; this is modeled by connections $\mathbf{CO}_{A,B,Y} \rightarrow \mathbf{RS}_{A,B,Y}$. This can be considered as B asking A for the information about him or herself, upon which A communicates this information. Similarly, activation of an observation control state $\mathbf{CO}_{A,B,Y}$ makes that the connection $\mathbf{S}_{A,Y} \rightarrow \mathbf{RS}_{A,B,Y}$ from A's state $\mathbf{S}_{A,Y}$ to B's state $\mathbf{RS}_{A,B,Y}$ gets a high value (1 or close to 1) so that the transfer of information by observation takes place; this is modeled by connections $\mathbf{CO}_{A,B,Y} \rightarrow \mathbf{RS}_{A,B,Y}$. As an example used in the case study in Sect. 4, the control states $\mathbf{CC}_{A,B,Y}$ and $\mathbf{CO}_{A,B,Y}$ themselves may become active depending on B's state $\mathbf{RS}_{B,B,Y}$; this is modeled by connections $\mathbf{RS}_{B,B,Y} \rightarrow \mathbf{CO}_{A,B,Y}$ and $\mathbf{RS}_{B,B,Y} \rightarrow \mathbf{CO}_{A,B,Y}$. But this can be addressed in many other ways as well, including externally determined control, for example, by enabling or allowing observation or communication (only) at specific time slots.

4 Simulation for a Tetradic Relationship Example

In this section, a simulation of an example scenario will be discussed to illustrate the introduced second-order adaptive social network model. The example scenario describes an adaptive tetradic relationship configuration with initially two couples all four of which are friends: Mark and Dion, and Ann and Jenny. After the process described in the scenario they find themselves in a slightly changed configuration, where Mark and Jenny, and Dion and Ann have the stronger connections; see Fig. 2. This adaptation process takes place because Mark and Jenny realize that they have more in common with as an example used here their preference to watch Netflix series. Similarly, Dion and Ann realize that they also have more in common, in their case disliking watching Netflix series (and instead a preference for outdoor activities).

To specify a network model according to the approach described in [27], as discussed in Sect. 2, three types of network characteristics are to be addressed: *connectivity*, *aggregation* and *timing* characteristics. They have been specified in role matrix format as shown in the Appendix Sect. 7 and used for the simulation discussed after. For the sake of simplicity, the subscript Y (which for the example stands for a preference to watch Netflix series) has been left out here. Role matrices indicate in rows successively for all network states, the factors that affect them from different roles. In role matrix **mb** (see Sect. 7), for each state it is indicated from which other states it has incoming connections from the same or a lower level. In the same box in role matrix **mcw**, it is indicated what are the connection weights for the connected states indicated in **mb**. If the connection weights are static, their static value is indicated in matrix **mb**, but if the connection weight is adaptive, the self-model state representing this weight is indicated, as in that case at each time point this is where the (dynamic) connection weight value can be found. This can be seen for all incoming connections for the first four states X_1 to X_4, and for all incoming connections for the state representation states X_9 to X_{20}. Indicating these adaptive value representations, defines the downward connections of Fig. 1. Also the speed factors are shown in Sect. 7 (role matrix **ms**, which actually is a vector).

In the second box in Sect. 7, showing the aggregation characteristics, it can be seen which states use which combination functions (role matrix **mcfw**) and which parameter values for them (role matrix **mcfp**); also the initial values for the example simulation are shown here.

Figures 3 and 4 show the simulation for the example scenario. In Fig. 3 the co-evolution of changing states and connection (self-model) representations is shown

Fig. 2 Example scenario for a tetradic relationship configuration where initially M and D and J and A have strong connections and in the end M and J and D and A have the strong connections

without showing the underlying personal state representations. Here the states $\mathbf{S}_A$ are slowly changing whereas the connection representations $\mathbf{RW}_{A,B}$ are changing faster. It indeed can be seen that for Mark and Jenny both directional connection representations $\mathbf{RW}_{\mathrm{M,J}}$ and $\mathbf{RW}_{\mathrm{J,M}}$ start to increase from timepoint 5 resp. 10 on to finally end up at a value (close to) 1. Similarly, the connection representations $\mathbf{RW}_{\mathrm{D,A}}$ and $\mathbf{RW}_{\mathrm{A,D}}$ between Dion and Ann start to increase after time 5. In the same time period, the connection representations $\mathbf{RW}_{\mathrm{M,D}}$ and $\mathbf{RW}_{\mathrm{D,M}}$ between Mark and Dion and $\mathbf{RW}_{\mathrm{J,A}}$ and $\mathbf{RW}_{\mathrm{A,J}}$ between Jenny and Ann decrease to (close to) 0. All these changes are a consequence of the homophily principle, as the state values $\mathbf{S}_{\mathrm{M}}$ and $\mathbf{S}_{\mathrm{J}}$ for Mark and Jenny are close to each other (0.8 and 0.9), and $\mathbf{S}_{\mathrm{D}}$ and $\mathbf{S}_{\mathrm{A}}$ for Dion and Ann also (0.2 and 0.3); note that the tipping point for similarity set was 0.25, so (only) a difference < 0.25 is strengthening a relationship. In contrast, the values for Mark and Dion differ a lot (0.8 versus 0.3), which is much higher than the tipping point 0.25 and therefore has a decreasing effect on their relationship; the same pattern holds for Jenny and Ann.

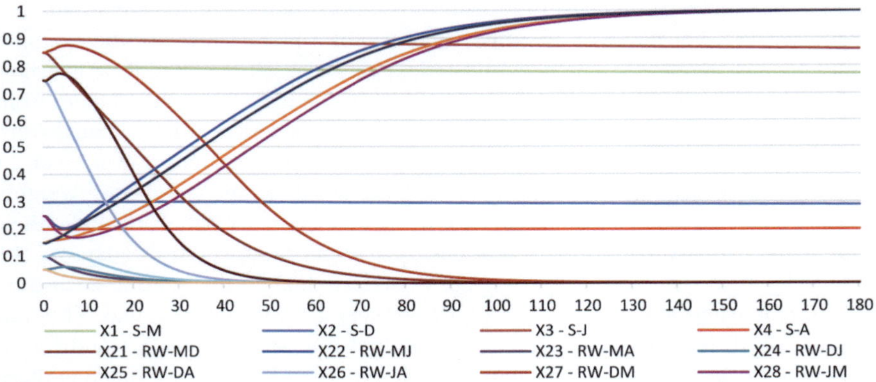

Fig. 3 Outcomes for the example scenario simulation: the changes in all relationships

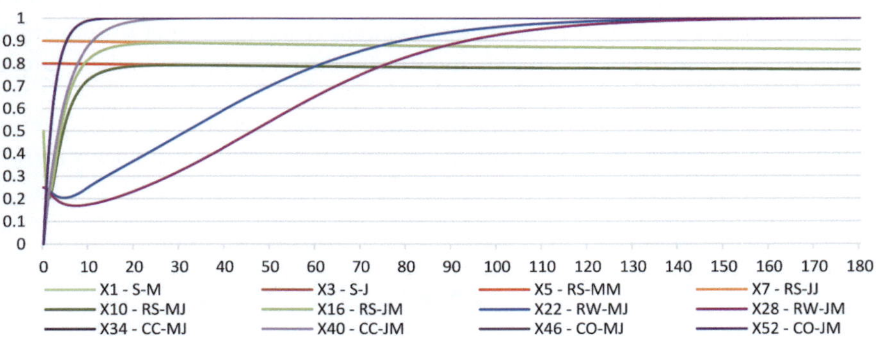

Fig. 4 The role of control and subjective states for the relationship between Mark and Jenny

Finally, in the right lower corner it can be seen that the other connection representations (for example, for Jenny and Dion) were already low and still became lower because of big differences in their states. It can be noted that all connection representations converge to 0 or 1, which shows that clustering (and seggregation) takes place, where the emerging clusters are Mark-Jenny and Dion-Ann, whereas the initial configuration (approximately) had clusters Mark-Dion and Jenny-Ann (also see Fig. 2).

Figure 4, the focus is on the development of the connections between Mark and Jenny and in particular it zooms in on the role that is played by the control states $\mathbf{CC}_{A,B,Y}$ and $\mathbf{CO}_{A,B,Y}$ and the subjective representation states $\mathbf{RS}_{A,B,Y}$. The dark purple line that gets close to 1 before time 10 indicates the control states $\mathbf{CC}_{A,B,Y}$ for the communication between them, which makes that at that time their mutual communication channels $\mathbf{RS}_{A,A,Y} \rightarrow \mathbf{RS}_{A,B,Y}$ get weights close to 1. This implies that before time 10 they indeed both communicate to each other that they like watching Netflix series. These control states for the communication are triggered in this example scenario because each of them observes his or her own behaviour and therefore they form representations $\mathbf{RS}_{A,A}$ of their own states $\mathbf{S}_A$ concerning watching the series. Next, around time point 20 the control states $\mathbf{CO}_{A,B,Y}$ for observation (the grey line) get close to 1, triggered in a similar way (but just a bit slower) as the control states for communication. This gives the relevant observation channel $\mathbf{S}_{A,Y} \rightarrow \mathbf{RS}_{A,B,Y}$ a weight close to 1. Due to that, mutual observation takes place.

Because of these communication and observation actions, the mutual subjective representations $\mathbf{RS}_{\mathrm{M,J},Y}$ of Jenny about Mark (the dark green line) and $\mathbf{RS}_{\mathrm{J,M},Y}$ of Mark about Jenny (the light green line) are formed and around time 20 reach levels around 0.8 (Jenny representing Mark) and 0.9 (Mark representing Jenny), respectively; these representations are close to the actual values, as are the representations $\mathbf{RS}_{A,A}$ of their own states, so all of them achieve faithful representations. Only now these subjective representations have been formed in a controlled manner, the homophily principle can start to work: the bonding works through the (subjective) representation states $\mathbf{RS}_{A,B,Y}$, not through the (objective) states $\mathbf{S}_{A,Y}$ themselves. More specifically, from the moment on that the subjective representations of Jenny about Mark and Jenny's own subjective representation about herself get closer than 0.25, (which is somewhere before time point 10 but not earlier), her self-model representation $\mathbf{RW}_{\mathrm{J,M},Y}$ of her connection to Mark (the pink line) starts to increase from 0.2 or lower to finally becoming very close to 1. Similarly, the effect of the subjective representations of Mark for Jenny and Mark's own subjective self-model representation about himself, on the subsequent increase of his representation $\mathbf{RW}_{\mathrm{M,J},Y}$ of his connection to Jenny (the blue line) can be noted. Before that point in time their connections were not increasing, but instead go slightly downward; this illustrates the effect of the control via the subjective self-model representation states on the adaptation.

5 Equilibrium Analysis

Moreover, based on mathematical analysis, following [31], from formula (4) for the homophily function it can be predicted that when the model reaches an equilibrium, it holds:

$W = 0$ or $W = 1$ or $|V_1 - V_2| = \boldsymbol{\tau}_{\mathbf{homo}}$

This is indeed the case, as can also be seen in the case study simulation in Fig. 3 where all **W**-states for connection weight representations end up in 0 or 1. This illustrates that the process leads to a form of community formation.

6 Discussion

The material for this chapter is based on [32]. In this chapter, a computational analysis was made of the role of subjective elements and control in social network adaptation. It was analysed: (1) how the coevolution of social contagion and bonding by homophily may be controlled by the persons involved, and (2) how subjective representation states (e.g., what they know about themselves and each other and about their connections) can play a role in this coevolution and its control. To address this, a second-order adaptive social network model was presented in which persons do have a form of control over the coevolution process, and, in relation to this, their bonding depends on their subjective representation states about themselves and about each other, and social contagion depends on their subjective representation states about their connections.

Postulating such a form of subjective homophily detection is an application of the more general philosophical principle of temporal factorization underlying dynamical systems as introduced in [22, 23]. This principle describes that when some temporal pattern *a* in the past entails some temporal pattern *b* in the future, then a state *p* exists (called mediating state) that occurs in the present such that pattern *a* entails state *p* and state *p* entails pattern *b*. Temporal factorization for the brain in particular has been formulated in [37] and called criterial causation, see also [30]. Indeed, in the model introduced here, the subjective respresentation states used for homophily detection in conjunction form a mediating state *p* between synchrony patterns and the (adapted) future behavioral patterns they entail. This shows how the designed network model successfully exploits the temporal factorization principle to realize adaptivity of social interaction. In [33] it has been shown how in such a way the temporal factorisation principle can be applied in a multilevel manner to model higher-order adaptivity for any adaptive dynamical system as a multi-order self-modeling network model. It is also discussed in [33] how evolution often seems to apply the same temporal factorisation principle in the world's emerging evolutionary dynamics, for example, using the structure of 'frozen metabolic accidents' as described in literature such as [12, 14] as mediating states.

Concerning evaluation, the model behaviour is as expected from the mentioned literature. Moreover, note that a basic design choice for the model is that the subjective representations of the connections determine the actual social contagion in the objective social world. This is based on the assumption that persons socially behave according to what they know or believe about their connections. Also here misrepresentation can be modeled easily by introducing some deviations in the mental homophily representations and connection representations underlying the subjective bonding by homophily mechanism within the model. Then the social behaviour leading to social contagion will (falsely) take place based on these misrepresentations. On the other hand, it may as well be assumed that the subjective representations of the connections do not play an exclusive role in the social behaviour, but also a more objective form of connections may have influence. To cover this, the model can easily be extended by also adding (in parallel) a more standard objective mechanism for bonding by homophily based on the objective states and then combine (according to some chosen ratio) both the objective and subjective connection representations to jointly make social contagion work.

The proposed computational network model where mental states are modeled as a basis for social mechanisms also roughly relates to (noncomputational) literature in Social Science such as [6, 13, 41] which addresses more in general the role of cognitive interpretation and cultural influence on social interactions. Such literature may provide inspiration to design computational network models for other situations where mental states and social dynamics interact. Other possible extensions may consider the integration of different adaptation principles, such as addressed (without subjectivity or control) in [26].

As an extension of the work presented in this chapter, in [4] it has been shown how bonding based on faked homophily can be modeled, see also this volume [8] Part III, Chap. (**Connecting Social Media Platforms and Users through Bonding by Faked Homophily: A Second-Order Adaptive Network Model**) [5]. Also in Chaps. (**The Interplay of Bonding based on Subjective Homophily and Learning During a Varied Rhythmic Action: An Adaptive Network Model**, **Human Learning, Attunement and Social Bonding based on Adaptive Homophily: A Multi-Adaptive Network Model**) [18, 19] of Part III of this volume this subjective form of homophily-induced connectedness adaptivity is applied. Moreover, in Part IV, Chap. (**Growing Connectedness through Synchrony: The Role of Subjective Analysis of Intrapersonal Synchrony and Interpersonal Synchrony within an Adaptive Agent Model**) [10] of this volume, an analogous concept of subjective synchrony detection is introduced as a basis for synchrony-induced connectedness adaptivity. The latter concept will also be used in Parts IV, V and VI of this volume.

Table 1 Types of states in the introduced controlled adaptive social network model

$\mathbf{S}_{B,Y}$	Objective state *Y* of person *B*
$\mathbf{RS}_{A,B,Y}$	Subjective representation of person *B* for state *Y* of person *A*
$\mathbf{RW}_{A,B,Y}$	Subjective representation of person *A* for the connection weight from person *A* to person *B*
$\mathbf{CC}_{A,B,Y}$	Control state for communication from *A* to *B*: representation of the weight of the connection from $\mathbf{RS}_{A,A,Y}$ to $\mathbf{RS}_{A,B,Y}$
$\mathbf{CO}_{A,B,Y}$	Control state for observation by *B*: representation of the weight of the connection from $\mathbf{S}_{A,Y}$ to $\mathbf{RS}_{A,B,Y}$ for the state *Y* of *A* observed by *B*

Table 2 Types of connections in the controlled adaptive social network model

<table>
<tr><th colspan="3">Intralevel connections</th></tr>
<tr><td>$\mathbf{S}_{A,Y} \rightarrow \mathbf{S}_{B,Y}$</td><td colspan="2">Social contagion from A to B for state Y</td></tr>
<tr><td>$\mathbf{RS}_{A,A,Y} \rightarrow \mathbf{RS}_{A,B,Y}$</td><td colspan="2">Communication of $\mathbf{S}_{A,Y}$ from A to B</td></tr>
<tr><td>$\mathbf{RS}_{A,A,Y} \rightarrow \mathbf{RW}_{A,B,Y}$</td><td colspan="2">Effect of state Y of A on bonding by homophily from A to B</td></tr>
<tr><td>$\mathbf{RS}_{B,A,Y} \rightarrow \mathbf{RW}_{A,B,Y}$</td><td colspan="2">Effect of state Y of B on bonding by homophily from A to B</td></tr>
<tr><th colspan="3">Interlevel connections</th></tr>
<tr><td>$\mathbf{S}_{A,Y} \rightarrow \mathbf{RS}_{A,B,Y}$</td><td>Observation by B of A's state Y</td><td>Upward from base level to first reification level</td></tr>
<tr><td>$\mathbf{RW}_{A,B,Y} \rightarrow \mathbf{S}_{B,Y}$</td><td>Effectuation of base connection weights for social contagion from A to B</td><td>Downward from first reification level to base level</td></tr>
<tr><td>$\mathbf{RS}_{B,B,Y} \rightarrow \mathbf{CO}_{A,B,Y}$</td><td>Observation control monitoring connections for B</td><td rowspan="2">Upward from first to second reification level</td></tr>
<tr><td>$\mathbf{RS}_{B,B,Y} \rightarrow \mathbf{CC}_{A,B,Y}$</td><td>Communication control monitoring connections for B</td></tr>
<tr><td>$\mathbf{CO}_{A,B,Y} \rightarrow \mathbf{RS}_{A,B,Y}$</td><td>Effectuation of control of observation of A by B</td><td rowspan="2">Downward from second to first reification level</td></tr>
<tr><td>$\mathbf{CC}_{A,B,Y} \rightarrow \mathbf{RS}_{A,B,Y}$</td><td>Effectuation of control of communication from A by B</td></tr>
</table>

Appendix

First box: Role matrices **mb** and **mcw** for the connectivity and **ms** for timing characteristics of the network model.

Second box: Role matrices **mcfw** and **mcfp** for the aggregation characteristics and the initial values of the network model.

mb	base connectivity	1	2	3
X_1	$\mathbf{S}_{M}$	X_2	X_3	X_4
X_2	$\mathbf{S}_{D}$	X_1	X_3	X_4
X_3	$\mathbf{S}_{J}$	X_1	X_2	X_4
X_4	$\mathbf{S}_{A}$	X_1	X_2	X_3
X_5	$\mathbf{RS}_{M,M}$	X_1		
X_6	$\mathbf{RS}_{D,D}$	X_2		
X_7	$\mathbf{RS}_{J,J}$	X_3		
X_8	$\mathbf{RS}_{A,A}$	X_4		
X_9	$\mathbf{RS}_{M,D}$	X_1	X_5	
X_{10}	$\mathbf{RS}_{M,J}$	X_1	X_5	
X_{11}	$\mathbf{RS}_{M,A}$	X_1	X_5	
X_{12}	$\mathbf{RS}_{D,J}$	X_2	X_6	
X_{13}	$\mathbf{RS}_{D,A}$	X_2	X_6	
X_{14}	$\mathbf{RS}_{J,A}$	X_3	X_7	
X_{15}	$\mathbf{RS}_{D,M}$	X_2	X_6	
X_{16}	$\mathbf{RS}_{J,M}$	X_3	X_7	
X_{17}	$\mathbf{RS}_{A,M}$	X_4	X_8	
X_{18}	$\mathbf{RS}_{J,D}$	X_3	X_7	
X_{19}	$\mathbf{RS}_{A,D}$	X_4	X_8	
X_{20}	$\mathbf{RS}_{A,J}$	X_4	X_8	
X_{21}	$\mathbf{RW}_{M,D}$	X_5	X_{15}	X_{21}
X_{22}	$\mathbf{RW}_{M,J}$	X_5	X_{16}	X_{22}
X_{23}	$\mathbf{RW}_{M,A}$	X_5	X_{17}	X_{23}
X_{24}	$\mathbf{RW}_{D,J}$	X_6	X_{18}	X_{24}
X_{25}	$\mathbf{RW}_{D,A}$	X_6	X_{19}	X_{25}
X_{26}	$\mathbf{RW}_{J,A}$	X_7	X_{20}	X_{26}
X_{27}	$\mathbf{RW}_{D,M}$	X_6	X_9	X_{27}
X_{28}	$\mathbf{RW}_{J,M}$	X_7	X_{10}	X_{28}
X_{29}	$\mathbf{RW}_{A,M}$	X_8	X_{11}	X_{29}
X_{30}	$\mathbf{RW}_{J,D}$	X_7	X_{12}	X_{30}
X_{31}	$\mathbf{RW}_{A,D}$	X_8	X_{13}	X_{31}
X_{32}	$\mathbf{RW}_{A,J}$	X_8	X_{14}	X_{32}
X_{33}	$\mathbf{CC}_{M,D}$	X_6		
X_{34}	$\mathbf{CC}_{M,J}$	X_7		
X_{35}	$\mathbf{CC}_{M,A}$	X_8		
X_{36}	$\mathbf{CC}_{D,J}$	X_7		
X_{37}	$\mathbf{CC}_{D,A}$	X_8		
X_{38}	$\mathbf{CC}_{J,A}$	X_8		
X_{39}	$\mathbf{CC}_{D,M}$	X_5		
X_{40}	$\mathbf{CC}_{J,M}$	X_5		
X_{41}	$\mathbf{CC}_{A,M}$	X_5		
X_{42}	$\mathbf{CC}_{J,D}$	X_6		
X_{43}	$\mathbf{CC}_{A,D}$	X_6		
X_{44}	$\mathbf{CC}_{A,J}$	X_7		
X_{45}	$\mathbf{CO}_{M,D}$	X_6		
X_{46}	$\mathbf{CO}_{M,J}$	X_7		
X_{47}	$\mathbf{CO}_{M,A}$	X_8		
X_{48}	$\mathbf{CO}_{D,J}$	X_7		
X_{49}	$\mathbf{CO}_{D,A}$	X_8		
X_{50}	$\mathbf{CO}_{J,A}$	X_8		
X_{51}	$\mathbf{CO}_{D,M}$	X_5		
X_{52}	$\mathbf{CO}_{J,M}$	X_5		
X_{53}	$\mathbf{CO}_{A,M}$	X_5		
X_{54}	$\mathbf{CO}_{J,D}$	X_6		
X_{55}	$\mathbf{CO}_{A,D}$	X_6		
X_{56}	$\mathbf{CO}_{A,J}$	X_7		

mcw	connection weights	1	2	3
X_1	$\mathbf{S}_{M}$	X_{27}	X_{28}	X_{29}
X_2	$\mathbf{S}_{D}$	X_{21}	X_{30}	X_{31}
X_3	$\mathbf{S}_{J}$	X_{22}	X_{24}	X_{32}
X_4	$\mathbf{S}_{A}$	X_{23}	X_{25}	X_{26}
X_5	$\mathbf{RS}_{M,M}$	1		
X_6	$\mathbf{RS}_{D,D}$	1		
X_7	$\mathbf{RS}_{J,J}$	1		
X_8	$\mathbf{RS}_{A,A}$	1		
X_9	$\mathbf{RS}_{M,D}$	X_{45}	X_{33}	
X_{10}	$\mathbf{RS}_{M,J}$	X_{46}	X_{34}	
X_{11}	$\mathbf{RS}_{M,A}$	X_{47}	X_{35}	
X_{12}	$\mathbf{RS}_{D,J}$	X_{48}	X_{36}	
X_{13}	$\mathbf{RS}_{D,A}$	X_{49}	X_{37}	
X_{14}	$\mathbf{RS}_{J,A}$	X_{50}	X_{38}	
X_{15}	$\mathbf{RS}_{D,M}$	X_{51}	X_{39}	
X_{16}	$\mathbf{RS}_{J,M}$	X_{52}	X_{40}	
X_{17}	$\mathbf{RS}_{A,M}$	X_{53}	X_{41}	
X_{18}	$\mathbf{RS}_{J,D}$	X_{54}	X_{42}	
X_{19}	$\mathbf{RS}_{A,D}$	X_{55}	X_{43}	
X_{20}	$\mathbf{RS}_{A,J}$	X_{56}	X_{44}	
X_{21}	$\mathbf{RW}_{M,D}$	1	1	1
X_{22}	$\mathbf{RW}_{M,J}$	1	1	1
X_{23}	$\mathbf{RW}_{M,A}$	1	1	1
X_{24}	$\mathbf{RW}_{D,J}$	1	1	1
X_{25}	$\mathbf{RW}_{D,A}$	1	1	1
X_{26}	$\mathbf{RW}_{J,A}$	1	1	1
X_{27}	$\mathbf{RW}_{D,M}$	1	1	1
X_{28}	$\mathbf{RW}_{J,M}$	1	1	1
X_{29}	$\mathbf{RW}_{A,M}$	1	1	1
X_{30}	$\mathbf{RW}_{J,D}$	1	1	1
X_{31}	$\mathbf{RW}_{A,D}$	1	1	1
X_{32}	$\mathbf{RW}_{A,J}$	1	1	1
X_{33}	$\mathbf{CC}_{M,D}$	1		
X_{34}	$\mathbf{CC}_{M,J}$	1		
X_{35}	$\mathbf{CC}_{M,A}$	1		
X_{36}	$\mathbf{CC}_{D,J}$	1		
X_{37}	$\mathbf{CC}_{D,A}$	1		
X_{38}	$\mathbf{CC}_{J,A}$	1		
X_{39}	$\mathbf{CC}_{D,M}$	1		
X_{40}	$\mathbf{CC}_{J,M}$	1		
X_{41}	$\mathbf{CC}_{A,M}$	1		
X_{42}	$\mathbf{CC}_{J,D}$	1		
X_{43}	$\mathbf{CC}_{A,D}$	1		
X_{44}	$\mathbf{CC}_{A,J}$	1		
X_{45}	$\mathbf{CO}_{M,D}$	1		
X_{46}	$\mathbf{CO}_{M,J}$	1		
X_{47}	$\mathbf{CO}_{M,A}$	1		
X_{48}	$\mathbf{CO}_{D,J}$	1		
X_{49}	$\mathbf{CO}_{D,A}$	1		
X_{50}	$\mathbf{CO}_{J,A}$	1		
X_{51}	$\mathbf{CO}_{D,M}$	1		
X_{52}	$\mathbf{CO}_{J,M}$	1		
X_{53}	$\mathbf{CO}_{A,M}$	1		
X_{54}	$\mathbf{CO}_{J,D}$	1		
X_{55}	$\mathbf{CO}_{A,D}$	1		
X_{56}	$\mathbf{CO}_{A,J}$	1		

ms	speed factors	1
X_1	$\mathbf{S}_{M}$	0.0005
X_2	$\mathbf{S}_{D}$	0.0005
X_3	$\mathbf{S}_{J}$	0.0005
X_4	$\mathbf{S}_{A}$	0.0005
X_5	$\mathbf{RS}_{M,M}$	0.9
X_6	$\mathbf{RS}_{D,D}$	0.9
X_7	$\mathbf{RS}_{J,J}$	0.9
X_8	$\mathbf{RS}_{A,A}$	0.9
X_9	$\mathbf{RS}_{M,D}$	0.9
X_{10}	$\mathbf{RS}_{M,J}$	0.9
X_{11}	$\mathbf{RS}_{M,A}$	0.9
X_{12}	$\mathbf{RS}_{D,J}$	0.9
X_{13}	$\mathbf{RS}_{D,A}$	0.9
X_{14}	$\mathbf{RS}_{J,A}$	0.9
X_{15}	$\mathbf{RS}_{D,M}$	0.9
X_{16}	$\mathbf{RS}_{J,M}$	0.9
X_{17}	$\mathbf{RS}_{A,M}$	0.9
X_{18}	$\mathbf{RS}_{J,D}$	0.9
X_{19}	$\mathbf{RS}_{A,D}$	0.9
X_{20}	$\mathbf{RS}_{A,J}$	0.9
X_{21}	$\mathbf{RW}_{M,D}$	0.1
X_{22}	$\mathbf{RW}_{M,J}$	0.1
X_{23}	$\mathbf{RW}_{M,A}$	0.1
X_{24}	$\mathbf{RW}_{D,J}$	0.1
X_{25}	$\mathbf{RW}_{D,A}$	0.1
X_{26}	$\mathbf{RW}_{J,A}$	0.1
X_{27}	$\mathbf{RW}_{D,M}$	0.1
X_{28}	$\mathbf{RW}_{J,M}$	0.1
X_{29}	$\mathbf{RW}_{A,M}$	0.1
X_{30}	$\mathbf{RW}_{J,D}$	0.1
X_{31}	$\mathbf{RW}_{A,D}$	0.1
X_{32}	$\mathbf{RW}_{A,J}$	0.1
X_{33}	$\mathbf{CC}_{M,D}$	0.2
X_{34}	$\mathbf{CC}_{M,J}$	0.2
X_{35}	$\mathbf{CC}_{M,A}$	0.2
X_{36}	$\mathbf{CC}_{D,J}$	0.2
X_{37}	$\mathbf{CC}_{D,A}$	0.2
X_{38}	$\mathbf{CC}_{J,A}$	0.2
X_{39}	$\mathbf{CC}_{D,M}$	0.2
X_{40}	$\mathbf{CC}_{J,M}$	0.2
X_{41}	$\mathbf{CC}_{A,M}$	0.2
X_{42}	$\mathbf{CC}_{J,D}$	0.2
X_{43}	$\mathbf{CC}_{A,D}$	0.2
X_{44}	$\mathbf{CC}_{A,J}$	0.2
X_{45}	$\mathbf{CO}_{M,D}$	0.4
X_{46}	$\mathbf{CO}_{M,J}$	0.4
X_{47}	$\mathbf{CO}_{M,A}$	0.4
X_{48}	$\mathbf{CO}_{D,J}$	0.4
X_{49}	$\mathbf{CO}_{D,A}$	0.4
X_{50}	$\mathbf{CO}_{J,A}$	0.4
X_{51}	$\mathbf{CO}_{D,M}$	0.4
X_{52}	$\mathbf{CO}_{J,M}$	0.4
X_{53}	$\mathbf{CO}_{A,M}$	0.4
X_{54}	$\mathbf{CO}_{J,D}$	0.4
X_{55}	$\mathbf{CO}_{A,D}$	0.4
X_{56}	$\mathbf{CO}_{A,J}$	0.4

mcfw combination function weights		1 eucl	2 slhomo	3 alogistic
X_{1}	$\mathbf{S}_{M}$	1		
X_{2}	$\mathbf{S}_{D}$	1		
X_{3}	$\mathbf{S}_{J}$	1		
X_{4}	$\mathbf{S}_{A}$	1		
X_{5}	$\mathbf{RS}_{M,M}$	1		
X_{6}	$\mathbf{RS}_{D,D}$	1		
X_{7}	$\mathbf{RS}_{J,J}$	1		
X_{8}	$\mathbf{RS}_{A,A}$	1		
X_{9}	$\mathbf{RS}_{M,D}$	1		
X_{10}	$\mathbf{RS}_{M,J}$	1		
X_{11}	$\mathbf{RS}_{M,A}$	1		
X_{12}	$\mathbf{RS}_{D,J}$	1		
X_{13}	$\mathbf{RS}_{D,A}$	1		
X_{14}	$\mathbf{RS}_{J,A}$	1		
X_{15}	$\mathbf{RS}_{D,M}$	1		
X_{16}	$\mathbf{RS}_{J,M}$	1		
X_{17}	$\mathbf{RS}_{A,M}$	1		
X_{18}	$\mathbf{RS}_{J,D}$	1		
X_{19}	$\mathbf{RS}_{A,D}$	1		
X_{20}	$\mathbf{RS}_{A,J}$	1		
X_{21}	$\mathbf{RW}_{M,D}$		1	
X_{22}	$\mathbf{RW}_{M,J}$		1	
X_{23}	$\mathbf{RW}_{M,A}$		1	
X_{24}	$\mathbf{RW}_{D,J}$		1	
X_{25}	$\mathbf{RW}_{D,A}$		1	
X_{26}	$\mathbf{RW}_{J,A}$		1	
X_{27}	$\mathbf{RW}_{D,M}$		1	
X_{28}	$\mathbf{RW}_{J,M}$		1	
X_{29}	$\mathbf{RW}_{A,M}$		1	
X_{30}	$\mathbf{RW}_{J,D}$		1	
X_{31}	$\mathbf{RW}_{A,D}$		1	
X_{32}	$\mathbf{RW}_{A,J}$		1	
X_{33}	$\mathbf{CC}_{M,D}$			1
X_{34}	$\mathbf{CC}_{M,J}$			1
X_{35}	$\mathbf{CC}_{M,A}$			1
X_{36}	$\mathbf{CC}_{D,J}$			1
X_{37}	$\mathbf{CC}_{D,A}$			1
X_{38}	$\mathbf{CC}_{J,A}$			1
X_{39}	$\mathbf{CC}_{D,M}$			1
X_{40}	$\mathbf{CC}_{J,M}$			1
X_{41}	$\mathbf{CC}_{A,M}$			1
X_{42}	$\mathbf{CC}_{J,D}$			1
X_{43}	$\mathbf{CC}_{A,D}$			1
X_{44}	$\mathbf{CC}_{A,J}$			1
X_{45}	$\mathbf{CO}_{M,D}$			1
X_{46}	$\mathbf{CO}_{M,J}$			1
X_{47}	$\mathbf{CO}_{M,A}$			1
X_{48}	$\mathbf{CO}_{D,J}$			1
X_{49}	$\mathbf{CO}_{D,A}$			1
X_{50}	$\mathbf{CO}_{J,A}$			1
X_{51}	$\mathbf{CO}_{D,M}$			1
X_{52}	$\mathbf{CO}_{J,M}$			1
X_{53}	$\mathbf{CO}_{A,M}$			1
X_{54}	$\mathbf{CO}_{J,D}$			1
X_{55}	$\mathbf{CO}_{A,D}$			1
X_{56}	$\mathbf{CO}_{A,J}$			1

mcfp combination function parameters		1 eucl		2 slhomo		3 alogistic	
		1	2	1	2	1	2
		n	λ	α	τ_{homo}	σ	τ_{log}
X_{1}	$\mathbf{S}_{M}$	1	1.5				
X_{2}	$\mathbf{S}_{D}$	1	1.5				
X_{3}	$\mathbf{S}_{J}$	1	1.5				
X_{4}	$\mathbf{S}_{A}$	1	1.5				
X_{5}	$\mathbf{RS}_{M,M}$	1	1				
X_{6}	$\mathbf{RS}_{D,D}$	1	1				
X_{7}	$\mathbf{RS}_{J,J}$	1	1				
X_{8}	$\mathbf{RS}_{A,A}$	1	1				
X_{9}	$\mathbf{RS}_{M,D}$	1	2				
X_{10}	$\mathbf{RS}_{M,J}$	1	2				
X_{11}	$\mathbf{RS}_{M,A}$	1	2				
X_{12}	$\mathbf{RS}_{D,J}$	1	2				
X_{13}	$\mathbf{RS}_{D,A}$	1	2				
X_{14}	$\mathbf{RS}_{J,A}$	1	2				
X_{15}	$\mathbf{RS}_{D,M}$	1	2				
X_{16}	$\mathbf{RS}_{J,M}$	1	2				
X_{17}	$\mathbf{RS}_{A,M}$	1	2				
X_{18}	$\mathbf{RS}_{J,D}$	1	2				
X_{19}	$\mathbf{RS}_{A,D}$	1	2				
X_{20}	$\mathbf{RS}_{A,J}$	1	2				
X_{21}	$\mathbf{RW}_{M,D}$			3	0.25		
X_{22}	$\mathbf{RW}_{M,J}$			3	0.25		
X_{23}	$\mathbf{RW}_{M,A}$			3	0.25		
X_{24}	$\mathbf{RW}_{D,J}$			3	0.25		
X_{25}	$\mathbf{RW}_{D,A}$			3	0.25		
X_{26}	$\mathbf{RW}_{J,A}$			3	0.25		
X_{27}	$\mathbf{RW}_{D,M}$			3	0.25		
X_{28}	$\mathbf{RW}_{J,M}$			3	0.25		
X_{29}	$\mathbf{RW}_{A,M}$			3	0.25		
X_{30}	$\mathbf{RW}_{J,D}$			3	0.25		
X_{31}	$\mathbf{RW}_{A,D}$			3	0.25		
X_{32}	$\mathbf{RW}_{A,J}$			3	0.25		
X_{33}	$\mathbf{CC}_{M,D}$					50	0.1
X_{34}	$\mathbf{CC}_{M,J}$					50	0.1
X_{35}	$\mathbf{CC}_{M,A}$					50	0.1
X_{36}	$\mathbf{CC}_{D,J}$					50	0.1
X_{37}	$\mathbf{CC}_{D,A}$					50	0.1
X_{38}	$\mathbf{CC}_{J,A}$					50	0.1
X_{39}	$\mathbf{CC}_{D,M}$					50	0.1
X_{40}	$\mathbf{CC}_{J,M}$					50	0.1
X_{41}	$\mathbf{CC}_{A,M}$					50	0.1
X_{42}	$\mathbf{CC}_{J,D}$					50	0.1
X_{43}	$\mathbf{CC}_{A,D}$					50	0.1
X_{44}	$\mathbf{CC}_{A,J}$					50	0.1
X_{45}	$\mathbf{CO}_{M,D}$					50	0.1
X_{46}	$\mathbf{CO}_{M,J}$					50	0.1
X_{47}	$\mathbf{CO}_{M,A}$					50	0.1
X_{48}	$\mathbf{CO}_{D,J}$					50	0.1
X_{49}	$\mathbf{CO}_{D,A}$					50	0.1
X_{50}	$\mathbf{CO}_{J,A}$					50	0.1
X_{51}	$\mathbf{CO}_{D,M}$					50	0.1
X_{52}	$\mathbf{CO}_{J,M}$					50	0.1
X_{53}	$\mathbf{CO}_{A,M}$					50	0.1
X_{54}	$\mathbf{CO}_{J,D}$					50	0.1
X_{55}	$\mathbf{CO}_{A,D}$					50	0.1
X_{56}	$\mathbf{CO}_{A,J}$					50	0.1

iv initial values		
X_{1}	$\mathbf{S}_{M}$	0.8
X_{2}	$\mathbf{S}_{D}$	0.3
X_{3}	$\mathbf{S}_{J}$	0.9
X_{4}	$\mathbf{S}_{A}$	0.2
X_{5}	$\mathbf{RS}_{M,M}$	0.8
X_{6}	$\mathbf{RS}_{D,D}$	0.3
X_{7}	$\mathbf{RS}_{J,J}$	0.9
X_{8}	$\mathbf{RS}_{A,A}$	0.2
X_{9}	$\mathbf{RS}_{M,D}$	0.5
X_{10}	$\mathbf{RS}_{M,J}$	0.5
X_{11}	$\mathbf{RS}_{M,A}$	0.5
X_{12}	$\mathbf{RS}_{D,J}$	0.5
X_{13}	$\mathbf{RS}_{D,A}$	0.5
X_{14}	$\mathbf{RS}_{J,A}$	0.5
X_{15}	$\mathbf{RS}_{D,M}$	0.5
X_{16}	$\mathbf{RS}_{J,M}$	0.5
X_{17}	$\mathbf{RS}_{A,M}$	0.5
X_{18}	$\mathbf{RS}_{J,D}$	0.5
X_{19}	$\mathbf{RS}_{A,D}$	0.5
X_{20}	$\mathbf{RS}_{A,J}$	0.5
X_{21}	$\mathbf{RW}_{M,D}$	0.85
X_{22}	$\mathbf{RW}_{M,J}$	0.25
X_{23}	$\mathbf{RW}_{M,A}$	0.1
X_{24}	$\mathbf{RW}_{D,J}$	0.05
X_{25}	$\mathbf{RW}_{D,A}$	0.15
X_{26}	$\mathbf{RW}_{J,A}$	0.75
X_{27}	$\mathbf{RW}_{D,M}$	0.85
X_{28}	$\mathbf{RW}_{J,M}$	0.25
X_{29}	$\mathbf{RW}_{A,M}$	0.1
X_{30}	$\mathbf{RW}_{J,D}$	0.05
X_{31}	$\mathbf{RW}_{A,D}$	0.15
X_{32}	$\mathbf{RW}_{A,J}$	0.75
X_{33}	$\mathbf{CC}_{M,D}$	0
X_{34}	$\mathbf{CC}_{M,J}$	0
X_{35}	$\mathbf{CC}_{M,A}$	0
X_{36}	$\mathbf{CC}_{D,J}$	0
X_{37}	$\mathbf{CC}_{D,A}$	0
X_{38}	$\mathbf{CC}_{J,A}$	0
X_{39}	$\mathbf{CC}_{D,M}$	0
X_{40}	$\mathbf{CC}_{J,M}$	0
X_{41}	$\mathbf{CC}_{A,M}$	0
X_{42}	$\mathbf{CC}_{J,D}$	0
X_{43}	$\mathbf{CC}_{A,D}$	0
X_{44}	$\mathbf{CC}_{A,J}$	0
X_{45}	$\mathbf{CO}_{M,D}$	0
X_{46}	$\mathbf{CO}_{M,J}$	0
X_{47}	$\mathbf{CO}_{M,A}$	0
X_{48}	$\mathbf{CO}_{D,J}$	0
X_{49}	$\mathbf{CO}_{D,A}$	0
X_{50}	$\mathbf{CO}_{J,A}$	0
X_{51}	$\mathbf{CO}_{D,M}$	0
X_{52}	$\mathbf{CO}_{J,M}$	0
X_{53}	$\mathbf{CO}_{A,M}$	0
X_{54}	$\mathbf{CO}_{J,D}$	0
X_{55}	$\mathbf{CO}_{A,D}$	0
X_{56}	$\mathbf{CO}_{A,J}$	0

References

1. Abraham, W.C., Bear, M.F.: Metaplasticity: the plasticity of synaptic plasticity. Trends Neurosci. **19**(4), 126–130 (1996)
2. Blankendaal, R., Parinussa, S., Treur, J.: A temporal-causal modelling approach to integrated contagion and network change in social networks. In: Proceedings of the 22nd European Conference on Artificial Intelligence, ECAI'16, pp. 1388–1396. IOS Press, Frontiers in Artificial Intelligence and Applications, vol. 285 (2016)
3. Boomgaard, G., Lavitt, F., Treur, J.: Computational analysis of social contagion and homophily based on an adaptive social network model. In: Proceedings of the 10th International Conference on Social Informatics, SocInfo'18. Lecture Notes in Computer Science, **11185**, 86–101. Springer Publishers (2018)
4. Bössenecker, G., Mreijen, P., Treur, J: A second-order adaptive network model for connecting social media platforms and users through bonding by faked homophily. Procedia Comput. Sci. **213**, 32–40 (2022)
5. Bössenecker, G., Mreijen, P., Treur, J: Connecting social media platforms and users through bonding by faked homophily: a second-order adaptive network model. In: Hen-drikse, S.C.F., Treur, J., Koole, S.L. (eds.) New Analysis and Modeling Directions in Social Interaction Science: Emergent Multimodal Interpersonal Synchrony, Affiliation, and Bonding, Ch 9. Springer Nature (2025)
6. Casciaro, T., Carley, K.M., Krackhardt, D.: Positive affectivity and accuracy in social network perception. Motiv. Emot. **23**(4), 285–305 (1999)
7. Hendrikse, S.C.F., Treur, J., Koole, S.L.: Modeling emerging interpersonal synchrony and its related adaptive short-term affliation and long-term bonding: a second-order multi-adaptive neural agent model. Inter. J. Neural Syst **33**(7), 2350038 (2023)
8. Hendrikse, S.C.F., Treur, J., Koole, S.L. (eds.), New Analysis and Modeling Directions in Social Interaction Science: Emergent Multimodal Interpersonal Synchrony, Affiliation, and Bonding. Springer Nature (2025)
9. Hendrikse, S.C.F., Treur, J., Koole, S.L.: The adaptive interplay of interpersonal synchrony, short-term affiliation and long-term bonding: a multi-timescale multi-adaptive agent model. In: Hendrikse, S.C.F., Treur, J., Koole, S.L. (eds.) New Analysis and Modeling Directions in Social Interaction Science: Emergent Multimodal Interpersonal Synchrony, Affiliation, and Bonding, Ch 12. Springer Nature (2025)
10. Hendrikse, S.C.F., Treur, J., Wilderjans, T.F., Dikker, S., Koole, S.L.: Growing connectedness through synchrony: the role of subjective analysis of intrapersonal synchrony and interpersonal synchrony within an adaptive agent model. In: Hendrikse, S.C.F., Treur, J., Koole, S.L. (eds.) New Analysis and Modeling Directions in Social Interaction Science: Emergent Multimodal Interpersonal Synchrony, Affiliation, and Bonding, Ch 11. Springer Nature (2025)
11. Holme, P., Newman, M.E.J.: Nonequilibrium phase transition in the coevolution of networks and opinions. Phy. Rev. E **74**(5), e056108 (2006)
12. Koonin, E.V.: Frozen accident pushing 50: stereochemistry, expansion, and chance in the evolution of the genetic code. Life **7**(2), E22 (2017)
13. Krackhardt, D.: Cognitive social structures. Soc. Netw. **9**, 109–134 (1987)
14. Leister, D.: Thawing out frozen metabolic accidents. BMC Biol. **17**(1), 8–18 (2019). https://doi.org/10.1186/s12915-018-0621-5
15. Levy, D.A., Nail, P.R.: Contagion: a theoretical and empirical review and reconceptualization. Genet. Soc. Gen. Psychol. Monogr. **119**(2), 233–284 (1993)
16. Magerl, W., Hansen, N., Treede, R.D., Klein, T.: The human pain system exhibits higher-order plasticity (metaplasticity). Neurobiol. Learn. Mem. **154**, 112–120 (2018)
17. McPherson, M., Smith-Lovin, L., Cook, J.M.: Birds of a feather: homophily in social networks. Ann. Rev. Sociol. **27**(1), 415–444 (2001)
18. Mukeriia, Y., Treur, J., Hendrikse, S.C.F.: The interplay of bonding based on subjective homophily and learning during a varied rhythmic action: an adaptive network model. In:

Hendrikse, S.C.F., Treur, J., Koole, S.L. (eds.) New Analysis and Modeling Directions in Social Interaction Science: Emergent Multimodal Interpersonal Synchrony, Affiliation, and Bonding, Ch 8. Springer Nature (2025)

19. Mukeriia, Y., Treur, J., Hendrikse, S.C.F.: Human learning, attunement and social bonding based on adaptive homophily: a multi-adaptive network model. In: Hendrikse, S.C.F., Treur, J., Koole, S.L. (eds.) New Analysis and Modeling Directions in Social Interaction Science: Emergent Multimodal Interpersonal Synchrony, Affiliation, and Bonding, Ch 10. Springer Nature (2025)
20. Pearson, M., Steglich, C., Snijders, T.: Homophily and assimilation among sport-active adolescent substance users. Connections **27**(1), 47–63 (2006)
21. Sharpanskykh, A., Treur, J.: Modelling and analysis of social contagion in dynamic networks. Neurocomputing **146**, 140–150 (2014)
22. Treur, J.: Temporal factorisation: a unifying principle for dynamics of the world and of mental states. Cog. Syst. Res. **8**(2), 57–74 (2007a)
23. Treur, J.: Temporal factorisation: realisation of mediating state properties for dynamics. Cog. Syst. Res. **8**(2), 75–88 (2007b)
24. Treur, J.: Network-oriented modeling: addressing complexity of cognitive, affective and social interactions. Springer Publishers (2016)
25. Treur, J.: Modelling and analysis of the dynamics of adaptive temporal-causal network models for evolving social interactions. Comput. Soc. Netw. **4**, e4 (2017)
26. Treur, J.: Mathematical analysis of the emergence of communities based on coevolution of social contagion and bonding by homophily. Appl. Netw. Sci. **4**, e1 (2019)
27. Treur, J.: A modeling environment for reified temporal-causal networks: modeling plasticity and metaplasticity in cognitive agent models. In Proc. of the International Conference on Principles and Practice of Multi-Agent Systems, PRIMA'19. Lecture Notes in Computer Science, vol 11873. pp. 487–495. Cham: Springer International Publishing (2019)
28. Treur, J.: Network-oriented modeling for adaptive networks: Designing Higher-order Adaptive Biological, Mental and Social Network Models. Springer Nature Publishing, Cham, Switzerland (2020)
29. Treur, J.: On the dynamics and adaptivity of mental processes: relating adaptive dynamical systems and self-modeling network models by mathematical analysis. Cogn. Syst. Res. **70**, 93–100 (2021)
30. Treur, J.: Modeling the emergence of informational content by adaptive networks for temporal factorisation and criterial causation. Cogn. Syst. Res. **68**, 34–52 (2021)
31. Treur, J.: Where is this leading me: stationary point and equilibrium analysis of self-modeling network models. In: Treur, J., and van Ments, L. (eds.) Mental Models and their Dynamics, Adaptation and Control: a Self-Modeling Network Modeling Approach, Ch 18. Springer Nature Switzerland (2022)
32. Treur, J.: Controlled social network adaptation: subjective elements in an objective social world. In: Yang, X.S., Sherratt, S., Dey, N., Joshi, A. (eds) Proceedings of Sixth International Congress on Information and Communication Technology, ICICT'21. Lecture Notes in Networks and Systems, vol. 235, pp. 263–274. Springer Nature (2022)
33. Treur, J.: Using multilevel temporal factorisation to analyse structure and dynamics for higher-order adaptive and evolutionary processes. In: Nguyen, N.T., et al. Computational Collective Intelligence. Proceedings of ICCCI 2024. Lecture Notes in Computer Science, vol. 14811, pp. 378–392. Springer Nature, Cham (2024)
34. Treur, J.: Modeling coevolution of dynamics, adaptivity and control of adaptivity by self-modeling networks. In: Hendrikse, S.C.F., Treur, J., Koole, S.L. (eds.) New Analysis and Modeling Directions in Social Interaction Science: Emergent Multimodal Interpersonal Synchrony, Affiliation, and Bonding, Ch 2. Springer Nature (2025)
35. Treur, J., Glas, G.: A multi-level cognitive architecture for self-referencing, self-awareness and self-interpretation. Cogn. Syst. Res. **68**, 125–142 (2021)
36. Treur, J., Glas, G.: Who am i really: an adaptive network model addressing mental models for self-referencing, self-awareness and self-interpretation. In: Treur, J., Van Ments, L. (eds.)

Mental Models and Their Dynamics, Adaptation, and Control: a Self-Modeling Network Modeling Approach, Ch 8, pp. 175–206. Springer Nature, Cham (2022)

37. Tse, P.: The neural basis of free will: Criterial causation. MIT Press (2013)
38. Van Beukel, S., Goos, S., Treur, J.: An adaptive temporal-causal network model for social networks based on the homophily and more-becomes-more principle. Neurocomputing **338**, 361–371 (2019)
39. Vazquez, F.: Opinion dynamics on coevolving networks. In: A. Mukherjee et al (eds): Dynamics On and Of Complex Networks, Volume 2, Modeling and Simulation in Science, Engineering and Technology, pp. 89–107. Springer New York (2013)
40. Vazquez, F., Gonzalez-Avella, J.C., Egu´ıluz, V.M., and San Miguel, M.: Time-scale competition leading to fragmentation and recombination transitions in the coevolution of network and states. Phys. Rev. E **76**, e046120 (2007)
41. Vaisey, S., Lizardo, O.: Can cultural worldviews influence network composition? Soc. Forces **88**(4), 1595–1618 (2010)

The Interplay of Bonding Based on Subjective Homophily and Learning During a Varied Rhythmic Action: An Adaptive Network Model

Yelyzaveta Mukeriia, Jan Treur, and Sophie Hendrikse

Abstract This chapter presents an adaptive network model in the context of joint action and social bonding. Exploration of mechanisms for mental and social network models are presented, specifically focusing on adaptation by bonding based on homophily and Hebbian learning during joint rhythmic action. The chapter provides a comprehensive explanation of these concepts and their role in controlled adaptation within illustrative scenarios.

1 Introduction

The phenomenon of human synchronization and social bonding has received much interest for researchers from various scientific disciplines for decades. From an early age, humans exhibit a remarkable ability to synchronize their actions, thoughts, and emotions with others, resulting in formation of strong social bonds that shape our interactions and relationships. This innate capacity for synchronization and social bonding plays a vital role in our social development, cooperation, and overall well-being. To model these various dynamics of human interaction in real-world scenarios, it is essential to consider both cognitive processes in the brain and interpersonal relationships within social networks. The adopted network modeling approach used in the chapter, serves as a conceptual tool for understanding the structure and dynamics of complex adaptive processes [19, 20]. Like [2] the proposed network model integrates two fundamental principles: Hebbian learning for progressive changes in connections between mental states (modeling plasticity in the brain) and a bonding by homophily

Y. Mukeriia · J. Treur (✉)
Social AI Group, Department of Computer Science, Vrije Universiteit Amsterdam, Amsterdam, The Netherlands
e-mail: j.treur@vu.nl

S. Hendrikse
Department of Methodology and Statistics, Tilburg School of Social and Behavioral Sciences, Tilburg University, Tilburg, The Netherlands

S. C. F. Hendrikse et al. (eds.), *New Analysis and Modeling Directions in Social Interaction Science*, Studies in Systems, Decision and Control 614,
https://doi.org/10.1007/978-3-031-99968-0_8

principle. However, this time the subjective homophily principle for adaptive interpersonal connectedness introduced in [22] will be used, see also this volume [9] Part 3, Chapter (A Second-Order Adaptive Network Model for Bonding Based on Subjectively Perceived Homophily) [23].

The focus of this chapter is on modeling the adaptation and bonding process of two individuals who are participating in a tapping activity, in which case, the adaptation happens through the motor control skill, named sensorimotor synchronization (SMS). Broadly speaking, SMS is the temporal coordination of an action with a predictable external event, for instance an external rhythm such as a sequence of musical beats [16]. It commonly occurs within social contexts, where individuals synchronize their movements with sequences produced by others. Instances of synchronized movement in response to live musical performances, such as nodding, clapping, or dancing, offer numerous examples of sensorimotor synchronization (SMS). When individuals engage with music, they naturally develop temporal expectations shaped by the underlying structural regularities, notably the musical beat, which represents a recurring pulse. These regularities often evoke a strong inclination to generate movements that synchronize with the rhythmic patterns [24].

During such joint actions the adaptations taking place involve both connections within internal mental networks and connections within social networks between persons. Within the introduced adaptive network model, the former type of adaptation will be addressed by Hebbian learning and the latter by bonding based on homophily. Hebbian learning, a synaptic plasticity principle proposed by [3] and also put forward in [5], posits that when a neuron A consistently and causally contributes to activating neuron B, their connection strengthens. In simpler terms, Hebbian learning explains how neurons enhance their communication when they frequently fire together, emphasizing the importance of one neuron's consistent and causal involvement in the firing of another. The homophily adaptation principle in interpersonal relationships highlights that the level of bonding between individuals is influenced by their degree of similarity. People who share more similarities tend to form stronger connections, as observed, for instance, in children's interactions, where shared demographic characteristics often lead to more frequent friendships and playgroup engagements [13].

To apply these adaptation principles effectively, it is essential to grasp the model's structure, which involves multiple levels of adaptation. The foundational mental network model operates at the lowest level, initially with given weights of connections between states. To make it more realistic, we introduce a first-order adaptation level, incorporating principles like Hebbian learning and subjective homophily-induced bonding. These principles themselves have default parameters, which can be made dynamic by including second-order adaptation. In this chapter, the second-order adaptation is applied to model adaptive adaptation speed for the adaptation speed both of the Hebbian learning and of the bonding. Such second-order adaptation mechanisms are also referred to by metaplasticity or 'plasticity of the plasticity' [1].

2 The Self-Modeling Network Modeling Approach Used

The modeling approach used is network-oriented and addresses adaptation of networks by self-modeling networks, also called reified networks [20]. Within a network model, at each time point *t*, the impact of state *X* on state *Y* is depicted through their connections, as $\mathbf{impact}_{X,Y}(t) = \boldsymbol{\omega}_{X,Y}\, X(t)$ where $\boldsymbol{\omega}_{X,Y}$ is the weight of the connection from *X* to *Y* and *X(t)* the activation value of X at time *t*. There could be one or multiple connections impacting a single state, and in order to aggregate them, combination functions $\mathbf{c}_Y(..)$ are used,

$$\begin{aligned}\mathbf{aggimpact}_Y(t) &= \mathbf{c}_Y\big(\mathbf{impact}_{X_1,Y}(t),\ \ldots,\ \mathbf{impact}_{X_k,Y}(t)\big)\\ &= \mathbf{c}_Y\big(\boldsymbol{\omega}_{X_1,Y}X_1(t),\ \ldots,\ \boldsymbol{\omega}_{X_k,Y}X_k(t)\big)\end{aligned}$$

Lastly, the speed factor $\boldsymbol{\eta}_Y$ is what determines the rate at which the $\mathbf{aggimpact}_Y(t)$ is exerted on state *Y*. Thus, the full difference and differential equations are the following:

$$Y(t + \Delta t) = Y(t) + \boldsymbol{\eta}_Y\big(\mathbf{aggimpact}_Y(t) - Y(t)\big)\Delta t$$

$$\mathbf{d}Y(t)/\mathbf{d}t = \boldsymbol{\eta}_Y\big(\mathbf{aggimpact}_Y(t) - Y(t)\big)$$

The three types of network characteristics $\boldsymbol{\omega}_{X,Y}$, $\mathbf{c}_Y(..)$, $\boldsymbol{\eta}_Y$ defining a *temporal-causal network* can be constant, in which case they model a nonadaptive network. However, the realistic case is that they also change over time and therefore the network is adaptive. Self-modelling networks are an easy way to address adaptivity by adding states (called *self-model states*) to the network that represent these characteristics. For example, a self-model state $\mathbf{W}_{X,Y}$ can be added to represent a connection weight $\boldsymbol{\omega}_{X,Y}$, or a self-model state $\mathbf{H}_Y$ can be used to represent a speed factor $\boldsymbol{\eta}_Y$. This can also be done in a higher-order manner, for example, a second-order self-model state $\mathbf{H}_{\mathbf{W}_{X,Y}}$ can be used to represent the adaptation speed of $\mathbf{W}_{X,Y}$; see [20] for more details.

The following combination functions were used. The identity function **id(.)** is the combination function that is usually used for the states that have an impacting link from only one state. The numerical representation of the function is $\mathbf{c}_Y(V) = \mathbf{id}(V) = V$. The scaled maximum function $\mathbf{smax}_\lambda(\ldots)$ is used for a smooth transition from one value to another. In this chapter, it is only applied to a self-modeling state for a combination function. Numerically, it is represented as $\mathbf{c}_Y(V_1, \ldots, V_k) = \mathbf{smax}_\lambda(V_1, \ldots, V_k) = \max(V_1, \ldots, V_k)/\boldsymbol{\lambda}$ where $\boldsymbol{\lambda}$ is the scaling factor.

The advanced logistic sum function $\mathbf{alogistic}_{\sigma,\tau}(\ldots)$, is usually used for the states that have impacting connections from multiple states, and can be expressed as

$$\begin{aligned}\mathbf{c}_Y(V_1, \ldots, V_k) &= \mathbf{alogistic}_{\sigma,\tau}(V_1, \ldots, V_k)\\ &= [\frac{1}{1+e^{-\sigma(V_1+\cdots+V_k-\tau)}} - \frac{1}{1+e^{\sigma\tau}})](1+e^{-\sigma\tau})\end{aligned}$$

where σ is a steepness parameter and τ a threshold parameter.

The Hebbian function $\mathbf{hebb}_{\mu}(\ldots)$, is used for adaptation based on Hebbian learning and is expressed for self-model state $\mathbf{W}_{X,Y}$ as

$$\mathbf{c}_{\mathbf{W}_{X,Y}}(V_1, V_2, W) = \mathbf{hebb}_{\mu}(V_1, V_2, W) = V_1 V_2(1 - W) + \mu W$$

where μ in [0, 1] is the persistence factor. The simple linear homophily function $\mathbf{slhomo}_{\alpha,\tau}(\ldots)$, is applied to the homophily adaptation states $\mathbf{W}_{X,Y}$. This function has two parameters: amplification factor α and tipping point parameter τ, and can be expressed as

$$\mathbf{c}_{\mathbf{W}_{X,Y}}(V_1, V_2, W) = \mathbf{slhomo}_{\alpha,\tau}(V_1, V_2, W) = W + \alpha W(1 - W)(\tau - |V_1 - V_2|)$$

The step-once function $\mathbf{steponce}_{\alpha,\beta}(\ldots)$ is used to model changing context factors, and is defined by two parameters: the start point α and the end point β for the simulation time t, during which the state needs to be active. This function is applied in cases when a state is not needed throughout the whole simulation. In mathematical terms, the function yields a value of 1 if $\alpha \leq t \leq \beta$, else 0. The step-modulo function $\mathbf{stepmod}_{\rho,\delta}(\ldots)$ is used to model behavior with constant intervals, using two parameters: the repeated time duration ρ and the transition point δ. The function can be expressed as formula which yields 0 if mod(t, ρ) < δ, else 1 for time t. Lastly, the beats per minute function $\mathbf{bpm}_{\gamma,\varphi}(\ldots)$ generates a pulse-like pattern based on the given values for the amount of beats per minute γ, and the duration of each beat φ. Numerically, it can be represented as: 1 if mod(t, (400/γ)) < φ, else 0, where 400 is used as the determinant for the amount of beats that are in a unit of time in the model. The higher the value, the greater the amount of time allocated for adaptation and learning processes to occur and unfold effectively.

3 Design of the Adaptive Network Model

In this section, the various components of the network model are explained. In the graphical representation of the model shown in Fig. 1, the base level contains an individual mental network for each agent, where the individual nodes represent the various states of the mental process, which are defined in Table 1. The black arrows represent the connections indicating the causal relationship between the states. The

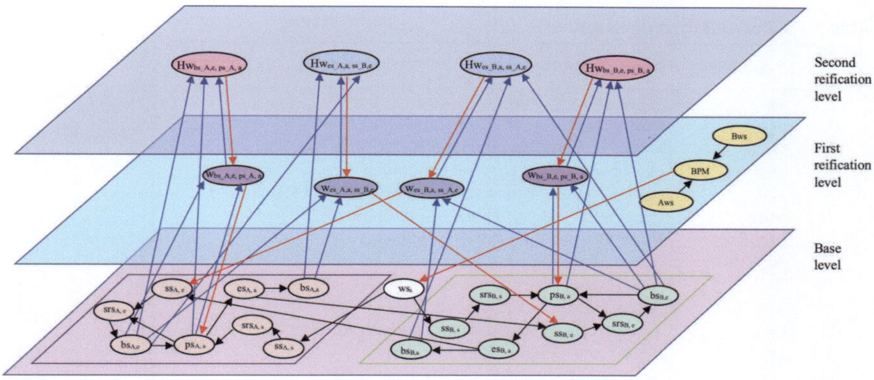

Fig. 1 Graphical representation of the adaptive network model

starting point is the world state ws_s, which is the beat. An agent senses the beat ss_s, which leads to sensory representation srs_s. At the same time, it senses a beat made by the other agent ss_e and has a sensory representation srs_e of it too. The belief state bs_e serves as a (subjective) interpretation the agent has of the perceived stimuli e from the other agent. Taking into account both perceived stimuli, the agent activates preparation state ps_a for its own action a and related execution state (the drumming beat) es_a, and then again has a belief state bs_a for being (also subjectively) aware of the own action. The same process unfolds for the second agent. It is important to note that all of the network connections within the mental network model of each agent, depicted in the model, are involved in standard mental processes that occur within any agent, regardless of the scenario at hand.

The combination functions and their parameters, are the determining factors for the way the connections leading to each state are aggregated. For both of the agents, the ss_s, srs_s, ss_e, es_a, bs_e, and bs_a states the identity function is used. The srs_e and ps_a states are modeled using the **alogistic** function, with the steepness $\boldsymbol{\sigma}$ set at 20 for both, and the threshold $\boldsymbol{\tau}$ set at 0.6 and 0.7 respectively.

The first-order self-modeling level addresses the Bonding based on subjective homophily and Hebbian learning adaptation for the weights of the different types of connections, with the blue arrow representing the sources of the learning. For Hebbian learning, these sources are the subjective belief state bs_e of the perceived action and preparation state ps_a for the own action. For Bonding based on subjective homophily the sources are the subjective belief states bs_e for the perceived action and bs_a for the own action. The red downward arrows point towards the base states that are affected by the learning (ps_a for Hebbian; the other agent's ss_e for Bonding based on subjective homophily). In order to achieve a varying rhythm, beat for the world state, a first-order self-model state BPM for the combination function **bpm** (further explained in the next section) was added, with input from two self-model states, $\mathbf{A}_{\text{ws}}$ and $\mathbf{B}_{\text{ws}}$.

Table 1 States and their descriptions

State nr	State name	Explanation	Level
X_1	ws_s	World state for stimulus s	Base level
X_2	$ss_{A,s}$	Sensor state of agent A for stimulus s	
X_3	$srs_{A,s}$	Sensory representation state of agent A for stimulus s	
X_4	$ss_{A,e}$	Sensor state of agent A for stimulus e	
X_5	$srs_{A,e}$	Sensory representation state of agent A for stimulus e	
X_6	$ps_{A,a}$	Preparation state of agent A for response a	
X_7	$es_{A,a}$	Execution state of agent A for stimulus a	
X_8	$bs_{A,e}$	Belief state of agent A for response e	
X_9	$bs_{A,a}$	Belief state of agent A for stimulus a	
X_{10}	$ss_{B,s}$	Sensor state of agent B for stimulus s	
X_{11}	$srs_{B,s}$	Sensory representation state of agent B for stimulus s	
X_{12}	$ss_{B,e}$	Sensor state of agent B for stimulus e	
X_{13}	$srs_{B,e}$	Sensory representation state of agent B for stimulus e	
X_{14}	$ps_{B,a}$	Preparation state of agent B for response a	
X_{15}	$es_{B,a}$	Execution state of agent B for stimulus a	
X_{16}	$bs_{B,e}$	Belief state of agent B for response e	
X_{17}	$bs_{B,a}$	Belief state of agent B for stimulus a	
X_{18}	$\mathbf{W}_{bs_{A,e}, ps_{A,a}}$	First-order self-model state for connection weight $\boldsymbol{\omega}_{bs_{A,e}, ps_{A,a}}$	First-order self-model level
X_{19}	$\mathbf{W}_{bs_{B,e}, ps_{B,a}}$	First-order self-model state for connection weight $\boldsymbol{\omega}_{bs_{B,e}, ps_{B,a}}$	
X_{20}	$\mathbf{W}_{es_{A,a}, ss_{B,e}}$	First-order self-model state for connection weight $\boldsymbol{\omega}_{es_{A,a}, ss_{B,e}}$	
X_{21}	$\mathbf{W}_{es_{B,a}, ss_{A,e}}$	First-order self-model state for connection weight $\boldsymbol{\omega}_{es_{B,a}, bs_{A,e}}$	
X_{22}	**BPM**	First-order self-model state for the frequency parameter of the bpm combination function	
X_{23}	$\mathbf{A}_{ws}$	First-order self-model state for phases of the world state ws	
X_{24}	$\mathbf{B}_{ws}$	First-order self-model state for phases of the world state ws	
X_{25}	$\mathbf{H}_{\mathbf{W}_{bs_{A,e}, ps_{A,a}}}$	Second-order self-model state for speed factor for first-order self-model state $\mathbf{W}_{srs_{A,e}, ps_{A,a}}$	Second-order self-model level
X_{26}	$\mathbf{H}_{\mathbf{W}_{bs_{B,e}, ps_{B,a}}}$	Second-order self-model state for speed factor for first-order self-model state $\mathbf{W}_{srs_{B,e}, ps_{B,a}}$	
X_{27}	$\mathbf{H}_{\mathbf{W}_{es_{A,a}, ss_{B,e}}}$	Second-order self-model state for speed factor for first-order self-model state $\mathbf{W}_{es_{A,a}, ss_{B,e}}$	
X_{28}	$\mathbf{H}_{\mathbf{W}_{es_{B,a}, ss_{A,e}}}$	Second-order self-model state for speed factor for first-order self-model state $\mathbf{W}_{es_{B,a}, bs_{A,e}}$	

The second-order self-model level addresses the adaptation $\mathbf{H_W}$-states for the speed factors of the first-order adaptation **W**-states. This is done to have a varying speed of Hebbian learning and Homophily adaptation, since it is not a constant or linear process. These second-order self-model states are based on the adaptation principle formulated by [17]: 'Adaptation accelerates with increasing stimulus exposure'. Stimuli exposure is detectable in activation of the related base states,therefore, the second-order speed adaptation $\mathbf{H_W}$-states receive information from the same base level states as the first-order learning **W**-states do, plus from the first-order learning **W**-states themselves.

On both of the adaptation levels states have persistence links, which connect the state to itself. It is a necessary link for the learning process, as it assures that previously attained (learnt) values for the states still persist, although often only partially. The speed adaptation $\mathbf{H_W}$-states, however, must have only very low persistence link as otherwise it would make it more difficult to attain a context-sensitive effect for the speed adaptation.

Next, information about the connection weights between the states (for the strength of the impact), the type of combination function used (to aggregate multiple impacts on the state), and the states' speed factors (for the timing of the effect of the impact) will be discussed. The numerical values presented in the section are in line with the scenario of an interaction between an adult and a child. In such a scenario, it is expected that the adult would have a bigger influence on the child in the learning and adaptation process, rather than the other way around. This effect is established using the function parameters for Hebbian and Homophily learning, discussed in the aggregation characteristics section.

As a default all of the connection weights are set at 1, to later be adjusted in accordance with specific scenarios. Additionally, the various links between the states have different purposes and effects in the learning. There are within-agent and inter-agent links. The former includes the prediction links ($ps_{A,a} \rightarrow srs_{A,e}$, $ps_{B,a} \rightarrow srs_{B,e}$), which model the strength of anticipating the stimuli of another agent, perception links ($ss_{A,e} \rightarrow srs_{A,e}$, $ss_{B,e} \rightarrow srs_{B,e}$), which determine how sensitive the agent is to perceiving another agent's stimuli, belief of perceived stimulus links ($srs_{A,e} \rightarrow bs_{A,e}$, $srs_{B,e} \rightarrow bs_{B,e}$), which are responsible for how aware the agent is of the perceived stimulus, and belief of own execution links ($es_{A,a} \rightarrow bs_{A,a}$, $es_{B,a} \rightarrow bs_{B,a}$) which determine how the agent is aware of its own actions. The expression links ($ps_{A,a} \rightarrow es_{A,a}$, $ps_{B,a} \rightarrow es_{B,a}$) signify how expressive the agent is in its own actions, for the other agent to be able to perceive it. The various settings of these links and their effects are explored in the evaluation section of the chapter.

The Hebbian learning states are modeled using their own combination function, **hebb**, whose only parameter, the persistence factor $\boldsymbol{\mu}$, is set to 0.85 and 0.9. The adult's persistence factor is slightly lower, as their learning persistence is expected to be lower.

The bonding by homophily adaptation states use the simple logistic homophily **slhomo** function, where the amplification parameter $\boldsymbol{\alpha}$ is set at 0.35 for the agent A affecting B, and 0.3 for the agent B affecting A. The tipping point $\boldsymbol{\tau}$ set at 0.7 and 0.6, respectively. When the dissimilarity between the individuals is less than $\boldsymbol{\tau}$, the strength of their connection will become higher. The speed adaptation states are modeled using the **alogistic** combination function with the steepness $\boldsymbol{\sigma}$ and threshold $\boldsymbol{\tau}$ being 2 and 2.5 for the Hebbian learning, and 1.5 and 2.5 for the bonding by homophily adaptation.

As briefly mentioned in the previous section, the network model has three self-modeling states which contribute to varying beat frequencies. The $\mathbf{A}_{ws}$ and $\mathbf{B}_{ws}$ self-modeling states for the world state use the **steponce** function to determine when in the simulation they happen. The former happens the first 200 units of time, and the latter the last 200 units of time. These two states have connections to the self-model state **BPM**, with the weight of these connections determining the frequency of how often the beat happens; the weight of the link from $\mathbf{A}_{ws}$ and $\mathbf{B}_{ws}$, are 30 and 20, respectively. In order for the world state to be affected by the self-model state **BPM**, the **bpm** function that is used to model the *ws* state, uses **BPM** as an adaptive frequency parameter. Thus, the world state *ws* has an adaptive number of beats per unit of time, and for the specific scenario the duration of the gap between the beats is 1 unit of time.

Finally, each state has a speed factor that defines how fast it changes for the given aggregated impact. As default all the states are set to a speed factor of 10. The red cells for the learning states are the adaptive speed factors from the second-order self-model level. These values are not constant; they vary based on the behavior of the $\mathbf{H_W}$-states (X_{25} to X_{28}). The $\mathbf{H_W}$-states themselves have a constant speed factor, set to a much lower speed factor of 1, in order to provide time for the learning aspect of the network model to happen.

The full specification of the network model in role matrix format can be found in the Appendix in Sect. 6.

4 Simulation Results

The subsequent section showcases example simulation graphs. Within such a graph, each state is depicted by an individual line, portraying its behavior over the time axis. Figure 2 illustrates the base level states for both of the agents and the world state *ws*, along with its fluctuating beat, noticeable by an apparent change in rhythm at time point 200.

It is noteworthy that both agents' mental states exhibit parallel behavior to the world state, indicating the synchronization of their perception, preparation and execution with the beat. This observation signifies that there is no loss of information

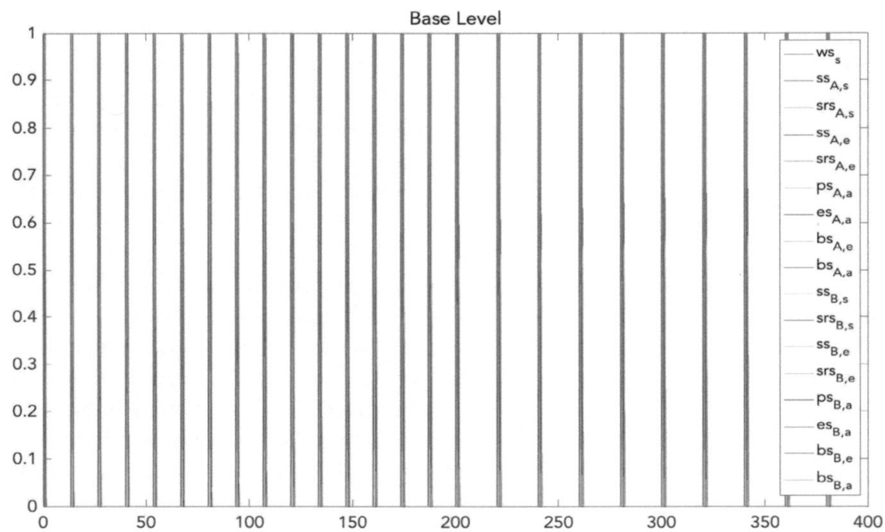

Fig. 2 Simulation of all of the base level states

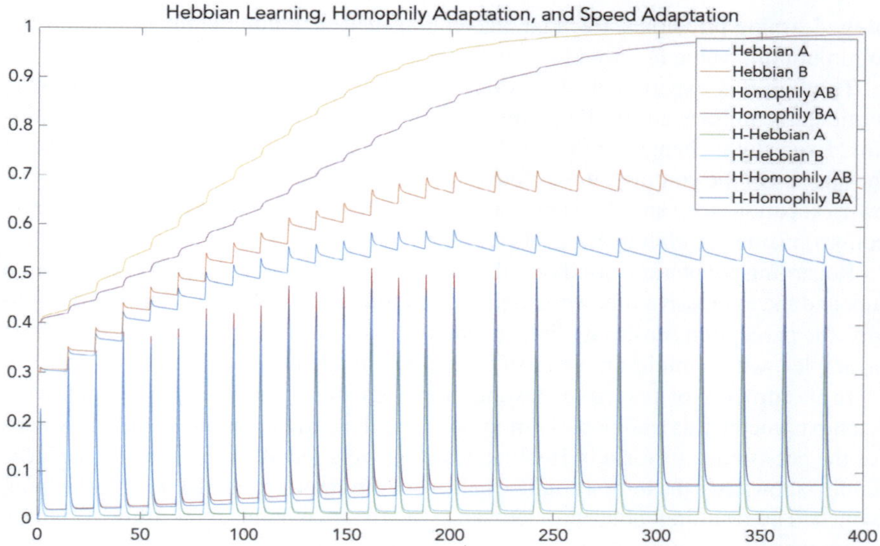

Fig. 3 Simulation of the learning, adaptation, and adaptive speed

between the agents and their stimuli perception, ensuring a solid basis for further adaptation.

Figure 3 displays the outcomes of Hebbian learning and homophily adaptation, accompanied by their respective speed adaptation states. The Hebbian states initiate at 0.3 and progressively grow, eventually reaching a final level of approximately 0.7 for agent B and 0.55 for agent A. Notably, the learning process exhibits slightly greater strength, higher values, and reduced variation for agent B (the student), as it was given more robust learning parameters. As for homophily adaptation, the curves demonstrate a more gradual progression, steadily approaching a value of 1 by the end of the simulation. Importantly, these curves do not decline, indicating a consistent growth in adaptation for both the adult and the student. Moreover, considering that the adult exerts a stronger influence on B than vice versa, the Hebbian AB curve appears higher and more pronounced. The speed adaptation states exhibit patterns with high amplitude fluctuations, characterized by alternating increments and decrements in speed values.

5 Model Evaluation and Discussion

The material for this chapter came from [14]. As part of the model evaluation, a series of tests were conducted to explore the capabilities of the system, seen in Table 2. These tests involved manipulating the connection weights and parameters of the learning principles. Each parameter was modified individually to identify the minimum link value required for satisfactory learning and adaptation.

Through this experiment, it was observed that the Hebbian learning process is particularly influenced by the prediction link and the belief of perceived stimulus link. Even slight changes in these values had a substantial impact on the behavior of the states. On the contrary, it was found that the perception sensitivity link, belief of own execution link, and the expressive link could be set to 0 without significantly compromising the effectiveness of Hebbian learning for both agents.

Regarding adaptation based on subjective homophily, the belief of own execution link and the expressive link were identified as influential sensitivity links. Additionally, the perception sensitivity link appeared to have the least impact on the learning principles, with a minimum value of 0 resulting in adequate performance.

In the process of parameter testing, a systematic methodology was employed. Each parameter was individually modified, and the corresponding minimum values for the persistence factor $\boldsymbol{\mu}$ in Hebbian learning were recorded in Table 3. Similarly, Table 4 showcases the lowest attained values for the amplification factor $\boldsymbol{\alpha}$ and tipping point $\boldsymbol{\tau}$. The resulting tables provide valuable reference points, presenting a concise summary of the minimum values attained for each parameter. It is worth noting that in the table, a value of 0* signifies that there is an observable impact on learning, however, is not substantial enough to hinder the learning process significantly, and could even be as low as 0.

These findings highlight the varying degrees of importance and sensitivity associated with different connection links, learning principles, and function parameters. This comprehensive analysis provides valuable insights into optimizing the model's performance and achieving optimal settings for enhanced functionality. It's worth noting that the explored model is inherently universal, meaning it can be applied across various scenarios. What distinguishes its performance and adaptability in different contexts are the specific network characteristics under which the combination function parameters. These network characteristics allow for tailoring the model to suit a wide range of real-world situations and personal characteristics, making it a versatile tool for understanding and simulating complex adaptive real-world systems. Furthermore, the model's complexity, while substantial, is essential for capturing the intricate dynamics of sensorimotor synchronization and achieving adaptability across various scenarios. For example, to distinguish a person who only slowly adapt from a person B that adapt fast, the weights of the incoming connections of the **H**-states of A representing adaptation speed can be set lower for connection weights. As another example, for a person who initially adapts very slowly but after some tipping point adapts fast can be modeled by using a high steepness parameter for the related **H**-states.

Regarding the notion of bonding based on subjectively perceived homophily, this study builds upon our work in [22]. In the context of the illustrative scenario presented in this chapter, our model extends beyond the simplified version discussed in [2]. Notably, this chapter's model deviates from the one in [2] through the incorporation of additional base-level states, the introduction of belief states for subjectively perceived homophily instead of objective homophily, integration of actual rhythm

Table 2 Connection weight model evaluation values

	Agent sensitivity		Hebbian learning			Homophily adaptation		
			Initial weight	Lowest weight	Median weight	Initial weight	Lowest weight	Median weight
Within-agent sensitivity	Prediction link	$ps_{A,a} \rightarrow srs_{A,e}$	1	0.3	0.65	1	0	0.5
		$ps_{B,a} \rightarrow srs_{B,e}$	1	0.2	0.6	1	0	0.5
	Perception sensitivity link	$ss_{A,e} \rightarrow srs_{A,e}$	1	0	0.5	1	0	0.5
		$ss_{B,e} \rightarrow srs_{B,e}$	1	0	0.5	1	0	0.5
	Belief of perceived stimulus link	$srs_{A,e} \rightarrow bs_{A,e}$	1	0.5	0.75	1	0	0.5
		$srs_{B,e} \rightarrow bs_{B,e}$	1	0.45	0.73	1	0	0.5
	Belief of Own execution link	$es_{A,a} \rightarrow bs_{A,a}$	1	0*	0.5	1	0	0.5
		$es_{B,a} \rightarrow bs_{B,a}$	1	0*	0.5	1	0	0.5
Inter-agent sensitivity	Expressivity link	$ps_{A,a} \rightarrow es_{A,a}$	1	0	0.5	1	0*	0.5
		$ps_{B,a} \rightarrow es_{B,a}$	1	0	0.5	1	0*	0.5

Table 3 Hebbian function parameter model evaluation values

Hebbian learning: Function parameter alterations			
		Initial value	Lowest value
Hebbian A	Persistence **μ**	0.85	0.65
Hebbian B	Persistence **μ**	0.9	0.65

Table 4 Homophily function parameter model evaluation values

Bonding by homophily: Function parameter alterations				
		Initial value	Lowest **α** value	Lowest **τ** value
Homophily AB	Amplif. factor **α**	0.35	0.05	0.35
	Tipping point **τ**	0.7	0.7	0.1
Homophily BA	Amplif. factor **α**	0.3	0.01	0.3
	Tipping point **τ**	0.6	0.6	0.15

with extended time pulses, adaptive irregular beat, and the inclusion of adaptive speed for the adaptation process. Regarding the model itself, this study builds upon our previous work in [15], with a focus on a related but distinct aspect of the multi-adaptive network model, delving into the dynamics of homophily adaptation within the context of social interactions among multiple individuals. In [18] also a combination of Hebbian learning and homophily was addressed, see also this volume [9], Chapter 6 (Bonding by Homophily and Adaptive Mirroring in Joint Decision Making: a Second-Order Adaptive Network Model). However, in that case no subjectively perceived homophily and rythmic action was addressed like the focus was in the current chapter.

In other earlier work, in [6] it was analysed how persons can synchronize in the context of adaptive joint decision making, see also this volume [9], Chapter (Learning to Synchronize: Adaptive Joint Decision-Making by Virtual Agents) [7]. Furthermore, in [6] it is analysed how subjective detection of synchrony can play a causal role for bonding, see also this volume Chapter (Growing Connectedness through Synchrony: The Role of Subjective Analysis of Intrapersonal Synchrony and Interpersonal Synchrony within an Adaptive Agent Model) [12]. Moreover, in [6] the distinction for different timescales between short-term affiliation and long-term bonding and their relation to synchronisation was introduced, see also this volume [9] Part 4, Chapter (The Adaptive Interplay of Interpersonal Synchrony, Short-Term Affiliation and Long-Term Bonding: A Multi-Timescale Multi-Adaptive Agent Model) [10]. These three papers do not consider subjective homophily as a basis for bonding like the current chapter does, but focus on subjectively deteced multimodal synchrony instead.

Table 5 Role matrix specifications for connectivity, connection weights, and initial values

	mb base connectivity					mcw connection weights						ms speed factors	iv initial values
		1	2	3	4		1	2	3	4			
ws_s	X_1					X_1					X_1	10	0
$ss_{A,s}$	X_2	X_1				X_2	1				X_2	10	0
$srs_{A,s}$	X_3	X_2				X_3	1				X_3	10	0
$ss_{A,e}$	X_4	X_{15}				X_4	X_{21}				X_4	10	0
$srs_{A,e}$	X_5	X_4	X_6			X_5	1	1			X_5	10	0
$ps_{A,a}$	X_6	X_3	X_8			X_6	0.7	X_{18}			X_6	10	0
$es_{A,a}$	X_7	X_6				X_7	1				X_7	10	0
$bs_{A,e}$	X_8	X_5				X_8	1				X_8	10	0
$bs_{A,a}$	X_9	X_7				X_9	1				X_9	10	0
$ss_{B,s}$	X_{10}	X_1				X_{10}	1				X_{10}	10	0
$srs_{B,s}$	X_{11}	X_{10}				X_{11}	1				X_{11}	10	0
$ss_{B,e}$	X_{12}	X_7				X_{12}	X_{20}				X_{12}	10	0
$srs_{B,e}$	X_{13}	X_{12}	X_{14}			X_{13}	1	1			X_{13}	10	0
$ps_{B,a}$	X_{14}	X_{11}	X_{16}			X_{14}	0.7	X_{19}			X_{14}	10	0
$es_{B,a}$	X_{15}	X_{14}				X_{15}	1				X_{15}	10	0
$bs_{B,e}$	X_{16}	X_{13}				X_{16}	1				X_{16}	10	0
$bs_{B,a}$	X_{17}	X_{15}				X_{17}	1				X_{17}	10	0
$\mathbf{W}_{bsA,e\,,\,psA,a}$	X_{18}	X_8	X_6	X_{18}		X_{18}	1	1	1		X_{18}	X_{25}	0.3
$\mathbf{W}_{bsB,e\,,\,psB,a}$	X_{19}	X_{16}	X_{14}	X_{19}		X_{19}	1	1	1		X_{19}	X_{26}	0.3
$\mathbf{W}_{esA,a,\,ssB,e}$	X_{20}	X_9	X_8	X_{20}		X_{20}	1	1	1		X_{20}	X_{27}	0.4
$\mathbf{W}_{esB,a,\,ssA,e}$	X_{21}	X_{17}	X_{16}	X_{21}		X_{21}	1	1	1		X_{21}	X_{28}	0.4
BPM	X_{22}	X_{23}	X_{24}			X_{22}	30	20			X_{22}	10	0
$\mathbf{A}_{ws}$	X_{23}	X_{23}				X_{23}	1				X_{23}	10	0
$\mathbf{B}_{ws}$	X_{24}	X_{24}				X_{24}	1				X_{24}	10	0
$\mathbf{H}_{W_{bsA,e\,,\,psA,a}}$	X_{25}	X_8	X_6	X_{18}	X_{25}	X_{25}	1	1	1	0.2	X_{25}	1.5	0
$\mathbf{H}_{W_{bsB,e\,,\,psB,a}}$	X_{26}	X_{16}	X_{14}	X_{19}	X_{26}	X_{26}	1	1	1	0.2	X_{26}	1.5	0
$\mathbf{H}_{W_{esA,a,\,ssB,e}}$	X_{27}	X_9	X_8	X_{20}	X_{27}	X_{27}	1	1	1	0.2	X_{27}	1.5	0
$\mathbf{H}_{W_{esB,a,\,ssA,e}}$	X_{28}	X_{17}	X_{16}	X_{21}	X_{28}	X_{28}	1	1	1	0.2	X_{28}	1.5	0

6 Appendix: Role Matrices

This section provides the full specification of the introduced model by role matrices. Within Table 5. the cells marked in red correspond to the red downward arrows in the graphical representation of the model and define the adaptive connections weights for the adaptive connections, and hence do not have any predetermined weights values. Their weights values are adapting to the Hebbian and Homophily learning behavior. All of the values for the weights of the links can be found in Table 5. Moreover, the initial values necessary for accurate model replication in future studies can also be found in Table 5.

References

1. Abraham, W.C., Bear, M.F.: Metaplasticity: the plasticity of synaptic plasticity. Trends Neurosci. **19**(4), 126–130 (1996). https://doi.org/10.1016/s0166-2236(96)80018-x.PMID:8658594
2. Accetto, M., Treur, J., Villa, V.: An adaptive cognitive-social model for mirroring and social bonding during synchronous joint action. In: Proceedings BICA 2018. Procedia Comput. Sci. **145**, 3–12 (2018)

3. Bain, A.: Mind and body: The theories of their relation. HS King & Company 1873 Farnborough Eng. Gregg International (1873)
4. Bössenecker, G., Mreijen, P., Treur, J: A second-order adaptive network model for connecting social media platforms and users through bonding by faked homophily. Procedia Comput. Sci. **213**, 32–40 (2022). https://doi.org/10.1016/j.procs.2022.11.035
5. Hebb, D.: The Organisation of Behavior. Wiley (1949)
6. Hendrikse, S.C.F., Kluiver, S., Treur, J., Wilderjans, T.F., Dikker, S., Koole, S.L.: How virtual agents can learn to synchronize: an adaptive joint decision-making model of psychotherapy. Cogn. Syst. Res. **79**, 138–155 (2023)
7. Hendrikse, S.C.F., Kluiver, S., Treur, J., Wilderjans, T.F., Dikker, S., Koole, S.L. (2025). Learning to synchronize: adaptive joint decision-making by virtual agents. In: Hendrikse, S.C.F., Treur, J., Koole, S.L. (eds.) New Analysis and Modeling Directions in Social Interaction Science: Emergent Multimodal Interpersonal Synchrony, Affiliation, and Bonding, Ch 5. Springer Nature (2025)
8. Hendrikse, S.C.F., Treur, J., Koole, S.L.: Modeling emerging interpersonal synchrony and its related adaptive short-term affiliation and long-term bonding: a second-order multi-adaptive neural agent model. Int. J. Neural Syst. **33**(7), 2350038 (2023)
9. Hendrikse, S.C.F., Treur, J., Koole, S.L. (eds.) New Analysis and Modeling Directions in Social Interaction Science: Emergent Multimodal Interpersonal Synchrony, Affiliation, and Bonding. Springer Nature (2025)
10. Hendrikse, S.C.F., Treur, J., Koole, S.L.: The adaptive interplay of interpersonal synchrony, short-term affiliation and long-term bonding: a multi-timescale multi-adaptive agent model. In: Hendrikse, S.C.F., Treur, J., Koole, S.L. (eds.) New Analysis and Modeling Directions in Social Interaction Science: Emergent Multimodal Interpersonal Synchrony, Affiliation, and Bonding, Ch 12. Springer Nature (2025)
11. Hendrikse, S.C.F., Treur, J., Wilderjans, T.F., Dikker, S., Koole, S.L.: On becoming in sync with yourself and others: an adaptive agent model for how persons connect by detecting intra- and interpersonal synchrony. Hum.-Centric Intell. Syst. **3**, 123–146 (2023)
12. Hendrikse, S.C.F., Treur, J., Wilderjans, T.F., Dikker, S., Koole, S.L.: Growing connectedness through synchrony: the role of subjective analysis of intrapersonal synchrony and interpersonal synchrony within an adaptive agent model. In: Hendrikse, S.C.F., Treur, J., Koole, S.L. (eds.) New Analysis and Modeling Directions in Social Interaction Science: Emergent Multimodal Interpersonal Synchrony, Affiliation, and Bonding, Ch 11. Springer Nature (2025)
13. McPherson, M., Smith-Lovin, L., Cook, J.M.: Birds of a feather: homophily in social networks. Ann. Rev. Sociol. **27**, 425–444 (2001)
14. Mukeriia, Y., Treur, J., & Hendrikse, S.C.F.: An adaptive network model for learning and bonding during a varying in rhythm synchronous joint action. In: Proceedings of the 12th International Conference on Complex Networks and Their Applications. Studies in Computational Intelligence, vol. 1144, pp. 41–52. Cham: Springer Nature Switzerland (2023)
15. Mukeriia, Y., Treur, J., Hendrikse, S.C.F.: A multi-adaptive network model for human Hebbian learning, synchronization and social bonding based on adaptive homophily. Cogn. Syst. Res. **84**, 101187 (2024)
16. Repp, B.H.: Sensorimotor synchronization: a review of the tapping literature. Psychon. Bull. Rev. **12**, 969–992 (2005). https://doi.org/10.3758/BF03206433
17. Robinson, B.L., Harper, N.S., McAlpine, D.: Meta-adaptation in the auditory midbrain under cortical influence. Nat. Commun. **7**, e13442 (2016)
18. Tichelaar, C., Treur, J.: Network-oriented modeling of the interaction of adaptive joint decision making, bonding and mirroring. In: Fagan, D., Martín-Vide, C., O'Neill, M., Vega-Rodríguez, M.A. (eds.) Proceedings of the 7th International Conference on Theory and Practice of Natural Computing, TPNC'18. Lecture Notes in Computer Science, vol. 11324, pp. 328–343. Springer Publishers (2018)
19. Treur, J.: Dynamic modeling based on a temporal-causal network modeling approach. Biol Inspired Cogn Arch **16**, 131–168 (2016)

20. Treur, J.: Network-oriented modeling for adaptive networks: designing higher-order adaptive biological, mental and social network models. Springer Nature (2020). https://doi.org/10.1007-978-3-030-31445-3
21. Treur, J.: Relating a reified adaptive network's structure to its emerging behaviour for bonding by homophily. In: Network-Oriented Modeling for Adaptive Networks: Designing Higher-Order Adaptive Biological, Mental and Social Network Models, Ch 13. Springer Nature, Cham (2020). https://doi.org/10.1007/978-3-030-31445-3_13
22. Treur, J.: Controlled social network adaptation: subjective elements in an objective social world. In: Yang, X.S., Sherratt, S., Dey, N., Joshi, A. (eds.) Proceedings of Sixth International Congress on Information and Communication Technology, ICICT'21. Lecture Notes in Networks and Systems, vol. 235, pp. 263–274. Springer Nature (2022)
23. Treur, J.: A second-order adaptive network model for bonding based on subjectively perceived homophily. In: Hendrikse, S.C.F., Treur, J., Koole, S.L. (eds.) New Analysis and Modeling Directions in Social Interaction Science: Emergent Multimodal Interpersonal Synchrony, Affiliation, and Bonding, Ch. 7. Springer Nature (2025)
24. van der Steen, M.C., Keller, P.E.: The adaptation and anticipation model (ADAM) of sensorimotor synchronization. Front. Hum. Neurosci. **7**, 253 (2013). https://doi.org/10.3389/fnhum.2013.00253.PMID:23772211;PMCID:PMC3677131

Connecting Social Media Platforms and Users Through Bonding by Faked Homophily: A Second-Order Adaptive Network Model

Giulia Bössenecker, Pim Mreijen, and Jan Treur

Abstract In this chapter a second-order adaptive network model is introduced for presenting a user with content they like on a social media platform. The platform's method is using a fake state to control the user's bonding with the platform. Simulation results have been performed with different scenarios for different starting values of the user and the platform. In all scenarios the platform's method can move towards the state of the user in order to achieve a form of bonding through faked homophily. After the user indeed connects to the platform it becomes more easy for the platform to move the user's preferences in a desired direction so that the platform can offer more adequate content more efficiently.

Keyword Second-order adaptive · Bonding · Homophily · Faked

1 Introduction

Similarity leads to connection, often expressed by 'Birds of a feather, flock together'; e.g., McPherson et al. [8]. We can find this principle everywhere in daily life. We see it in friendship, colleague bonds, marriage, school, social media and so on. A feature of social networks is that social bonds seem to be more persistent in similar persons than in non similar ones. This phenomenon of similarity, known as homophily, applies to a wide range of social interaction [2]. It has been widely researched that individuals bond with other individuals based on similar demographics and interests [4]. Homophily is of great value to many popular platforms nowadays, such as Amazon, Instagram and Netflix, who apply similarity of watching, click and buying patterns to

G. Bössenecker · J. Treur (✉)
Social AI Group, Department of Computer Science, Vrije Universititeit Amsterdam, Amsterdam, Netherlands
e-mail: j.treur@vu.nl

P. Mreijen
Management, Policy Analysis and Entrepreneurship in Health and Life Science, Vrije Universiteit Amsterdam, Amsterdam, Netherlands

S. C. F. Hendrikse et al. (eds.), *New Analysis and Modeling Directions in Social Interaction Science*, Studies in Systems, Decision and Control 614,
https://doi.org/10.1007/978-3-031-99968-0_9

recommend new patterns and predict future behavior of their clients [9]. Individuals with similar behavior patterns might also have the same future interests, hence, the platform can recommend things to one that the other has interest in.

Streaming platforms like Netflix and Amazon prime do not offer 'visible' social connections, it is not possible to add other users as is possible on e.g. Instagram or Facebook. However, the user still does make a connection, now to the platform. Netflix, for example, uses phrases such as 'Because you watched…, you might like…' and 'Because you added … to your list, we think you might like…', creating some sort of familiarity or connection between the user and the platform [10]. We consider this connection to be developing through bonding by homophily: a connection is created through similarity. Platforms want this connection to be as strong as possible to be able to create user engagement, getting the user to stay on, use and recommend the platform to others as much as possible. And through all of this possibly more profit for the platform. A streaming platform might want to produce and offer its user certain content that the user is not currently interested in. It may find pathways to get the user's interest changed over time so that such interest in the available content develops in a direction that is good for the platform.

Next to bonding by homophily another principle also plays an important role, namely social contagion. This entails that the stronger the connection between two individuals is, the more they will behave similarly [6]. This principle can be of use to platforms, i.e., this implies that a platform can influence a users behavior if the connection to this user is strong enough.

This leads to an assumption that it may be advantageous for a platform if the bonding by homophily could be controlled in a certain way instead of the bonding only happening through actual mutual interests. Between humans such a form of control often happens by faking similarity. Bonding by homophily through faking means that bonding does not depend on the objective states for two persons, but rather on how these states are perceived and represented by the persons involved. As far as the authors know, no other models in the literature address this quite realistic issue of 'bonding by faked homophily', except [11] which was used as inspiration for the model introduced here, see also this volume [5], Chap. 7 [15].

It would also be beneficial to a platform if many users on the platform have the same interests, making it easier and cheaper to produce content that many users will like. Hence, according to the social contagion principle, by first creating a strong connection between a platform and its user based on faked homophily, the platform might be able to influence the users interests and push these interests towards the platform interests.

In the current chapter, a second-order adaptive network model is introduced that incorporates bonding by faked homophily. First, the faked homophily principle is discussed in relation to the approach of modeling based on self-modeling networks. Hereafter, example simulations are performed with the model to explore the effect of different network characteristics incorporated in the model.

2 Modeling by Self-Modeling Networks

In the current chapter, the Network-Oriented Modeling approach described in [11] is applied, and its supporting software environment described in [11] implemented in MATLAB [7]. This modeling approach based on temporal-causal networks distinguishes itself through the incorporation of an adaptive and dynamic temporal perspective, on both nodes (also called states) and their connections (used to model causal relations between states). This approach uses the following network structure characteristics:

(a) **Connectivity of the network**

- Each connection from a node (or state) X has a connection weight $\boldsymbol{\omega}_{X,Y}$ to a state Y. This weight represents the strength of the connection.

(b) **Aggregation of multiple connections in the network**

- For each state Y there is a combination function $\mathbf{c}_Y$ to determine the aggregation of causal impacts that are incoming and combine these impacts.

(c) **Timing in the network**

- For each state Y there is a speed factor $\boldsymbol{\eta}_Y$. This defines how fast a node or state will adjust upon aggregated impact.

A numerical representation of the dynamics based on these network characteristics for connectivity, aggregation and timing is as follows [11]:

1. State values over time t
   Each state Y in the network model has a real number value $Y(t)$ at each time point t.
2. Single causal impact
   State X with a connection to state Y has an impact $\mathbf{impact}_{X,Y}(t) = \boldsymbol{\omega}_{X,Y}\, X(t)$ on Y at t.
3. Aggregating multiple impacts
   The aggregated causal impact of multiple states X_i on Y at t (those from which Y gets its incoming connections) is determined by applying combination function $\mathbf{c}_Y$ to the k single causal impacts on Y:

$$\begin{aligned}\mathbf{aggimpact}_Y(t) &= \mathbf{c}_Y\left(\mathbf{impact}_{X_1,Y}(t), ..., \mathbf{impact}_{X_k,Y}(t)\right)\\ &= \mathbf{c}_Y\left(\boldsymbol{\omega}_{X_1,Y}X_1(t),\ \ldots,\ \boldsymbol{\omega}_{X_k,Y}X_k(t)\right)\end{aligned} \tag{1}$$

4. Timing of the causal effect
   The causal impact on Y is exerted over time gradually, using speed factor $\boldsymbol{\eta}_Y$, where the X_i are all states from which state Y has incoming connections:

$$\begin{aligned}Y(t+\Delta t) &= Y(t) + \boldsymbol{\eta}_Y\left[\mathbf{aggimpact}_Y(t) - Y(t)\right]\Delta t\\ &= Y(t) + \boldsymbol{\eta}_Y[\mathbf{c}_Y(\boldsymbol{\omega}_{X_1,Y}X_1(t),\ \ldots,\ \boldsymbol{\omega}_{X_k,Y}X_k(t)) - Y(t)]\Delta t\end{aligned} \tag{2}$$

Table 1 Combination functions used [11]

Name	Formula	Parameters	Used for
$\mathbf{eucl}_{n,\lambda}$ $(V_1, \ldots, V_k)$	$\sqrt[n]{\frac{V_1{}^n+\cdots+V_k{}^n}{\lambda}}$	Order $\boldsymbol{n}$ Scaling factor $\boldsymbol{\lambda}$	X_3-X_7, X_9-X_{11}, X_{13}-X_{14}, X_{16}
$\mathbf{alogistic}_{\sigma,\tau}$ $(V_1, \ldots, V_k)$	$[\frac{1}{1+e^{-\sigma(V_1+\cdots+V_k-\tau)}} - \frac{1}{1+e^{\sigma\tau}}]\,(1+e^{-\sigma\tau})$	Steepness $\boldsymbol{\sigma}$ Threshold $\boldsymbol{\tau}$	X_1-X_2
$\mathbf{slhomo}_{\alpha,\tau}$ (V_1, V_2, W)	$W + \alpha\,(\tau - \lvert V_1 - V_2\rvert)(1-W)W$	Modulation factor $\boldsymbol{\alpha}$ Tipping point $\boldsymbol{\tau}$	X_8, X_{12}
Inverseid $(V_1, \ldots, V_k)$	$1/V_1$	–	X_{15}

The choice of which combination function to used for a given network model is usually problem-dependent. The available software environment provides a combination function library with more then 65 different combination functions that can be applied in a model. In this chapter the combination functions were used as shown in Table 1.

It is common for network models to be adaptive when the model should be as realistic as possible. When adaptive, some of the network characteristics change over time. Through the usage of 'self-modeling networks' or 'network reification', a similar network-oriented conceptualisation can be applied to adaptive networks to be able to realize a declarative description by mathematically defined relations and functions. This can be obtained by adding new states to the network that are called self-model states.

These states represent the network characteristics through network states and during processing their values are used for the represented network characteristic. When these self-model states are dynamic, they are able to describe the network characteristics that are adaptive [11]. A main example of this used in this chapter is that for states *A*, *B* of two persons or agents, self-model states $\mathbf{RW}_{A,B}$ are used representing the weights $\boldsymbol{\omega}_{A,B}$ of their connections. By changing the values of these self-model states (due to the perceived extent of similarity), the connections adapt over time. The option of self-modeling turned out very useful to model what is needed here to represent perceived states and connections. For mor details on the modeling approach, see also this volume [5], Chap. 2 [14].

3 The Adaptive Network Model for Faked Bonding by Homophily

This section describes how the adaptive temporal-causal network model was designed, creating bonding by faked homophily. In this section in Fig. 1, these states (and all other states) are presented in a graphical 3D format where the self-model

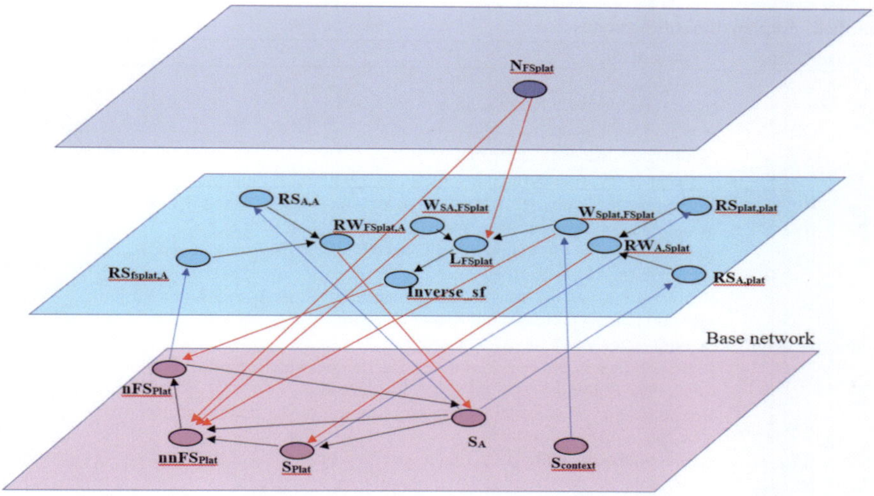

Fig. 1 The connectivity of the adaptive network. Lower level (pink plane): base states of user and platform. Middle level (blue plane): first-order self-model states for perceived base states **RS** and connection weights **W** and **RW**. Upper level (purple plane): Second-order self-model state N for the platform. For an explanation of the states, see Table 2

states are depicted in the next-level plane. In Table 2 an overview can be found of all states and their explanations.

The proposed network consists of 16 different states of which 5 states are at the base level, 10 states at the second level (the first-order self-model level) and 1 state at the third level (the second-order self-model level). As stated earlier, bonding by faked homophily and social contagion are the primary characteristics of the network model that was designed. This bonding by homophily is not assumed to be based on two objective world states, but rather on subjective representations of world states modeling how these world states are perceived by the persons or agents. From these subjective representations, one state that is a false representation, as the related world state is faked by the platform. Therefore, the user does not directly perceive the true intentions of the platform and only interacts with the faked state. The platform, on the other hand, does perceive the user's state since the user interacts with the platform. If the user's preferred content is far off from the (hidden) state of the platform, it is possible for the platform to move itself towards the user. This can be seen as a platform acquiring or producing new types of content that matches the user more. It must be noted that this platform changing its state is a far slower process than the platform's fake state. This should be only necessary if the platform really lacks the content the user likes. Eventually it is assumed that the platform state, the platform fake state and the user can converge to the same point.

The model's connectivity can be seen in Fig. 1, where the platform's fake state tries to approximately match the users' interests. The base level consists of a network of four states and one context state, see the pink plane. The base network has a state

Table 2 Explanation of the states

State	Name	Explanation	Level
X_1	S_A	User state for A	Base level
X_2	S_{Plat}	State of the platform (remains hidden from user)	
X_3	$nnFS_{Plat}$	Non-normalised fake state of the platform	
X_4	nFS_{Plat}	Normalised fake state of the platform	
X_5	$S_{Context}$	Distributes the weight of the **W**-states	
X_6	$\mathbf{RS}_{FSPlat,A}$	Representation state user A has of the fake state $nnFS_{Plat}$ of the platform	First-order self-model level
X_7	$\mathbf{RS}_{A,A}$	Representation state that shows what user A thinks of state S_A	
X_8	$\mathbf{RW}_{FSPlat,SA}$	Representation of the weight of the connection from the fake state $nnFS_{Plat}$ of the platform to state S_A of A	
X_9	$\mathbf{W}_{SA,FSPlat}$	Representation of the weight of the connection from user state S_A to the (nonnormalised) fake state $nnFS_{Plat}$ of the platform	
X_{10}	$\mathbf{L}_{FSPlat}$	Scaling factor for the faked state $nnFS_{Plat}$	
X_{11}	$\mathbf{W}_{SPlat,FSPlat}$	Representation of the weight of the connection from platform state S_{Plat} to the (nonnormalised) fake state $nnFS_{Plat}$ of the platform	
X_{12}	$\mathbf{RW}_{SA,Splat}$	Representation of the weight of the connection from the state S_A of user A to state S_{Plat} of the platform	
X_{13}	$\mathbf{RS}_{Plat,Plat}$	Representation state for platform thinks of own state S_{Plat}	
X_{14}	$\mathbf{RS}_{A,Plat}$	Representation state for platform thinks of user A's state S_A	
X_{15}	$Inverse_{sf}$	Inverse of the scaling factor represented by $\mathbf{L}_{FSplat}$	
X_{16}	$\mathbf{N}_{FSPlat}$	Order ***n*** of the Euclidean function used for aggregation by nFS_{Plat}	Second-order self-model level

for the user S_A. The user can only perceive the platform's (normalised) fake state nFS_{Plat}. This fake state is the state that can quickly change and move towards the user, after which bonding through homophily with the user can occur. The $Inverse_{sf}$ state in the second layer with its accompanying components is there to normalise the output from $nnFS_{Plat}$. The platform's objective state S_{Plat} cannot be perceived by the user, hence the user cannot elucidate the platform's true intentions, since the user only perceives the fake state. The platform on the other hand, does receive input from the user S_A, since it can register what kind of content the user likes. If needed, the platform can adjust its state to represent one that is closer to the user. It is important to notice that this platform adjusting has a lower speed factor, and therefore reacts and moves slower than the platform's fake state.

At the first-order self-model level, subjective representations of the other states are modelled, see the middle (blue) plane. For example, user A has an internal representation state $\mathbf{RS}_{FS\_plat,A}$ for the fake state of the platform. User A also has a subjective representation of the connection from the platform's fake state modelled through the adaptive weight representation state $\mathbf{RW}_{FSPlat,SA}$. Based on this weight, by the social contagion at the base level, A's state S_A is affected by the platform's fake state $nnFS_{Plat}$. The other **RW**-state $\mathbf{RW}_{SA,Splat}$ works via social contagion in the other direction: from A's state S_A to the platform's state S_{Plat}. The values of both **RW**-states are determined by the function **slhomo** for bonding by homophily, thereby using input from the **RS**-states involved: $\mathbf{RS}_{FSPlat,A}$ and $\mathbf{RS}_{A,A}$ for $\mathbf{RW}_{FSPlat,SA}$ by A, respectively $\mathbf{RS}_{A,Plat}$ and $\mathbf{RS}_{Plat,Plat}$ for $\mathbf{RW}_{SA,Splat}$ by the platform. At the second-order self-model level in the upper (purple) plane there is only one state that represents the order ***n*** of the euclidean combination function used for states $\mathbf{L}_{FSPlat}$ in the middle plane and $nnFS_{Plat}$ in the base plane.

For more details and a full specification of the adaptive network model, see the Appendix in Sect. 7.

4 Simulation Results

The available dedicated software environment developed in MATLAB was used for simulation [11]; see URL https://www.researchgate.net/publication/368775720.

It takes input from role matrices: tables in a standard format that specify the network structure characteristics for all the different types of states. The environment is able to run the model based on these role matrices. Three different simulation experiments were conducted with varying initial values on the adaptive network model described above.

Experiment 1

First, a simulation was created with the initial values shown in Table 3.

Here S_A and nFS_{Plat} are at the same initial value (0.5), while S_{Plat} has a higher initial value (0.8). The goal is to bring S_A closer to S_{Plat} through the fake state nFS_{Plat}, mimicking a situation where faked bonding by homophily is created by fake state nFS_{Plat}, whereafter social contagion makes it possible for S_A to move towards S_{Plat}. Indeed, it can be seen in Fig. 2 that the bonding is successful; see the blue and green line starting at 0.3 and 0.5 and ending up at 1 around times 200 and 400, respectively.

In Fig. 3, states S_A and nFS_{Plat} start out at the same initial value 0.5, whereafter S_A is pulled towards S_{Plat}. At the same time, S_{Plat} also moves slowly towards S_A, which is caused by the input it receives from S_A. We can conclude from this figure that the model is working as intended for this specific situation. Note that nFS_{Plat} first drops in value because it waits for input from both S_A and S_{Plat}.

Experiment 2

Next, a second simulation was created where user state S_A has a relatively high initial value while S_{Plat} and nFS_{Plat} have relatively low initial values to be able to check if the model will still bring S_A and S_{Plat} together. The initial values are as shown in Table 4.

For clarity, we only display S_A, S_{Plat} and nFS_{Plat} in Fig. 4.

In Fig. 4 we can see user state S_A still being pulled towards S_{Plat} through nFS_{Plat}, meaning that the model is working correctly.

Table 3 Initial values for Experiment 1

X_1	X_2	X_3	X_4	X_5	X_6	X_7	X_8	X_9	X_{10}	X_{11}	X_{12}	X_{13}	X_{14}	X_{15}	X_{16}
0.5	0.8	0.5	0.5	0.5	0.1	0.1	0.3	1	0.5	0.2	0.5	0.1	0.1	0.1	1

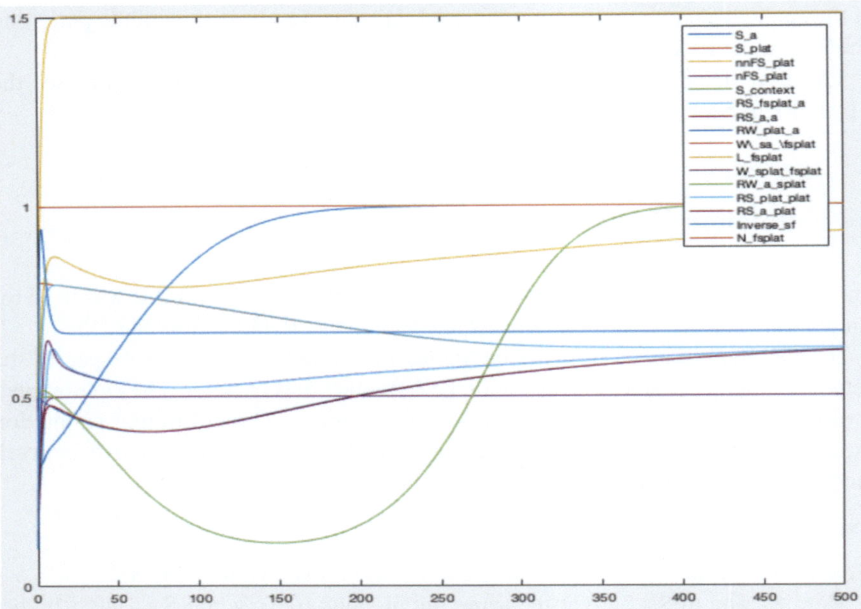

Fig. 2 Simulation results for the first experiment: bonding takes place as both **RW**-states (green and blue line starting at 0.3 and 0.5) representing the connections increase to 1; as a result social contagion becomes stronger

Experiment 3

Lastly a simulation was created where the platform state S_{Plat} itself would only converge very slowly, simulating a situation where the platform does not want to adjust its content to the user S_A. This was done by adjusting the speed factor of the platform S_{Plat} to 0.0001.

In Fig. 5 we can still see the expected result, user state S_A now moves towards the platform more quickly than the platform is moving towards S_A.

5 Mathematical Verification of the Model

Verification of an AI model by mathematical analysis can be used to check correctness of a model; e.g., [3]. In order to analyse whether the implemented model is correct given its design, it is useful to analyse stationary points or equilibria; we follow [12]. A stationary point for state Y occurs if $\mathbf{d}Y(t)/\mathbf{d}t = 0$. For the case of network models described by (1) and (2) from Sect. 2, for nonzero speed factors, the condition $\mathbf{d}Y(t)/\mathbf{d}t = 0$ is equivalent to the following criterion in terms of the network characteristics:

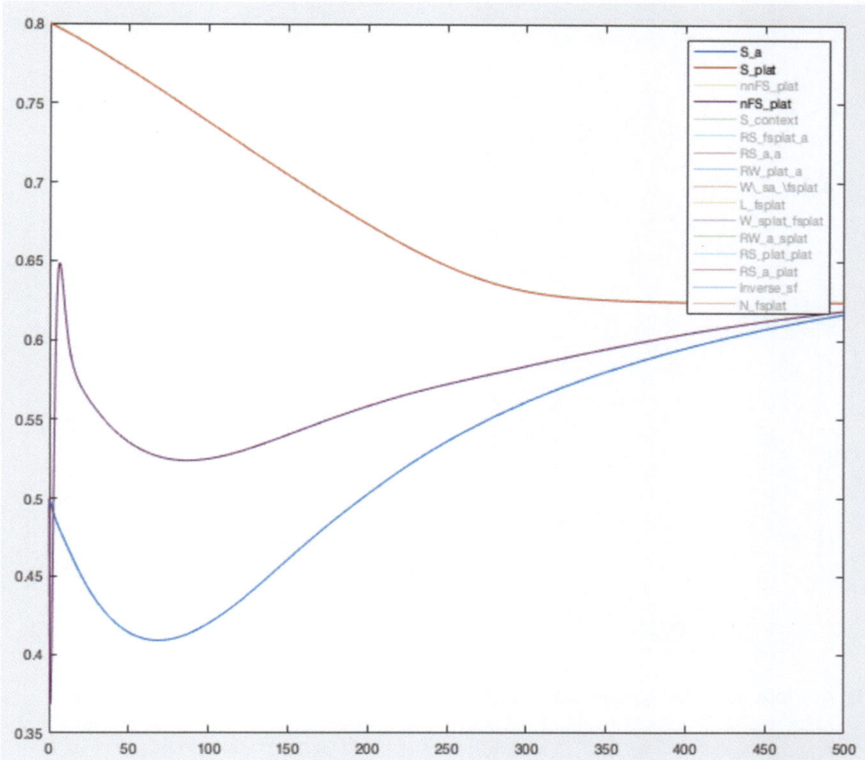

Fig. 3 Close-up of the relevant base states in Experiment 1

Table 4 Initial values for Experiment 2

X_1	X_2	X_3	X_4	X_5	X_6	X_7	X_8	X_9	X_{10}	X_{11}	X_{12}	X_{13}	X_{14}	X_{15}	X_{16}
0.8	0.2	0.5	0.2	0.5	0.1	0.1	0.3	1	0.5	0.2	0.5	0.1	0.1	0.1	1

$$\begin{aligned} \mathbf{aggimpact}_Y(t) &= Y(t) \\ \text{where } \mathbf{aggimpact}_Y(t) &= \mathbf{c}_Y\big(\boldsymbol{\omega}_{X_1,Y}X_1(t), \ldots, \ \boldsymbol{\omega}_{X_k,Y}X_k(t)\big) \end{aligned} \tag{3}$$

The model was mathematically verified by measuring the deviation between the value of a state at a given time point *Y(t)* and the corresponding aggregated impact $\mathbf{aggimpact}_Y\,(t)$. The time point $t = 425$ was chosen due to the fact that the simulation gave stable horizontal lines at that given time, so is expected to be stationary (in equilibrium). The states indicated in the first row of Table 5 were analysed. The following values are shown:

1. The time point $t = 425$

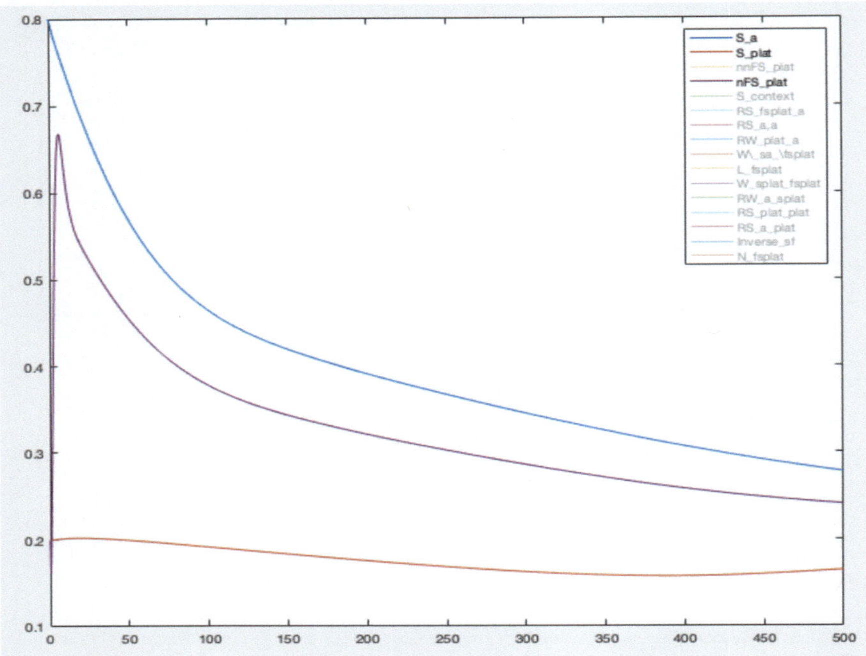

Fig. 4 Close-up of the relevant states in Experiment 2

2. The state value $Y(t)$ at that specific time point
3. The aggregated impact $\mathbf{aggimpact}_Y\ (t)$
4. The deviation $|\mathbf{aggimpact}_Y\ (t) - Y(t)|$

The deviation of the sampled states is at value that is close to 0, which provides evidence that the implemented model is mathematically correct with respect to its design.

6 Discussion

The material used in this chapter was taken from [1]. In this chapter an adaptive network model has been discussed for presenting on a platform a user with content they like to a certain extent. This enables bonding between user and platform. Simulation results have been performed with different scenarios for different starting values of the user and the platform. In all scenarios the platform's so-called fake state, can move towards the state of the user. After the user connects to the platform through bonding by faked homophily it becomes possible to move the user in a desired direction. This conceptual representation of bonding through homophily with a fake state

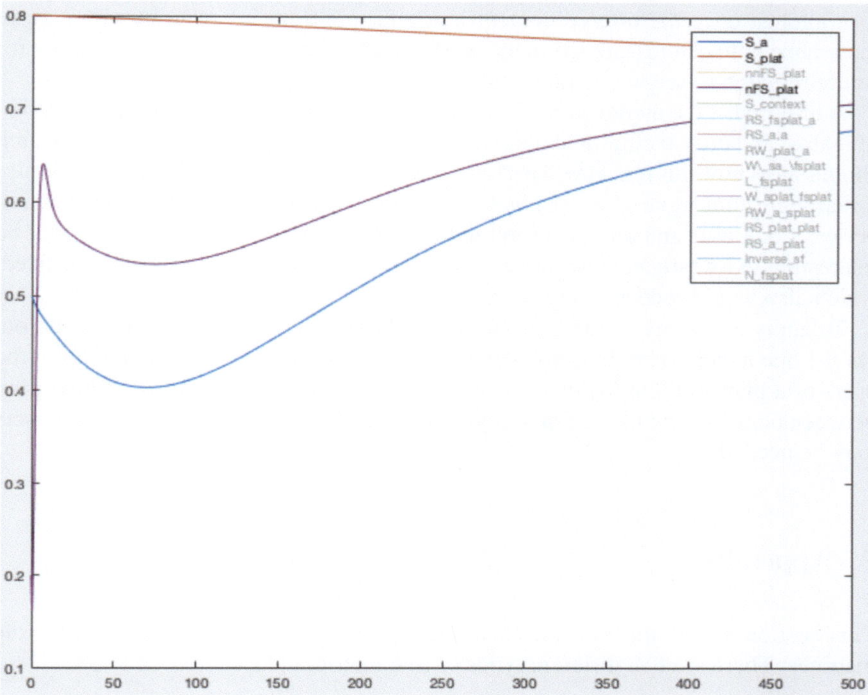

Fig. 5 Close-up of the relevant states in Experiment 3

Table 5 Verification results

State Name Y	S_A	S_{Plat}	nFS_{plat}	$RS_{FSPlat,A}$	$RW_{Plat,A}$
Function	**alogistic**	**alogistic**	**eucl**	**eucl**	**slhom**
Time point	425	425	425	425	425
$Y(t)$	0.601483	0.623936	0.608331	0.608008	1
$\text{Aggimpact}_y(t)$	0.625349	0.618527	0.608650	0.608331	1
Deviation	0.023866	0.005409	0.000320	0.000322	0

shows that bonding is possible even when the created connection is based on fake assumptions. It is not necessary for all parties involved to know what *exactly* made the connection happen, only that similarities are present.

The simulations that were performed all showed user state S_A being lured in by fake platform state nFS_{Plat} towards platform state S_{Plat}. This information can be useful for platforms that want to incorporate a mechanism alike. However, it would also be interesting to research how far off an algorithm can be before a user loses interest and the created bond by faked homophily is broken. Is it possible for a platform to push the user in any direction it wants when a connection is made or will the user lose interest once the offered content is too far off from his own interests? Since these have not been researched in the work reported here, it would be interesting for future research to see where this limit lies.

Additionally, the model created for this chapter necessarily is a simplified version of reality, and no real empirical data were used to validate the model. Future research should show how this model will perform with real data with an algorithm that mimics a real algorithm more closely. Lastly it should be noted that only the connection between one user and an oversimplified mechanism was investigated. It would be interesting to investigate how the model would work when more users are involved, which also would reflect reality more.

To apply a strategy of the type as analysed may need further ethical discussion. On the one hand, it may be considered a form of misleading or manipulating of the users of a platform. On the other hand, it might be considered a form of brokering between platform and users. For applying it, at least a form of transparency for users may be needed.

7 Appendix

This section shows the full specification of the adaptive network model by role matrices. There are five different types of role matrices [11]:

- **mb** for base connectivity (Fig. 6).This specifies on each row for a given state from which states it has incoming connections from the same or lower level
- **mcw** for the role of the connection weights (Fig. 7). This specifies on each row for a given state what the connection weights are for the states indicated in the corresponding cells in the **mb** matrix
- **ms** for the role of the speed factors (Fig. 8). This specifies the speed factor for each state
- **mcfw** for the role of the combination function weights (Fig. 9). This specifies which basic combination functions $\mathbf{c}_Y(..)$ are used for each state Y and with what weights. A nonzero weight indicates that the combination function in question is used for that specific state
- **mcfp** for the role of the combination function parameters (Fig. 10). This specifies for each state Y and each combination function, the parameters of this combination function for state Y.

Note that in the first four rows of the **mcw** matrix an adaptive connection weight is indicated by the label of the self-model state that represents this connection weight shown. For example, the X_8 in the first row for X_1 of role matrix mcw indicates that the weight of the connection as indicated by role matrix **mb** (i.e., from X_4 to X_1 here) for that cell has an adaptive weight which is at time t the value $X_8(t)$ of state X_8 at time t.

base connectivity	mb	1	2	
x1	S_a	x4		
x2	S_plat	x1		
x3	nnFS_plat	x1	x2	
x4	nFS_plat	x3		
x5	S_context	x5		
x6	RS_fsplat_a	x4		
x7	RS_a,a	x1		
x8	RW_plat_a	x6	x7	x8
x9	W_sa_fsplat	x9		
x10	L_fsplat	x9	x11	
x11	W_splat_fsplat	x5		
x12	RW_a_splat	x13	x14	x12
x13	RS_plat_plat	x2		
x14	RS_a_plat	x1		
x15	Inverse_sf	x10		
x16	N_fsplat	x16		

Fig. 6 Role matrix **mb**

mcw	connection weights	1	2	3
x1	S_a	x8		
x2	S_plat	x12		
x3	nnFS_plat	x9	x11	
x4	nFS_plat	x15		
x5	S_context	1		
x6	RS_fsplat_a	1		
x7	RS_a,a	1		
x8	RW_plat_a	1	1	1
x9	W_sa_fsplat	1		
x10	L_fsplat	1	1	
x11	W_splat_fsplat	1		
x12	RW_a_splat	1	1	1
x13	RS_plat_plat	1		
x14	RS_a_plat	1		
x15	Inverse_sf	1		
x16	N_fsplat	1		

Fig. 7 Role matrix **mcw**

msv	speed factor	1
x1	S_a	0.01
x2	S_plat	0.0001
x3	nnFS_plat	0.5
x4	nFS_plat	0.5
x5	S_context	0.5
x6	RS_fsplat_a	0.5
x7	RS_a,a	0.5
x8	RW_plat_a	0.1
x9	W_sa_fsplat	0.5
x10	L_fsplat	0.5
x11	W_splat_fsplat	0.5
x12	RW_a_splat	0.1
x13	RS_plat_plat	0.5
x14	RS_a_plat	0.5
x15	Inverse_sf	0.5
x16	N_fsplat	0

Fig. 8 Role matrix **ms**

mcfw	combination function weights	1 eucl.	2 alogistic	5 homophily	49 inverse identity
x1	S_a		1		
x2	S_plat		1		
x3	nnFS_plat	1			
x4	nFS_plat	1			
x5	S_context	1			
x6	RS_fsplat_a	1			
x7	RS_a,a	1			
x8	RW_plat_a			1	
x9	W_sa_fsplat	1			
x10	L_fsplat	1			
x11	W_splat_fsplat	1			
x12	RW_a_splat			1	
x13	RS_plat_plat	1			
x14	RS_a_plat	1			
x15	Inverse_sf				1
x16	N_fsplat	1			

Fig. 9 Role matrix **mcfw**

mcfp	combination function parameters	1 eucl		2 alogistic		homophily		inverse identity	
				steepness	threshold	modulatie factor	tipping point		
x1	S_a			3	0.25				
x2	S_plat			3	0.25				
x3	nnFS_plat	x16	1						
x4	nFS_plat	1	1						
x5	S_context	1	1						
x6	RS_fsplat_a	1	1						
x7	RS_a,a	1	1						
x8	RW_plat_a					2	0.25		
x9	W_sa_fsplat	1	1						
x10	L_fsplat	x16	1						
x11	W_splat_fsplat	1	1						
x12	RW_a_splat					2	0.25		
x13	RS_plat_plat	1	1						
x14	RS_a_plat	1	1						
x15	Inverse_sf								
x16	N_fsplat	1	1						

Fig. 10 Role matrix **mcfp**

References

1. Bössenecker, G., Mreijen, P., Treur, J.: A second-order adaptive network model for connecting social media platforms and users through bonding by faked homophily. Proced. Comput. Sci. **213**, 32–40 (2022)
2. Currarini, S., Matheson, J., Vega-Redondo, F.: A simple model of homophily in social networks. Eur. Econ. Rev. **90**, 18–39 (2016)
3. Cornelissen, F., Jonker, C.M., Treur, J.: Compositional verification of knowledge-based systems: a case study for diagnostic reasoning. In: Plaza, E., Benjamins, R. (eds.) Knowledge Acquisition, Modeling and Management. Proceedings EKAW 1997. Lecture Notes in Computer Science, vol. 1319, pp. 65–80. Springer, Berlin, Heidelberg (1997)
4. Dehghani, M., Johnson, K., Hoover, J., Sagi, E., Garten, J., Parmar, N.J., Graham, J.: Purity homophily in social networks. J. Exp. Psychol. Gen. **145**(3), 366 (2016)
5. Hendrikse, S.C.F., Treur, J., Koole, S.L. (eds.).: New Analysis and Modeling Directions in Social Interaction Science: Emergent Multimodal Interpersonal Synchrony, Affiliation, and Bonding. Springer Nature (this volume) (2025)

6. Levy, D.A., Nail, P.R.: Contagion: A theoretical and empirical review and reconceptualization. Genetic, social, and general psychology monographs (1993)
7. MATLAB. version 7.10.0 (R2017a).: Natick, Massachusetts: The MathWorks Inc (2017)
8. McPherson, M., Smith-Lovin, L., Cook, J.M.: Birds of a feather: homophily in social networks. Ann. Rev. Sociol. **27**(1), 415–444 (2001)
9. Noë, N., Whitaker, R.M., Chorley, M.J., Pollet, T.V.: Birds of a feather locate together? Foursquare checkins and personality homophily. Comput. Hum. Behav. **58**, 343–353 (2016)
10. Pilipets, E.: From Netflix streaming to Netflix and chill: the (Dis) connected body of serial binge-viewer. Soc Media Soc **5**(4), 2056305119883426 (2019)
11. Treur, J.: Network-oriented modeling for adaptive networks: designing higher-order adaptive biological, mental and social network models. Springer Nature, Cham (2020)
12. Treur, J.: Where is this leading me: stationary point and equilibrium analysis of self-modeling network models. In: Treur, J., and van Ments, L. (eds.), Mental Models and their Dynamics, Adaptation and Control: a Self-Modeling Network Modeling Approach, Ch 18. Springer Nature Switzerland (2022a)
13. Treur, J.: Controlled social network adaptation: subjective elements in an objective social world. In: Yang, XS., Sherratt, S., Dey, N., Joshi, A. (eds.) Proceedings of Sixth International Congress on Information and Communication Technology, ICICT'21. Lecture Notes in Networks and Systems, vol 235, pp 263–274. Springer Nature (2022b)
14. Treur, J.: Modeling coevolution of dynamics, adaptivity and control of adaptivity by self-modeling networks. In: Hendrikse, S.C.F., Treur, J., Koole, S.L. (eds.), New Analysis and Modeling Directions in Social Interaction Science: Emergent Multimodal Interpersonal Synchrony, Affiliation, and Bonding, Ch 2, this volume. Springer Nature (2025a)
15. Treur, J.: A second-order adaptive network model for bonding based on subjectively perceived homophily. In: Hendrikse, S.C.F., Treur, J., Koole, S.L. (eds.), New Analysis and Modeling Directions in Social Interaction Science: Emergent Multimodal Interpersonal Synchrony, Affiliation, and Bonding, Ch 7, this volume. Springer Nature (2025b)

Human Learning, Attunement and Social Bonding Based on Adaptive Homophily: A Multi-Adaptive Network Model

Yelyzaveta Mukeriia, Jan Treur, and Sophie Hendrikse

Abstract This chapter present a multi-adaptive network model integrating multiple adaptation mechanisms, specifically focusing on five types of adaptation. Two of them address first-order adaptation by learning of responding on others and first-order adaptation by bonding with others based on homophily. Three other adaptation mechanisms addressed are second-order adaptation of the speed of both Hebbian learning and bonding by homophily, and second-order adaptation of the homophily tipping point. The latter enables to control the bonding based on the contextual relational situation of the persons. The chapter provides a comprehensive explanation of these concepts and their role in controlled learning within the diverse context scenarios of the chapter.

1 Introduction

The phenomenon of human attunement and social bonding has been a curious topic for researchers from various scientific disciplines for decades. From an early age, humans exhibit a remarkable ability to attune their actions, thoughts, and emotions with others, forming strong social bonds that shape our interactions and relationships. This innate capacity for synchronization and social bonding plays a vital role in our social development, cooperation, and overall well-being. An interesting phenomenon regarding human interaction in daily life activities is our ability to effectively coordinate our actions with each other, especially given that it must occur in a dynamic environment, as nothing around us is constant [37].

Y. Mukeriia · J. Treur (✉)
Social AI Group, Department of Computer Science, Vrije Universiteit Amsterdam, Amsterdam, The Netherlands
e-mail: j.treur@vu.nl

S. Hendrikse
Department of Methodology and Statistics, Tilburg School of Social and Behavioral Sciences, Tilburg University, Tilburg, The Netherlands

S. C. F. Hendrikse et al. (eds.), *New Analysis and Modeling Directions in Social Interaction Science*, Studies in Systems, Decision and Control 614,
https://doi.org/10.1007/978-3-031-99968-0_10

To model these various dynamics of human interaction in real-world scenarios, it is essential to consider both cognitive processes in the brain and interpersonal relationships within social networks. The adopted network-oriented modeling approach in the chapter, serves as a conceptual tool for understanding the structure and dynamics of complex adaptive processes [31, 33]. The presented approach integrates in a multi-level manner two fundamental principles: Hebbian learning for progressive changes in connections between mental states modeling plasticity in the brain [7] and the subjective variant of bonding by homophily principle for adaptive interpersonal connectedness in modeling adaptive social networks [26], see this volume, Chapter 7 [35]. For both fundamental principles a variant for first-order adaptation and one for second-order adaptation is addressed.

The chapter examines how homophily-based adaptation operates in different multi-agent interaction scenarios using a social network model with multiple agents. The objective is on the one hand to understand how the strength of connections between individuals changes based on their similarities, but on the other hand how this can depend on circumstances, in particular on how many or how strong relationships someone already has.

The approach followed is that when individuals are less dissimilar than a tipping point τ, their connection strengthens and otherwise it becomes weaker, as applied in Sharpanskykh and Treur [28], see also Accetto et al. [2], Blankendaal et al. [5], Kozyreva et al. [21], Van den Beukel et al. [36]. Similar to Mukeriia et al. [23], a form of subjective homophily is used, based on perceived and interpreted actions of both agents, see also this volume [13], Chap. 8.

But in contrast this time second-order adaptivity for the homophily criterion is incorporated. To address this, in the current chapter an adaptive tipping point for the homophily is used, which is a form of second-order adaptation over the first-order adaptation described as bonding by homophily. By making the homophily tipping point adaptive, it ensures that it fluctuates in response to changes in how many or how strong relations already exist for the person. Two approaches are explored to model these adaptive tipping points: a simple linear tipping point function that focuses on the average weight of the existing connections of a person (indicating how strong are the connections), and a logistic tipping point function that emphasizes the total sum of the connection weights (indicating how many are they).

2 Background Knowledge

In this section we introduce the main concepts used in this chapter, both about the human processes addressed and about the modelling approach followed. First the two first-order adaptation principles are briefly explained.

2.1 First-Order Adaptation Principles

Two first-order adaptation principles are addressed:

Hebbian Learning Adaptation Principle

Hebbian learning is a type of synaptic plasticity that occurs when there is consistent and correlated activation between two neurons [6]. This learning principle, first presented by Bain [3] and Hebb [7], suggests that if a presynaptic neuron repeatedly activates a postsynaptic neuron while both neurons are active, the connection between them becomes stronger [7, 29]. In simpler form, Hebbian learning describes how neurons strengthen their communication pathway when they fire together frequently. An important element is the causality involved in the simultaneous firing. It is the repeated participation (consistency) of one neuron in the firing activity (causality) of the other that leads to strengthened connectivity [19]. In other words, the focus is on the sustained and causal engagement of one neuron with the firing of another.

Homophily Adaptation Principle

In regards to interpersonal relations, a key feature that has an influence on the level of adaptation is the extent of similarity between the individuals in question [26]. People who are more 'alike' tend to develop stronger connections, rather than the ones who are less 'alike' [33], Chaps. 6 and 13. Such behavior can be seen, for example, in children's interaction. Researchers in the field of social science have observed that school children exhibit a tendency to form friendships and engage in play groups more frequently when they share similarities in demographic characteristics, reinforcing their position within the social space [26]. This type of adaptation can be explained using the homophily principle. According to this principle, the more alike the states of two connected individuals are, the stronger their connection becomes.

2.2 Second-Order Adaptation Principles

In order to apply the above mentioned adaptation principles, it is important to understand the overall structure of the entire model. Adaptive network models can be depicted in a form of a multi-level construction, containing different orders of adaptations on different levels. The base mental network model is depicted at the lower level, with connections between different mental states. As default, these base states have static parameters for weights of connections between each other, which is not a realistic set up if the goal is to mimic a real-world scenario. Thus, in order to make them adaptive, a 'first-order adaptation level' is introduced, modeling first-order connection adaptation principles such as Hebbian learning and bonding by homophily, briefly explained in the Sect. 2.1. Now, these first-order adaptation principles themselves also have parameters by default, and these parameters can in reality also be dynamic, in this case by second-order adaptation. In this chapter, second-order adaptation is applied in three ways: to the adaptation speed both of the Hebbian learning and of the bonding by homophily, and to the tipping point of the homophily adaptation. Such second-order adaptation mechanisms are sometimes described as metaplasticity or 'plasticity of the plasticity' [1, 27, 30].

2.3 The Self-Modeling Modeling Approach Used

The modeling approach used is network-oriented and addresses adaptation of networks by self-modeling networks [31, 33].

Network models

Within a network model, like in neural network models at each time point t, the impact of state X on state Y is depicted through their connections, as

$$\mathbf{impact}_{X,Y}(t) = \boldsymbol{\omega}_{X,Y} X(t) \tag{1}$$

where $\boldsymbol{\omega}_{X,Y}$ is the weight of the connection from X to Y and $X(t)$ the activation value of X at time t. There could be one or multiple connections impacting a single state, and in order to aggregate them, combination functions $\mathbf{c}_Y(..)$ are used

$$\begin{aligned}\mathbf{aggimpact}_Y(t) &= \mathbf{c}_Y(\mathbf{impact}_{X_1,Y}(t), \ldots, \mathbf{impact}_{X_k,Y}(t)) \\ &= \mathbf{c}_Y(\boldsymbol{\omega}_{X_1,Y} X_1(t), \ldots, \boldsymbol{\omega}_{X_k,Y} X_k(t))\end{aligned} \tag{2}$$

This corresponds to activation functions in neural network models.

Lastly, the speed factor $\boldsymbol{\eta}_Y$ is what determines the timing with which the aggegrated impact $\mathbf{aggimpact}_Y(t)$ is exerted on state Y. Thus, the full difference and differential equations are the following [31]:

$$Y(t + \Delta t) = Y(t) + \boldsymbol{\eta}_Y[\mathbf{aggimpact}_Y(t) - Y(t)]\Delta t \tag{3}$$

$$dY(t)/dt = \boldsymbol{\eta}_Y[\mathbf{aggimpact}_Y(t) - Y(t)] \tag{4}$$

The three types of network characteristics $\boldsymbol{\omega}_{X,Y}$, $\mathbf{c}_Y(..)$, $\boldsymbol{\eta}_Y$ defining a *temporal-causal network* in principle are constant, in which case they model a nonadaptive network. However, the realistic case is that they also change over time and therefore the network is adaptive. Self-modelling networks are an easy way to address adaptivity by adding states (called *self-model states*) to the network that represent these adaptive network characteristics. For example, a self-model state $\mathbf{W}_{X,Y}$ can be added to represent a connection weight $\boldsymbol{\omega}_{X,Y}$, or a self-model state $\mathbf{H}_Y$ can be used to represent a speed factor $\boldsymbol{\eta}_Y$; such types of self-model states are also called **W**-states and **H**-states, respectively. In such a case in Eqs. (2), (3) and (4) for the adaptive network characteristics $\boldsymbol{\omega}$ and $\boldsymbol{\eta}$ the values of their corresponding self-model **W**-states and **H**-states are used:

$$\mathbf{aggimpact}_Y(t) = \mathbf{c}_Y(\mathbf{W}_{X_1,Y}(t)X_1(t), \ldots, \mathbf{W}_{X_k,Y}(t)X_k(t)) \tag{5}$$

$$Y(t + \Delta t) = Y(t) + \mathbf{H}_Y(t)[\mathbf{c}_Y(\mathbf{W}_{X_1,Y}(t)X_1(t), \ldots, \mathbf{W}_{X_k,Y}(t)X_k(t)) - Y(t)]\Delta t \tag{6}$$

$$dY(t)/dt = \mathbf{H}_Y(t)[\mathbf{c}_Y(\mathbf{W}_{X_1,Y}(t)X_1(t), \ldots, \mathbf{W}_{X_k,Y}(t)X_k(t)) - Y(t)] \tag{7}$$

This can also be done in a higher-order manner, for example, a second-order self-model state $\mathbf{H}_{\mathbf{W}_Y}$ (also called a $\mathbf{H}_{\mathbf{W}}$-state) can be used to represent the adaptation speed of $\mathbf{W}_{X,Y}$; see Treur [33], Chap. 4 for more details.

Combination functions and their parameters

Next, we will introduce the combination functions used in the chapter, and the individual parts of the chapter will then delve into the application of the functions that are relevant to their models (Treur [33]). The identity function **id(.)** is the combination function that is usually used for the states that have an impacting link from only one state. The numerical representation of the function is

$$\mathbf{c}_Y(V_1, \ldots, V_k) = \mathbf{id}(V_1, \ldots, V_k) = V_1 \tag{8}$$

The scaled maximum function $\mathbf{smax}_{\boldsymbol{\lambda}}$ is used to ensure a smooth transition from one value to another. In this chapter, it is only applied to a self-modeling state for a combination function. Numerically, it is represented as

$$\mathbf{c}_Y(V_1, \ldots, V_k) = \mathbf{smax}_{\boldsymbol{\lambda}}(V_1, \ldots, V_k) = \mathbf{max}(V_1, \ldots, V_k)/\boldsymbol{\lambda} \tag{9}$$

where $\boldsymbol{\lambda}$ is the scaling factor.

The advanced logistic sum function $\mathbf{alogistic}_{\boldsymbol{\sigma},\boldsymbol{\tau}}$, is usually used for the states that have impacting connections from multiple states, and can be expressed as

$$\begin{aligned}\mathbf{c}_Y(V_1, \ldots, V_k) &= \mathbf{alogistic}_{\boldsymbol{\sigma},\boldsymbol{\tau}}(V_1, \ldots, V_k)\\ &= [\frac{1}{1 + \mathrm{e}^{-\sigma(V_1 + \cdots + V_k - \tau)}} - \frac{1}{1 + \mathrm{e}^{\sigma\tau}}](1 + \mathrm{e}^{-\sigma\tau})\end{aligned} \tag{10}$$

where $\boldsymbol{\sigma}$ is a steepness parameter and $\boldsymbol{\tau}$ the excitability threshold parameter.

The Hebbian function $\mathbf{hebb}_{\boldsymbol{\mu}}$**(...)**, is used for adaptation based on Hebbian learning and is expressed for self-model state $\mathbf{W}_{X,Y}$ as

$$\mathbf{c}_{\mathbf{W}_{X,Y}}(V_1, V_2, W) = \mathbf{hebb}_{\boldsymbol{\mu}}(V_1, V_2, W) = V_1 V_2(1 - W) + \boldsymbol{\mu} W \tag{11}$$

where $\mathbf{\mu}$ in [0, 1] is the persistence factor (1 full persistence, 0 no persistence at all). The simple linear homophily function $\mathbf{slhomo}_{\alpha,\tau}(\ldots)$, is applied to the homophily adaptation states $\mathbf{W}_{X,Y}$. This function has two parameters: amplification factor $\boldsymbol{\alpha}$ and the tipping point value $\boldsymbol{\tau}$, and can be expressed as

$$\mathbf{c}_{\mathbf{W}_{X,Y}}(V_1, V_2, W) = \mathbf{slhomo}_{\alpha,\tau}(V_1, V_2, W) = \alpha(\tau - |V_1 - V_2|)W(1 - W) + W \tag{12}$$

The step-once function $\mathbf{steponce}_{\alpha,\beta}(\ldots)$ is used to model changing context factors, and is defined by two parameters: the start point $\boldsymbol{\alpha}$ and the end point $\boldsymbol{\beta}$ with-in the simulation time t, during which the state needs to be active. This function is applied in cases when a state is not needed throughout the whole simulation. In mathematical terms, the function yields a value of 1 if $\boldsymbol{\alpha} \leq t \leq \boldsymbol{\beta}$, else 0. The step-modulo function $\mathbf{stepmod}_{\rho,\delta}(\ldots)$ is used to model behavior with constant intervals, using two parameters: the repeated time duration $\boldsymbol{\rho}$ and the tipping point $\boldsymbol{\delta}$. The function can be expressed as a formula which yields 0 if mod(t, $\boldsymbol{\rho}$) $< \boldsymbol{\delta}$, else 1.

The simple linear tipping point function $\mathbf{sltip}_{\nu,\alpha}(\ldots)$, is used for the homophily adaptive tipping point self-model **TP**-states. It can be expressed as

$$\begin{aligned}\mathbf{c}_{\mathbf{TP}_{\mathbf{W}_{X,Y}}}(W_1, \ldots, W_k, T) &= \mathbf{sltip}_{\nu,\alpha}(W_1, \ldots, W_k, T)\\ &= T + \alpha T(1 - T)\left(\nu - \frac{W_1 + \cdots + W_k}{k}\right)\end{aligned} \tag{13}$$

where $\boldsymbol{\alpha}$ is the modulation factor for the tipping point $\mathbf{TP}_{\mathbf{W}_{X_i,Y}}$ and $\boldsymbol{\nu}$ is the norm for the average incoming connection weights from X_i [33], Chaps. 6 and 13. The logistic tipping point function $\mathbf{alogtip}_{\nu,\alpha}(\ldots)$, is a varied form of the $\mathbf{sltip}_{\nu,\alpha}(\ldots)$ function, also used for the homophily adaptive tipping point self-model **TP**-states. The difference lies in the way the norm $\boldsymbol{\nu}$ is approximated. In this function, the norm $\boldsymbol{\nu}$ is approximated using the function, $\boldsymbol{\nu}$ - $\mathbf{alogistic}_{\sigma,\tau}(V_1, \ldots, V_k)$. Hence, the full function is

$$\begin{aligned}\mathbf{c}_{\mathbf{TP}_{\mathbf{W}_{X,Y}}}(W_1, \ldots, W_k, T) = \mathbf{alogtip}_{\nu,\alpha}(W_1, \ldots, W_k, T) = T + \alpha T(1 - T)\\ \left(\nu - \mathbf{alogistic}_{\sigma,\tau}(W_1, \ldots, W_k)\right)\end{aligned} \tag{14}$$

The parameters for the $\mathbf{alogistic}_{\sigma,\tau}(\ldots)$ function within the $\mathbf{alogtip}_{\nu,\alpha}(\ldots)$ function, are set at 5 for the steepness $\boldsymbol{\sigma}$ and 1.5 for the excitability threshold $\boldsymbol{\tau}$, as they are default values for a sum of W's that is in the range of [0, 3].

For more details of the modeling approach, see this volume, Chapter 2 [34].

3 Design of the Multi-adaptive Network Model

The designed complex multi-adaptive network model is shown in Fig. 1 and explained in Table 1 (base level), Table 2 (first-order self-model level), and Table 3 (second-order self-model level).

Base level processes

The base level comprises mental network models for four distinct individuals, and the social interactions begins upon activation of the world state ws_s, which serves as the triggering signal. When the agents sense the beginning of the social interaction ss_s, they receive a sensory representation of the action srs_a. At the same time, agents initiate their mental models and start interacting with each other individually. Agent A detects interactions from each individual agent via sensor states ss_{e_B}, ss_{e_C}, ss_{e_D}, generates sensory representation states srs_{e_B}, srs_{e_C}, srs_{e_D} for them via states and forms beliefs bs_{e_B}, bs_{e_C}, bs_{e_D} about them. Via a preparation state ps_a, executes the

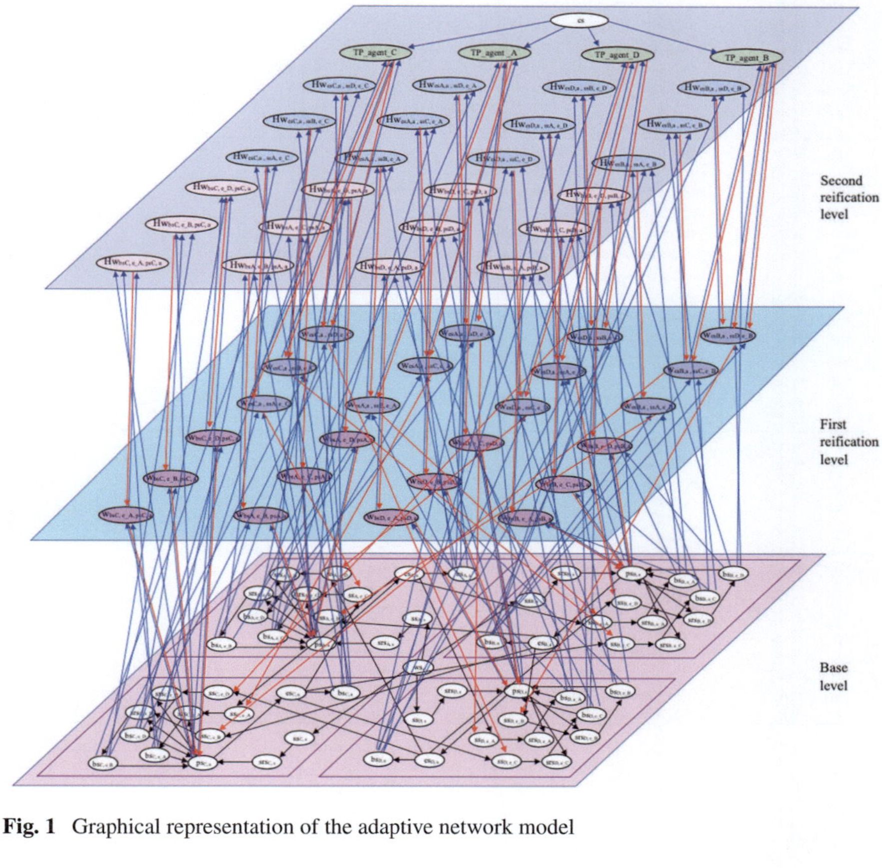

Fig. 1 Graphical representation of the adaptive network model

Table 1 Base level states and their explanation

State nr	State name	Explanation	Level
X_1	ws_s	World state for stimulus s	Base level
X_2	$ss_{A,s}$	Sensor state of agent A for stimulus s	
X_3	$srs_{A,s}$	Sensory representation state of agent A for stimulus s	
X_4	$ps_{A,a}$	Preparation state of agent A for response a	
X_5	ss_{A,e_B}	Sensor state of agent A for stimulus e from Agent B	
X_6	srs_{A,e_B}	Sensory representation state of agent A for stimulus e from Agent B	
X_7	bs_{A,e_B}	Belief state of agent A for stimulus e from Agent B	
X_8	ss_{A,e_C}	Sensor state of agent A for stimulus e from Agent C	
X_9	srs_{A,e_C}	Sensory representation state of agent A for stimulus e from Agent C	
X_{10}	bs_{A,e_C}	Belief state of agent A for stimulus e from Agent C	
X_{11}	ss_{A,e_D}	Sensor state of agent A for stimulus e from Agent D	
X_{12}	srs_{A,e_D}	Sensory representation state of agent A for stimulus e from Agent D	
X_{13}	bs_{A,e_D}	Belief state of agent A for stimulus e from Agent D	
X_{14}	$es_{A,a}$	Execution state of agent A for response a	
X_{15}	$bs_{A,a}$	Belief state of agent A for response a	
X_{16}	$ss_{B,s}$	Sensor state of agent B for stimulus s	
X_{17}	$srs_{B,s}$	Sensory representation state of agent B for stimulus s	
X_{18}	$ps_{B,a}$	Preparation state of agent B for response a	
X_{19}	ss_{B,e_A}	Sensor state of agent B for stimulus e from Agent A	
X_{20}	srs_{B,e_A}	Sensory representation state of agent B for stimulus e from Agent A	
X_{21}	bs_{B,e_A}	Belief state of agent B for stimulus e from Agent B	
X_{22}	ss_{B,e_C}	Sensor state of agent B for stimulus e from Agent C	
X_{23}	srs_{B,e_C}	Sensory representation state of agent B for stimulus e from Agent C	
X_{24}	bs_{B,e_C}	Belief state of agent B for stimulus e from Agent C	
X_{25}	ss_{B,e_D}	Sensor state of agent B for stimulus e from Agent D	
X_{26}	srs_{B,e_D}	Sensory representation state of agent B for stimulus e from Agent D	
X_{27}	bs_{B,e_D}	Belief state of agent B for stimulus e from Agent D	
X_{28}	$es_{B,a}$	Execution state of agent B for response a	
X_{29}	$bs_{B,a}$	Belief state of agent B for response a	
X_{30}	$ss_{C,s}$	World state for stimulus s	
X_{31}	$srs_{C,s}$	Sensor state of agent C for stimulus s	
X_{32}	$ps_{C,a}$	Sensory representation state of agent C for stimulus s	
X_{33}	ss_{C,e_A}	Preparation state of agent C for response a	
X_{34}	srs_{C,e_A}	Sensor state of agent C for stimulus e from Agent A	
X_{35}	bs_{C,e_A}	Sensory representation state of agent C for stimulus e from Agent A	
X_{36}	ss_{C,e_B}	Belief state of agent C for stimulus e from Agent A	
X_{37}	srs_{C,e_B}	Sensor state of agent C for stimulus e from Agent B	
X_{38}	bs_{C,e_B}	Sensory representation state of agent C for stimulus e from Agent B	
X_{39}	ss_{C,e_D}	Belief state of agent C for stimulus e from Agent B	
X_{40}	srs_{C,e_D}	Sensor state of agent C for stimulus e from Agent D	
X_{41}	bs_{C,e_D}	Sensory representation state of agent C for stimulus e from Agent D	
X_{42}	$es_{C,a}$	Execution state of agent C for response a	
X_{43}	$bs_{C,a}$	Belief state of agent C for response a	
X_{44}	$ss_{D,s}$	World state for response s	
X_{45}	$srs_{D,s}$	Sensor state of agent D for stimulus s	
X_{46}	$ps_{D,a}$	Sensory representation state of agent D for stimulus s	
X_{47}	ss_{D,e_A}	Preparation state of agent D for response a	
X_{48}	srs_{D,e_A}	Sensor state of agent D for stimulus e from Agent A	
X_{49}	bs_{D,e_A}	Sensory representation state of agent D for stimulus e from Agent A	
X_{50}	ss_{D,e_B}	Belief state of agent D for response e from Agent A	
X_{51}	srs_{D,e_B}	Sensor state of agent D for stimulus e from Agent B	
X_{52}	bs_{D,e_B}	Sensory representation state of agent D for stimulus e from Agent B	
X_{53}	ss_{D,e_C}	Belief state of agent D for stimulus e from Agent B	
X_{54}	srs_{D,e_C}	Sensor state of agent D for stimulus e from Agent C	
X_{55}	bs_{D,e_C}	Sensory representation state of agent D for stimulus e from Agent C	
X_{56}	$es_{D,a}$	Execution state of agent D for response a	
X_{57}	$bs_{D,a}$	Belief state of agent D for response a	

Table 2 First-order self-model states and their explanation

State nr	State name	Explanation	Level
X_{58}	$\mathbf{W}_{bs\_A,e\_B,ps\_A,a}$	First-order self-model state for weight for $bs_{A,e\_B} \to ps_{A,a}$	First-order self-model level
X_{59}	$\mathbf{W}_{bs\_A,e\_C,ps\_A,a}$	First-order self-model state for weight for $bs_{A,e\_C} \to ps_{A,a}$	
X_{60}	$\mathbf{W}_{bs\_A,e\_D,ps\_A,a}$	First-order self-model state for weight for $bs_{A,e\_D} \to ps_{A,a}$	
X_{61}	$\mathbf{W}_{bs\_B,e\_A,ps\_B,a}$	First-order self-model state for weight for $bs_{B,e\_A} \to ps_{B,a}$	
X_{62}	$\mathbf{W}_{bs\_B,e\_C,ps\_B,a}$	First-order self-model state for weight for $bs_{B,e\_C} \to ps_{B,a}$	
X_{63}	$\mathbf{W}_{bs\_B,e\_D,ps\_B,a}$	First-order self-model state for weight for $bs_{B,e\_D} \to ps_{B,a}$	
X_{64}	$\mathbf{W}_{bs\_C,e\_A,ps\_C,a}$	First-order self-model state for weight for $bs_{C,e\_A} \to ps_{C,a}$	
X_{65}	$\mathbf{W}_{bs\_C,e\_B,ps\_C,a}$	First-order self-model state for weight for $bs_{C,e\_B} \to ps_{C,a}$	
X_{66}	$\mathbf{W}_{bs\_C,e\_D,ps\_C,a}$	First-order self-model state for weight for $bs_{C,e\_D} \to ps_{C,a}$	
X_{67}	$\mathbf{W}_{bs\_D,e\_A,ps\_D,a}$	First-order self-model state for weight for $bs_{D,e\_A} \to ps_{D,a}$	
X_{68}	$\mathbf{W}_{bs\_D,e\_B,ps\_D,a}$	First-order self-model state for weight for $bs_{D,e\_B} \to ps_{D,a}$	
X_{69}	$\mathbf{W}_{bs\_D,e\_C,ps\_D,a}$	First-order self-model state for weight for $bs_{D,e\_C} \to ps_{D,a}$	
X_{70}	$\mathbf{W}_{es\_A,a,ss\_B,e\_A}$	First-order self-model state for weight for $es_{A,a} \to ss_{B,e\_A}$	
X_{71}	$\mathbf{W}_{es\_A,a,ss\_C,e\_A}$	First-order self-model state for weight for $es_{A,a} \to ss_{C,e\_A}$	
X_{72}	$\mathbf{W}_{es\_A,a,ss\_D,e\_A}$	First-order self-model state for weight for $es_{A,a} \to ss_{D,e\_A}$	
X_{73}	$\mathbf{W}_{es\_B,a,ss\_A,e\_B}$	First-order self-model state for weight for $es_{B,a} \to ss_{A,e\_B}$	
X_{74}	$\mathbf{W}_{es\_B,a,ss\_C,e\_B}$	First-order self-model state for weight for $es_{B,a} \to ss_{C,e\_B}$	
X_{75}	$\mathbf{W}_{es\_B,a,ss\_D,e\_B}$	First-order self-model state for weight for $es_{B,a} \to ss_{D,e\_B}$	
X_{76}	$\mathbf{W}_{es\_C,a,ss\_A,e\_C}$	First-order self-model state for weight for $es_{C,a} \to ss_{A,e\_C}$	
X_{77}	$\mathbf{W}_{es\_C,a,ss\_B,e\_C}$	First-order self-model state for weight for $es_{C,a} \to ss_{B,e\_C}$	
X_{78}	$\mathbf{W}_{es\_C,a,ss\_D,e\_C}$	First-order self-model state for weight for $es_{C,a} \to ss_{D,e\_C}$	
X_{79}	$\mathbf{W}_{es\_D,a,ss\_A,e\_D}$	First-order self-model state for weight for $es_{D,a} \to ss_{A,e\_D}$	
X_{80}	$\mathbf{W}_{es\_D,a,ss\_B,e\_D}$	First-order self-model state for weight for $es_{D,a} \to ss_{B,e\_D}$	
X_{81}	$\mathbf{W}_{es\_D,a,ss\_C,e\_D}$	First-order self-model state for weight for $es_{D,a} \to ss_{C,e\_D}$	

(inter)action es_a, and forms a belief state bs_a regarding the level of its own action. The same procedure is followed by the other three agents. All base level states use combination function **id** (when they only have one incoming connection) or **alogistic** (otherwise), see Eqs. (8) and (10), respectivily.

First-order self-model level processes

The first-order self-model level (also called first reification level) models the first-order adaptation processes for the base level processes; see Table 2. Adaptation takes place in two ways: individual adaptivity by Hebbian learning and social adaptivity through bonding by homophily. To model this, for each agent or pair of agents self-model **W**-states X_{58} to X_{69} for Hebbian learning adaptation and self-model **W**-states X_{70} to X_{81} for bonding by homophily adaptation are included. For the Hebbian learning **W**-states, the combination function **hebb** is used and for the bonding by homophily **W**-states the combination function **slhom,** see Eqs. (11) and (12), respectively.

The Hebbian learning states gather information from the bs_e and ps_a states, indicated by blue arrows, while the homophily states receive input from the bs_e and bs_a states. Note that by using belief states as input for the homophily **W**-states, a form of subjective homophily is modeled in contrast to the objective homophily based on execution states that usually is applied in literature such as [2, 5, 21, 28, 36].

Table 3 Second-order self-model states and their explanation

State nr	State name	Explanation	Level
X_{82}	$\mathbf{H}_{\mathbf{W}bs\_A,e\_B,ps\_A,a}$	Second-order self-model state for speed factor for first-order self-model state $\mathbf{W}_{bs\_A,e\_B,ps\_A,a}$	
X_{83}	$\mathbf{H}_{\mathbf{W}bs\_A,e\_C,ps\_A,a}$	Second-order self-model state for speed factor for first-order self-model state $\mathbf{W}_{bs\_A,e\_C,ps\_A,a}$	
X_{84}	$\mathbf{H}_{\mathbf{W}bs\_A,e\_D,ps\_A,a}$	Second-order self-model state for speed factor for first-order self-model state $\mathbf{W}_{bs\_A,e\_D,ps\_A,a}$	
X_{85}	$\mathbf{H}_{\mathbf{W}bs\_B,e\_A,ps\_B,a}$	Second-order self-model state for speed factor for first-order self-model state $\mathbf{W}_{bs\_B,e\_A,ps\_B,a}$	
X_{86}	$\mathbf{H}_{\mathbf{W}bs\_B,e\_C,ps\_B,a}$	Second-order self-model state for speed factor for first-order self-model state $\mathbf{W}_{bs\_B,e\_C,ps\_B,a}$	
X_{87}	$\mathbf{H}_{\mathbf{W}bs\_B,e\_D,ps\_B,a}$	Second-order self-model state for speed factor for first-order self-model state $\mathbf{W}_{bs\_B,e\_D,ps\_B,a}$	
X_{88}	$\mathbf{H}_{\mathbf{W}bs\_C,e\_A,ps\_C,a}$	Second-order self-model state for speed factor for first-order self-model state $\mathbf{W}_{bs\_C,e\_A,ps\_C,a}$	
X_{89}	$\mathbf{H}_{\mathbf{W}bs\_C,e\_B,ps\_C,a}$	Second-order self-model state for speed factor for first-order self-model state $\mathbf{W}_{bs\_C,e\_B,ps\_C,a}$	
X_{90}	$\mathbf{H}_{\mathbf{W}bs\_C,e\_D,ps\_C,a}$	Second-order self-model state for speed factor for first-order self-model state $\mathbf{W}_{bs\_C,e\_D,ps\_C,a}$	
X_{91}	$\mathbf{H}_{\mathbf{W}bs\_D,e\_A,ps\_D,a}$	Second-order self-model state for speed factor for first-order self-model state $\mathbf{W}_{bs\_D,e\_A,ps\_D,a}$	
X_{92}	$\mathbf{H}_{\mathbf{W}bs\_D,e\_B,ps\_D,a}$	Second-order self-model state for speed factor for first-order self-model state $\mathbf{W}_{bs\_D,e\_B,ps\_D,a}$	
X_{93}	$\mathbf{H}_{\mathbf{W}bs\_D,e\_C,ps\_D,a}$	Second-order self-model state for speed factor for first-order self-model state $\mathbf{W}_{bs\_D,e\_C,ps\_D,a}$	First-order self-model level
X_{94}	$\mathbf{H}_{\mathbf{W}es\_A,a,ss\_B,e\_A}$	Second-order self-model state for speed factor for first-order self-model state $\mathbf{W}_{es\_A,a,ss\_B,e\_A}$	
X_{95}	$\mathbf{H}_{\mathbf{W}es\_A,a,ss\_C,e\_A}$	Second-order self-model state for speed factor for first-order self-model state $\mathbf{W}_{es\_A,a,ss\_C,e\_A}$	
X_{96}	$\mathbf{H}_{\mathbf{W}es\_A,a,ss\_D,e\_A}$	Second-order self-model state for speed factor for first-order self-model state $\mathbf{W}_{es\_A,a,ss\_D,e\_A}$	
X_{97}	$\mathbf{H}_{\mathbf{W}es\_B,a,ss\_A,e\_B}$	Second-order self-model state for speed factor for first-order self-model state $\mathbf{W}_{es\_B,a,ss\_A,e\_B}$	
X_{98}	$\mathbf{H}_{\mathbf{W}es\_B,a,ss\_C,e\_B}$	Second-order self-model state for speed factor for first-order self-model state $\mathbf{W}_{es\_B,a,ss\_C,e\_B}$	
X_{99}	$\mathbf{H}_{\mathbf{W}es\_B,a,ss\_D,e\_B}$	Second-order self-model state for speed factor for first-order self-model state $\mathbf{W}_{es\_B,a,ss\_D,e\_B}$	
X_{100}	$\mathbf{H}_{\mathbf{W}es\_C,a,ss\_A,e\_C}$	Second-order self-model state for speed factor for first-order self-model state $\mathbf{W}_{es\_C,a,ss\_A,e\_C}$	
X_{101}	$\mathbf{H}_{\mathbf{W}es\_C,a,ss\_B,e\_C}$	Second-order self-model state for speed factor for first-order self-model state $\mathbf{W}_{es\_C,a,ss\_B,e\_C}$	
X_{102}	$\mathbf{H}_{\mathbf{W}es\_C,a,ss\_D,e\_C}$	Second-order self-model state for speed factor for first-order self-model state $\mathbf{W}_{es\_C,a,ss\_D,e\_C}$	
X_{103}	$\mathbf{H}_{\mathbf{W}es\_D,a,ss\_A,e\_D}$	Second-order self-model state for speed factor for first-order self-model state $\mathbf{W}_{es\_D,a,ss\_A,e\_D}$	
X_{104}	$\mathbf{H}_{\mathbf{W}es\_D,a,ss\_B,e\_D}$	Second-order self-model state for speed factor for first-order self-model state $\mathbf{W}_{es\_D,a,ss\_B,e\_D}$	
X_{105}	$\mathbf{H}_{\mathbf{W}es\_D,a,ss\_C,e\_D}$	Second-order self-model state for speed factor for first-order self-model state Wes_D,a , ss_C, e_D	
X_{106}	$\mathbf{TP}_{Agent\_A}$	Second-order self-model state for the tipping point for first-order self-model **W**-states for bonding by homophily of A	
X_{107}	$\mathbf{TP}_{Agent\_B}$	Second-order self-model state for the tipping point for first-order self-model **W**-states for bonding by homophily of B	
X_{108}	$\mathbf{TP}_{Agent\_C}$	Second-order self-model state for the tipping point for first-order self-model **W**-states for bonding by homophily of C	
X_{109}	$\mathbf{TP}_{Agent\_D}$	Second-order self-model state for the tipping point for first-order self-model **W**-states for bonding by homophily of D	
X_{110}	cs	Context state for how far someone prefers to have bonding	

The impact of the learning states on their related base level states is indicated by red downward arrows: toward ps_a for Hebbian learning and toward ss_e for bonding by homophily.

Second-order self-model level processes

The second-order self-model level (second reification level) models adaptivity of the adaptivity of the base level; see Table 3. It makes the first-order adaptation level itself adaptive as well: second-order adaptation. This is specified by self-model states for network characteristics used for the first-order self-model level. In particular this level contains a context state and the second-order self-model $\mathbf{H_W}$-states for the speed factors for all first-order adaptation **W**-states, as well as the second-order self-model $\mathbf{TP_W}$-states for the adaptive tipping points for the adaptation **W**-states for bonding by homophily. The self-model $\mathbf{H_W}$-states for adaptive adaptation speed are unique to each pair of interactions and receive information from the same base-level states as the learning **W**-states, along with the learning **W**-states themselves. The $\mathbf{H_W}$-states use combination function **alogistic**, see Eq. (10). It expresses the second-order adaptation principle that the speed of adaptation is higher when there is more stimulus exposure [27].

On the other hand, the adaptation principle for the tipping point $\mathbf{TP_W}$-states can be described in terms of the intensity of the existing connections of the individual, modeled by the adaptive **W**-states of the individual. Unlike for the adaptive speed states, in the case of a tipping point self-model state, the base level states are not involved. It receives information only from the **W**-states and from itself, by the means of a persistence link. Lastly, the context state represents how far someone prefers to have bonding, as a personality characteristic. Its link to the adaptive tipping point states will amplify or reduce the tolerance of the agent when adapting to others. Depending on the addressed scenario, the $\mathbf{TP_W}$-states for the tipping points use combination function **sltip** or **alogtip**, see Eqs. (13) and (14), respectively.

3.1 Scenario A: Exclusion of an Agent

The first scenario explored is concerning a social interaction where initially one agent's, in our case agent D's, connection with other agents is too weak, causing an eventual loss of connections. As a result, the agent is separated from the others. This section covers the processes through which such separation effect is achieved.

Connectivity characteristics

As previously mentioned, various connection weights contribute to interactions. Among them are two types of links: within-agent and inter-agent. Within-agent links encompass prediction link, perception link, belief of perceived stimulus link, and belief of own execution link.In the scenario separating agent D, agents A, B, and C possess strong prediction, perception, and belief of perceived stimulus links towards each other, while having very weak such links towards agent D. Specifically, in the model, the links between agents A, B, and C fall within the range of [0.7, 0.85], whereas for agent D, they fall within the range of [0.3, 0.4]. Hence, the links from all agents to agent D are intentionally weakened, posing a challenge for agent D to perceive any incoming information from them. Moreover, the expressive link from D to other agents is also weakened, further hindering the adaptation of other agents towards D. The aim of this scenario is to make agent D a less favorable candidate for connection. Detailed numerical information can be found in Table 4, while the complete connection weight matrix is presented in Table 8.

Aggregation characteristics

The aggregation of connections leading to any given state is shaped by the combination functions and their parameters, serving as the decisive factors. The combination functions used for this scenario include the identity function, alogistic function, simple linear homophily function, Hebbian function, steponce function, and the simple linear tipping point function. Within the combination functions, their parameters are determining factors for the way the causal impacts on a given state are aggregated. For all of the agents, for the ss_s, srs_s, ss_e, es_a, bs_e, and bs_a states the identity function is used, while the srs_e and ps_a states are modeled using the **alogistic** function, with the same steepness $\boldsymbol{\sigma}$ and excitability threshold $\boldsymbol{\tau}$ parameters; the steepness is 20 for all, and the threshold is 0.6 for srs_e and 0.7 for ps_a. The world state ws is modeled using the **stepmod** function, where the repeated time duration $\boldsymbol{\rho}$ of the interaction is set at 50, and the transition point $\boldsymbol{\delta}$ which serves as a breaking moment between interactions is set at 20. These parameters result in 30 units of time of interaction and 20 units of time of break.

The Hebbian learning **W**-states are modeled using their own combination function, **hebb**, and its only parameter, the amplification factor $\boldsymbol{\mu}$, is set to values in range of [0.94, 0.995]. The values are all relatively close to 1, yet are not all the

Table 4 Connection weights for Scenario 1

Agent connection weight			Agent A			Agent B			Agent C			Agent D		
			B	C	D	A	C	D	A	B	D	A	B	C
Within-Agent	Prediction Link	$ps_{X,a} \rightarrow srs_{X,e}$	0.7	0.75	0.4	0.8	0.85	0.3	0.85	0.7	0.35	0.8	0.85	0.7
	Perception Sensitivity Link	$ss_{X,e} \rightarrow srs_{X,e}$	0.7	0.75	0.4	0.8	0.85	0.3	0.85	0.7	0.35	0.8	0.85	0.7
	Belief of Perceived Stimulus Link	$srs_{X,e} \rightarrow bs_{X,e}$	0.7	0.75	0.4	0.8	0.85	0.3	0.85	0.7	0.35	0.8	0.85	0.7
	Belief of Own Execution Link	$es_{X,a} \rightarrow bs_{X,a}$	1			0.9			0.9			0.9		
Inter-Agent	Expressive Link	$ps_{X,a} \rightarrow es_{X,a}$	0.95			0.9			1			0.4		

same to ensure that differences in learning would be visible. The homophily bonding **W**-states use the simple logistic homophily function **slhomo**, where the amplification parameter $\boldsymbol{\sigma}$ is set at 0.2 for all of the homophily **W**-states. The tipping point parameter $\boldsymbol{\tau}$ of the homophily function is adaptive and thus the tipping points follow the $\mathbf{TP_W}$-states. When the tipping point is increased, more similarity is perceived and as a consequence more connections are strengthened by the homophily adaptation, so the average connection weight and total sum of connection weights will become higher compared to the norm $\mathbf{v}$, and vice versa. The speed adaptation $\mathbf{H_W}$-states are modeled using the **alogistic** combination function with the steepness $\boldsymbol{\sigma}$ and excitability threshold $\boldsymbol{\tau}$ being 1.5 and 1 for both the Hebbian and homophily speed adaptation $\mathbf{H_W}$-states.

Lastly, the tipping point adaptation $\mathbf{TP_W}$-states are modeled using two combination functions, simple linear tipping point function and the alogistic tipping point function. The simple linear tipping point function focuses on the average weight of connections in comparison to the norm $\mathbf{v}$ and has parameters 0.6 and 0.4 for the modulation factor $\boldsymbol{\sigma}$ and the norm $\mathbf{v}$, respectively. In how far the agents achieve being close to the precise value for their norm $\mathbf{v}$, can be found by graphing the average trend of homophily learning for each agent, and then taking the mean of each. As can be seen from Fig. 2, the norm lies at around 0.4, for all four agents.

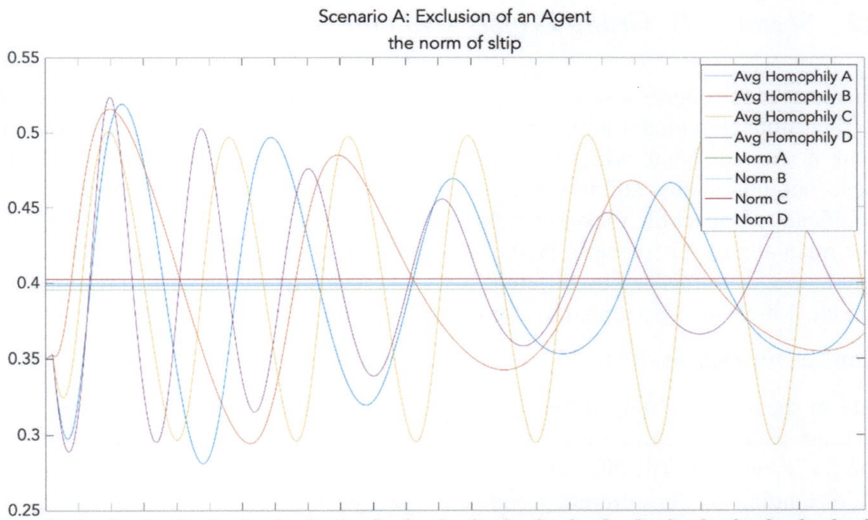

Fig. 2 Scenario A: Norm of **sltip**

The **alogtip** tipping point function, on the other hand, has the same parameter values (0.6 and 0.4) for the modulation factor $\boldsymbol{\alpha}$ and the norm **v**, respectively. The norm **v** for this function, was found using the **alogistic** function, where the function parameters (5 for the steepness $\boldsymbol{\sigma}$ and 1.5 for the excitability threshold $\boldsymbol{\tau}$). The step-size of the simulation was 0.1. All of the values for the function parameters for both functions can be found in Figs. 8, 9, 10, 11.

Timing characteristics

Each state is set at a default speed factor of 10, which defines the speed at which the state changes for the given aggregated impact. In the same manner, for the learning **W**-states X_{58} to X_{81} the adaptive speed factors are indicated (in red in Table 5) from the second-order self-model level $\mathbf{H_W}$-states X_{82} to X_{105}, whose own speed factors are set at 1.3. The speeds of the tipping point adaptation $\mathbf{TP_W}$-states are nonadaptive and have a range of numbers between 0.05 and 1, to explore the naturally occurring varying tipping point outcomes for the agents. Due to the identical timing characteristic shared by both scenarios A and B, it is not necessary to include a dedicated section about the speed factor in Scenario B. This omission is deliberate to avoid redundant repetition in the subsequent section.

3.2 *Scenario B: Group Project Dynamics*

This section introduces a second scenario centered around the dynamic interactions of four students collaborating on a group project. Initially, individuals A and B share a stronger bond, while C and D also possess a closer connection. Each of these four individuals exhibits unique personalities and communication styles, which significantly impact their level of expressiveness and openness to receiving stimuli. The main aim of this scenario is to explore the extent of adaptation throughout the simulation, as influenced by different ways and levels of willingness to respond to stimuli, which are represented by the varying tipping point.

Connectivity characteristics

Due to the scenario only differing in the values of the connection weights, and function parameters, role matrix **mb** for base connectivity remains identical to the one for Scenario 1. The difference lies in the weights of these connections. A clear understanding of the different weights can be found in Table 6. Since agents A and B have a closer bond, they have stronger prediction, perception sensitivity, and belief of perceived link towards each other, while a bit weaker towards the other two agents. The same applies to the pair of agents C and D. The selected values for weights of connections are intentionally non-patterned and exhibit a degree of

Table 5 Role matrix ms specifying the both the adaptive (red cells) and nonadaptive (green cells) speed factors

ms speed factors	
X_{1}	10
X_{2}	10
X_{3}	10
X_{4}	10
X_{5}	10
X_{6}	10
X_{7}	10
X_{8}	10
X_{9}	10
X_{10}	10
X_{11}	10
X_{12}	10
X_{13}	10
X_{14}	10
X_{15}	10
X_{16}	10
X_{17}	10
X_{18}	10
X_{19}	10
X_{20}	10
X_{21}	10
X_{22}	10
X_{23}	10
X_{24}	10
X_{25}	10
X_{26}	10
X_{27}	10
X_{28}	10
X_{29}	10
X_{30}	10
X_{31}	10
X_{32}	10
X_{33}	10
X_{34}	10
X_{35}	10
X_{36}	10
X_{37}	10
X_{38}	10
X_{39}	10
X_{40}	10
X_{41}	10
X_{42}	10
X_{43}	10
X_{44}	10
X_{45}	10
X_{46}	10
X_{47}	10
X_{48}	10
X_{49}	10
X_{50}	10
X_{51}	10
X_{52}	10
X_{53}	10
X_{54}	10
X_{55}	10
X_{56}	10
X_{57}	10
X_{58}	X_{82}
X_{59}	X_{83}
X_{60}	X_{84}
X_{61}	X_{85}
X_{62}	X_{86}
X_{63}	X_{87}
X_{64}	X_{88}
X_{65}	X_{89}
X_{66}	X_{90}
X_{67}	X_{91}
X_{68}	X_{92}
X_{69}	X_{93}
X_{70}	X_{94}
X_{71}	X_{95}
X_{72}	X_{96}
X_{73}	X_{97}
X_{74}	X_{98}
X_{75}	X_{99}
X_{76}	X_{100}
X_{77}	X_{101}
X_{78}	X_{102}
X_{79}	X_{103}
X_{80}	X_{104}
X_{81}	X_{105}
X_{82}	1.3
X_{83}	1.3
X_{84}	1.3
X_{85}	1.3
X_{86}	1.3
X_{87}	1.3
X_{88}	1.3
X_{89}	1.3
X_{90}	1.3
X_{91}	1.3
X_{92}	1.3
X_{93}	1.3
X_{94}	1.3
X_{95}	1.3
X_{96}	1.3
X_{97}	1.3
X_{98}	1.3
X_{99}	1.3
X_{100}	1.3
X_{101}	1.3
X_{102}	1.3
X_{103}	1.3
X_{104}	1.3
X_{105}	1.3
X_{106}	1
X_{107}	0.5
X_{108}	0.1
X_{109}	0.05
X_{110}	10

randomness, as they reflect the natural variations in the strengths of connections that individuals naturally possess. Regarding the belief in their own action links, agents B and D receive stronger values, while agents A and C have slightly weaker values. To ensure a balanced distribution of advantages and disadvantages across all connection links, agents A and C now possess stronger values for the expressive link. The full connection weights matrix can be seen in Tables 7 and 10.

Aggregation characteristics

In contrast to the previous scenario, where the combination function parameters (the modulation factor $\boldsymbol{\alpha}$ and the norm $\mathbf{v}$) remained consistent for both functions (**sltip**

Table 6 Connection weights for Scenario 2

Agent connection weight			Agent A			Agent B			Agent C			Agent D		
			B	C	D	A	C	D	A	B	D	A	B	C
Within-Agent	Prediction Link	$ps_{X,a} \rightarrow srs_{X,e}$	0.95	0.6	0.4	0.9	0.5	0.4	0.35	0.5	0.8	0.2	0.35	0.9
	Perception Sensitivity Link	$ss_{X,e} \rightarrow srs_{X,e}$	0.9	0.7	0.2	0.95	0.6	0.45	0.4	0.2	0.85	0.25	0.6	0.85
	Belief of Perceived Stimulus Link	$srs_{X,e} \rightarrow bs\_{X,e}$	0.85	0.75	0.35	0.9	0.65	0.3	0.5	0.4	0.8	0.4	0.55	0.9
	Belief of Own Execution Link	$es_{X,a} \rightarrow bs_{X,a}$	0.7			0.9			0.6			0.8		
Inter-Agent	Expressive Link	$ps_{X,a} \rightarrow es_{X,a}$	0.9			0.5			0.8			0.4		

and **alogtip**) and across all agents, this scenario introduces a variation of the modulation factor **α** among agents and functions. This distinction aims to further explore the various outcomes that can arise from the varying tipping points. However, it is important to note that the norm **ν** should remain unchanged, as it is either predetermined based on the parameters of the alogistic function or manually determined. For the **sltip** function, the modulation factor **α** is set at 0.8 for agents A and D, and 0.5 for agents B and C. Similarly, for the **alogtip** function, the modulation factor **α** is set at 0.7 for agents A and D, and 0.4 for agents B and C. From the results gathered in Fig. 3, to ensure uniformity and simplicity, a value of 0.4 is assigned as the norm **ν** for all agents. All of the values for the function parameters for both functions can be found in Figs. 8, 10, 12, 13.

4 Simulation Results for Scenario A: Separation of an Agent

This section showcases the simulation results in graphical form, specifically focusing on Scenario A described earlier. The section is divided into two parts: one presents the results obtained using the **sltip** function for the tipping point of homophily, while the other focuses on the **alogtip** function. While the primary emphasis is on the homophily adaptation, the Hebbian learning graphs will not be extensively analyzed here, but can be found in the appendices for reference. As the two tipping point functions model different adaptation principles (average connection vs total sum of connections), the objective is not to compare the effects of the two functions, but rather to highlight the distinct outputs they produce. Therefore, the results of each function will be presented separately, providing a comprehensive understanding of their individual impact.

In this scenario, specific adjustments were made to the connection weight links of the base level states, aiming to discourage Agent D from connecting to others. By gradually weakening selected links, the strength of homophily adaptation between all agents and agent D was significantly reduced. The impact of these weight alterations is visualized in Figs. 4, 5, 6, 7, 8, 9, 10 and 11, where the individual states of each agent exhibit similar patterns across the two functions. However, variations arise in the intensity and frequency of these states, attributable to the distinct characteristics of the **sltip** and **alogtip** functions. The introduction of weak weights for prediction, perception sensitivity, belief of perceived stimulus, and expressive links with agent

Table 7 Role matrix mcw for both adaptive and nonadaptive connection weights at the base level

mcw connection weights					
	1	2	3	4	5
X_1	1				
X_2	1				
X_3	1				
X_4	0.7	X_{58}	X_{59}	X_{60}	
X_5	X_{73}				
X_6	0.9	0.95			
X_7	0.85				
X_8	X_{76}				
X_9	0.7	0.6			
X_{10}	0.75				
X_{11}	X_{79}				
X_{12}	0.2	0.4			
X_{13}	0.35				
X_{14}	0.9				
X_{15}	0.7				
X_{16}	1				
X_{17}	1				
X_{18}	0.7	X_{61}	X_{62}	X_{63}	
X_{19}	X_{70}				
X_{20}	0.95	0.9			
X_{21}	0.95				
X_{22}	X_{77}				
X_{23}	0.6	0.5			
X_{24}	0.65				
X_{25}	X_{80}				
X_{26}	0.45	0.4			
X_{27}	0.3				
X_{28}	0.5				
X_{29}	0.9				

(continued)

Table 7 (continued)

X_{30}	1			
X_{31}	1			
X_{32}	0.7	X_{64}	X_{65}	X_{66}
X_{33}	X_{71}			
X_{34}	0.4	0.35		
X_{35}	0.5			
X_{36}	X_{74}			
X_{37}	0.2	0.5		
X_{38}	0.4			
X_{39}	X_{81}			
X_{40}	0.85	0.8		
X_{41}	0.85			
X_{42}	0.8			
X_{43}	0.6			
X_{44}	1			
X_{45}	1			
X_{46}	0.7	X_{67}	X_{68}	X_{69}
X_{47}	X_{72}			
X_{48}	0.25	0.2		
X_{49}	0.4			
X_{50}	X_{75}			
X_{51}	0.6	0.35		
X_{52}	0.55			
X_{53}	X_{78}			
X_{54}	0.85	0.9		
X_{55}	0.9			
X_{56}	0.4			
X_{57}	0.8			

D results in the agent displaying weak sensory, sensory representation, and belief states from the other agents, as indicated by near-zero or zero curves. Thus, in the base level graphs for agent D, the sensory states drop fairly quickly from the start of the simulation as the agent struggles to attain information from others.

The weakened performance of the base level states poses challenges for achieving effective homophily adaptation between agents, seen in Figs. 12, 13. For the homophily adaptation to have a high performance, the agent is expected to have strong inter-agent links with the agent they are adapting to, and strong expressive link from the agent it wants to receive stimuli from. It is important to note that in the graph, the label "Homophily AB" signifies A's influence on B, or in simpler terms, B adapting to A. When using the **sltip** function, the strongest adaptation occurred in cases where C was adapting to A, C adapting to D, and A adapting to B. This can be attributed to the stronger within-agent link weights and more prominent expressive

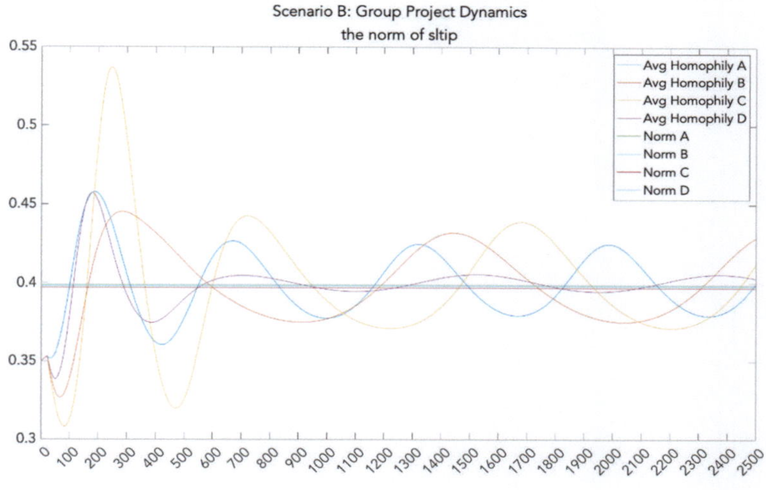

Fig. 3 Scenario B: Norm of sltip

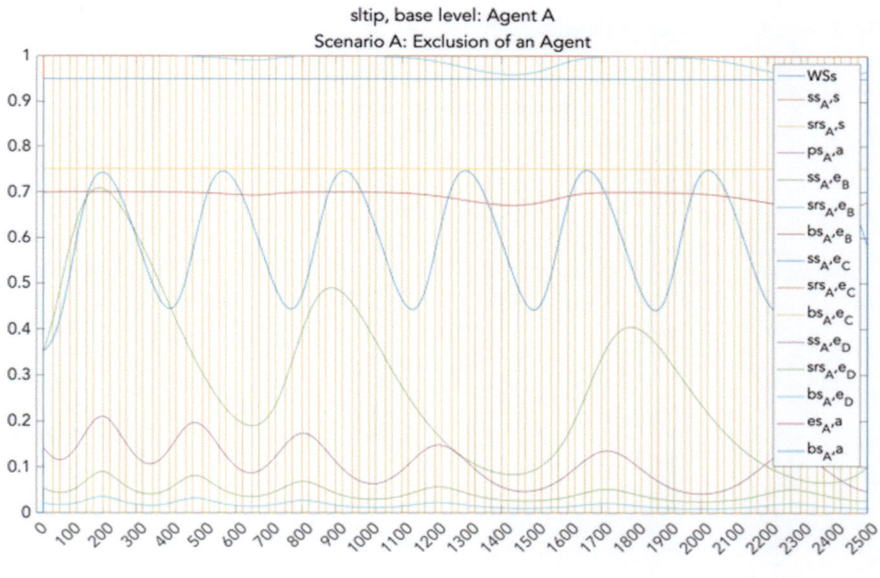

Fig. 4 Base level of Agent A (**sltip**)

link of agent C. Conversely, the weakest adaptation was observed in cases where D was adapting to C, D adapting to B, D adapting to A, and A adapting to B. When using the **alogistic** function, the strongest adaptation was observed in cases where C was adapting to A, C adapting to D, and C adapting to A. On the other hand, the weakest adaptation occurred in cases where D was adapting to C, D adapting to A,

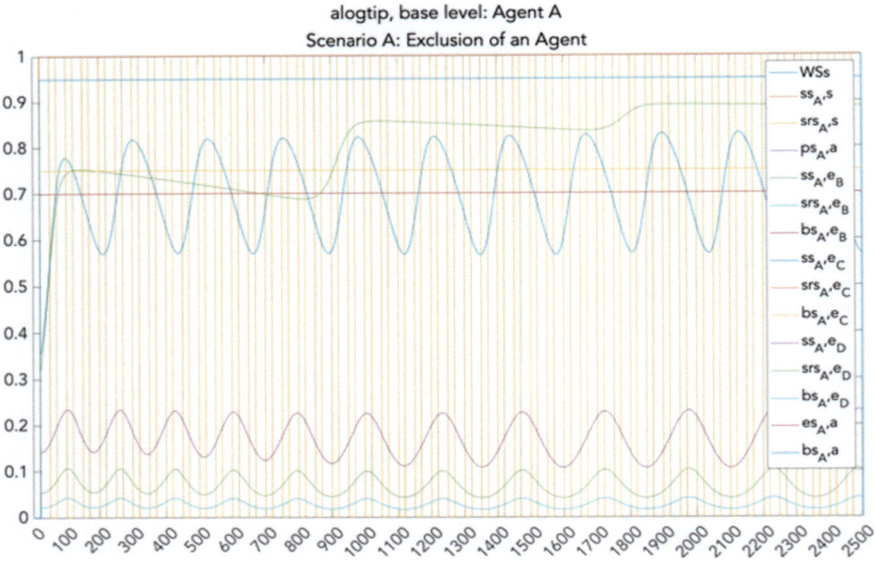

Fig. 5 Base level of Agent A (**alogtip**)

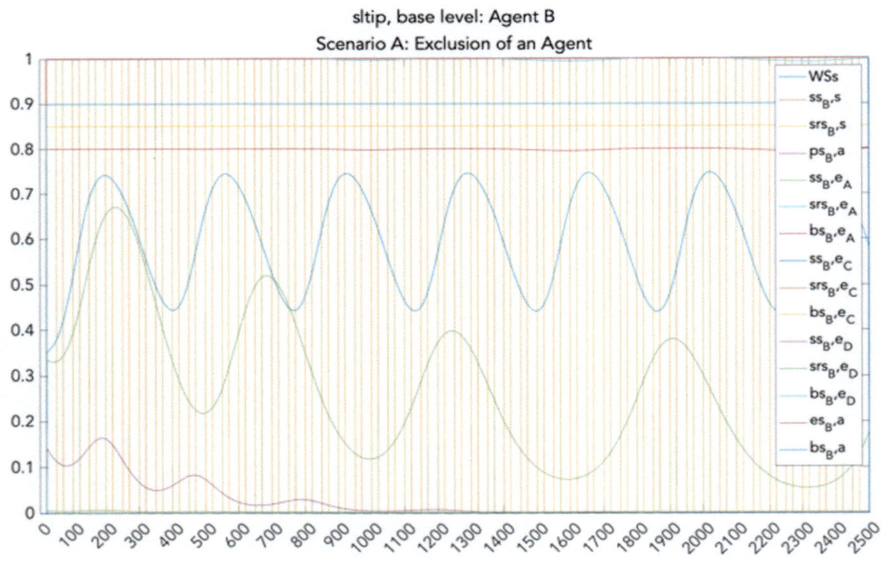

Fig. 6 Base level of Agent B (**sltip**)

Fig. 7 Base level of Agent B (**alogtip**)

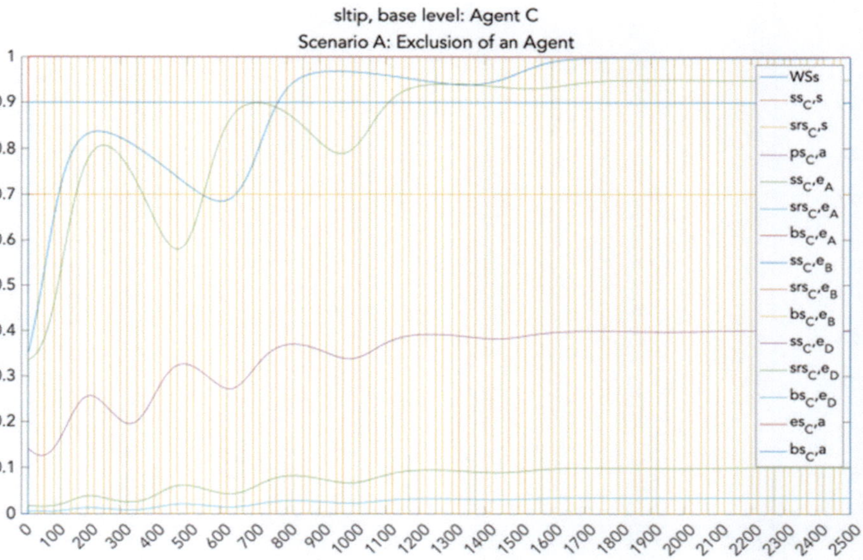

Fig. 8 Base level of Agent C (**sltip**)

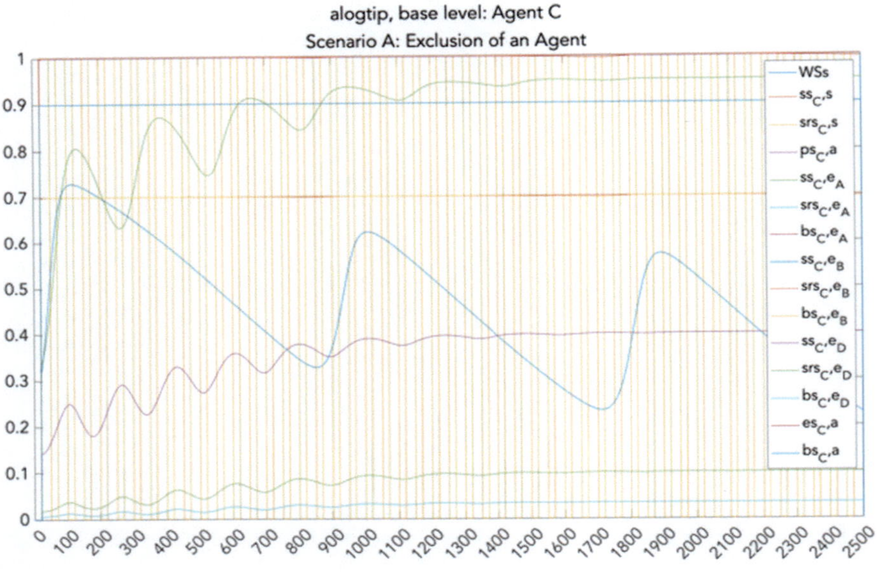

Fig. 9 Base level of Agent C (**alogtip**)

Fig. 10 Base level of Agent D (**sltip**)

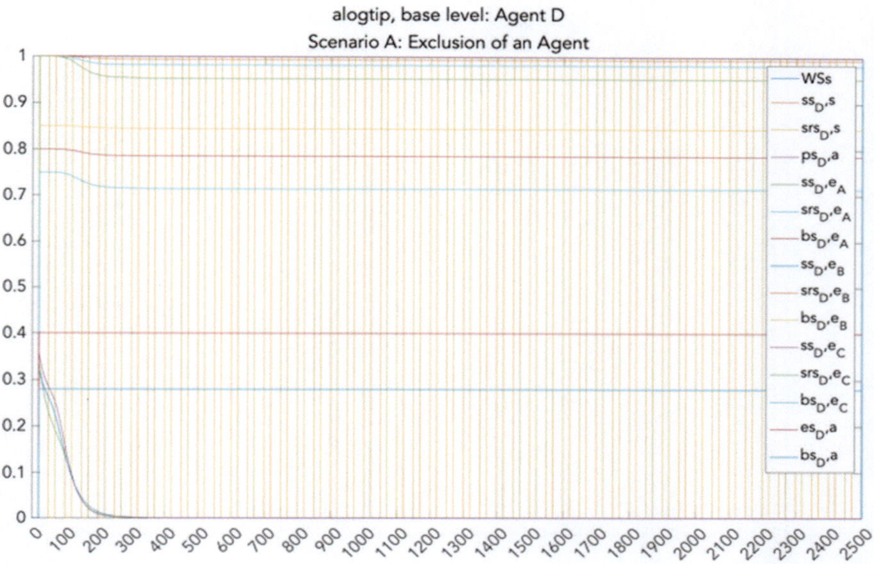

Fig. 11 Base level of Agent D (**alogtip**)

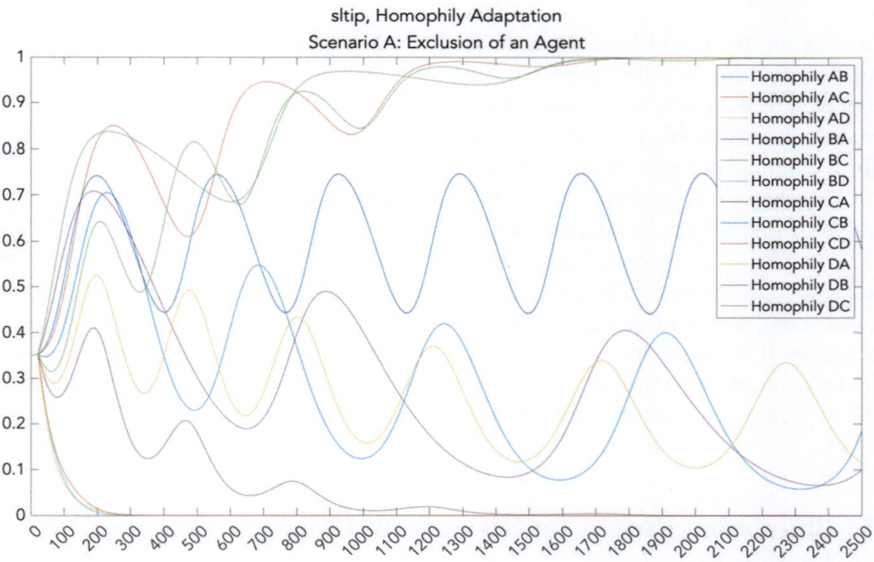

Fig. 12 Homophily adaptation with **sltip**

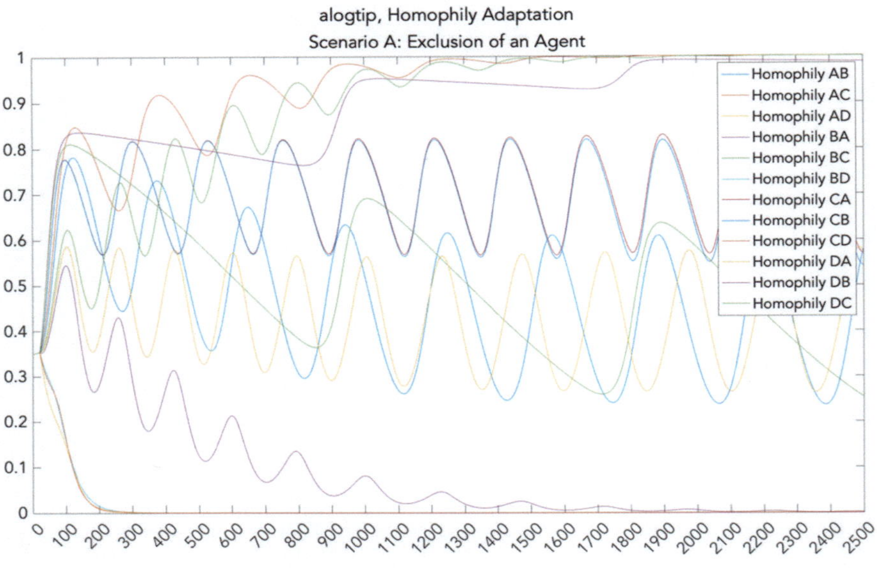

Fig. 13 Homophily adaptation with **alogtip**

D adapting to B, and B adapting to D. In both functions, agent D displayed zero adaptation with all other agents, highlighting how variation in connection weights can lead to the exclusion of an individual. Although agent D could share its information, it was unable to receive any meaningful input due to the weak connections from other agents, preventing it from adapting.

This shows that separation of a person during an initial phase can lead to long-term isolation due to the two adaptation mechanisms considered here.

The $\mathbf{TP_W}$-states corresponding to the tipping point and the speed of adaptations are also represented in a graphical form and are essential in confirming that they are functioning as intended. These graphs are used to validate that the expected behavior is observed, ensuring the reliability of the adaptation graphs' results. The graphs are included in the appendix for readers to reference and gain a better understanding of the expected behavior and effectiveness of the adaptation process.

5 Simulation Results for Scenario B: Group Project Dynamics

The aim of this simulation was to investigate the dynamics of a project group comprising four individuals with distinct connection links. Upon analyzing the base level states graphs in Figs. 14, 15, 16, 17, 18, 19, 20, 21, it is observed that agent A demonstrates a weak sensitivity towards stimuli from agent B, despite the expectation of a strong connection between them. This weak sensitivity can be attributed to agent B's limited expressiveness. Conversely, although the links with agents C and D were intentionally weakened, agent A still obtains information from agent C due to its strong expressive links. On the other hand, Agent B demonstrates a high level of connection strength with Agent A, attributed to their strong connection links and Agent A's high expressiveness. Additionally, Agent B exhibits moderate connection strengths with the other two agents in the group. Interestingly, agent C, who is expected to have a stronger bond with agent D, demonstrates a weak sensory state in response to stimuli from agent D. This can be attributed to the weak expressive link of agent D, which hampers stimulus detection. As for agent D, its sensory state initially rises rapidly but then quickly drops in response to agent C's stimuli, possibly due to the weak connection links established from agent D's weak expressive link. Consequently, as both agents are more interconnected and one of them has limited expressiveness, their ability to share knowledge gradually weakens over time, as evident in the graph for agent D.

The graphs in Figs. 22 and 23, present the final adaptation of each agent based on the given scenario and individual function parameters. Interestingly, notable adaptations were observed in cases where the connection strengths were relatively weak. For example, the adaptation of Agent C to Agent B, despite B having a weak expressiveness and their links being among the weakest, showed significant results. A similar pattern emerged in the adaptation of Agent B to Agent D, where their connections were weak yet exhibited considerable adaptation.

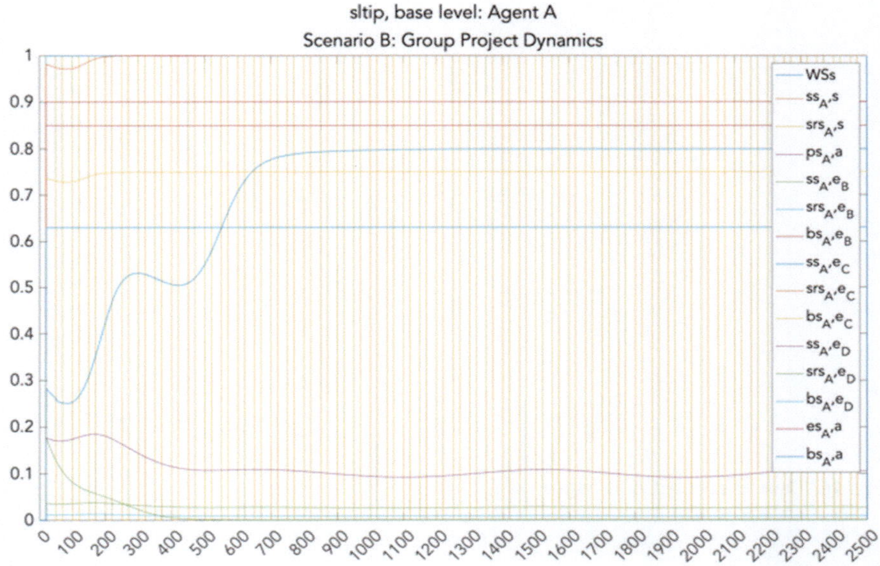

Fig. 14 Base level of Agent A (**sltip**)

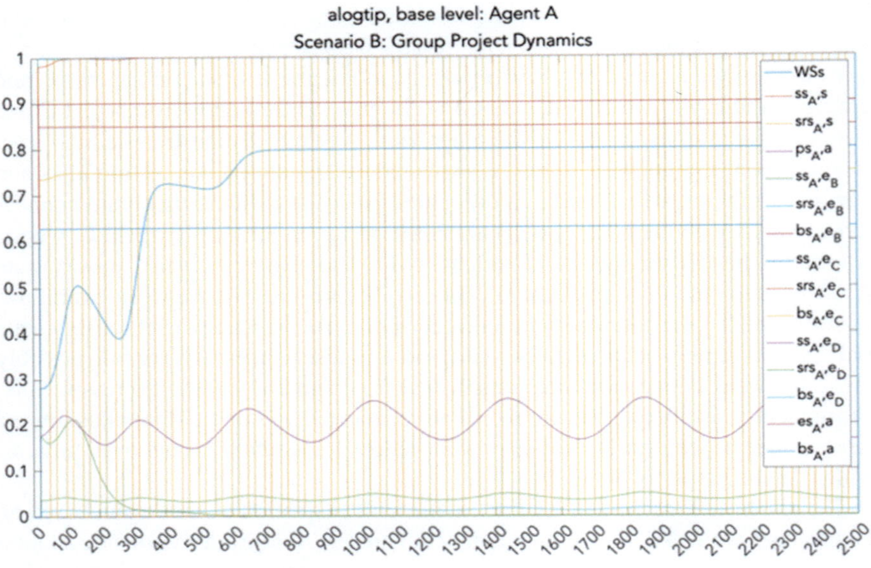

Fig. 15 Base level of Agent A (**alogtip**)

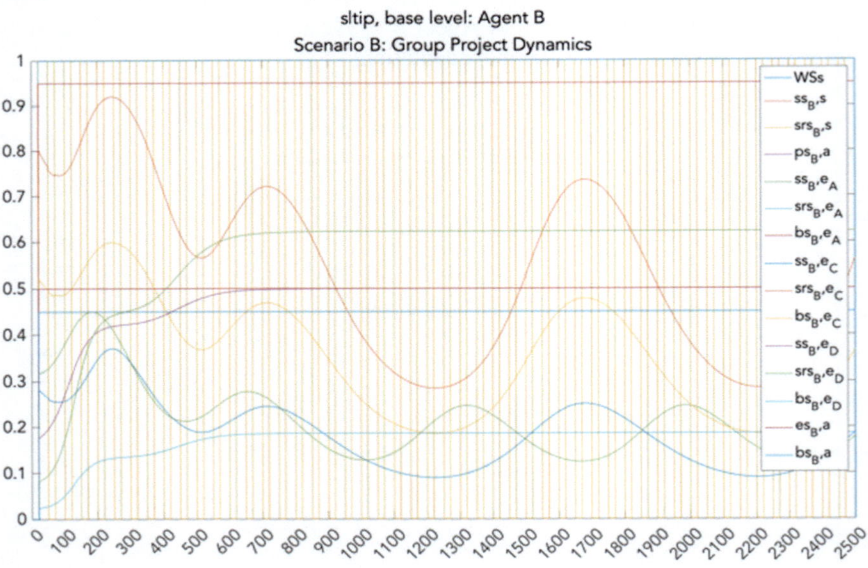

Fig. 16 Base level of Agent B (**sltip**)

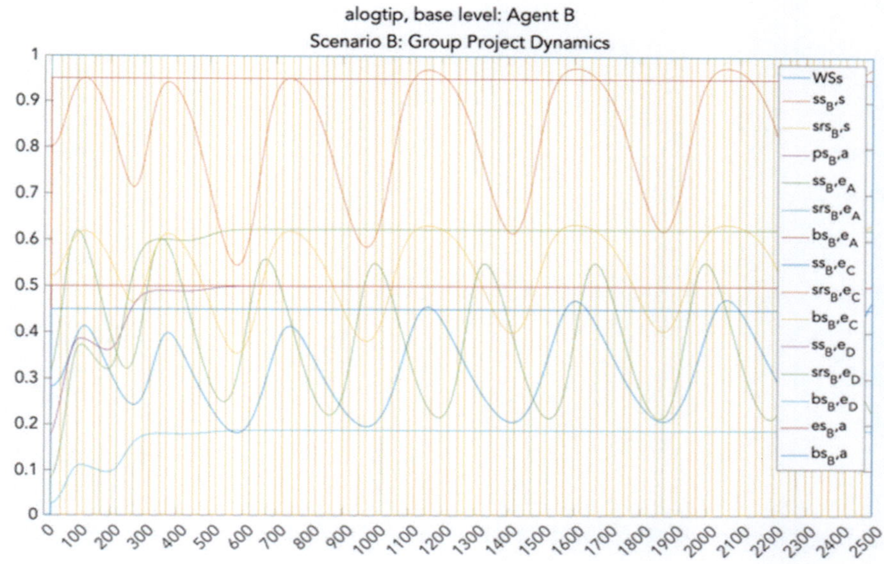

Fig. 17 Base level of Agent B (**alogtip**)

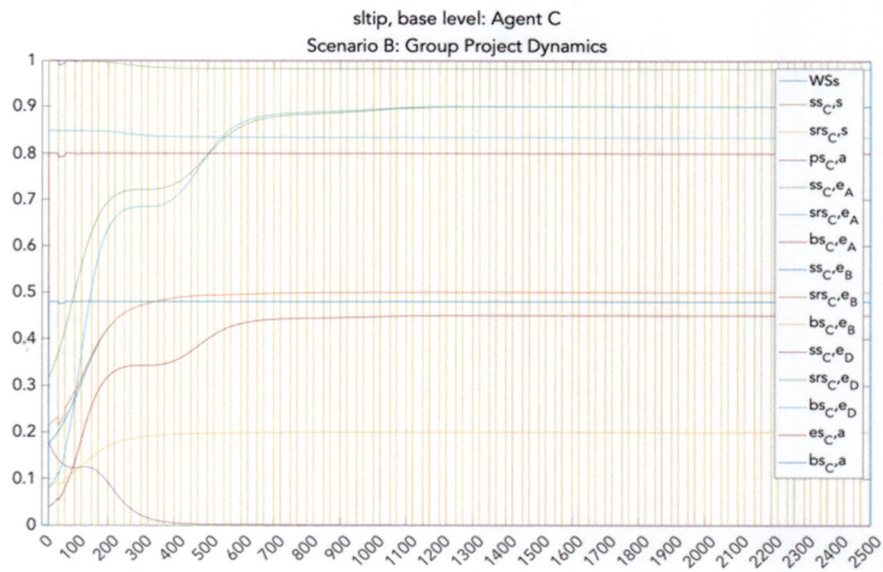

Fig. 18 Base level of Agent C (**sltip**)

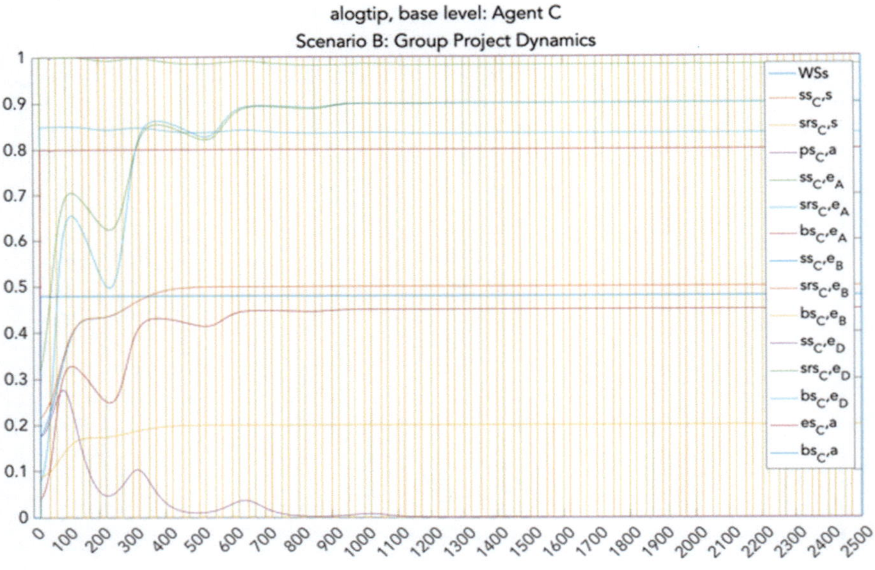

Fig. 19 Base level of Agent C (**alogtip**)

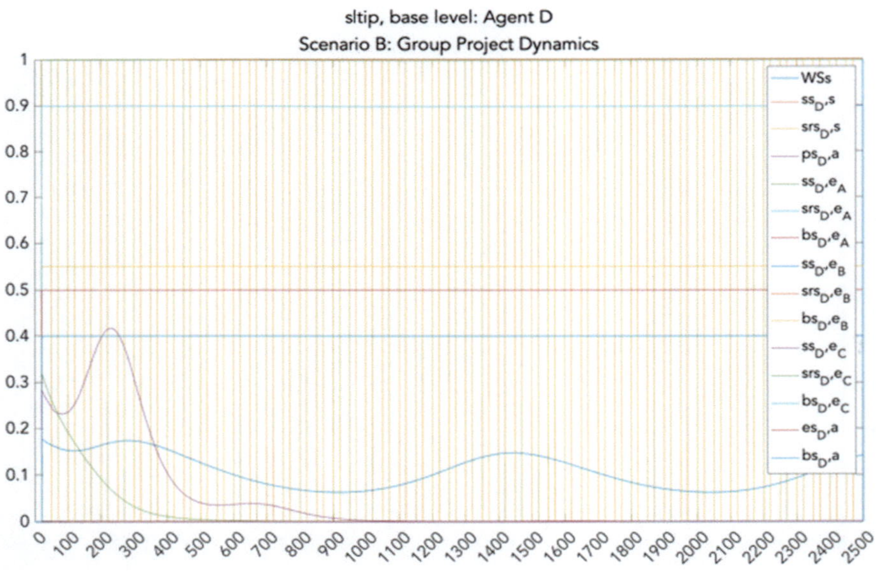

Fig. 20 Base level of Agent D (**sltip**)

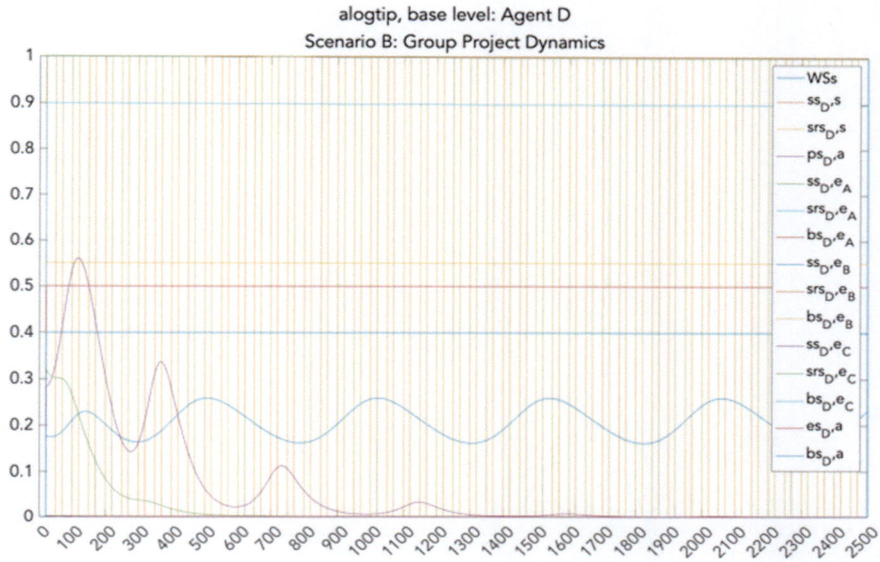

Fig. 21 Base level of Agent D (**alogtip**)

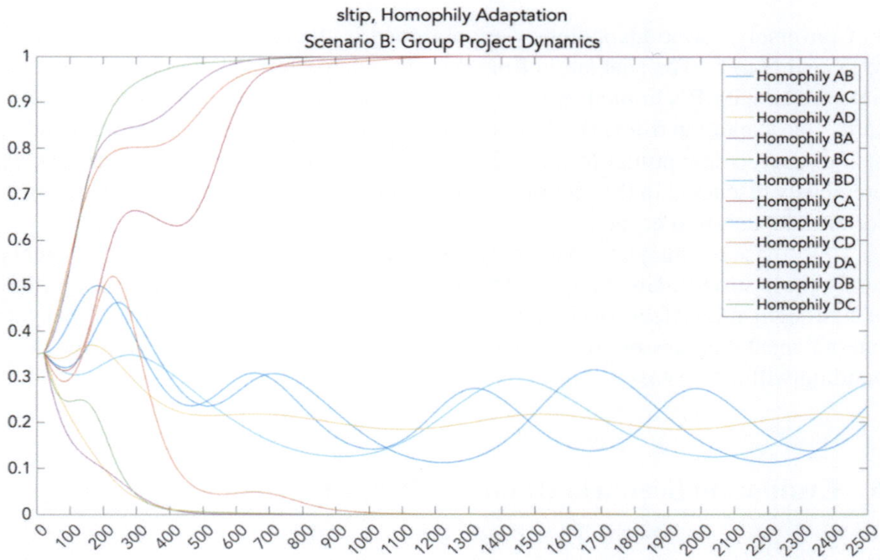

Fig. 22 Homophily adaptation with **sltip**

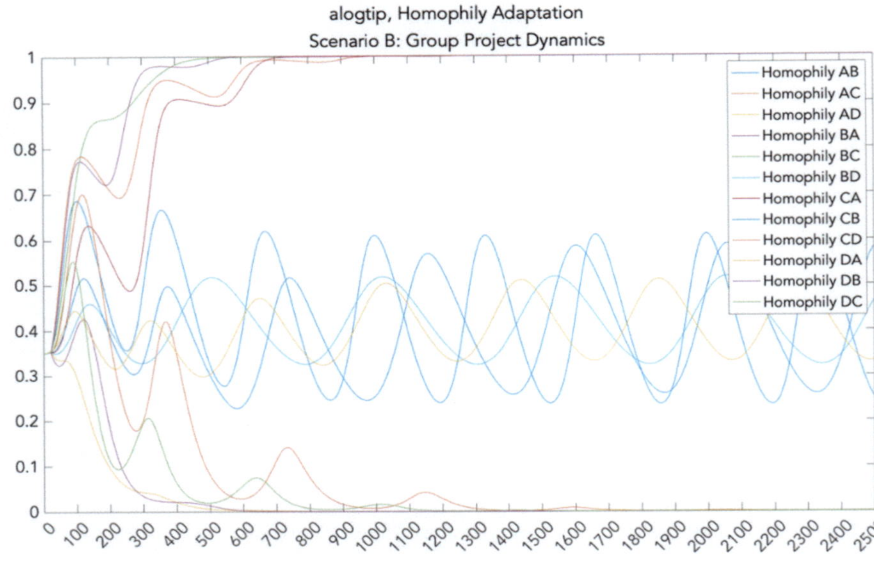

Fig. 23 Homophily adaptation with **alogtip**

Conversely, lower adaptations were observed in the cases of A to B, D to A, C to D, and D to C. The adaptation from A to B underperformed, despite their strong links. Even with B's limited expressiveness, higher adaptation was expected in this case. The adaptation from D to A, where A had strong expressiveness but weak links, also displayed suboptimal results. The weakened connection between C and D, as previously discussed in the section on base level states, posed significant challenges for the adaptation to occur.

Thus, these results yield that strong connection weights of links between agents rarely cause strong adaptation, regardless of how strong the expressive link is, which at first sight might feel as counter intuitive. However, an explanation is that the internal mental processes are decisive for bonding as well. Without their adaptation, bonding will not be easy.

6 Evaluation Scenario B: Group Project Dynamics

Searching for the cause of such unexpected adaptation, the amplification factor function parameter for the tipping point was the first factor that was presumed to have been the cause. Agents A and D, with high amplification factors, exhibited greater variation in tolerance levels. This could explain why Agent A only displayed strong adaptation with Agent C, who had strong expressiveness, while Agent D did not exhibit strong adaptations to any other agent. In order to assess such hypothesis, the

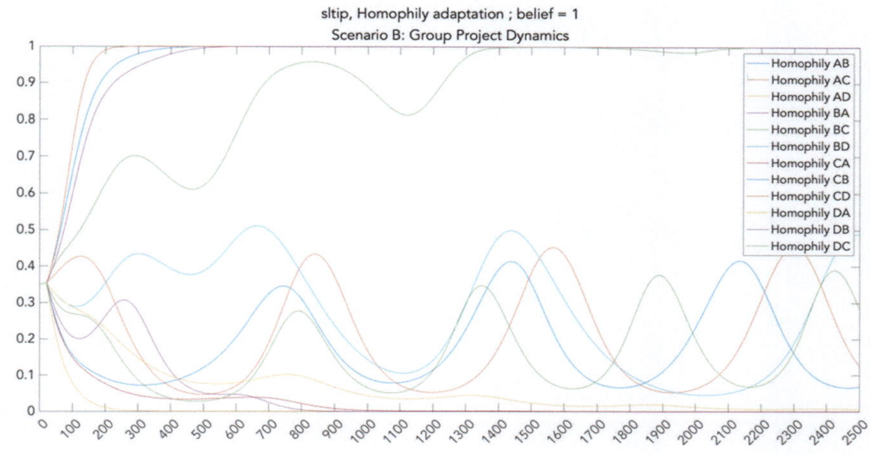

Fig. 24 Homophily adaptation with **alogtip;** belief = 1

alpha parameter for the tipping point of each agent was set to 0.6 for both the **sltip** and **alogtip** functions. Remarkably, this adjustment produced equivalent adaptation performance compared to when different alpha values were used. The sole distinction observed was in the wavelength of the tipping point.

Upon further investigation, it was discovered that the link to the belief of own execution for agents A and C was the factor that played the crucial role in the observed adaptation. Initially lowered to 0.7 and 0.6 in scenario B, these links were subsequently adjusted to a value of 1 for evaluation purposes. As anticipated, this adjustment resulted in the desired outcome, indicating that agents with strong within-agent links and expressive links exhibited the most favorable performance (Figs. 24 and 25).

7 Discussion

This chapter is based on material from Mukeriia et al. [24]. The model presented in this chapter explores two scenarios to investigate the dynamics of bonding by homophily by manipulating connection weights and function parameters. Taking inspiration from Treur [33], Chap. 6 the model in this chapter expands upon it. In addition to incorporating the homophily adaptation and the adaptive tipping point functions, the model contains adaptive speed and Hebbian learning, while exploring both the **sltip** function and the **alogtip** function. This extension enhances the capabilities of the model from Treur [33], Chap. 6, enabling a more comprehensive analysis of the dynamics of homophily adaptation.

In the first scenario, which involved agent exclusion, the findings revealed successful adaptation among agents who possessed strong within-agent links and

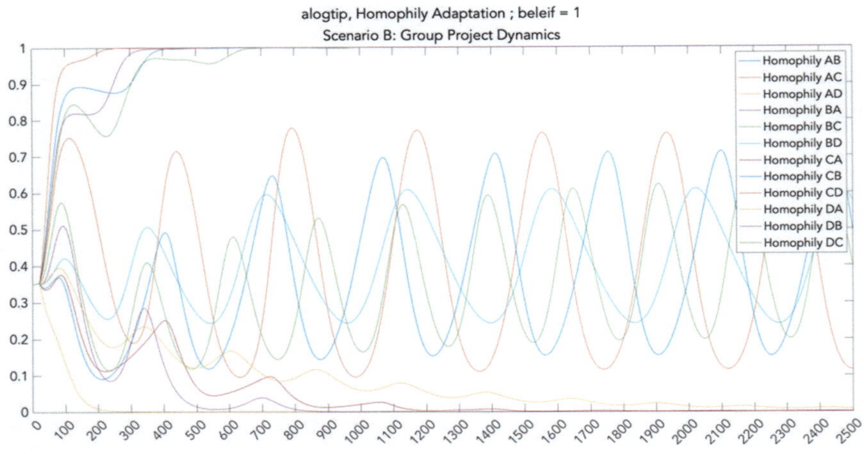

Fig. 25 Homophily adaptation with **alogtip;** belief = 1

expressive links. This supports the understanding that robust connections between agents facilitate effective homophily adaptation. However, the second scenario introduced additional variations, particularly in the amplification factor of the tipping point, with-in agent links and the expressive links between agents. This scenario initially challenged the previously established notion. Contrary to expectations, originally the combination of strong connection links and expressive links did not lead to strong adaptation outcomes in the context of homophily. Initially, it was presumed that the amplification factor was responsible for the unexpected adaptation observed among certain agents. However, upon evaluation, it became evident that this was not the underlying cause. Surprisingly, it was the belief of own execution link that effected the intensity of adaptation. A slightly weakened belief in an agent's own execution hindered the adaptation of the other agent towards it. Consequently, it can be inferred that the most robust adaptation occurs when an agent possesses a strong within-agent link towards another agent with strong expressive links.

These findings highlight the complex nature of bonding by homophily and its ability to manifest in diverse ways, even when certain link configurations appear advantageous. The adaptability of homophily suggests the existence of underlying mechanisms that go beyond simple connectivity patterns, emphasizing the importance of considering multiple factors in understanding and predicting the dynamics of social interactions.

Other work om bonding by homophily can be found in Acetto et al. [2]; Blankendaal et al. [5], Kozyreva et al. [21], Sharpanskykh and Treur [28], Van den Beukel et al. [36]. Here, the second reference combines bonding by homophily with another principle for bonding: the more becomes more principle. The third one considers multicriteria similarity for the homophily. All of these four references do not address the second-order adaptivity of the homophily tipping point and do not consider subjective homophily as is the focus in the current chapter. Moreover, the last three do not integrate bonding by homophily with hebbian learning.

Other related research also addresses bonding in social interaction but not in relation to homophily but to synchronisation (or mimicry) between persons, e.g., empirically focused research such as [4, 20, 22, 25] and analysis and modeling research such as [8, 11, 12, 17], see also this volume [13], Chapters 5, 11, 12, 13 [9, 15, 16, 18]. As also discussed in this volume [13], Chapter 1 and 20 [10, 14], this area has some similarity to the topic of the current chapter, but does not address the approach based on homophily addressed here in relation to literature such as [26]. An important difference is that synchronisation is more complex as it is based on temporal patterns (time series) for activation levels of states that become attuned whereas homophily is based on states with activation levels that become attuned at single time points. It would be an interesting question whether second-order adaptivity in relation to subjective synchrony detection can be modeled more or less similar to the adaptive tipping point for homophily in the current chapter.

A limitation of the current work is that it only considers bonding by homophily and does not integrate other adaptation principles such as more becomes more as addressed in Blankendaal et al. [5], Van den Beukel et al. [36]. This can be work for future research. Another limitation is that homophily was based on single time point comparison of activation levels for similarity. The work of Hendrikse et al. [8, 11, 12, 17] mentioned above points to a (more complex) way to how such similarity can alternatively be based on temporal patterns of activation levels (time series) and synchrony of them.

In future studies using the presented model, an interesting aspect to explore is adjusting the unique values of the link between the context state and the tipping point. This link represents a personal characteristic that reflects an individual's preference for bonding. In this chapter, the value for this link was set to 1 for all agents, ensuring they had the same preference. However, varying this value could create more differences among the agents, allowing a better understanding of how individual preferences influence bonding dynamics in the model.

8 Appendix: Role Matrices and Further Graphs

8.1 Role Matrices

To support reproducibility, in Tables 8, 9, 10, 11, 12, 13 and 14 the role matrices are included used to run the scenarios. They include the settings of all network characteristics used: $\boldsymbol{\omega}$ for connectivity in **mb** and **mcw**, **c** for aggregation in **mcfw** and **mcfp**, and $\boldsymbol{\eta}$ for timing in **ms**.

Table 8 Scenario A: role matrices with connections, connection weights, and initial values

mb base connectivity	1	2	3	4	5
X_1					
X_2	X_1				
X_3	X_2				
X_4	X_3	X_7	X_{10}	X_{13}	
X_5	X_{28}				
X_6	X_5	X_4			
X_7	X_6				
X_8	X_{42}				
X_9	X_8	X_4			
X_{10}	X_9				
X_{11}	X_{56}				
X_{12}	X_{11}	X_4			
X_{13}	X_{12}				
X_{14}	X_4				
X_{15}	X_{14}				
X_{16}	X_1				
X_{17}	X_{16}				
X_{18}	X_{17}	X_{21}	X_{24}	X_{27}	
X_{19}	X_{14}				
X_{20}	X_{19}	X_{18}			
X_{21}	X_{20}				
X_{22}	X_{42}				
X_{23}	X_{22}	X_{18}			
X_{24}	X_{23}				
X_{25}	X_{56}				
X_{26}	X_{25}	X_{18}			
X_{27}	X_{26}				
X_{28}	X_{18}				
X_{29}	X_{28}				
X_{30}	X_1				
X_{31}	X_{30}				
X_{32}	X_{31}	X_{35}	X_{38}	X_{41}	
X_{33}	X_{14}				
X_{34}	X_{33}	X_{32}			
X_{35}	X_{34}				
X_{36}	X_{28}				
X_{37}	X_{36}	X_{32}			
X_{38}	X_{37}				
X_{39}	X_{56}				
X_{40}	X_{39}	X_{32}			
X_{41}	X_{40}				
X_{42}	X_{32}				
X_{43}	X_{42}				
X_{44}	X_1				
X_{45}	X_{44}				
X_{46}	X_{45}	X_{49}	X_{52}	X_{55}	
X_{47}	X_{14}				
X_{48}	X_{47}	X_{46}			
X_{49}	X_{48}				
X_{50}	X_{28}				
X_{51}	X_{50}	X_{46}			
X_{52}	X_{51}				
X_{53}	X_{42}				
X_{54}	X_{53}	X_{46}			

mcwv connection weights	1	2	3	4	5
X_1					
X_2	1				
X_3	1				
X_4	0.7	X_{58}	X_{59}	X_{60}	
X_5	X_{73}				
X_6	0.7	0.7			
X_7	0.7				
X_8	X_{76}				
X_9	0.75	0.75			
X_{10}	0.75				
X_{11}	X_{79}				
X_{12}	0.4	0.4			
X_{13}	0.4				
X_{14}	0.95				
X_{15}	1				
X_{16}	1				
X_{17}	1				
X_{18}	0.7	X_{61}	X_{62}	X_{63}	
X_{19}	X_{70}				
X_{20}	0.8	0.8			
X_{21}	0.8				
X_{22}	X_{77}				
X_{23}	0.85	0.85			
X_{24}	0.85				
X_{25}	X_{80}				
X_{26}	0.3	0.3			
X_{27}	0.3				
X_{28}	0.9				
X_{29}	0.9				
X_{30}	1				
X_{31}	1				
X_{32}	0.7	X_{64}	X_{65}	X_{66}	
X_{33}	X_{71}				
X_{34}	0.85	0.85			
X_{35}	0.7				
X_{36}	X_{74}				
X_{37}	0.7	0.7			
X_{38}	0.7				
X_{39}	X_{81}				
X_{40}	0.35	0.35			
X_{41}	0.35				
X_{42}	1				
X_{43}	0.9				
X_{44}	1				
X_{45}	1				
X_{46}	0.7	X_{67}	X_{68}	X_{69}	
X_{47}	X_{72}				
X_{48}	0.8	0.8			
X_{49}	0.8				
X_{50}	X_{75}				
X_{51}	0.85	0.85			
X_{52}	0.85				
X_{53}	X_{78}				
X_{54}	0.75	0.75			

iv initial values	
X_1	0
X_2	0
X_3	0
X_4	0
X_5	0
X_6	0
X_7	0
X_8	0
X_9	0
X_{10}	0
X_{11}	0
X_{12}	0
X_{13}	0
X_{14}	0
X_{15}	0
X_{16}	0
X_{17}	0
X_{18}	0
X_{19}	0
X_{20}	0
X_{21}	0
X_{22}	0
X_{23}	0
X_{24}	0
X_{25}	0
X_{26}	0
X_{27}	0
X_{28}	0
X_{29}	0
X_{30}	0
X_{31}	0
X_{32}	0
X_{33}	0
X_{34}	0
X_{35}	0
X_{36}	0
X_{37}	0
X_{38}	0
X_{39}	0
X_{40}	0
X_{41}	0
X_{42}	0
X_{43}	0
X_{44}	0
X_{45}	0
X_{46}	0
X_{47}	0
X_{48}	0
X_{49}	0
X_{50}	0
X_{51}	0
X_{52}	0
X_{53}	0
X_{54}	0

(continued)

Table 8 (continued)

X_{55}	X_{54}				
X_{56}	X_{46}				
X_{57}	X_{56}				
X_{58}	X_{7}	X_{4}	X_{58}		
X_{59}	X_{10}	X_{4}	X_{59}		
X_{60}	X_{13}	X_{4}	X_{60}		
X_{61}	X_{21}	X_{18}	X_{61}		
X_{62}	X_{24}	X_{18}	X_{62}		
X_{63}	X_{27}	X_{18}	X_{63}		
X_{64}	X_{35}	X_{32}	X_{64}		
X_{65}	X_{38}	X_{32}	X_{65}		
X_{66}	X_{41}	X_{32}	X_{66}		
X_{67}	X_{49}	X_{46}	X_{67}		
X_{68}	X_{52}	X_{46}	X_{68}		
X_{69}	X_{55}	X_{46}	X_{69}		
X_{70}	X_{7}	X_{15}	X_{70}		
X_{71}	X_{10}	X_{15}	X_{71}		
X_{72}	X_{13}	X_{15}	X_{72}		
X_{73}	X_{21}	X_{29}	X_{73}		
X_{74}	X_{24}	X_{29}	X_{74}		
X_{75}	X_{27}	X_{29}	X_{75}		
X_{76}	X_{35}	X_{43}	X_{76}		
X_{77}	X_{38}	X_{43}	X_{77}		
X_{78}	X_{41}	X_{43}	X_{78}		
X_{79}	X_{49}	X_{57}	X_{79}		
X_{80}	X_{52}	X_{57}	X_{80}		
X_{81}	X_{55}	X_{57}	X_{81}		
X_{82}	X_{7}	X_{4}	X_{58}		
X_{83}	X_{10}	X_{4}	X_{59}		
X_{84}	X_{13}	X_{4}	X_{60}		
X_{85}	X_{21}	X_{18}	X_{61}		
X_{86}	X_{24}	X_{18}	X_{62}		
X_{87}	X_{27}	X_{18}	X_{63}		
X_{88}	X_{35}	X_{32}	X_{64}		
X_{89}	X_{38}	X_{32}	X_{65}		
X_{90}	X_{41}	X_{32}	X_{66}		
X_{91}	X_{49}	X_{46}	X_{67}		
X_{92}	X_{52}	X_{46}	X_{68}		
X_{93}	X_{55}	X_{46}	X_{69}		
X_{94}	X_{7}	X_{15}	X_{70}		
X_{95}	X_{10}	X_{15}	X_{71}		
X_{96}	X_{13}	X_{15}	X_{72}		
X_{97}	X_{21}	X_{29}	X_{73}		
X_{98}	X_{24}	X_{29}	X_{74}		
X_{99}	X_{27}	X_{29}	X_{75}		
X_{100}	X_{35}	X_{43}	X_{76}		
X_{101}	X_{38}	X_{43}	X_{77}		
X_{102}	X_{41}	X_{43}	X_{78}		
X_{103}	X_{49}	X_{57}	X_{79}		
X_{104}	X_{52}	X_{57}	X_{80}		
X_{105}	X_{55}	X_{57}	X_{81}		
X_{106}	X_{70}	X_{71}	X_{72}	X_{106}	X_{110}
X_{107}	X_{73}	X_{74}	X_{75}	X_{107}	X_{110}
X_{108}	X_{76}	X_{77}	X_{78}	X_{108}	X_{110}
X_{109}	X_{79}	X_{80}	X_{81}	X_{109}	X_{110}
X_{110}	X_{110}				

X_{55}	0.75				
X_{56}	0.4				
X_{57}	0.9				
X_{58}	1	1	1		
X_{59}	1	1	1		
X_{60}	1	1	1		
X_{61}	1	1	1		
X_{62}	1	1	1		
X_{63}	1	1	1		
X_{64}	1	1	1		
X_{65}	1	1	1		
X_{66}	1	1	1		
X_{67}	1	1	1		
X_{68}	1	1	1		
X_{69}	1	1	1		
X_{70}	1	1	1		
X_{71}	1	1	1		
X_{72}	1	1	1		
X_{73}	1	1	1		
X_{74}	1	1	1		
X_{75}	1	1	1		
X_{76}	1	1	1		
X_{77}	1	1	1		
X_{78}	1	1	1		
X_{79}	1	1	1		
X_{80}	1	1	1		
X_{81}	1	1	1		
X_{82}	1	1	0.3		
X_{83}	1	1	0.3		
X_{84}	1	1	0.3		
X_{85}	1	1	0.3		
X_{86}	1	1	0.3		
X_{87}	1	1	0.3		
X_{88}	1	1	0.3		
X_{89}	1	1	0.3		
X_{90}	1	1	0.3		
X_{91}	1	1	0.3		
X_{92}	1	1	0.3		
X_{93}	1	1	0.3		
X_{94}	1	1	0.3		
X_{95}	1	1	0.3		
X_{96}	1	1	0.3		
X_{97}	1	1	0.3		
X_{98}	1	1	0.3		
X_{99}	1	1	0.3		
X_{100}	1	1	0.3		
X_{101}	1	1	0.3		
X_{102}	1	1	0.3		
X_{103}	1	1	0.3		
X_{104}	1	1	0.3		
X_{105}	1	1	0.3		
X_{106}	1	1	1	1	1
X_{107}	1	1	1	1	1
X_{108}	1	1	1	1	1
X_{109}	1	1	1	1	1
X_{110}	1				

X_{55}	0
X_{56}	0
X_{57}	0
X_{58}	0.6
X_{59}	0.6
X_{60}	0.6
X_{61}	0.6
X_{62}	0.6
X_{63}	0.6
X_{64}	0.6
X_{65}	0.6
X_{66}	0.6
X_{67}	0.6
X_{68}	0.6
X_{69}	0.6
X_{70}	0.35
X_{71}	0.35
X_{72}	0.35
X_{73}	0.35
X_{74}	0.35
X_{75}	0.35
X_{76}	0.35
X_{77}	0.35
X_{78}	0.35
X_{79}	0.35
X_{80}	0.35
X_{81}	0.35
X_{82}	0
X_{83}	0
X_{84}	0
X_{85}	0
X_{86}	0
X_{87}	0
X_{88}	0
X_{89}	0
X_{90}	0
X_{91}	0
X_{92}	0
X_{93}	0
X_{94}	0
X_{95}	0
X_{96}	0
X_{97}	0
X_{98}	0
X_{99}	0
X_{100}	0
X_{101}	0
X_{102}	0
X_{103}	0
X_{104}	0
X_{105}	0
X_{106}	0.2
X_{107}	0.2
X_{108}	0.2
X_{109}	0.2
X_{110}	1

Table 9 Scenario A and B: role matrices with combination functions with sltip

sltip	mcfwv combination function weights					
	21 id	2 alogistic	5 slhomo	3 hebb	35 stepmod	51 sltip
X_1					1	
X_2	1					
X_3	1					
X_4		1				
X_5	1					
X_6		1				
X_7	1					
X_8	1					
X_9		1				
X_{10}	1					
X_{11}	1					
X_{12}		1				
X_{13}	1					
X_{14}	1					
X_{15}	1					
X_{16}	1					
X_{17}	1					
X_{18}		1				
X_{19}	1					
X_{20}		1				
X_{21}	1					
X_{22}	1					
X_{23}		1				
X_{24}	1					
X_{25}	1					
X_{26}		1				
X_{27}	1					
X_{28}	1					
X_{29}	1					
X_{30}	1					
X_{31}	1					
X_{32}		1				
X_{33}	1					
X_{34}		1				
X_{35}	1					
X_{36}	1					
X_{37}		1				
X_{38}	1					
X_{39}	1					
X_{40}		1				

(continued)

Table 9 (continued)

X_{41}	1			
X_{42}	1			
X_{43}	1			
X_{44}	1			
X_{45}	1			
X_{46}		1		
X_{47}	1			
X_{48}		1		
X_{49}	1			
X_{50}	1			
X_{51}		1		
X_{52}	1			
X_{53}	1			
X_{54}		1		
X_{55}	1			
X_{56}	1			
X_{57}	1			
X_{58}				1
X_{59}				1
X_{60}				1
X_{61}				1
X_{62}				1
X_{63}				1
X_{64}				1
X_{65}				1
X_{66}				1
X_{67}				1
X_{68}				1
X_{69}				1
X_{70}			1	
X_{71}			1	
X_{72}			1	
X_{73}			1	
X_{74}			1	
X_{75}			1	
X_{76}			1	
X_{77}			1	
X_{78}			1	
X_{79}			1	
X_{80}			1	
X_{81}			1	
X_{82}		1		
X_{83}		1		

(continued)

Table 9 (continued)

X_{84}		1				
X_{85}		1				
X_{86}		1				
X_{87}		1				
X_{88}		1				
X_{89}		1				
X_{90}		1				
X_{91}		1				
X_{92}		1				
X_{93}		1				
X_{94}		1				
X_{95}		1				
X_{96}		1				
X_{97}		1				
X_{98}		1				
X_{99}		1				
X_{100}		1				
X_{101}		1				
X_{102}		1				
X_{103}		1				
X_{104}		1				
X_{105}		1				
X_{106}						1
X_{107}						1
X_{108}						1
X_{109}						1
X_{110}	1					

9 Additional Simulation Graphs

For the sake of reproducibility, more detailed graphs are included.

Scenario A Graphs (Figs. 26, 27, 28, 29, 30 and 31).

Scenario B Graphs (Figs. 32, 33, 34, 35, 36 and 37).

Table 10 Scenario A: role matrices with combination function parameters with sltip

sltip	mcfwv combination function weights											
	21 id		2 alogistic		5 slhomo		3 hebb		35 stepmod		51 sltip	
	-	-	σ	τ	α	τ	μ	-	α	β	α	ν
X_1									50	20		
X_2												
X_3												
X_4			20	0.7								
X_5												
X_6			20	0.6								
X_7												
X_8												
X_9			20	0.6								
X_{10}												
X_{11}												
X_{12}			20	0.6								
X_{13}												
X_{14}												
X_{15}												
X_{16}												
X_{17}												
X_{18}			20	0.7								
X_{19}												
X_{20}			20	0.6								
X_{21}												
X_{22}												
X_{23}			20	0.6								
X_{24}												
X_{25}												
X_{26}			20	0.6								
X_{27}												
X_{28}												
X_{29}												
X_{30}												
X_{31}												
X_{32}			20	0.7								
X_{33}												
X_{34}			20	0.6								
X_{35}												
X_{36}												
X_{37}			20	0.6								
X_{38}												
X_{39}												
X_{40}			20	0.6								
X_{41}												

(continued)

Table 10 (continued)

X_{42}											
X_{43}											
X_{44}											
X_{45}											
X_{46}		20	0.7								
X_{47}											
X_{48}		20	0.6								
X_{49}											
X_{50}											
X_{51}		20	0.6								
X_{52}											
X_{53}											
X_{54}		20	0.6								
X_{55}											
X_{56}											
X_{57}											
X_{58}						0.94					
X_{59}						0.945					
X_{60}						0.95					
X_{61}						0.955					
X_{62}						0.96					
X_{63}						0.965					
X_{64}						0.97					
X_{65}						0.975					
X_{66}						0.98					
X_{67}						0.985					
X_{68}						0.99					
X_{69}						0.995					
X_{70}				0.1	X_{106}						
X_{71}				0.1	X_{106}						
X_{72}				0.1	X_{106}						
X_{73}				0.1	X_{107}						
X_{74}				0.1	X_{107}						
X_{75}				0.1	X_{107}						
X_{76}				0.1	X_{108}						
X_{77}				0.1	X_{108}						
X_{78}				0.1	X_{108}						
X_{79}				0.1	X_{109}						
X_{80}				0.1	X_{109}						
X_{81}				0.1	X_{109}						
X_{82}		1.5	1								
X_{83}		1.5	1								
X_{84}		1.5	1								
X_{85}		1.5	1								
X_{86}		1.5	1								

(continued)

Table 10 (continued)

X_{87}			1.5	1								
X_{88}			1.5	1								
X_{89}			1.5	1								
X_{90}			1.5	1								
X_{91}			1.5	1								
X_{92}			1.5	1								
X_{93}			1.5	1								
X_{94}			1.5	1								
X_{95}			1.5	1								
X_{96}			1.5	1								
X_{97}			1.5	1								
X_{98}			1.5	1								
X_{99}			1.5	1								
X_{100}			1.5	1								
X_{101}			1.5	1								
X_{102}			1.5	1								
X_{103}			1.5	1								
X_{104}			1.5	1								
X_{105}			1.5	1								
X_{106}											0.6	0.4
X_{107}											0.6	0.4
X_{108}											0.6	0.4
X_{109}											0.6	0.4
X_{110}												

Table 11 Scenario A and B: role matrices with combination functions for alogtip

alogtip	mcfwv combination function weights					
	21 id	2 alogistic	5 slhomo	3 hebb	35 step-mod	52 alogtip
X_{106}						1
X_{107}						1
X_{108}						1
X_{109}						1

Table 12 Scenario A: role matrices with combination function parameters for alogtip

alogtip	mcfwv combination function weights											
	21 id		2 alogistic		5 slhomo		3 hebb		25 stepmod		52 alogtip	
	-	-	σ	τ	α	τ	μ	-	α	β	α	ν
X_{106}											0.6	0.4
X_{107}											0.6	0.4
X_{108}											0.6	0.4
X_{109}											0.6	0.4

Table 13 Scenario B: role matrices with combination function weight for sltip

sltip	**mcfwv combination function weights**											
	21 id		**2 alogistic**		**5 slhomo**		**3 hebb**		**35 stepmod**		**51 sltip**	
	-	-	σ	τ	α	τ	μ	-	α	β	α	ν
X_{106}											0.8	0.4
X_{107}											0.5	0.4
X_{108}											0.5	0.4
X_{109}											0.8	0.4

Table 14 Scenario B: role matrices with combination function weight for alogtip

alogtip	**mcfwv combination function weights**											
	21 id		**2 alogistic**		**5 slhomo**		**3 hebb**		**35 stepmod**		**52 alogtip**	
	-	-	σ	τ	α	τ	μ	-	α	β	α	ν
X_{106}											0.7	0.4
X_{107}											0.4	0.4
X_{108}											0.4	0.4
X_{109}											0.7	0.4

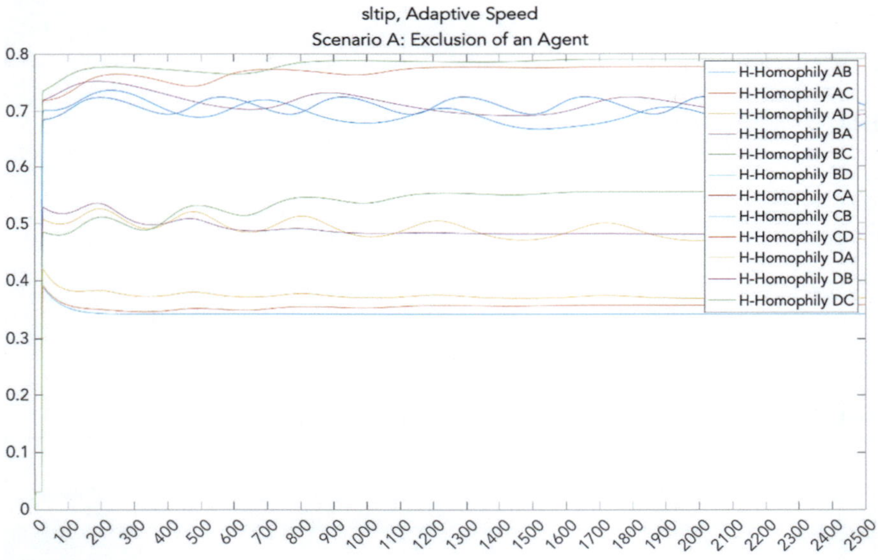

Fig. 26 Scenario A: Homophily adaptive speed with **sltip**

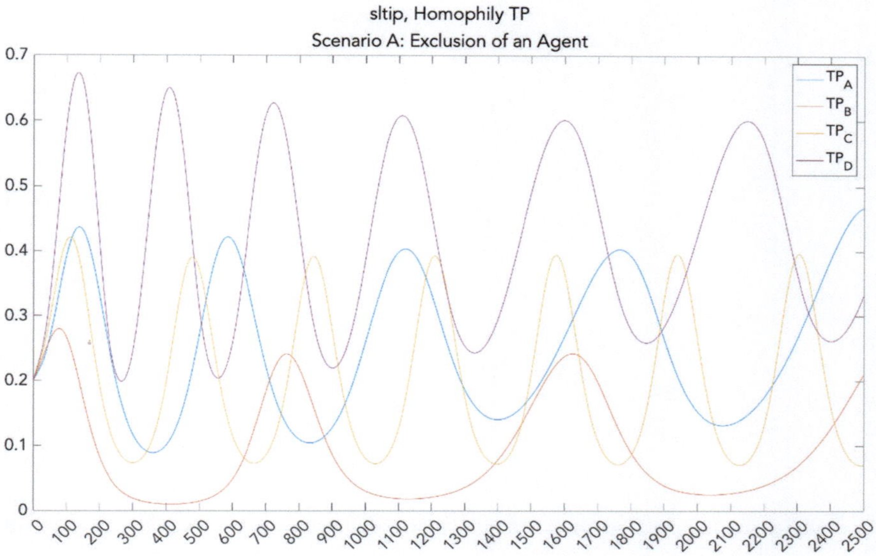

Fig. 27 Scenario A: Homophily Tipping Point with **sltip**

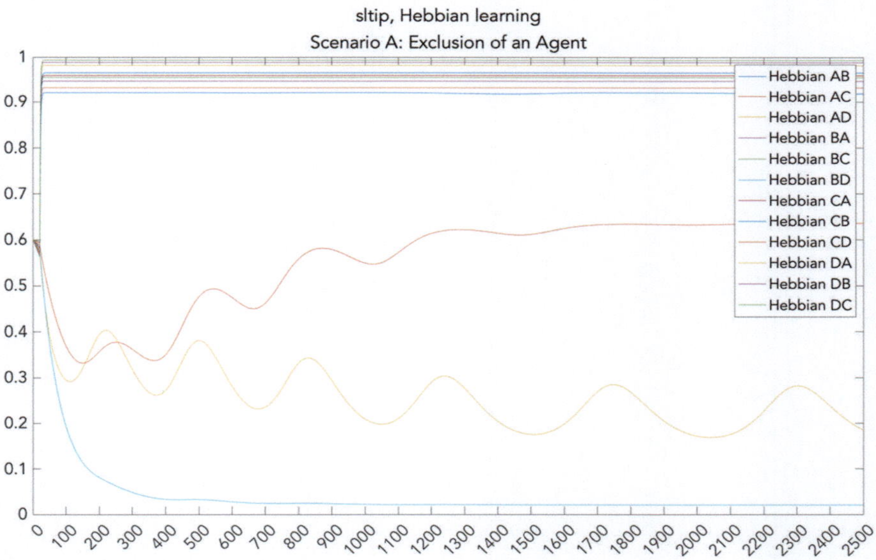

Fig. 28 Scenario A: Hebbian learning with **sltip**

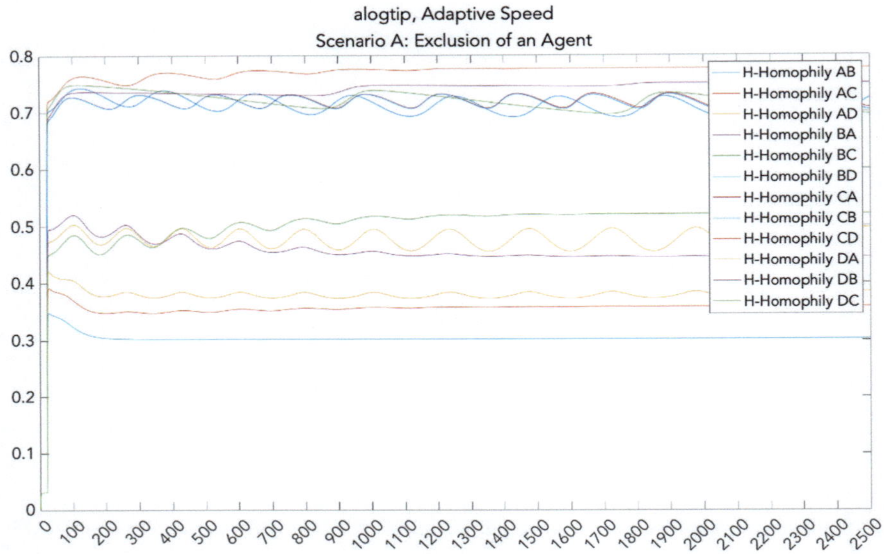

Fig. 29 Scenario A: Homophily adaptive speed with **alogtip**

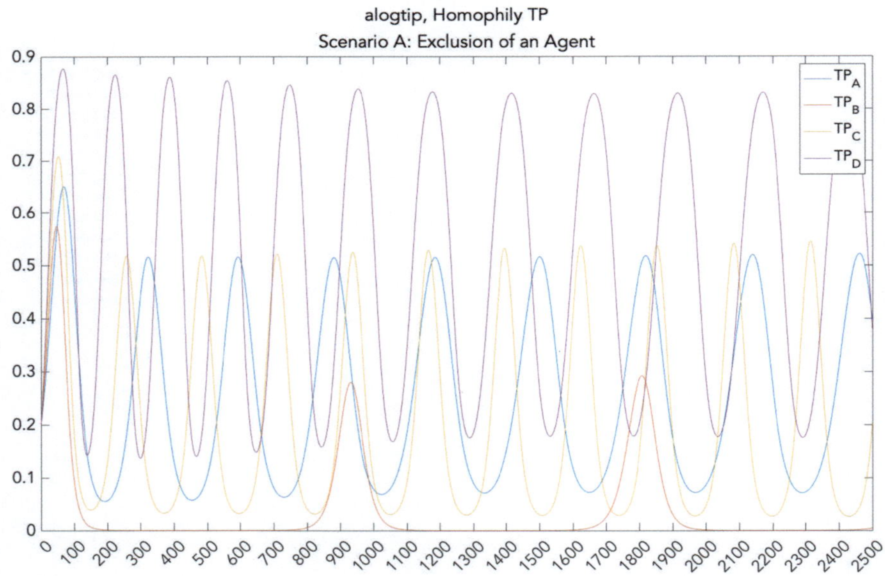

Fig. 30 Scenario A: Homophily Tipping Point with **sltip**

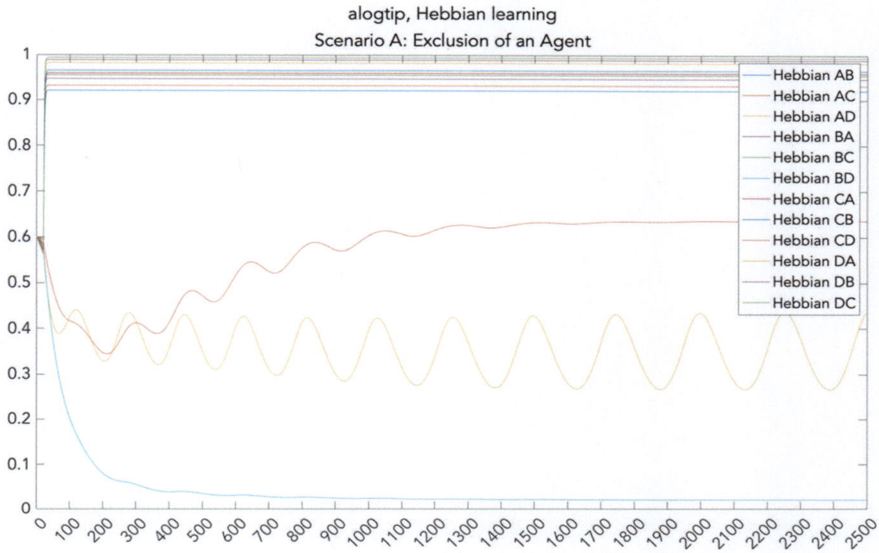

Fig. 31 Scenario A: Hebbian learning with **alogtip**

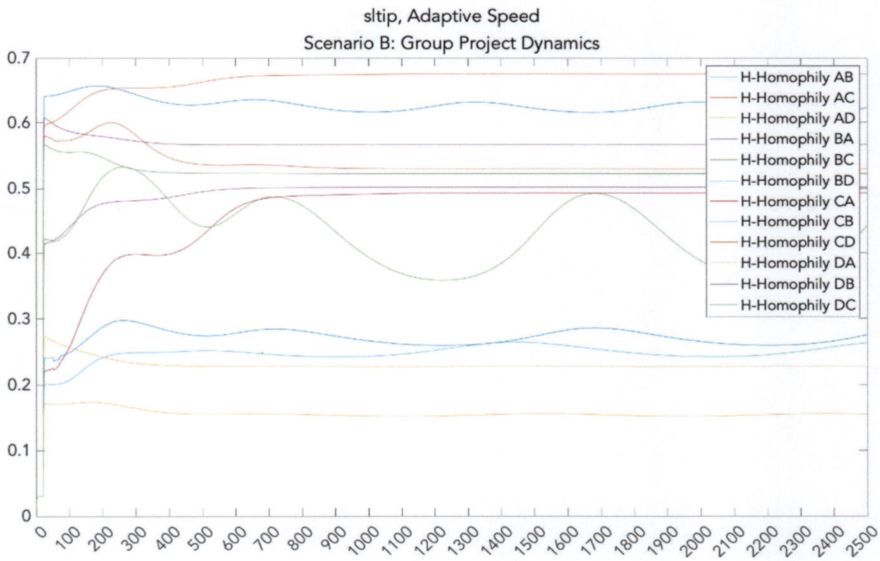

Fig. 32 Scenario B: Homophily adaptive speed with **sltip**

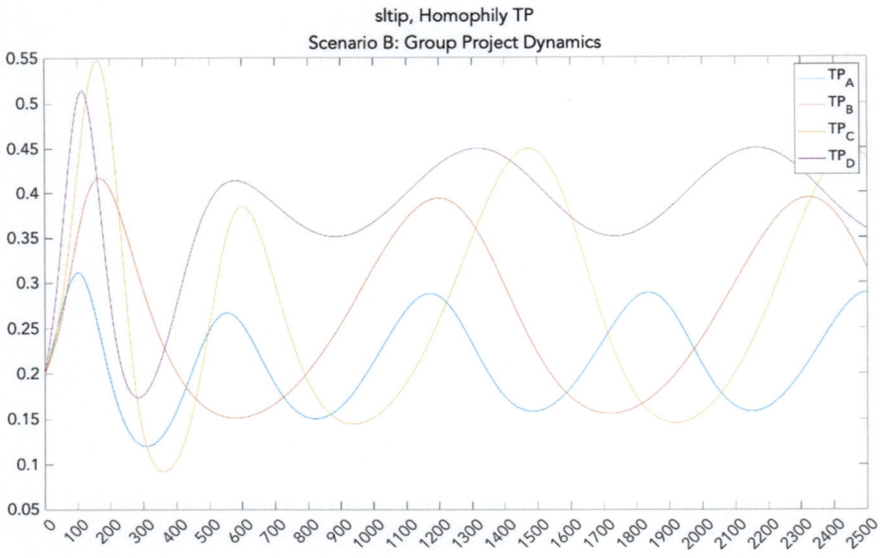

Fig. 33 Scenario B: Homophily Tipping Point with **sltip**

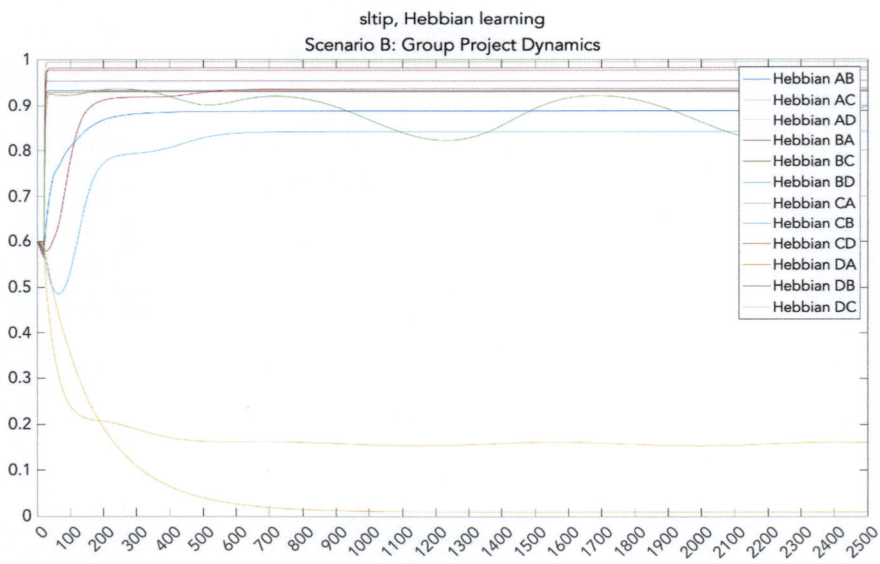

Fig. 34 Scenario B: Hebbian learning with **sltip**

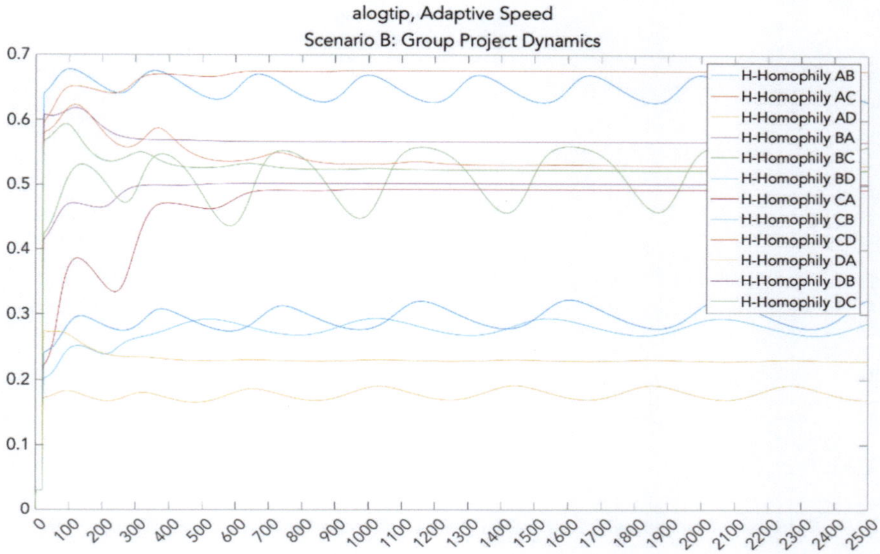

Fig. 35 Scenario B: Homophily adaptive speed with **alogtip**

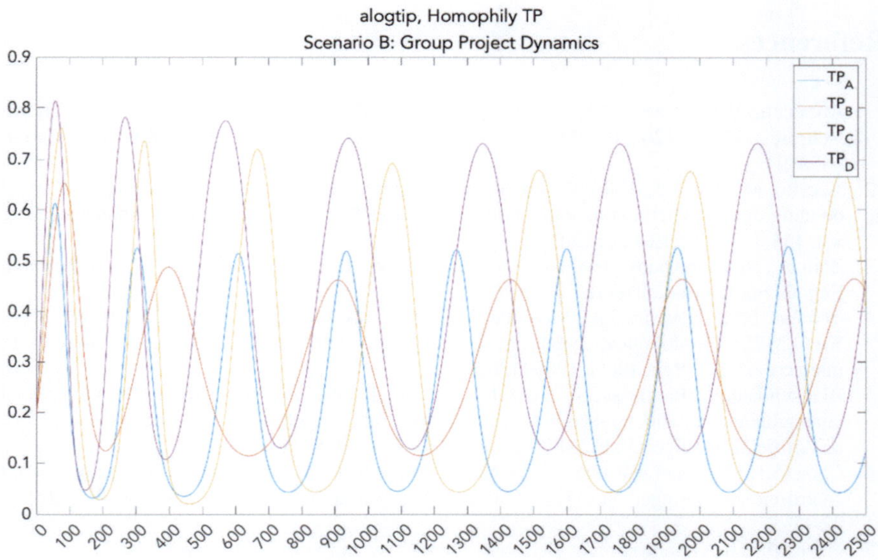

Fig. 36 Scenario B: Homophily Tipping Point with **alogtip**

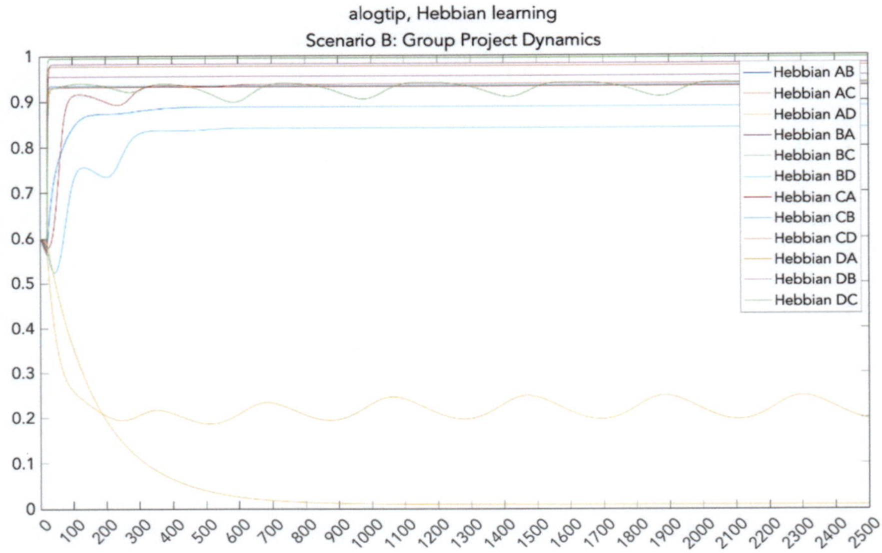

Fig. 37 Scenario B: Hebbian learning with **alogtip**

References

1. Abraham, W.C., Bear, M.F.: Metaplasticity: the plasticity of synaptic plasticity. Trends Neurosci. **19**(4), 126–30 (1996). https://doi.org/10.1016/s0166-2236(96)80018-x. PMID: 8658594
2. Accetto, M., Treur, J., Villa, V.: An adaptive cognitive-social model for mirroring and social bonding during synchronous joint action. In: Proc. BICA 2018. Procedia Computer Science, vol. 145, pp. 3–12. Elsevier (2018)
3. Bain, A.: Mind and body: the theories of their relation. HS King & Company 1873 Farnborough Eng. Gregg International (1873)
4. Behrens, F., Snijdewint, J.A., Moulder, R.G., Prochazkova, E., Sjak-Shie, E.E., Boker, S.M., Kret, M.E.: Physiological synchrony is associated with cooperative success in real-life interactions. Sci. Rep. **10**(1), 19609 (2020)
5. Blankendaal, R., Parinussa, S., Treur, J.: A temporal-causal modelling approach to integrated contagion and network change in social networks. In: Proc. of the 22nd European Conference on Artificial Intelligence, ECAI 2016, pp. 1388–1396. IOS Press (2016)
6. Choe, Y.L.: Hebbian learning. In: Jaeger, D., Jung, R. (eds) Encyclopedia of Computational Neuroscience. Springer, New York, NY (2014). https://doi.org/10.1007/978-1-4614-7320-6_672-1
7. Hebb, D.: The Organisation of Behavior. John Wiley and Sons (1949)
8. Hendrikse, S.C.F., Kluiver, S., Treur, J., Wilderjans, T.F., Dikker, S., Koole, S.L.: How virtual agents can learn to synchronize: an adaptive joint decision-making model of psychotherapy. Cogn. Syst. Res. **79**, 138–155 (2023)
9. Hendrikse, S.C.F., Kluiver, S., Treur, J., Wilderjans, T.F., Dikker, S., Koole, S.L.: Learning to synchronize: adaptive joint decision-making by virtual agents. In: Hendrikse, S.C.F., Treur, J., Koole, S.L. (eds.), New Analysis and Modeling Directions in Social Interaction Science: Emergent Multimodal Interpersonal Synchrony, Affiliation, and Bond-ing, Ch 5. Springer Nature (2025)

10. Hendrikse, S.C.F., Treur, J.: From a multilevel multi-timescale computational architecture for synchrony-induced and homophily-induced attunement to designing bonding bots. In: Hendrikse, S.C.F., Treur, J., Koole, S.L. (eds.), New Analysis and Modeling Directions in Social Interaction Science: Emergent Multimodal Interpersonal Synchrony, Affiliation, and Bonding, Ch 20. Springer Nature (2025)
11. Hendrikse, S.C.F., Treur, J., Koole, S.L.: Modeling emerging interpersonal synchrony and its related adaptive short-term affiliation and long-term bonding: a second-order multi-adaptive neural agent model. Int. J. Neural Syst. **33**(7), 2350038 (2023)
12. Hendrikse, S.C.F., Treur, J., Koole, S.L.: Relationship-specific and relationship-independent behavioural adaptivity in affiliation and bonding: a multi-adaptive dynamical systems approach. Cogn. Syst. Res. **84**, 101182 (2024)
13. Hendrikse, S.C.F., Treur, J., Koole, S. (eds.): New Analysis and Modeling Directions in Social Interaction Science: Emergent Multimodal Interpersonal Synchrony, Affiliation, and Bonding, this volume. Springer Nature (2025a)
14. Hendrikse, S.C.F., Treur, J., Koole, S.L.: Introduction to the role of adaptivity in social interaction science. In: Hendrikse, S.C.F., Treur, J., Koole, S.L. (eds.), New Analysis and Modeling Directions in Social Interaction Science: Emergent Multimodal Interpersonal Synchrony, Affiliation, and Bonding, Ch 1, this volume. Springer Nature (2025b)
15. Hendrikse, S.C.F., Treur, J., Koole, S.L.: The adaptive interplay of interpersonal synchrony, short-term affiliation and long-term bonding: a multi-timescale multi-adaptive agent model. In: Hendrikse, S.C.F., Treur, J., Koole, S.L. (eds.), New Analysis and Modeling Directions in Social Interaction Science: Emergent Multimodal Interpersonal Synchrony, Affiliation, and Bonding, Ch 12, this volume. Springer Nature (2025c)
16. Hendrikse, S.C.F., Treur, J., Koole, S.L.: A multi-timescale multi-adaptive dynamical system approach for relationship-specific and relationship-independent adaptivity in affiliation and bonding. In: Hendrikse, S.C.F., Treur, J., Koole, S.L. (eds.), New Analysis and Modeling Directions in Social Interaction Science: Emergent Multimodal Interpersonal Synchrony, Affiliation, and Bonding, Ch 13, this volume. Springer Nature (2025d)
17. Hendrikse, S.C.F., Treur, J., Wilderjans, T.F., Dikker, S., Koole, S.L.: On becoming in sync with yourself and others: an adaptive agent model for how persons connect by detecting intrapersonal and interpersonal synchrony. Hum.Centric Intell. Syst. **3**, 123–146 (2023)
18. Hendrikse, S.C.F., Treur, J., Wilderjans, T.F., Dikker, S., Koole, S.L.: Growing connectedness through synchrony: the role of subjective analysis of intrapersonal synchrony and interpersonal synchrony within an adaptive agent model. In: Hendrikse, S.C.F., Treur, J., Koole, S.L. (eds.), New Analysis and Modeling Directions in Social Interaction Science: Emergent Multimodal Interpersonal Synchrony, Affiliation, and Bonding, Ch 11, this volume. Springer Nature (2025)
19. Keysers, C., Gazzola, V.: Hebbian learning and predictive mirror neurons for actions, sensations and emotions. Philos. Trans. R. Soc. Lond. B Biol. Sci. **369**(1644), 20130175 (2014). https://doi.org/10.1098/rstb.2013.0175.PMID:24778372;PMCID:PMC4006178
20. Koole, S.L., Tschacher, W.: Synchrony in psychotherapy: a review and an integrative framework for the therapeutic alliance. Front. Psychol. **7**, 862 (2016)
21. Kozyreva, O., Pechina, A., Treur, J.: Network-oriented modeling of multi-criteria homophily and opinion dynamics in social media. In: Social Informatics, Proceedings of the 10th International Conference SocInfo'18, Part I. Lecture Notes in Computer Science, vol 11185, pp. 322–335. Springer International Publishing (2018)
22. Kret, M.E., AkyR.: Mimicry eases prediction and thereby smoothens social interactions. Cogn. Emot. **36**(5), 794–798 (2022).ated with cooperative success in real-life interactions. Sci. Rep. **10**(1), 19609 (2022)
23. Mukeriia, Y., Treur, J., Hendrikse, S.C.F.: An adaptive network model for learning and bonding during a varying in rhythm synchronous joint action. In: Proceedings of the 12th International Conference on Complex Networks and their Applications. Studies in Computational Intelligence, vol. 1144, pp. 41–52. Springer Nature Switzerland, Cham (2023)
24. Mukeriia, Y., Treur, J., Hendrikse, S.C.F.: A multi-adaptive network model for human hebbian learning, synchronization and social bonding based on adaptive homophily. Cogn. Syst. Res. **84**, 101187 (2024)

25. Prochazkova, E., Sjak-Shie, E., Behrens, F., Lindh, D., Kret, M.E.: Physiological synchrony is associated with attraction in a blind date setting. Nat. Hum. Behav. **6**(2), 269–278 (2022)
26. McPherson, M., Smith-Lovin, L., Cook, J.M.: Birds of a feather: homophily in social networks. Ann. Rev. Sociol. **27**, 425–444 (2001)
27. Robinson, B.L., Harper, N.S., McAlpine, D.: Meta-adaptation in the auditory midbrain under cortical influence. Nat. Commun. **7**, e13442 (2016)
28. Sharpanskykh, A., Treur, J.: Modelling and analysis of social contagion in dynamic networks. Neurocomputing **146**, 140–150 (2014)
29. Shatz, C.J.: The developing brain. Sci. Am. **267**(3), 60–67 (1992)
30. SjP.J., Rancz, E.A., Roth, A., Hausser, M.: Dendritic excitability and synaptic plasticity. Physiol. Rev. **88**, 769–840 (2008)
31. Treur, J.: Dynamic modeling based on a temporal-causal network modeling approach. Biol. Inspired Cogn. Arch. **16**, 131–168 (2016)
32. Treur, J.: Network reification as a unified approach to represent network adaptation principles within a network. In: Fagan, D., Martide, C., O'Neill, M., Vega-Rodrz, M.A. (eds) Theory and Practice of Natural Computing. Proc. TPNC 2018. Lecture Notes in Computer Science, vol 11324. Springer, Cham (2018). https://doi.org/10.1007/978-3-030-04070-3_27
33. Treur, J.: Network-Oriented Modeling for Adaptive Networks: Designing Higher-Order Adaptive Biological, Mental and Social Network Models. Springer Nature, Cham (2020). 10.1007-978-3-030-31445-3
34. Treur, J.: Modeling coevolution of dynamics, adaptivity and control of adaptivity by self-modeling networks. In: Hendrikse, S.C.F., Treur, J., Koole, S.L. (eds.), New Analysis and Modeling Directions in Social Interaction Science: Emergent Multimodal Interpersonal Synchrony, Affiliation, and Bonding, Ch 2. Springer Nature (2025a)
35. Treur, J.: A second-order adaptive network model for bonding based on subjectively perceived homophily. In: Hendrikse, S.C.F., Treur, J., Koole, S.L. (eds.), New Analysis and Modeling Directions in Social Interaction Science: Emergent Multimodal Interpersonal Synchrony, Affiliation, and Bonding, Ch 7. Springer Nature (2025b)
36. Van den Beukel, S., Goos, S.H., Treur, J.: An adaptive temporal-causal network model for social networks based on the homophily and more-becomes-more principle. Neurocomputing **338**, 361–371 (2019)
37. Van der Steen, M.C., Keller, P.E.: The adaptation and anticipation model (ADAM) of sensorimotor synchronization. Front. Hum. Neurosci. **7**, 253 (2013). https://doi.org/10.3389/fnhum.2013.00253. PMID: 23772211; PMCID: PMC3677131

Computational Analysis of Social Interaction Adaptivity Based on Subjective Synchrony Detection

In this part, in "Growing Connectedness through Synchrony: The Role of Subjective Analysis of Intrapersonal Synchrony and Interpersonal Synchrony within an Adaptive Agent Model" it is computationally explored in more detail how (1) emerging intrapersonal and interpersonal synchrony during multimodal social interaction can subjectively be detected and analysed by an agent and (2) how this internal subjective analysis can lead to adaptivity of the social interaction process. In "The Adaptive Interplay of Interpersonal Synchrony, Short-Term Affiliation and Long-Term Bonding: A Multi-Timescale Multi-Adaptive Agent Model" it is explored how adaptivity during social interaction can happen on different timescales and, for example, lead to affiliation on a short term and bonding on a longer term. This chapter addresses how such short-term and long-term adaptivity induced by synchronization can be modeled computationally by a second-order multi-adaptive agent model. It addresses movement, affect and verbal modalities and both intrapersonal synchrony and interpersonal synchrony. Moreover, in this chapter, mathematical analysis is addressed for adaptive network models and their positioning within the landscape of adaptive dynamical systems. The first type of analysis addressed shows that any smooth adaptive dynamical system has a canonical representation by a self-modeling network. This implies theoretically that the self-modeling network format is widely applicable, which also has been found in many practical applications using this approach. Furthermore, stationary point and equilibrium analysis was addressed and applied to the introduced self-modeling network model. It was used to obtain verification of the model providing evidence that the implemented model is correct with respect to its design specifications. In "A Multi-Timescale Multi-Adaptive Dynamical System Approach for Relationship-Specific and Relationship-Independent Adaptivity in Affiliation and Bonding" it is analysed how behavioural adaptivity during multimodal social interaction can partly be relationship-specific and partly relationship-independent. In theories such as attachment theory, it is claimed that some adaptations acquired in one relationship also have their effects in other relationships. Such transference phenomena are demonstrated and analysed here by a multi-adaptive second-order network model.

Growing Connectedness Through Synchrony: The Role of Subjective Analysis of Intrapersonal Synchrony and Interpersonal Synchrony Within an Adaptive Agent Model

Sophie C. F. Hendrikse, Jan Treur, Tom F. Wilderjans, Suzanne Dikker, and Sander L. Koole

Abstract Research indicates that interpersonal synchronization leads to more closeness, mutual coordination, alliance, or affiliation between the synchronized persons. There is a pathway leading from interpersonal interaction to interpersonal synchronisation, and then to interpersonal affiliation. If persons act on temporal patterns of synchrony, this suggests that they possess a facility to detect such patterns. Therefore, we assume here that persons indeed detect when temporal patterns of synchrony occur and that a stronger affiliation or connection may grow from this detection. We developed a multi-adaptive agent model that contains detector states for both intrapersonal

---

S. C. F. Hendrikse
Department of Methodology and Statistics, Tilburg School of Social and Behavioral Sciences, Tilburg University, Tilburg, The Netherlands
e-mail: S.C.F.Hendrikse@tilburguniversity.edu

S. L. Koole
Amsterdam Emotion Regulation Lab, Department of Clinical Psychology, Vrije Universiteit Amsterdam, Amsterdam, the Netherlands
e-mail: s.l.koole@vu.nl

J. Treur (✉)
Social AI Group, Department of Computer Science, Vrije Universiteit Amsterdam, Amsterdam, the Netherlands
e-mail: j.treur@vu.nl

T. F. Wilderjans
Methodology and Statistics Research Unit, Institute of Psychology, Faculty of Social and Behavioural Sciences, Leiden University, Leiden, the Netherlands

Research Group of Quantitative Psychology and Individual Differences, Faculty of Psychology and Educational Sciences, Katholieke Universiteit (KU) Leuven, Leuven, Belgium
e-mail: t.f.wilderjans@leidenuniv.nl

S. Dikker
NYU—Max Planck Center for Language, Music and Emotion, New York University, New York, USA

S. C. F. Hendrikse et al. (eds.), *New Analysis and Modeling Directions in Social Interaction Science*, Studies in Systems, Decision and Control 614,
https://doi.org/10.1007/978-3-031-99968-0_11

and interpersonal synchrony, which in turn dynamically affect interpersonal affiliation. We evaluated the behavior of two agents in multiple simulation experiments in which stochastic stimuli were manipulated. Several expected types of patterns were reproduced computationally, and our multi-adaptive agent model may serve as base for the development of virtual agents that can synchronize with their users. By multiple simulation experiments for stochastic stimuli from the environment, it was found that indeed several expected types of patterns are reproduced computationally.

Keywords Social interaction · Interpersonal synchrony · Intrapersonal synchrony · Connecting · Behavioural adaptivity · Adaptive agent model

1 Introduction

Connection or affiliation between persons often stems from experiencing some forms of emerging synchrony with each other [2, 26, 30, 40]. Such synchrony can occur in different forms. Intrapersonal synchrony happens when persons synchronize different modalities with each other, such as movements, different brain regions, emotions and verbal actions. Such intrapersonal synchrony can lead to conscious emotions [11]. Interpersonal synchrony occurs when persons synchronize corresponding modalities, such as their movements, or their emotions. Intrapersonal and interpersonal synchrony are often explored separately, although they are potentially related to each other and other outcomes (e.g., [4]).

To account in more detail for the interplay between intrapersonal and interpersonal synchrony and their link to affiliation, a computational agent model with both built-in intrapersonal and interpersonal synchrony detectors was designed. These built-in detectors are one of the main innovations of this chapter, along with their role in the agents' behavioral adaptivity concerning affiliation. The intrapersonal and interpersonal synchrony covers movement, affective and verbal regions.

An experimental scenario was designed, in which two periods alternate: one with a common stochastic stimulus and one with two different stochastic stimuli for two agents. Although, this scenario was kept abstract, it was inspired by the intended application on how synchrony between therapist and client and related behavioral adaptivity within counseling or therapy sessions plays an important role in appreciation and success rate of the sessions, see [30–32, 36, 38].

The adaptive agent model was designed using the self-modeling network modeling approach from [44, 45], this is briefly described in Sect. 3. The simulation experiments described in Sect. 5 were conducted using the available dedicated software environment. In Sect. 2 some background literature is discussed and in Sect. 4 the designed adaptive agent model is explained in more detail. Finally, Sect. 6 is a discussion.

2 Background Knowledge

It has often been found that when synchrony originates, this is accompanied with a changed connection or affiliation between persons, e.g., [2, 26, 29, 30, 35, 40, 55, 57]. More specifically, number of experiments have analyzed manipulated psychotherapy sessions whose outcome was that therapists were rated more favorably and as more empathic when they were instructed in advance to synchronize their movements with those of the client [32, 38, 51]. Additionally, a clinical study reported in [36] found that initial movement synchrony between client and therapist, was predictive of the client's appraisal of the quality of the alliance at the end of each session. Therefore, there is much support for the idea that movement synchrony has a positive effect on the alliance, see also [30]. Furthermore, according to various researchers [10, 53, 57], synchrony in face-to-face interactions turns out to play a key role in interpersonal bonding. For some more background about synchrony, see, for example, this volume [15], Chapter 1 [16], Sects. 1.3 and 1.4.

If persons notice or experience synchrony and change in some way as a result, this may suggest that they possess some facility to detect it. Therefore, it does not seem strange to assume that persons do (perhaps unconsciously) detect in some way synchrony and from that make a stronger connection. This is indeed the first main assumption made here. Note that the relationship between synchrony and being connected works in two ways, as not only may synchrony strengthen the connection between persons, in turn for more strongly connected persons, due to a stronger social contagion it may also be easier to reach synchrony when they interact. In this way an interplay seems to exist between synchrony and connectedness such as (short-term) affiliation.

A subsequent question is how connecting actually occurs. For example, is it a form of mentalizing or is it a form of adaptation of behavior? In their metareview, [35], contend that the defining feature of what they call partnered interaction:

> '…is not mentalizing, but instead real-time behavioral adaptivity between the partners, as studied in a large literature devoted to joint action and joint agency (e.g., [2, 9, 27], [34]).'

Here, as a second main assumption, we follow the latter position of behavioral adaptivity within social interaction as a main mechanism for connecting. Therefore, in combination with the first main assumption mentioned above we analyze two persons that are connecting through synchronization with mechanisms of circular pathways having roughly the following form.

interaction patterns → synchrony patterns → detected synchrony →

adaptation of connections → (changed) interaction patterns

Notice that here 'detected synchrony' acts as a specific information-carrying state p mediating between (1) interaction or synchrony patterns a that occur in the past before p's activation and (2) interaction patterns b (entailed by a) that will occur in the future after p's activation. In [41] such a state p is called a mediating state for the effect of past pattern a on future pattern b entailed by pattern a; similarly, in

[52] such a (brain) state is referred to as describing 'informational criteria' for future activation, see also [46].

In addition to interpersonal synchrony, intrapersonal synchrony for different modalities also often occurs. It has been found that intrapersonal (response) synchrony often is at the basis of a conscious emotion, e.g., [11]. In return, from an emotion integration viewpoint, such a conscious emotion can be thought of as having an effect on responses as well. However, from the perspective of causality it might be plausible that these response effects are causally driven not by the 'being conscious' of the emotion, but rather the underlying mechanisms and the awareness is more like monitoring such underlying processes. These observations form the basis for our third main assumption, namely that the detected intrapersonal synchrony underlying a conscious emotion has a causal effect on the responses and the related behavioral adaptivity.

3 The Self-Modeling Network Modeling Approach Used

In this section, we briefly introduce the network-oriented modeling approach used. Following [44, 45], a *temporal-causal network* model is characterized by the following characteristics (with X and Y denoting nodes of the network, also called states):

- Connectivity characteristics
  Connections from a state X to a state Y and their weights $\boldsymbol{\omega}_{X,Y}$
- Aggregation characteristics
  For any state Y, some combination function $\mathbf{c}_Y(..)$ defines the aggregation that is applied to the impacts $\boldsymbol{\omega}_{X,Y}X(t)$ on Y from its incoming connections from states X
- Timing characteristics
  Each state Y has a speed factor $\boldsymbol{\eta}_Y$ defining how fast it changes for given causal impact

The following difference (or related differential) equations that are used for simulation purposes as well as for analysis of temporal-causal networks, incorporate these network characteristics $\boldsymbol{\omega}_{X,Y}$, $\mathbf{c}_Y(..)$ and $\boldsymbol{\eta}_Y$ in a standard numerical format:

$$Y(t + \Delta t) = Y(t) + \boldsymbol{\eta}_Y[\mathbf{c}_Y(\boldsymbol{\omega}_{X_1,Y}X_1(t), \ldots, \boldsymbol{\omega}_{X_k,Y}X_k(t)) - Y(t)]\Delta t \quad (1)$$

for any state Y and where X_1 to X_k are the states from which Y gets its incoming connections. Note that (1) has a format similar to that of recurrent neural networks. Within the software environment described in [44], Ch. 9, a large number of useful basic combination functions (around 60) are included in a combination function library.

The above concepts make it possible to design network models and their dynamics in a declarative manner, based on mathematically defined functions and relations. The examples of combination functions that are applied in the model introduced here can be found in Table 1. Here, for the third and fourth function rand (1, 1) draws a random number from [0, 1] in a uniform manner and a is a persistence factor (with value 0.5 used in the simulations).

Realistic network models are usually adaptive: often not only their states but also some of their network characteristics change over time. By using a *self-modeling network* (also called a *reified* network), a similar network-oriented conceptualization can also be applied to *adaptive* networks thus producing a declarative description using mathematically defined functions and relations for them as well; see [44, 45]. This process is enabled by the addition of new states to the network (called *self-model states*) which represent (adaptive) network characteristics. In the graphical 3D-format as shown in Sect. 4, such additional states are depicted at a next level (called *self-model level* or *reification level*), while the original network is at the *base level*.

As an example, the weight $\boldsymbol{\omega}_{X,Y}$ of a connection from state X to state Y can be represented (at a next self-model level) by a self-model state named $\mathbf{W}_{X,Y}$. Similarly, all other network characteristics from $\boldsymbol{\omega}_{X,Y}$, $\mathbf{c}_Y(..)$ and $\boldsymbol{\eta}_Y$ can be made adaptive by including self-model states for them. For example, an adaptive speed factor $\boldsymbol{\eta}_Y$ can be represented by a self-model state named $\mathbf{H}_Y$. Using this concept, adaptation processes can be modeled in a declarative form via specification of proper combination functions in the library that are applied to self-model states, for example, a combination function for Hebbian learning. This enables us to model adaptivity in

Table 1 The combination functions used in the introduced network model

Name	Formula	Parameters	Used for
Advanced logistic sum $\mathbf{alogistic}_{\sigma,\tau}(V_1, \ldots, V_k)$	$[\frac{1}{1+e^{-\sigma(V_1+\cdots+V_k-\tau)}} - \frac{1}{1+e^{\sigma\tau})}]\ (1 + e^{-\sigma\tau})$	Steepness $\boldsymbol{\sigma}$ Excitability threshold $\boldsymbol{\tau}$	X_4-X_{15}, X_{22}-X_{36}, X_{43}-X_{77}
Complemental difference $\mathbf{compdiff}(V_1, V_2)$	$1 - \frac{\|V_1-V_2\|}{\max(V_1,V_2)}$ if V_1 or $V_2 > 0$ 0 if $V_1 = V_2 = 0$	–	X_{16}-X_{21}, X_{37}-X_{42} (synchrony detectors)
Random Stepmod $\mathbf{randstepmod}_{\rho,\delta}(V)$	0 if $0 \leq$ time t mod $\boldsymbol{\rho} \leq \boldsymbol{\delta}$ aV + (1-a)rand(1,1) else	Repitition $\boldsymbol{\rho}$ Step time $\boldsymbol{\delta}$	X_3 (common stimulus)
Random Stepmodopp $\mathbf{randstepmodopp}_{\rho,\delta}(V)$	0 if $\boldsymbol{\delta} \leq$ time t mod $\boldsymbol{\rho} \leq \boldsymbol{\rho}$ aV + (1-a)rand(1,1) else	Repitition $\boldsymbol{\rho}$ Step time $\boldsymbol{\delta}$	X_1-X_2 (individual stimuli)

a declarative format and not by procedural descriptions (e.g., learning algorithms) often used in the literature on learning.

As the outcome of this process of network reification is also a temporal-causal network model in itself, as has been shown in [44], Ch 10, this self-modeling network construction can easily be applied iteratively to obtain multiple orders of self-models at multiple (first-order, second-order, ...) self-model levels. For example, a second-order self-model may include a second-order self-model state $\mathbf{H}_{\mathbf{W}_{X,Y}}$ representing the speed factor $\boldsymbol{\eta}_{\mathbf{W}_{X,Y}}$ for the dynamics of first-order self-model state $\mathbf{W}_{X,Y}$ which in turn represents the adaptation of connection weight $\boldsymbol{\omega}_{X,Y}$.

Due to the library of combination functions, the modeling approach can also be seen as a generalized form of neural networks, as for any activation function used in neural networks a related combination function in this library can be included and used [47]. Even more, in [47] it has been shown more generally how (1) any (smooth) dynamical system can be modeled as a (temporal-causal) network model and (2) any (smooth) adaptive dynamical system can be modeled as a self-modeling network model. This makes it a widely applicable modeling approach.

In the current chapter, this multi-level self-modeling network perspective will be applied to obtain a second-order adaptive network architecture addressing controlled adaptation induced by detected synchrony. In this self-modeling network architecture, the first-order self-model models the adaptation of the base level network, while the second-order self-model level exerts control over this; see Fig. 1. In this picture, the upward arrows indicate how the adaptation depends on contextual circumstances at the base level and how control of the adaptation depends on the state of adaptation. The downward arrows indicate the effect of the adaptation on the mental processes and of the control on the adaptation processes.

As an example, the control level can be used to make the adaptation speed context-sensitive as addressed by metaplasticity literature such as [1, 37]. For instance, the metaplasticity principle 'Adaptation accelerates with increasing stimulus exposure' formulated by [37] can easily be modeled by using second-order self-model states; this has actually been done for the introduced model, as will be discussed in Sect. 4.

For more details of this modeling approach, see [44] or this volume [15], Chapter 2 [49].

Fig. 1 Three-level self-modeling network architecture for controlled adaptation

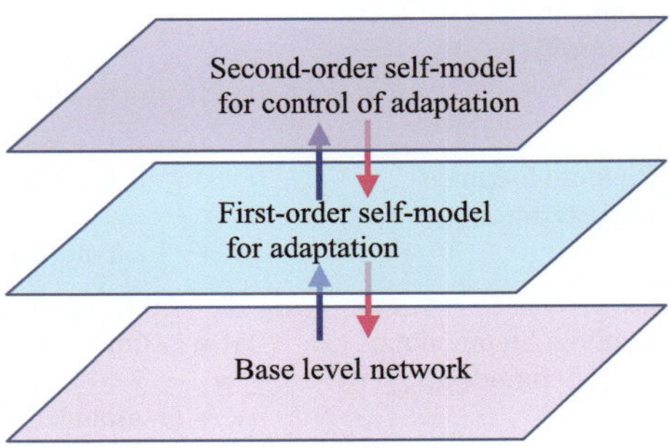

4 The Introduced Controlled Adaptive Agent Model

In this section, the introduced controlled adaptive agent model is explained in some detail. The controlled adaptive agent design uses a self-modeling network architecture of three levels as discussed in Sect. 3: a base level, a first-order self-model level, and a second-order self-model level. Here the (middle) first-order self-model level models how connections of the base level are adapted over time, and the (upper) second-order self-model level models the control over the adaptation.

4.1 Base Level

Figure 2 shows a graphical overview of the base level of the agent model (agents are indicated by the big boxes) and Table 2 provides explanations for all its base states. This figure and table explain more details about the agent models and the interaction between the agents, such as how agents interact, which mental states they have, and the causal relations between their states.

For each agent, a number of interaction states were modeled: states involved in sensing (indicated by sense) are on the left-hand side of each box, and states involved in execution or expression of actions (move, exp_affect, talk) on the right-hand side. In between these interaction states, within a box are the agent's internal mental states; outside the boxes are the world states. Note that it is assumed that each agent also senses its own actions, modeled by the arrows from right to left outside the box.

4.1.1 Internal Mental States and Their Connections

For each agent there are internal mental states such as sensory representation states (rep) and preparation states (prep) for each of the three modalities: movement *m*, expression of affect *b*, and verbal action *v*. These *m*, *b*, and *v* are kept abstract and generic in the model. Examples that can be kept in mind are moving backward or forward for movement *m*, a happy or sad emotion for affect *b*, and a verbal utterance expressing happiness for *v*. Furthermore, each agent has a conscious emotion state for affective response *b* (cons_emotion). Each of the mentioned states is depicted in Fig. 2 by a light pink circle shape. For each modality, its representation state has an outgoing (response) connection to the corresponding preparation state and it has an incoming (prediction) connection back from the preparation state to model internal mental simulation of predicted effects of actions [3, 25].

Last but not least, there are the six synchrony detector states (depicted in Fig. 2 by the darker pink diamond shapes) which are introduced here. There are three *intrapersonal synchrony detection states* for the three pairs of the three modalities:

movement—emotion (*m*-*b*)

movement—verbal action (*m*-*v*)

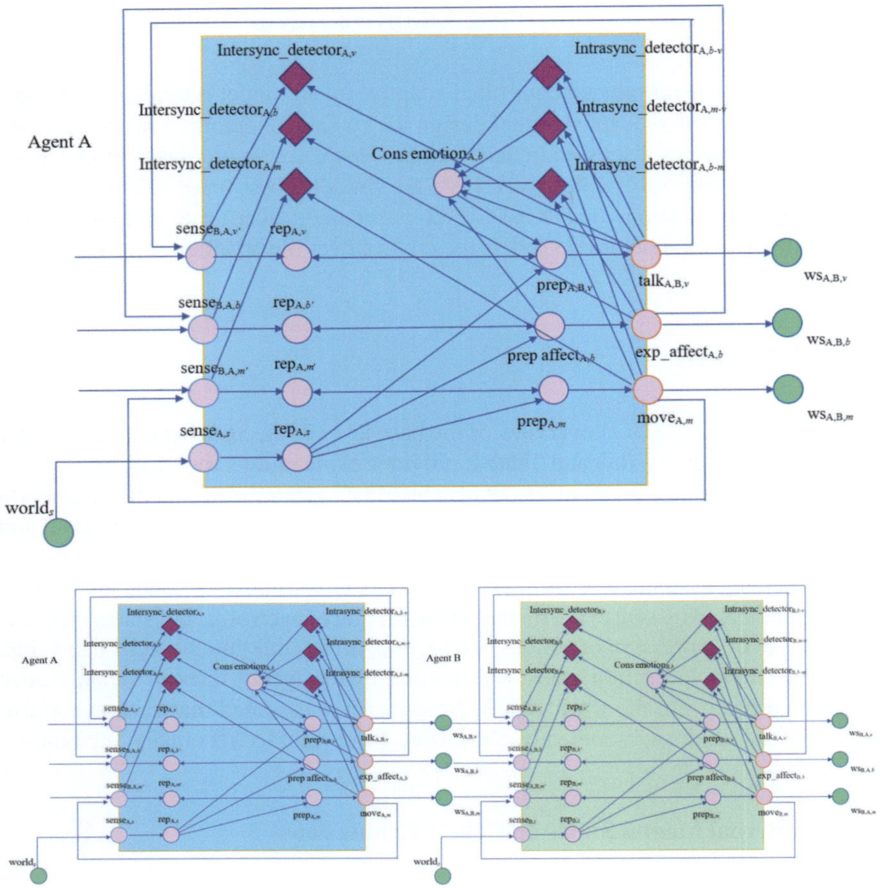

Fig. 2 Base level of (1) the introduced adaptive agent model (upper picture) with three modalities and (in dark pink) six synchony detection states for intrapersonal and interpersonal synchrony and (2) how the agents interact (lower picture) according to the three modalities

emotion—verbal action (*b*-*v*)

These intrapersonal synchrony detection states have incoming connections from the two execution states for the modalities they address.

The conscious emotion state is triggered by incoming connections from the preparation state for affective response *b* together with the three intrapersonal synchrony detection states (following [11]). In addition, the conscious emotion state has an incoming connection from the verbal action execution state (for noticing the emotion in the verbal utterance) and an outgoing connection to the preparation of the verbal action (for emotion integration in the verbal action preparation).

There are three *interpersonal synchrony detection states* for the three modalities *m*, *b*, and *v*. Each of them has two incoming connections: from the sensing state

Table 2 Base states of the computational network model

state	name	Explanation
X_1	$ws_{s,A}$	World state for stimulus *s* for *A*
X_2	$ws_{s,B}$	World state for stimulus *s* for *B*
X_3	$ws_{s,AB}$	World state for stimulus *s* both for *A* and *B*
X_4	$sense_{s,A}$	Sensor state of *A* for stimulus *s* for *A*
X_5	$sense_{B,m,A}$	Sensor state of *A* for movement *m* of *B*
X_6	$sense_{B,b,A}$	Sensor state of *A* for expressed affective response *b* of *B*
X_7	$sense_{B,v,A}$	Sensor state of *A* for verbal action *v* of *B*
X_8	$rep_{s,A}$	Sensory representation state of *A* for stimulus *s* for *A*
X_9	$rep_{B,m,A}$	Sensory representation state of *A* for movement *m* of *B*
X_{10}	$rep_{B,b,A}$	Sensory representation state of *A* for expressed affective response *b* of *B*
X_{11}	$rep_{B,v,A}$	Sensory representation state of *A* for verbal action *v* of *B*
X_{12}	$prep_{m,A}$	Preparation state for movement *m* of *A*
X_{13}	$prep_{b,A}$	Preparation state for affective response *b* of *A*
X_{14}	$prep_{v,A}$	Preparation state for verbal action *v* of *A*
X_{15}	$cons\_emotion_{b,A}$	Conscious emotion state for *b* of *A*
X_{16}	$intrasyncdet_{A,b\text{-}m}$	Intrapersonal synchrony detection of *A* for executing *b* and *m* by *A*
X_{17}	$intrasyncdet_{A,m\text{-}v}$	Intrapersonal synchrony detection of *A* for executing *m* and *v* by *A*
X_{18}	$intrasyncdet_{A,b\text{-}v}$	Intrapersonal synchrony detection of *A* for executing *b* and *v* by *A*
X_{19}	$intersyncdet_{B,A,m}$	Interpersonal synchrony detection of *A* for executing *m* by *B* and *A*
X_{20}	$intersyncdet_{B,A,b}$	Interpersonal synchrony detection of *A* for executing *b* by *B* and *A*
X_{21}	$intersyncdet_{B,A,v}$	Interpersonal synchrony detection of *A* for executing *v* by *B* and *A*
X_{22}	$move_{m,A}$	Executing movement *m* by *A*
X_{23}	$exp\_affect_{b,A}$	Executing expression of *b* by *A*
X_{24}	$talk_{A,B,v}$	Executing verbal action *v* by *A*
X_{25}	$sense_{s,B}$	Sensor state of *B* for stimulus *s* for *B*
X_{26}	$sense_{A,m,B}$	Sensor state of *B* for movement *m* of *A*
X_{27}	$sense_{A,b,B}$	Sensor state of *B* for expressed affective response *b* of *A*
X_{28}	$sense_{A,v,B}$	Sensor state of *B* for verbal action *v* of *A*
X_{29}	$rep_{s,B}$	Sensory representation state of *B* for stimulus *s* for *B*
X_{30}	$rep_{A,m,B}$	Sensory representation state of *B* for movement *m* of *A*
X_{31}	$rep_{A,b,B}$	Sensory representation state of *B* for expressed affective response *b* of *A*
X_{32}	$rep_{A,v,B}$	Sensory representation state of *B* for verbal action *v* of *A*
X_{33}	$prep_{m,B}$	Preparation state for movement *m* of *B*
X_{34}	$prep_{b,B}$	Preparation state for affective response *b* of *B*
X_{35}	$prep_{v,B}$	Preparation state for verbal action *v* of *B*
X_{36}	$cons\_emotion_{b,B}$	Conscious emotion state for *b* of *B*
X_{37}	$intrasyncdet_{B,b\text{-}m}$	Intrapersonal synchrony detection of *B* for executing *b* and *m* by *B*
X_{38}	$intrasyncdet_{B,m\text{-}v}$	Intrapersonal synchrony detection of *B* for executing *m* and *v* by *B*
X_{39}	$intrasyncdet_{B,b\text{-}v}$	Intrapersonal synchrony detection of *B* for executing *b* and *v* by *B*
X_{40}	$intersyncdet_{A,B,m}$	Interpersonal synchrony detection of *B* for executing *m* by *A* and *B*
X_{41}	$intersyncdet_{A,B,b}$	Interpersonal synchrony detection of *B* for executing *b* by *A* and *B*
X_{42}	$intersyncdet_{A,B,v}$	Interpersonal synchrony detection of *B* for executing *v* by *A* and *B*
X_{43}	$move_{m,B}$	Executing movement *m* by *B*
X_{44}	$exp\_affect_{b,B}$	Executing expression of *b* by *B*
X_{45}	$talk_{B,A,v}$	Executing verbal action *v* by *B*
X_{46}	$ws_{m,A,B}$	World state for transmitting movement *m* of *A* to *B*
X_{47}	$ws_{b,A,B}$	World state for transmitting affective response *b* of *A* to *B*
X_{48}	$ws_{v,A,B}$	World state for transmitting verbal action *v* of *A* to *B*
X_{49}	$ws_{m,B,A}$	World state for transmitting movement *m* of *B* to *A*
X_{50}	$ws_{b,B,A}$	World state for transmitting affective response *b* of *B* to *A*
X_{51}	$ws_{v,B,A}$	World state for transmitting verbal action *v* of *B* to *A*

(representing the action of the other person) and the execution state (representing the own action) of the modality addressed.

4.1.2 Interaction States and World States

All connections discussed this far are assumed to be nonadaptive. However, a few other connections are adaptive depending on detected synchrony: more synchrony detected will lead to connecting stronger to the other person, which is modeled by making the weights of such connections stronger. This applies to four types of connections, all playing an important role in the interaction between two agents:

- **Adaptive internal connections**
  - the (representing) connections from sensing to representation states for each of the three modalities
  - the (executing) connections from preparation to execution states for each of the three modalities

- **Adaptive external connections**
  - the (observing) connections from world states to sensing states
  - the (effectuating) connections from execution states to world states.

So, more synchrony detected will lead to these types of connections becoming stronger; each type of them contributes in its own way to the interaction between the persons. A stronger observing connection will lead to sensing the other person better (e.g., turning sensors in the right direction and bending or getting closer to the other), and a stronger representing connection will make the sensed signals better available and accessible for the brain. From the other side, a stronger executing connection will lead to stronger expression and acting toward the other and a stronger effectuating connection to better availability (for the other) of the action effects in the world (e.g., more visible, better hearable by directing and positioning in the right direction and bending or getting closer to the other). In Sect. 4.2, it will be discussed in more detail how this adaptivity and its control were modeled using the notion of self-modeling of the network model.

Finally, at the base level some world states are modeled for stimuli s that are sensed by the agents. In the simulations, they have stochastic activation levels. In some episodes one common stimulus is observed by both agents (for example when they physically meet and therefore are in the same environment), but in other episodes the agents receive different stimuli.

4.2 Modeling the Adaptation and Its Control

The adaptation and control needed in this agent model were modeled by the notion of self-modeling network from [44, 45]: by adding first-order self-model states for adaptation of the adaptive base connections discussed in Sect. 4.1 and second-order self-model states for control of the adaptation. In Fig. 3 the overall design of the network model is shown; here, the first-order self-model states are in the middle (blue) plane and the second-order self-model states in the upper (purple) plane. See Table 3 for explanations of all self-model states.

The first-order states are all **W**-states representing the weights of the adaptive base connections. By changing the activation values of these **W**-states, the corresponding connection weights change accordingly. This change occurs due to the influences from the detected synchronies, modeled by the upward (blue) arrows in Fig. 3 from the synchrony detection states in the base plane to the **W**-states in the middle plane.

To keep the model a bit simpler, there are only two second-order self-model states $\mathbf{H}_{\mathbf{W}_A}$ and $\mathbf{H}_{\mathbf{W}_B}$, one for each agent. They represent the adaptation speed (learning rate) for the adaptive connections for the concerning agent. They model the second-order adaptation principle ‘Adaptation accelerates with stimulus exposure’ (Robinson et al. 2016). To this end they have incoming connections (blue upward arrows from base plane to upper plane) from the stimulus representation states at the base level.

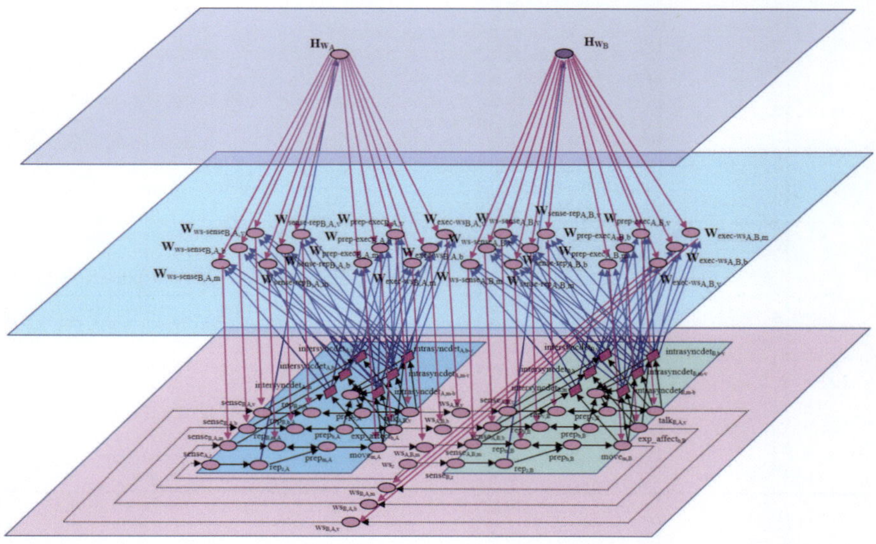

Fig. 3 Overview of the overall controlled adaptive network model

Table 3 First-order self-model **W**-states for connection weights and second-order self-model $\mathbf{H_W}$-states for the adaptation speed of the **W**-states of the computational network model

state	name	Explanation
X_{52}	$\mathbf{W}_{\text{sense-rep\_}m,A}$	First-order self-model state for the weight of *A*'s internal connection from sensing to representing movement *m*
X_{53}	$\mathbf{W}_{\text{sense-rep\_}b,A}$	First-order self-model state for the weight of *A*'s internal connection from sensing to representing affective response *b*
X_{54}	$\mathbf{W}_{\text{sense-rep\_}v,A}$	First-order self-model state for the weight of *A*'s internal connection from sensing to representing verbal action *v*
X_{55}	$\mathbf{W}_{\text{prep-exec\_}m,A}$	First-order self-model state for the weight of *A*'s internal connection from preparing to executing movement *m*
X_{56}	$\mathbf{W}_{\text{prep-exec\_}b,A}$	First-order self-model state for the weight of *A*'s internal connection from preparing to expressing affective response *b*
X_{57}	$\mathbf{W}_{\text{prep-exec\_}v,A}$	First-order self-model state for the weight of *A*'s internal connection from preparing to executing verbal action *v*
X_{58}	$\mathbf{W}_{\text{exec-ws\_}m,A,B}$	First-order self-model state for the weight of *A*'s external connection from executing to world state for movement *m* for *B*
X_{59}	$\mathbf{W}_{\text{exec-ws\_}b,A,B}$	First-order self-model state for the weight of *A*'s external connection from executing to world state for afective response *b* for *B*
X_{60}	$\mathbf{W}_{\text{exec-ws\_}v,A,B}$	First-order self-model state for the weight of *A*'s external connection from executing to world state for verbal action *v* for *B*
X_{61}	$\mathbf{W}_{\text{ws-sense\_}m,B,A}$	First-order self-model state for the weight of *A*'s external connection from world state to sensing for movement *m* from *B*
X_{62}	$\mathbf{W}_{\text{ws-sense\_}b,B,A}$	First-order self-model state for the weight of *A*'s external connection from world state to sensing for affective response *b* from *B*
X_{63}	$\mathbf{W}_{\text{ws-sense\_}v,B,A}$	First-order self-model state for the weight of *A*'s external connection from world state to sensing for verbal action *v* from *B*
X_{64}	$\mathbf{W}_{\text{sense-rep\_}m,B}$	First-order self-model state for the weight of *B*'s internal connection from sensing to representing movement *m*
X_{65}	$\mathbf{W}_{\text{sense-rep\_}b,B}$	First-order self-model state for the weight of *B*'s internal connection from sensing to representing affective response *b*
X_{66}	$\mathbf{W}_{\text{sense-rep\_}v,B}$	First-order self-model state for the weight of *B*'s internal connection from sensing to representing verbal action *v*
X_{67}	$\mathbf{W}_{\text{prep-exec\_}m,B}$	First-order self-model state for the weight of *B*'s internal connection from preparing to executing movement *m*
X_{68}	$\mathbf{W}_{\text{prep-exec\_}b,B}$	First-order self-model state for the weight of *B*'s internal connection from preparing to expressing affective response *b*
X_{69}	$\mathbf{W}_{\text{prep-exec\_}v,B}$	First-order self-model state for the weight of *B*'s internal connection from preparing to executing verbal action *v*
X_{70}	$\mathbf{W}_{\text{exec-ws\_}m,B,A}$	First-order self-model state for the weight of *B*'s external connection from executing to world state for movement *m* for *A*
X_{71}	$\mathbf{W}_{\text{exec-ws\_}b,B,A}$	First-order self-model state for the weight of *B*'s external connection from executing to world state for affective response *b* for *A*
X_{72}	$\mathbf{W}_{\text{exec-ws\_}v,B,A}$	First-order self-model state for the weight of *B*'s external connection from executing to world state for verbal action *v* for *A*
X_{73}	$\mathbf{W}_{\text{ws-sense\_}m,A,B}$	First-order self-model state for the weight of *B*'s external connection from world state to sensing for movement *m* from *A*
X_{74}	$\mathbf{W}_{\text{ws-sense\_}b,A,B}$	First-order self-model state for the weight of *B*'s external connection from world state to sensing for affective response *b* from *A*
X_{75}	$\mathbf{W}_{\text{ws-sense\_}v,A,B}$	First-order self-model state for the weight of *B*'s external connection from world state to sensing for verbal action *v* from *A*
X_{76}	$\mathbf{H}_{\mathbf{W_A}}$	Second-order self-model state for the speed factor of the first-order self-model **W**-states for *A*
X_{77}	$\mathbf{H}_{\mathbf{W_B}}$	Second-order self-model state for the speed factor of the first-order self-model **W**-states for *B*

5 Simulation Results

Simulation experiments have been conducted for a scenario that was inspired by how synchrony between therapist and client within counseling or therapy sessions can occur and actually plays an important role in appreciation and success rate of such sessions, as also discussed in [30–32, 36, 38]. We have run 20 repetitions of the simulation scenario for our two agents. The total time duration of each simulation equaled 360 time units and the step size (Δt) regarded 0.5, meaning 720 computational steps were made in total in each of our experiments. All 20 simulations had the same structure concerning the type and duration of stochastic stimuli: an episode of 40 time units with different stimuli between the two agents, followed by an episode of equal size (40 time units) with a common stimulus for the two agents. This same cycle of 80 time units in total is repeated four times, after which a final episode of 40 time units with two different stimuli is presented. The episodes with the different stochastic stimuli reflect an environment where different influences affect each agent or the two agents concentrate on different foci, whereas the agents experience a shared environment or concentrate on the same focus in the episodes with the same stochastic stimulus. Within a therapeutical context, these episodes can be interpreted as periods in which a therapist and a client concentrate on the same focus and periods in which their foci differ. In all episodes, regardless the environment, the agents are able to interact with each other. We will discuss two illustrative simulations that displayed distinct patterns that were most often present during the 20 repetitions.

5.1 From Intrapersonal Synchrony to Conscious Emotions

Regarding the first illustrative simulation (Fig. 4), the intrapersonal synchrony between movement and affect of agent A and B (dark green solid line and light green dashed line) were steeply activated first and followed exactly the same trajectory. Thereafter, the intrapersonal synchrony of agent B's affect and verbal actions and movement and verbal actions followed exactly the same pathway, and a bit slower these corresponding states of agent A. After the activation of the three types of intrapersonal synchrony, the conscious emotions of both agent B and A started to occur (blue lines).

In the second example simulation (Fig. 5), the same overlaps are present as in the first illustrative simulation. Agent A reached high levels of conscious emotion before agent B reached all three intrapersonal synchronies. It also shows that within each agent, all three intrapersonal synchronies are required before the conscious emotion can elevate, independently of the intrapersonal synchrony and conscious emotion of the other agent.

The patterns described above cover the first 40 time units. After this period, the agents start to interact with each other in both simulations. Then, the levels of the

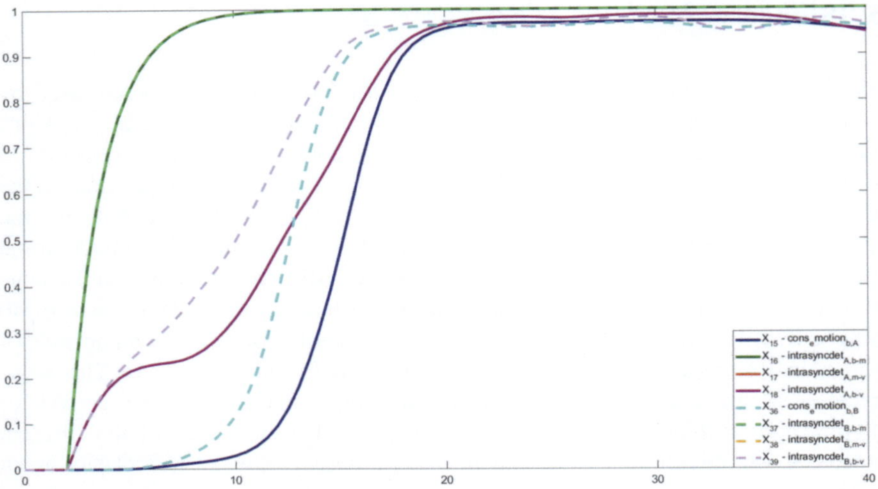

Fig. 4 The intrapersonal synchrony and conscious emotions activation levels from 0 to 40 time units in illustrative simulation 1

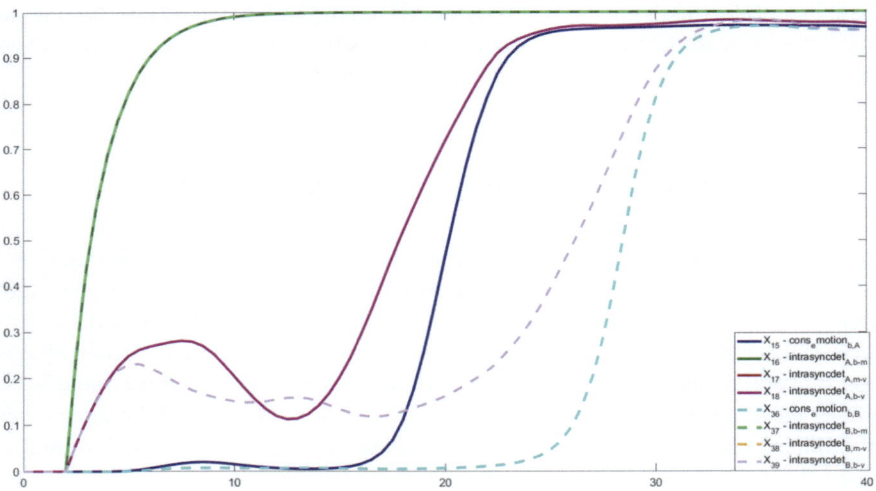

Fig. 5 The intrapersonal synchrony and conscious emotions activation levels from 0 to 40 time units in illustrative simulation 2

intrapersonal synchrony detectors and the conscious emotion remain high, although they fluctuate slightly under influence of the stochastic stimuli (Figs. 6 and 7).

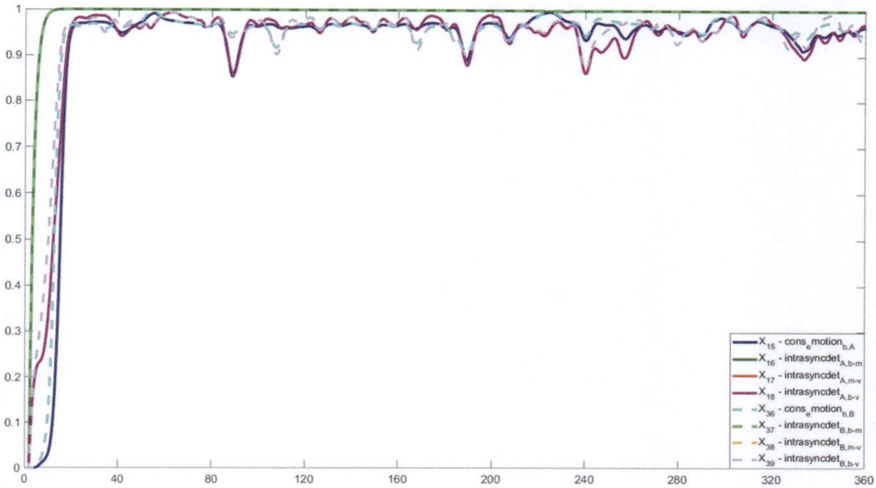

Fig. 6 The intrapersonal synchrony and conscious emotions activation levels from 0 to 360 time units in illustrative simulation 1

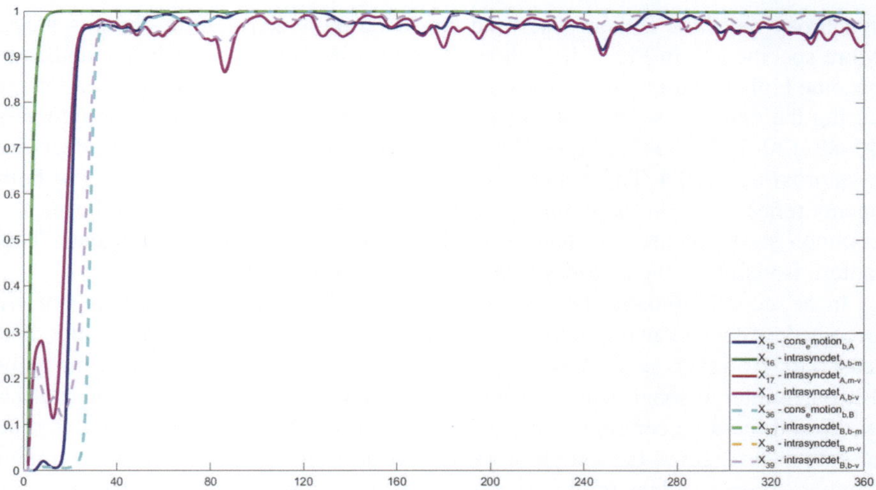

Fig. 7 The intrapersonal synchrony and conscious emotions activation levels from 0 to 360 time units in illustrative simulation 2

5.2 Intrapersonal Versus Interpersonal Synchrony

In Fig. 8 and 9, the intrapersonal and interpersonal synchrony detectors are compared. For the first illustrative simulation (Fig. 8), it can be seen that while the intrapersonal ones activate early and remain high (as also noticed in Sect. 5.1), the interpersonal

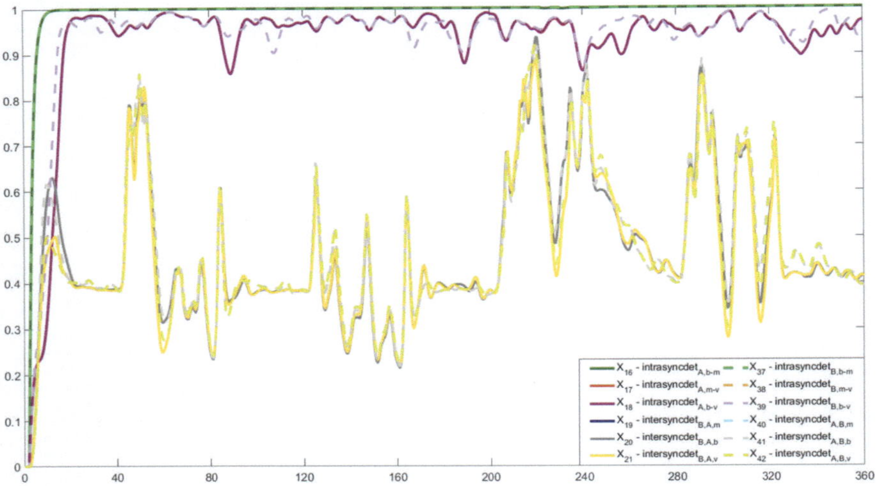

Fig. 8 The intrapersonal and interpersonal synchrony activation levels from 0 to 360 time units in illustrative simulation 1

synchrony detectors activate later and are much more sensitive to circumstances. More specifically, in Fig. 8 it is shown that the interpersonal synchrony detectors become higher (with values up to 0.8 and more) and more fluctuating in particular during the episodes where both agents receive a common stimulus (time intervals 40–80, 120–160, 200–240, 280–320). In between these episodes the levels decrease to approximately 0.4. This result suggests that the interpersonal synchrony of both agents relies more on the common stimulus (in-phase synchronization during the common stimulus interval) than on the interactions between the two agents. This pattern is visible in the majority of the 20 simulations run.

In the second simulation (Fig. 9), for agent B the levels of interpersonal synchrony become less dependent on a common stimulus and stay high all the time (values of at least 0.9). However, agent A shows a completely different pattern. From time point 120 on, its interpersonal synchrony detectors display lower levels of synchrony during the episodes with a common stimulus. In some of the 20 run simulations, this type of pattern in agent A (an antiphase synchronization in episodes with the common stimulus) occurs for one of the two agents, in contrast to the pattern of the agents A and B displayed (first illustrative simulation, Fig. 8).

5.3 *Raise of Affiliation After Synchronization*

A second important aim of the computational neural model introduced here was to study how synchrony leads to certain forms of affiliation or connecting. In Fig. 10 it is shown for the first presented simulation how the two agents connect stronger when

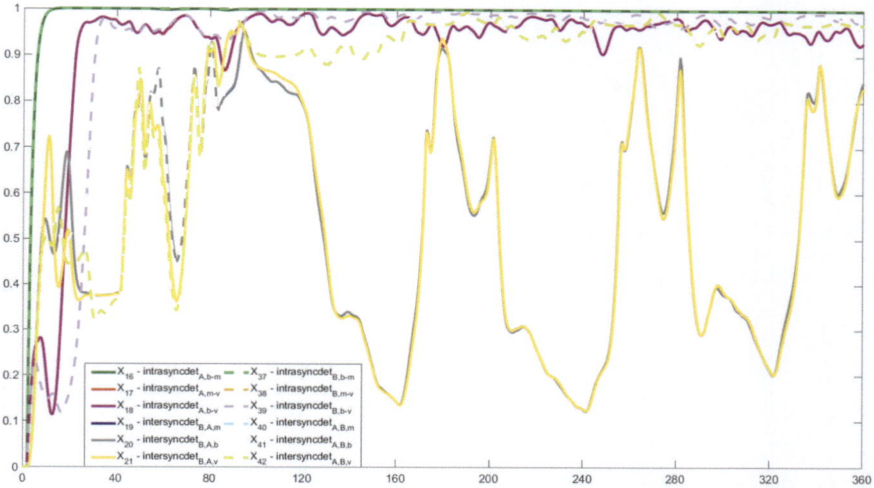

Fig. 9 The intrapersonal and interpersonal synchrony activation levels from 0 to 360 time units in illustrative simulation 2

the detected synchrony levels are higher. The type of connecting modelled here is instantaneous, which means that it reacts in a flexible way to the circumstances, in particular to the detected interpersonal synchrony, as can be seen in Fig. 10.

Also, in Fig. 11 for agent B the connecting goes hand in hand with the synchrony. However, in this case, agent B ends up in a constant high level of both detected interpersonal synchrony and affiliation. Therefore, agent B is not acting in a regular manner in accordance with the different stimulus intervals like happened in the first simulation example. Consequently, agent A also does not behave in accordance

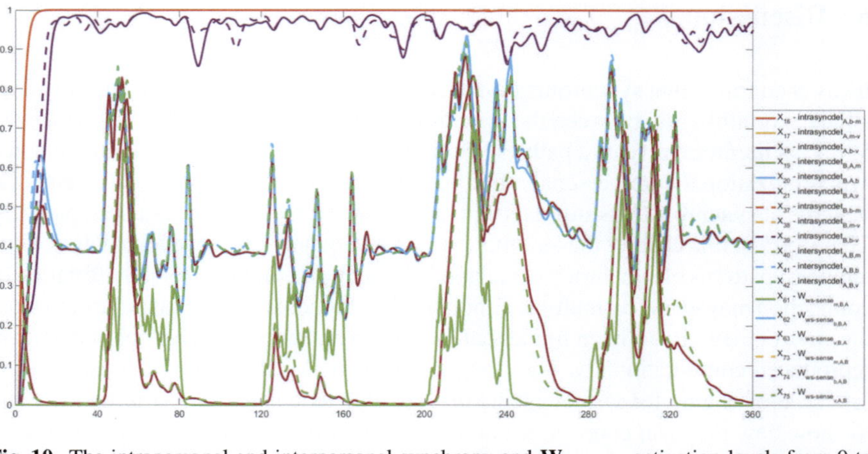

Fig. 10 The intrapersonal and interpersonal synchrony and $\mathbf{W}_{\text{ws-sense}}$ activation levels from 0 to 360 time units in illustrative simulation 1

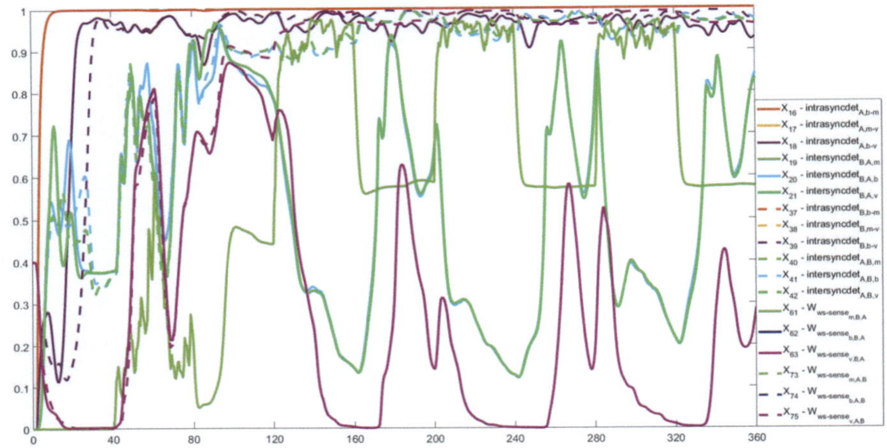

Fig. 11 The intrapersonal and interpersonal synchrony and $\mathbf{W}_{\text{ws-sense}}$ activation levels from 0 to 360 time units in illustrative simulation 2

with the different stimulus intervals, due to their interaction. Therefore, in this case some patterns occur that are counterintuitive from the perspective of the patterns for simulation example 1. From time 120 on, for all modalities the detected interpersonal synchrony of agent A becomes high in the intervals with different stimuli. The same holds for the affiliation (the $\mathbf{W}_{\text{ws-sense}}$ states) concerning affective response *b* and verbal response *v*, but not for movement *m*. This shows that due to environmental randomness also patterns can occur that may feel unexpected. Graphs for the other **W**-states can be found in the Appendix section.

6 Discussion

It has been found that synchronization leads to more closeness, mutual coordination, alliance, or affiliation between the synchronized persons; e.g., [2, 26, 30, 31, 35, 40, 55]. This literature reveals a pathway from interpersonal interaction to interpersonal synchronization to interpersonal affiliation or connection. If persons act on temporal patterns of synchrony, this suggests that they possess a facility to detect such patterns. Therefore, in this chapter the assumption was made that persons indeed detect when temporal patterns of synchrony occur and from such detection a stronger affiliation or connection may grow. By multiple simulation experiments for stochastic stimuli from the environment, it has been found that indeed several expected types of patterns are reproduced computationally. The designed computational agent model can be used as a kind of engine for virtual agents that are able to synchronize with humans and to show how they can connect or bond in a human-like manner. The material from this chapter is based on Hendrikse and colleagues [22].

Postulating such synchrony detector states can actually be seen as an application of the more general philosophical principle of temporal factorization introduced in [41]. This principle describes that when some temporal pattern a in the past entails some temporal pattern b in the future, then a state p exists (called mediating state) that occurs in the present such that pattern a entails state p and state p entails pattern b. An example from [41, 46] illustrates this principle for animal behavior. In many experiments it has been observed that when an animal has seen food at some position in the past, it shows delayed response behavior in the future: going to that position, even when nothing is visible. By the temporal factorization principle, it can be postulated that some kind of memory state exists in the present. A principle similar to temporal factorization has been formulated in a specialized form (called criterial causation) for the brain in particular in [52], see also [46]. Indeed, in the model introduced here, each synchrony detection state serves as such a mediating state p between synchrony patterns and the (adapted) future behavioral patterns they entail. In this way, the designed network model successfully exploits the temporal factorization principle to develop adaptivity of the interaction. A similar step to postulate subjective homophily detection for homophily-induced social interaction adaptivity was introduced in [48] and applied in [5], see also this volume [15] Part III, Chap. 7 [50] and Chap. 9 [6]).

For the synchrony detection states, it is easy to use any desired combination function, based on any method for this; see, for example, this volume [15], Chapter 1 [16], Sect. 1.4. Both heuristic and statistical functions may be used. The specific function used here for the synchrony detection states can be considered a heuristic function. As an alternative, also any statistical function can be used, for example Auguste Bravais's well-known statistical correlation coefficient introduced in [7], which sometimes is attributed to Karl Pearson and accordingly termed the Pearson correlation coefficient. It can be argued, however, whether such a statistical function describes what the brain actually does. Probably even the brains of Bravais (or Pearson) never actually used this statistical function to detect synchronies when they interacted with others in their personal life.

Brains usually apply heuristic approximations as these are much more efficient. Therefore, to design human-like agent models, heuristic functions such as the one used here based on comparing two different signals are probably more realistic. This function actually has some similarity to the well-known and often used comparator model to determine whether an action should be attributed to the self or to another person; e.g., [58, 59]. In [43], following [33, 39, 56], it has been shown how the comparator model can be remodeled more like an integration model where the two signals are not compared but aggregated over time, which makes it still more biologically plausible. This may also be done for the function used here for detection of synchrony. Another future research direction is to distinguish different types of affiliation, for example according to different time scales.

The research work on the network model introducing synchrony detectors and related behavioral adaptivity as presented here has created a strong basis for further research. Although the simulated scenario was kept abstract here, when data on particular therapeutical sessions are acquired, the model can be used to simulate such sessions in a more specific manner. This is meant for further research, see also [30–32, 36, 38]. In the meantime, since the work presented here was finished, a few refinements and variations on the introduced model have been and are being addressed.

One refinement addressed in this further research is a distinction in the timescales of the adaptation processes: between short-term and long-term behavioral adaptivity, for example, affiliation (e.g., within 45 min therapy episodes) versus bonding (e.g., over weeks or months). This can be found in [13], see also this volume [15], Chapter 12 [17].

Another extension addressed in further work is using more than two agents and distinguishing (1) relationship-specific behavioral adaptivity from (2) relationship-independent behavioral adaptivity. This can be found in [14], see also this volume [15], Chapter 13 [18].

Yet two other issues addressed as further work based on the work presented in this chapter are considering time lags between the incoming signals for the synchrony detectors [20] and adding detectors for switching in and out of synchrony [21]. See also this volume [15] Chapter 14 [23] and Chapter 15 [24].

Thus, the network model introducing synchrony detectors and related behavioral adaptivity as presented here has already played a very fruitful role in further research.

7. Appendix

In this appendix, first some further simulation pictures are shown for the affiliation patterns represented by the **W**-states in relation to the detected intrapersonal and interpersonal synchronies. Next, the full specification of the introduced adaptive network model is shown in terms of role matrices which are tables with the network characteristics in standardized table format. These tables are readable by the dedicated software environment available at https://www.researchgate.net/publication/368775720 and then can generate simulations. In this way reproducibility is supported (Figs. 12, 13, 14, 15, 16 and 17).

In Tables 4 to 8 the full specification of the adaptive network model by role matrices is shown. Each role matrix has 77 rows for all states X_1 to X_{77} of the model.

The connectivity characteristics are specified by role matrices **mb** and **mcw** shown in Tables 4 and 5. Role matrix **mb** lists for each state the states (at the same or lower level) from which the state gets its incoming connections, while in role matrix **mcw** the connection weights are listed for these connections.

Nonadaptive connection weights are indicated in **mcw** (in Table 5) by a number (in a green shaded cell), but adaptive connection weights are indicated by a reference to the (self-model) state representing the adaptive value (in a peach-red shaded cell). This can be seen for states X_5 to X_7 (with self-model states X_{61} to X_{63}), states X_9 to X_{11} (with self-model states X_{52} to X_{54}), X_{22} to X_{24} (with self-model states X_{55} to X_{57}), X_{26} to X_{28} (with self-model states X_{73} to X_{75}), X_{30} to X_{32} (with self-model

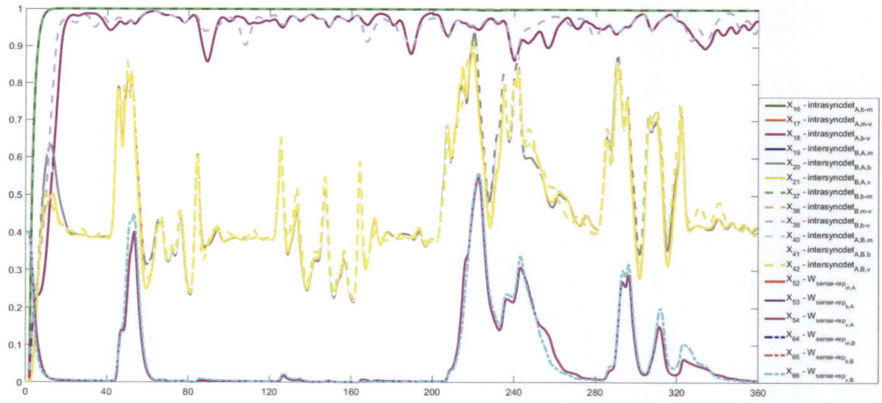

Fig. 12 The intrapersonal and interpersonal synchrony and $\mathbf{W}_{\text{sense-rep}}$ activation levels from 0 to 360 time units in illustrative simulation 1

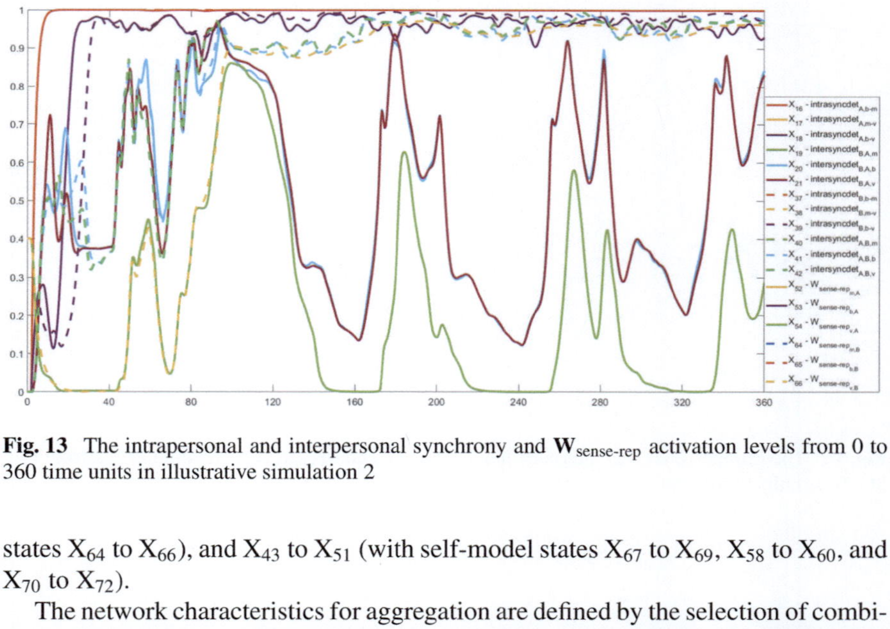

Fig. 13 The intrapersonal and interpersonal synchrony and $\mathbf{W}_{\text{sense-rep}}$ activation levels from 0 to 360 time units in illustrative simulation 2

states X_{64} to X_{66}), and X_{43} to X_{51} (with self-model states X_{67} to X_{69}, X_{58} to X_{60}, and X_{70} to X_{72}).

The network characteristics for aggregation are defined by the selection of combination functions from the library and values for their parameters. In role matrix **mcfw** it is specified by weights which state uses which combination function; see Table 6.

In role matrix **mcfp** (see Table 7) it is indicated what the parameter values are for the chosen combination functions. None of them are adaptive.

In Table 8, the role matrix **ms** for speed factors is shown, which lists all speed factors. Next to it, the list of initial values can be found. Also for **ms** some entries are adaptive: the speed factors of X_{52} to X_{75} are represented by (second-order) self-model states X_{76} and X_{77}.

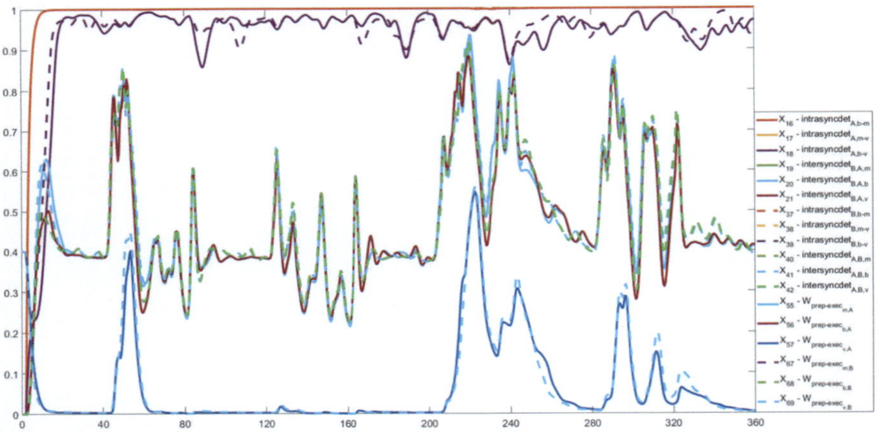

Fig. 14 The intrapersonal and interpersonal synchrony and $\mathbf{W}_{\text{prep-exec}}$ activation levels from 0 to 360 time units in illustrative simulation 1

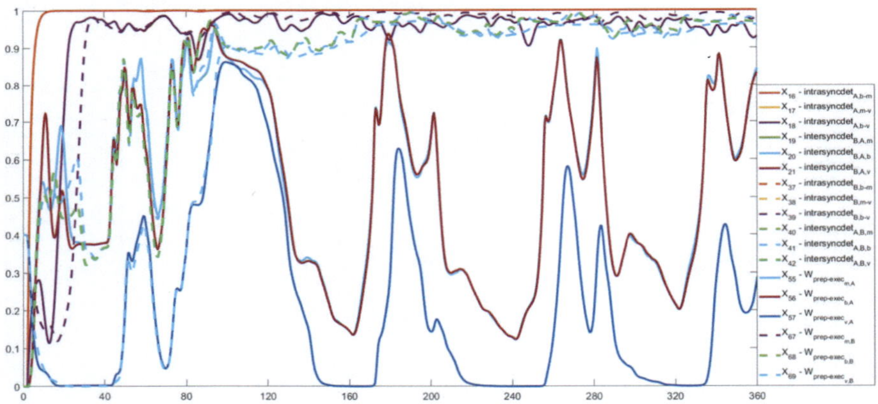

Fig. 15 The intrapersonal and interpersonal synchrony and $\mathbf{W}_{\text{prep-exec}}$ activation levels from 0 to 360 time units in illustrative simulation 2

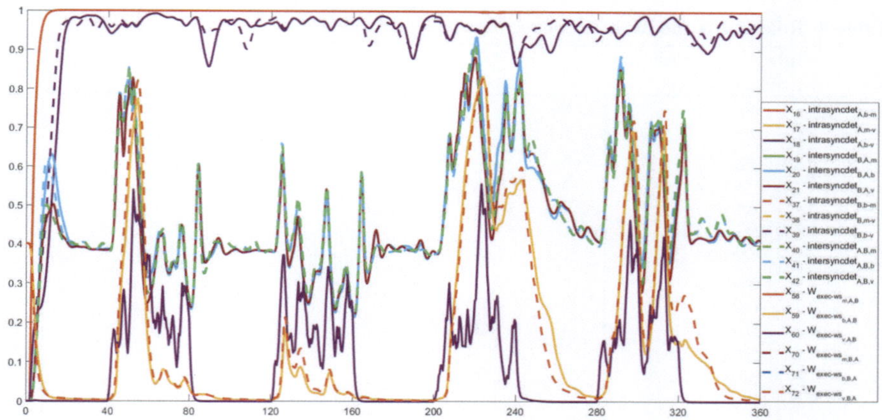

Fig. 16 The intrapersonal and interpersonal synchrony and $\mathbf{W}_{\text{exec-ws}}$ activation levels from 0 to 360 time units in illustrative simulation 1

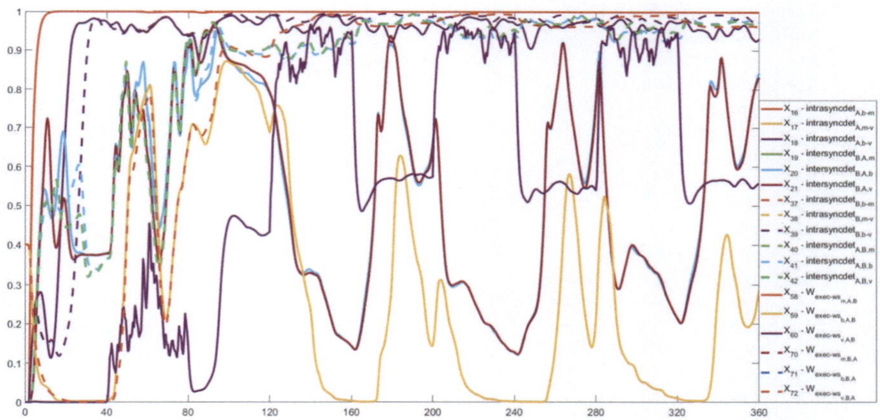

Fig. 17 The intrapersonal and interpersonal synchrony and $\mathbf{W}_{\text{exec-ws}}$ activation levels from 0 to 360 time units in illustrative simulation 2

Table 4 Role matrix **mb** for base connectivity

mb	base connectivity	1	2	3	4	5	6	7
X_1	$ws_{s,A}$	X_1						
X_2	$ws_{s,B}$	X_2						
X_3	$ws_{s,AB}$	X_3						
X_4	$sense_{s,A}$	X_3	X_1					
X_5	$sense_{B,m,A}$	X_{22}	X_{49}					
X_6	$sense_{B,b,A}$	X_{23}	X_{50}					
X_7	$sense_{B,v,A}$	X_{24}	X_{51}					
X_8	$rep_{s,A}$	X_4						
X_9	$rep_{B,m,A}$	X_5	X_{12}					
X_{10}	$rep_{B,b,A}$	X_6	X_{13}					
X_{11}	$rep_{B,v,A}$	X_7	X_{14}					
X_{12}	$prep_{m,A}$	X_8	X_9					
X_{13}	$prep_{b,A}$	X_8	X_{10}					
X_{14}	$prep_{v,A}$	X_8	X_{11}	X_{15}				
X_{15}	$cons\_emotion_{b,A}$	X_{13}	X_{24}	X_{16}	X_{17}	X_{18}		
X_{16}	$intrasyncdet_{A,\text{b-m}}$	X_{22}	X_{23}	X_{16}				
X_{17}	$intrasyncdet_{A,\text{m-v}}$	X_{22}	X_{24}	X_{17}				
X_{18}	$intrasyncdet_{A,\text{b-v}}$	X_{23}	X_{24}	X_{18}				
X_{19}	$intersyncdet_{B,A,\text{m}}$	X_{22}	X_5	X_{19}				
X_{20}	$intersyncdet_{B,A,\text{b}}$	X_{23}	X_6	X_{20}				
X_{21}	$intersyncdet_{B,A,\text{v}}$	X_{24}	X_7	X_{21}				
X_{22}	$move_{m,A}$	X_{12}						
X_{23}	$exp\_affect_{b,A}$	X_{13}						
X_{24}	$talk_{A,B,v}$	X_{14}						
X_{25}	$sense_{s,B}$	X_3	X_2					
X_{26}	$sense_{A,m,B}$	X_{43}	X_{46}					
X_{27}	$sense_{A,b,B}$	X_{44}	X_{47}					
X_{28}	$sense_{A,v,B}$	X_{45}	X_{48}					
X_{29}	$rep_{s,B}$	X_{25}						
X_{30}	$rep_{A,m,B}$	X_{26}	X_{33}					
X_{31}	$rep_{A,b,B}$	X_{27}	X_{34}					
X_{32}	$rep_{A,v,B}$	X_{28}	X_{35}					
X_{33}	$prep_{m,B}$	X_{29}	X_{30}					
X_{34}	$prep_{b,B}$	X_{29}	X_{31}					
X_{35}	$prep_{v,B}$	X_{29}	X_{32}	X_{36}				
X_{36}	$cons\_emotion_{b,B}$	X_{34}	X_{45}	X_{37}	X_{38}	X_{39}		
X_{37}	$intrasyncdet_{B,\text{b-m}}$	X_{43}	X_{44}	X_{37}				
X_{38}	$intrasyncdet_{B,\text{m-v}}$	X_{43}	X_{45}	X_{38}				
X_{39}	$intrasyncdet_{B,\text{b-v}}$	X_{44}	X_{45}	X_{39}				
X_{40}	$intersyncdet_{A,B,\text{m}}$	X_{43}	X_{26}	X_{40}				
X_{41}	$intersyncdet_{A,\text{B,b}}$	X_{44}	X_{27}	X_{41}				
X_{42}	$intersyncdet_{A,\text{B,v}}$	X_{45}	X_{28}	X_{42}				
X_{43}	$move_{m,B}$	X_{33}						
X_{44}	$exp\_affect_{b,B}$	X_{34}						
X_{45}	$talk_{B,A,v}$	X_{35}						
X_{46}	$ws_{m,A,B}$	X_{22}	X_3					
X_{47}	$ws_{b,A,B}$	X_{23}	X_3					
X_{48}	$ws_{v,A,B}$	X_{24}	X_3					
X_{49}	$ws_{m,B,A}$	X_{43}	X_3					
X_{50}	$ws_{b,B,A}$	X_{44}	X_3					

(continued)

Table 4 (continued)

X_{51}	$ws_{v,B,A}$	X_{45}	X_3					
X_{52}	$\mathbf{W}_{\text{sense-rep\_}m,A}$	X_{16}	X_{17}	X_{18}	X_{19}	X_{20}	X_{21}	
X_{53}	$\mathbf{W}_{\text{sense-rep\_}b,A}$	X_{16}	X_{17}	X_{18}	X_{19}	X_{20}	X_{21}	
X_{54}	$\mathbf{W}_{\text{sense-rep\_}v,A}$	X_{16}	X_{17}	X_{18}	X_{19}	X_{20}	X_{21}	
X_{55}	$\mathbf{W}_{\text{prep-exec\_}m,A}$	X_{16}	X_{17}	X_{18}	X_{19}	X_{20}	X_{21}	
X_{56}	$\mathbf{W}_{\text{prep-exec\_}b,A}$	X_{16}	X_{17}	X_{18}	X_{19}	X_{20}	X_{21}	
X_{57}	$\mathbf{W}_{\text{prep-exec\_}v,A}$	X_{16}	X_{17}	X_{18}	X_{19}	X_{20}	X_{21}	
X_{58}	$\mathbf{W}_{\text{exec-ws\_}m,A,B}$	X_{16}	X_{17}	X_{18}	X_{19}	X_{20}	X_{21}	X_3
X_{59}	$\mathbf{W}_{\text{exec-ws\_}b,A,B}$	X_{16}	X_{17}	X_{18}	X_{19}	X_{20}	X_{21}	X_3
X_{60}	$\mathbf{W}_{\text{exec-ws\_}v,A,B}$	X_{16}	X_{17}	X_{18}	X_{19}	X_{20}	X_{21}	X_3
X_{61}	$\mathbf{W}_{\text{ws-sense\_}m,B,A}$	X_{16}	X_{17}	X_{18}	X_{19}	X_{20}	X_{21}	X_3
X_{62}	$\mathbf{W}_{\text{ws-sense\_}b,B,A}$	X_{16}	X_{17}	X_{18}	X_{19}	X_{20}	X_{21}	X_3
X_{63}	$\mathbf{W}_{\text{ws-sense\_}v,B,A}$	X_{16}	X_{17}	X_{18}	X_{19}	X_{20}	X_{21}	X_3
X_{64}	$\mathbf{W}_{\text{sense-rep\_}m,B}$	X_{37}	X_{38}	X_{39}	X_{40}	X_{41}	X_{42}	
X_{65}	$\mathbf{W}_{\text{sense-rep\_}b,B}$	X_{37}	X_{38}	X_{39}	X_{40}	X_{41}	X_{42}	
X_{66}	$\mathbf{W}_{\text{sense-rep\_}v,B}$	X_{37}	X_{38}	X_{39}	X_{40}	X_{41}	X_{42}	
X_{67}	$\mathbf{W}_{\text{prep-exec\_}m,B}$	X_{37}	X_{38}	X_{39}	X_{40}	X_{41}	X_{42}	
X_{68}	$\mathbf{W}_{\text{prep-exec\_}b,B}$	X_{37}	X_{38}	X_{39}	X_{40}	X_{41}	X_{42}	
X_{69}	$\mathbf{W}_{\text{prep-exec\_}v,B}$	X_{37}	X_{38}	X_{39}	X_{40}	X_{41}	X_{42}	
X_{70}	$\mathbf{W}_{\text{exec-ws\_}m,B,A}$	X_{37}	X_{38}	X_{39}	X_{40}	X_{41}	X_{42}	X_3
X_{71}	$\mathbf{W}_{\text{exec-ws\_}b,B,A}$	X_{37}	X_{38}	X_{39}	X_{40}	X_{41}	X_{42}	X_3
X_{72}	$\mathbf{W}_{\text{exec-ws\_}v,B,A}$	X_{37}	X_{38}	X_{39}	X_{40}	X_{41}	X_{42}	X_3
X_{73}	$\mathbf{W}_{\text{ws-sense\_}m,A,B}$	X_{37}	X_{38}	X_{39}	X_{40}	X_{41}	X_{42}	X_3
X_{74}	$\mathbf{W}_{\text{ws-sense\_}b,A,B}$	X_{37}	X_{38}	X_{39}	X_{40}	X_{41}	X_{42}	X_3
X_{75}	$\mathbf{W}_{\text{ws-sense\_}v,A,B}$	X_{37}	X_{38}	X_{39}	X_{40}	X_{41}	X_{42}	X_3
X_{76}	$\mathbf{H}_{\mathbf{W_A}}$	X_8						
X_{77}	$\mathbf{H}_{\mathbf{W_B}}$	X_{29}						

Table 5 Role matrix **mcw** for connection weights

mcw	connection weights	1	2	3	4	5	6	7
X_1	$ws_{s,A}$	1						
X_2	$ws_{s,B}$	1						
X_3	$ws_{s,AB}$	1						
X_4	$sense_{s,A}$	1	1					
X_5	$sense_{B,m,A}$	1	X_{61}					
X_6	$sense_{B,b,A}$	1	X_{62}					
X_7	$sense_{B,v,A}$	1	X_{63}					
X_8	$rep_{s,A}$	1						
X_9	$rep_{B,m,A}$	X_{52}	1					
X_{10}	$rep_{B,b,A}$	X_{53}	1					
X_{11}	$rep_{B,v,A}$	X_{54}	1					
X_{12}	$prep_{m,A}$	1	1					
X_{13}	$prep_{b,A}$	1	1					
X_{14}	$prep_{v,A}$	1	1	1				
X_{15}	$cons\_emotion_{b,A}$	1	1	1	1	1		
X_{16}	$intrasyncdet_{A,\text{b-m}}$	1	1	1				
X_{17}	$intrasyncdet_{A,\text{m-v}}$	1	1	1				
X_{18}	$intrasyncdet_{A,\text{b-v}}$	1	1	1				
X_{19}	$intersyncdet_{B,A,\text{m}}$	1	1	1				
X_{20}	$intersyncdet_{B,A,\text{b}}$	1	1	1				
X_{21}	$intersyncdet_{B,A,\text{v}}$	1	1	1				
X_{22}	$move_{m,A}$	X_{55}						
X_{23}	$exp\_affect_{b,A}$	X_{56}						
X_{24}	$talk_{A,B,v}$	X_{57}						
X_{25}	$sense_{s,B}$	1	1					
X_{26}	$sense_{A,m,B}$	1	X_{73}					
X_{27}	$sense_{A,b,B}$	1	X_{74}					
X_{28}	$sense_{A,v,B}$	1	X_{75}					
X_{29}	$rep_{s,B}$	1						
X_{30}	$rep_{A,m,B}$	X_{64}	1					
X_{31}	$rep_{A,b,B}$	X_{65}	1					
X_{32}	$rep_{A,v,B}$	X_{66}	1					
X_{33}	$prep_{m,B}$	1	1					
X_{34}	$prep_{b,B}$	1	1					
X_{35}	$prep_{v,B}$	1	1	1				
X_{36}	$cons\_emotion_{b,B}$	1	1	1	1	1		
X_{37}	$intrasyncdet_{B,\text{b-m}}$	1	1	1				
X_{38}	$intrasyncdet_{B,\text{m-v}}$	1	1	1				
X_{39}	$intrasyncdet_{B,\text{b-v}}$	1	1	1				
X_{40}	$intersyncdet_{A,B,\text{m}}$	1	1	1				
X_{41}	$intersyncdet_{A,\text{B,b}}$	1	1	1				
X_{42}	$intersyncdet_{A,\text{B,v}}$	1	1	1				

(continued)

Table 5 (continued)

X_{43}	$\text{move}_{m,B}$	X_{67}						
X_{44}	$\text{exp\_affect}_{b,B}$	X_{68}						
X_{45}	$\text{talk}_{B,A,v}$	X_{69}						
X_{46}	$\text{ws}_{m,A,B}$	X_{58}	1					
X_{47}	$\text{ws}_{b,A,B}$	X_{59}	1					
X_{48}	$\text{ws}_{v,A,B}$	X_{60}	1					
X_{49}	$\text{ws}_{m,B,A}$	X_{70}	1					
X_{50}	$\text{ws}_{b,B,A}$	X_{71}	1					
X_{51}	$\text{ws}_{v,B,A}$	X_{72}	1					
X_{52}	$\mathbf{W}_{\text{sense-rep\_}m,A}$	1	1	1	1	1	1	
X_{53}	$\mathbf{W}_{\text{sense-rep\_}b,A}$	1	1	1	1	1	1	
X_{54}	$\mathbf{W}_{\text{sense-rep\_}v,A}$	1	1	1	1	1	1	
X_{55}	$\mathbf{W}_{\text{prep-exec\_}m,A}$	1	1	1	1	1	1	
X_{56}	$\mathbf{W}_{\text{prep-exec\_}b,A}$	1	1	1	1	1	1	
X_{57}	$\mathbf{W}_{\text{prep-exec\_}v,A}$	1	1	1	1	1	1	
X_{58}	$\mathbf{W}_{\text{exec-ws\_}m,A,B}$	1	1	1	1	1	1	1
X_{59}	$\mathbf{W}_{\text{exec-ws\_}b,A,B}$	1	1	1	1	1	1	1
X_{60}	$\mathbf{W}_{\text{exec-ws\_}v,A,B}$	1	1	1	1	1	1	1
X_{61}	$\mathbf{W}_{\text{ws-sense\_}m,B,A}$	1	1	1	1	1	1	1
X_{62}	$\mathbf{W}_{\text{ws-sense\_}b,B,A}$	1	1	1	1	1	1	1
X_{63}	$\mathbf{W}_{\text{ws-sense\_}v,B,A}$	1	1	1	1	1	1	1
X_{64}	$\mathbf{W}_{\text{sense-rep\_}m,B}$	1	1	1	1	1	1	
X_{65}	$\mathbf{W}_{\text{sense-rep\_}b,B}$	1	1	1	1	1	1	
X_{66}	$\mathbf{W}_{\text{sense-rep\_}v,B}$	1	1	1	1	1	1	
X_{67}	$\mathbf{W}_{\text{prep-exec\_}m,B}$	1	1	1	1	1	1	
X_{68}	$\mathbf{W}_{\text{prep-exec\_}b,B}$	1	1	1	1	1	1	
X_{69}	$\mathbf{W}_{\text{prep-exec\_}v,B}$	1	1	1	1	1	1	
X_{70}	$\mathbf{W}_{\text{exec-ws\_}m,B,A}$	1	1	1	1	1	1	1
X_{71}	$\mathbf{W}_{\text{exec-ws\_}b,B,A}$	1	1	1	1	1	1	1
X_{72}	$\mathbf{W}_{\text{exec-ws\_}v,B,A}$	1	1	1	1	1	1	1
X_{73}	$\mathbf{W}_{\text{ws-sense\_}m,A,B}$	1	1	1	1	1	1	1
X_{74}	$\mathbf{W}_{\text{ws-sense\_}b,A,B}$	1	1	1	1	1	1	1
X_{75}	$\mathbf{W}_{\text{ws-sense\_}v,A,B}$	1	1	1	1	1	1	1
X_{76}	$\mathbf{H}_{\mathbf{W}_\mathbf{A}}$	1						
X_{77}	$\mathbf{H}_{\mathbf{W}_\mathbf{B}}$	1						

Table 6 Role matrix **mcfw** for combination function weights

mcfw combination function weights		**1 alogistic**	**2 compdiff**	**3 randstepmod**	**4 randstepmo-dopp**
X_1	$ws_{s,A}$				1
X_2	$ws_{s,B}$				1
X_3	$ws_{s,AB}$			1	
X_4	$sense_{s,A}$	1			
X_5	$sense_{B,m,A}$	1			
X_6	$sense_{B,b,A}$	1			
X_7	$sense_{B,v,A}$	1			
X_8	$rep_{s,A}$	1			
X_9	$rep_{B,m,A}$	1			
X_{10}	$rep_{B,b,A}$	1			
X_{11}	$rep_{B,v,A}$	1			
X_{12}	$prep_{m,A}$	1			
X_{13}	$prep_{b,A}$	1			
X_{14}	$prep_{v,A}$	1			
X_{15}	$cons\_emotion_{b,A}$	1			
X_{16}	$intrasyncdet_{A,b\text{-}m}$		1		
X_{17}	$intrasyncdet_{A,m\text{-}v}$		1		
X_{18}	$intrasyncdet_{A,b\text{-}v}$		1		
X_{19}	$intersyncdet_{B,A,m}$		1		
X_{20}	$intersyncdet_{B,A,b}$		1		
X_{21}	$intersyncdet_{B,A,v}$		1		
X_{22}	$move_{m,A}$	1			
X_{23}	$exp\_affect_{b,A}$	1			
X_{24}	$talk_{A,B,v}$	1			
X_{25}	$sense_{s,B}$	1			
X_{26}	$sense_{A,m,B}$	1			
X_{27}	$sense_{A,b,B}$	1			
X_{28}	$sense_{A,v,B}$	1			
X_{29}	$rep_{s,B}$	1			
X_{30}	$rep_{A,m,B}$	1			
X_{31}	$rep_{A,b,B}$	1			
X_{32}	$rep_{A,v,B}$	1			
X_{33}	$prep_{m,B}$	1			
X_{34}	$prep_{b,B}$	1			
X_{35}	$prep_{v,B}$	1			
X_{36}	$cons\_emotion_{b,B}$	1			
X_{37}	$intrasyncdet_{B,b\text{-}m}$		1		
X_{38}	$intrasyncdet_{B,m\text{-}v}$		1		
X_{39}	$intrasyncdet_{B,b\text{-}v}$		1		
X_{40}	$intersyncdet_{A,B,m}$		1		
X_{41}	$intersyncdet_{A,B,b}$		1		
X_{42}	$intersyncdet_{A,B,v}$		1		
X_{43}	$move_{m,B}$	1			
X_{44}	$exp\_affect_{b,B}$	1			
X_{45}	$talk_{B,A,v}$	1			
X_{46}	$ws_{m,A,B}$	1			
X_{47}	$ws_{b,A,B}$	1			
X_{48}	$ws_{v,A,B}$	1			
X_{49}	$ws_{m,B,A}$	1			
X_{50}	$ws_{b,B,A}$	1			
X_{51}	$ws_{v,B,A}$	1			

(continued)

Table 6 (continued)

X_{52}	$\mathbf{W}_{\text{sense-rep\_}m,A}$	1	
X_{53}	$\mathbf{W}_{\text{sense-rep\_}b,A}$	1	
X_{54}	$\mathbf{W}_{\text{sense-rep\_}v,A}$	1	
X_{55}	$\mathbf{W}_{\text{prep-exec\_}m,A}$	1	
X_{56}	$\mathbf{W}_{\text{prep-exec\_}b,A}$	1	
X_{57}	$\mathbf{W}_{\text{prep-exec\_}v,A}$	1	
X_{58}	$\mathbf{W}_{\text{exec-ws\_}m,A,B}$	1	
X_{59}	$\mathbf{W}_{\text{exec-ws\_}b,A,B}$	1	
X_{60}	$\mathbf{W}_{\text{exec-ws\_}v,A,B}$	1	
X_{61}	$\mathbf{W}_{\text{ws-sense\_}m,B,A}$	1	
X_{62}	$\mathbf{W}_{\text{ws-sense\_}b,B,A}$	1	
X_{63}	$\mathbf{W}_{\text{ws-sense\_}v,B,A}$	1	
X_{64}	$\mathbf{W}_{\text{sense-rep\_}m,B}$	1	
X_{65}	$\mathbf{W}_{\text{sense-rep\_}b,B}$	1	
X_{66}	$\mathbf{W}_{\text{sense-rep\_}v,B}$	1	
X_{67}	$\mathbf{W}_{\text{prep-exec\_}m,B}$	1	
X_{68}	$\mathbf{W}_{\text{prep-exec\_}b,B}$	1	
X_{69}	$\mathbf{W}_{\text{prep-exec\_}v,B}$	1	
X_{70}	$\mathbf{W}_{\text{exec-ws\_}m,B,A}$	1	
X_{71}	$\mathbf{W}_{\text{exec-ws\_}b,B,A}$	1	
X_{72}	$\mathbf{W}_{\text{exec-ws\_}v,B,A}$	1	
X_{73}	$\mathbf{W}_{\text{ws-sense\_}m,A,B}$	1	
X_{74}	$\mathbf{W}_{\text{ws-sense\_}b,A,B}$	1	
X_{75}	$\mathbf{W}_{\text{ws-sense\_}v,A,B}$	1	
X_{76}	$\mathbf{H}\mathbf{w}_{\mathbf{A}}$	1	
X_{77}	$\mathbf{H}\mathbf{w}_{\mathbf{B}}$	1	

Table 7 Role matrix **mcfp** for combination function parameters

mcfp combination function parameters	1 alogistic		2 compdiff		3 randstepmod		4 randstepmo-dopp	
X_1 ws$_{s,A}$							80	40
X_2 ws$_{s,B}$							80	40
X_3 ws$_{s,AB}$					80	40		
X_4 sense$_{s,A}$	5	0.5						
X_5 sense$_{B,m,A}$	5	0.5						
X_6 sense$_{B,b,A}$	5	0.5						
X_7 sense$_{B,v,A}$	5	0.5						
X_8 rep$_{s,A}$	5	0.5						
X_9 rep$_{B,m,A}$	5	0.5						
X_{10} rep$_{B,b,A}$	5	0.5						
X_{11} rep$_{B,v,A}$	5	0.5						
X_{12} prep$_{m,A}$	5	0.6						
X_{13} prep$_{b,A}$	5	0.6						
X_{14} prep$_{v,A}$	5	0.9						
X_{15} cons_emotion$_{b,A}$	5	3.25						
X_{16} intrasyncdet$_{A,\text{b-m}}$								
X_{17} intrasyncdet$_{A,\text{m-v}}$								
X_{18} intrasyncdet$_{A,\text{b-v}}$								
X_{19} intersyncdet$_{B,A,\text{m}}$								
X_{20} intersyncdet$_{B,A,\text{b}}$								
X_{21} intersyncdet$_{B,A,\text{v}}$								
X_{22} move$_{m,A}$	5	0.5						
X_{23} exp_affect$_{b,A}$	5	0.5						
X_{24} talk$_{A,B,v}$	5	0.5						
X_{25} sense$_{s,B}$	5	0.5						
X_{26} sense$_{A,m,B}$	5	0.5						
X_{27} sense$_{A,b,B}$	5	0.5						
X_{28} sense$_{A,v,B}$	5	0.5						
X_{29} rep$_{s,B}$	5	0.5						
X_{30} rep$_{A,m,B}$	5	0.5						
X_{31} rep$_{A,b,B}$	5	0.5						
X_{32} rep$_{A,v,B}$	5	0.5						
X_{33} prep$_{m,B}$	5	0.6						
X_{34} prep$_{b,B}$	5	0.6						
X_{35} prep$_{v,B}$	5	0.9						
X_{36} cons_emotion$_{b,B}$	5	3.25						
X_{37} intrasyncdet$_{B,\text{b-m}}$								
X_{38} intrasyncdet$_{B,\text{m-v}}$								
X_{39} intrasyncdet$_{B,\text{b-v}}$								
X_{40} intersyncdet$_{A,B,\text{m}}$								
X_{41} intersyncdet$_{A,B,\text{b}}$								
X_{42} intersyncdet$_{A,B,\text{v}}$								
X_{43} move$_{m,B}$	5	0.5						
X_{44} exp_affect$_{b,B}$	5	0.5						
X_{45} talk$_{B,A,v}$	5	0.5						

(continued)

Table 7 (continued)

X_{46}	$\text{ws}_{m,A,B}$	5	0.8	
X_{47}	$\text{ws}_{b,A,B}$	5	0.8	
X_{48}	$\text{ws}_{v,A,B}$	5	0.8	
X_{49}	$\text{ws}_{m,B,A}$	5	0.8	
X_{50}	$\text{ws}_{b,B,A}$	5	0.8	
X_{51}	$\text{ws}_{v,B,A}$	5	0.8	
X_{52}	$\mathbf{W}_{\text{sense-rep}\_m,A}$	5	5.25	
X_{53}	$\mathbf{W}_{\text{sense-rep}\_b,A}$	5	5.25	
X_{54}	$\mathbf{W}_{\text{sense-rep}\_v,A}$	5	5.25	
X_{55}	$\mathbf{W}_{\text{prep-exec}\_m,A}$	5	5.25	
X_{56}	$\mathbf{W}_{\text{prep-exec}\_b,A}$	5	5.25	
X_{57}	$\mathbf{W}_{\text{prep-exec}\_v,A}$	5	5.25	
X_{58}	$\mathbf{W}_{\text{exec-ws}\_m,A,B}$	5	5.25	
X_{59}	$\mathbf{W}_{\text{exec-ws}\_b,A,B}$	5	5.25	
X_{60}	$\mathbf{W}_{\text{exec-ws}\_v,A,B}$	5	5.25	
X_{61}	$\mathbf{W}_{\text{ws-sense}\_m,B,A}$	5	5.25	
X_{62}	$\mathbf{W}_{\text{ws-sense}\_b,B,A}$	5	5.25	
X_{63}	$\mathbf{W}_{\text{ws-sense}\_v,B,A}$	5	5.25	
X_{64}	$\mathbf{W}_{\text{sense-rep}\_m,B}$	5	5.25	
X_{65}	$\mathbf{W}_{\text{sense-rep}\_b,B}$	5	5.25	
X_{66}	$\mathbf{W}_{\text{sense-rep}\_v,B}$	5	5.25	
X_{67}	$\mathbf{W}_{\text{prep-exec}\_m,B}$	5	5.25	
X_{68}	$\mathbf{W}_{\text{prep-exec}\_b,B}$	5	5.25	
X_{69}	$\mathbf{W}_{\text{prep-exec}\_v,B}$	5	5.25	
X_{70}	$\mathbf{W}_{\text{exec-ws}\_m,B,A}$	5	5.25	
X_{71}	$\mathbf{W}_{\text{exec-ws}\_b,B,A}$	5	5.25	
X_{72}	$\mathbf{W}_{\text{exec-ws}\_v,B,A}$	5	5.25	
X_{73}	$\mathbf{W}_{\text{ws-sense}\_m,A,B}$	5	5.25	
X_{74}	$\mathbf{W}_{\text{ws-sense}\_b,A,B}$	5	5.25	
X_{75}	$\mathbf{W}_{\text{ws-sense}\_v,A,B}$	5	5.25	
X_{76}	$\mathbf{Hw_A}$	5	0.7	
X_{77}	$\mathbf{Hw_B}$	5	0.7	

Table 8 Role matrix **ms** for speed factors and **iv** for initial values

ms	**speed factors**	**1**	**iv initial values**	**1**
X_1	$ws_{s,A}$	1		1
X_2	$ws_{s,B}$	1		1
X_3	$ws_{s,AB}$	1		0
X_4	$sense_{s,A}$	1		0
X_5	$sense_{B,m,A}$	1		0
X_6	$sense_{B,b,A}$	1		0
X_7	$sense_{B,v,A}$	1		0
X_8	$rep_{s,A}$	1		0
X_9	$rep_{B,m,A}$	1		0
X_{10}	$rep_{B,b,A}$	1		0
X_{11}	$rep_{B,v,A}$	1		0
X_{12}	$prep_{m,A}$	1		0
X_{13}	$prep_{b,A}$	1		0
X_{14}	$prep_{v,A}$	1		0
X_{15}	$cons\_emotion_{b,A}$	1		0
X_{16}	$intrasyncdet_{A,\text{b-m}}$	0.5		0
X_{17}	$intrasyncdet_{A,\text{m-v}}$	0.5		0
X_{18}	$intrasyncdet_{A,\text{b-v}}$	0.5		0
X_{19}	$intersyncdet_{B,A,\text{m}}$	0.5		0
X_{20}	$intersyncdet_{B,A,\text{b}}$	0.5		0
X_{21}	$intersyncdet_{B,A,\text{v}}$	0.5		0
X_{22}	$move_{m,A}$	1		0
X_{23}	$exp\_affect_{b,A}$	1		0
X_{24}	$talk_{A,B,v}$	1		0
X_{25}	$sense_{s,B}$	1		0
X_{26}	$sense_{A,m,B}$	1		0
X_{27}	$sense_{A,b,B}$	1		0
X_{28}	$sense_{A,v,B}$	1		0
X_{29}	$rep_{s,B}$	1		0
X_{30}	$rep_{A,m,B}$	1		0
X_{31}	$rep_{A,b,B}$	1		0
X_{32}	$rep_{A,v,B}$	1		0
X_{33}	$prep_{m,B}$	1		0
X_{34}	$prep_{b,B}$	1		0
X_{35}	$prep_{v,B}$	1		0
X_{36}	$cons\_emotion_{b,B}$	1		0
X_{37}	$intrasyncdet_{B,\text{b-m}}$	0.5		0
X_{38}	$intrasyncdet_{B,\text{m-v}}$	0.5		0
X_{39}	$intrasyncdet_{B,\text{b-v}}$	0.5		0
X_{40}	$intersyncdet_{A,B,\text{m}}$	0.5		0
X_{41}	$intersyncdet_{A,\text{B,b}}$	0.5		0
X_{42}	$intersyncdet_{A,\text{B,v}}$	0.5		0
X_{43}	$move_{m,B}$	1		0
X_{44}	$exp\_affect_{b,B}$	1		0
X_{45}	$talk_{B,A,v}$	1		0
X_{46}	$ws_{m,A,B}$	1		0
X_{47}	$ws_{b,A,B}$	1		0
X_{48}	$ws_{v,A,B}$	1		0
X_{49}	$ws_{m,B,A}$	1		0

(continued)

Table 8 (continued)

X_{50}	$ws_{b,B,A}$	1		0
X_{51}	$ws_{v,B,A}$	1		0
X_{52}	$\mathbf{W}_{\text{sense-rep\_}m,A}$	X_{76}		0.4
X_{53}	$\mathbf{W}_{\text{sense-rep\_}b,A}$	X_{76}		0.4
X_{54}	$\mathbf{W}_{\text{sense-rep\_}v,A}$	X_{76}		0.4
X_{55}	$\mathbf{W}_{\text{prep-exec\_}m,A}$	X_{76}		0.4
X_{56}	$\mathbf{W}_{\text{prep-exec\_}b,A}$	X_{76}		0.4
X_{57}	$\mathbf{W}_{\text{prep-exec\_}v,A}$	X_{76}		0.4
X_{58}	$\mathbf{W}_{\text{exec-ws\_}m,A,B}$	X_{76}		0.4
X_{59}	$\mathbf{W}_{\text{exec-ws\_}b,A,B}$	X_{76}		0.4
X_{60}	$\mathbf{W}_{\text{exec-ws\_}v,A,B}$	X_{76}		0.4
X_{61}	$\mathbf{W}_{\text{ws-sense\_}m,B,A}$	X_{76}		0.4
X_{62}	$\mathbf{W}_{\text{ws-sense\_}b,B,A}$	X_{76}		0.4
X_{63}	$\mathbf{W}_{\text{ws-sense\_}v,B,A}$	X_{76}		0.4
X_{64}	$\mathbf{W}_{\text{sense-rep\_}m,B}$	X_{77}		0.4
X_{65}	$\mathbf{W}_{\text{sense-rep\_}b,B}$	X_{77}		0.4
X_{66}	$\mathbf{W}_{\text{sense-rep\_}v,B}$	X_{77}		0.4
X_{67}	$\mathbf{W}_{\text{prep-exec\_}m,B}$	X_{77}		0.4
X_{68}	$\mathbf{W}_{\text{prep-exec\_}b,B}$	X_{77}		0.4
X_{69}	$\mathbf{W}_{\text{prep-exec\_}v,B}$	X_{77}		0.4
X_{70}	$\mathbf{W}_{\text{exec-ws\_}m,B,A}$	X_{77}		0.4
X_{71}	$\mathbf{W}_{\text{exec-ws\_}b,B,A}$	X_{77}		0.4
X_{72}	$\mathbf{W}_{\text{exec-ws\_}v,B,A}$	X_{77}		0.4
X_{73}	$\mathbf{W}_{\text{ws-sense\_}m,A,B}$	X_{77}		0.4
X_{74}	$\mathbf{W}_{\text{ws-sense\_}b,A,B}$	X_{77}		0.4
X_{75}	$\mathbf{W}_{\text{ws-sense\_}v,A,B}$	X_{77}		0.4
X_{76}	$\mathbf{H}_{\mathbf{W_A}}$	1		0
X_{77}	$\mathbf{H}_{\mathbf{W_B}}$	1		0

References

1. Abraham, W. C., Bear, M. F. (1996). Metaplasticity: the plasticity of synaptic plasticity. Trends in neurosciences, 19(4), 126-130.
2. Accetto, M., Treur, J., Villa, V.: An adaptive cognitive-social model for mirroring and social bonding during synchronous joint action. Procedia Comput. Sci. **145**, 3–12 (2018). https://doi.org/10.1016/j.procs.2018.11.002
3. Antonio Damasio (1999). The Feeling of What Happens: Body and Emotion in the Making of Consciousness. New York: Harcourt Brace
4. Bloch, C., Vogeley, K., Georgescu, A. L., Falter-Wagner, C. M.: INTRApersonal synchrony as constituent of INTERpersonal synchrony and its relevance for autism spectrum disorder. Front Robot AI **73** (2019)
5. Bössenecker, G., Mreijen, P., Treur, J: A second-order adaptive network model for connecting social media platforms and users through bonding by faked homophily. Procedia Comput. Sci. **213**, 32–40 (2022). https://doi.org/10.1016/j.procs.2022.11.035
6. Bössenecker, G., Mreijen, P., Treur, J: Connecting social media platforms and users through bonding by faked homophily: a second-order adaptive network model. In: Hendrikse, S.C.F., Treur, J., Koole, S.L. (eds.) New Analysis and Modeling Directions in Social Interaction Science: Emergent Multimodal Interpersonal Synchrony, Affiliation, and Bonding, Ch 9. Springer Nature (2025)
7. Bravais, A.: Analyse Mathematique. Sur les probabilités des erreurs de situation d'un point. Mem. Acad. Roy. Sei. Inst. France, Sci. Math, et Phys. **9**, 255–332 (1844)
8. Collins, D.F., Cameron, T., Gillard, D.M., Prochazka, A.: Muscular sense is attenuated when humans move. J. Physiol. **508**, 635–643 (1998)
9. Fairhurst, M.T., Janata, P., Keller, P.E.: Being and feeling in sync with an adaptive virtual partner: Brain mechanisms underlying dynamic cooperativity. Cereb. Cortex **23**(11), 2592–2600 (2013). https://doi.org/10.1093/cercor/bhs243
10. Feldman, R.: Parent–infant synchrony biological foundations and developmental outcomes. Curr. Dir. Psychol. Sci. **16**, 340–345 (2007). https://doi.org/10.1111/j.1467-8721.2007.00532.x
11. Grandjean, D., Sander, D., Scherer, K. R. (2008). Conscious emotional experience emerges as a function of multilevel, appraisal-driven response synchronization. Consciousness and cognition, 17(2), 484-495.
12. Hendrikse, S.C.F., Kluiver, S., Treur, J., Wilderjans, T.F., Dikker, S., Koole, S.L.: How virtual agents can learn to synchronize: an adaptive joint decision-making model of psychotherapy. Cogn. Syst. Res. **79**, 138–155 (2022)
13. Hendrikse, S.C.F., Treur, J., Koole, S.L.: Modeling emerging interpersonal synchrony and its related adaptive short-term affliation and long-term bonding: a second-order multi-adaptive neural agent model. Inter. J. Neural Syst **33**(7), 2350038 (2023)
14. Hendrikse, S.C.F., Treur, J., Koole, S.L.: Relationship-specific and relationship-independent behavioural adaptivity in affiliation and bonding: a multi-adaptive dynamical systems approach. Cogn. Syst. Res. **84**, 101182 (2024)
15. Hendrikse, S.C.F., Treur, J., Koole, S.L. (eds.), New analysis and modeling directions in social interaction science: emergent multimodal interpersonal synchrony, affiliation, and bonding. Springer Nature (2025)
16. Hendrikse, S.C.F., Treur, J., Koole, S.L.: Introduction to the role of adaptivity in social interaction science. In: Hendrikse, S.C.F., Treur, J., Koole, S.L. (eds.) New Analysis and Modeling Directions in Social Interaction Science: Emergent Multimodal Interpersonal Synchrony, Affiliation, and Bonding, Ch 1. Springer Nature (2025)
17. Hendrikse, S.C.F., Treur, J., Koole, S.L.: The adaptive interplay of interpersonal synchrony, short-term affiliation and long-term bonding: a multi-timescale multi-adaptive agent model. In: Hendrikse, S.C.F., Treur, J., Koole, S.L. (eds.), New Analysis and Modeling Directions in Social Interaction Science: Emergent Multimodal Interpersonal Synchrony, Affiliation, and Bonding, Ch 12. Springer Nature (2025)

18. Hendrikse, S.C.F., Treur, J., Koole, S.L.: A multi-timescale multi-adaptive dynamical system approach for relationship-specific and relationship-independent adaptivity in affiliation and bonding. In: Hendrikse, S.C.F., Treur, J., Koole, S.L. (eds.) New Analysis and Modeling Directions in Social Interaction Science: Emergent Multimodal Interpersonal Synchrony, Affiliation, and Bonding, Ch 13. Springer Nature (2025)
19. Hendrikse, S., Treur, J., Wilderjans, T., Dikker, S., Koole, S.L.: On the same wavelengths: emergence of multiple synchronies among multiple agents. In: Proceedings of the 22nd International Workshop on Multi-Agent-Based Simulation, MABS'21. Lecture Notes in AI, vol. 13128, pp. 57–71. Springer Nature (2022)
20. Hendrikse, S.C.F., Treur, J., Wilderjans, T.F., Dikker, S., Koole, S.L.: Becoming attuned to each other over time: a computational neural agent model for the role of time lags in subjective synchrony detection and related behavioral adaptivity. In: Proceedings of the 15th International Conference on Brain Informatics, BI'22. Lecture Notes in AI, vol. 13406, pp 369–383. Springer Nature (2022)
21. Hendrikse, S.C.F., Treur, J., Wilderjans, T.F., Dikker, S., Koole, S.L.: Switching in and out of sync: a controlled adaptive network model of transition dynamics in the effects of interpersonal synchrony on affiliation. In: Proceedings of the 11th International Conference on Complex Networks and their Applications. Studies in Computational Intelligence, vol. 1078, pp. 81–95. Cham: Springer Nature Switzerland (2022)
22. Hendrikse, S.C.F., Treur, J., Wilderjans, T.F., Dikker, S., Koole, S.L.: On becoming in sync with yourself and others: an adaptive agent model for how persons connect by detecting intrapersonal and interpersonal synchrony. Hum-Centric Intell Syst **3**(2), 123–146 (2023)
23. Hendrikse, S.C.F., Treur, J., Wilderjans, T.F., Dikker, S., Koole, S.L.: The role of timing in synchrony detection and social interaction adaptivity: including time lags in a second-order adaptive agent model. In: Hendrikse, S.C.F., Treur, J., Koole, S.L. (eds.) New Analysis and Modeling Directions in Social Interaction Science: Emergent Multimodal Interpersonal Synchrony, Affiliation, and Bonding, Ch 14. Springer Nature (2025)
24. Hendrikse, S.C.F., Treur, J., Wilderjans, T.F., Dikker, S., Koole, S.L.: The effect of transition dynamics in synchrony on social interaction adaptivity: a multi-adaptive network model. In: Hendrikse, S.C.F., Treur, J., Koole, S.L. (eds.) New Analysis and Modeling Directions in Social Interaction Science: Emergent Multimodal Interpersonal Synchrony, Affiliation, and Bonding, Ch 15. Springer Nature (2025)
25. Hesslow, G. (2002). Conscious thought as simulation of behaviour and perception. Trends in cognitive sciences, 6(6), 242-247.
26. Hove, M.J., Risen, J.L.: It's all in the timing: interpersonal synchrony increases affiliation. Soc. Cogn. **27**(6), 949–960 (2009)
27. Izawa, J., Rane, T., Donchin, O., Shadmehr, R.: Motor adaptation as a process of reoptimization. J. Neurosci.: Off. J. Soc. Neurosci. **28**(11), 2883–2891 (2008).https://doi.org/10.1523/JNEUROSCI.5359-07.2008
28. Keller, P.E., Novembre, G., Hove, M.J.: Rhythm in joint action: psychological and neurophysiological mechanisms for real-time interpersonal coordination. Philos. Trans. R. Soc. Lond. Ser. B: Biol. Sci **369**(1658), Article 20130394 (2014). https://doi.org/10.1098/rstb.2013.0394
29. Kirschner, S., Tomasello, M.: Joint music making promotes prosocial behavior in 4-year-old children. Evol. Hum. Behav. **31**, 354–364 (2010). https://doi.org/10.1016/j.evolhumbehav.2010.04.004
30. Koole, S.L., Tschacher, W.: Synchrony in psychotherapy: a review and an integrative framework for the therapeutic alliance. Front. Psychol. **7**, 862 (2016)
31. Koole, S.L., Tschacher, W., Butler, E., Dikker, S., Wilderjans, T.F.: In sync with your shrink. Forgas, J.P., Crano, W.D., Fiedler, K. (eds.) Applications of Social Psychology. Milton Park: Taylor and Francis, pp. 161–184 (2020)
32. Maurer, R.E., Tindall, J. H.: Effect of postural congruence on client's perception of counselor empathy. J. Counseling Psychol. **30**:158 (1983). https://doi.org/10.1037/0022-0167.30.2.158
33. Moore, J., Haggard, P.: Awareness of action: inference and prediction. Conscious. Cogn. **17**, 136–144 (2008)

34. Pacherie, E.: The phenomenology of joint action: Self-agency vs. joint-agency. In: Seeman, A (ed.) Joint Attention: New Developments. MIT Press, pp. 343–389 (2012)
35. Prince, K., Brown, S.: Neural correlates of partnered interaction as revealed by cross-domain ALE meta-analysis. Psychol. Neurosci. (2022). https://doi.org/10.1037/pne0000282
36. Ramseyer, F., Tschacher, W.: Nonverbal synchrony in psychotherapy: coordinated body movement reflects relationship quality and outcome. J. Consult. Clin. Psychol. **79**, 284–295 (2011). https://doi.org/10.1037/a0023419a
37. Robinson, B.L., Harper, N.S., McAlpine, D.: Meta-adaptation in the auditory midbrain under cortical influence. Nat. Commun. 7, e13442 (2016)
38. Sharpley, C.F., Halat, J., Rabinowicz, T., Weiland, B., Stafford, J.: Standard posture, postural mirroring and client-perceived rapport. Couns. Psychol. Q. **14**, 267–280 (2001). https://doi.org/10.1080/09515070110088843
39. Synofzik, M., Thier, P., Leube, D.T., Schlotterbeck, P., Lindner, A.: Misattributions of agency in schizophrenia are based on imprecise predictions about the sensory consequences of one's actions. Brain **133**, 262–271 (2010)
40. Tarr, B., Launay, J., Dunbar, R.I.M.: Silent disco: dancing in synchrony leads to elevated pain thresholds and social closeness. Evol. Hum. Behav. **37**(5), 343–349 (2016)
41. Treur, J.: Temporal factorisation: a unifying principle for dynamics of the world and of mental states. Cogn. Syst. Res. **8**(2), 57–74 (2007)
42. Treur, J.: Temporal factorisation: realisation of mediating state properties for dynamics. Cogn. Syst. Res. **8**(2), 75–88 (2007)
43. Treur, J.: A cognitive agent model incorporating prior and retrospective ownership states for actions. In: Walsh, T. (ed.) Proceedings of the Twenty-Second International Joint Conference on Artificial Intelligence, IJCAI'11, pp. 1743–1749 (2011)
44. Treur, J.: Network-oriented modeling for adaptive networks: designing higher-order adaptive biological, mental and social network models. Springer Nature (2020)
45. Treur, J.: Modeling Multi-order adaptive processes by self-modeling networks. In: Antonio J. Tallón-Ballesteros and Chi-Hua Chen (eds.) Proceedings of the 2nd International Conference on Machine Learning and Intelligent Systems, MLIS'20. Frontiers in Artificial Intelligence and Applications, vol. 332, pp. 206–217. IOS Press (2020)
46. Treur, J.: Modeling the emergence of informational content by adaptive networks for temporal factorisation and criterial causation. Cogn. Syst. Res. **68**, 34–52 (2021)
47. Treur, J.: On the dynamics and adaptivity of mental processes: relating adaptive dynamical systems and self-modeling network models by mathematical analysis. Cogn. Syst. Res. **70**, 93–100 (2021)
48. Treur, J.: Controlled social network adaptation: subjective elements in an objective social world. In: Yang, X.S., Sherratt, S., Dey, N., Joshi, A. (eds.) Proceedings of Sixth International Congress on Information and Communication Technology, ICICT'21. Lecture Notes in Networks and Systems, vol. 235, pp. 263–274. Springer Nature (2022). https://doi.org/10.1007/978-981-16-2377-6_26
49. Treur, J.: Modeling coevolution of dynamics, adaptivity and control of adaptivity by self-modeling networks. In: Hendrikse, S.C.F., Treur, J., Koole, S.L. (eds.) New Analysis and Modeling Directions in Social Interaction Science: Emergent Multimodal Interpersonal Synchrony, Affiliation, and Bonding, Ch 2. Springer Nature (2025)
50. Treur, J.: A second-order adaptive network model for bonding based on subjectively perceived homophily. In: Hendrikse, S.C.F., Treur, J., Koole, S.L. (eds.) New Analysis and Modeling Directions in Social Interaction Science: Emergent Multimodal Interpersonal Synchrony, Affiliation, and Bonding, Ch 7. Springer Nature (2025)
51. Trout, D.L., Rosenfeld, H.M.: The effect of postural lean and body congruence on the judgment of psychotherapeutic rapport. J. Nonverbal Behav. **4**,176–190 (1980). https://doi.org/10.1007/BF00986818
52. Tse, P.U.: The Neural Basis of Free Will: Criterial Causation. MIT Press, Cambridge (2013)
53. Vacharkulksemsuk, T., Fredrickson, B. L.: Strangers in sync: achieving embodied rapport through shared movements. J. Exp. Soc. Psychol. **48**(1), 399-402 (2012)

54. Valdesolo, P., DeSteno, D.: Synchrony and the social tuning of compassion. Emotion **11**, 262 (2011). https://doi.org/10.1037/a0021302
55. Valdesolo, P., Ouyang, J., DeSteno, D.: The rhythm of joint action: synchrony promotes cooperative ability. J. Exp. Soc. Psychol. **46**(4), 693–695 (2010)
56. Voss, M., Moore, J., Hauser, M., Gallinat, J., Heinz, A., Haggard, P.: Altered awareness of action in schizophrenia: a specific deficit in predicting action consequences. Brain **133**, 3104–3112 (2010)
57. Wiltermuth, S. S., Heath, C.: Synchrony and cooperation. Psychological science, 20(1), 1-5 (2009)
58. Wolpert, D.M.: Computational approaches to motor control. Trends Cogn. Sci. **1**, 209–216 (1997)
59. Zaadnoordijk, L., Besold, T.R., Hunnius, S.: A match does not make a sense: on the sufficiency of the comparator model for explaining the sense of agency. Neurosci. Conscious. **5**(1), niz006 (2019). https://doi.org/10.1093/nc/niz006

The Adaptive Interplay of Interpersonal Synchrony, Short-Term Affiliation and Long-Term Bonding: A Multi-timescale Multi-adaptive Agent Model

Sophie C. F. Hendrikse, Jan Treur, and Sander L. Koole

Abstract When people interact, their behavior tends to become synchronized, a mutual coordination process that fosters short-term adaptations, like increased affiliation, and long-term adaptations, like increased bonding. This chapter addresses how such short-term and long-term adaptivity induced by synchronization can be modeled computationally by a second-order multi-adaptive agent model. It addresses movement, affect and verbal modalities and both intrapersonal synchrony and interpersonal synchrony. The behavior of the introduced agent model was evaluated in a simulation paradigm with different stimuli and communication enabling conditions. Moreover, in this chapter, mathematical analysis is also addressed for adaptive network models and their positioning within the landscape of adaptive dynamical systems. The first type of analysis addressed shows that any smooth adaptive dynamical system has a canonical representation by a self-modeling network. This implies theoretically that the self-modeling network format is widely applicable, which also has been found in many practical applications using this approach. Furthermore, stationary point and equilibrium analysis was addressed and applied to the introduced self-modeling network model. It was used to obtain verification of the model providing evidence that the implemented model is correct with respect to its design specifications.

S. C. F. Hendrikse
Department of Methodology and Statistics, Tilburg School of Social and Behavioral Sciences, Tilburg University, Tilburg, The Netherlands
e-mail: s.c.f.hendrikse@tilburguniversity.edu

J. Treur (✉)
Social AI Group, Department of Computer Science, Vrije Universiteit Amsterdam, Amsterdam, The Netherlands
e-mail: j.treur@vu.nl

S. L. Koole
Amsterdam Emotion Regulation Lab, Department of Clinical Psychology, Vrije Universiteit Amsterdam, Amsterdam, The Netherlands
e-mail: s.l.koole@vu.nl

S. C. F. Hendrikse et al. (eds.), *New Analysis and Modeling Directions in Social Interaction Science*, Studies in Systems, Decision and Control 614,
https://doi.org/10.1007/978-3-031-99968-0_12

Keywords Multi-adaptive agent model · Social interaction · Interpersonal synchrony · Intrapersonal synchrony · Short-term affiliation · Long-term bonding

1 Introduction

Whenever people interact, their behavior tends to become mutually coordinated in time, or synchronized. Interpersonal synchrony has been found to enhance relationship functioning, for example, by inducing greater levels of closeness, concentration, coordination, cooperation, affiliation, alliance, connection, or bonding [2, 19, 20, 23, 27, 35, 59, 65, 87]. For some more background about synchrony, see, for example, this volume [39], Chap. 1 [40], Sects. "The Multidisciplinary Field of Synchrony" and "Methods for Detecting Interpersonal Synchrony". In literature [29, 85] it was suggested that to model the complex, cyclical types of dynamics that occur, a dynamical systems modeling approach is needed. Notably, the benefits of interpersonal synchrony include patterns of mutual adaptation both in the short term and in the long term. For instance, in the context of psychotherapy, a patient and therapist who synchronize their movements [59] may experience a stronger sense of sharing the present moment during a therapeutic session [83]. Over multiple sessions, this increased social presence may strengthen the therapeutic bond, which allows the patient and therapist to work together more effectively [49].

This chapter has as *main goal* to address computational and mathematical analyses of the complex adaptive dynamics at different timescales of such forms of short-term and long-term adaptivity of interaction behavior related to interpersonal synchronization and to verify the *hypothesis* that the underlying mechanisms put forward in the literature indeed generate these social phenomena by an emerging and adaptive interactive process. More specifically, these analyses cover three different but closely related levels:

- Analysis of *mechanisms* from the literature in *psychology and neuroscience* that are suggested to play a role in these complex multi-adaptive dynamics. These include mechanisms for different forms of (synaptic and non-synaptic) plasticity and control over plasticity by meta-plasticity (also called second-order plasticity or second-order adaptivity). Here the *underlying hypothesis* in the literature is that these mechanisms put forward are sufficient and able to generate the emerging and adaptive patterns of synchronization and adaptation of the interaction behavior. This hypothesis is tested 'in silico' in the current chapter by computational simulation based on these mechanisms.
- Analysis of *mathematical formalization* of these *mechanisms* behind the considered complex adaptive dynamics by an agent-based second-order multi-adaptive dynamical system. This covers analysis of conducted simulation experiments based on such formalization and includes analysis of stationary points and equilibria for the occurring dynamics and adaptivity.

- Analysis of how the specific *representation* of such an adaptive dynamical system by *self-modeling networks* used here is positioned in the wider landscape of adaptive dynamical systems. It is analyzed how the self-modeling network format can be used to provide a canonical representation for any smooth adaptive dynamical system, which also covers most neural system models.

So, more specifically, the agent model that is a central focus here is an adaptive dynamical systems model based on a number of mechanisms in the literatures on cognitive, behavioral, and affective neuroscience. A neural basis for short-term behavioral adaptivity can be found in the recent work on the (non-synaptic, intrinsic) adaptive excitability of (neural) states [24, 26, 86, 89]. By contrast, a neural basis for long-term adaptivity can be found in the classical notion of synaptic plasticity [9, 10, 34, 63, 64]. Together, these two fundamentally different forms of adaptation yield a model of a multi-adaptive agent. The two forms of adaptation also interact with each other where the nonsynaptic adaptation is more short-term and synaptic adaptation more long-term, e.g., [6, 33, 51], see also this volume [39], Chap. 2 "Modeling Coevolution of Dynamics, Adaptivity and Control of Adaptivity by Self-modeling Networks" [80].

The extent of adaptation that an agent requires may vary from situation to situation. The capacity to adjust plasticity to the demands of the situation relates to metaplasticity [1, 60]. The present model of a agent models meta-plasticity as a second-order form of plasticity that controls plasticity in a context-sensitive manner. The resulting model yields a second-order multi-adaptive agent, which is human-like in the sense that it incorporates an interplay of three major mechanisms for adaptivity that according to the neuroscientific literature characterise human agents.

Note that it is not claimed that the model is human-like in the sense that it would cover the applied neural mechanisms at a physiological level. Addressing that level was not within the scope of this research. Instead, these neural mechanisms were considered and modeled in a more abstract manner at a functional level. Investigating physiological scalability would be another, next enterprise to be addressed.

The present model of a agent further includes intrapersonal synchrony and interpersonal synchrony and their links to short-term and long-term behavioural adaptivity. To model the pathway from synchrony patterns to this behavioural adaptivity, we included both built-in intrapersonal synchrony and interpersonal synchrony detectors. Here, intrapersonal synchrony means that within an agent, actions for the different modalities occur in a coordinated manner. Interpersonal synchrony means that for each modality the actions of the two agents occur in a coordinated manner. The addressed modalities are movement, affect and verbal modalities. We included these three modalities because they have each been shown to be influential in interpersonal behavior [49].

We evaluated the agent model in a series of simulation experiments for two agents with a setup in which a number of stochastic circumstances were covered in different (time) episodes. The simulations included not only episodes with a stochastic common stimulus for the two agents, but also episodes with different stochastic stimuli for the agents. Moreover, to analyse the role of communication, stochastic circumstances were also included for episodes when communication was enabled by the environment and episodes when communication was not enabled.

Next, as part of further analysis of the self-modeling network modeling approach it is shown how any (smooth) adaptive dynamical system can be modeled in a canonical way as a self-modeling network model. In this way, any adaptive dynamical system has its canonical representation as a self-modeling network model and can be analysed based on this canonical representation. On the one hand, this shows that the chosen modeling approach does not introduce biases or limitations if adaptive dynamical systems are modeled using it. In particular, it shows also that the approach generalises most common neural system models. On the other hand, this was a basis to show how stationary point analysis and equilibrium analysis for adaptive dynamical systems can be performed by using the self-modeling network representation for the specific adaptive dynamical system model introduced here.

2 Main Assumptions and Background Knowledge

In this section, we present the main assumptions behind the introduced adaptive agent model and relate them to the relevant neuroscience literatures. This grounding in neuroscience is based on pathways for a circular interplay of synchrony with both non-synaptic plasticity [26] and synaptic plasticity [9, 10, 34, 63, 64], thereby covering an interplay of both short-term time scales and long-term time scales and their interaction [6, 33, 51]. More specifically, the following underlying assumptions are made for the pathways involved; for a conceptual overview, see Fig. 1. Note that the main example used in the presentation concerns two agents A and B and their interaction.

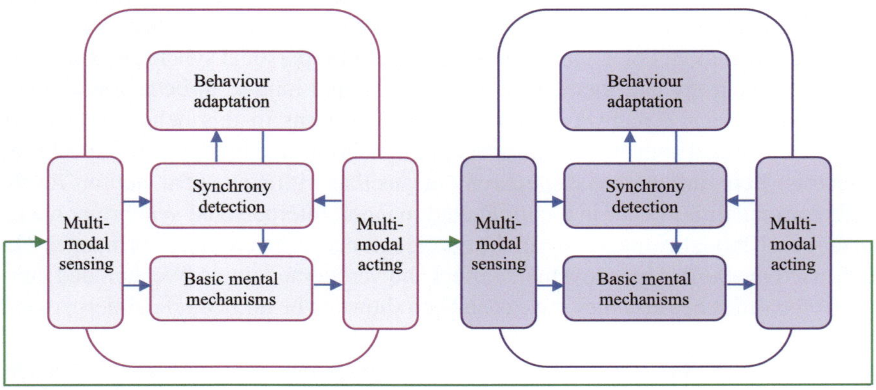

Fig. 1 Conceptual overview of the processes involved in multimodal (intra- and interpersonal) synchrony and social interaction adaptivity; see also [30]

2.1 Interpersonal Synchrony Leads to Adaptation of Interaction Behaviour

Interpersonal synchrony is often followed by a behavioural change or adaptation of mutual behaviour [2, 19, 20, 23, 27, 35, 48, 65, 87]. This adaptive shift in mutual behavioral coordination has been observed, for instance, in psychotherapy sessions. Research has shown that therapists were rated more favorably and as more empathic when, beforehand, they were instructed to make their movements more synchronized with the client [53, 62]. Similarly, [59] found that initial movement synchrony between client and therapist was predictive of the client's experience of the quality of the alliance at the end of each session. Furthermore, Koole and Tschacher [49] reviewed converging evidence that movement synchrony has a positive effect on the working alliance between patient and therapist. More generally, synchrony in face-to-face interactions has been found to promote interpersonal affiliation [28, 87]. More background about synchrony can be found in this volume [39], Chap. "Introduction to the Role of Adaptivity in Social Interaction Science" [40], Sects. The Multidisciplinary Field of Synchrony and Methods for Detecting Interpersonal Synchrony.

Behavioural adaptation after interpersonal synchrony occurs both in the form of short-term adaptation and long-term adaptation. Much research on interpersonal synchrony has focused on short-term adaptive changes in interpersonal coordination [2, 35, 65, 87]. However, several lines of research have observed effects of interpersonal synchrony on long-term adaptation as well. First, developmental research has observed that movement synchrony between infant and caregivers predict social interaction patterns of the child several years later [28]. Second, research on close relationships suggests that early patterns of interpersonal synchrony predict subsequent indicators of relationship functioning, for instance, one study found that spouses' patterns of cortisol variation converged over a period of years, indicating long-term shifts in interpersonal coordination [50].

Third and last, research on psychotherapy processes has found that markers of interpersonal synchrony in early sessions can predict the development of the therapeutic relationship and therapeutic outcomes [48]. Long-term adaptation processes remain less well-studied than short-term adaptation processes. Nevertheless, the convergence of evidence is sufficient to conclude that interpersonal synchrony is likely to promote both short-term and long-term adaptation in interpersonal relationships.

2.2 The Behavioural Adaptation Relies on Different Neural Mechanisms: Synaptic Plasticity of Connections and Non-synaptic Plasticity of Intrinsic Excitability

In the neuroscientific literature, a distinction is made between synaptic and non-synaptic (intrinsic) adaptation. The classical notion of synaptic plasticity [9, 10, 34, 63, 64] has been used to explain long-term behavioural adaptation. This addresses how the strength of a connection between different states is adapted over time due to simultaneous activation of the connected states. By contrast, the non-synaptic adaptation of intrinsic excitability of (neural) states has been addressed in more detail more recently [24, 26, 86, 89]. The latter form of adaptation has been related, for example, to homeostatic regulation [86] and also to how deviant dopamine levels during sleep make that dreams can use more associations due to easier excitable neurons [17]. Moreover, both (synaptic and non-synaptic) forms of adaptation can easily work together [6, 33, 51].

In the agent model these two adaptation mechanisms and their interaction have been used to model behavioural adaptivity: the former (synaptic) for long-term adaptation and the latter (nonsynaptic) for short-term adaptation. Here an interplay of two types of adaptivity occurs. Synchrony does not only lead to short-term adaptation, but short-term adaptation itself also intensifies interaction which can lead to more synchrony which in turn can strengthen the long-term adaptation. Besides, also long-term adaptivity strengthens interaction which leads to more synchrony and consequently stronger short-term adaptivity. In this way, via multiple circular pathways a dynamic interplay occurs between synchrony, short-term adaptivity and long-term adaptivity.

Plasticity is not a constant feature, as it often is highly context-dependent according to what is called metaplasticity [1, 60]. For example, 'Adaptation accelerates with increasing stimulus exposure' [60]. To enable such context-sensitive control of plasticity, second-order adaptation (i.e., adaptation of the adaptation) has been included in the agent model, which makes the model more realistic.

2.3 The Pathways from Synchrony to Behavioural Adaptation Involve Synchrony Detection States

If synchrony occurs for an agent and due to this the agent adapts the interaction behaviour, this suggests that agents possess a facility to notice or experience synchrony patterns for the different modalities. Indeed, the assumption is made that agents do in some way (perhaps unconsciously) detect synchrony and from there may trigger behavioural adaptation for their interaction behaviour. In the pathway from synchrony patterns to changed interaction behaviour patterns, such synchrony detection states can be considered as specific mediating mental states. Such a state p in general has been called a mediating state for the effect of a past pattern a on a future pattern *b* entailed by pattern *a* [67, 68]; similarly, such a (brain) state is referred to as describing 'informational criteria' for future activation [74, 84]. In line with previous research, it is assumed that not only the detected interpersonal synchrony but also the detected intrapersonal synchrony relating to a conscious emotion has a causal effect on the behavioural adaptivity: [43], see also this volume, Chap. 11 [45].

3 Self-modeling Network Modeling

The presented agent model is an adaptive dynamical system model designed and specified based on network-oriented modeling. The network-oriented modeling approach used here is basically a causal network modeling approach where nodes model states that have activation values that change over time and connections between these states model causal relations that have their effects in a temporal, dynamic manner on the state activations. Thus, dynamical systems are modeled. Moreover, by enabling these causal relations and characteristics for their effects on state activations to change as well, also adaptive dynamical systems are covered. This is also done in a network-oriented manner, using a so-called self-modeling network architecture. In Sect. 6, it will be shown that any smooth adaptive dynamical system can be modeled in this way. In the current section, this modeling approach is briefly introduced; see also this volume [39], Chap. 2 "Modeling Coevolution of Dynamics, Adaptivity and Control of Adaptivity by Self-modeling Networks" [80].

Following the adaptive network-oriented modeling approach [69–73] used here, a temporal-causal network model is characterized by (here *X* and *Y* denote nodes of the network, also called states, and *X(t)* and *Y(t)* denote their activation values at time *t*):

- *Connectivity characteristics*

  Connections from a state *X* to a state *Y* and their weights $\boldsymbol{\omega}_{X,Y}$
- *Aggregation characteristics*

  For any state *Y*, some combination function

  $\mathbf{c}_{\boldsymbol{\pi}_Y,Y}(V_1, .., V_k)$ with vector of parameter values $\boldsymbol{\pi}_Y = (\boldsymbol{\pi}_{1,Y}, \ldots, \boldsymbol{\pi}_{m,Y})$ defines the aggregation that is applied to the impacts $V_i = \boldsymbol{\omega}_{X_i,Y} X_i(t)$ on *Y* from its incoming connections from states X_i.
- *Timing characteristics*

  Each state *Y* has a speed factor $\boldsymbol{\eta}_Y$ defining how fast it changes for given causal impact.

Note that for the sake of notational simplicity, often $\mathbf{c}_{\boldsymbol{\pi}_Y,Y}$ will be denoted by $\mathbf{c}_Y$, omitting the subscript $\boldsymbol{\pi}_Y$; this will not mean that there are no parameters, they are just left implicit.

These network characteristics $\boldsymbol{\omega}_{X,Y}$, $\mathbf{c}_{\boldsymbol{\pi}_Y,Y}$, $\boldsymbol{\pi}_Y$ and $\boldsymbol{\eta}_Y$ for a given network model serve as a (formal) design specification of this network model. The following *canonical difference and differential equations for temporal-causal network models* used for simulation and analysis of such network models, incorporate these network characteristics $\boldsymbol{\omega}_{X,Y}$, $\mathbf{c}_Y$, $\boldsymbol{\pi}_{i,Y}$ and $\boldsymbol{\eta}_Y$ in a standard numerical format:

$$Y(t+\Delta t) = Y(t) + \quad \boldsymbol{\eta}_Y\big[\mathbf{c}_{\boldsymbol{\pi}_Y,Y}\big(\boldsymbol{\omega}_{X_1,Y}X_1(t), \ldots, \boldsymbol{\omega}_{X_k,Y}X_k(t)\big) - Y(t)\big]\Delta t$$
$$\frac{dY(t)}{dt} = \boldsymbol{\eta}_Y\big[\mathbf{c}_{\boldsymbol{\pi}_Y,Y}\big(\boldsymbol{\omega}_{X_1,Y}X_1(t), \ldots, \boldsymbol{\omega}_{X_k,Y}X_k(t)\big) - Y(t)\big] \tag{1}$$

for any state Y and where X_1 to X_k are the states from which Y gets its incoming connections. Note that (1) has a format similar to that of recurrent neural networks. Within the dedicated software environment implemented in MATLAB, a large number of currently around 60 useful basic combination functions are included in a combination function library [72].

The above concepts enable us to design network models and their dynamics in a declarative manner, based on mathematically defined functions and relations and specified in a standard table format covering all network characteristics (called role matrices, see the Appendix B in Sect. 10). The examples of combination functions that are applied in the model introduced here can be found in Table 1. Here, for the third and fourth function, rand (1, 1) draws a random number from [0, 1] in a uniform manner and a is a persistence factor (with value 0.5 used in the simulations).

Realistic network models are usually adaptive: often not only their states but also some of their network characteristics change over time. By using a *self-modeling network* (also called a *reified* network), a similar network-oriented conceptualization can also be applied to adaptive networks to obtain a declarative description using mathematically defined functions and relations for them as well [71–73]. This works through the addition of new states to the network (called *self-model states*) which represent (adaptive) network characteristics. In the graphical 3D-format as shown in Sect. 4, such additional states are depicted at a next level (called *self-model level* or *reification level*), where the original network is at the *base level*.

As an example, the weight $\boldsymbol{\omega}_{X,Y}$ of a connection from state X to state Y can be represented (at a next self-model level) by a (connectivity) self-model state named $\mathbf{W}_{X,Y}$, which can be used to model synaptic plasticity [71–73]. Similarly, all other network characteristics from $\boldsymbol{\omega}_{X,Y}$, $\mathbf{c}_Y(..)$ and $\boldsymbol{\eta}_Y$ can be made adaptive by including self-model states for them. For example, for adaptive excitability [24, 26, 86, 89] the threshold $\boldsymbol{\tau}_Y$ (for a logistic combination function) for a state Y can be represented by a (aggregation) self-model state named $\mathbf{T}_Y$ and an adaptive speed factor $\boldsymbol{\eta}_Y$ can be represented by a (timing) self-model state named $\mathbf{H}_Y$. Dynamics for the activation values for these self-model states are modeled by adding their own network characteristics (thus integrating them in the network structure) and applying Eq. (1) for them.

Table 1 The combination functions used in the introduced network model

Name	Specification	Parameters	Used for
Advanced logistic sum **alogistic**$_{\sigma,\tau}(V_1, \ldots, V_k)$	$\left[\frac{1}{1+e^{-\sigma(V_1+\ldots+V_k-\tau)}} - \frac{1}{(1+e^{\sigma\tau})}\right](1+e^{-\sigma\tau})$	π_1: steepness $\boldsymbol{\sigma}$ π_2: excitability Threshold $\boldsymbol{\tau}$	$X_4 - X_5,\ X_{10} - X_{16},$ $X_{24} - X_{26},\ X_{31} - X_{38},$ $X_{45} - X_{47},\ X_{54} - X_{59},$ $X_{63} - X_{71},\ X_{75} - X_{93}$
Complemental difference **compdiff**(V_1, V_2)	0 if $V_1 = V_2 = 0$ $1 - \frac{\lvert V_1-V_2\rvert}{\max(V_1,V_2)}$ else	None	$X_{18} - X_{23},\ X_{39} - X_{44}$ (synchrony detectors)
Random Stepmod **randstepmod**$_{\rho,\delta}(V)$	0 if $0 \leq$ time t mod $\boldsymbol{\rho} \leq \boldsymbol{\delta}$ $aV + (1-a)$rand(1, 1) else	π_1: repetition $\boldsymbol{\rho}$ π_2: step time $\boldsymbol{\delta}$	X_3 (common stimulus) $X_{60} - X_{62},\ X_{72} - X_{74}$ (communication enablers)
Random Stepmodopp **randstepmodopp**$_{\rho,\delta}(V)$	0 if $\boldsymbol{\delta} \leq$ time t mod $\boldsymbol{\rho} \leq \boldsymbol{\rho}$ $aV + (1-a)$rand(1, 1) else	π_1: repetition $\boldsymbol{\rho}$ π_2: step time $\boldsymbol{\delta}$	$X_1 - X_2$ (individual stimuli)
Euclidean **eucl**$_{n,\lambda}(V_1, \ldots, V_k)$	$\sqrt[n]{\frac{V_1^n+\cdots+V_k^n}{\lambda}}$	π_1: order $\boldsymbol{n}$ π_2: scaling factor $\boldsymbol{\lambda}$	$X_6 - X_9,\ X_{27} - X_{30}$ (sensing) $X_{48} - X_{53}$ (communication)

If for all network characteristics $\boldsymbol{\omega}$, $\boldsymbol{\pi}$, $\boldsymbol{\eta}$ for all base level states, respective self-model states **W**, **P**, **H** are introduced representing these network characteristics, then the *canonical difference and differential equations* for the base level states of the self-modeling network model are:

$$Y(t + \Delta t) = Y(t) + \mathbf{H}_Y(t)[\mathbf{c}_{\mathbf{P}_Y(t),Y}(\mathbf{W}_{X_1,Y}(t)X_1(t), \ldots, \mathbf{W}_{X_k,Y}(t)X_k(t)) - Y(t)]\Delta t$$
$$dY(t)/dt) = \mathbf{H}_Y(t)[\mathbf{c}_{\mathbf{P}_Y(t),Y}(\mathbf{W}_{X_1,Y}(t)X_1(t), \ldots, \mathbf{W}_{X_k,Y}(t)X_k(t)) - Y(t)] \quad (2)$$

where $\mathbf{P}_Y(t) = \left(\mathbf{P}_{1,Y}(t), \ldots, \mathbf{P}_{m,Y}(t)\right)$.

This canonical difference equation is incorporated in the dedicated software environment. By instantiating this general difference Eq. (2) by proper values for the network characteristics for all base states Y and similarly instantiating Eq. (1) for all self-model states, the software environment runs a system of n difference equations where n is the number of (base and self-model) states in the network.

When Eq. (2) are compared to Eq. (1), it can be noticed that at each point in time t, for the value of each network characteristic the activation value of its corresponding self-model state is used: for $\boldsymbol{\eta}_Y$ the value $\mathbf{H}_Y(t)$ is used, for $\boldsymbol{\omega}_{X_i,Y}$ the value $\mathbf{W}_{X_i,Y}(t)$, etc. In this way, each of these self-model states is assigned the functional role of the specific network characteristic it represents. In particular, when the activation values of these **W**-states, **P**-states and **H**-states change, the corresponding network characteristics change accordingly. This makes these network characteristics adaptive.

More mathematical background of this self-modeling network architecture construction and Eq. (2) has been described in [71, 72].

Usually not for all network characteristics self-model states are introduced but only for part of them. For example, in case only self-model states **P** for the combination function parameters $\boldsymbol{\pi}$ are introduced, then the canonical difference and differential equation are:

$$Y(t + \Delta t) = Y(t) + \boldsymbol{\eta}_Y\left[\mathbf{c}_{\mathbf{P}_Y(t),Y}\left(\boldsymbol{\omega}_{X_1,Y}X_1(t), \ldots, \boldsymbol{\omega}_{X_k,Y}X_k(t)\right) - Y(t)\right]\Delta t$$
$$\frac{dY(t)}{dt} = \boldsymbol{\eta}_Y\left[\mathbf{c}_{\mathbf{P}_Y(t),Y}\left(\boldsymbol{\omega}_{X_1,Y}X_1(t), \ldots, \boldsymbol{\omega}_{X_k,Y}X_k(t)\right) - Y(t)\right] \quad (3)$$

where $\mathbf{P}_Y(t) = \left(\mathbf{P}_{1,Y}(t), \ldots, \mathbf{P}_{m,Y}(t)\right)$.

This specific case will come back in the mathematical analysis addressed in Sect. 6.2 (to establish Theorem 2 there).

Note that difference and differential Eq. (2) are not exactly in the standard format of a temporal-causal network, as $\mathbf{H}_Y$ is not a constant speed factor and also the **P**- and **W**-values are not constant. However, it can be rewritten into the temporal-causal network format when the following combination function $\mathbf{c}^*_Y(..)$ is defined:

$$\begin{aligned} &\mathbf{c}^*_Y(H, P_1, \ldots, P_m, W_1, \ldots, W_k, V_1, \ldots, V_k, V) \\ &= H\mathbf{c}_{P,Y}(W_1V_1, \ldots, W_kV_k) + (1 - H)V \end{aligned} \quad (4)$$

where $P = (P_1, \ldots, P_m)$.

Based on this combination function, consider the following difference equation:

$$\begin{aligned} &Y(t+\Delta t) = Y(t) + \\ &[\mathbf{c}^*_Y(\mathbf{H}_Y(t), \mathbf{P}_{1,Y}(t), \ldots, \mathbf{P}_{m,Y}(t), \mathbf{W}_{X_1,Y}(t), \ldots, \mathbf{W}_{X_k,Y}(t), \\ &X_1(t), \ldots, X_k(t), Y(t)) - Y(t)]\Delta t \end{aligned} \tag{5}$$

This is indeed in temporal-causal network format (1) (with speed factor 1). Now note that using (4), Eq. (5) can be rewritten as follows:

$$\begin{aligned} Y(t+\Delta t) &= Y(t) + \Big[\mathbf{H}_Y(t)\mathbf{c}_{\mathbf{P}_Y(t),Y}\big(\mathbf{W}_{X_1,Y}(t)X_1(t), \ldots, \mathbf{W}_{X_k,Y}(t)X_k(t)\big) + (1-\mathbf{H}_Y(t))Y(t) - Y(t)\Big]\Delta t \\ &= Y(t) + \mathbf{H}_Y(t)\mathbf{c}_{\mathbf{P}_Y(t),Y}\big(\mathbf{W}_{X_1,Y}(t)X_1(t), \ldots, \mathbf{W}_{X_k,Y}(t)X_k(t)\big) - \mathbf{H}_Y(t)Y(t)\Big]\Delta t \\ &= Y(t) + \mathbf{H}_Y(t)\Big[\mathbf{c}_{\mathbf{P}_Y(t),Y}\big(\mathbf{W}_{X_1,Y}(t)X_1(t), \ldots, \mathbf{W}_{X_k,Y}(t)X_k(t)\big) - Y(t)\Big]\Delta t \end{aligned} \tag{6}$$

This (6) shows exactly difference Eq. (2) above; this confirms that the chosen combination function $\mathbf{c}^*_Y(..)$ in (4) to show that the self-modeling network has a temporal-causal network format (1) works.

As the outcome of a process of network reification is also a temporal-causal network model itself, as has been shown above, this self-modeling network construction can easily be applied iteratively to obtain multiple orders of self-models at multiple (first-order, second-order, …) self-model levels. For example, a second-order self-model may include a second-order (timing) self-model state $\mathbf{H}_{\mathbf{W}_{X,Y}}$ representing the speed factor $\boldsymbol{\eta}_{\mathbf{W}_{X,Y}}$ for the dynamics of first-order self-model state $\mathbf{W}_{X,Y}$ which in turn represents the adaptation of connection weight $\boldsymbol{\omega}_{X,Y}$. Similarly, a second-order self-model may include a second-order (timing) self-model state $\mathbf{H}_{\mathbf{T}_Y}$ representing the speed factor $\boldsymbol{\eta}_{\mathbf{T}_Y}$ for the dynamics of first-order self-model state $\mathbf{T}_Y$ which in turn represents the adaptation of excitability threshold $\boldsymbol{\tau}_Y$ for Y.

In the current chapter, this multi-level self-modeling network modelling perspective will be applied to obtain a second-order adaptive network architecture addressing controlled behavioural adaptation induced by detected synchrony. In this self-modeling network architecture, the first-order self-model models the adaptation of the base level network that models behaviour, and the second-order self-model level the control over this adaptation. As an example, the control level can be used to make the adaptation speed context-sensitive as addressed by meta-plasticity literature [1, 60]. For instance, the meta-plasticity principle 'Adaptation accelerates with increasing stimulus exposure' formulated by [60] can easily be modeled by using second-order self-model states for timing; this actually has been done for the introduced model, as will be discussed in Sect. 4.

4 The Adaptive Agent Model

In this section, our adaptive agent model is explained in some detail. The controlled adaptive agent design uses a self-modeling network architecture of three levels as discussed in Sect. 3: a base level, a first-order self-model level, and a second-order self-model level. Here the (middle) first-order self-model level models how connections and excitability thresholds of the base level are adapted over time, and the (upper) second-order self-model level models the context-sensitive control over the adaptations. The Appendix B in Sect. 10 at the end of this aricle provides explanations for all of its states and a full specification of the model.

4.1 *Base Level*

Figure 2 shows a graphic overview of the base level. For each agent, interaction states were modeled: states involved in sensing (indicated by sense) are on the left-hand side of each box, and states involved in execution or expression of actions (move, exp_affect, talk) on the right-hand side. In between these interaction states, within a box are the agent's internal mental states; outside the boxes are the world states. Note that we assume that each agent also senses its own actions, modeled by the arrows from right to left outside the box.

For each agent, we modeled a few internal mental states such as sensory representation states (rep) and preparation states (prep) for each of the three modalities: movement *m*, expression of affect *b*, and verbal action *v*.

Furthermore, each agent has a conscious emotion state for affective response *b* (cons_emotion). Each of the mentioned states is depicted in Fig. 2 by a light pink circle shape. For each modality, its representation state has an outgoing (response) connection to the corresponding preparation state and it has an incoming (prediction) connection back from the preparation state to model internal mental simulation [25, 46].

Finally, there are the six synchrony detector states (depicted in Fig. 2 by the darker pink diamond shapes) which are introduced here. As in previous research [43] (see also this volume [39], Chap. 11 [45]) we cover three *intrapersonal synchrony detection states* for the three pairs of the three modalities:

$$\begin{aligned}
&\text{movement} - \text{emotion } (m - b)\\
&\text{movement} - \text{verbal action } (m - v)\\
&\text{emotion} - \text{verbal action } (b - v).
\end{aligned}$$

These intrapersonal synchrony detection states have incoming connections from the two execution states for the modalities they address. The conscious emotion state is triggered by incoming connections from the preparation state for affective response *b* together with the three intrapersonal synchrony detection states [31]. In addition, the

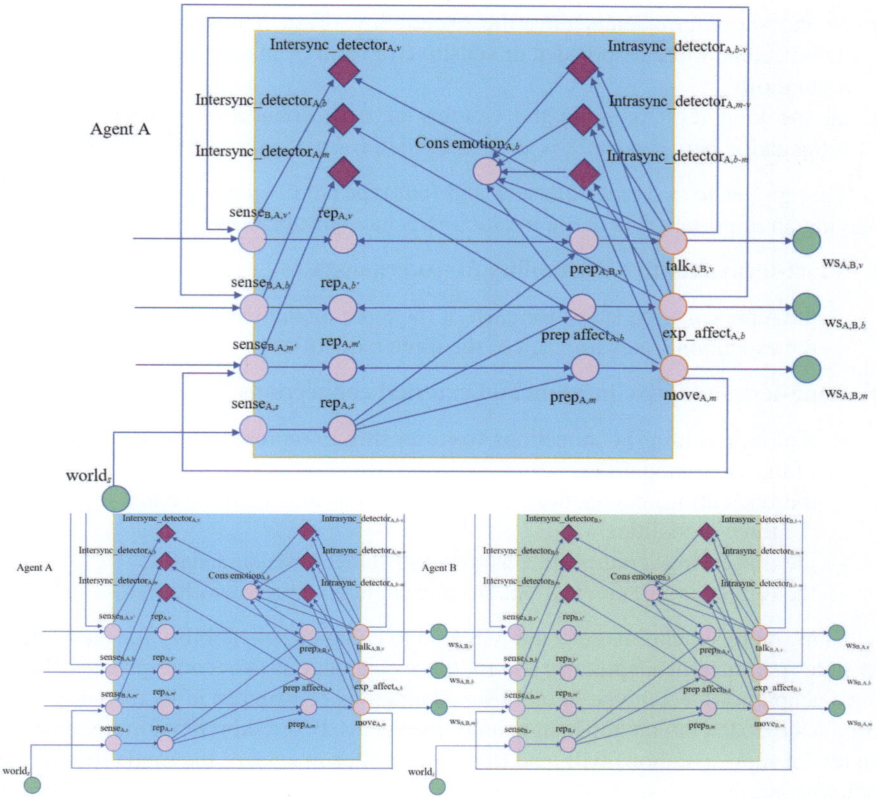

Fig. 2 Base level of the introduced adaptive agent model (upper picture) with three modalities and (in dark pink) six synchrony detection states for intrapersonal and interpersonal synchrony and how the agents interact (lower picture) according to the three modalities

conscious emotion state has an incoming connection from the verbal action execution state (for noticing the emotion in the verbal utterance) and an outgoing connection to the preparation of the verbal action (for emotion integration in the verbal action preparation).

There are three *interpersonal synchrony detection states* for the three modalities *m*, *b*, and *v*. Each of them has two incoming connections: from the sensing state (representing the action of the other agent) and the execution state (representing the own action) of the modality addressed.

For a few states and connections, their excitability and connection weights are adaptive depending on detected synchrony: detected synchrony leads to becoming more sensitive to sensing an agent and expressing to that agent (short-term effect) and to connecting stronger to the agent (long-term effect). Here, two different time scales for the adaptations are considered:

- on the short term, enhancing the excitability of such internal states, so that they become more responsive or sensitive (a form of instantaneous homeostatic regulation)
- on the long term, making the weights of such connections stronger so that propagation between states is strengthened (a form of a more endurable bonding)

This applies to two types of states and four types of connections in particular, all playing an important role in the interaction behaviour of the two agents:

- **Short-term adaptive excitability for internal states**
  - the representation states for each of the three modalities
  - the execution states for each of the three modalities
- **Long-term adaptive internal and external connections**
  - the (representing) connections from sensing to representation states for each of the three modalities
  - the (executing) connections from preparation to execution states for each of the three modalities
  - the (observing) connections from world states to sensing states
  - the (effectuating) connections from execution states to world states.

Thus, more synchrony detected will lead to enhanced excitability for these types of states (short-term adaptation) and for these connections to become stronger (long-term adaptation); each type of all these adaptations contributes in its own way (and time scale) to the interaction behaviour of the agents. In the short term, more sensitive states for representations will lead to gaining better images of the modalities of the other agent; this will make the sensed signals better available and accessible for the agent. More sensitive states for execution will lead to better expressed own modalities, so that the other agent can sense them better.

Over time and repeated interactions, a stronger (external) observing connection will lead to sensing the other agent better (e.g., turning sensors in the right direction and bending or getting closer to the other), and a stronger representing connection again (but now in a more endurable manner) will make the sensed signals better available and accessible for the agent. Conversely, a stronger executing connection will also contribute (in an endurable manner) to stronger expression and acting toward the other and a stronger effectuating connection to better availability (for the other) of the action effects in the world (e.g., more visible, better hearable by directing and positioning in the right direction and bending or getting closer to the other). In Sect. 4.2, we discuss in more detail how we modeled these forms of adaptivity and their control using the principle of self-modeling of the network model.

Finally, at the base level some world states are modeled for stimuli s that are sensed by the agents. In the simulations, they have stochastic activation levels. In some episodes one common stimulus is observed by both agents (for example when they physically meet and therefore are in the same environment), but in other episodes the agents receive different stimuli. Furthermore, also the world situation's suitability for enabling communication between the two agents is modeled by similar stochastic fluctuations. Moreover, two context states are included to model the conditions to maintain excitability thresholds well.

4.2 Modeling Adaptation and Its Control

We modeled adaptation and its control needed in the agent model using a self-modeling network [73], see also Sect. 3. Following what has been described in Sect. 4.1, for a number of states Y adaptive excitability has been modeled via the excitability threshold $\boldsymbol{\tau}_Y$ of the logistic function used for these states (see Table 1). Moreover, the strengthening of connections from X to Y has been modeled via adaptive connection weights $\boldsymbol{\omega}_{X,Y}$. Following Sect. 3, these adaptations have been modeled in particular through self-modeling for these $\boldsymbol{\tau}_Y$ and $\boldsymbol{\omega}_{X,Y}$ by adding the following first- and second-order self-model states:

- First-order self-model **T**-states $\mathbf{T}_Y$ are used for short-term adaptation of the adaptive base excitability thresholds $\boldsymbol{\tau}_Y$ for the internal representation states and execution states Y for the three considered modalities (movement, affective response and verbal action). For each agent there are 6 of these **T**-states, both for the 3 representation base states and the 3 execution base states (both for the three modalities).
- First-order self-model **W**-states $\mathbf{W}_{X,Y}$ are used for adaptation of the adaptive base connection weights $\boldsymbol{\omega}_{X,Y}$ for both internal and external connections for the three considered modalities; internal connections at the base level from sense states to representation states and from preparation states to execution states, and external connections from execution states to world states and from world states to sense states. For each agent there are 12 of these **W**-states, for the connections from world states to sensor states, from sensor states to representation states, from representation states to execution states and from execution states to world states (all for the three modalities).
- Second-order self-model $\mathbf{H_T}$-states are used for control of the **T**-states for adaptation of the adaptive excitability thresholds $\boldsymbol{\tau}_Y$ for the internal representation states and execution states Y. For each agent there is one of these states.
- Second-order self-model $\mathbf{H_W}$-states are used for control of the **W**-states for the adaptation of the adaptive base connection weights $\boldsymbol{\omega}_{X,Y}$. For each agent there is one of these states.

Figure 3 shows the overall design of the network model; here, the first-order self-model states are in the middle (blue) plane and the second-order self-model states in the upper (purple) plane. The first-order states include **T**-states representing the excitability thresholds of representation and execution states and **W**-states representing the weights of the different types of adaptive connections addressed.

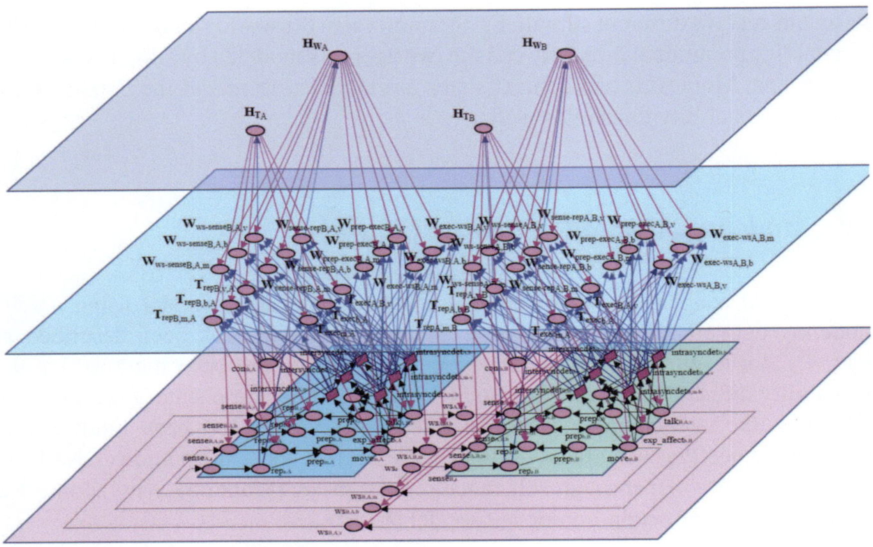

Fig. 3 Overview of the overall second-order adaptive network model

Recall from Sect. 3 the canonical difference and differential Eq. (2) for a self-modeling network. In this equation it is shown that at each time point for the values of the network characteristics the values of these self-model **T**-states and **W**-states are used. Based on this equation, by changing the activation values of these **T**-states and **W**-states over time *t*, the corresponding excitability thresholds and connection weights change accordingly which makes them adaptive. Such change of the values of the **T**-states and **W**-states occurs due to the influences from the detected synchronies, modeled by the upward (blue) arrows in Fig. 3 from the synchrony detection states in the base plane to the **T**-states and **W**-states in the middle plane.

For most states the combination function **alogistic** is used which has an excitability threshold parameter that can be made adaptive; see the last column in Table 1. The synchrony detections states, however, have a different function called **compdiff** to measure the extent of synchrony. For further details, see Table 1 and Appendix B in Sect. 10.

There are four second-order self-model states to control the adaptation: two second-order self-model states $\mathbf{H}_{\mathbf{T}_A}$ and $\mathbf{H}_{\mathbf{T}_B}$ for excitability adaptation control, one for each agent, and two second-order self-model states $\mathbf{H}_{\mathbf{W}_A}$ and $\mathbf{H}_{\mathbf{W}_B}$ for connection weight adaptation control, also one for each agent. These second-order self-model states are used to represent the adaptation speed (learning rate) for the adaptive excitability threshold **T**-states and connection weight **W**-states for the concerning agents *A* and *B*. Based on the canonical difference and differential Eq. (2) for a self-modeling network from Sect. 3, for each time point *t* the activation values of the second-order self-model states $\mathbf{H}_{\mathbf{T}_A}$, $\mathbf{H}_{\mathbf{T}_B}$, $\mathbf{H}_{\mathbf{W}_A}$, and $\mathbf{H}_{\mathbf{W}_B}$ at *t* are used as the values for these network characteristics of the first-order self-model **T**-states and **W**-states for

A and *B*. These second-order self-model states $\mathbf{H}_{\mathbf{T}_A}$, $\mathbf{H}_{\mathbf{T}_B}$, $\mathbf{H}_{\mathbf{W}_A}$, and $\mathbf{H}_{\mathbf{W}_B}$ model the second-order adaptation (or meta-plasticity) principle 'Adaptation accelerates with stimulus exposure' [60]. To this end they have incoming connections (blue arrows from base plane to upper plane) from the stimulus representation states at the base level.

5 Simulation Results

Appendix B in Sect. 10 at the end of this chapter provides a full specification of the model as used in our simulations. As can be seen, in general the values have been chosen in a standard manner. For example, all positive connection weights are 1, except for the long-term adaptation speed self-model states $\mathbf{H}_{\mathbf{W}_A}$ and $\mathbf{H}_{\mathbf{W}_B}$ which are 0.01, see Table 9. The negative connection weights for the **T**-states are –0.12 so that they together add up to a value –0.72, which compensates the positive weight 1 enough to get effect. The values for the steepness parameter $\boldsymbol{\sigma}$ (see Table 11) for the combination function **alogistic** were all set on 5 which also is a kind of default value. For values of the threshold parameter $\boldsymbol{\tau}$, they often have been set at 0.5 but accordingly higher for states with multiple incoming connections.

For all states, when the Euclidean combination function **eucl** is used, the order parameter $\boldsymbol{n}$ is 1 which makes it linear and the scaling factor $\boldsymbol{\lambda}$ is the sum of the weights of the incoming connections which normalizes it. Note that the time unit is kept abstract. Depending on the application context one might think of minutes, for example for therapeutical sessions.

5.1 Design of the Simulation Experiments

In this section, we evaluate our agent model in an experimental simulation paradigm. Our paradigm was set up in such a way that we could evaluate the behaviour of our two agents during four different types of consecutive episodes (see Table 2 and Fig. 4) which are explained below. Each of these types of episodes lasts for 30 time units, so that a cycle of four episodes equals 120 time units. Our total simulation run had a duration of 840 time units and the step size (Δt) was 0.5, resulting in 1680 computational steps in total for each simulation run. This means that each cycle of four episodes was repeated 7 times in each simulation. As it concerns a partly stochastic simulation, we ran 20 repetitions of each simulation with the same episodic paradigm and parameter settings, to get a sense of the robustness of the agent model's behaviour. It turned out that general patterns were approximately similar across all independent simulations. Therefore, we selected one simulation to discuss in the upcoming subsections.

Regarding the four different types of episodes in this simulation, they manipulate both whether or not the two agents received the same or a different stochastic stimulus

Table 2 Simulation paradigm of each run with the agent model: the pattern of stimuli and communication enables repeating every 120 time units

Time	Episode	Different stimuli $ws_{s,A}$ $ws_{s,B}$	Common stimulus ws_s	Communication enabled $\mathbf{W}_{exec-wsx,A,B}$ $\mathbf{W}_{exec-wsx,B,A}$
0–30	Episode 1	Yes	No	No
30–60	Episode 2	Yes	No	Yes
60–90	Episode 3	No	Yes	No
90–120	Episode 4	No	Yes	Yes
120–150	Episode 5	Yes	No	No
150–180	Episode 6	Yes	No	Yes
180–210	Episode 7	No	Yes	No
210–240	Episode 8	No	Yes	Yes
240–270	Episode 9	Etc.	Etc.	Etc.
Etc.	Etc.	Etc.	Etc.	Etc.

and whether or not they were able to communicate (with some stochastic variations in enabling conditions, due to environmental changes and noise) with each other (Table 2). The specific episodes for the considered example simulation are shown in Fig. 4.

The world states $ws_{s,A}$ and $ws_{s,B}$ indicate the different stimuli for agent A and B from the world (activated from time 0 to 60 and then repeated every 120 time units; see the dark solid and dashed blue lines for A, resp. B). Similarly, world state ws_s indicates the common stimulus (activated from time 60–120 and then repeated every 120 time units; see the purple line). These three states have values stochastically fluctuating approximately between 0.7 and 0.9. Furthermore, the self-model states $\mathbf{W}_{exec-wsx,A,B}$ (from A to B) and the states $\mathbf{W}_{exec-wsx,B,A}$ (from B to A) indicate the communication enabling conditions in the environment. They are activated from time 30 to time 60 thereby fluctuating stochastically roughly between 0.45 and 0.65 and then repeated every 60 time units.

All these stochastic activation patterns indeed follow the pattern shown in Table 2 with repetition every 120 time units until end time 840.

5.2 Behaviour of the Base States of the Agent Model

For the base states, in the first phase for time 0–10 the representations (states $rep_{s,A}$ and $rep_{s,B}$) for the stimulus are activated (the curves fluctuating around 0.8) and preparations (states $prep_{x,A}$) for actions are triggered (curves going to 1); see the upper graph in Fig. 5. This leads, together with the intrapersonal synchrony detection activation (see Figs. 5 and 6), to the conscious emotion around time 10 (red

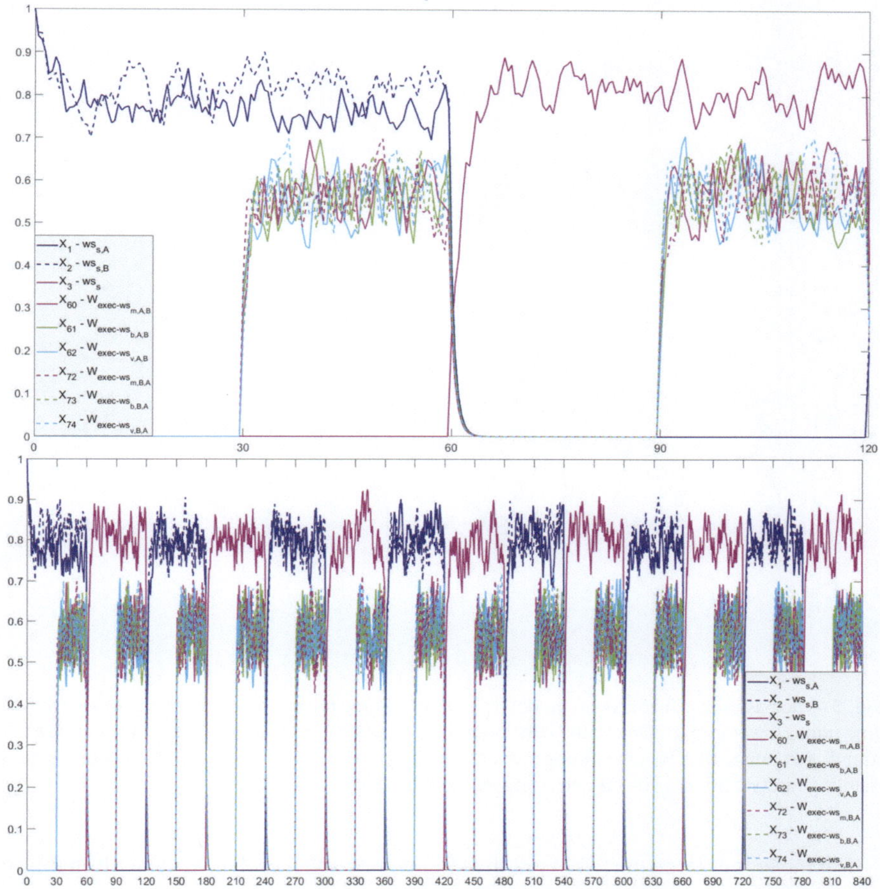

Fig. 4 The stimuli and interaction enabling states in the agent model from 0 to 120 time units (upper graph) and from 0 to 840 time units (lower graph): interaction enabling (multi-colour) from 30 to 60, 90 to 120, etc., different stimuli (blue) 0–60, 120–180, etc., common stimulus (purple) 60–120, 180–240, etc. (see also Table 2)

curve going to 1), but this still is only internal processing as no executions of actions take place yet. The action executions (states $\text{move}_{m,\text{A}}$, $\text{exp\_affect}_{b,\text{A}}$, and $\text{talk}_{\text{A,B},v}$) for both agents start to come up after time 10 (e.g., the purple line); this also depends on the short-term adaptations that will be discussed in Sect. 5.4. The curves immediately under these executions concern the sensing of the other agent's actions (the $\text{sense}_{\text{A},x,\text{B}}$ and $\text{sense}_{\text{B},x,\text{A}}$ states); in some periods they are slightly fluctuating due to environmental noise on the communication channels. The actual communication level (the $\text{ws}_{x,\text{A,B}}$ and $\text{ws}_{x,\text{B,A}}$ states) is seen below it from 30 to 60 and from 90 to 120.

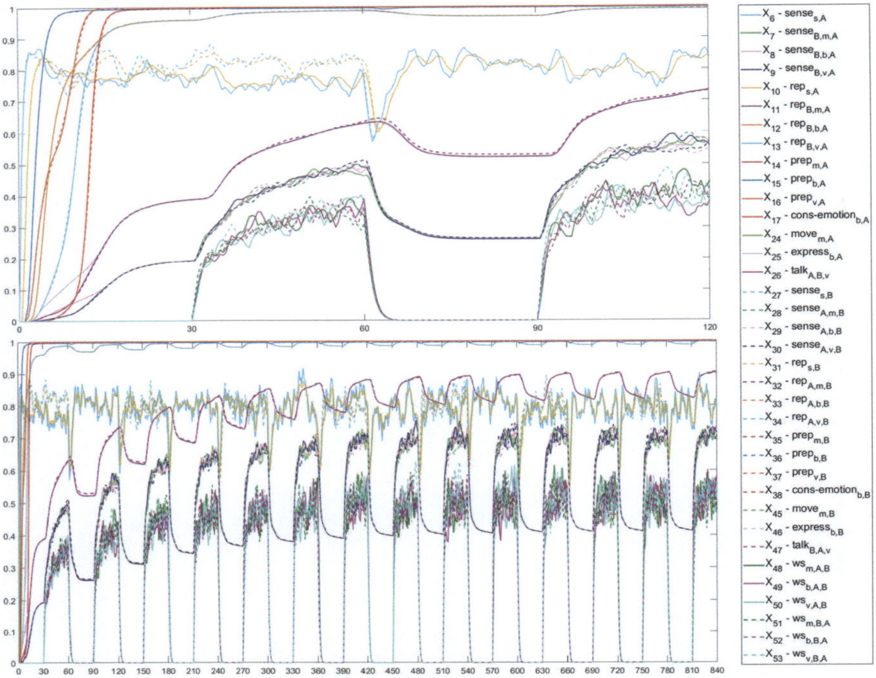

Fig. 5 The base states in the agent model from 0 to 120 time units (upper graph) and from 0 to 840 time units (lower graph). Due to the behavioural adaptivity the activation levels in response to the stimuli and interaction become stronger over time, both on the short term (within each interaction enabling interval 30–60, 90–120, etc.) and on the long term

For the longer term, the lower graph in Fig. 5 shows that each interval with enabling conditions for communication leads to higher activations of the action executions (the purple line) until values around 0.8 are reached. This is due to a long-term behavioural adaptation that is discussed in Sect. 5.5. Accordingly, the sensing states become higher as well over this longer term, but not as high as the action executions, due to a communication bias incorporated in the model. This overall pattern shows that the enabling conditions for communication have a stronger adaptive effect on the actions than having a common stimulus.

5.3 Behaviour of the Intrapersonal Synchrony and Interpersonal Synchrony Detector States

The curves that the graphs in Figs. 6 and 7 have in common depict the detected intrapersonal synchrony and interpersonal synchrony. Here:

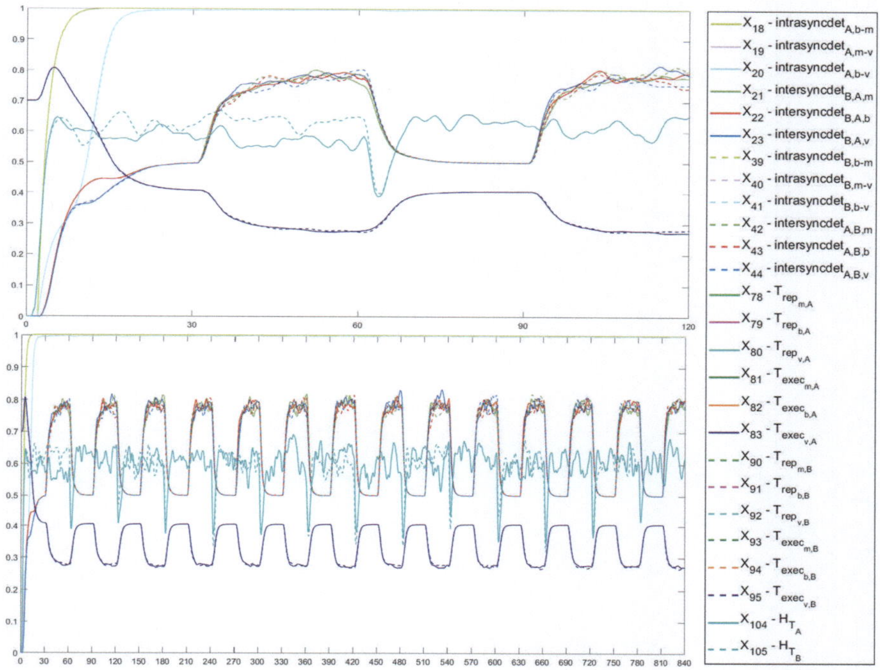

Fig. 6 The detected intrapersonal synchrony, interpersonal synchrony and short-term adaptation **T**-states in the agent model from 0 to 120 time units (upper graph) and from 0 to 840 time units (lower graph). As a form of short-term behavioural adaptivity, during each interaction interval the **T**-states become lower: 30–60, 90–120, etc. This makes in an adaptive manner the thresholds of the base states lower and therefore the activation values of them higher within each of these intervals, as also shows in Fig. 5

- the detected intrapersonal synchrony detection is represented by the states $\text{intrasyncdet}_{A,\, x-y}$ and $\text{intrasyncdet}_{B,\, x-y}$ shown as the light green and light blue curves going to 1 from time 0 to 15.
- the interpersonal synchrony detection is represented by the states $\text{intersyncdet}_{A,\, B,\, x}$ and $\text{intersyncdet}_{B,\, A,\, x}$ shown as the red and blue curves going to 0.4 from time 0 to 30 and further to 0.8 from time 30 to 60.

Here it can be observed that the detection of intrapersonal synchrony takes place already in the first episode from time 0 to 30, meaning no common stimulus or communication is required. In contrast, the detection of interpersonal synchrony strongly depends on the interaction between the two agents. Note also that the former type of detected synchrony reaches a perfect level of 1, due to the coherent internal makeup of the agents, while the latter type does not get higher than around 0.8. At first sight this may look strange, given that the actual executions of actions of both agents are practically the same, as discussed above (see Sect. 5.2 and Fig. 5). However, this is due to the communication bias that was also noted above in Sect. 5.2. This

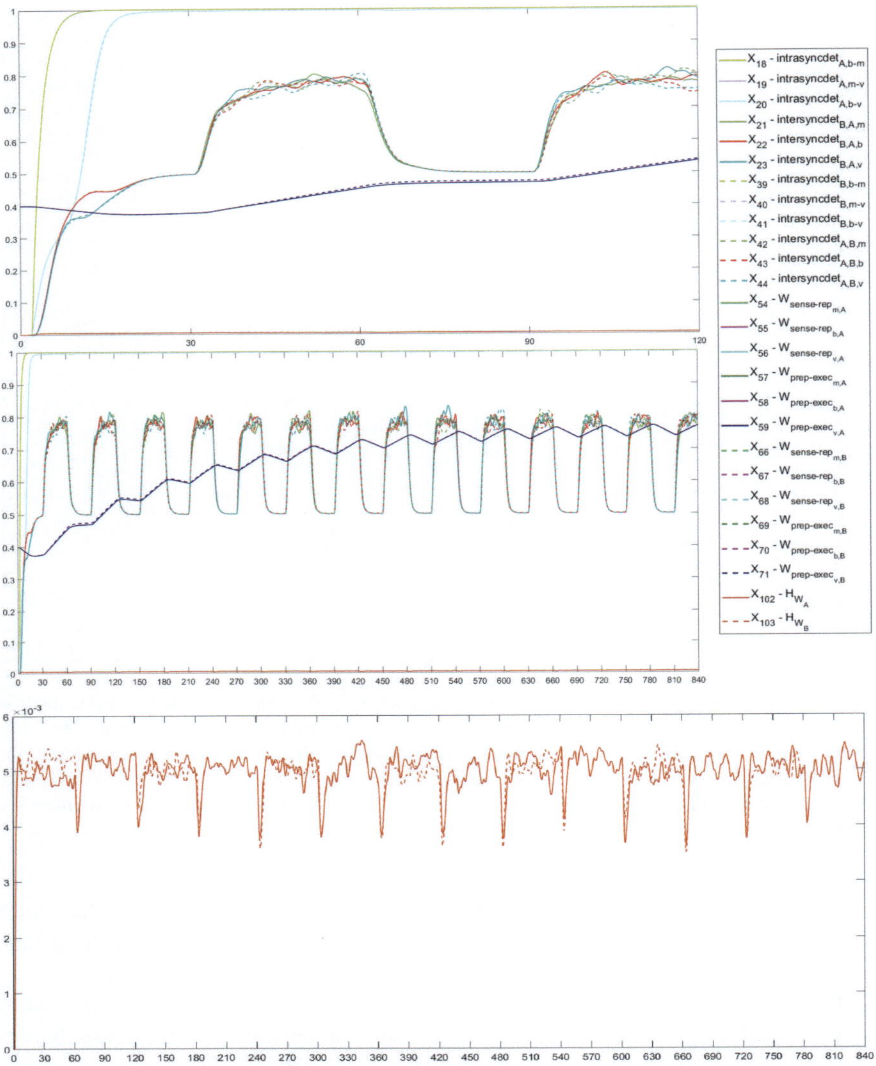

Fig. 7 The detected intrapersonal synchrony, interpersonal synchrony and long-term adaptation **W**-states and $\mathbf{H_W}$-states of their adaptation speed in the agent model from 0 to 120 time units (upper graph) and from 0 to 840 time units (middle graph). The lower graph depicts the $\mathbf{H_W}$-states with a different vertical scale (times 10^{-3})

demonstrates the capability of the model that it is able to distinguish a subjective personally detected interpersonal synchrony from an objective form of interpersonal synchrony detection as might be assigned by an external observer but not by the agent itself.

5.4 The Interplay Between Synchrony and Short-Term Adaptation

In Fig. 6 the synchrony detection states are shown together with the states involved in the short-term adaptation: the first-order self-model **T**-states that represent the adaptive excitability thresholds for representation and execution states and the second-order self-model $\mathbf{H_T}$-states that represent the **T**-states' speed factors (adaptive learning rates). Except the synchrony detection states already discussed in Sect. 5.3, the graphs show two light green curves fluctuating around 0.6 for the $\mathbf{H_T}$-states and a blue curve going down to below 0.3 for the **T**-states.

According to the meta-plasticity principle 'Adaptation accelerates with increasing stimulus exposure' [60], the $\mathbf{H_T}$-states indeed fluctuate with the stimuli. Furthermore, in accordance with this, when one stimulus period is in transition to another stimulus period, it can be seen that there is a short dip in the values of the $\mathbf{H_T}$-states, as stimuli start from 0, so there is a very short period of a lower level, as can also be seen in Fig. 4. Moreover, it is clear that the **T**-states (e.g., the blue curve) show an opposite pattern compared to the pattern of the interpersonal synchrony detection states. In particular, in the episodes from 30 to 60 and from 90 to 120 (and so on), where the detected interpersonal synchrony is the highest, the **T**-states for the excitability thresholds are the lowest. This is a short-term adaptation that makes that these agent states related to the communication with the other agent have a higher excitability due to the detected interpersonal synchrony, which will have an intensifying effect on their communication. Not coincidentally, the mentioned periods are also the periods with good enabling conditions for communication (see also Sect. 5.2). It can also be noted that this tendency is a short-term effect and is reversible: the **T**-states get higher again when the detected interpersonal synchrony gets lower.

5.5 The Interplay Between Synchrony and Long-Term Adaptation

In Fig. 7 the synchrony detection states are shown together with the states involved in the long-term adaptation: the first-order self-model **W**-states that represent the adaptive weights for the connections to the representation and execution states and the second-order self-model $\mathbf{H_W}$-states that represent the **W**-states' speed factors (adaptive learning rates). Here, except the synchrony detection states already discussed in Sect. 5.3, the graphs show the **W**-states (e.g., a blue curve) slowly and gradually going up to above 0.5 at time 120 and further up to about 0.8 at time 840.

Moreover, at a very low level, the curves for the $\mathbf{H_W}$-states can be seen. They also fluctuate according to the meta-plasticity principle 'Adaptation accelerates with increasing stimulus exposure' [60], but at a very low level around 0.005 (see the lower graph in Fig. 7). Again, following the same principle, when one stimulus period is in transition to another stimulus period, it can be seen that there is a short dip in the values of the $\mathbf{H_W}$-states. This happens because stimuli start from 0, so there is a very short period of a lower level (see Fig. 4). The pattern of the **W**-states indeed shows a long-term adaptation effect. It highlights that they get a repeated boost in the time intervals 30–60, 90–120, and so on, and show a form of persistency. These boosts occur specifically in these intervals for a reason. These intervals are when there are communication enabling conditions and as discussed in Sect. 5.4 that induces synchrony and the short-term adaptation via the **T**-states, which in turn add to synchrony. Therefore, these two effects are at the basis of these boosts for the long-term adaptation. In this way there is a form of interaction between short-term and long-term adaptation.

6 Modeling and Analysis of Adaptive Dynamical Systems via Their Canonical Self-modeling Network Representation

This section discusses how any smooth adaptive dynamical system can be modeled by a self-modeling network model, following and further detailing [7, 75]. It is shown in particular that any adaptive dynamical system has a canonical representation as a self-modeling network defined by network characteristics for connectivity, aggregation, and timing. The network concepts of this canonical representation of an adaptive dynamical system provide useful tools both for designing a model and for formal analysis of the dynamics of the adaptive dynamical system addressed.

With this idea in mind, as a next step equilibrium analysis of self-modeling network models is addressed. Dynamics in network models are described by node states that change over time (for example, for individuals' opinions, intentions, emotions, beliefs, …). Such dynamics depend on network characteristics for the connectivity between nodes, the aggregation of impacts from different nodes on a given node, and the timing of the node activation updates [72]. For example, whether within a well-connected group in the end a common opinion, intention, emotion or belief is reached (a common value for all node states) depends on all these network characteristics. Sometimes silent assumptions are made about the aggregation and timing characteristics. For timing characteristics, often it is silently assumed that the nodes are updated in a synchronous manner, although in application domains this assumption is usually not fulfilled. For aggregation, in social network models usually linear functions are applied, which means that it is often not investigated how a variation of this choice of aggregation would affect the dynamics.

In the modeling and analysis approach used in this chapter, a more diverse landscape is covered which is not limited by the fixed conditions on connectivity, aggregation or timing as are so often imposed. For connectivity, both acyclic and cyclic networks are covered here. For aggregation, both networks with linear and nonlinear aggregation are considered and for networks with nonlinear aggregation, networks with logistic aggregation are addressed but also networks with other forms of nonlinear aggregation. Finally, both synchronous and asynchronous timing are provided. The often-occurring use of linear functions for aggregation for social network models may be based on a more general belief that dynamical system models can be analysed better for linear functions than for nonlinear functions. Although there may be some truth in this if specifically logistic nonlinear functions are compared to linear functions, such a belief is not correct in general. It has been found that also classes of nonlinear functions exist that enable good analysis possibilities when it comes to the emerging dynamics within a network model, thereby among others not using any conditions on the connectivity but instead exploiting for any network its structure of strongly connected components.

In Sect. 6.1 it is shown that for the nonadaptive case this network-oriented modelling approach is equivalent to any dynamical systems modelling approach (Theorem 1 and Corollary 1), and in Sect. 6.2 that for the adaptive case self-modeling networks are equivalent to any adaptive dynamical systems approach (Theorem 2 and Corollary 2). In Sect. 6.3 equilibrium analysis of network models is provided and applied to the model introduced earlier.

6.1 Dynamical Systems and Their Canonical Network Representation

Dynamical systems are usually specified in certain mathematical formats; see [7], pp. 241–252 for some details. In the first place a finite set of states (or state variables) $X_1, \ldots, X_n$ is assumed describing how the system changes over time via functions $X_1(t), \ldots, X_n(t)$ of time t. As discussed by [7, 58], a dynamical system is a state-determined system which can be formalized in a numerical manner by a relation (rule of evolution) that expresses how for each time point t the future value of each state X_i at time $t + s$ uniquely depends on s and on $X_1(t), \ldots, X_n(t)$. Therefore, a dynamical system can be described via n functions $F_j(V_1, \ldots, V_n, s)$ for each X_j in the following manner, see also [7], pp. 243–244:

$$X_j(t+s) = F_j(X_1(t), \ldots, X_n(t), s) \text{ for } s > 0 \tag{7}$$

If these functions F_j and the X_j are continuously differentiable (which also implies they are continuous), we call the dynamical system smooth. Suppose such a smooth dynamical system is given. It turns out that it can always be described in a canonical manner by a temporal-causal network model; the argument is as follows. Consider (7) where the functions F_i and X_i are continuously differentiable. In the particular case of s approaching 0 it holds

$$\lim_{s\downarrow 0} X_j(t+s) = \lim_{s\downarrow 0} F_j(X_l(t), \ldots, X_n(t), s)$$

which due to continuity of the involved functions implies

$$X_j(t) = F_j(X_l(t), \ldots, X_n(t), 0) \quad (8)$$

So, Eq. (7) also holds for $s = 0$.

Let $X_j{}'$ denote the derivative of X_j with respect to time. Apply the partial derivative $\partial(..)/\partial s$ on both sides of Eq. (1)

$$X_j(t+s) = F_j(X_1(t), \ldots, X_n(t), s)$$

Then it follows

$$\partial X_j(t+s)/\partial s = \partial F_j(X_l(t), \ldots, X_n(t), s)/\partial s$$

Here for the left hand side, by the chain rule for function composition it holds

$$\partial X_j(t+s)/\partial s = X_j'(t+s)\partial(t+s)/\partial s = X_j'(t+s)$$

So, it is found that for all t and s it holds

$$X_j'(t+s) = \partial F_j(X_l(t), \ldots, X_n(t), s)/\partial s \quad (9)$$

In particular, this holds for $s = 0$; therefore

$$X_j'(t) = \left[\partial F_j(X_1(t), \ldots, X_n(t), s)/\partial s\right]_{s=0} \quad (10)$$

For a more detailed explanation of the argument for (10), see the Appendix A in Sect. 9, where also the differences and relations with [7]'s approach are discussed.

Now define the (combination) function $g_j(V_1, \ldots, V_n)$ by

$$g_j(V_1, \ldots, V_n) = V_j + \left[\partial F_j(V_1, \ldots, V_n, s)/\partial s\right]_{s=0} \quad (11)$$

Then it holds

$$dX_j(t)/dt = \left[\partial F_j(X_1(t), \ldots, X_n(t), s)/\partial s\right]_{s=0} = g_j(X_1(t), \ldots, X_n(t)) - X_i(t) \quad (12)$$

Comparing Eq. (12) to the canonical format (1) in Sect. 3 that defines dynamics of temporal-causal networks, it immediately follows that this matches to each other as long as the speed factors $\boldsymbol{\eta}$ and connection weights $\boldsymbol{\omega}$ are set at 1, i.e., from (12) it follows:

$$dX_j(t)/dt = \boldsymbol{\eta}_{X_j}\left[\mathbf{c}_{X_j}\left(\boldsymbol{\omega}_{X_1,X_j}X_1(t), \ldots, \boldsymbol{\omega}_{X_n,X_j}X_n(t)\right) - X_j(t)\right] \quad (13)$$

with $\boldsymbol{\eta}_{X_j} = 1$ and $\mathbf{c}_{X_j} = g_j$ for all j and $\boldsymbol{\omega}_{X_i,X_j} = 1$ for all i and j. For an example of this, see Box 1.

Box 1 Example of the canonical transformation of any smooth dynamical system into temporal-causal network format

Consider the following example dynamical system from [7], p. 244:

$$X_1(t+s) = X_1(t) + X_2(t)s + s^2$$
$$X_2(t+s) = X_2(t) + 2s$$

This can be formalised in the format of (1) by

$$X_1(t+s) = F_1(X_1(t), X_2(t), s)$$
$$X_2(t+s) = F_2(X_1(t), X_2(t), s)$$

where the functions F_1 and F_2 are defined by

$$F_1(V_1, V_2, s) = V_1 + V_2 s + s^2$$
$$F_2(V_1, V_2, s) = V_2 + 2s$$

Then

$$[\partial F_1(V_1, V_2, s)/\partial s]_{s=0} = [V_2 + 2s]_{s=0} = V_2$$
$$[\partial F_2(V_1, V_2, s)/\partial s]_{s=0} = [2]_{s=0} = 2$$

This leads to the differential equations

$$dX_1(t)/dt = X_2(t)$$
$$dX_2(t)/dt = 2$$

When the (combination) functions g_1 and g_2 are defined by

$$g_1(V_1, V_2) = V_1 + [\partial F_1(V_1, V_2, s)/\partial s]_{s=0} = V_1 + V_2$$
$$g_2(V_1, V_2) = V_2 + [\partial F_2(V_1, V_2, s)/\partial s]_{s=0} = V_2 + 2$$

then the following is obtained:

$$dX_1(t)/dt = g_1(X_1(t), X_2(t)) - X_1(t)$$
$$dX_2(t)/dt = g_2(X_1(t), X_2(t)) - X_2(t)$$

This is equivalent to

$$dX_1(t)/dt = \boldsymbol{\eta}_{X_1}\left[\mathbf{c}_{X_1}\left(\boldsymbol{\omega}_{X_1,X_1}X_1(t), \boldsymbol{\omega}_{X_2,X_1}X_2(t)\right) - X_1(t)\right]$$
$$dX_2(t)/dt = \boldsymbol{\eta}_{X_2}\left[\mathbf{c}_{X_2}\left(\boldsymbol{\omega}_{X_1,X_2}X_1(t), \boldsymbol{\omega}_{X_2,X_2}X_2(t)\right) - X_2(t)\right]$$

with $\boldsymbol{\eta}_{X_j} = 1$ and $\mathbf{c}_{X_j} = g_j$ for all j and $\boldsymbol{\omega}_{X_i,X_j} = 1$ for all i and j

This shows that any given smooth dynamical system can be formalised in this canonical manner by a representation in the temporal-causal network format; this notion is described in more detail in Definition 1 and Theorem 1. Note that this also shows theoretically that the use of specific values for speed factors and connection weights is not essential, as they can all be set to 1. However, they still are convenient instruments in the practice of modeling real-world processes.

Definition 1 (*Canonical Network Representation of a Smooth Dynamical System*)

Let any smooth dynamical system be given by.

$$X_j(t+s) = F_j(X_1(t), \ldots, X_n(t), s) \text{ for } s \geq 0, j = 1, \ldots, n$$

where the functions F_i are continuously differentiable. Then the *canonical temporal-causal network representation* of it is defined by network characteristics

$\boldsymbol{\omega}_{X_i,X_j}$, $\mathbf{c}_{X_j}$, $\boldsymbol{\eta}_{X_j}$ for all i and j.
with $\boldsymbol{\omega}_{X_i,X_j} = 1$ for all i and j

$$\mathbf{c}_{X_j}(V_1, \ldots, V_n) = V_j + \left[\partial F_j(V_1, \ldots, V_n, s)/\partial s\right]_{s=0}$$

$\boldsymbol{\eta}_{X_j} = 1$ for all j.
This network representation has dynamics induced by the following canonical differential equations for temporal-causal networks

$$dX_j(t)/dt = \boldsymbol{\eta}_{X_j}\left[\mathbf{c}_{X_j}\left(\boldsymbol{\omega}_{X_1,X_j}X_1(t), \ldots, \boldsymbol{\omega}_{X_n,X_j}X_n(t)\right) - X_j(t)\right]$$

So, by the argument above, the following theorem is obtained:

Theorem 1 (the Canonical Network Representation of a Smooth Dynamical System)

Any smooth dynamical system can be formalised in a canonical manner by a temporal-causal network model called its canonical network representation. Conversely, any temporal-causal network model is a dynamical system model. ■

As a corollary from Theorem 1 the following well-known result immediately follows.

Corollary 1 (from Smooth Dynamical System to First-Order Differential Equations)

Any smooth dynamical system can be formalised as a system of first-order differential equations. ■

The latter result was also proven in a different way in [7], pp. 241–252. See Appendix A in Sect. 9 for some more details.

6.2 Adaptive Dynamical Systems and Their Canonical Self-modeling Network Representation

In this section, it is shown how the approach described in Sect. 6.1 can be extended to obtain a transformation of any smooth *adaptive* dynamical system into a self-modeling network model. Adaptive dynamical systems are usually modeled by two levels of dynamical systems (see Fig. 8). Here the higher level dynamical system models the dynamics of the parameters $P_{i,j}$ of the lower level dynamical system (the lower level component in Fig. 8) that describes the dynamics of variables X_i, for example by

$$X_j(t+s) = F_j\big(P_{j,1}, \ldots, P_{j,k}, X_1(t), \ldots, X_n(t), s\big) \text{ for } s > 0 \tag{14}$$

In addition, for the dynamics of the $P_{i,j}$ there will also be a dynamical system (the upper level component in Fig. 8) for $s \geq 0$:

$$P_{i,j}(t+s) = G_{i,j}\big(P_{1,1}(t), \ldots, P_{n,k}(t), X_1(t), \ldots, X_n(t), s\big) \tag{15}$$

By applying the argument from Sect. 6.2 to both levels, the following differential equations are obtained covering the entire adaptive dynamical system:

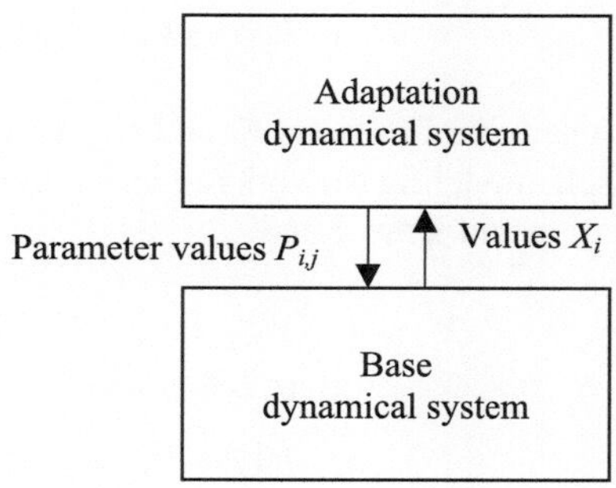

Fig. 8 Overall picture of an adaptive dynamical system

$$
\begin{aligned}
&\mathrm{d}X_i(t)/\mathrm{d}t \\
&= \boldsymbol{\eta}_{X_i}\big[\mathbf{c}_i\big(\boldsymbol{\omega}_{P_{i,1},X_i}P_{i,1}(t), \ldots, \boldsymbol{\omega}_{P_{i,k}X_i}P_{i,k}(t), \boldsymbol{\omega}_{X_1,X_i}X_1(t), \ldots, \boldsymbol{\omega}_{X_nX_i}X_n(t)\big) - X_i(t)\big] \\
&\mathrm{d}P_{i,j}(t)/\mathrm{d}t \\
&= \boldsymbol{\eta}_{P_{i,j}}\big[\mathbf{c}_{P_{i,j}}\big(\boldsymbol{\omega}_{P_{i,1}P_{i,j}}P_{i,1}(t), \ldots, \boldsymbol{\omega}_{P_{i,k}P_{i,j}}P_{i,k}(t), \\
&\boldsymbol{\omega}_{X_1,P_{i,j}}X_1(t), \ldots, \boldsymbol{\omega}_{X_nP_{i,j}}X_n(t)\big) - P_{i,j}(t)\big]
\end{aligned}
\tag{16}
$$

where all $\boldsymbol{\eta}$ and $\boldsymbol{\omega}$ are 1. Recall from Sect. 3 the canonical differential Eq. (3) that defines a self-modeling network model for the case when self-model states $\mathbf{P}$ are introduced for the combination function parameters $\boldsymbol{\pi}_Y$ for all base level states Y:

$$
dY(t)/dt = \boldsymbol{\eta}_Y\big[\mathbf{c}_{\mathbf{P}_Y(t),Y}\big(\boldsymbol{\omega}_{X_1,Y}X_1(t), \ldots, \boldsymbol{\omega}_{X_k,Y}X_k(t)\big) - Y(t)\big] \tag{3}
$$

where $\mathbf{P}_Y(t) = \big(\mathbf{P}_{1,Y}(t), \ldots, \mathbf{P}_{m,Y}(t)\big)$.

The first equation of (16) is (although in a slightly different mathematical notation) equal to Eq. (3), which shows that the former equation defines a self-modeling temporal-causal network model. In this self-modeling network model, the parameters $P_{i,\,j}$ from the adaptive dynamical system are modeled by (aggregation) self-model **P**-states within the self-modeling network model for parameters in the combination functions used for states X_i in the base network defined by the states X_i. In this way a canonical self-modeling network representation is obtained for the considered smooth adaptive dynamical system; this notion is defined by Definition 2.

Definition 2 (*Canonical Self-Modeling Network Representation of a Smooth Adaptive Dynamical System*)

Let any smooth adaptive dynamical system for $s \geq 0$, $j = 1, \ldots, n$, and for $i = 1, \ldots, k$ be given by

$$
\begin{aligned}
X_j(t+s) &= F_j\big(P_{j,1}, \ldots, P_{j,k}, X_1(t), \ldots, X_n(t), s\big) \\
P_{i,j}(t+s) &= G_{i,j}\big(P_{1,1}(t), \ldots, P_{n,k}(t), X_1(t), \ldots, X_n(t), s\big)
\end{aligned}
$$

where the functions F_j and $G_{i,j}$ are continuously differentiable. Then the canonical self-modeling network representation of it is defined by characteristics $\boldsymbol{\omega}$, $\boldsymbol{\pi}$, $\boldsymbol{\eta}$ where all $\boldsymbol{\omega}$ and $\boldsymbol{\eta}$ are 1 and

$$
\begin{aligned}
&\mathbf{c}_{X_j}\big(W_{i,1}, \ldots, W_{i,k}, V_1, \ldots, V_n\big) \\
&= V_i + \big[\partial F_i\big(W_{i,1}, \ldots, W_{i,k}, V_1, \ldots, V_n, s\big)/\partial s\big]_{s=0} \\
&\mathbf{c}_{P_{i,j}}\big(W_{i,1}, \ldots, W_{i,k}, V_1, \ldots, V_n\big) \\
&= W_{i,j} + \big[\partial G_i\big(W_{i,1}, \ldots, W_{i,k}, V_1, \ldots, V_n, s\big)/\partial s\big]_{s=0}
\end{aligned}
$$

This self-modeling network representation has dynamics induced by the following canonical differential equations

$$dX_j(t)/dt$$
$$= \eta_{X_j}\Big[\mathbf{c}_{X_j}\Big(\omega_{P_{i,1},X_j}P_{i,1}(t),\ldots,\omega_{P_{i,k},X_j}P_{i,k}(t),\omega_{X_1,X_j}X_1(t),\ldots,\omega_{X_n,X_j}X_n(t)\Big)-X_j(t)\Big]$$
$$dP_{i,j}(t)/dt$$
$$= \eta_{P_{i,j}}\Big[\mathbf{c}_{P_{i,j}}\Big(\omega_{P_{i,1},P_{i,j}}P_{i,1}(t),\ldots,\omega_{P_{i,k},P_{i,j}}P_{i,k}(t),\omega_{X_1,P_{i,j}}X_1(t),\ldots,\omega_{X_n,P_{i,j}}X_n(t)\Big)-P_{i,j}(t)\Big]$$

■

Thus, by the argument preceeding Definition 2, the following theorem was obtained:

Theorem 2 (the Canonical Self-Modeling Network Representation of an Adaptive Dynamical System)

Any adaptive smooth dynamical system model can be transformed in a canonical manner into a self-modeling network model called its canonical self-modeling network representation described by Definition 2 *above. Conversely, any self-modeling network model is an adaptive dynamical system model. These also apply to higher-order adaptive dynamical systems in relation to higher-order self-modeling networks.* ■

As a corollary it now follows that any adaptive dynamical system can be described by first-order differential equations:

Corollary 2 (from a Smooth Adaptive Dynamical System to First-Order Differential Equations)

Any smooth adaptive dynamical system can be formalised as a system of first-order differential equations. ■

Theorems 1 and 2 demonstrate that a modeling approach based on the self-modeling network format is at least as general as any other adaptive dynamical system modeling approach. Therefore the choice for this format to model adaptive dynamical systems does not introduce any limitation. In particular, it also can be viewed as generalising the most common types of neural network models.

6.3 *Stationary Point and Equilibrium Analysis for Self-modeling Networks*

In this section, following [76], it is shown how stationary point and equilibrium analysis can be performed for self-modeling networks (Sect. 7.1) and how this can be applied to verify the correctness of the implemented self-modeling network model introduced in Sect. 4 compared to its design specifications (Sect. 7.2).

The General Analysis Approach

The following types of properties are often considered for equilibrium analysis of dynamical systems in general.

Definition 3 (*Stationary Point, Increasing, Decreasing, Equilibrium*)

Let Y be a network state.

- Y has a *stationary point* at t if $\mathbf{d}Y(t)/\mathbf{d}t = 0$
- Y is *increasing* at t if $\mathbf{d}Y(t)/\mathbf{d}t > 0$
- Y is *decreasing* at t if $\mathbf{d}Y(t)/\mathbf{d}t < 0$
- The network model is in *equilibrium* at t if every state Y of the model has a stationary point at t.

Note that for mathematical analysis of dynamical system models, usually there is an emphasis on equilibrium analysis. However, in many cases no equilibria occur, for example in cases of oscillatory limit cycle behaviour. In such cases, still stationary points can be analysed. In the case considered in this chapter no equilibria occur when the environment is changing all the time.

By considering the canonical network representation of a dynamical system, the above criteria are formulated in terms of the network characteristics: for network models, the following criteria in terms of the network characteristics $\boldsymbol{\omega}_{X,Y}$, $\mathbf{c}_Y$, $\boldsymbol{\eta}_Y$ can be derived from the generic difference Eq. (1). Let Y be a state and $X_1, \ldots, X_k$ the states connected toward Y. For non-zero speed factors $\boldsymbol{\eta}_Y$, the following criteria in terms of network characteristics for connectivity and aggregation apply; here $\mathbf{aggimpact}_Y(t) = \mathbf{c}_Y(\boldsymbol{\omega}_{X_1,Y}X_1(t), \ldots, \boldsymbol{\omega}_{X_k,Y}X_k(t))$.

- Y has a stationary point at $t \Leftrightarrow \mathbf{aggimpact}_Y(t) = Y(t)$
- Y is increasing at $t \Leftrightarrow \mathbf{aggimpact}_Y(t) > Y(t)$
- Y is decreasing at $t \Leftrightarrow \mathbf{aggimpact}_Y(t) < Y(t)$
- The network model is in equilibrium at $t \Leftrightarrow \mathbf{aggimpact}_Y(t) = Y(t)$ for every state Y.

The above criteria for a network being in an equilibrium (assuming non-zero speed factors) depend both on the connections weights $\boldsymbol{\omega}_{X,Y}$ used for connectivity and on the combination function $\mathbf{c}_Y$ used for aggregation. Note that in a self-modeling network, these criteria can be applied not only to base states but also to self-model states. In the latter case they can be used for equilibrium analysis of learning or adaptation processes.

In particular, a network model with states $X_1, \ldots, X_n$ is in equilibrium if and only if the following n equations (called *equilibrium equations*) are satisfied:

$$\mathbf{aggimpact}_{X_1}(t) = X_1(t)$$

$$\ldots\ldots\ldots\ldots$$

$$\ldots\ldots\ldots\ldots$$

$$\mathbf{aggimpact}_{X_n}(t) = X_n(t)$$

These equations express relations between values of the states in an equilibrium: they indicate how values in an equilibrium relate to each other and contain as parameters network characteristics $\boldsymbol{\omega}_{X_i,X_j}$ and $\mathbf{c}_{X_j}$. Sometimes it is possible to solve these equations, for example, when they are linear, or when they are nonlinear Euclidean or geometric equations. When there is no equilibrium, still stationary points for a given state X_i at some time point t can be analysed based on the above criteria, for example, (local) maxima or minima of the function $X_i(t)$.

The above criteria can be used to verify correctness of (the implementation of) a network model based on inspection of stationary points or equilibria in the following manner.

Verification by Checking the Criteria Through Substitution

1. Generate a simulation.
2. For a sample of states X_j identify stationary points with their time points t and state values $X_j(t)$.
3. For each of these stationary points for a state X_j from the chosen sample at time t, identify the values $X_i(t)$ at time t of states X_i among $X_1, \ldots, X_n$ that are connected toward X_j.
4. Substitute all these values $X_i(t)$ in the criterion $\mathbf{aggimpact}_{X_j}(t) = X_j(t)$.
5. If the equation holds (for example, with absolute deviation < 10^{-2}), then this test succeeds, otherwise it fails.
6. If this test fails, then it should be explored what error is causing this failure and how this error can be corrected.
7. If the test succeeds, this contributes to evidence that the implemented network model is correct in comparison with its design specification.

In Sect. 7.2, it is shown in detail how this form of verification by substitution can be applied for an example network.

Analysis of the Introduced Agent Model

The procedure for testing on correctness described in Sect. 7.1 has been applied to the agent model introduced in Sect. 4 of this chapter. It has been applied for two different scenarios, in each of them for a sample of states covering all levels of the model. One scenario used for this type of test is a scenario where an equilibrium occurs, see Fig. 9. This was achieved by setting the external factors for stimuli and communication enabling to constant values 1 instead of random values as applied in the scenarios in Sect. 5 where no equilibria occur due to this randomness.

The analysis focuses on time $t = 1000$ and the chosen sample consists of the following states X_j: the second-order self-model state $\mathbf{H}_{\mathbf{W}_B}$ (X_{105}, green line in Fig. 9

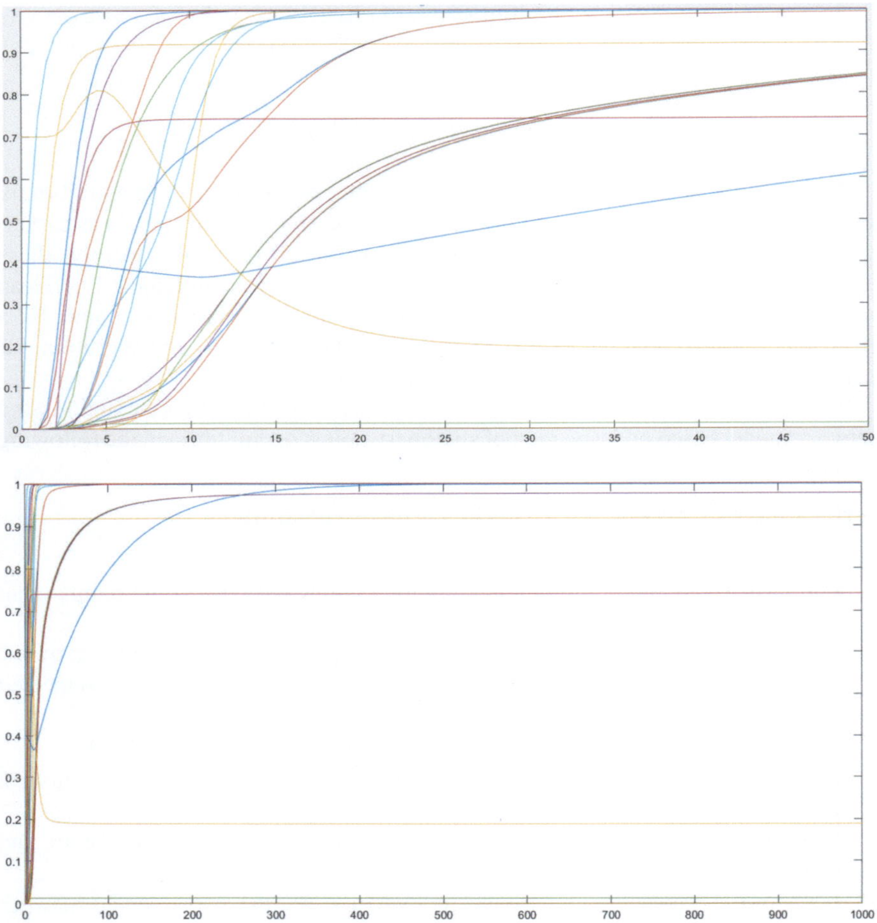

Fig. 9 Simulation scenario for a constant environment leading to an equilibrium. Upper graph: initial phase time 0 to 50. Lower graph: up to time 1000. For a complete legend for the colours, see Fig. 10

ending up around 0.01), first-order self-model state $\mathbf{T}_{\text{exec\_v, B}}$ (X_{95}, orange line ending up around 0.2), second-order self-model state $\mathbf{H}_{\mathbf{T}_B}$ (X_{103}, red line ending up around 0.75), base states $\text{rep}_{\text{s, B}}$ (X_{31}, orange line ending up around 0.92), $\text{ws}_{\text{v, B, A}}$ (X_{53}, purple line ending up around 0.97), $\text{intrasyncdet}_{\text{A, m–v}}$ (X_{19}, blue line ending up at time 20 around 1), $\text{intersyncdet}_{\text{B, A, m}}$ (X_{21}, blue line ending up at time 50 around 1), and first-order self-model state $\mathbf{W}_{\text{prep-exec\_v, B}}$ (X_{71}, blue line ending up at time 500 around 1). To calculate

$$\mathbf{aggimpact}_{X_j}(t) = \mathbf{c}_Y\big(\boldsymbol{\omega}_{X_1,Y}X_1(t), \ldots, \boldsymbol{\omega}_{X_k,Y}X_k(t)\big) \tag{18}$$

for each of these chosen states X_j, from the network characteristics the states X_i are determined with connections to X_j, (see Table 3, second column) and for each of these X_i the weights $\boldsymbol{\omega}_{X_i,X_j}(t)$ of the connection from X_i to X_j (see Table 3, middle part), and from the simulation data the values $X_i(t)$ of the X_i at time t (see Table 3, right-hand part).

Based on these connection weights $\boldsymbol{\omega}_{X_i,X_j}(t)$ and state values $X_i(t)$ in Table 3, the products $\boldsymbol{\omega}_{X_i,X_j}(t)X_i(t)$ are determined (see Table 4, sixth column). Then by applying the combination function from Table 1 on these products, via Eq. (10) $\mathbf{aggimpact}_{X_j}(t)$ is determined for each X_j (see one but last column in Table 4). Finally, this value is compared to the state value $X_j(t)$ of X_j to obtain the deviation = $\mathbf{aggimpact}_{X_j}(t) - X_j$ in the last column of Table 4. In Table 4, all absolute deviations are smaller than 10^{-6}. This provides evidence that the implemented model is correct with respect to its design specifications. Note that this provides evidence for correctness of the model in general, not only for this special scenario, as if there were errors, they would most probably also show their effects in this example scenario.

Yet, for still more evidence another scenario has been analysed as well, where the environmental factors for stimuli and communication enabling do change but in a nonrandom manner; see Fig. 11.

Here, for both agents, stimulus s occurs from time 0 to time 450 and disappears from 450 to 500; this is repeated every 500 time units. Moreover, interaction is not enabled from time 0 to time 50 and is enabled from time 50 to time 400, which is repeated every 400 time units. Due to the changing environment, no equilibrium occurs here. However, there are many cases of (approximate) stationary points. In particular, stationary points have been analysed as above for a sample of states and time points indicated in the left three columns in Table 5. Here most absolute deviations are <0.01. However, there are three of them in the order of 0.02 which is larger than expected for a stationary point. The graph indeed shows that they actually are not approximately stationary points. All in all, also these results provide evidence that the implemented model is correct with respect to its design specifications.

7 Discussion

The material presented in this chapter is mostly based on Hendrikse et al. [38]. In this chapter, an agent model was introduced for the way intrapersonal synchrony and interpersonal synchrony induce behavioural adaptivity between the synchronized persons [2, 19, 20, 23, 27, 35, 48, 59, 65, 87]. In literature, it was advocated to use a dynamical systems modeling approach to model the complex, cyclical types of dynamics that occur [29, 85]. The model presented here is indeed a dynamical system model; moreover, it is multi-adaptive in that the behavioural adaptivity covers both short-term adaptations and long-term adaptations, reflecting short-term affiliation and long-term bonding. The former type of adaptation was modeled using (non-synaptic) adaptive excitability [24, 26, 86, 89], whereas for the latter type a more classical synaptic type of adaptation [9, 10, 34, 63, 64] was used. Following the aforementioned

Table 3 Weights $\boldsymbol{\omega}_{X_i,X_j}(t)$ and simulation values $X_i(t)$ for incoming connections from states X_i to X_j used for equilibrium analysis for the scenario with constant environment depicted in Fig. 9

X_j	Incoming X_i	Connection weights $\boldsymbol{\omega}_{X_i,X_j}(t)$ at t							State values $X_i(t)$ at t						
X_{19}	X_{24}, X_{26}	1	1						0.976171	0.976334					
X_{21}	X_{24}, X_7	1	1						0.976171	0.976171					
X_{31}	X_{27}	1							1						
X_{53}	X_3, X_{47}	1	0						0.976334	0					
X_{71}	X_{39}–X_{44}	1	1	1	1	1	1		1	0.999833	0.999833	1	1	1	
X_{95}	X_{39}–X_{44}, X_5	−0.12	−0.12	−0.12	−0.12	−0.12	−0.12	1	1	0.999833	0.999833	1	1	1	1
X_{103}	X_{31}	0.08							0.917915						
X_{105}	X_{31}	1							0.917915						

Table 4 Equilibrium analysis for the scenario with constant environment depicted in Fig. 9

State	X_j	Time point t	State value $X_j(t)$	Incoming states X_i	Impact values $\boldsymbol{\omega}_{X_i,X_j}(t)X_i(t)$	$\mathbf{aggimpact}_{X_j}(t)$	Deviation
$\text{intrasyncdet}_{A,m-v}$	X_{19}	1000	0.999832697	X_{24}	0.97617082	0.999832697	$3.3\ 10^{-11}$
				X_{26}	0.976334164		
$\text{intersyncdet}_{B,A,m}$	X_{21}	1000	0.999999995	X_{24}	0.97617082	0.999999996	$1.2\ 10^{-10}$
				X_7	0.976170816		
$\text{rep}_{s,B}$	X_{31}	1000	0.917915001	X_{27}	1	0.917915001	$< 10^{-17}$
$\text{ws}_{v,B,A}$	X_{53}	1000	0.976334161	X_3	0.976334164	0.976334164	$2.9\ 10^{-9}$
				X_{47}	0		
$\mathbf{W}_{\text{prep-exec\_v},B}$	X_{71}	1000	0.99944441	X_{39}	1	0.999446296	$1.9\ 10^{-6}$
				X_{40}	0.999832697		
				X_{41}	0.999832697		
				X_{42}	0.999999995		
				X_{43}	0.999999995		
				X_{44}	0.999999995		
$\mathbf{T}_{\text{exec\_v},B}$	X_{95}	1000	0.188195503	X_{39}	–0.120000000	0.188195503	$-3.5\ 10^{-11}$
				X_{40}	–0.119979924		
				X_{41}	–0.119979924		
				X_{42}	–0.119999999		
				X_{43}	–0.119999999		
				X_{44}	–0.119999999		
				X_5	1		
$\mathbf{H}_{WB}$	X_{103}	1000	0.012837073	X_{31}	0.0734332	0.012837073	$2.9\ 10^{-17}$
$\mathbf{H}_{TB}$	X_{105}	1000	0.740701054	X_{31}	0.917915001	0.740701054	$< 10^{-17}$

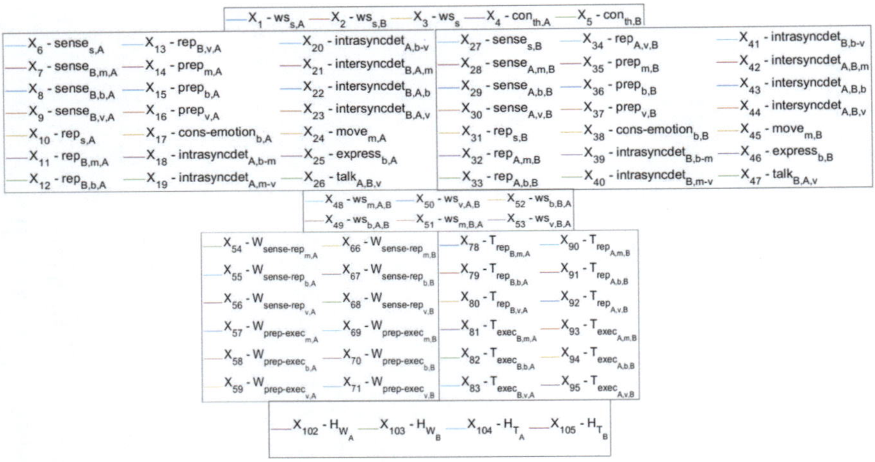

Fig. 10 Legend for the colours in Figs. 9 and 11

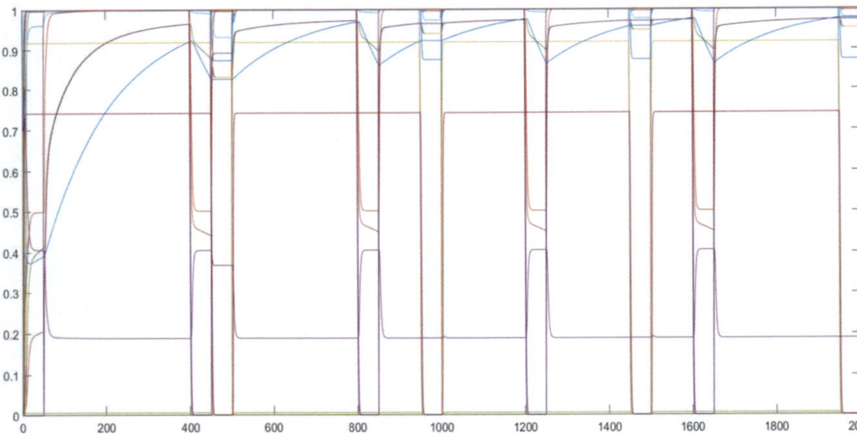

Fig. 11 Simulation scenario for a nonrandomly changing environment. For a legend for the colours, see Fig. 10

literature on synchrony, both types of adaptivity were modeled as driven by the (internally detected) intrapersonal synchrony and interpersonal synchrony for the agent. By also including metaplasticity [1] in the model to control the adaptations in a context-sensitive manner, the agent model became second-order adaptive.

For more background literature on the use of an architecture with multiple adaptation or control levels from biological and philosophical perspectives, see [11, 12, 14, 15, 78, 79]. Although in the current chapter only two adaptation or control levels were used, in other research where the role of environment-induced epigenetic changes

Table 5 Stationary point analysis for the scenario with non-random non-constant environment depicted in Fig. 11

State X_j		Time point t	State value $X_j(t)$	Incoming states X_i	Impact values $\boldsymbol{\omega}_{X_i,X_j}(t)X_i(t)$	**aggimpact**$_{Xj}(t)$	Deviation
$\text{sense}_{A,v,B}$	X_{30}	1999	0.972980783	X_{47}	0.972980783	0.972980783	$<10^{-17}$
				X_{50}	0.972980783		
$\text{rep}_{s,B}$	X_{31}	1949	0.917915001	X_{27}	1	0.917915001	$<10^{-17}$
$\text{intrasyncdet}_{B,b-v}$	X_{41}	1949	0.999816153	X_{46}	0.973237336	0.999816356	$2.0\ 10^{-7}$
				X_{47}	0.973416098		
				X_{41}	0.999816153		
$\text{intrasyncdet}_{B,b-v}$	X_{41}	1999	0.978349541	X_{46}	0.951915303	0.978349541	$-5.7\ 10^{-13}$
				X_{47}	0.972980783		
				X_{41}	0.978349541		
$\text{intersyncdet}_{A,B,v}$	X_{44}	1949	0.999971441	X_{47}	0.973416098	0.999971816	$3.8\ 10^{-7}$
				X_{30}	0.973388663		
				X_{44}	0.999971441		
$\text{intersyncdet}_{A,B,v}$	X_{44}	1999	1	X_{47}	0.973416098	0.999552797	–0.000447203
				X_{30}	0.972980783		
				X_{44}	1		
$\text{ws}_{b,B,A}$	X_{52}	1949	0.973218991	X_{46}	0.947347128	0.947347128	–0.025871863
				X_3	0		
$\text{ws}_{b,B,A}$	X_{52}	1999	0.951915303	X_{46}	0.926195296	0.926195296	–0.025720006
				X_3	0		
$\text{ws}_{v,B,A}$	X_{53}	1999	0.972980783	X_{47}	0.972980783	0.972980783	$<10^{-17}$
				X_3	0		
$\mathbf{W}_{\text{prep-exec\_v,B}}$	X_{71}	1999	0.976505645	X_{39}	1	0.999313691	0.022808046

(continued)

Table 5 (continued)

State X_j		Time point t	State value $X_j(t)$	Incoming states X_i	Impact values $\boldsymbol{\omega}_{X_i,X_j}(t)X_i(t)$	$\mathbf{aggimpact}_{X_j}(t)$	Deviation
				X_{40}	0.978349541		
				X_{41}	0.978349541		
				X_{42}	1		
				X_{43}	1		
				X_{44}	1		
$\mathbf{T}_{exec\_v,B}$	X_{95}	1649	0.404971073	X_{39}	–0.12	0.404971648	5.8 10^{-7}
				X_{40}	–0.119924731		
				X_{41}	–0.119924731		
				X_{42}	–0.060065522		
				X_{43}	–0.060065522		
				X_{44}	–0.060065218		
				X_5	1		
$\mathbf{T}_{exec\_v,B}$	X_{95}	1999	0.188288623	X_{39}	–0.12	0.193456528	0.005167905
				X_{40}	–0.117401945		
				X_{41}	–0.117401945		
				X_{42}	–0.12		
				X_{43}	–0.12		
				X_{44}	–0.12		
				X_5	1		
$\mathbf{H}_{WB}$	X_{103}	1949	0.005872159	X_{31}	0.0367166	0.005872159	$<10^{-17}$
$\mathbf{H}_{WB}$	X_{103}	1999	1.4 10^{-14}	X_{31}	0.039982777	0.006444919	0.006444919

(continued)

Table 5 (continued)

State X_j		Time point t	State value $X_j(t)$	Incoming states X_i	Impact values $\boldsymbol{\omega}_{X_i,X_j}(t)X_i(t)$	$\mathbf{aggimpact}_{X_j}(t)$	Deviation
$\mathbf{H_{TB}}$	X_{105}	1949	0.740701054	X_{31}	0.917915001	0.740701054	$<10^{-17}$
$\mathbf{H_{TB}}$	X_{105}	1999	$4.36807\ 10^{-13}$	X_{31}	0.999569427	0.006444919	0.006444919

for the development of mental disorders such as ASPD, Binge eating, Insomnia, Psychopathy, SUD is analysed, a fifth-order control architecture turned out useful [16, 21, 32, 47, 61, 78, 79].

The simulations of the model have been performed using the dedicated software environment developed in MATLAB on HP Intel Core i5 and Apple Macbook Pro Intel Core i9 laptops. Execution times per run were less than a minute, for example, on the HP Intel Core i5 with MATLAB 2017a between 40 and 50 s. The software environment used can be downloaded via URL https://www.researchgate.net/publication/368775720.

Thus, this chapter has focused on the emerging and adaptive effects of human social interaction, concerning emerging synchronisation and related adaptive affiliation and bonding, with the therapist-client interaction as a central application option. By such simulations, for example, a therapist can get insight about how they can improve the way they interact with their clients and make therapy or counselling more effective. From the scientific perspective, the modeling also contributes formalisation to this area of psychology which is almost always addressed in informal manners. The contributed approach can also be used as a solid basis for development of supporting virtual agents in that context.

Synchrony and related patterns in the brain are also analysed, for example, for atypical brain conditions of subjects such as for PTSD [88], epilepsy [52, 57], ADHD [3, 4], or autism [5]. The work presented in the current chapter distinguishes itself from this in four ways: (1) it abstracts from the specific brain processes and instead focuses on the level of mental processes, (2) it addresses not only the emergence of synchrony but also the causal effects of synchronisation on adaptivity of interaction behaviour such as affiliation and bonding, (3) it does not address the emergence of synchrony from an objective external observer perspective but from a subjective perspective from the agents themselves, and (4) it focusses on typical instead of atypical conditions of subjects.

We already engaged in computational modeling of synchrony between agents in earlier work, see this volume Chap. 3 [44] and Chap. 5 [37]. However, in the models described there, no (subjective) internal detection of synchrony takes place. Moreover, in the first one no adaptivity was covered, and in the second one another type of adaptivity was incorporated, namely of internal connections from representation states to preparation states. In this volume Chap. 5 [40] describes another computational agent model where subjective synchrony detection is addressed. However, by that model no long-term behavioural adaptivity is covered and also no adaptive intrinsic excitability is addressed, whereas both are included in the current model.

Earlier work [2, 66] addressing behavioural adaptation due to coordinated actions used a dynamic form of the 'bonding based on homophily' principle [54] was used to model the effect of coordination of emotions and actions on behavioural adaptivity but no (subjective) detection of synchrony was used. However, now also bonding by subjective homophily has been addressed in this volume Chaps. 7, 8, 9 and 10 [18, 55, 56, 80].

In this chapter, also mathematical analysis was addressed for the modeling approach applied. The first type of analysis shows that any smooth adaptive dynamical system has a canonical representation as a self-modeling network. This implies theoretically that the self-modeling network format is widely applicable and that no biases or limitations are introduced by choosing a network modeling approach to design adaptive dynamical system models. In particular, it also generalises the most common neural system models.

This finding also has been shown in many practical applications varying from biological, cognitive, affective to social processes and their interaction. It is illustrated by many examples in particular in a book [72] introducing the self-modeling network modeling approach and its applications, a book [81] focusing on the use of self-modeling network models to handle dynamics, adaptation and control of internal mental models, and a book [22] focusing on the use of self-modeling network models to model organizational learning processes. Furthermore, stationary point and equilibrium analysis was addressed and applied to the introduced self-modeling network model. These analyses were used as a form of verification of the model, which provided evidence that the implemented model is correct with respect to its design specifications.

Thus, a flexible human-like second-order multi-adaptive agent model was obtained for the way in which detected synchrony leads to different types of behavioural adaptivity concerning the short-term affiliation and long-term bonding between the two agents.

A number of aspects that still may be considered relevant are not covered by the model introduced here. One of these aspects is the use of time lags in the process of synchrony detection. This has not been addressed here, but can be a relevant extension of the work reported here. More in general, different combination functions describing methods for synchrony detection may be tried out either with time lags or without time lags. Another relevant aspect that is not addressed in the current chapter is the role of interruptions or transitions in synchrony and their effect on behavioural adaptivity.

For further work, many more simulation experiments can be designed and conducted, for example to explore the question which types of short-term synchrony are most likely to become translated into long-term benefits for a relationship, or to explore in more detail the roles of intrapersonal synchrony and interpersonal synchrony.

Also, the model can easily be extended to cover interaction between more than two agents. To achieve that, for each additional other agent, within the agent model the number of sensing states can be extended by three of them (for each modality) and similarly three additional representation states and three interpersonal synchrony detection states can be added for each additional other agent. Accordingly, also additional first- and second-order self-model states can be added. This will add complexity to the agent model.

Alternatively, if the model abstracts from the differences between the other agents, then the current agent model without any additional states can be directly applied by using the current sensing, representation, and synchrony detection states and also the self-model states as states aggregating all other agents. This keeps the complexity of the agent model the same, but of course then the model is less context-sensitive as different adaptations to specific other agents are not possible. So, as more often happens, there is a trade-off between the complexity of the agent model and the extent of context-sensitivity here: more context sensitivity comes with more complexity.

Considered from a wider scientific perspective, the model can provide a basis to develop adaptive virtual agents that are able to concentrate on each other by short-term behavioural adaptivity and bond with each other by long-term behavioural adaptivity in a human-like manner. For example, in other work [13] the focus is on virtual conversational agents and how they can adapt to their human users. In that work classical learning techniques from AI are used to optimize the agent's behaviour with respect to a given user such as Q-learning, which is not directly inspired or justified by neuroscience.

In contrast, the current chapter offers approaches such as synaptic plasticity by adaptive connection weights [9, 10, 34, 63, 64] and non-synaptic plasticity by adaptive excitability thresholds [24, 26, 86, 89], and in addition metaplasticity [1] to control both types of plasticity. As all of these forms of adaptivity are justified in neuroscience literature, this will in principle lead to a more human-like agent model. Nevertheless, it will be interesting to explore in further work how these two different perspectives can benefit from each other.

Concerning the relation of the considered agent model to mechanisms from neuroscience, note that these mechanisms have been incorporated only from an abstracted functional perspective. This means that it cannot be claimed that the model is human-like for the (neuro)physiological level. The latter has been left out of consideration here and would require another research project.

Validation of the model has only been done based on qualitative empirical information from the psychological literature. Dynamic and adaptive patterns have been obtained that are in accordance with that type of empirical information. Due to the lack of quantitative (numerical) empirical information, no quantitative validation has been performed yet. For future research, it is considered to try to acquire such numerical data and then perform quantitative numerical validation by parameter tuning. The way how that can be done is described in Chap. 19 of [81].

Part of the work presented in the current chapter was presented in a preliminary form in the AIAI'22 conference and published in their proceedings [42]. That paper was limited to the design of the model and an example simulation. In contrast, the fundamental mathematical analysis of the positioning of the modeling approach based on self-modeling temporal-causal networks in the landscape of adaptive dynamical systems (described in Sects. 6 and 10) is new compared to Hendrikse et al. [42]. It has been shown here that any smooth adaptive dynamical system has a canonical representation as a self-modeling temporal-causal network, which means that the applied modeling approach is universal for smooth adaptive dynamical systems. Moreover, the in-depth verification of the introduced model (described in Sect. 7) is also new. Here, substantial evidence was added that the implemented model is correct with respect to its design specification. Finally, the full specification of the model in Appendix B in Sect. 10 is new as well.

8 Conclusion

All in all, we achieved the following summarised findings and discoveries:

- Formalisation of the informal domain of human social interaction involving emerging and multiple types of adaptive dynamical system effects is possible.
- Unifying bridges between causal modeling, network modeling, and dynamical systems modeling are possible.
- A systematic approach to (higher-order) adaptivity in these different modeling perspectives is possible.

More specifically, the following has been achieved:

- The notion of canonical temporal-causal network representation for any smooth dynamical system is introduced; it is shown how any smooth dynamical system can be assigned this canonical network representation. This also applies to most common neural network approaches. This creates a bridge between different subdisciplines that are usually kept separate: causal modeling in AI, neural networks in AI, (multidisciplinary) network science, computational science. This enables simulation and analysis of any smooth dynamical system in terms of network concepts.
- The notion of canonical temporal-causal self-modeling network representation for any smooth adaptive dynamical system is introduced; it is shown how any smooth adaptive dynamical system can be assigned this canonical network representation. This again creates a bridge between different subdisciplines that are usually kept separate and provides a clear way of addressing higher-order adaptivity to any of these: meta-level architectures in AI, causal modeling in AI, neural networks in AI, (multidisciplinary) network science, computational science. This enables simulation and analysis of any smooth higher-order adaptive dynamical system in terms of network concepts.
- It is shown by mathematical stationary point and equilibrium analysis how verification of correctness of the implemented agent model introduced (in comparison to its design specifications) provides more evidence of its correctness.
- By explaining the full specification of the introduced agent model reproducibility is obtained.

Appendix A: More Details for Sect. 6

In this section, a more detailed explanation for the main argument in Sect. 6.1 for Theorem 1 can be found (see Box 2) and it is discussed how the chosen approach differs from and relates to [7]'s approach.

Box 2 More detailed explanation of the argument for Theorem 1
Assume $s > 0$. Subtracting Eq. (8) from Eq. (7) (see Sect. 6.1) and dividing by s provides:

$$\begin{aligned}&\left[X_j(t+s) - X_j(t)\right]/s = \\ &\quad\left[F_j(X_1(t), \ldots, X_n(t), s) - F_j(X_1(t), \ldots, X_n(t), 0)\right]/s\end{aligned}$$

When for both sides of this equation the limit for s approaching 0 is taken, the left-hand side of (3) becomes (renaming s to Δt to get the familiar expression)

$$\begin{aligned}\lim_{s\downarrow 0}\left(\left[X_j(t+s) - X_i(t)\right]/s\right) &= \lim_{\Delta t\downarrow 0}\left(\left[X_j(t+\Delta t) - X_j(t)\right]/\Delta t\right) \\ &= dX_j(t)/dt = X_j'(t)\end{aligned}$$

and the right-hand side (here renaming s to Δs to get the familiar expression)

$$\begin{aligned}&\lim_{s\downarrow 0}\left(\left[F_j(X_1(t), \ldots, X_n(t), s) - F_j(X_1(t), \ldots, X_n(t), 0)\right]/s\right) = \\ &\lim_{\Delta s\downarrow 0}\left(\left[F_j(X_1(t), \ldots, X_n(t), \Delta s) - F_j(X_1(t), \ldots, X_n(t), 0)\right]/\Delta s\right) \\ &= \left[\partial F_j(X_1(t), \ldots, X_n(t), s)/\partial s\right]_{s=0}\end{aligned}$$

Therefore, it is obtained:

$$dX_j(t)/dt = \left[\partial F_j(X_1(t), \ldots, X_n(t), s)/\partial s\right]_{s=0} \tag{19}$$

The differences and relationships with the approach by [7] are as follows. Instead of Eq. (7), Ashby [7], pp. 243–245 uses the special case of (7) for $t = 0$ as indication for a state-determined system:

$$X_j(s) = F_j(X_1(0), \ldots, X_n(0), s) \text{ for } s > 0 \tag{20}$$

As this special case by itself is not enough to characterise a state-determined system, he furthermore also uses a second condition for a state-determined system that can be called *transitivity*:

$$\begin{aligned}&F_i\left(X_1(t), \ldots, X_n(t), s' + s''\right) = \\ &F_i\left(F_1\left(X_1(t), \ldots, X_n(t), s'\right), \ldots, F_n\left(X_1(t), \ldots, X_n(t), s'\right), s''\right)\end{aligned} \tag{21}$$

So, in the end he characterises a state-determined system by the conjunction of conditions (20) and (21). It turns out that condition (7) that we use here to characterise a state-determined system is not equivalent to (20) but to this conjunction of (20) and (21), in other words, the following holds:

Theorem 3
(Characterising State-Determined Systems)

The following are equivalent:

(i) $X_j(t+s) = F_j(X_1(t), \ldots, X_n(t), s)$ *for* $s > 0$
(ii) $X_j(s) = F_j(X_1(0), \ldots, X_n(0), s)$ *f*or $s > 0$ *and the system is transitive*

Proof (i) ⇒ (ii) That (22) implies transitivity follows from.

$$X_i(t + (s' + s'')) = X_i((t + s') + s'')$$

and working out both sides of this:
$X_i(t + (s' + s'')) = F_i(X_1(t), \ldots, X_n(t), s' + s'')$ for $s', s'' > 0$

$$X_i((t + s') + s'') = F_i(X_1(t + s'), \ldots, X_n(t + s'), s'')$$

where

$$X_i(t + s') = F_i(X_1(t), \ldots, X_n(t), s')$$

So

$$X_i((t + s') + s'') =$$

$F_i(F_1(X_1(t), \ldots, X_n(t), s'), \ldots, F_n(X_1(t), \ldots, X_n(t), s'), s'')$ This proves transitivity from (22).

(ii) ⇒ (i) To be proven

$$X_j(t+s) = F_j(X_1(t), \ldots, X_n(t), s) \tag{22}$$

Given.
$X_j(s) = F_j(X_1(0), \ldots, X_n(0), s)$ for all s.
So also

$$X_i(t+s) = F_i(X_1(0), \ldots, X_n(0), t+s)$$

Now by transitivity it holds

$$\begin{aligned} X_j(t+s) &= F_j(X_1(0), \ldots, X_n(0), t+s) \\ &= F_j(F_1(X_1(0), \ldots, X_n(0), t), \ldots, F_n(X_1(0), \ldots, X_n(0), t), s) \\ &= F_j(X_1(t), \ldots, X_n(t), s) \end{aligned}$$

This proves (22). ■

This explains how the approach used here differs from but still relates to Ashby's [7] approach. Note that another main difference is that Ashby did not analyse how (adaptive) dynamical systems can be related to (adaptive) network models, what is our main focus here.

Appendix B: Further Details of the Introduced Model

Overview of All States of the Model

Tables 6 (base states) and 7 (first- and second-order self-model states) provide explanations of all states of the introduced model.

Full Specification of the Model in Role Matrices Format

In this section, first some further simulation pictures are shown for the affilation patterns represented by the **W**-states in relation to the detected intra- and interpersonal synchronies. Next, the full specification of the introduced adaptive network model is shown in terms of role matrices which are tables with the network characteristics in standardized table format. These tables are readable by the dedicated software environment available via. https://www.researchgate.net/publication/368775720_Network-Oriented_Modeling_Software.

And then can generate simulations. In this way reproducibility is supported. In Tables 8, 9, 10, 11 and 12 the full specification of the adaptive network model by role matrices is shown. Each role matrix has 93 rows for all states X_1 to X_{93} of the model.

The connectivity characteristics are specified by role matrices **mb** and **mcw** shown in Tables 8 and 9. Role matrix **mb** lists for each state the states (at the same or lower level) from which the state gets its incoming connections, while in role matrix **mcw** the connection weights are listed for these connections.

Nonadaptive connection weights are indicated in **mcw** (in Table 9) by a number (in a green shaded cell), but adaptive connection weights are indicated by a reference to the (self-model) **W**-state representing the adaptive value (in a peach-red shaded cell). This can be seen for states X_7 to X_9 (with self-model **W**-states X_{63} to X_{65}), states X_{11} to X_{13} (with self-model **W**-states X_{54} to X_{56}), X_{24} to X_{26} (with self-model **W**-states X_{57} to X_{59}), X_{28} to X_{30} (with self-model **W**-states X_{75} to X_{77}), X_{32} to X_{34} (with self-model **W**-states X_{66} to X_{68}), and X_{45} to X_{53} (with self-model **W**-states X_{69} to X_{71}, X_{60} to X_{62}, and X_{72} to X_{74}).

Table 6 Base states of the computational network model

state	name	Explanation
X_1	$ws_{s,A}$	World state for stimulus *s* for *A*
X_2	$ws_{s,B}$	World state for stimulus *s* for *B*
X_3	$ws_{s,AB}$	World state for stimulus *s* both for *A* and *B*
X_4	$con_{th,A}$	Context state for excitability threshold for *A*
X_5	$con_{th,B}$	Context state for excitability threshold for *B*
X_6	$sense_{s,A}$	Sensor state of *A* for stimulus *s* for *A*
X_7	$sense_{B,m,A}$	Sensor state of *A* for movement *m* of *B*
X_8	$sense_{B,b,A}$	Sensor state of *A* for expressed affective response *b* of *B*
X_9	$sense_{B,v,A}$	Sensor state of *A* for verbal action *v* of *B*
X_{10}	$rep_{s,A}$	Sensory representation state of *A* for stimulus *s* for *A*
X_{11}	$rep_{B,m,A}$	Sensory representation state of *A* for movement *m* of *B*
X_{12}	$rep_{B,b,A}$	Sensory representation state of *A* for expressed affective response *b* of *B*
X_{13}	$rep_{B,v,A}$	Sensory representation state of *A* for verbal action *v* of *B*
X_{14}	$prep_{m,A}$	Preparation state for movement *m* of *A*
X_{15}	$prep_{b,A}$	Preparation state for affective response *b* of *A*
X_{16}	$prep_{v,A}$	Preparation state for verbal action *v* of *A*
X_{17}	$cons\_emotion_{b,A}$	Conscious emotion state for *b* of *A*
X_{18}	$intrasyncdet_{A,\text{b-m}}$	Intrapersonal synchrony detection of *A* for executing *b* and *m* by *A*
X_{19}	$intrasyncdet_{A,\text{m-v}}$	Intrapersonal synchrony detection of *A* for executing *m* and *v* by *A*
X_{20}	$intrasyncdet_{A,\text{b-v}}$	Intrapersonal synchrony detection of *A* for executing *b* and *v* by *A*
X_{21}	$intersyncdet_{B,A,\text{m}}$	Interpersonal synchrony detection of *A* for executing *m* by *B* and *A*
X_{22}	$intersyncdet_{B,A,\text{b}}$	Interpersonal synchrony detection of *A* for executing *b* by *B* and *A*
X_{23}	$intersyncdet_{B,A,\text{v}}$	Interpersonal synchrony detection of *A* for executing *v* by *B* and *A*
X_{24}	$move_{m,A}$	Executing movement *m* by *A*
X_{25}	$exp\_affect_{b,A}$	Executing expression of *b* by *A*
X_{26}	$talk_{A,B,v}$	Executing verbal action *v* by *A*
X_{27}	$sense_{s,B}$	Sensor state of *B* for stimulus *s* for *B*
X_{28}	$sense_{A,m,B}$	Sensor state of *B* for movement *m* of *A*
X_{29}	$sense_{A,b,B}$	Sensor state of *B* for expressed affective response *b* of *A*
X_{30}	$sense_{A,v,B}$	Sensor state of *B* for verbal action *v* of *A*
X_{31}	$rep_{s,B}$	Sensory representation state of *B* for stimulus *s* for *B*
X_{32}	$rep_{A,m,B}$	Sensory representation state of *B* for movement *m* of *A*
X_{33}	$rep_{A,b,B}$	Sensory representation state of *B* for expressed affective response *b* of *A*
X_{34}	$rep_{A,v,B}$	Sensory representation state of *B* for verbal action *v* of *A*
X_{35}	$prep_{m,B}$	Preparation state for movement *m* of *B*
X_{36}	$prep_{b,B}$	Preparation state for affective response *b* of *B*
X_{37}	$prep_{v,B}$	Preparation state for verbal action *v* of *B*
X_{38}	$cons\_emotion_{b,B}$	Conscious emotion state for *b* of *B*
X_{39}	$intrasyncdet_{B,\text{b-m}}$	Intrapersonal synchrony detection of *B* for executing *b* and *m* by *B*
X_{40}	$intrasyncdet_{B,\text{m-v}}$	Intrapersonal synchrony detection of *B* for executing *m* and *v* by *B*
X_{41}	$intrasyncdet_{B,\text{b-v}}$	Intrapersonal synchrony detection of *B* for executing *b* and *v* by *B*
X_{42}	$intersyncdet_{A,B,\text{m}}$	Interpersonal synchrony detection of *B* for executing *m* by *A* and *B*
X_{43}	$intersyncdet_{A,\text{B},\text{b}}$	Interpersonal synchrony detection of *B* for executing *b* by *A* and *B*
X_{44}	$intersyncdet_{A,\text{B},\text{v}}$	Interpersonal synchrony detection of *B* for executing *v* by *A* and *B*
X_{45}	$move_{m,B}$	Executing movement *m* by *B*
X_{46}	$exp\_affect_{b,B}$	Executing expression of *b* by *B*
X_{47}	$talk_{B,A,v}$	Executing verbal action *v* by *B*
X_{48}	$ws_{m,A,B}$	World state for transmitting movement *m* of *A* to *B*
X_{49}	$ws_{b,A,B}$	World state for transmitting affective response *b* of *A* to *B*
X_{50}	$ws_{v,A,B}$	World state for transmitting verbal action *v* of *A* to *B*
X_{51}	$ws_{m,B,A}$	World state for transmitting movement *m* of *B* to *A*
X_{52}	$ws_{b,B,A}$	World state for transmitting affective response *b* of *B* to *A*
X_{53}	$ws_{v,B,A}$	World state for transmitting verbal action *v* of *B* to *A*

Table 7 First-order self-model **T**-states and **W**-states for excitability thresholds and connection weights and second-order self-model $\mathbf{H_T}$-states and $\mathbf{H_W}$-states for the adaptation speed of the **T**-states and **W**-states of the computational network model

state	name	Explanation
X_{54}	$\mathbf{W}_{\text{sense-rep\_}m,A}$	First-order self-model state for the weight of *A*'s internal connection from sensing to representing movement *m*
X_{55}	$\mathbf{W}_{\text{sense-rep\_}b,A}$	First-order self-model state for the weight of *A*'s internal connection from sensing to representing affective response *b*
X_{56}	$\mathbf{W}_{\text{sense-rep\_}v,A}$	First-order self-model state for the weight of *A*'s internal connection from sensing to representing verbal action *v*
X_{57}	$\mathbf{W}_{\text{prep-exec\_}m,A}$	First-order self-model state for the weight of *A*'s internal connection from preparing to executing movement *m*
X_{58}	$\mathbf{W}_{\text{prep-exec\_}b,A}$	First-order self-model state for the weight of *A*'s internal connection from preparing to expressing affective response *b*
X_{59}	$\mathbf{W}_{\text{prep-exec\_}v,A}$	First-order self-model state for the weight of *A*'s internal connection from preparing to executing verbal action *v*
X_{60}	$\mathbf{W}_{\text{exec-ws\_}m,A,B}$	First-order self-model state for the weight of *A*'s external connection from executing to world state for movement *m* for *B* (enabling interaction from *A* to *B*)
X_{61}	$\mathbf{W}_{\text{exec-ws\_}b,A,B}$	First-order self-model state for the weight of *A*'s external connection from executing to world state for affective response *b* f *B* (enabling interaction from *A* to *B*)
X_{62}	$\mathbf{W}_{\text{exec-ws\_}v,A,B}$	First-order self-model state for the weight of *A*'s external connection from executing to world state for verbal action *v* for *B* (enabling interaction from *A* to *B*)
X_{63}	$\mathbf{W}_{\text{ws-sense\_}m,B,A}$	First-order self-model state for the weight of *A*'s external connection from world state to *A*'s sensing of movement *m* from *B*
X_{64}	$\mathbf{W}_{\text{ws-sense\_}b,B,A}$	First-order self-model state for the weight of *A*'s external connection from world state to *A*'s sensing of affective response *b* from *B*
X_{65}	$\mathbf{W}_{\text{ws-sense\_}v,B,A}$	First-order self-model state for the weight of *A*'s external connection from world state to *A*'s sensing of verbal action *v* from
X_{66}	$\mathbf{W}_{\text{sense-rep\_}m,B}$	First-order self-model state for the weight of *B*'s internal connection from sensing to representing movement *m*
X_{67}	$\mathbf{W}_{\text{sense-rep\_}b,B}$	First-order self-model state for the weight of *B*'s internal connection from sensing to representing affective response *b*
X_{68}	$\mathbf{W}_{\text{sense-rep\_}v,B}$	First-order self-model state for the weight of *B*'s internal connection from sensing to representing verbal action *v*
X_{69}	$\mathbf{W}_{\text{prep-exec\_}m,B}$	First-order self-model state for the weight of *B*'s internal connection from preparing to executing movement *m*
X_{70}	$\mathbf{W}_{\text{prep-exec\_}b,B}$	First-order self-model state for the weight of *B*'s internal connection from preparing to expressing affective response *b*
X_{71}	$\mathbf{W}_{\text{prep-exec\_}v,B}$	First-order self-model state for the weight of *B*'s internal connection from preparing to executing verbal action *v*
X_{72}	$\mathbf{W}_{\text{exec-ws\_}m,B,A}$	First-order self-model state for the weight of *B*'s external connection from executing to world state for movement *m* for *A* (enabling interaction from *B* to *A*)
X_{73}	$\mathbf{W}_{\text{exec-ws\_}b,B,A}$	First-order self-model state for the weight of *B*'s external connection from executing to world state for affective response *b* f *A* (enabling interaction from *B* to *A*)
X_{74}	$\mathbf{W}_{\text{exec-ws\_}v,B,A}$	First-order self-model state for the weight of *B*'s external connection from executing to world state for verbal action *v* for *A* (enabling interaction from *B* to *A*)
X_{75}	$\mathbf{W}_{\text{ws-sense\_}m,A,B}$	First-order self-model state for the weight of *B*'s external connection from world state to *B*'s sensing of movement *m* from *A*
X_{76}	$\mathbf{W}_{\text{ws-sense\_}b,A,B}$	First-order self-model state for the weight of *B*'s external connection from world state to *B*'s sensing of affective response *b* from *A*
X_{77}	$\mathbf{W}_{\text{ws-sense\_}v,A,B}$	First-order self-model state for the weight of *B*'s external connection from world state to *B*'s sensing of verbal action *v* from
X_{78}	$\mathbf{T}_{\text{rep\_}m,A}$	First-order self-model state for the excitability threshold of *A*'s sensory representation state $\text{rep}_{m,\text{A}}$ for movement *m*
X_{79}	$\mathbf{T}_{\text{rep\_}b,A}$	First-order self-model state for the excitability threshold of *A*'s sensory representation state $\text{rep}_{b,\text{A}}$ for affective response *b*
X_{80}	$\mathbf{T}_{\text{rep\_}v,A}$	First-order self-model state for the excitability threshold of *A*'s sensory representation state $\text{rep}_{v,\text{A}}$ for verbal response *v*
X_{81}	$\mathbf{T}_{\text{exec\_}m,A}$	First-order self-model state for the excitability threshold of *A*'s execution state state $\text{move}_{m,\text{A}}$ for movement *m*
X_{82}	$\mathbf{T}_{\text{exec\_}b,A}$	First-order self-model state for the excitability threshold of *A*'s execution state state $\text{exp\_affect}_{b,\text{A}}$ for affective response *b*
X_{83}	$\mathbf{T}_{\text{exec\_}v,A}$	First-order self-model state for the excitability threshold of *A*'s execution state state $\text{talk}_{v,\text{A}}$ for verbal response *v*
X_{84}	$\mathbf{T}_{\text{rep\_}m,B}$	First-order self-model state for the excitability threshold of B's sensory representation state $\text{rep}_{m,B}$ for movement *m*
X_{85}	$\mathbf{T}_{\text{rep\_}b,B}$	First-order self-model state for the excitability threshold of *B*'s sensory representation state $\text{rep}_{b,\text{B}}$ for affective response *b*
X_{86}	$\mathbf{T}_{\text{rep\_}v,B}$	First-order self-model state for the excitability threshold of *B*'s sensory representation state $\text{rep}_{v,\text{B}}$ for verbal response *v*
X_{87}	$\mathbf{T}_{\text{exec\_}m,B}$	First-order self-model state for the excitability threshold of *B*'s execution state state $\text{move}_{m,\text{B}}$ for movement *m*
X_{88}	$\mathbf{T}_{\text{exec\_}b,B}$	First-order self-model state for the excitability threshold of *B*'s execution state state $\text{exp\_affect}_{b,\text{B}}$ for affective response *b*
X_{89}	$\mathbf{T}_{\text{exec\_}v,B}$	First-order self-model state for the excitability threshold of *B*'s execution state state $\text{talk}_{v,\text{B}}$ for verbal response *v*
X_{90}	$\mathbf{H}_{\mathbf{W}_\text{A}}$	Second-order self-model state for the speed factor of the first-order self-model **W**-states for *A*
X_{91}	$\mathbf{H}_{\mathbf{W}_\text{B}}$	Second-order self-model state for the speed factor of the first-order self-model **W**-states for *B*
X_{92}	$\mathbf{H}_{\mathbf{T}_\text{A}}$	Second-order self-model state for the speed factor of the first-order self-model **T**-states for *A*
X_{93}	$\mathbf{H}_{\mathbf{T}_\text{B}}$	Second-order self-model state for the speed factor of the first-order self-model **T**-states for *B*

The network characteristics for aggregation are defined by the selection of combination functions from the library and values for their parameters. In role matrix **mcfw** it is specified by weights which state uses which combination function; see Table 10.

In role matrix **mcfp** (see Table 11) it is indicated what the parameter values are for the chosen combination functions. A number of them are adaptive: their adaptive excitability thresholds are represented by self-model **T**-states. These concern for agent A states X_{11} to X_{13} (with excitability threshold self-model **T**-states X_{78} to X_{80}) and X_{24} to X_{26} (with self-model **T**-states X_{81} to X_{83}), and for agent B states X_{32} to X_{34} (with excitability threshold self-model **T**-states X_{84} to X_{86}) and X_{35} to X_{47} (with self-model **T**-states X_{87} to X_{89}).

In Table12, the role matrix **ms** for speed factors is shown, which lists all speed factors. Next to it, the list of initial values can be found. For **ms** some entries are

Table 8 Role matrix **mb** for base connectivity

mb base connectivity	1	2	3	4	5	6	7
X_1 $ws_{s,A}$	X_1						
X_2 $ws_{s,B}$	X_2						
X_3 $ws_{s,AB}$	X_3						
X_4 $con_{th,A}$	X_4						
X_5 $con_{th,B}$	X_5						
X_6 $sense_{s,A}$	X_3	X_1					
X_7 $sense_{B,m,A}$	X_{24}	X_{51}					
X_8 $sense_{B,b,A}$	X_{25}	X_{52}					
X_9 $sense_{B,v,A}$	X_{26}	X_{53}					
X_{10} $rep_{s,A}$	X_6						
X_{11} $rep_{B,m,A}$	X_7	X_{14}					
X_{12} $rep_{B,b,A}$	X_8	X_{15}					
X_{13} $rep_{B,v,A}$	X_9	X_{16}					
X_{14} $prep_{m,A}$	X_{10}	X_{11}					
X_{15} $prep_{b,A}$	X_{10}	X_{12}					
X_{16} $prep_{v,A}$	X_{10}	X_{13}	X_{17}				
X_{17} $cons\_emotion_{b,A}$	X_{15}	X_{26}	X_{18}	X_{19}	X_{20}		
X_{18} $intrasyncdet_{A,\text{b-m}}$	X_{24}	X_{25}					
X_{19} $intrasyncdet_{A,\text{m-v}}$	X_{24}	X_{26}					
X_{20} $intrasyncdet_{A,\text{b-v}}$	X_{25}	X_{26}					
X_{21} $intersyncdet_{B,A,\text{m}}$	X_{24}	X_7					
X_{22} $intersyncdet_{B,A,\text{b}}$	X_{25}	X_8					
X_{23} $intersyncdet_{B,A,\text{v}}$	X_{26}	X_9					
X_{24} $move_{m,A}$	X_{14}						
X_{25} $exp\_affect_{b,A}$	X_{15}						
X_{26} $talk_{A,B,v}$	X_{16}						
X_{27} $sense_{s,B}$	X_3	X_2					
X_{28} $sense_{A,m,B}$	X_{45}	X_{48}					
X_{29} $sense_{A,b,B}$	X_{46}	X_{49}					
X_{30} $sense_{A,v,B}$	X_{47}	X_{50}					
X_{31} $rep_{s,B}$	X_{27}						
X_{32} $rep_{A,m,B}$	X_{28}	X_{35}					
X_{33} $rep_{A,b,B}$	X_{29}	X_{36}					
X_{34} $rep_{A,v,B}$	X_{30}	X_{37}					
X_{35} $prep_{m,B}$	X_{31}	X_{32}					
X_{36} $prep_{b,B}$	X_{31}	X_{33}					
X_{37} $prep_{v,B}$	X_{31}	X_{34}	X_{38}				
X_{38} $cons\_emotion_{b,B}$	X_{36}	X_{47}	X_{39}	X_{40}	X_{41}		
X_{39} $intrasyncdet_{B,\text{b-m}}$	X_{45}	X_{46}					
X_{40} $intrasyncdet_{B,\text{m-v}}$	X_{45}	X_{47}					
X_{41} $intrasyncdet_{B,\text{b-v}}$	X_{46}	X_{47}					
X_{42} $intersyncdet_{A,B,\text{m}}$	X_{45}	X_{28}					
X_{43} $intersyncdet_{A,\text{B,b}}$	X_{46}	X_{29}					
X_{44} $intersyncdet_{A,\text{B,v}}$	X_{47}	X_{30}					
X_{45} $move_{m,B}$	X_{35}						
X_{46} $exp\_affect_{b,B}$	X_{36}						
X_{47} $talk_{B,A,v}$	X_{37}						
X_{48} $ws_{m,A,B}$	X_{24}						
X_{49} $ws_{b,A,B}$	X_{25}						
X_{50} $ws_{v,A,B}$	X_{26}						
X_{51} $ws_{m,B,A}$	X_{45}						
X_{52} $ws_{b,B,A}$	X_{46}						
X_{53} $ws_{v,B,A}$	X_{47}						

(continued)

Table 8 (continued)

X_{54} $\mathbf{W}_{sense\text{-}rep\_m,A}$	X_{18}	X_{19}	X_{20}	X_{21}	X_{22}	X_{23}	
X_{55} $\mathbf{W}_{sense\text{-}rep\_b,A}$	X_{18}	X_{19}	X_{20}	X_{21}	X_{22}	X_{23}	
X_{56} $\mathbf{W}_{sense\text{-}rep\_v,A}$	X_{18}	X_{19}	X_{20}	X_{21}	X_{22}	X_{23}	
X_{57} $\mathbf{W}_{prep\text{-}exec\_m,A}$	X_{18}	X_{19}	X_{20}	X_{21}	X_{22}	X_{23}	
X_{58} $\mathbf{W}_{prep\text{-}exec\_b,A}$	X_{18}	X_{19}	X_{20}	X_{21}	X_{22}	X_{23}	
X_{59} $\mathbf{W}_{prep\text{-}exec\_v,A}$	X_{18}	X_{19}	X_{20}	X_{21}	X_{22}	X_{23}	
X_{60} $\mathbf{W}_{exec\text{-}ws\_m,A,B}$	X_{60}						
X_{61} $\mathbf{W}_{exec\text{-}ws\_b,A,B}$	X_{61}						
X_{62} $\mathbf{W}_{exec\text{-}ws\_v,A,B}$	X_{62}						
X_{63} $\mathbf{W}_{ws\text{-}sense\_m,B,A}$	X_{63}						
X_{64} $\mathbf{W}_{ws\text{-}sense\_b,B,A}$	X_{64}						
X_{65} $\mathbf{W}_{ws\text{-}sense\_v,B,A}$	X_{65}						
X_{66} $\mathbf{W}_{sense\text{-}rep\_m,B}$	X_{39}	X_{40}	X_{41}	X_{42}	X_{43}	X_{44}	
X_{67} $\mathbf{W}_{sense\text{-}rep\_b,B}$	X_{39}	X_{40}	X_{41}	X_{42}	X_{43}	X_{44}	
X_{68} $\mathbf{W}_{sense\text{-}rep\_v,B}$	X_{39}	X_{40}	X_{41}	X_{42}	X_{43}	X_{44}	
X_{69} $\mathbf{W}_{prep\text{-}exec\_m,B}$	X_{39}	X_{40}	X_{41}	X_{42}	X_{43}	X_{44}	
X_{70} $\mathbf{W}_{prep\text{-}exec\_b,B}$	X_{39}	X_{40}	X_{41}	X_{42}	X_{43}	X_{44}	
X_{71} $\mathbf{W}_{prep\text{-}exec\_v,B}$	X_{39}	X_{40}	X_{41}	X_{42}	X_{43}	X_{44}	
X_{72} $\mathbf{W}_{exec\text{-}ws\_m,B,A}$	X_{72}						
X_{73} $\mathbf{W}_{exec\text{-}ws\_b,B,A}$	X_{73}						
X_{74} $\mathbf{W}_{exec\text{-}ws\_v,B,A}$	X_{74}						
X_{75} $\mathbf{W}_{ws\text{-}sense\_m,A,B}$	X_{75}						
X_{76} $\mathbf{W}_{ws\text{-}sense\_b,A,B}$	X_{76}						
X_{77} $\mathbf{W}_{ws\text{-}sense\_v,A,B}$	X_{77}						
X_{78} $\mathbf{T}_{rep\_m,A}$	X_{18}	X_{19}	X_{20}	X_{21}	X_{22}	X_{23}	X_{4}
X_{79} $\mathbf{T}_{rep\_b,A}$	X_{18}	X_{19}	X_{20}	X_{21}	X_{22}	X_{23}	X_{4}
X_{80} $\mathbf{T}_{rep\_v,A}$	X_{18}	X_{19}	X_{20}	X_{21}	X_{22}	X_{23}	X_{4}
X_{81} $\mathbf{T}_{exec\_m,A}$	X_{18}	X_{19}	X_{20}	X_{21}	X_{22}	X_{23}	X_{4}
X_{82} $\mathbf{T}_{exec\_b,A}$	X_{18}	X_{19}	X_{20}	X_{21}	X_{22}	X_{23}	X_{4}
X_{83} $\mathbf{T}_{exec\_v,A}$	X_{18}	X_{19}	X_{20}	X_{21}	X_{22}	X_{23}	X_{4}
X_{84} $\mathbf{T}_{rep\_m,B}$	X_{39}	X_{40}	X_{41}	X_{42}	X_{43}	X_{44}	X_{5}
X_{85} $\mathbf{T}_{rep\_b,B}$	X_{39}	X_{40}	X_{41}	X_{42}	X_{43}	X_{44}	X_{5}
X_{86} $\mathbf{T}_{rep\_v,B}$	X_{39}	X_{40}	X_{41}	X_{42}	X_{43}	X_{44}	X_{5}
X_{87} $\mathbf{T}_{exec\_m,B}$	X_{39}	X_{40}	X_{41}	X_{42}	X_{43}	X_{44}	X_{5}
X_{88} $\mathbf{T}_{exec\_b,B}$	X_{39}	X_{40}	X_{41}	X_{42}	X_{43}	X_{44}	X_{5}
X_{89} $\mathbf{T}_{exec\_v,B}$	X_{39}	X_{40}	X_{41}	X_{42}	X_{43}	X_{44}	X_{5}
X_{90} $\mathbf{H}_{\mathbf{W}_A}$	X_{10}						
X_{91} $\mathbf{H}_{\mathbf{W}_B}$	X_{31}						
X_{92} $\mathbf{H}_{\mathbf{T}_A}$	X_{10}						
X_{93} $\mathbf{H}_{\mathbf{T}_B}$	X_{31}						

adaptive as well. For agent A the speed factors of **W**-states X_{54} to X_{65} are represented by (second-order) self-model $\mathbf{H}_\mathbf{W}$-state X_{90} and for agent B the speed factors of **W**-states X_{66} to X_{77} are represented by (second-order) self-model $\mathbf{H}_\mathbf{W}$-state X_{91}. Moreover, for agent A the speed factors for **T**-states X_{78} to X_{83} are represented by (second-order) self-model $\mathbf{H}_\mathbf{T}$-state X_{92} and for agent B the speed factors of **T**-states X_{84} to X_{89} are represented by (second-order) self-model $\mathbf{H}_\mathbf{T}$-state X_{93}.

Table 9 Role matrix **mcw** for connection weights

	mcw connection weights	1	2	3	4	5	6	7	8
X_1	$ws_{s,A}$	1							
X_2	$ws_{s,B}$	1							
X_3	$ws_{s,AB}$	1							
X_4	$con_{th,A}$	1							
X_5	$con_{th,B}$	1							
X_6	$sense_{s,A}$	1	1						
X_7	$sense_{B,m,A}$	1	X_{63}						
X_8	$sense_{B,b,A}$	1	X_{64}						
X_9	$sense_{B,v,A}$	1	X_{65}						
X_{10}	$rep_{s,A}$	1							
X_{11}	$rep_{B,m,A}$	X_{54}	1						
X_{12}	$rep_{B,b,A}$	X_{55}	1						
X_{13}	$rep_{B,v,A}$	X_{56}	1						
X_{14}	$prep_{m,A}$	1	1						
X_{15}	$prep_{b,A}$	1	1						
X_{16}	$prep_{v,A}$	1	1	1					
X_{17}	$cons\_emotion_{b,A}$	1	1	1	1	1			
X_{18}	$intrasyncdet_{A,\text{b-m}}$	1	1						
X_{19}	$intrasyncdet_{A,\text{m-v}}$	1	1						
X_{20}	$intrasyncdet_{A,\text{b-v}}$	1	1						
X_{21}	$intersyncdet_{B,A,\text{m}}$	1	1						
X_{22}	$intersyncdet_{B,A,\text{b}}$	1	1						
X_{23}	$intersyncdet_{B,A,\text{v}}$	1	1						
X_{24}	$move_{m,A}$	X_{57}							
X_{25}	$exp\_affect_{b,A}$	X_{58}							
X_{26}	$talk_{A,B,v}$	X_{59}							
X_{27}	$sense_{s,B}$	1	1						
X_{28}	$sense_{A,m,B}$	1	X_{75}						
X_{29}	$sense_{A,b,B}$	1	X_{76}						
X_{30}	$sense_{A,v,B}$	1	X_{77}						
X_{31}	$rep_{s,B}$	1							
X_{32}	$rep_{A,m,B}$	X_{66}	1						
X_{33}	$rep_{A,b,B}$	X_{67}	1						
X_{34}	$rep_{A,v,B}$	X_{68}	1						
X_{35}	$prep_{m,B}$	1	1						
X_{36}	$prep_{b,B}$	1	1						
X_{37}	$prep_{v,B}$	1	1	1					
X_{38}	$cons\_emotion_{b,B}$	1	1	1	1	1			
X_{39}	$intrasyncdet_{B,\text{b-m}}$	1	1						
X_{40}	$intrasyncdet_{B,\text{m-v}}$	1	1						
X_{41}	$intrasyncdet_{B,\text{b-v}}$	1	1						
X_{42}	$intersyncdet_{A,B,\text{m}}$	1	1						
X_{43}	$intersyncdet_{A,\text{B,b}}$	1	1						
X_{44}	$intersyncdet_{A,\text{B,v}}$	1	1						
X_{45}	$move_{m,B}$	X_{69}							
X_{46}	$exp\_affect_{b,B}$	X_{70}							
X_{47}	$talk_{B,A,v}$	X_{71}							
X_{48}	$ws_{m,A,B}$	X_{60}							
X_{49}	$ws_{b,A,B}$	X_{61}							
X_{50}	$ws_{v,A,B}$	X_{62}							
X_{51}	$ws_{m,B,A}$	X_{72}							
X_{52}	$ws_{b,B,A}$	X_{73}							
X_{53}	$ws_{v,B,A}$	X_{74}							

(continued)

Table 9 (continued)

X_{54}	$\mathbf{W}_{\text{sense-rep\_}m,A}$	1	1	1	1	1	1		
X_{55}	$\mathbf{W}_{\text{sense-rep\_}b,A}$	1	1	1	1	1	1		
X_{56}	$\mathbf{W}_{\text{sense-rep\_}v,A}$	1	1	1	1	1	1		
X_{57}	$\mathbf{W}_{\text{prep-exec\_}m,A}$	1	1	1	1	1	1		
X_{58}	$\mathbf{W}_{\text{prep-exec\_}b,A}$	1	1	1	1	1	1		
X_{59}	$\mathbf{W}_{\text{prep-exec\_}v,A}$	1	1	1	1	1	1		
X_{60}	$\mathbf{W}_{\text{exec-ws\_}m,A,B}$	1							
X_{61}	$\mathbf{W}_{\text{exec-ws\_}b,A,B}$	1							
X_{62}	$\mathbf{W}_{\text{exec-ws\_}v,A,B}$	1							
X_{63}	$\mathbf{W}_{\text{ws-sense\_}m,B,A}$	1							
X_{64}	$\mathbf{W}_{\text{ws-sense\_}b,B,A}$	1							
X_{65}	$\mathbf{W}_{\text{ws-sense\_}v,B,A}$	1							
X_{66}	$\mathbf{W}_{\text{sense-rep\_}m,B}$	1	1	1	1	1	1		
X_{67}	$\mathbf{W}_{\text{sense-rep\_}b,B}$	1	1	1	1	1	1		
X_{68}	$\mathbf{W}_{\text{sense-rep\_}v,B}$	1	1	1	1	1	1		
X_{69}	$\mathbf{W}_{\text{prep-exec\_}m,B}$	1	1	1	1	1	1		
X_{70}	$\mathbf{W}_{\text{prep-exec\_}b,B}$	1	1	1	1	1	1		
X_{71}	$\mathbf{W}_{\text{prep-exec\_}v,B}$	1	1	1	1	1	1		
X_{72}	$\mathbf{W}_{\text{exec-ws\_}m,B,A}$	1							
X_{73}	$\mathbf{W}_{\text{exec-ws\_}b,B,A}$	1							
X_{74}	$\mathbf{W}_{\text{exec-ws\_}v,B,A}$	1							
X_{75}	$\mathbf{W}_{\text{ws-sense\_}m,A,B}$	1							
X_{76}	$\mathbf{W}_{\text{ws-sense\_}b,A,B}$	1							
X_{77}	$\mathbf{W}_{\text{ws-sense\_}v,A,B}$	1							
X_{78}	$\mathbf{T}_{\text{rep\_}m,A}$	-0.12	-0.12	-0.12	-0.12	-0.12	-0.12		1
X_{79}	$\mathbf{T}_{\text{rep\_}b,A}$	-0.12	-0.12	-0.12	-0.12	-0.12	-0.12		1
X_{80}	$\mathbf{T}_{\text{rep\_}v,A}$	-0.12	-0.12	-0.12	-0.12	-0.12	-0.12		1
X_{81}	$\mathbf{T}_{\text{exec\_}m,A}$	-0.12	-0.12	-0.12	-0.12	-0.12	-0.12		1
X_{82}	$\mathbf{T}_{\text{exec\_}b,A}$	-0.12	-0.12	-0.12	-0.12	-0.12	-0.12		1
X_{83}	$\mathbf{T}_{\text{exec\_}v,A}$	-0.12	-0.12	-0.12	-0.12	-0.12	-0.12		1
X_{84}	$\mathbf{T}_{\text{rep\_}m,B}$	-0.12	-0.12	-0.12	-0.12	-0.12	-0.12		1
X_{85}	$\mathbf{T}_{\text{rep\_}b,B}$	-0.12	-0.12	-0.12	-0.12	-0.12	-0.12		1
X_{86}	$\mathbf{T}_{\text{rep\_}v,B}$	-0.12	-0.12	-0.12	-0.12	-0.12	-0.12		1
X_{87}	$\mathbf{T}_{\text{exec\_}m,B}$	-0.12	-0.12	-0.12	-0.12	-0.12	-0.12		1
X_{88}	$\mathbf{T}_{\text{exec\_}b,B}$	-0.12	-0.12	-0.12	-0.12	-0.12	-0.12		1
X_{89}	$\mathbf{T}_{\text{exec\_}v,B}$	-0.12	-0.12	-0.12	-0.12	-0.12	-0.12		1
X_{90}	$\mathbf{H}_{\mathbf{W_A}}$	0.01							
X_{91}	$\mathbf{H}_{\mathbf{W_B}}$	0.01							
X_{92}	$\mathbf{H}_{\mathbf{T_A}}$	1							
X_{93}	$\mathbf{H}_{\mathbf{T_B}}$	1							

Table 10 Role matrix **mcfw** for combination function weights

mcfw combination function weights	1 alogistic	2 compdiff	3 randstepmod	4 randstepmodopp	5 eucl
X_1 $ws_{s,A}$				1	
X_2 $ws_{s,B}$				1	
X_3 $ws_{s,AB}$			1		
X_4 $con_{th,A}$	1				
X_5 $con_{th,B}$	1				
X_6 $sense_{s,A}$					1
X_7 $sense_{B,m,A}$					1
X_8 $sense_{B,b,A}$					1
X_9 $sense_{B,v,A}$					1
X_{10} $rep_{s,A}$	1				
X_{11} $rep_{B,m,A}$	1				
X_{12} $rep_{B,b,A}$	1				
X_{13} $rep_{B,v,A}$	1				
X_{14} $prep_{m,A}$	1				
X_{15} $prep_{b,A}$	1				
X_{16} $prep_{v,A}$	1				
X_{17} $cons\_emotion_{b,A}$	1				
X_{18} $intrasyncdet_{A,\text{b-m}}$		1			
X_{19} $intrasyncdet_{A,\text{m-v}}$		1			
X_{20} $intrasyncdet_{A,\text{b-v}}$		1			
X_{21} $intersyncdet_{B,A,\text{m}}$		1			
X_{22} $intersyncdet_{B,A,\text{b}}$		1			
X_{23} $intersyncdet_{B,A,\text{v}}$		1			
X_{24} $move_{m,A}$	1				
X_{25} $exp\_affect_{b,A}$	1				
X_{26} $talk_{A,B,v}$	1				
X_{27} $sense_{s,B}$					1
X_{28} $sense_{A,m,B}$					1
X_{29} $sense_{A,b,B}$					1
X_{30} $sense_{A,v,B}$					1
X_{31} $rep_{s,B}$	1				
X_{32} $rep_{A,m,B}$	1				
X_{33} $rep_{A,b,B}$	1				
X_{34} $rep_{A,v,B}$	1				
X_{35} $prep_{m,B}$	1				
X_{36} $prep_{b,B}$	1				
X_{37} $prep_{v,B}$	1				
X_{38} $cons\_emotion_{b,B}$	1				
X_{39} $intrasyncdet_{B,\text{b-m}}$		1			
X_{40} $intrasyncdet_{B,\text{m-v}}$		1			
X_{41} $intrasyncdet_{B,\text{b-v}}$		1			
X_{42} $intersyncdet_{A,B,\text{m}}$		1			
X_{43} $intersyncdet_{A,\text{B,b}}$		1			
X_{44} $intersyncdet_{A,\text{B,v}}$		1			
X_{45} $move_{m,B}$	1				
X_{46} $exp\_affect_{b,B}$	1				
X_{47} $talk_{B,A,v}$	1				
X_{48} $ws_{m,A,B}$					1
X_{49} $ws_{b,A,B}$					1
X_{50} $ws_{v,A,B}$					1
X_{51} $ws_{m,B,A}$					1
X_{52} $ws_{b,B,A}$					1
X_{53} $ws_{v,B,A}$					1

(continued)

Table 10 (continued)

X_{54} $\mathbf{W}_{\text{sense-rep}\ m,A}$	1			
X_{55} $\mathbf{W}_{\text{sense-rep}\ b,A}$	1			
X_{56} $\mathbf{W}_{\text{sense-rep}\ v,A}$	1			
X_{57} $\mathbf{W}_{\text{prep-exec}\ m,A}$	1			
X_{58} $\mathbf{W}_{\text{prep-exec}\ b,A}$	1			
X_{59} $\mathbf{W}_{\text{prep-exec}\ v,A}$	1			
X_{60} $\mathbf{W}_{\text{exec-ws}\ m,A,B}$			1	
X_{61} $\mathbf{W}_{\text{exec-ws}\ b,A,B}$			1	
X_{62} $\mathbf{W}_{\text{exec-ws}\ v,A,B}$			1	
X_{63} $\mathbf{W}_{\text{ws-sense}\ m,B,A}$	1			
X_{64} $\mathbf{W}_{\text{ws-sense}\ b,B,A}$	1			
X_{65} $\mathbf{W}_{\text{ws-sense}\ v,B,A}$	1			
X_{66} $\mathbf{W}_{\text{sense-rep}\ m,B}$	1			
X_{67} $\mathbf{W}_{\text{sense-rep}\ b,B}$	1			
X_{68} $\mathbf{W}_{\text{sense-rep}\ v,B}$	1			
X_{69} $\mathbf{W}_{\text{prep-exec}\ m,B}$	1			
X_{70} $\mathbf{W}_{\text{prep-exec}\ b,B}$	1			
X_{71} $\mathbf{W}_{\text{prep-exec}\ v,B}$	1			
X_{72} $\mathbf{W}_{\text{exec-ws}\ m,B,A}$			1	
X_{73} $\mathbf{W}_{\text{exec-ws}\ b,B,A}$			1	
X_{74} $\mathbf{W}_{\text{exec-ws}\ v,B,A}$			1	
X_{75} $\mathbf{W}_{\text{ws-sense}\ m,A,B}$	1			
X_{76} $\mathbf{W}_{\text{ws-sense}\ b,A,B}$	1			
X_{77} $\mathbf{W}_{\text{ws-sense}\ v,A,B}$	1			
X_{78} $\mathbf{T}_{\text{rep}\ m,A}$	1			
X_{79} $\mathbf{T}_{\text{rep}\ b,A}$	1			
X_{80} $\mathbf{T}_{\text{rep}\ v,A}$	1			
X_{81} $\mathbf{T}_{\text{exec}\ m,A}$	1			
X_{82} $\mathbf{T}_{\text{exec}\ b,A}$	1			
X_{83} $\mathbf{T}_{\text{exec}\ v,A}$	1			
X_{84} $\mathbf{T}_{\text{rep}\ m,B}$	1			
X_{85} $\mathbf{T}_{\text{rep}\ b,B}$	1			
X_{86} $\mathbf{T}_{\text{rep}\ v,B}$	1			
X_{87} $\mathbf{T}_{\text{exec}\ m,B}$	1			
X_{88} $\mathbf{T}_{\text{exec}\ b,B}$	1			
X_{89} $\mathbf{T}_{\text{exec}\ v,B}$	1			
X_{90} $\mathbf{H}_{\mathbf{W}_\mathbf{A}}$	1			
X_{91} $\mathbf{H}_{\mathbf{W}_\mathbf{B}}$	1			
X_{92} $\mathbf{H}_{\mathbf{T}_\mathbf{A}}$	1			
X_{93} $\mathbf{H}_{\mathbf{T}_\mathbf{B}}$	1			

Table 11 Role matrix **mcfp** for combination function parameters

mcfp combination function parameters	1 alogistic		2 compdiff		3 randstepmod		4 randstepmodopp		5 eucl	
X_1 $ws_{s,A}$							120	60		
X_2 $ws_{s,B}$							120	60		
X_3 $ws_{s,AB}$					120	60				
X_4 $con_{th,A}$	5	0.5								
X_5 $con_{th,B}$	5	0.5								
X_6 $sense_{s,A}$									1	1
X_7 $sense_{B,m,A}$									1	2
X_8 $sense_{B,b,A}$									1	2
X_9 $sense_{B,v,A}$									1	2
X_{10} $rep_{s,A}$	5	0.5								
X_{11} $rep_{B,m,A}$	5	X_{78}								
X_{12} $rep_{B,b,A}$	5	X_{79}								
X_{13} $rep_{B,v,A}$	5	X_{80}								
X_{14} $prep_{m,A}$	5	0.6								
X_{15} $prep_{b,A}$	5	0.6								
X_{16} $prep_{v,A}$	5	0.9								
X_{17} $cons\_emotion_{b,A}$	5	3.25								
X_{18} $intrasyncdet_{A,b\text{-}m}$										
X_{19} $intrasyncdet_{A,m\text{-}v}$										
X_{20} $intrasyncdet_{A,b\text{-}v}$										
X_{21} $intersyncdet_{B,A,m}$										
X_{22} $intersyncdet_{B,A,b}$										
X_{23} $intersyncdet_{B,A,v}$										
X_{24} $move_{m,A}$	5	X_{81}								
X_{25} $exp\_affect_{b,A}$	5	X_{82}								
X_{26} $talk_{A,B,v}$	5	X_{83}								
X_{27} $sense_{s,B}$									1	1
X_{28} $sense_{A,m,B}$									1	2
X_{29} $sense_{A,b,B}$									1	2
X_{30} $sense_{A,v,B}$									1	2
X_{31} $rep_{s,B}$	5	0.5								
X_{32} $rep_{A,m,B}$	5	X_{84}								
X_{33} $rep_{A,b,B}$	5	X_{85}								
X_{34} $rep_{A,v,B}$	5	X_{86}								
X_{35} $prep_{m,B}$	5	0.6								
X_{36} $prep_{b,B}$	5	0.6								
X_{37} $prep_{v,B}$	5	0.9								
X_{38} $cons\_emotion_{b,B}$	5	3.25								
X_{39} $intrasyncdet_{B,b\text{-}m}$										
X_{40} $intrasyncdet_{B,m\text{-}v}$										
X_{41} $intrasyncdet_{B,b\text{-}v}$										
X_{42} $intersyncdet_{A,B,m}$										
X_{43} $intersyncdet_{A,B,b}$										
X_{44} $intersyncdet_{A,B,v}$										

(continued)

Table 11 (continued)

X_{45}	$\text{move}_{m,B}$	5	X_{87}				
X_{46}	$\text{exp\_affect}_{b,B}$	5	X_{88}				
X_{47}	$\text{talk}_{B,A,v}$	5	X_{89}				
X_{48}	$\text{ws}_{m,A,B}$					1	1
X_{49}	$\text{ws}_{b,A,B}$					1	1
X_{50}	$\text{ws}_{v,A,B}$					1	1
X_{51}	$\text{ws}_{m,B,A}$					1	1
X_{52}	$\text{ws}_{b,B,A}$					1	1
X_{53}	$\text{ws}_{v,B,A}$					1	1
X_{54}	$\mathbf{W}_{\text{sense-rep } m,A}$	5	4.5				
X_{55}	$\mathbf{W}_{\text{sense-rep } b,A}$	5	4.5				
X_{56}	$\mathbf{W}_{\text{sense-rep } v,A}$	5	4.5				
X_{57}	$\mathbf{W}_{\text{prep-exec } m,A}$	5	4.5				
X_{58}	$\mathbf{W}_{\text{prep-exec } b,A}$	5	4.5				
X_{59}	$\mathbf{W}_{\text{prep-exec } v,A}$	5	4.5				
X_{60}	$\mathbf{W}_{\text{exec-ws } m,A,B}$			60	30		
X_{61}	$\mathbf{W}_{\text{exec-ws } b,A,B}$			60	30		
X_{62}	$\mathbf{W}_{\text{exec-ws } v,A,B}$			60	30		
X_{63}	$\mathbf{W}_{\text{ws-sense } m,B,A}$	5	4.5				
X_{64}	$\mathbf{W}_{\text{ws-sense } b,B,A}$	5	4.5				
X_{65}	$\mathbf{W}_{\text{ws-sense } v,B,A}$	5	4.5				
X_{66}	$\mathbf{W}_{\text{sense-rep } m,B}$	5	4.5				
X_{67}	$\mathbf{W}_{\text{sense-rep } b,B}$	5	4.5				
X_{68}	$\mathbf{W}_{\text{sense-rep } v,B}$	5	4.5				
X_{69}	$\mathbf{W}_{\text{prep-exec } m,B}$	5	4.5				
X_{70}	$\mathbf{W}_{\text{prep-exec } b,B}$	5	4.5				
X_{71}	$\mathbf{W}_{\text{prep-exec } v,B}$	5	4.5				
X_{72}	$\mathbf{W}_{\text{exec-ws } m,B,A}$			60	30		
X_{73}	$\mathbf{W}_{\text{exec-ws } b,B,A}$			60	30		
X_{74}	$\mathbf{W}_{\text{exec-ws } v,B,A}$			60	30		
X_{75}	$\mathbf{W}_{\text{ws-sense } m,A,B}$	5	4.5				
X_{76}	$\mathbf{W}_{\text{ws-sense } b,A,B}$	5	4.5				
X_{77}	$\mathbf{W}_{\text{ws-sense } v,A,B}$	5	4.5				
X_{78}	$\mathbf{T}_{\text{rep } m,A}$	5	0.5				
X_{79}	$\mathbf{T}_{\text{rep } b,A}$	5	0.5				
X_{80}	$\mathbf{T}_{\text{rep } v,A}$	5	0.5				
X_{81}	$\mathbf{T}_{\text{exec } m,A}$	5	0.5				
X_{82}	$\mathbf{T}_{\text{exec } b,A}$	5	0.5				
X_{83}	$\mathbf{T}_{\text{exec } v,A}$	5	0.5				
X_{84}	$\mathbf{T}_{\text{rep } m,B}$	5	0.5				
X_{85}	$\mathbf{T}_{\text{rep } b,B}$	5	0.5				
X_{86}	$\mathbf{T}_{\text{rep } v,B}$	5	0.5				
X_{87}	$\mathbf{T}_{\text{exec } m,B}$	5	0.5				
X_{88}	$\mathbf{T}_{\text{exec } b,B}$	5	0.5				
X_{89}	$\mathbf{T}_{\text{exec } v,B}$	5	0.5				
X_{90}	$\mathbf{H}_{\mathbf{W_A}}$	5	0.7				
X_{91}	$\mathbf{H}_{\mathbf{W_B}}$	5	0.7				
X_{92}	$\mathbf{H}_{\mathbf{T_A}}$	5	0.7				
X_{93}	$\mathbf{H}_{\mathbf{T_B}}$	5	0.7				

Table 12 Role matrix **ms** for speed factors and **iv** for initial values

ms	speed factors	1	iv initial values	1
X_1	$ws_{s,A}$	1		1
X_2	$ws_{s,B}$	1		1
X_3	$ws_{s,AB}$	1		0
X_4	$con_{th,A}$	0		1
X_5	$con_{th,B}$	0		1
X_6	$sense_{s,A}$	1		0
X_7	$sense_{B,m,A}$	1		0
X_8	$sense_{B,b,A}$	1		0
X_9	$sense_{B,v,A}$	1		0
X_{10}	$rep_{s,A}$	1		0
X_{11}	$rep_{B,m,A}$	1		0
X_{12}	$rep_{B,b,A}$	1		0
X_{13}	$rep_{B,v,A}$	1		0
X_{14}	$prep_{m,A}$	1		0
X_{15}	$prep_{b,A}$	1		0
X_{16}	$prep_{v,A}$	1		0
X_{17}	$cons\_emotion_{b,A}$	1		0
X_{18}	$intrasyncdet_{A,\text{b-m}}$	0.5		0
X_{19}	$intrasyncdet_{A,\text{m-v}}$	0.5		0
X_{20}	$intrasyncdet_{A,\text{b-v}}$	0.5		0
X_{21}	$intersyncdet_{B,A,\text{m}}$	0.5		0
X_{22}	$intersyncdet_{B,A,\text{b}}$	0.5		0
X_{23}	$intersyncdet_{B,A,\text{v}}$	0.5		0
X_{24}	$move_{m,A}$	1		0
X_{25}	$exp\_affect_{b,A}$	1		0
X_{26}	$talk_{A,B,v}$	1		0
X_{27}	$sense_{s,B}$	1		0
X_{28}	$sense_{A,m,B}$	1		0
X_{29}	$sense_{A,b,B}$	1		0
X_{30}	$sense_{A,v,B}$	1		0
X_{31}	$rep_{s,B}$	1		0
X_{32}	$rep_{A,m,B}$	1		0
X_{33}	$rep_{A,b,B}$	1		0
X_{34}	$rep_{A,v,B}$	1		0
X_{35}	$prep_{m,B}$	1		0
X_{36}	$prep_{b,B}$	1		0
X_{37}	$prep_{v,B}$	1		0
X_{38}	$cons\_emotion_{b,B}$	1		0
X_{39}	$intrasyncdet_{B,\text{b-m}}$	0.5		0
X_{40}	$intrasyncdet_{B,\text{m-v}}$	0.5		0
X_{41}	$intrasyncdet_{B,\text{b-v}}$	0.5		0
X_{42}	$intersyncdet_{A,B,\text{m}}$	0.5		0
X_{43}	$intersyncdet_{A,\text{B},\text{b}}$	0.5		0
X_{44}	$intersyncdet_{A,\text{B},\text{v}}$	0.5		0
X_{45}	$move_{m,B}$	1		0
X_{46}	$exp\_affect_{b,B}$	1		0
X_{47}	$talk_{B,A,v}$	1		0
X_{48}	$ws_{m,A,B}$	1		0
X_{49}	$ws_{b,A,B}$	1		0
X_{50}	$ws_{v,A,B}$	1		0

(continued)

Table 12 (continued)

X_{51}	$ws_{m,B,A}$	1		0
X_{52}	$ws_{b,B,A}$	1		0
X_{53}	$ws_{v,B,A}$	1		0
X_{54}	$\mathbf{W}_{\text{sense-rep } m,A}$	X_{90}		0.4
X_{55}	$\mathbf{W}_{\text{sense-rep } b,A}$	X_{90}		0.4
X_{56}	$\mathbf{W}_{\text{sense-rep } v,A}$	X_{90}		0.4
X_{57}	$\mathbf{W}_{\text{prep-exec } m,A}$	X_{90}		0.4
X_{58}	$\mathbf{W}_{\text{prep-exec } b,A}$	X_{90}		0.4
X_{59}	$\mathbf{W}_{\text{prep-exec } v,A}$	X_{90}		0.4
X_{60}	$\mathbf{W}_{\text{exec-ws } m,A,B}$	1		0
X_{61}	$\mathbf{W}_{\text{exec-ws } b,A,B}$	1		0
X_{62}	$\mathbf{W}_{\text{exec-ws } v,A,B}$	1		0
X_{63}	$\mathbf{W}_{\text{ws-sense } m,B,A}$	0		1
X_{64}	$\mathbf{W}_{\text{ws-sense } b,B,A}$	0		1
X_{65}	$\mathbf{W}_{\text{ws-sense } v,B,A}$	0		1
X_{66}	$\mathbf{W}_{\text{sense-rep } m,B}$	X_{91}		0.4
X_{67}	$\mathbf{W}_{\text{sense-rep } b,B}$	X_{91}		0.4
X_{68}	$\mathbf{W}_{\text{sense-rep } v,B}$	X_{91}		0.4
X_{69}	$\mathbf{W}_{\text{prep-exec } m,B}$	X_{91}		0.4
X_{70}	$\mathbf{W}_{\text{prep-exec } b,B}$	X_{91}		0.4
X_{71}	$\mathbf{W}_{\text{prep-exec } v,B}$	X_{91}		0.4
X_{72}	$\mathbf{W}_{\text{exec-ws } m,B,A}$	1		0
X_{73}	$\mathbf{W}_{\text{exec-ws } b,B,A}$	1		0
X_{74}	$\mathbf{W}_{\text{exec-ws } v,B,A}$	1		0
X_{75}	$\mathbf{W}_{\text{ws-sense } m,A,B}$	0		1
X_{76}	$\mathbf{W}_{\text{ws-sense } b,A,B}$	0		1
X_{77}	$\mathbf{W}_{\text{ws-sense } v,A,B}$	0		1
X_{78}	$\mathbf{T}_{\text{rep } m,A}$	X_{92}		0.7
X_{79}	$\mathbf{T}_{\text{rep } b,A}$	X_{92}		0.7
X_{80}	$\mathbf{T}_{\text{rep } v,A}$	X_{92}		0.7
X_{81}	$\mathbf{T}_{\text{exec } m,A}$	X_{92}		0.7
X_{82}	$\mathbf{T}_{\text{exec } b,A}$	X_{92}		0.7
X_{83}	$\mathbf{T}_{\text{exec } v,A}$	X_{92}		0.7
X_{84}	$\mathbf{T}_{\text{rep } m,B}$	X_{93}		0.7
X_{85}	$\mathbf{T}_{\text{rep } b,B}$	X_{93}		0.7
X_{86}	$\mathbf{T}_{\text{rep } v,B}$	X_{93}		0.7
X_{87}	$\mathbf{T}_{\text{exec } m,B}$	X_{93}		0.7
X_{88}	$\mathbf{T}_{\text{exec } b,B}$	X_{93}		0.7
X_{89}	$\mathbf{T}_{\text{exec } v,B}$	X_{93}		0.7
X_{90}	$\mathbf{H}_{\mathbf{W_A}}$	1		0
X_{91}	$\mathbf{H}_{\mathbf{W_B}}$	1		0
X_{92}	$\mathbf{H}_{\mathbf{T_A}}$	1		0
X_{93}	$\mathbf{H}_{\mathbf{T_B}}$	1		0

References

1. Abraham, W.C., Bear, M.F.: Metaplasticity: the plasticity of synaptic plasticity. Trends Neurosci. **19**(4), 126–130 (1996)
2. Accetto, M., Treur, J., Villa, V.: An adaptive cognitive-social model for mirroring and social bonding during synchronous joint action. Procedia Computer Science 145, 3–12 (2018). https://doi.org/10.1016/j.procs.2018.11.002
3. Ahmadlou, M., Adeli, H.: Fuzzy synchronization likelihood with application to attention-deficit/hyperactivity disorder. Clin. EEG Neurosci. **42**(1), 6–13 (2011)
4. Ahmadlou, M., Adeli, H.: Visibility graph similarity: a new measure of generalized synchronization in coupled dynamic systems. Physica D **241**(4), 326–332 (2012)
5. Ahmadlou, M., Adeli, H., Adeli, A.: Fuzzy synchronization likelihood-wavelet methodology for diagnosis of autism spectrum disorder. J. Neurosci. Methods **211**(2), 203–209 (2012)
6. Alejandre-García, T., Kim, S., Pérez-Ortega, J., Yuste, R.: Intrinsic excitability mechanisms of neuronal ensemble formation. eLife **11**, e77470 (2022)
7. Ashby, W.R.: Design for a Brain. Chapman and Hall, London (second extended edition) (1960)
8. Augustinus, F., Hendrikse, S.C.F., Treur, J.: Psychopathic genes: a multi-level adaptive dynamic modelling approach. In: Proceedings of the 8th Computational Methods in Systems and Software Conference, CoMeSySo'24. Lecture Notes in Networks and Systems, vol 1471, pp 477–498. Springer Nature (2025)
9. Bain, A.: Mind and Body: The Theories of Their Relation. HS King & Company. Farnborough Eng. Gregg International (1873)
10. Bear, M.F., Malenka, R.C.: Synaptic plasticity: LTP and LTD. Curr. Opin. Neurobiol. **4**(3), 389–399 (1994)
11. Bechtel, W., Bich, L.: Organisms need mechanisms; mechanisms need organisms. In: Cordovil, J.L., Santos, G., Vecchi, D. (eds.) New Mechanism: Explanation, Emergence and Reduction. History, Philosophy and Theory of the Life Sciences, vol. 35, pp. 85–108. Springer Nature (2024)
12. Bevilacqua, A., Wilkinson, S. J., Dimelow, R., Murabito, E., Rehman, S., Nardelli, M., et al.: Vertical systems biology: from DNA to flux and back. In: Hetherington, A., Grierson, C. (eds.) Practical Systems Biology, pp. 79–106. Taylor & Francis (2008)
13. Biancardi, B., Dermouche, S., Pelachaud, C.: Adaptation mechanisms in human-agent interaction: effects on user's impressions and engagement. Front. Comput. Sci. **3**, 696682 (2021)
14. Bich, L., Bechtel, W.: Control mechanisms: explaining the integration and versatility of biological organisms. Adapt. Behav. **30**(5), 389–407 (2022)
15. Bich, L., Bechtel, W.: Organization needs organization: understanding integrated control in living organisms. Stud. Hist. Philos. Sci. **93**, 96–106 (2022)
16. Blei, Z.A., Hendrikse, S.C.F., Treur, J.: Modeling epigenetic modification due to childhood abuse and its relation to ASPD: a fifth-order adaptive network model. In: Proceedings of the 15th Biologically Inspired Cognitive Architectures for AI Conference, BICA*AI'24. Studies in Computational Intelligence, vol 477, pp 59–70. Springer Nature (2024)
17. Boot, N., Baas, M., Van Gaal, S., Cools, R., De Dreu, C.K.W.: Creative cognition and dopaminergic modulation of fronto-striatal networks: integrative review and research agenda. Neurosci. Biobehav. Rev. **78**, 13–23 (2017)
18. Bössenecker, G., Mreijen, P., Treur, J: Connecting social media platforms and users through bonding by faked homophily: a second-order adaptive network model. In: Hen-drikse, S.C.F., Treur, J., Koole, S.L. (eds.) New Analysis and Modeling Directions in Social Interaction Science: Emergent Multimodal Interpersonal Synchrony, Affiliation, and Bonding, Chap. 9. Springer Nature (2025)
19. Burgoon, J.K., Dillman, L., Stem, L.A.: Adaptation in dyadic interaction: defining and operationalizing patterns of reciprocity and compensation. Commun. Theory **3**(4), 295–316 (1993)

20. Burgoon, J.K., Stem, L.A, Dillman, L.: Interpersonal Adaptation: Dyadic Interaction Patterns. Cambridge University Press (1995). https://doi.org/10.1017/CBO9780511720314
21. Călin, R.S., Hendrikse, S.C.F., Treur, J.: Environmental and Epigenetic Influences in Binge Eating: An Adaptive Network Modeling Approach. To appear (2024)
22. Canbaloğlu, G., Treur, J., Wiewiora, A. (eds.): Computational Modeling of Multilevel Organisational Learning and its Control Using Self-Modeling Network Models. Springer Nature (2023)
23. Cappella, J.N.: Mutual influence in expressive behavior: adult–adult and infant–adult dyadic interaction. Psychol. Bull. **89**(1), 101 (1981)
24. Chandra, N., Barkai, E.: A non-synaptic mechanism of complex learning: modulation of intrinsic neuronal excitability. Neurobiol. Learn. Mem. **154**, 30–36 (2018)
25. Damasio, A.R.: The Feeling of What Happens: Body and Emotion in the Making of Consciousness. Houghton Mifflin Harcourt (1999)
26. Debanne, D., Inglebert, Y., Russier, M.: Plasticity of intrinsic neuronal excitability. Curr. Opin. Neurobiol. **54**, 73–82 (2019)
27. Dumas, G., Nadel, J., Soussignan, R., Martinerie, J., Garnero, L.: Inter-brain synchronization during social interaction. PLoS ONE **5**(8), e12166 (2010)
28. Feldman, R.: Parent–infant synchrony biological foundations and developmental outcomes. Curr. Dir. Psychol. Sci. **16**, 340–345 (2007). https://doi.org/10.1111/j.1467-8721.2007.00532.x
29. Ferrer, E., Helm, J.L.: Dynamical systems modeling of physiological coregulation in dyadic interactions. Int. J. Psychophysiol. **88**(3), 296–308 (2013)
30. Górriz, J.M., Álvarez-Illán, I., Álvarez-Marquina, A., Arco, J.E., Atzmueller, M., Ballarini, F., et al.: Computational approaches to explainable artificial intelligence: advances in theory, applications and trends. Inf. Fusion **100**, 101945 (2023)
31. Grandjean, D., Sander, D., Scherer, K.R.: Conscious emotional experience emerges as a function of multilevel, appraisal-driven response synchronization. Conscious. Cogn. **17**(2), 484–495 (2008)
32. Hassouna, S., Hendrikse, S.C.F., Treur, J.: The Dynamics of Epigenetic Influence in Insomnia: A Higher-Order Adaptive Modeling Perspective. In: Itthipuripat, S., Ascoli, G.A., Li, A., Pat, N., Kuai, H. (eds) Brain Informatics. BI 2024. Lecture Notes in Computer Science, vol 15541, pp 312–325. Springer Nature (2025)
33. Hansel, C., Yuste, R.: Neural ensembles: role of intrinsic excitability and its plasticity. Front. Cell. Neurosci. **18**, 1440588 (2024)
34. Hebb, D.O.: The Organization of Behavior: A Neuropsychological Theory. Wiley, New York (1949)
35. Hove, M.J., Risen, J.L.: It's all in the timing: Interpersonal synchrony increases affiliation. Soc. Cogn. **27**(6), 949–960 (2009)
36. Hendrikse, S.C.F., Kluiver, S., Treur, J., Wilderjans, T.F., Dikker, S., Koole, S.L.: How virtual agents can learn to synchronize: an adaptive joint decision-making model of psychotherapy. Cogn. Syst. Res. **79**, 138–155 (2023)
37. Hendrikse, S.C.F., Kluiver, S., Treur, J., Wilderjans, T.F., Dikker, S., Koole, S.L.: Learning to synchronize: adaptive joint decision-making by virtual agents. In: Hendrikse, S.C.F., Treur, J., Koole, S.L. (eds.) New Analysis and Modeling Directions in Social Interaction Science: Emergent Multimodal Interpersonal Synchrony, Affiliation, and Bonding, Chap. 5. Springer Nature (2025)
38. Hendrikse, S.C.F., Treur, J., Koole, S.L.: Modeling emerging interpersonal synchrony and its related adaptive short-term affliation and long-term bonding: a second-order multi-adaptive neural agent model. Int. J. Neural Syst. **33**(7), 2350038 (2023)
39. Hendrikse, S.C.F., Treur, J., Koole, S.L. (eds.): New Analysis and Modeling Directions in Social Interaction Science: Emergent Multimodal Interpersonal Synchrony, Affiliation, and Bonding. Springer Nature (2025a)
40. Hendrikse, S.C.F., Treur, J., Koole, S.L.: Introduction to the Role of Adaptivity in Social Interaction Science. In: Hendrikse, S.C.F., Treur, J., Koole, S.L. (eds.) New Analysis and Modeling Directions in Social Interaction Science: Emergent Multimodal Interpersonal Synchrony, Affiliation, and Bonding, Chap. 1. Springer Nature (2025b).

41. Hendrikse, S.C.F., Treur, J., Wilderjans, T.F., Dikker, S., Koole, S.L.: On the same wavelengths: emergence of multiple synchronies among multiple agents. In: Proceedings of the 22nd International Workshop on Multi-Agent-Based Simulation, MABS'21. Lecture Notes in AI, vol. 13128, pp. 57–71. Springer Nature (2022a)
42. Hendrikse, S.C.F., Treur, J., Wilderjans, T.F., Dikker, S., Koole, S.L.: On the interplay of interpersonal synchrony, short-term affiliation and long-term bonding: a second-order multi-adaptive neural agent model. In: Maglogiannis et al. (eds.) Proceedings of the 18th International Conference on Artificial Intelligence Applications and Innovations, AIAI'22. Advances in Information and Communication Technology, vol. 646, pp. 37–57. Springer Nature (2022b)
43. Hendrikse, S.C.F., Treur, J., Wilderjans, T.F., Dikker, S., Koole, S.L.: On becoming in sync with yourself and others: an adaptive agent model for how persons connect by detecting intrapersonal and interpersonal synchrony. Human-Centric Intell. Syst. **3**(2), 123–146 (2023). https://www.springer.com/journal/44230
44. Hendrikse, S.C.F., Treur, J., Wilderjans, T.F., Dikker, S., Koole, S.L.: The emergence of multimodal interpersonal synchrony during social interaction: a network model. In: Hendrikse, S.C.F., Treur, J., Koole, S.L. (eds.), New Analysis and Modeling Directions in Social Interaction Science: Emergent Multimodal Interpersonal Synchrony, Affiliation, and Bonding, Chap. 3. Springer Nature (2025a)
45. Hendrikse, S.C.F., Treur, J., Wilderjans, T.F., Dikker, S., Koole, S.L.: Growing connectedness through synchrony: the role of subjective analysis of intrapersonal synchrony and interpersonal synchrony within an adaptive agent model. In: Hendrikse, S.C.F., Treur, J., Koole, S.L. (eds.) New Analysis and Modeling Directions in Social Interaction Science: Emergent Multimodal Interpersonal Synchrony, Affiliation, and Bonding, Ch. 11. Springer Nature (2025b)
46. Hesslow, G.: Conscious thought as simulation of behaviour and perception. Trends Cogn. Sci. **6**, 242–247 (2002)
47. Jaganjac, I., Hendrikse, S.C.F., Treur, J.: An Adaptive dynamical system model for development of schizophrenia: epigenetics and false memories. Cogn. Syst. Res. 88, 101288 (2024)
48. Koole, S.L., Tschacher, W., Butler, E., Dikker, S., Wilderjans, T.F.: In sync with your shrink. In: Forgas, J.P., Crano, W.D., Fiedler, K. (eds.) Applications of Social Psychology, pp. 161–184. Milton Park, Taylor and Francis (2020)
49. Koole, S.L., Tschacher, W.: Synchrony in psychotherapy: a review and an integrative framework for the therapeutic alliance. Front. Psychol. **7**, 862 (2016)
50. Laws, H.B., Sayer, A.G., Pietromonaco, P.R., Powers, S.I.: Longitudinal changes in spouses' HPA responses: convergence in cortisol patterns during the early years of marriage. Health Psychol. **34**(11), 1076 (2015)
51. Lisman, J., Cooper, K., Sehgal, M., Silva, A.J.: Memory formation depends on both synapse-specific modifications of synaptic strength and cell-specific increases in excitability. Nat. Neurosci. **21**, 309–314 (2018)
52. Liu, G., Tian, L., Zhou, W.: Patient-independent seizure detection based on channel-perturbation convolutional neural network and bidirectional long short-term memory. Int. J. Neural Syst. **32**(6), 2150051 (2022)
53. Maurer, R.E., Tindall, J.H.: Effect of postural congruence on client's perception of counselor empathy. J. Counseling Psychol. **30**(158). https://doi.org/10.1037/0022-0167.30.2.158 (1983)
54. McPherson, M., Smith-Lovin, L., Cook, J.M.: Birds of a feather: homophily in social networks. Ann. Rev. Sociol. **27**(1), 415–444 (2001)
55. Mukeriia, Y., Treur, J., Hendrikse, S.C.F.: Learning and bonding based on subjective homophily during a varying in rhythm synchronous joint action: an adaptive network model. In: Hendrikse, S.C.F., Treur, J., Koole, S.L. (eds.), New Analysis and Modeling Directions in Social Interaction Science: Emergent Multimodal Interpersonal Synchrony, Affiliation, and Bonding, Chap. 8. Springer Nature (2025a).
56. Mukeriia, Y., Treur, J., Hendrikse, S.C.F.: Human learning, attunement and social bonding based on adaptive homophily: a multi-adaptive network model. In: Hendrikse, S.C.F., Treur, J., Koole, S.L. (eds.) New Analysis and Modeling Directions in Social Interaction Science: Emergent Multimodal Interpersonal Synchrony, Affiliation, and Bonding, Chap. 10. Springer Nature (2025b)

57. Olamat, A., Ozel, P., Akan, A.: Synchronization analysis in epileptic EEG signals via state transfer networks based on visibility graph technique. Int. J. Neural Syst. **32**(2), 2150041 (2022)
58. Port, R.F., van Gelder, T.: Mind as Motion: Explorations in the Dynamics of Cognition. MIT Press, Cambridge, MA (1995)
59. Ramseyer, F., Tschacher, W.: Nonverbal synchrony in psychotherapy: coordinated body movement reflects relationship quality and outcome. J. Consult. Clin. Psychol. **79**, 284–295 (2011). https://doi.org/10.1037/a0023419a
60. Robinson, B.L., Harper, N.S., McAlpine, D.: Meta-adaptation in the auditory midbrain under cortical influence. Nat. Commun. **7**, e13442 (2016)
61. Schlöder, D., Hendrikse, S.C.F., Treur, J.: Higher-order adaptive dynamical system modeling of the role of epigenetics in substance use disorder. In: Proceedings of the 15th Biologically Inspired Cognitive Architectures for AI Conference, BICA*AI'24. Studies in Computational Intelligence, vol 477, pp 358–368. Springer Nature (2024)
62. Sharpley, C.F., Halat, J., Rabinowicz, T., Weiland, B., Stafford, J.L." Standard posture, postural mirroring and client-perceived rapport. Couns. Psychol. Q. **14**, 267–280 (2001). https://doi.org/10.1080/09515070110088843
63. Shatz, C.J.: The developing brain. Sci. Am. **267**, 60–67 (1992)
64. Stanton, P.K.: LTD, LTP, and the sliding threshold for long-term synaptic plasticity. Hippocampus **6**(1), 35–42 (1996)
65. Tarr, B., Launay, J., Dunbar, R.I.M.: Silent disco: dancing in synchrony leads to elevated pain thresholds and social closeness. Evol. Hum. Behav. **37**(5), 343–349 (2016)
66. Tichelaar, C., Treur, J.: Network-oriented modeling of the interaction of adaptive joint decision making, bonding and mirroring. In: Proceedings of the 7th International Conference on Theory and Practice of Natural Computing, TPNC'18. Lecture Notes in Computer Science, vol. 11324, pp. 328–343. Springer Nature, Cham (2018)
67. Treur, J.: Temporal factorisation: a unifying principle for dynamics of the world and of mental states. Cogn. Syst. Res. **8**(2), 57–74 (2007)
68. Treur, J.: Temporal factorisation: realisation of mediating state properties for dynamics. Cogn. Syst. Res. **8**(2), 75–88 (2007)
69. Treur, J.: Network reification as a unified approach to represent network adaptation principles within a network. In: Theory and Practice of Natural Computing: Proceedings of the 7th International Conference TPNC'18. Lecture Notes in Computer Science, vol. 11324, pp. 344–358. Springer International Publishing (2018)
70. Treur, J.: Multilevel network reification: representing higher-order adaptivity in a network. In: Aiello, L., Cherifi, C., Cherifi, H., Lambiotte, R., Lió, P., Rocha, L. (eds.) Complex Networks and Their Applications VII: Volume 1 Proceedings The 7th International Conference on Complex Networks and Their Applications, COMPLEX NETWORKS'18. Studies in Computational Intelligence, vol. 812, pp. 635–651. Springer International Publishing (2019)
71. Treur, J.: Modeling higher-order adaptivity of a network by multilevel network reification. Netw. Sci. **8**, S110–S144 (2020)
72. Treur, J.: Network-Oriented Modeling for Adaptive Networks: Designing Higher-Order Adaptive Biological, Mental and Social Network Models. Springer Nature (2020b)
73. Treur, J.: Modeling multi-order adaptive processes by self-modeling networks. In: Tallón-Ballesteros, A.J., Chen, C.-H. (eds.) Proceedings of the 2nd International Conference on Machine Learning and Intelligent Systems, MLIS'20. Frontiers in Artificial Intelligence and Applications, vol. 332, pp. 206–217. IOS Press (2020c)
74. Treur, J.: Modeling the emergence of informational content by adaptive networks for temporal factorisation and criterial causation. Cogn. Syst. Res. **68**, 34–52 (2021)
75. Treur, J.: On the dynamics and adaptivity of mental processes: relating adaptive dynamical systems and self-modeling network models by mathematical analysis. Cogn. Syst. Res. **70**, 93–100 (2021)

76. Treur, J.: Where is this leading me: stationary point and equilibrium analysis of self-modeling network models. In: Treur, J., van Ments, L. (eds.), Mental Models and their Dynamics, Adaptation and Control: A Self-Modeling Network Modeling Approach, Chap. 18, pp. 491–536. Springer Nature Switzerland (2022a)
77. Treur, J.: Does This suit me? validation of self-modeling network models by parameter tuning. In: Treur, J., Van Ments, L. (eds.) Mental Models and their Dynamics, Adaptation, and Control: A Self-Modeling Network Modeling Approach, Chap. 19, pp. 537–565. Springer Nature (2022b)
78. Treur, J.: On structure, dynamics, and adaptivity for biological and mental processes: a higher-order adaptive dynamical system modeling perspective. In: Proceedings of the 46th Annual Meeting of the Cognitive Science Society: Dynamics of Cognition, CogSci'24 (2024a)
79. Treur, J.: Using multilevel temporal factorisation to analyse structure and dynamics for higher-order adaptive and evolutionary processes. In: Proceedings of the 16th International Conference on Computational Collective Intelligence, ICCCI'24. Lecture Notes in AI, vol. 14811, pp. 378–392. Springer Nature (2024b)
80. Treur, J.: Modeling coevolution of dynamics, adaptivity and control of adaptivity by self-modeling networks. In: Hendrikse, S.C.F., Treur, J., Koole, S.L. (eds.), New Analysis and Modeling Directions in Social Interaction Science: Emergent Multimodal Interpersonal Synchrony, Affiliation, and Bonding, Chap. 2. Springer Nature (2025)
81. Treur, J., Van Ments, L. (eds.): Mental Models and their Dynamics, Adaptation, and Control: a Self-Modeling Network Modeling Approach. Springer Nature (2022)
82. Trout, D.L., Rosenfeld, H.M.: The effect of postural lean and body congruence on the judgment of psychotherapeutic rapport. J. Nonverbal Behav. **4**, 176–190 (1980)
83. Tschacher, W., Ramseyer, F., Koole, S.L.: Sharing the now in the social present: duration of nonverbal synchrony is linked with personality. J. Pers. **86**(2), 129–138 (2018)
84. Tse, P.U.: The Neural Basis of Free Will: Criterial Causation. MIT Press, Cambridge (2013)
85. Warner, R.M.: Cyclicity of vocal activity increases during conversation: support for a nonlinear systems model of dyadic social interaction. Behav. Sci. **37**(2), 128–138 (1992)
86. Williams, A.H., O'Leary, T., Marder, E.: Homeostatic regulation of neuronal excitability. Scholarpedia **8**(1), 1656 (2013)
87. Wiltermuth, S.S., Heath, C.: Synchrony and cooperation. Psychol. Sci. **20**(1), 1–5 (2009)
88. Zweerings, J., Sarasjärvi, K., Mathiak, K.A., Iglesias-Fuster, J., Cong, F., Zvyagintsev, M., Mathiak, K.: Data-driven approach to the analysis of real-time fMRI neurofeedback data: disorder-specific brain synchrony in PTSD. Int. J. Neural Syst. **31**(11), 2150043 (2021)
89. Zhang, A., Li, X., Gao, Y., Niu, Y.: Event-driven intrinsic plasticity for spiking convolutional neural networks. IEEE Trans. Neural Netw. Learn. Syst. **33**(5), 1986–1995 (2021)

A Multi-timescale Multi-adaptive Dynamical System Approach for Relationship-Specific and Relationship-Independent Adaptivity in Affiliation and Bonding

Sophie C. F. Hendrikse, Jan Treur, and Sander L. Koole

Abstract Humans often adapt their behaviour toward each other when they interact. From a neuroscientific perspective, such adaptivity can involve mechanisms based on adaptive connections (synaptic plasticity) and adaptive excitability thresholds (nonsynaptic plasticity) within the mental or neural network concerned. It is, however, often left unaddressed which of the types of adaptation are specific for the relationship and which are more general for multiple relationships. We focus on this differentiation between relationship-specific and relationship-independent adaptation in social interactions. We analysed computationally how an interplay of adaptive relationship-specific and relationship-independent mechanisms occurs within the causal pathways for social interaction. As part of this, we cover also the context-sensitive control of these types of adaptation (adaptive speeds and strengths of adaptation), which is sometimes termed higher-order adaptation or metaplasticity. The model was evaluated by a number of explored runs where within a group of four agents each agent randomly has episodes of interaction with one of the three other agents. The outcomes of the analysis of the (stochastic) simulation results show a strong dependence of adaptation on the extent of social interaction: more social interaction leads to more adaptation of the interaction behaviour. This holds both for the short-term and long-term first-order adaptation and for the second-order adaptation, which is long-term.

S. C. F. Hendrikse
Department of Methodology and Statistics, Tilburg School of Social and Behavioral Sciences, Tilburg University, Tilburg, The Netherlands
e-mail: S.C.F.Hendrikse@tilburguniversity.edu

J. Treur (✉)
Social AI Group, Department of Computer Science, Vrije Universiteit Amsterdam, Amsterdam, The Netherlands
e-mail: j.treur@vu.nl

S. L. Koole
Amsterdam Emotion Regulation Lab, Department of Clinical Psychology, Vrije Universiteit Amsterdam, Amsterdam, The Netherlands
e-mail: s.l.koole@vu.nl

S. C. F. Hendrikse et al. (eds.), *New Analysis and Modeling Directions in Social Interaction Science*, Studies in Systems, Decision and Control 614,
https://doi.org/10.1007/978-3-031-99968-0_13

Keywords Relationship-specific · Relationship-independent · Behavioural adaptivity · Affiliation · Bonding · Multi-adaptive · Dynamical system

1 Introduction

During social interaction, humans often adapt their interaction behaviour toward each other. This behavioural adaptivity may concern short-term effects such as affiliation but also long-term effects such as bonding. Interaction can involve different modalities like movement, affect and verbal actions. Central mechanisms in the adaptive causal pathways for social interaction can have the form of synaptic plasticity based on adaptive connections [4] and nonsynaptic plasticity based on adaptive excitability thresholds [13]. An interesting issue is which of these mechanisms are specific for the other person and which are more general for multiple persons. In the latter case, transference takes place: what you learn in one relationship will also influence how you interact in another relationship. Attachment theory is a theory that claims that adaptations acquired in one relationship also have their effects in other relationships [9, 52–55]. In accordance with that claim, we assume that at least part of the considered behavioural adaptivity in social interaction is relationship-independent. With this assumption as a point of departure, we analysed computationally how an interplay of adaptive relationship-specific and relationship-independent mechanisms occurs within the causal pathways leading to behavioural adaptivity during social interaction.

As part of this, in the current chapter we analyse from a neuroscience perspective which learning or adaptation principles can apply to the mechanisms within the different types of causal pathways indicated above. We do not only cover both relationship-specific and relationship-independent adaptation as forms of first-order adaptation, but also second-order adaptation to control these types of first-order adaptation in a context-sensitive manner, in particular adaptive speeds and strengths of adaptation. As such causal pathways in general involve connections between states and excitability of states, from neuroscience both synaptic and nonsynaptic forms of plasticity are exploited: adaptive connection weights and adaptive excitability thresholds. In addition, metaplasticity (second-order adaptation) is incorporated to control these forms of plasticity depending on the context. We evaluated the model by a number of explored runs where within a group of four agents each agent randomly has episodes of interaction with one of the three other agents and due to these episodes displays both short-term and long-term behavioural adaptivity.

2 Background Knowledge

In this section, we elaborate on the main assumptions behind our multi-adaptive dynamical systems approach applied here. These main assumptions are grounded in the neuroscience literature. There are relationships between interpersonal synchrony and nonsynaptic plasticity [13] and synaptic plasticity [4] respectively. Such nonsynaptic plasticity functions on short-term time scales, whereas synaptic plasticity functions on the long-term time scales. More specifically, we assume the following mechanisms for the causal pathways leading to behavioural adaptivity in social interaction.

2.1 Emerging Interpersonal Synchrony and Related Adaptivity of Interaction Behaviour

Interpersonal synchrony has been found to be at the basis of mutual adaptation of behaviour, leading to changes in the interaction. Numerous studies have demonstrated this phenomenon, including (but not limited to) those conducted by Fairhurst et al. [14], Feldman [16], Hove and Risen [37], Kirschner and Tomasello [40], Koole and Tschacher [41], Koole et al. [42], Palumbo et al. [47], Prince and Brown [49], Tarr et al. [61], Valdesolo et al. [71], Valdesolo and DeSteno [70], and Wiltermuth and Heath [76]. The adaptive shift in mutual behavioural coordination has been observed in various contexts, including psychotherapy sessions [41, 46, 50, 59, 69].

Behavioural adaptation following interpersonal synchrony can take the form of both short-term changes and long-term changes in behaviour. While many studies have focused on short-term adaptive shifts in coordination [37, 61, 76], there is also evidence for long-term effects of interpersonal synchrony. For example, developmental research has shown that movement synchrony between infants and caregivers can predict social interaction patterns in the child several years later [16]. Similarly, research on close relationships has found that early patterns of interpersonal synchrony can predict subsequent indicators of relationship functioning, such as converging patterns of cortisol variation in spouses over a period of years [43]. The attachment theory also considers that behavioural adaptivity acquired in one relationship can influence interaction behaviour in other relationships too, which can be seen as a form of transference, and the attachment theory has been studied in numerous settings and has been commonly applied to therapeutic contexts [9, 15, 17–19, 39, 45, 52–55]. According to the attachment theory, the first attachment relationship between children and their primary caregiver significantly impacts children's future functioning in relationships. While the attachment theory has originally mainly been applied to intimate, romantic relationships, since more recently also friendship relationships are addressed, for example [12, 22, 73]. Moreover, the neuroscientific mechanisms underlying attachment theory have gathered attention [5, 6, 11, 74].

The current chapter analyses this in relation to basic mechanisms from neuroscience and their plasticity. To enable an individual to adapt behaviour upon being in synchrony, it is assumed that the individual has the ability to detect synchrony patterns across different modalities, like introduced in [30], see also this volume [26] Chap. 11. From an overall conceptual analysis perspective, mental states representing detected synchrony can be considered mediating mental states in the pathway from synchrony patterns to changed interaction behaviour [62, 63].

2.2 Using Mechanisms from Neuroscience: Synaptic and Nonsynaptic Plasticity, and Metaplasticity

The field of neuroscience distinguishes between synaptic plasticity and nonsynaptic plasticity (also called intrinsic plasticity). Synaptic plasticity is a classical concept explaining how the strength of a connection between different neural states adapts over time [4, 21, 58, 60]. In contrast, nonsynaptic adaptation of intrinsic excitability of neural states has been more recently addressed and linked to homeostatic regulation [8, 10, 13, 44, 75, 77]. Nonsynaptic plasticity and synaptic plasticity can work together, allowing for the modeling of simultaneously working mechanisms for different types of behavioural adaptivity in the multi-adaptive dynamical systems model for short-term adaptation and long-term adaptation. In this way, via multiple circular pathways an interplay between synchrony, short-term adaptivity, and long-term adaptivity occurs. Synchrony does not only lead to short-term adaptation but through this also intensifies interaction, leading to more synchrony and strengthening long-term adaptation. Conversely, long-term adaptivity strengthens interaction, leading to more synchrony and consequently stronger short-term adaptivity. Metaplasticity [1] controls the plasticity in a context-sensitive manner. Second-order adaptation (adaptation of the adaptation) has been included in our model to allow for more realistic and context-sensitive control of plasticity. This has been applied particularly to address adaptive speeds and strengths of the different types of first-order adaptation that form the plasticity.

2.3 Research Questions and Hypotheses

The general focus is on how interaction with other agents leads to adaptation of the interaction behaviour. The main research question considered is:

> How can an adaptive agent model for interacting agents be designed that captures how depending on the interaction durations with other agents that occur, an agent achieves relationship-specific and relationship-independent adaptivity concerning interaction behaviour?

In the experimental setup chosen, the interaction episodes and their durations form the independent variable (per simulation run chosen in a stochastic manner) whereas the agent's processes and adaptations depend on that. The main research question can be detailed by the following more specific hypotheses that will be verified through simulations:

A. Adaptation in basic interaction behaviour can be observed for
   (a) representing the other agent
   (b) responding to the representations of the other agent
   (c) executing interaction actions
   (d) emergence of synchrony shown by its detection
B. The adaptation can be considered for multiple modalities (movement *m*, affect *b*, verbal *v*).
C. Two types of adaptation will occur, (a) relationship-specific and (b) relationship-independent:
   (i) More experiences with interactions with a given agent A lead to faster and stronger adaptation in interactions with A in the future.
   (ii) More experiences with interactions with any agent A lead to faster and stronger adaptation in interactions with any agent B in the future (transference).
D. Such adaptation occurs both (i) in the short-term and (ii) in the long-term:
   (i) Interaction within episodes
   (ii) Interaction over multiple episodes
E. The relation between the extents of interaction and adaptation can be observed in two ways:
   (i) Within a given simulation run over time adaptation becomes stronger after more interaction has taken place
   (ii) In a comparative manner, simulation runs that show longer interaction durations will also show more adaptation compared to simulation runs with shorter interaction durations
F. More experiences with interactions will in general not only lead to adaptations in interaction behaviour (first-order adaptation effect) but also to faster and stronger adaptation in interactions in the future (second-order adaptation effect). This happens (i) within a given simulation and (ii) more in simulations where more interaction occurs.

3 Adaptive Dynamical Systems Modeled by Self-Modeling Networks

The dynamical systems view as a useful perspective on cognition has been put forward, for example, in [2, 48, 57]. Dynamical systems are usually considered as state-determined systems which are systems for which each (current) state at some time t determines the future states after t, e.g., [2, 72]. For example:

> the fact that the current state determines future behaviour implies the existence of some *rule of evolution* describing the behaviour of the system as a function of its current state. For systems we wish to understand we always hope that this rule can be specified in some reasonably succinct and useful fashion. [72, p. 6].

Concerning the issue how 'this rule can be specified in some reasonably succinct and useful fashion', often a first-order difference or differential equations format is suggested [2, 48]. However, also causally oriented views have been developed, which may provide more intuitive conceptualisations than mathematical difference or differential equations. For example, it has been shown by several applications how a causal format can be a useful format to model complex dynamic and adaptive processes, as long as dynamics of states [64] and adaptivity for causal relations and their characteristics [65, 66] are taken into account in that format, see also this volume [26], Chap. 2. Moreover, it has been mathematically proven in a general manner in [24, 68] that any smooth dynamical system has a canonical representation in network format based on temporal-causal networks and that any smooth adaptive dynamical system has a canonical representation in self-modeling temporal-causal network format, see also this volume [26], Chap. 12. It is this format that is used in the current chapter.

3.1 *Modeling Dynamical Systems by a Temporal-Causal Network Format*

A temporal-causal network model, as characterized by Treur [65, 66], is composed of nodes (also referred to as states) denoted by X and Y with real numbers (usually in the [0, 1] interval) as activation values $X(t)$ and $Y(t)$ over time t. A specific model is defined by the following network characteristics:

- **Connectivity characteristics**

There exist connections from states X to Y, each with a corresponding weight specified, denoted by $\boldsymbol{\omega}_{X,Y}$.

- **Aggregation characteristics**

For any state Y, a combination function $\mathbf{c}_{\boldsymbol{\pi}_Y,Y}\,(V_1, \ldots, V_k)$ with function parameter values $\boldsymbol{\pi}_Y = (\boldsymbol{\pi}_{1,Y}, \ldots, \boldsymbol{\pi}_{m,Y})$ aggregates the impacts $V_i = \boldsymbol{\omega}_{X_i,Y}X_i(t)$ on Y from its incoming connections from states X_i.

- **Timing characteristics**

Each state Y has a speed factor denoted by $\boldsymbol{\eta}_Y$, which determines the rate at which it changes for a given causal impact.

Note that sometimes the parameters in combination functions are left implicit and $\mathbf{c}_{\boldsymbol{\pi}_Y,Y}(V_1, \ldots, V_k)$ is simply denoted by $\mathbf{c}_Y(V_1, \ldots, V_k)$. For the network's dynamics, these network characteristics are incorporated into a canonical difference equation (or related differential equation) used for both simulation and analysis purposes:

$$Y(t + \Delta t) = Y(t) + \boldsymbol{\eta}_Y\left[\mathbf{c}_{\boldsymbol{\pi}_Y,Y}(\boldsymbol{\omega}_{X_1,Y}X_1(t), \ldots, \boldsymbol{\omega}_{X_k,Y}X_k(t)) - Y(t)\right]\Delta t \quad (1)$$

where X_1 to X_k represent states from which Y receives incoming connections. Equation (1) bears a resemblance to the format of recurrent neural networks.

The software environment outlined in Treur [65, Chap. 9] includes a library of approximately 70 basic combination functions for use in the model design process. Examples of these functions are listed in Table 1. Note that when the function $\mathbf{alogistic}_{\boldsymbol{\sigma},\boldsymbol{\tau}}(V_1, \ldots, V_k)$ is applied to intended positive values, it is cut off at 0 if its formula produces a negative value. However, this cut-off is not applied when the function is applied to intended negative values. Overall, these concepts enable the declarative design of network models and their dynamics based on mathematically defined functions and relations. By instantiating this general difference Eq. (1) by proper values for the network characteristics for all states Y, the software environment runs a system of n difference equations where n is the number of states in the network.

Table 1 The combination functions used in the introduced network model

	Notation	Formula	Parameters
Advanced logistic sum	$\mathbf{alogistic}_{\boldsymbol{\sigma},\boldsymbol{\tau}}(V_1, \ldots, V_k)$	$\left[\frac{1}{1+e^{-\sigma(V_1+\cdots+V_k-\tau)}} - \frac{1}{1+e^{\sigma\tau}}\right](1 + e^{-\sigma\tau})$	Steepness $\boldsymbol{\sigma}$ Excitability threshold $\boldsymbol{\tau}$
Complemental difference	$\mathbf{compdiff}(V_1, V_2)$	0 if $V_1 = V_2 = 0$ $1 - \frac{\lvert V_1 - V_2\rvert}{\max(V_1, V_2)}$ else	–
Stepmod	$\mathbf{stepmod}_{\boldsymbol{\rho},\boldsymbol{\delta}}(V)$	0 if $0 \leq t \;\mathbf{mod}\; \boldsymbol{\rho} \leq \boldsymbol{\delta}$ 1 else	Time t Repitition $\boldsymbol{\rho}$ Step time $\boldsymbol{\delta}$
Euclidean	$\mathbf{eucl}_{\boldsymbol{n},\boldsymbol{\lambda}}(V_1, \ldots, V_k)$	$\sqrt[n]{\frac{V_1^n+\cdots+V_k^n}{\lambda}}$	Order $\boldsymbol{n}$ Scaling factor $\boldsymbol{\lambda}$

The four functions in Table 1 are all used in the introduced model. To obtain stochastic generation of runs, in addition a fifth function is used, called **scenario-generation**. Over time this function determines in a stochastic manner environmental circumstances: which pairs of agents can interact with each other over which time periods and for which modalities.

3.2 *Modeling Adaptive Dynamical Systems by Self-Modeling Networks*

Dynamical systems for real-world cases often involve a number of parameters that for changing contextual circumstances have to be adaptive, resulting in an adaptive dynamical system. For network models representing adaptive dynamical systems, such parameters have the form of network characteristics: connection weights, combination functions and their parameters, and speed factors, like indicated in Sect. 3.1. To achieve a transparent declarative description for adaptive networks, self-modeling networks (also called reified networks) have been introduced [65, 66], see also this volume [26], Chap. 2. Self-model states are added to the (base) network to represent adaptive network characteristics, which are depicted at a next level known as the self-model level or reification level in the graphical 3D-format shown in Sect. 4. The original network is situated at the base level.

For example, the weight $\boldsymbol{\omega}_{X,Y}$ of a connection from state X to state Y can be represented by a self-model state named $\mathbf{W}_{X,Y}$ at the self-model level. The activation value of state $\mathbf{W}_{X,Y}$ is used for $\boldsymbol{\omega}_{X,Y}$ when updating the activation level of state Y over time using Eq. (1). If the activation level of the self-model state $\mathbf{W}_{X,Y}$ changes over time, this reflects the adaptivity of the weight of the connection from X to Y represented by $\mathbf{W}_{X,Y}$. This concept corresponds to the notion of synaptic plasticity in neuroscience, which, for example, is described by Bear and Malenka [4] and Stanton [60]. In contrast to this synaptic type of plasticity, an example of nonsynaptic plasticity [10, 13, 77] is modulation of excitability thresholds. This can be modeled for a given state Y by a self-model state $\mathbf{T}_Y$ which represents the excitability threshold $\boldsymbol{\tau}_Y$ of Y.

Similarly, the other network characteristics from $\boldsymbol{\omega}_{X,Y}$, $\mathbf{c}_{\boldsymbol{\pi}_Y,Y}(..)$ and $\boldsymbol{\eta}_Y$ can be made adaptive by including self-model states for them. For example, an adaptive speed factor $\boldsymbol{\eta}_Y$ can be represented by a self-model state named $\mathbf{H}_Y$ and an adaptive parameter $\boldsymbol{\pi}_{i,Y}$ can be represented by a self-model state $\mathbf{P}_{i,Y}$. If for all network characteristics $\boldsymbol{\omega}, \boldsymbol{\pi}, \boldsymbol{\eta}$ for all base level states, respective self-model states $\mathbf{W}, \mathbf{P}, \mathbf{H}$ are introduced representing these network characteristics, then the canonical difference equation for the base level states of the self-modeling network model is:

$$
\begin{aligned}
Y(t + \Delta t) &= Y(t) + \mathbf{H}_Y(t) \\
&\quad [\mathbf{c}_{\mathbf{P}_Y(t),Y}(\mathbf{W}_{X_1,Y}(t)X_1(t), \ldots, \mathbf{W}_{X_k,Y}(t)X_k(t)) - Y(t)]\Delta t
\end{aligned} \tag{2}
$$

where $\mathbf{P}_Y(t) = \left(\mathbf{P}_{1,Y}(t), \ldots, \mathbf{P}_{m,Y}(t)\right)$.

This universal difference equation is incorporated in the dedicated software environment. By instantiating this general difference Eq. (2) by proper values for the network characteristics for all base states Y and instantiating Eq. (1) for all self-model states, the software environment runs a system of $\boldsymbol{n}$ difference equations where $\boldsymbol{n}$ is the number of base states plus self-model states in the network. For the adaptive dynamical system model introduced in the current chapter, $\boldsymbol{n} = 357$.

Note that this difference Eq. (2) is not exactly in the standard format of a temporal-causal network, as $\mathbf{H}_Y$ is not a constant speed factor and also the **P**- and **W**-values are not constant. However, it can be rewritten into the temporal-causal network format when the following general combination function $\mathbf{c}^*_Y$**(..)** is defined:

$$\begin{aligned}&\mathbf{c}^*_Y\left(H, P_{1,Y}, \ldots, P_{m,Y}, W_1, \ldots, W_k, V_1, \ldots, V_k, V\right)\\&\quad= H\mathbf{c}_{P_Y,Y}(W_1V_1, \ldots, W_kV_k) + (1 - H)V\end{aligned} \tag{3}$$

where $P_Y = \left(P_{1,Y}, \ldots, P_{m,Y}\right)$ are variables for adaptive parameters of the combination function, H is a variable for adaptive speed factors, W_i are variables for adaptive connection weights, and V_i and V are variables for state activations. Based on this universal combination function, consider the following difference equation:

$$\begin{aligned}Y(t + \Delta t) = Y(t) + \big[&\mathbf{c}^*_Y\big(\mathbf{H}_Y(t), \mathbf{P}_{1,Y}(t), \ldots, \mathbf{P}_{m,Y}(t), \mathbf{W}_{X_1,Y}(t), \ldots,\\&\mathbf{W}_{X_k,Y}(t), X_1(t), \ldots, X_k(t), Y(t)\big) - Y(t)\big]\Delta t\end{aligned} \tag{4}$$

This is indeed in temporal-causal network format (with speed factor 1) defined by (1). Now note that using (3), Eq. (4) can be rewritten as follows:

$$\begin{aligned}Y(t + \Delta t) &= Y(t) + \big[\mathbf{H}_Y(t)\mathbf{c}_{\mathbf{P}_Y(t),Y}\big(\mathbf{W}_{X_1,Y}(t)X_1(t), \ldots, \mathbf{W}_{X_k,Y}(t)X_k(t)\big)\\&\quad+(1 - \mathbf{H}_Y(t))Y(t) - Y(t)]\Delta t\\&= Y(t) + \big[\mathbf{H}_Y(t)\mathbf{c}_{\mathbf{P}_Y(t),Y}\big(\mathbf{W}_{X_1,Y}(t)X_1(t), \ldots, \mathbf{W}_{X_k,Y}(t)X_k(t)\big)\\&\quad-\mathbf{H}_Y(t)Y(t)]\Delta t\\&= Y(t) + \mathbf{H}_Y(t)\big[\mathbf{c}_{\mathbf{P}_Y(t),Y}\big(\mathbf{W}_{X_1,Y}(t)X_1(t), \ldots, \mathbf{W}_{X_k,Y}(t)X_k(t)\big)\\&\quad-Y(t)]\Delta t\end{aligned} \tag{5}$$

where $\mathbf{P}_Y(t) = \left(\mathbf{P}_{1,Y}(t), \ldots, \mathbf{P}_{m,Y}(t)\right)$.

This (5) shows exactly difference Eq. (2) above; this confirms that the chosen combination function $\mathbf{c}^*_Y$**(..)** in (3) to show that the self-modeling network has a temporal-causal network format like (1) indeed works.

As the above shows that a self-modeling network is also a temporal-causal network model itself, this self-modeling network construction can easily be applied iteratively to obtain multiple orders of self-models at multiple (first-order, second-order, …) self-model levels. For example, a second-order self-model may include a second-order self-model state $\mathbf{H}_{\mathbf{W}_{X,Y}}$ representing the speed factor $\boldsymbol{\eta}_{\mathbf{W}_{X,Y}}$ for the dynamics of first-order self-model state $\mathbf{W}_{X,Y}$ which in turn represents the adaptation of connection weight $\boldsymbol{\omega}_{X,Y}$. So, this second-order self-model state $\mathbf{H}_{\mathbf{W}_{X,Y}}$ represents the adaptation speed (or learning rate) for that connection weight. Similarly, a second-order self-model may include a second-order self-model state $\mathbf{H}_{\mathbf{T}_Y}$ representing the speed factor $\boldsymbol{\eta}_{\mathbf{T}_Y}$ for the dynamics of first-order self-model state $\mathbf{T}_Y$ which in turn represents the adaptation of excitability threshold $\boldsymbol{\tau}_Y$ for Y, so this second-order self-model state $\mathbf{H}_{\mathbf{T}_Y}$ represents the adaptation speed for that excitability threshold. Moreover, also higher-order $\mathbf{W}$-states can be used of the form $\mathbf{W}_{Z,\mathbf{W}_{X,Y}}$ representing the weight of the connection from a given state Z to state $\mathbf{W}_{X,Y}$ or of the form $\mathbf{W}_{Z,\mathbf{T}_Y}$ representing the weight of the connection from a given state Z to state $\mathbf{T}_Y$. All such second-order self-model states can be used to modulate an adaptation process for $\mathbf{W}_{X,Y}$ or $\mathbf{T}_Y$ over time. Therefore, these second-order self-model states as indicated can be used to exert control over different aspects of first-order adaptation processes, in line with the notion of metaplasticity from neuroscience to control plasticity in a context-sensitive manner, e.g., [1]. In this way such second-order adaptation effects contribute to (the speed and strength of) first-order adaptation.

In the current chapter, this multi-level self-modeling network perspective will be applied to obtain a second-order adaptive network architecture addressing both first- and second-order adaptation. As an example, the adaptation control level can be used to make the adaptation speed context-sensitive as addressed by metaplasticity literature such as [1, 51]. This indeed has been done for the introduced model, as will be discussed in Sect. 4.

4 The Multi-adaptive Dynamical Systems Model

In this section, the agent model (modelled through multi-adaptive dynamical systems) is introduced. First, the conceptual assumptions underlying the model are discussed for the base behaviour (Sect. 4.1) and for the adaptivity (Sect. 4.2). Next, an overview of the model is presented (Sect. 4.3). After this, the states used in the model are discussed in more detail, together with the related network characteristics, subsequently for the base level (Sect. 4.4), the first-order self-model level (Sect. 4.5), and the second-order self-model level (Sect. 4.6).

4.1 The Base Multimodal Interaction Behaviour

All considered agents share the same design. For the base level of this design, see Fig. 1. An overview of explanations of all base states for agent A can be found in Table 2. Recall from Sect. 3 that all states have real numbers in the interval [0, 1] as activation values over time. Such states are depicted in Fig. 1 by circle or diamond shapes and there causal relations by arrows. All of the agents receive input from a stimulus $world_s$ regularly for certain time periods, meaning the activation of $world_s$ alternates between 0 (s absent) and 1 (s present), with s an instance of a stimulus (which is kept abstract). For the chosen scenario, this external stimulus influences any agent A by a causal relation to the sensing state indicated by $\text{sense}_{s,A}$ and subsequently by that to the sensory representation state of this stimulus indicated by $\text{rep}_{s,A}$ (a kind of mental image) which in turn affects the agent's own internal preparation states for the three considered modalities: $\text{prep}_{m,A}$, $\text{prep}_{b,A}$ and $\text{prep}_{v,A}$, with m, b and v instances of a movement, affective response and a verbal action of agent A, respectively (also kept abstract). For any other agent B with which agent A interacts, the model for agent A also includes sensing and sensory representation states for the three considered modalities for B: for B's verbal actions v (indicated by $\text{sense}_{B,v,A}$ and $\text{rep}_{B,v,A}$), emotions b (indicated by $\text{sense}_{B,b,A}$ and $\text{rep}_{B,b,A}$) and movements m (indicated by $\text{sense}_{B,m,A}$ and $\text{rep}_{B,m,A}$). The latter sensory representation states are directly connected in a reciprocal manner with the corresponding preparation states $\text{prep}_{m,A}$, $\text{prep}_{b,A}$ and $\text{prep}_{v,A}$ of agent A. These reciprocal causal links show (1) how the sensory representation states lead to mirroring of the same action by preparing for it [38], and (2) how internal simulation of the prepared action with a backward effect takes place [36].

As discussed in Sect. 2.1, during interaction often forms of interpersonal synchrony emerge and this affects the interaction behaviour in an adaptive manner. Therefore, in the model introduced here it is assumed that for each of the modalities and each other person such emerging synchrony can be detected. This is modeled by states such as $\text{intersyncdet}_{B,m,A}$, $\text{intersyncdet}_{B,b,A}$, $\text{intersyncdet}_{B,v,A}$ (depicted by diamond shapes in Fig. 1). These states have high activation values in the interval [0, 1] if the indicated synchrony is detected and low values when no synchrony is detected. In Fig. 2 it is shown how the agents interact with each other.

4.2 *Conceptual Assumptions Behind the Introduced Adaptive Dynamical System Model*

The aim of this chapter is to analyse in which ways agents change their behaviour during social interactions with multiple others over time and how these changes can be related to internal mental learning mechanisms. The chosen mental learning mechanisms are based on what is known from neuroscience, in particular on synaptic and nonsynaptic plasticity [4, 10, 13, 60] and metaplasticity [1]. Overall, the following assumptions were made on adaptive changes that can take place during social interaction:

(a) Changes in the way others are perceived
(b) Changes in the way of responding to other persons
(c) Changes in the way of executing or expressing certain behaviours to other persons

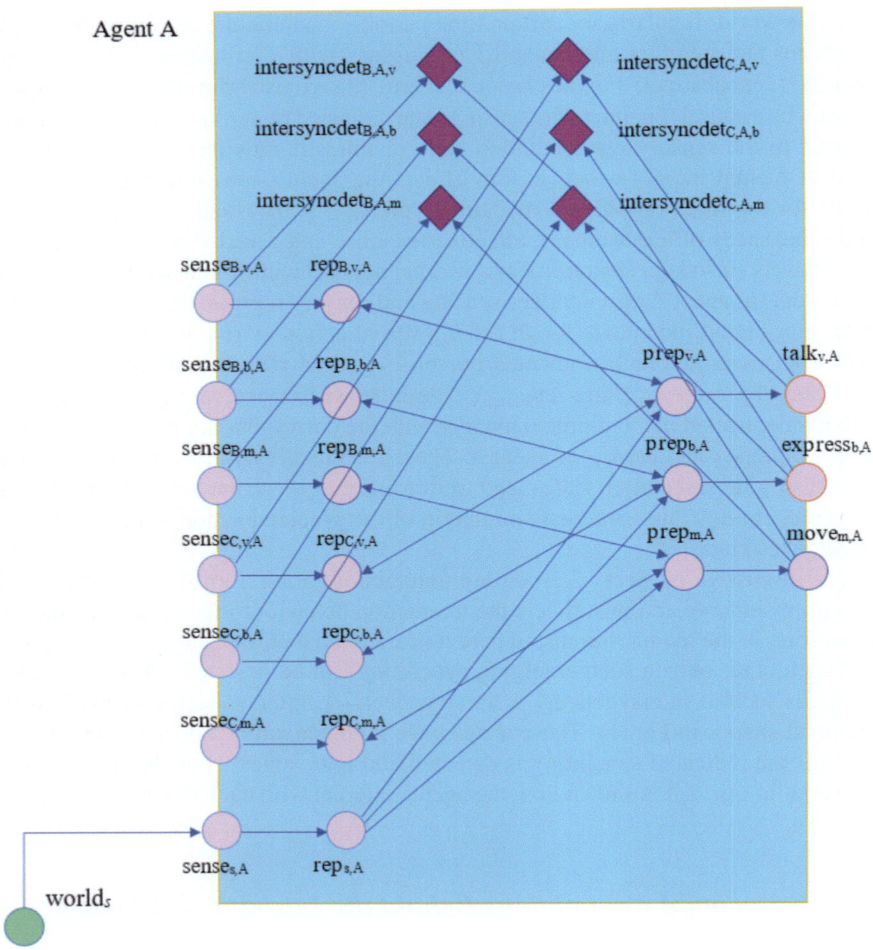

Fig. 1 Base level connectivity for agent A: states and within-agent connections. With three modalities and (in dark pink) six synchrony detection states for interpersonal synchrony

As discussed in Sect. 3, first-order self-model states can be used to model adaptation. In particular, $\mathbf{W}_{X,Y}$-states and $\mathbf{T}_Y$-states can be applied to model synaptic and nonsynaptic types of adaptation of connections from states X to Y and excitability thresholds for states Y used to model the mental processes perceiving, responding and executing or expressing (a) to (c) mentioned above. Thus, stronger and more sensitive activation of states used to model the three types of mental processes (a) to (c) can be obtained.

More specifically, if sensor states and sensory representation states for others are used to model the way how someone else is perceived, then strengthening the connections from sensor states and representation states and lowering the excitability

Table 2 Base states of the adaptive dynamical systems model for agent A. For the other agents, similar base states are used: X_{41} to X_{75} for B, X_{76} to X_{110} for C, X_{111} to X_{145} for D

state	name	Explanation
X_1	ws_s	World state for stimulus *s*
X_2	$con_{th,A}$	Context state for excitability thresholds for *A*
X_3	$con_{th,B}$	Context state for excitability thresholds for *B*
X_4	$con_{th,C}$	Context state for excitability thresholds for *C*
X_5	$con_{th,D}$	Context state for excitability thresholds for *D*
X_6	$sense_{s,A}$	Sensor state of *A* for stimulus *s* for *A*
X_7	$sense_{B,m,A}$	Sensor state of *A* for movement *m* of *B*
X_8	$sense_{B,b,A}$	Sensor state of *A* for expressed affective response *b* of *B*
X_9	$sense_{B,v,A}$	Sensor state of *A* for verbal action *v* of *B*
X_{10}	$sense_{C,m,A}$	Sensor state of *A* for movement *m* of *C*
X_{11}	$sense_{C,b,A}$	Sensor state of *A* for expressed affective response *b* of *C*
X_{12}	$sense_{C,v,A}$	Sensor state of *A* for verbal action *v* of *C*
X_{13}	$sense_{D,m,A}$	Sensor state of *A* for movement *m* of *D*
X_{14}	$sense_{D,b,A}$	Sensor state of *A* for expressed affective response *b* of *D*
X_{15}	$sense_{D,v,A}$	Sensor state of *A* for verbal action *v* of *D*
X_{16}	$rep_{s,A}$	Sensory representation state of *A* for stimulus *s* for *A*
X_{17}	$rep_{B,m,A}$	Sensory representation state of *A* for movement *m* of *B*
X_{18}	$rep_{B,b,A}$	Sensory representation state of *A* for expressed affective response *b* of *B*
X_{19}	$rep_{B,v,A}$	Sensory representation state of *A* for verbal action *v* of *B*
X_{20}	$rep_{C,m,A}$	Sensory representation state of *A* for movement *m* of *C*
X_{21}	$rep_{C,b,A}$	Sensory representation state of *A* for expressed affective response *b* of *C*
X_{22}	$rep_{C,v,A}$	Sensory representation state of *A* for verbal action *v* of *C*
X_{23}	$rep_{D,m,A}$	Sensory representation state of *A* for movement *m* of *D*
X_{24}	$rep_{D,b,A}$	Sensory representation state of *A* for expressed affective response *b* of *D*
X_{25}	$rep_{D,v,A}$	Sensory representation state of *A* for verbal action *v* of *D*
X_{26}	$prep_{m,A}$	Preparation state for movement *m* of *A*
X_{27}	$prep_{b,A}$	Preparation state for affective response *b* of *A*
X_{28}	$prep_{v,A}$	Preparation state for verbal action *v* of *A*
X_{29}	$intersyncdet_{B,A,m}$	Interpersonal synchrony detection of *A* for executing *m* by *B* and *A*
X_{30}	$intersyncdet_{B,A,b}$	Interpersonal synchrony detection of *A* for executing *b* by *B* and *A*
X_{31}	$intersyncdet_{B,A,v}$	Interpersonal synchrony detection of *A* for executing *v* by *B* and *A*
X_{32}	$intersyncdet_{C,A,m}$	Interpersonal synchrony detection of *A* for executing *m* by *C* and *A*
X_{33}	$intersyncdet_{C,A,b}$	Interpersonal synchrony detection of *A* for executing *b* by *C* and *A*
X_{34}	$intersyncdet_{C,A,v}$	Interpersonal synchrony detection of *A* for executing *v* by *C* and *A*
X_{35}	$intersyncdet_{D,A,m}$	Interpersonal synchrony detection of *A* for executing *m* by *D* and *A*
X_{36}	$intersyncdet_{D,A,b}$	Interpersonal synchrony detection of *A* for executing *b* by *D* and *A*
X_{37}	$intersyncdet_{D,A,v}$	Interpersonal synchrony detection of *A* for executing *v* by *D* and *A*
X_{38}	$move_{m,A}$	Executing movement *m* by *A*
X_{39}	$express_{b,A}$	Executing expression of *b* by *A*
X_{40}	$talk_{A,B,v}$	Executing verbal action *v* by *A*

thresholds for representation states can lead to stronger and more sensitive forms of perceiving (a). Furthermore, if in addition preparation states and connections from representation states to them are used to model the basis of the responding process, then strengthening these connections and lowering the excitability thresholds for these preparation states will lead to stronger and more sensitive responding (b). Finally, if in addition (action) execution states and connections from preparation states to them are used to model the execution process, then strengthening these

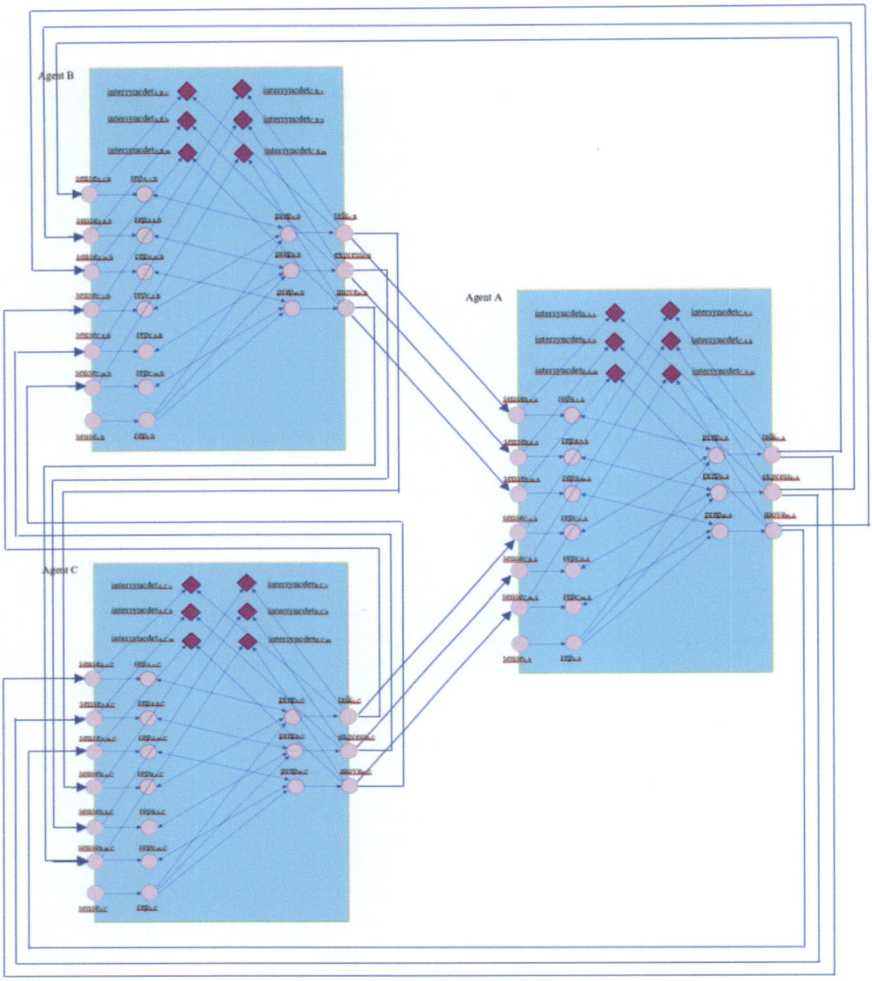

Fig. 2 Base level connectivity for three agents A, B, C: within-agent connections and between-agents connections

connections and lowering the excitability thresholds for these execution states will lead to stronger and more sensitive acting and expressing (c). We assume here that synaptic plasticity is used to model long-term changes (over multiple episodes) like in bonding, whereas nonsynaptic adaptation of excitability thresholds is used to model short-term changes like in affiliation (within one episode). For all of the above forms of adaptations, it has been be considered in how far they are relationship-specific or relationship-independent. For example, when an agent learns to respond stronger to an agent within an interaction, will that only change the behaviour in interaction with that specific agent, or also when interacting with other agents? In terms of the model to be designed, which (adaptive) state and connection characteristics play only a role when interacting with one and the same agent and which (adaptive) state and connection characteristics play a role when interacting with any other agent? Below we will make these distinctions for the different forms of learning addressed.

Still other assumptions are made about second-order adaptivity. Within neuroscience this is described by metaplasticity which can control plasticity in a context-sensitive manner [1]. In the first place, this concept is used to address adaptation of the adaptation strength for first-order self-model states $\mathbf{W}_{X,Y}$ and $\mathbf{T}_Y$. As discussed in Sect. 3, second-order self-model states $\mathbf{W}_{Z,\mathbf{W}_{X,Y}}$ and $\mathbf{W}_{Z,\mathbf{T}_Y}$ can be used to make the strength of adaptation of $\mathbf{W}_{X,Y}$ and $\mathbf{T}_Y$ adaptive. This can be used to model how over time a person learns to adapt stronger. Secondly, the concept of metaplasticity is used to address adaptation of the adaptation speed for $\mathbf{W}_{X,Y}$ and $\mathbf{T}_Y$. Again, as discussed in Sect. 3, second-order self-model states $\mathbf{H}_{\mathbf{W}_{X,Y}}$ and $\mathbf{H}_{\mathbf{T}_Y}$ can be used to make the speed of adaptation of $\mathbf{W}_{X,Y}$ and $\mathbf{T}_Y$ adaptive. This can be used to model how over time a person learns to adapt faster. In this way, in the model both stronger adaptation and faster adaptation over time are modelled by these second-order self-model states. This second-order adaptation effect can substantially contribute to first-order adaptation.

In total two first-order adaptation mechanisms have been identified, modelled through **W**-states for long-term adaptation and **T**-states for short-term adaptation. Moreover, these two learning mechanisms have been applied to the different types of mental processes (a) to (c) mentioned above, which makes six types of first-order adaptation. In addition, four second-order adaptation mechanisms have been identified modelled through $\mathbf{W}_{Z,\mathbf{W}_{X,Y}}$-, $\mathbf{W}_{Z,\mathbf{T}_Y}$-, $\mathbf{H}_{\mathbf{W}_{X,Y}}$-, and $\mathbf{H}_{\mathbf{T}_Y}$-states. Further assumptions have been made about which of these 10 learning types work within a given episode and which ones over multiple episodes, and which are other-agent-specific and which relationship-independent, see Table 3 for an overview of this and Figs. 3 and 4 for a 3D graphical overview of the model. Here, in the first four rows (in purple) the assumptions are indicated that all four types of second-order adaptivity are relationship-independent and long-term. So, their influence on the first-order adaptation is a long-term relationship-independent influence. Furthermore, in rows 5 to 7 (in blue) the assumptions are indicated that synaptic plasticity is used as a form of long-term learning, for representing and responding relationship-specific, and for executing actions relationship-independent. In contrast, in the last three rows (in blue) the assumptions are indicated that nonsynaptic plasticity of excitability thresholds is used as a form of short-term learning, again for representing and responding relationship-specific, and for executing actions relationship-independent. Note that the second-order adaptation also provides long-term relationship-independent effects, even on the relationship-specific short-term first-order adaptation modeled by the **T**-states for the representation states.

Table 3 Overview of the assumptions on first-order and second-order relationship-specific and relationship-independent adaptivity and short-term and long-term learning adaptivity

	Relationship-independent	Long-term	Order and type of adaptivity
$\mathbf{W}_{\mathbf{W}}$-states for connectivity of **W**-states	+	+	Second-order, relationship-independent adaptivity
$\mathbf{W}_{\mathbf{T}}$-states for connectivity of **T**-states	+	+	Second-order, relationship-independent adaptivity
$\mathbf{H}_{\mathbf{W}}$-states for timing of **W**-states	+	+	Second-order, relationship-independent adaptivity
$\mathbf{H}_{\mathbf{T}}$-states for timing of **T**-states	+	+	Second-order, relationship-independent adaptivity
W-states for representing	-	+	First-order, relationship-specific adaptivity
W-states for responding	-	+	First-order, relationship-specific adaptivity
W-states for executing	+	+	First-order, relationship-independent adaptivity
T-states for representing	-	-	First-order, relationship-specific adaptivity
T-states for responding	-	-	First-order, relationship-specific adaptivity
T-states for executing	+	-	First-order, relationship-independent adaptivity

4.3 Overview of the Model

The designed adaptive dynamical system model takes into account actions of agents for three different modalities: for moving *m*, expressing affect *b* and talking *v*. In total, the model covers four agents and their interactions and adaptivity, modelled according to a second-order adaptive dynamical system and represented in a network-oriented format; see Figs. 3 and 4 for a graphical overview and Table 4 for explanations of all types of states. Overall, it has 357 states X_i, which in view of Sect. 3 means that its dynamics is based on a system of 357 difference or differential equations which are instantiations of (1) for the chosen network characteristics including the specific combination functions shown in Table 1. Within the overall model, each of the four agents A to D is modelled by 79 states:

- From these 79 states per agent, 35 states are at the base level and model the base processes of perceiving other persons, responding to them and executing actions. These 35 states are sensor and execution states, representation and preparation states and synchrony detector states, all for different modalities.
- Furthermore, 33 states are at the first-order self-model level and model the agent's first-order adaptivity: **W**-states and **T**-states representing connectivity and aggregation characteristics of the base-level network. They model respectively synaptic

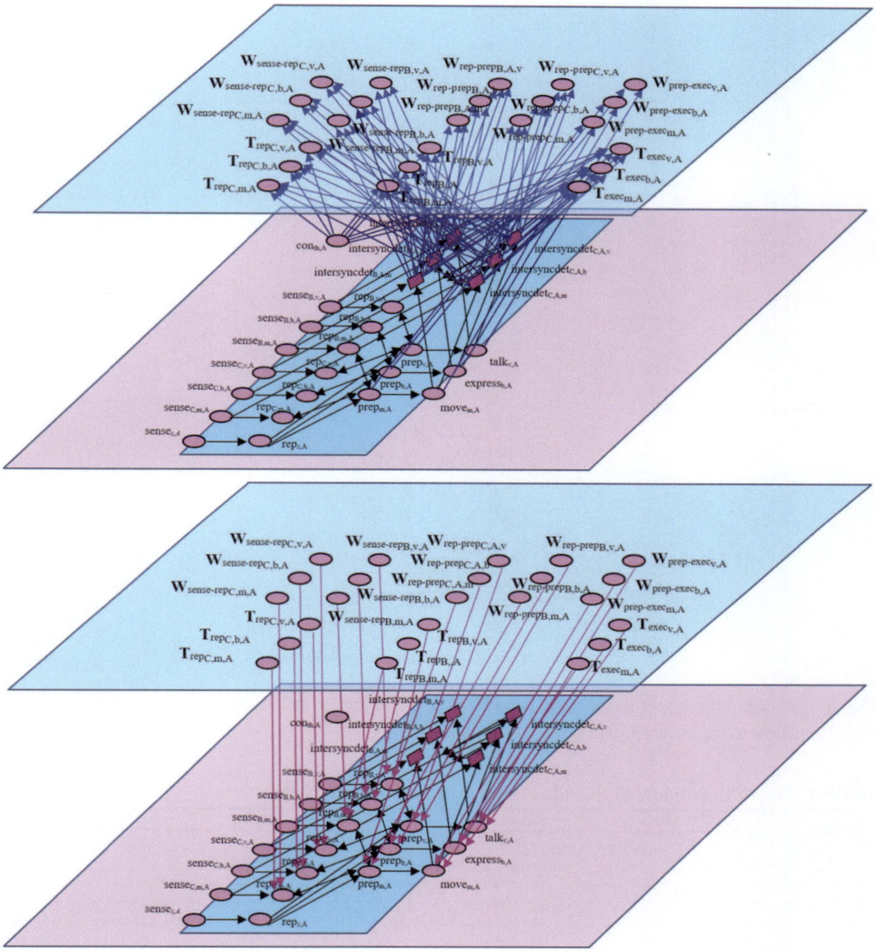

Fig. 3 The adaptive agent model: the base level and first-order self-model level with in the upper picture their upward interaction links (in blue) and in the lower picture their downward interaction links (in pink)

plasticity (adaptivity of connection weights) and nonsynaptic plasticity (adaptivity of excitability thresholds)

- Finally, 11 states are at the second-order self-model level and model the agent's second-order adaptivity for control over its adaptivity: aggregation states, $\mathbf{H_W}$- and $\mathbf{H_T}$-states, $\mathbf{W_W}$-, and $\mathbf{W_T}$-states representing timing and connectivity network characteristics for first-order self-model **W**- and **T**-states. Here, the $\mathbf{H_W}$- and $\mathbf{H_T}$-states represent the adaptive speed factors for the **W**-states and **T**-states and the $\mathbf{W_W}$- and $\mathbf{W_T}$-states represent the weights of the incoming connections for the **W**-states and **T**-states.

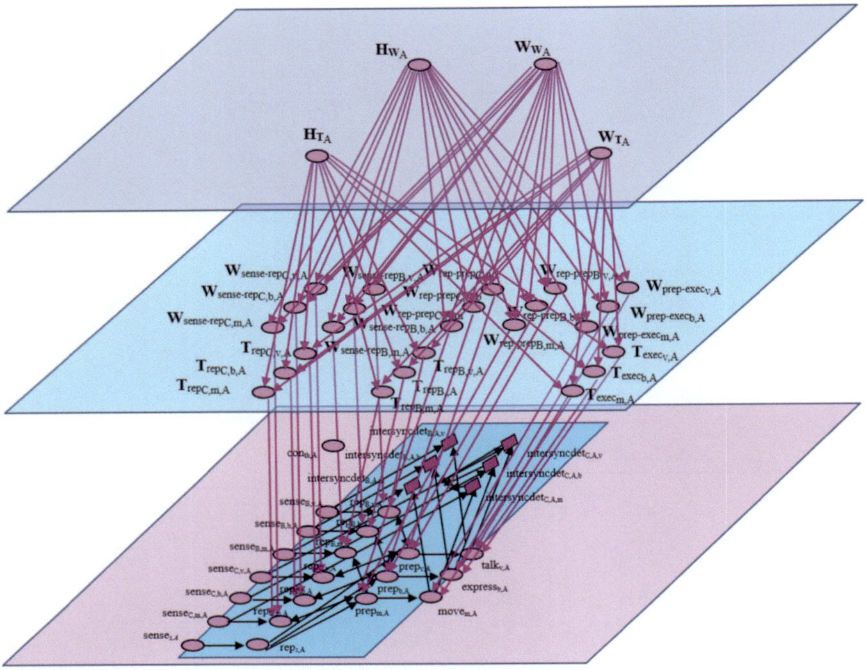

Fig. 4 The adaptive agent model: the base level and first-order and second-order self-model levels with their downward interaction links (in pink)

Table 4 Overall overview of the adaptive dynamical system model

Type	States	Explanation
Base world states: 5	X_1 X_2-X_5	• World state for stimulus *s* • Context states for excitability thresholds for the agent states
Base agent states: 35 base states per agent	agent A X_6-X_{40} agent B X_{41}-X_{75} agent C X_{76}-X_{110} Agent D X_{111}-X_{145}	• Sensor states $sense_{.,.,.}$ for stimulus *s* and observed multimodal actions • Representation states $rep_{.,.,.}$ for stimulus and actions in different modalities • Preparation states $prep_{.,.}$ for actions move, express, talk in different modalities. • Interpersonal synchrony detection states $intersyncdet_{.,.,.}$ for different modalities • Execution states move, express, talk for actions in different modalities.
First-order self-model agent states: 33 per agent	agent A X_{146}-X_{178} agent B X_{179}-X_{211} agent C X_{212}-X_{244} agent D X_{245}-X_{278}	• First-order self-model **W**-states for weights of connections *within* agents: ○ States $\mathbf{W}_{sense\text{-}rep.,.,.}$ for connections from sensing to representation states ○ States $\mathbf{W}_{rep\text{-}prep.,.,.}$ for connections from representation to preparation states ○ States $\mathbf{W}_{prep\text{-}exec.,.,.}$ for connections from preparation to execution states • First-order self model **T**-states for excitability thresholds: ○ States $\mathbf{T}_{rep.,.,.}$ for excitability thresholds of representation states ○ States $\mathbf{T}_{exec.,.,.}$ for excitability thresholds of execution states
First-order self-model interaction states: 35	X_{279}-X_{313}	• First-order self-model **W**-states for weights of connections *between* agents: ○ States $\mathbf{W}_{z,X,Y}$ for weights of connections from agent *X* to agent *Y* for modality *z*
Second-order self-model agent aggregation states: 7 per agent	agent A X_{314}-X_{320} agent B X_{321}-X_{327} agent C X_{328}-X_{334} agent D X_{335}-X_{341}	• Aggregation states $agg_{.,.,.}$ for aggregation of: ○ Sensing, execution, and for synchrony states ○ **W**-states for weights of connections from sensing to representation, from representation to preparation, and from preparation to execution
Second-order self-model agent states for adaptation Speed and adaptation strength: 4 per agent	agent A X_{342}-X_{345} agent B X_{346}-X_{329} agent C X_{350}-X_{353} agent D X_{354}-X_{357}	• Second-order self-model $\mathbf{H}_W$- and $\mathbf{H}_T$-states representing the speed factors of the first-order self-model **W**-states and **T**-states of the agent. • Second-order self-model $\mathbf{W}_W$- and $\mathbf{W}_T$-states representing the weights of the incoming connections of the first-order self-model **W**-states and **T**-states of the agent.

4.4 Network Characteristics for the Base Level

At this point, some more details are discussed of the network model's (connectivity, aggregation, timing) characteristics, starting with the base level states.

Connectivity characteristics for base states

- Connectivity for base agent states:
  - Within-agent connections for representing, responding, and executing have adaptive weights modelled by within-agent first-order self-model **W**-states.
  - Within-agent connections for representing stimulus *s* and from representation states of *s* to preparation states have weights 1.
  - Within-agent connections to synchrony detection have weights 1.
  - Within-agent connections from preparation states to representation states for predicting have weights 1.
  - Between-agents connections from execution states of one agent *X* to sensing states of another agent *Y* have adaptive weights modelled by between-agents first-order self-model **W**-states.
  - Connections from world state ws_s to agent sensor states for *s* have weights 1.
- Connectivity for base world states:
  - Base world states have no connections from other states. Instead, they have circular persistence connections to themselves (not depicted in the model pictures) with weight 1 which makes them persistent over time.

Aggregation characteristics for base states

- Aggregation for base agent states:
  - Sensing states use the Euclidean function **eucl** from Table 1 of order $\boldsymbol{n} = 1$ and scaling factor $\boldsymbol{\lambda} = 1.1$.
  - Interpersonal synchrony detector states use the synchrony detection function **compdiff** from Table 1.
  - All other base agent states use the logistic function **alogistic** from Table 1 with steepness $\boldsymbol{\sigma} = 5$ and adaptive thresholds modelled by within-agent first-order self-model **T**-states.
- Aggregation for base world states:
  - World state ws_s uses the stimulus repetition function **stepmod** from Table 1 with repetition $\boldsymbol{\rho} = 80$ time units and step time $\boldsymbol{\delta} = 40$ time units.
  - Context states use the identity function via Euclidean function **eucl** from Table 1 of order $\boldsymbol{n} = 1$ and scaling factor $\boldsymbol{\lambda} = 1$.

Timing characteristics for base states

- Timing for base agent states:
  - All interpersonal synchrony detector states have speed factor 0.5.

  - All other agent states have speed factor 1.
- Timing for base world states:
  - World state ws_s has speed factor 0.5.
  - The context states have speed factor 0, which makes them constant.

4.5 Network Characteristics for the First-Order Self-Model Level

Next, some more details are discussed of the network model's (connectivity, aggregation, and timing) characteristics for the first-order self-model level states.

Connectivity characteristics for first-order self-model states

- **Connectivity for within-agent first-order self-model states**
  - Within-agent connections from synchrony detector states to **T**-states for representation, preparation and execution states have (negative) adaptive weights that are represented by second-order self-model states $\mathbf{W_T}$.
  - Within-agent connections from context states to all **T**-states have weight 1.
  - Within-agent connections from synchrony detector states to **W**-states for representing and responding have adaptive weights represented by second-order self-model states $\mathbf{W_W}$. In addition, these **W**-states have circular persistence connections to themselves with weight 1.
  - Within-agent connections from preparation and execution states to **W**-states for execution have adaptive weights represented by second-order self-model states $\mathbf{W_W}$. In addition, these **W**-states have circular persistence connections to themselves with weight 1.
- **Connectivity for between-agents first-order self-model states**
  - Between-agents **W**-states have no connections from other states. However, these **W**-states have circular persistence connections to themselves with weight 1.

Aggregation characteristics for first-order self-model states

- **Aggregation for within-agent first-order self-model states**
  - All first-order self-model within-agent **W**-states use the logistic function **alogistic** from Table 1 with steepness $\boldsymbol{\sigma} = 5$ and threshold $\boldsymbol{\tau} = 0.5$ and **T**-states with steepness $\boldsymbol{\sigma} = 5$ and threshold $\boldsymbol{\tau} = 0.6$.
- **Aggregation for between-agents first-order self-model states**
  - All first-order self-model between-agents **W**-states use stochastic generation of interaction episodes for dyads via combination function **scenario-generation**.

Timing characteristics for first-order self-model states

- **Timing for within-agent first-order self-model states**
  - All first-order self-model within-agent **W**-states and **T**-states use adaptive speed factors represented by second-order self-model states $\mathbf{H_W}$ and $\mathbf{H_T}$, respectively.
- **Timing for between-agents first-order self-model states**
  - All first-order self-model between-agents **W**-states have timing specified by speed factor 2.

An overview of explanations of all first-order self-model states for agent A can be found in Table 5.

4.6 Network Characteristics for the Second-Order Self-Model Level

Finally, more details are discussed of the network model's (connectivity, aggregation, timing) characteristics for the second-order self-model level states.

Connectivity characteristics for second-order self-model states

- **Connectivity for within-agent second-order self-model states**
  - Within-agent connections from base-level within-agent states for sensing, execution and synchrony detection to respective aggregation states have weights 0.1.
  - Within-agent connections from within-agent first-order self-model **W**-states for representing, responding and executing to respective aggregation states have weight 0.1.
  - Within-agent connections from second-order within-agent self-model aggregation states for representing, responding and executing to overall **W**-aggregation states have weight 0.1.
  - Within-agent connections from aggregation states for sensing, execution and synchrony detection and from overall **W**-aggregation states to $\mathbf{H_W}$-, $\mathbf{H_T}$-, $\mathbf{W_W}$-, $\mathbf{W_T}$-states have weight 1 for the first three $\mathbf{H_W}$, $\mathbf{H_T}$, $\mathbf{W_W}$, and weight -0.35 for the $\mathbf{W_T}$-states. In addition, these $\mathbf{H_W}$-, $\mathbf{H_T}$-, $\mathbf{W_W}$-, $\mathbf{W_T}$-states have circular persistence connections to themselves with weight 1 for $\mathbf{H_W}$- and $\mathbf{H_T}$-states, 0.15 for $\mathbf{W_W}$-states, and -1 for $\mathbf{W_T}$-states.

Table 5 First-order self-model **T**-states and **W**-states of agent A in the adaptive dynamical systems model: modelling excitability thresholds and connection weights. For the other agents, similar first-order self-model states are used: X_{179} to X_{211} for B, X_{212} to X_{244} for C, X_{245} to X_{278} for D

State	Name	Explanation
X_{146}	$\mathbf{W}_{sense\text{-}rep\_m,B,A}$	First-order self-model state for the weight of *A*'s internal connection from sensing to representing movement *m* of *B*
X_{147}	$\mathbf{W}_{sense\text{-}rep\_b,B,A}$	First-order self-model state for the weight of *A*'s internal connection from sensing to representing affective response *b* of *B*
X_{148}	$\mathbf{W}_{sense\text{-}rep\_v,B,A}$	First-order self-model state for the weight of *A*'s internal connection from sensing to representing verbal action *v* of *B*
X_{149}	$\mathbf{W}_{sense\text{-}rep\_m,C,A}$	First-order self-model state for the weight of *A*'s internal connection from sensing to representing movement *m* of *B*
X_{150}	$\mathbf{W}_{sense\text{-}rep\_b,C,A}$	First-order self-model state for the weight of *A*'s internal connection from sensing to representing affective response *b* of *B*
X_{151}	$\mathbf{W}_{sense\text{-}rep\_v,C,A}$	First-order self-model state for the weight of *A*'s internal connection from sensing to representing verbal action *v* of *B*
X_{152}	$\mathbf{W}_{sense\text{-}rep\_m,D,A}$	First-order self-model state for the weight of *A*'s internal connection from sensing to representing movement *m* of *B*
X_{153}	$\mathbf{W}_{sense\text{-}rep\_b,D,A}$	First-order self-model state for the weight of *A*'s internal connection from sensing to representing affective response *b* of *B*
X_{154}	$\mathbf{W}_{sense\text{-}rep\_v,D,A}$	First-order self-model state for the weight of *A*'s internal connection from sensing to representing verbal action *v* of *B*
X_{155}	$\mathbf{W}_{rep\text{-}prep\_m,B,A}$	First-order self-model state for the weight of *A*'s internal connection from representing movement *m* of *B* to preparing for movement *m*
X_{156}	$\mathbf{W}_{rep\text{-}prep\_b,B,A}$	First-order self-model state for the weight of *A*'s internal connection from representing affective response *b* of *B* to preparing for affective response *b*
X_{157}	$\mathbf{W}_{rep\text{-}prep\_v,B,A}$	First-order self-model state for the weight of *A*'s internal connection from representing verbal action *v* of *B* to preparing for verbal action *v*
X_{158}	$\mathbf{W}_{rep\text{-}prep\_m,C,A}$	First-order self-model state for the weight of *A*'s internal connection from representing movement *m* of *C* to preparing for movement *m*
X_{159}	$\mathbf{W}_{rep\text{-}prep\_b,C,A}$	First-order self-model state for the weight of *A*'s internal connection from representing affective response *b* of *C* to preparing for affective response *b*
X_{160}	$\mathbf{W}_{rep\text{-}prep\_v,C,A}$	First-order self-model state for the weight of *A*'s internal connection from representing verbal action *v* of *C* to preparing for verbal action *v*
X_{161}	$\mathbf{W}_{rep\text{-}prep\_m,D,A}$	First-order self-model state for the weight of *A*'s internal connection from representing movement *m* of *D* to preparing for movement *m*
X_{162}	$\mathbf{W}_{rep\text{-}prep\_b,D,A}$	First-order self-model state for the weight of *A*'s internal connection from representing affective response *b* of *D* to preparing for affective response *b*
X_{163}	$\mathbf{W}_{rep\text{-}prep\_v,D,A}$	First-order self-model state for the weight of *A*'s internal connection from representing verbal action *v* of *D* to preparing for verbal action *v*
X_{164}	$\mathbf{W}_{prep\text{-}exec\_m,A}$	First-order self-model state for the weight of *A*'s internal connection from preparing to executing movement *m*
X_{165}	$\mathbf{W}_{prep\text{-}exec\_b,A}$	First-order self-model state for the weight of *A*'s internal connection from preparing to expressing affective response *b*
X_{166}	$\mathbf{W}_{prep\text{-}exec\_v,A}$	First-order self-model state for the weight of *A*'s internal connection from preparing to executing verbal action *v*
X_{167}	$\mathbf{T}_{rep\_B,m,A}$	First-order self-model state for the excitability threshold of *A*'s sensory representation state $rep_{m,A}$ for movement *m* of *B*
X_{168}	$\mathbf{T}_{rep\_B,b,A}$	First-order self-model state for the excitability threshold of *A*'s sensory representation state $rep_{b,A}$ for affective response *b* of *B*
X_{169}	$\mathbf{T}_{rep\_B,v,A}$	First-order self-model state for the excitability threshold of *A*'s sensory representation state $rep_{v,A}$ for verbal response *v* of *B*
X_{170}	$\mathbf{T}_{rep\_C,m,A}$	First-order self-model state for the excitability threshold of *A*'s sensory representation state $rep_{m,A}$ for movement *m* of *C*
X_{171}	$\mathbf{T}_{rep\_C,b,A}$	First-order self-model state for the excitability threshold of *A*'s sensory representation state $rep_{b,A}$ for affective response *b* of *C*
X_{172}	$\mathbf{T}_{rep\_C,v,A}$	First-order self-model state for the excitability threshold of *A*'s sensory representation state $rep_{v,A}$ for verbal response *v* of *C*
X_{173}	$\mathbf{T}_{rep\_D,m,A}$	First-order self-model state for the excitability threshold of *A*'s sensory representation state $rep_{m,A}$ for movement *m* of *D*
X_{174}	$\mathbf{T}_{rep\_D,b,A}$	First-order self-model state for the excitability threshold of *A*'s sensory representation state $rep_{b,A}$ for affective response *b* of *D*
X_{175}	$\mathbf{T}_{rep\_D,v,A}$	First-order self-model state for the excitability threshold of *A*'s sensory representation state $rep_{v,A}$ for verbal response *v* of *D*
X_{176}	$\mathbf{T}_{exec\_m,A}$	First-order self-model state for the excitability threshold of *A*'s execution state $move_{m,A}$ for movement *m*
X_{177}	$\mathbf{T}_{exec\_b,A}$	First-order self-model state for the excitability threshold of *A*'s execution state state $express_{b,A}$ for affective response *b*
X_{178}	$\mathbf{T}_{exec\_v,A}$	First-order self-model state for the excitability threshold of *A*'s execution state state $talk_{v,A}$ for verbal response *v*

Aggregation characteristics for second-order self-model states

- **Aggregation for within-agent second-order self-model states**
  - All second-order self-model aggregation states use the Euclidean function **eucl** from Table 1 of order $\boldsymbol{n} = 1$ with a normalising scaling factor $\boldsymbol{\lambda}$ = the sum of the weights of the incoming connections.
  - All second-order self-model states $\mathbf{H_W}$-, $\mathbf{H_T}$-, $\mathbf{W_W}$-states use the logistic function **alogistic** from Table 1 with steepness $\boldsymbol{\sigma} = 5$, threshold $\boldsymbol{\tau} = 0.8$ for $\mathbf{H_W}$- and $\mathbf{H_T}$-states and threshold $\boldsymbol{\tau} = 0.2$ for $\mathbf{W_W}$-states.
  - All second-order self-model states $\mathbf{W_T}$-states use the logistic function **alogistic** for negative values with steepness $\boldsymbol{\sigma} = 2$ and threshold $\boldsymbol{\tau} = 0$.

Timing characteristics for second-order self-model states

- **Timing for within-agent second-order self-model states**
  - All second-order self-model aggregation states use speed factor 1.
  - All second-order self-model $\mathbf{H_W}$-, $\mathbf{H_T}$-, $\mathbf{W_W}$-, $\mathbf{W_T}$-states use speed factors 0.005, 0.1, 0.005, and 0.009, respectively.

An overview of explanations of all second-order self-model states for agent A can be found in Table 6.

For more details of the model specification, see Linked Data at https://www.researchgate.net/publication/363802843.

Table 6 Second-order self-model $\mathbf{H_W}$-, $\mathbf{H_T}$-, $\mathbf{W_W}$- and $\mathbf{W_T}$-states for agent A in the adaptive dynamical systems model: modelling the adaptation speed of the **T**-states and **W**-states and the connection weights of their incoming connections. For the other agents, similar second-order self-model states are used: X_{346} to X_{349} for B, X_{350} to X_{353} for C, X_{354} to X_{357} for D

State	Name	Explanation
X_{342}	$\mathbf{H_{W_A}}$	Second-order self-model state for the speed factors of the first-order self-model **W**-states for *A*
X_{343}	$\mathbf{H_{T_A}}$	Second-order self-model state for the speed factors of the first-order self-model **T**-states for *A*
X_{344}	$\mathbf{W_{W_A}}$	Second-order self-model state for the weights of the incoming connections of the first-order self-model **W**-states for *A* (from the respective synchrony detector states for the $\mathbf{W}_{\text{sense-rep}}$-states and $\mathbf{W}_{\text{rep-prep}}$-states and from the preparation and execution states for the $\mathbf{W}_{\text{prep-exec}}$-states)
X_{345}	$\mathbf{W_{T_A}}$	Second-order self-model state for the weights of the incoming connections of the first-order self-model **T**-states for *A* (from the respective synchrony detector states)

5 Research Question and Hypotheses Operationalised for the Introduced Model

Recall the main research question and hypotheses A. to F. introduced in Sect. 2.3. Now the model has been described, these hypotheses can be operationalised by relating them to the states of the model. We go through them one by one.

A. Adaptation in basic interaction behaviour can be observed for

   (a) Representing the other agent
   (b) Responding to the representations of the other agent
   (c) Expressing (executing) interaction actions
   (d) Emergence of synchrony shown by its detection

B. The adaptation can be considered for multiple modalities (movement m, express b, verbal v).

The base states for representing, responding and executing are the rep-, prep-, and exec-states (note that the (inter)action execution states move_m, express_b, talk_v are also indicated by exec_m, exec_b, exec_v). Their activation values over time will be analysed to verify to what extent behaviour is adapted; see Table 7.

C. Two types of adaptation will occur, (i) other-agent specific and (ii) other-agent independent:

   (i) More experiences with interactions with a given agent A lead to faster and stronger adaptation in interactions with A in the future.
   (ii) More experiences with interactions with any agent A lead to faster and stronger adaptation in interactions with any agent B in the future (transference).

D. Such adaptation occurs both (i) in the short-term and (ii) in the long-term:

   (i) Interaction within episodes
   (ii) Interaction over multiple episodes

E. The relation between the extents of interaction and adaptation can be observed in two ways:

Table 7 The interaction-related base states that are examined to analyse behaviour adaptation

Interaction-related base processes		Modalities		
Type of processes	Type of states	Movement m	Affect b	Verbal action v
Representing	rep-states	rep_m	rep_b	rep_v
Responding	prep-states	prep_m	prep_b	prep_v
Expressing	exec-states	move_m	express_b	talk_v

(i) Within a given simulation run over time adaptation becomes stronger after more interaction has taken place
(ii) In a comparative manner, simulation runs that show longer interaction durations will also show more adaptation compared to simulation runs with shorter interaction durations

F. More experiences with interactions will in general not only lead to adaptations in interaction behaviour (first-order adaptation effect) but also to faster and stronger adaptation in interactions in the future (second-order adaptation effect). This happens (i) within a given simulation and (ii) more in simulations where more interaction occurs.

First-order self-model **T**- and **W**-states and second-order self-model $\mathbf{W_T}$-, $\mathbf{H_T}$-, $\mathbf{W_W}$-, $\mathbf{H_W}$-states are indicative for the different types of adaptations that occur. The above hypotheses can be observed at the first-order and second-order self-model level as follows (see also Table 8).

First-order self-model **T**-states represent the adaptive excitability thresholds for certain base states within the model: the rep-, prep-, and exec-states. Lower thresholds imply that a state can get stronger activation (more sensitive, higher extent of excitability). First-order self-model **W**-states represent adaptive connection weights between two states, e.g., between sense and representation states. Higher connection weights imply that a connection between two states is stronger, resulting in a stronger activation of the target state. Globally, for these first-order self-model states we expect that

- Within a given simulation run, the activation values of the **T**-states will become lower within each interaction episode, and higher when no interaction episode for a specific relationship occurs; in contrast, the activation values of the **W**-states will become higher over time in general. For the **W**-states for prep-exec connections this will be irrespective of which are the relationships that have interactions. For the other **W**-states this will be relationship-specific.

Table 8 The first-order and second-order self-model states that are examined to analyse adaptation

	First-order adaptation		Second-order adaptation
	Relationship-specific	Relationship-independent	Relationship-independent
Short term	**T**-states for all representation and preparation states for each dyad separately for 3 modalities *m*, *b*, *v*	**T**-states for all dyads and all execution states for 3 modalities *m*, *b*, *v*	
Long term	**W**-states for all sense-rep and rep-prep connections for each dyad separately for 3 modalities *m*, *b*, *v*	**W**-states for all dyads and all prep-exec connections for 3 modalities *m*, *b*, *v*	$\mathbf{W_T}$- and $\mathbf{W_W}$-states for the strengths of **T**- and **W**-states activation $\mathbf{H_T}$- and $\mathbf{H_W}$-states for the speeds of **T**- and **W**-states activation

- In a comparative manner, on the average, the activation values of the **T**-states will be lower and of the **W**-states will be higher in simulation runs with longer interaction durations. For the **W**-states for prep-exec connections this will be irrespective of which are the relationships that have interactions. For the other **W**-states this will be relationship-specific.

Second-order self-model $\mathbf{H_T}$- and $\mathbf{H_W}$-states control the speed of adaptivity for the **T**-states and **W**-states. Second-order self-model $\mathbf{W_W}$-states and $\mathbf{W_T}$-states control the strength of the incoming connections to the **T**-states and **W**-states and through that the strengths of the activations of the **T**-states and **W**-states. For the second-order self-model states we expect that

- Within a given simulation run, for all relationships the activation values of all second-order self-model states will increase over time, irrespective of which are the relationships that have interactions, except those of the $\mathbf{W_T}$-states which instead will decrease over time.
- In a comparative manner, on the average the activation values of all second-order self-model states will be higher in simulation runs with longer interaction durations irrespective of which relationships it concerns, except those of the $\mathbf{W_T}$-states which instead will be lower.

6 Simulation Setup and Example

In this section, the setup of the simulation experiments and one illustrative example simulation run is discussed. In Sect. 7 an extensive analysis is presented for a collection of 20 simulation runs performed according to this setup. For all settings for the simulations, see the full specification of the model available as supplementary material and as Linked Data at https://www.researchgate.net/publication/363802843.

6.1 Design of the Simulation Experiments

A stimulus occurs regularly during certain periods: 40 time units without the stimulus followed by 40 time units with the stimulus and this pattern is repeated every 80 time units. Moreover, in a random manner interaction-enabled periods happen for certain modalities for randomly chosen pairs of agents. After each interaction-enabled period for two agents, a new pair of agents is chosen at random from A, B, C and D for the next interaction-enabled period (see Fig. 5). In addition, the duration until the next interaction-enabled period (interaction break length: the durations of the intervals between the purple boxes in Fig. 5) and the duration of the next interaction-enabled period (interaction length: the lengths of the purple boxes in Fig. 5) are both chosen at random from the interval [0, 50].

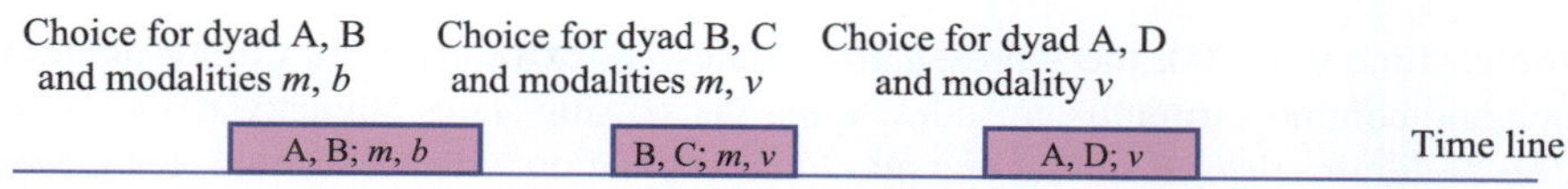

Fig. 5 Example of a timeline with interaction episodes based on randomly chosen dyads from {A, B, C, D}, up to three modalities from {*m*, *b*, *v*}, interaction durations from the interval [0, 50] and durations between interactions from the interval [0, 50]

Furthermore, the enabled modalities for the interaction in the next interaction-enabled period are chosen. Each modality has a 5/6 = 0.83 independent probability to be available. In other words, for each interaction-enabled period there is 0.58 chance that all three modalities are available and 0.42 chance that only one or two modalities are available (and 0.005 chance that none is available, in which case the agents in principle would be able to interact during some time, but actually do not have any of the three modalities available for it). The formulas within MATLAB to randomly select each of the three interaction modalities *m* (cmm), *b* (cma) and *v* (cmv) are:

$$\begin{aligned}\text{cmm} &= \text{round}(0.4 + 0.6 * \text{rand}(1, 1));\\ \text{cma} &= \text{round}(0.4 + 0.6 * \text{rand}(1, 1));\\ \text{cmv} &= \text{round}(0.4 + 0.6 * \text{rand}(1, 1));\end{aligned}$$

We have conducted 20 independent runs, each of them with a total time duration of 4000 time units and step size $\Delta t = 0.5$. We chose this approach because of the stochastic set up of the experiment, to evaluate the consistency of behavior under multiple circumstances.

6.2 Evaluation of a Simulation Run

First, we zoom in on how agents' states develop over time. Therefore, we evaluate the patterns of the activation values of the states of an agent A from simulation 1. In the other 19 simulations (available on request), we have seen that the patterns of adaptivity are roughly the same over simulations. This means that, overall, the described findings from simulation 1 are representative for all simulations.

Base level states

The sensing states are only activated during interaction episodes, and their activation values become generally higher within interaction episodes later in time (Fig. 6). The representation states get already activated within the stimulus intervals, regardless interaction is enabled (Fig. 6). During the interaction episodes, when no stimulus is present, the representation states do not become higher, because both agents do not have enough input to trigger their actions then. However, later in time, for example around time unit 800, the representation states are extra activated during interaction and common stimulus episodes, when the sensing states are activated as well. This highlights the role interaction plays in combination with the stimulus regarding increased activation of the representation states. The patterns of the preparation and the execution states tend in general to be the same: they align with the activations of the common stimulus and when on top of this common stimulus interaction between agents A and C happens, their activation levels elevate further (Fig. 7). The interpersonal synchrony states only get activated through the interaction episodes and seem to achieve higher peaks over time, indicating agents A and C become more attuned towards each other. These findings are in line with hypotheses A and B: adaptation of representing (rep-states), responding (prep-states) and expressing (exec-states) occurs for the three modalities. This adaptation is shown through the elevated activation values of the relevant states over the simulation when interaction between agents A and C (in combination with a common stimulus) happened.

First-order adaptation: T-states

The two types of first-order **T**-states are roughly following the same patterns for all modalities: they show downwards jumps during the interaction episodes of agents A and C and these jumps are becoming larger over time until time 1600 (Fig. 8). Not similarly, the $\mathbf{T}_{exec}$ states sometimes display small downwards jumps when no interaction between A and C occurs, for example around time 1950. An explanation for this might be that not only the interpersonal synchrony states of agents C and A influence the $\mathbf{T}_{exec}$ states of agent A, but also the detected interpersonal synchrony states from agent A towards agents B and D ($\mathbf{T}_{\mathbf{exec}}$-states are other-agent independent). These findings are in line with the expectations that within simulations the activation values of **T**-states will become lower during interaction episodes with

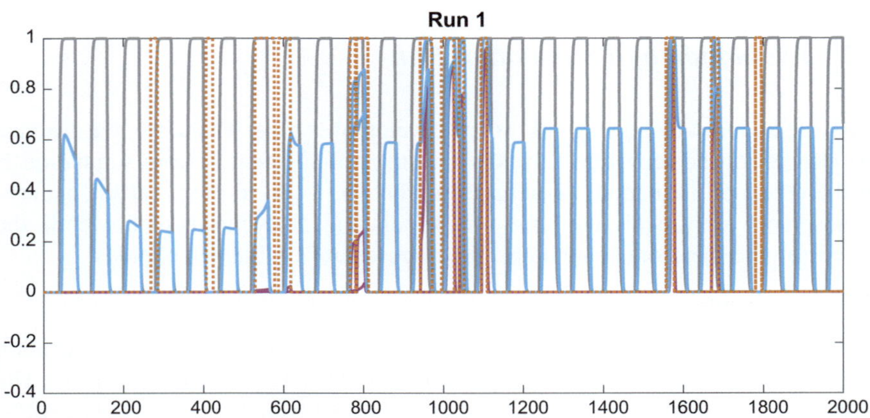

Fig. 6 The grey solid line is the common stimulus. The orange dashed lines are interaction episodes of agent A and C. The pink lines are the sensing states from agent A of agent's C *m*, *b* and *v*. The blue lines are agent A's representation states of agent C's actions

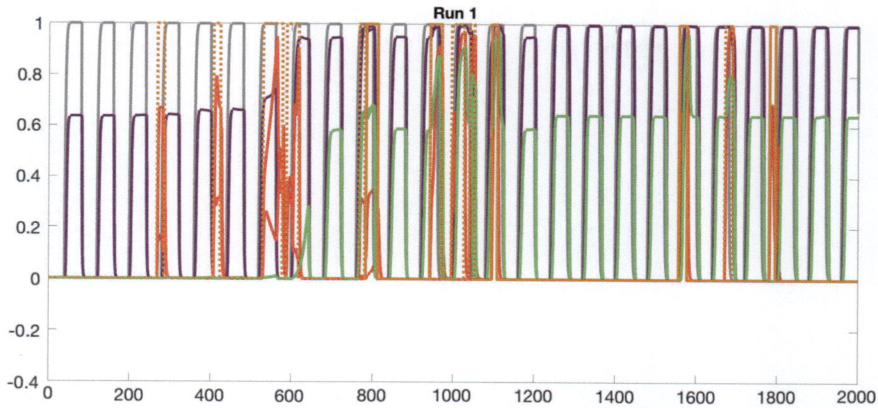

Fig. 7 The grey solid line is the common stimulus. The orange dashed lines are interaction episodes of agent A and C. The purple lines are the preparation states, the green lines the execution states and the red lines the interpersonal synchrony detection states for the three modalities (*m*, *b* and *v*) of A towards agent C

a specific agent and higher when the interaction episode is finished, meaning that during interaction with a specific agent the relationship-specific adaptivity emerges. This is in accordance with hypotheses D(i) and E(i) from Sect. 5.

First-order adaptation: W-states

All first-order self-model **W**-states show a kind of breakthrough around time 600: a sharp increase, when two relatively long interaction episodes follow closely on each other (Fig. 9). After that sharp increase, the $\mathbf{W}_{\text{sense-rep}_{x,C,A}}$ and $\mathbf{W}_{\text{rep-prep}_{x,C,A}}$ states

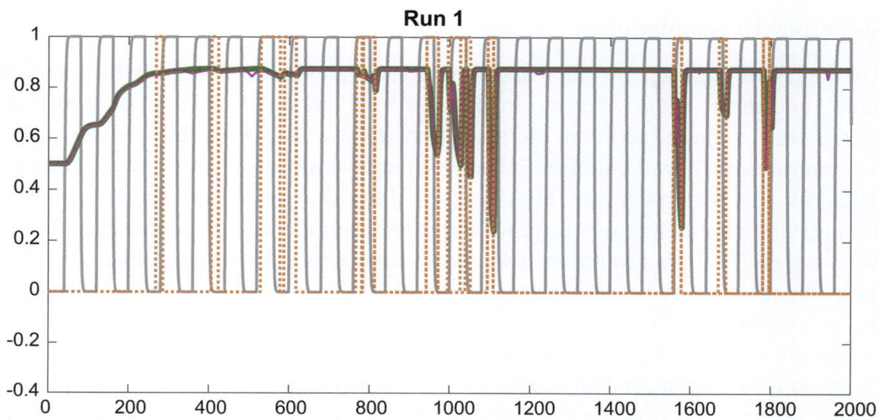

Fig. 8 The grey solid line is the common stimulus. The orange dashed lines are interaction episodes of agents A and C. The pink lines are the $\mathbf{T}_{\text{exec}}$ states and the green lines the $\mathbf{T}_{\text{rep}_{C,x,A}}$ states

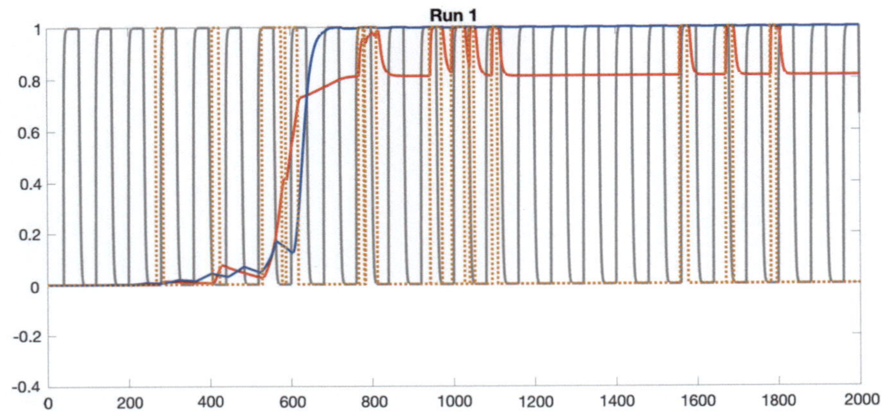

Fig. 9 The grey solid line is the common stimulus. The orange dashed lines are interaction episodes of agent A and C. The overlapping red lines are the $\mathbf{W}_{\text{sense-rep}_{x,C,A}}$ and $\mathbf{W}_{\text{rep-prep}_{x,C,A}}$ states and the overlapping blue lines the $\mathbf{W}_{\text{prep-exec}_{x,C,A}}$

balance around an activation level of 0.8, with jumps towards 1 during interaction episodes. More extremely, the $\mathbf{W}_{\text{prep-exec}_{x,C,A}}$-states already reach an activation level around 1 at time 700 at the end of the sharp increase, meaning its maximum is already achieved and effects of interaction episodes cannot cause an extra effect anymore. These results are in line with the hypotheses C, D(ii), and E(i). But note that this is not a gradual monotonous development in which the **W**-states become higher over time, but the main increase is rather suddenly (tipping points that are reached around time 600) and the $\mathbf{W}_{\text{sense-rep}_{x,C,A}}$ and $\mathbf{W}_{\text{rep-prep}_{x,C,A}}$ states are not completely monotonous during the non-interaction episodes, as observed by the drops in activation values from the moment interaction ends (especially from time 800 onwards).

Second-order adaptation

The $\mathbf{H}_{\mathbf{T}_A}$-state from agent A increases during each interaction episode with any other agent, and drops again when no interaction with any agent occurs. Around time 650, there is a sharp incline in the activation values, and thereafter there are again fluctuations that show the same trend between interaction and no interaction episodes (Fig. 10). The patterns for the $\mathbf{H}_{\mathbf{W}_A}$-states and the $\mathbf{W}_{\mathbf{W}_A}$-states are roughly similar, although the oscillations are less pronounced and their inclines are more gradual. The $\mathbf{W}_{\mathbf{T}_A}$-state is declining over time, towards negative activation levels around −0.2. All these patterns are in line with the hypothesis F(i) that within a given simulation run, the activation values of all second-order self-model states will increase over time, except those of the $\mathbf{W}_{\mathbf{T}}$-states which instead will decrease over time. This indicates that relationship-independent adaptivity can emerge from interactions with specific agents. Namely, each of these second-order self-model states from agent A adapt during the simulation, regardless the other agent with whom agent A interacts and all second-order self-model states have their effect in a relationship-independent manner. Moreover, these second-order effects induce relationship-independent influences on first-order adaptivity.

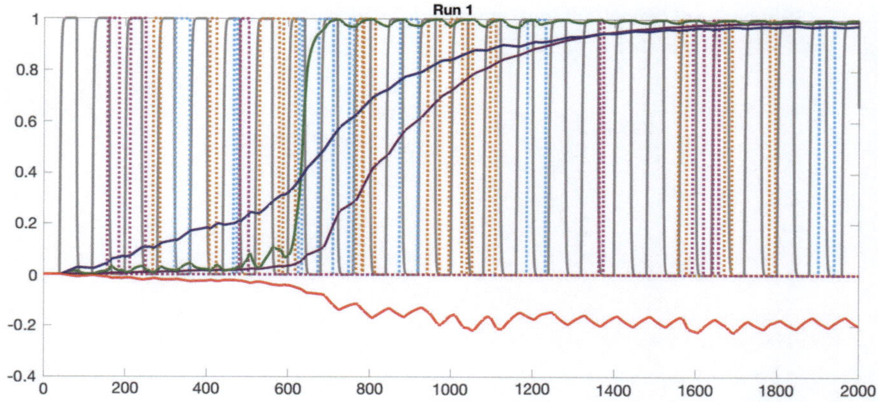

Fig. 10 The grey solid line is the common stimulus. The pink dotted lines are the interactions between A and D, the light blue dotted lines these between A and B and the orange dotted lines these between A and C. $\mathbf{H}_{\mathbf{W}_A}$ is the purple line, $\mathbf{H}_{\mathbf{T}_A}$ the dark green line, $\mathbf{W}_{\mathbf{W}_A}$ the dark blue line and $\mathbf{W}_{\mathbf{T}_A}$ the red line

7 Statistical Analysis of the Simulation Outcomes

In Sect. 6, we have evaluated the adaptation patterns of one typical simulation run at the base level, first-order level and second-order level. Within that simulation run it can be observed how over time more and more adaptations take place, based on the social interaction episodes and emerging synchronies within them. Although this run was claimed to be typical, a single run cannot show in how far the adaptations indeed depend on the extent to which social interaction takes place as that extent is fixed. Therefore, as an additional step, in the current section we quantify the main patterns for all 20 generated simulation runs in a statistical manner. These 20 runs do have variation in the extent of social interaction. In this way some more evidence is obtained about the behaviour of the model in relation to the hypotheses formulated in Sect. 2 and related to the model in Sect. 5, in particular, for example, also E(ii) and F(ii). This can be done especially in a comparative manner by comparing runs with more interaction to runs with less interaction, so that the extent of interaction can be considered an independent variable over the 20 runs and it can be analysed how other factors depend on this.

7.1 Variation in Total Interaction Durations

The 80 agents in the 20 simulation runs had on average a total individual interaction duration of 2512 time units (diameter: 1580 time units), with a standard deviation of 345 time units, and their individual interaction durations seemed to be normally distributed, see Fig. 11. The smallest and largest interaction durations from an agent

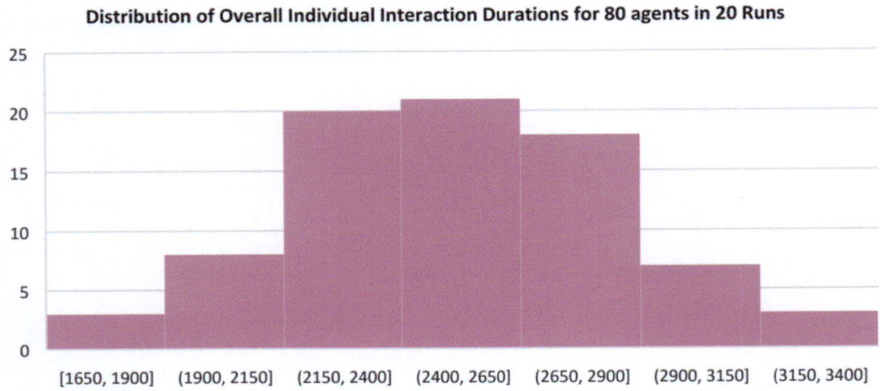

Fig. 11 The distribution of individual total interaction durations over the whole population of 80 agents

from the 80 agents equaled 1650 and 3230 time units, respectively. These results indicate that there was enough variation in the interaction durations, and that the average of 2512 of their total interaction durations over the whole simulation runs of 4000 time units was only slightly above the 2000 time units. These results enable a further evaluation of the agents' mechanisms performances in relation to their total interaction durations (see Sect. 7.3).

7.2 Averaged Learning Effects Between Two Phases

To evaluate the hypotheses, in all runs we zoom in on the first half (time 0–2000, Phase 1) and second half (time 2000–4000, Phase 2), see Fig. 12 and Table 9. All activation levels of the different types of states were averaged over time and over the three modalities *m*, *b* and *v*. Additionally, the synchrony detection activation levels are averaged over all interpersonal synchrony states. Regarding the states (rep, prep, exec and sync) from the base level, it appears that their activation levels have on average increased from Phase 1 to Phase 2 by a factor more than 2.5 (see Table 9). Although the average activation values of synchrony detection states are generally low in both phases (because each dyad only interacts only a small part of the time) and the differences are hardly seen in Fig. 12, the synchrony activation levels still have increased with a factor around 2, from 0.026 to 0.054. These results at the base level are in line with hypothesis A from Sects. 2 and 5 that the activation values (between 0 and 1) of the base states will increase over time within simulations.

Concerning the first-order self-model **W**- and **T**-states, the activation levels of all **W**-states display an increasing pattern and those of all **T**-states a decreasing pattern from Phase 1 to 2, see Fig. 12. The overall factors are approximately 3 and 0.9, respectively, see Table 9. These results are overall in line with our expectations

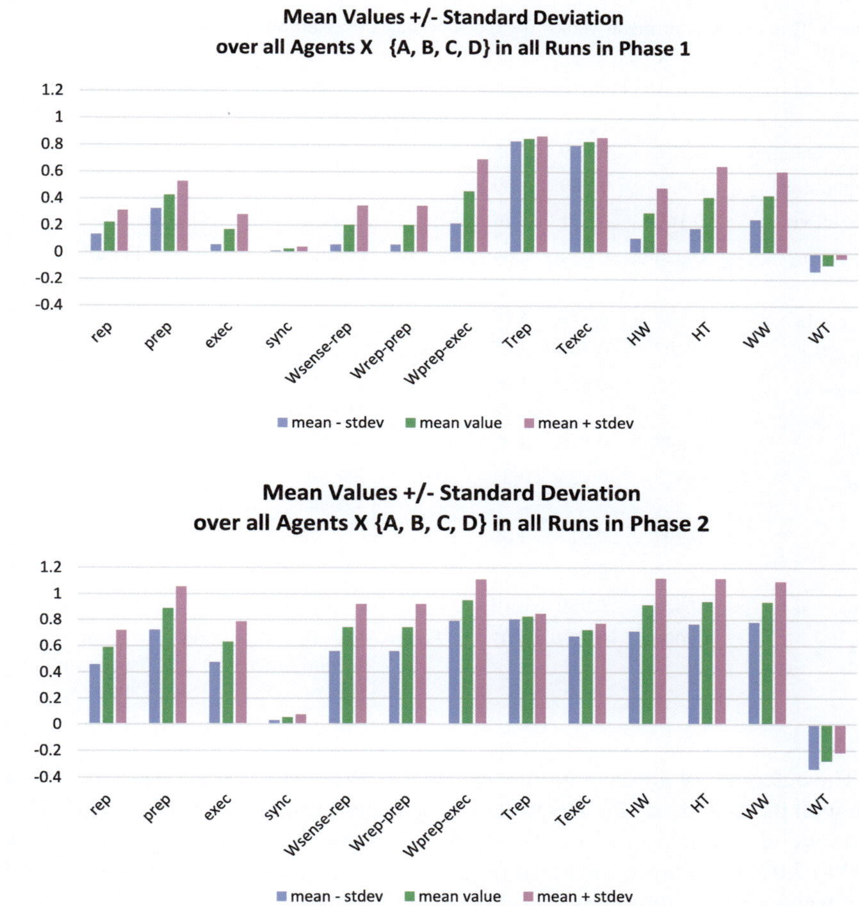

Fig. 12 Mean values and standard deviations for all types of states in comparison for Phase 1 (upper graph) for time 0–2000 and Phase 2 for time 2000–4000 (lower graph)

formulated in hypotheses C to E. Higher activation levels of **W**-states demonstrate that the adaptive connection weights became stronger, and lower activation levels of **T**-states that the target states can get a stronger activation (more sensitive target states: enhanced excitability). Although the overall decrease with a factor of 0.92 for the **T**-states does not seem that high, it fits perfectly within the hypothesis. We expected that the values of the **T**-states of given agents would decline when they interact, but would become higher when they are not interacting, because this type of adaptation works in the short-term only within interaction episodes. Since the values of the **T**-states are computed over all episodes (interaction episodes and no interaction episodes) and in the non-interaction periods they stay high as they should, in the averaging process the adaptation effects are flattened out.

Table 9 Comparisons of mean values for types of states over all 80 individuals for Phase 1 and Phase 2

	Phase 1	Phase 2	Factor	Geometric mean of factors		Arithmetic mean of factors	
Representation	0.222	0.589	2.65				
Preparation	0.426	0.889	2.09				
Execution	0.169	0.631	3.74				
Synchrony	0.026	0.054	2.04	2.55	For all base level states	2.63	For all base level states
$\mathbf{W}_{\text{sense-rep}}$	0.203	0.742	3.67				
$\mathbf{W}_{\text{rep-prep}}$	0.203	0.742	3.67				
$\mathbf{W}_{\text{prep-exec}}$	0.455	0.952	2.09	3.04	For all **W**-states	3.14	For all **W**-states
$\mathbf{T}_{\text{rep}}$	0.848	0.826	0.97				
$\mathbf{T}_{\text{exec}}$	0.828	0.723	0.87	0.92	For all **T**-states	0.92	For all **T**-states
$\mathbf{H}_{\mathbf{W}}$	0.297	0.917	3.09				
$\mathbf{H}_{\mathbf{T}}$	0.415	0.944	2.27				
$\mathbf{W}_{\mathbf{W}}$	0.428	0.939	2.19				
$\mathbf{W}_{\mathbf{T}}$	−0.090	−0.277	3.07	2.62	For all second-order states	2.66	For all second-order states

The outcomes of all second-order self-model $\mathbf{W}_{\mathbf{W}}$-, $\mathbf{H}_{\mathbf{W}}$-, $\mathbf{W}_{\mathbf{T}}$-, $\mathbf{H}_{\mathbf{T}}$-states are in line with the hypothesis F(ii) as well. The activation values of the $\mathbf{W}_{\mathbf{W}}$-, $\mathbf{H}_{\mathbf{W}}$-, $\mathbf{H}_{\mathbf{T}}$-states are all elevated in Phase 2 compared to Phase 1, with factors ranging from 2.19 to 3.09. In contrast, the mean (negative) values of the $\mathbf{W}_{\mathbf{T}}$-states are dropped by a factor of 3.07 from Phase 1 to Phase 2, so decreasing over time. These results for the second-order self-model states have the second-order adaptive effect that independent of the relationship:

- the speeds of adaptivity for the **W**-states and **T**-states increase over time,
- the strengths of the connections from the synchrony detection states to the **W**-states increase, and
- the strengths of the connections from the synchrony detection states to the **T**-states decrease over time.

This shows both second-order relationship-independent adaptivity and the relationship-independent influences on first-order adaptivity induced by it.

Note that all described differences between activation levels of states between Phase 1 and 2 do not only hold for the mean values, but as well for the mean values minus and plus the standard deviation (see Fig. 12). This is an extra indication that the hypotheses are confirmed.

7.3 Averaged Learning Effects Versus Overall Interaction Duration

Next, we want to more explicitly relate the adaptive effects to the extent of interaction during a simulation run. Therefore, we created scatterplots for the 80 agents of the 20 runs with for each of them their total interaction duration on the horizontal axis and for the different types of states their average activation levels on the vertical axis. Within each of the scatterplots we added the trendline and determined its slope and the R^2 coefficient.

Effects of the learning on the base states

First, we analyse how the adaptive effects on the base level states relate to the interaction duration. For each of the four considered types of base states, scatterplots were created: in Fig. 13 for representation and preparation states and in Fig. 14 for execution and interpersonal synchrony states. Indeed, they all show positive slopes for the trendlines. The R^2 coefficients are between 0.19 and 0.27. Moreover, as listed in Table 6, Pearson correlation coefficients were determined which varied from 0.44 to 0.51. This shows that the way how adaptivity results in changes for the base states strongly depends on the extent of interaction.

First-order adaptation: W-states

Next, the adaptivity shown in the first-order self-model states were addressed: see the scatterplots for the **W**-states in Fig. 15 and for the **T**-states in Fig. 16. All **W**-states show increasing trendlines. The R^2 coefficients in Table 10 are around 0.25–0.27. Note that de results for the $\mathbf{W}_{\text{sense-rep}}$-states and $\mathbf{W}_{\text{rep-prep}}$-states are the same because for all four agents we have used a uniform structure for this with only slight differences in the way they sense as indicated in Sect. 8. The correlation coefficients are around 0.51–0.52, see Table 6. This shows also a clear dependence of the adaptations of the **W**-states on the extent of interaction.

First-order adaptation: T-states

Similarly, we analysed the **T**-states. Here the trendlines in Fig. 16 have negative slopes. But these trends should indeed be negative as the adaptive effect concerns lowering the **T**-values during interactions so that more sensitive responses can be generated. In this case the R^2 values in Table 10 vary from 0.17 to 0.31. Moreover, the correlation coefficients vary from −0.41 to −0.56. So, also here a strong dependence of the adaptivity on the extent of interaction is found.

Second-order adaptive effects

Finally, we analysed how the states for second-order adaptation depend on interaction duration. For the scatterplots of the $\mathbf{H}_{\mathbf{W}}$- and $\mathbf{H}_{\mathbf{T}}$-states, see Fig. 17, for the $\mathbf{W}_{\mathbf{W}}$- and $\mathbf{W}_{\mathbf{T}}$-states, see Fig. 18. It is also found here that for the $\mathbf{H}_{\mathbf{W}}$-, $\mathbf{H}_{\mathbf{T}}$- and $\mathbf{W}_{\mathbf{W}}$- states the trendlines are positive. The R^2 coefficients are around 0.28 and the correlation coefficients are around 0.53, see Table 6. Indeed, for these second-order adaptivity

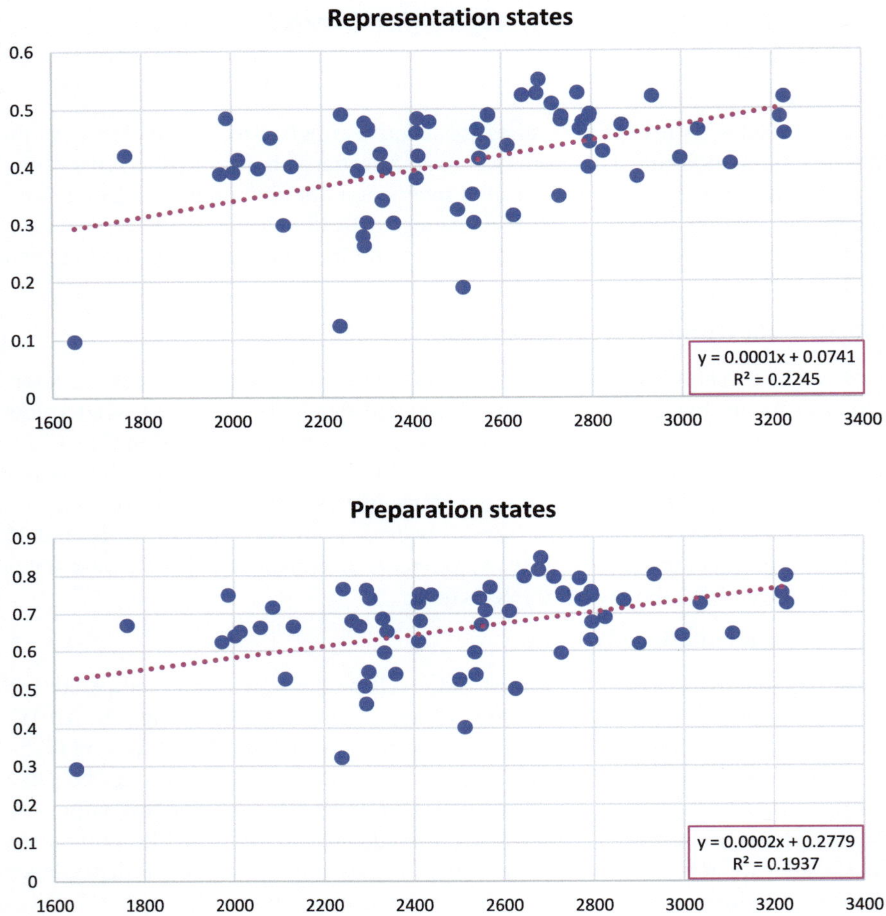

Fig. 13 The base level effects for the mean values for representation states (upper graph) and preparation states (lower graph) against the interaction durations (both for times 0–4000) over all 80 agents

states, a strong dependence is found on the extent of interaction as well. However, regarding the fourth type of states, the $\mathbf{W}_{\mathbf{T}}$-states show a negative trend, which is also what it is supposed to display as these activation values are negative and adaptation makes them more negative. Here the R^2-coefficient is 0.28 and the correlation coefficient is -0.53. These findings highlight how second-order relationship-independent adaptivity depends on the extent of social interaction and thus also the relationship-independent influences on first-order adaptivity induced by it depend on this extent of social interaction.

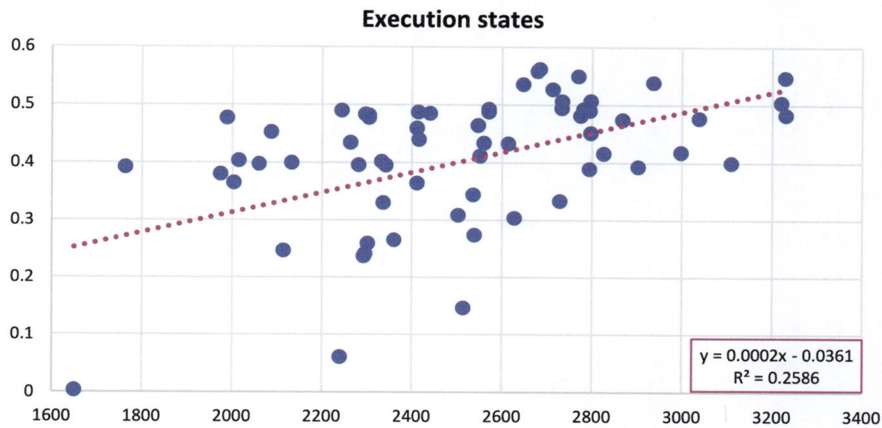

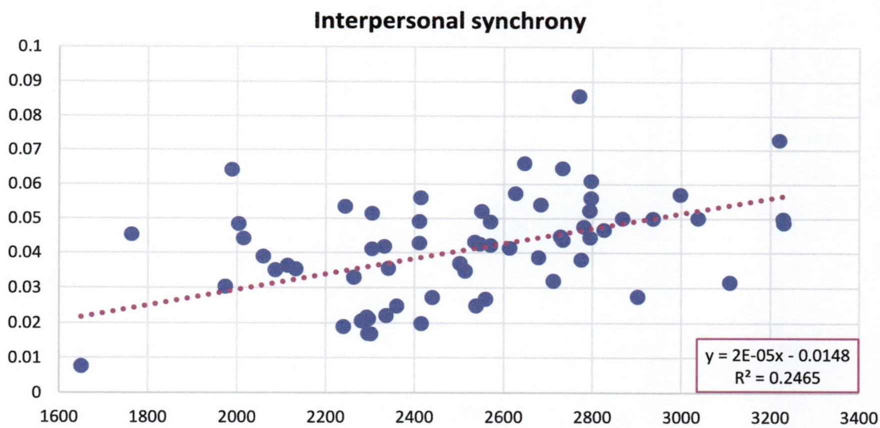

Fig. 14 The base level effects for the mean values for execution states (upper graph) and interpersonal synchrony detection states (lower graph) against the interaction durations (both for times 0–4000) over all 80 individuals

Overall findings

We conclude that the adaptive changes in activation values for the **T**-states for first-order adaptation and corresponding $\mathbf{W}_{\mathbf{T}}$-states for second-order adaptation may seem relatively low as the **T**-states concern short-term adaptation to contextual circumstances that only occasionally occur. This implies that there are long periods covered in the averages in which no adaptation takes place as the context does not ask for that. Moreover, for threshold values (represented by these **T**-states) small differences often already have a substantial effect. Therefore, most indicative are the three **W**-states for first-order adaptation and the first three states $\mathbf{H}_{\mathbf{W}}$, $\mathbf{H}_{\mathbf{T}}$, $\mathbf{W}_{\mathbf{W}}$ for second-order adaptation. For the Pearson correlation coefficients, these correlation coefficients

Fig. 15 The first-order learning effects for the mean values for the **W**-states against the interaction durations (both for times 0–4000) over all 80 individuals: $\mathbf{W}_{\text{sense-rep}}$ (upper graph), $\mathbf{W}_{\text{rep-prep}}$ (middle graph), $\mathbf{W}_{\text{prep-exec}}$ (lower graph)

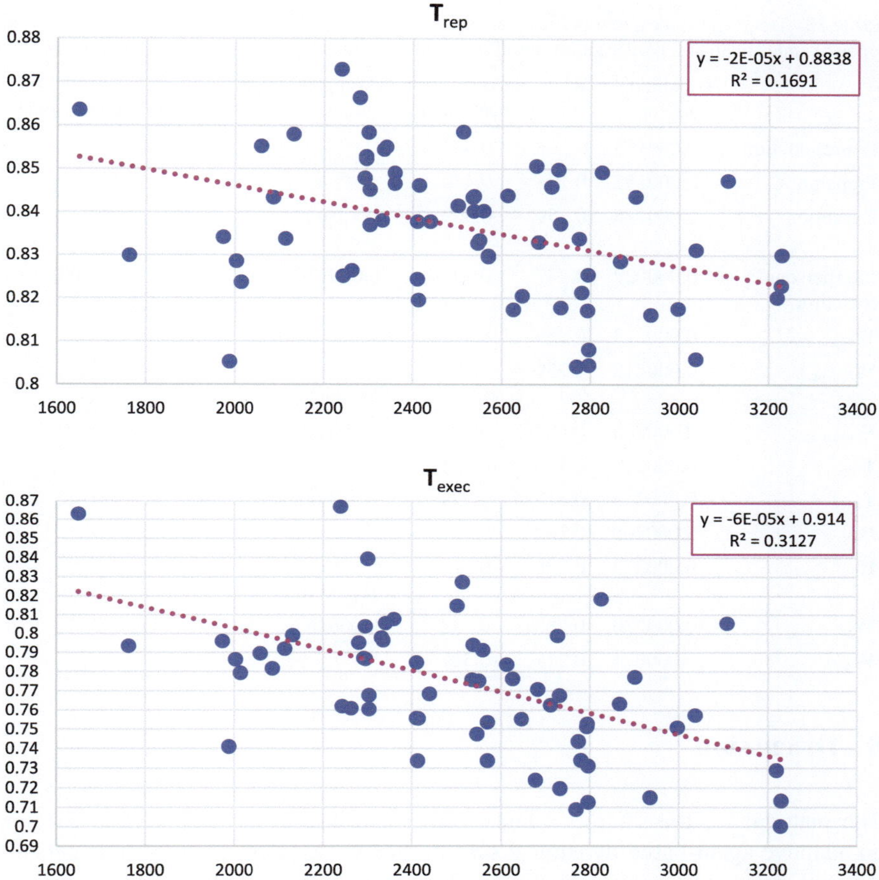

Fig. 16 The first-order learning effects for the mean values for the **T**-states against the interaction durations (both for times 0–4000) over all 80 individuals: $\mathbf{T}_{rep}$ (upper graph), $\mathbf{T}_{exec}$ (lower graph)

show convincing numbers around 0.51 for the first-order adaptation **W**-states, around −0.49 for the first-order adaptation **T**-states, and around 0.53 resp. −0.53 for the second-order adaptation states $\mathbf{H_W}$, $\mathbf{H_T}$, $\mathbf{W_W}$, $\mathbf{W_T}$. Similarly, the R^2 numbers are around 0.26 for the first-order adaptation **W**-states, around 0.24 for the first-order adaptation **T**-states, and around 0.28 resp. −0.28 for the second-order adaptation states $\mathbf{H_W}$, $\mathbf{H_T}$, $\mathbf{W_W}$, $\mathbf{W_T}$. All these data indicate a strong dependence of the different forms of adaptivity on the durations of the social interaction episodes. In particular this holds for the second-order adaptation states which have their effects on first-order adaptation in a relationship-independent manner.

Table 10 Trendline slopes, R^2 coefficients, and correlation coefficients for the mean values for all types of states against the interaction durations (both for times 0–4000) over all 80 individuals

	Trendline slope	R^2	Correlation coefficient	Averages slopes	R^2 coefficients	Correlation coefficients
Representation	0.000133	0.225	0.474			
Preparation	0.000152	0.194	0.440			
Execution	0.000175	0.259	0.509	For base states		
Interpersonal Synchrony	0.000022	0.247	0.497	0.000120	0.231	0.480
$\mathbf{W}_{\text{sense-rep}}$	0.000213	0.259	0.509			
$\mathbf{W}_{\text{rep-prep}}$	0.000213	0.259	0.509	For **W**-states		
$\mathbf{W}_{\text{prep-exec}}$	0.000260	0.267	0.517	0.000229	0.261	0.511
$\mathbf{T}_{\text{rep}}$	−0.000019	0.169	−0.411	For **T**-states		
$\mathbf{T}_{\text{exec}}$	−0.000056	0.313	−0.559	−0.000037	0.241	−0.485
$\mathbf{H}_{\mathbf{W}}$	0.000268	0.282	0.531			
$\mathbf{H}_{\mathbf{T}}$	0.000273	0.278	0.528	For $\mathbf{H}_{\mathbf{W}}$, $\mathbf{H}_{\mathbf{T}}$, $\mathbf{W}_{\mathbf{W}}$		
$\mathbf{W}_{\mathbf{W}}$	0.000230	0.283	0.532	0.000257	0.280	0.529
$\mathbf{W}_{\mathbf{T}}$	−0.000074	0.278	−0.527			

8 Discussion

The material for this chapter comes from Hendrikse et al. [25]. In this chapter, an adaptive agent-based dynamical system model was introduced for how persons develop during the social interaction they have. It incorporates how interaction behaviour changes on the short term during interaction episodes and on the long term over multiple interaction episodes. Furthermore, it addresses social interaction in multiple relationships and transference between them: how behaviour learned in one relationship can also be carried over to other relationships, like described, for example, in attachment theory developed by Mary Salter Ainsworth and John Bowlby [9, 52–55]. Moreover, it distinguishes first-order adaptation from second-order adaptation and to model them applies mechanisms known from neuroscience: synaptic plasticity [4] for long-term first-order adaptation, nonsynaptic plasticity [13] for short-term first-order adaptation and metaplasticity [1] for second-order adaptation.

The model was developed for four agents as an adaptive dynamical system specified by its canonical self-modeling network representation as described in [24, 68], see also this volume [26], Chap. 13. The four agents do not interact all the time but only during episodes separated by periods without interaction. In each interaction episode only one dyad interacts, selected at random, while also the modalities used, the durations of the interaction episodes, and the times between episodes are chosen at random. In this way, we generated 20 simulation runs and statistically

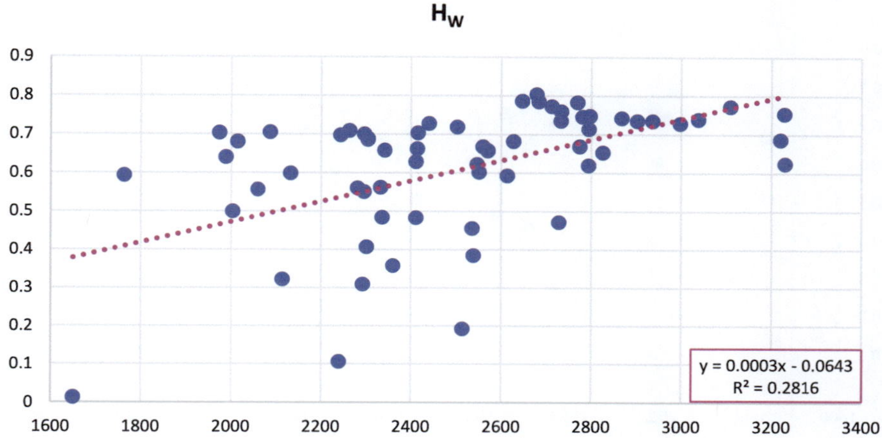

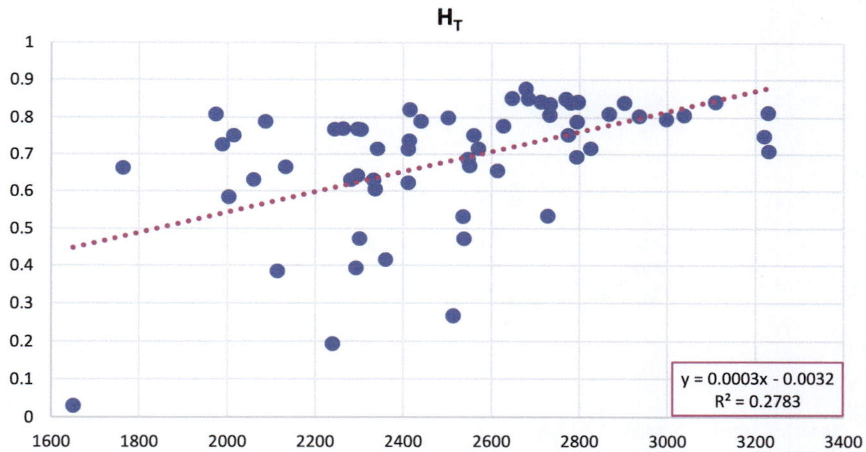

Fig. 17 The second-order learning effects for the mean values for the $\mathbf{H_W}$ and $\mathbf{H_T}$-states against the interaction durations (both for times 0–4000) over all 80 individuals: $\mathbf{H_W}$ (upper graph), $\mathbf{H_T}$ (lower graph)

analysed, among others, the dependence of the different types of adaptivity on the extent of social interaction. The outcomes of the analysis indeed show a strong dependence: more social interaction leads to more adaptation of the interaction behaviour, both for the short-term and long-term first-order adaptation and for the second-order adaptation, which is long-term.

Many other modeling approaches in mental domains and beyond such as Samsonovich [56], use some form of dynamical system modeling. The modeling approach used in the current chapter based on self-modeling networks can cover any smooth adaptive dynamical system as is shown in [24, 68], see also this volume [26], Chapter "The Adaptive Interplay of Interpersonal Synchrony, Short-Term Affiliation and Long-Term Bonding: A Multi-timescale Multi-adaptive Agent Model" [27]. As an illustrative example, consider Eq. (3) from [56, p. 67], which is:

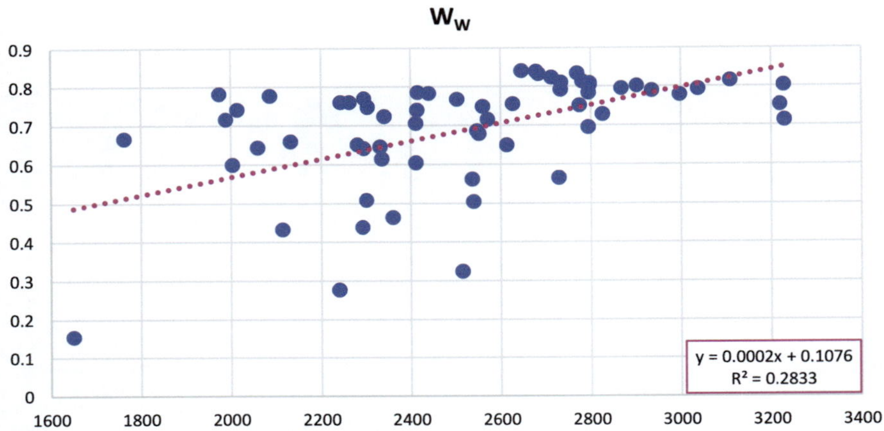

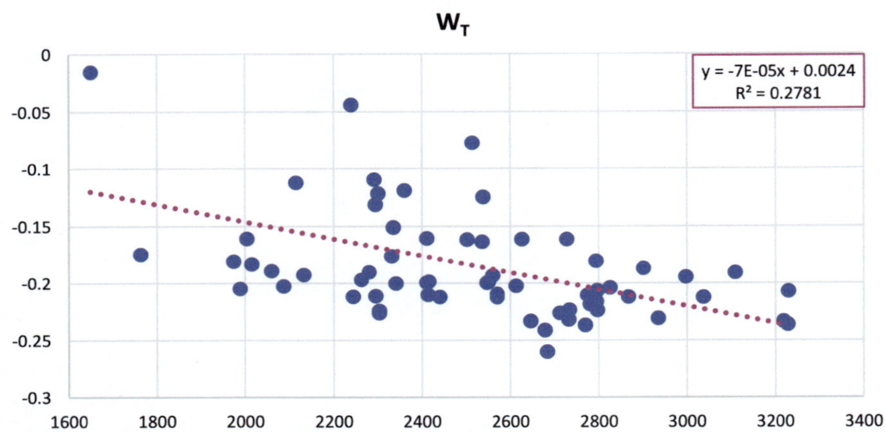

Fig. 18 The second-order learning effects for the mean values for the $\mathbf{W_W}$- and $\mathbf{W_T}$-states against the interaction durations (both for times 0–4000) over all 80 individuals: $\mathbf{W_W}$ (upper graph), $\mathbf{W_T}$ (lower graph)

$$A(actors,\ t + 1) = (1 - rW_1)A(actors,\ t) + rW_2A(actions,\ t)$$

This equation can be easily rewritten as follows:

$$A(actors,\ t + 1) = A(actors,\ t) + [-rW_1A(actors,\ t) + rW_2A(actions,\ t)]$$

$$A(actors,\ t + \Delta t) = A(actors,\ t) + [rW_2A(actions,\ t) - rW_1A(actors,\ t)]\Delta t$$

The last equation fits in our canonical Eq. (1)

$$Y(t + \Delta t) = Y(t) + \eta_Y[\mathbf{c}_{\pi_Y,Y}(\omega_{X_1,Y}X_1(t), \ldots, \omega_{X_k,Y}X_k(t)) - Y(t)]\Delta t$$

by setting

$$k = 2, Y = A(actors, \ldots), X_1 = A(actions, \ldots), X_2 = A(actors, \ldots)$$

and

$$\eta_Y = 1, \omega_{X_1,Y} = -rW_2, \ \omega_{X_2,Y} = -rW_1, \ \text{and } \mathbf{c}(V_1, V_2) = V_1 + V_2$$

In this way Eq. (3) from [56] can be modeled in the network-oriented modeling approach used in the current chapter.

The extent of adaptation of the agents can be observed in a most clear manner in their three **W**-states for first-order adaptation and the first three states $\mathbf{H_W}$, $\mathbf{H_T}$, $\mathbf{W_W}$ for second-order adaptation. It was found that the degree of adaptation of an agent depends in a significant manner on the overall duration of the interaction episodes of this agent. In more detail, the collected simulation data indicate a strong dependence of the different forms of adaptivity on the durations of the social interaction episodes. This does not only hold for the first-order adaptation states but also for the second-order adaptation states which have their effects on first-order adaptation in a relationship-independent manner.

The work presented here has adopted some elements of earlier work. For example, modeling of the emergence of synchrony during social interaction between agents was addressed in earlier work such as [23, 29], see also this volume [26], Chapters "The Emergence of Multimodal Interpersonal Synchrony During Social Interaction: A Network Model" and "Learning to Synchronize: Adaptive Joint Decision-Making by Virtual Agents" [32]. However, in these models no (subjective) internal detection of synchrony was incorporated and in [29] no adaptivity was modeled, whereas in [23] another type of adaptivity was captured: of internal responding connections from representation to preparation. The idea of subjective synchrony detection in an agent-based model was introduced in [30] and subsequently the distinction between short-term and long-term behavioural adaptivity was introduced in [24, 31], see also this volume [26], Chapters "Growing Connectedness through Synchrony: The Role of Subjective Analysis of Intrapersonal Synchrony and Interpersonal Synchrony within an Adaptive Agent Model" and "The Adaptive Interplay of Interpersonal Synchrony, Short-Term Affiliation and Long-Term Bonding: A Multi-timescale Multi-adaptive Agent Model" [28, 33]. However, these papers did not distinguish between relationship-specific and relationship-independent adaptation and were limited to a fixed dyad as they did not include a context of interaction with multiple agents as studied here.

A specific computational model for attachment theory has been contributed in [34, 35]. This model is based on the internal working models for the self and the other, following [3]. It does not address the distinction between short-term and long-term adaptivity and also not the differentiation between relationship-specific and relationship-independent adaptivity as in the current chapter. Moreover, there the second-order adaptation is limited to the speed of adaptation, whereas here also second-order adaptation for the strength of adaptation is addressed.

For further work, note that in the model only few variations for individual differences between the agents have been addressed. As follow-up research it can be interesting to study such differences in much more detail. Furthermore, as mentioned, mechanisms for plasticity and metaplasticity from neuroscience have been used as a basis for the adaptive agent model. However, more work on the neuroscientific mechanisms behind attachment theory exists, for example, in [5, 6, 11, 74]. This work can provide input for refinement of the presented models. As an example, Coan [11] mentions that topics concerning neural systems supporting emotion and motivation and emotion-regulation, filial bonding, familiarity, proximity seeking, and individual differences are important. Moreover, Beckes and Coan [5] put forward processes such as person perception, familiarity, anticipatory motivation, behavioral organisation, consummatory behaviour, emotion regulation, and aversive motivation. Another perspective that might be interesting for further work is the multidimensional model for attachment proposed by Gagliardi [20]. So, these literatures can provide many forms of inspiration to extend the current model. Our current adaptive agent model provides a solid base to model further refinements of relationship-specific and relationship-independent social behaviour development.

9 Conclusion

Based on 20 runs, the outcomes of the analysis of the (stochastic) simulation results show a strong dependence of adaptation on the extent of social interaction: more social interaction leads to more adaptation of the interaction behaviour. This holds both for the short-term and long-term first-order adaptation and for the second-order adaptation, which is long-term. The modeling approach used is a general dynamical system modeling approach which means that any other dynamical system model can be remodeled or interpreted using the network concepts used here.

References

1. Abraham, W.C., Bear, M.F.: Metaplasticity: the plasticity of synaptic plasticity. Trends Neurosci. **19**(4), 126–130 (1996)
2. Ashby, W.R.: Design for a Brain (Second Extended Edition). Chapman and Hall, London (1960)
3. Bartholomew, K., Horowitz, L.M.: Attachment styles among young adults: a test of a four-category model. J. Pers. Soc. Psychol. **61**(2), 226–244 (1991)
4. Bear, M.F., Malenka, R.C.: Synaptic plasticity: LTP and LTD. Curr. Opin. Neurobiol. **4**(3), 389–399 (1994)
5. Beckes, L., Coan, J.A.: Relationship neuroscience. In: Mikulincer, M.E., Shaver, P.R., Simpson, J.A., Dovidio, J.F. (eds.) APA Handbook of Personality and Social Psychology, vol. 3: Interpersonal relations, Ch 5, pp. 119–150. American Psychological Association (2015)

6. Beckes, L., Ijzerman, H., Tops, M.: Toward a radically embodied neuroscience of attachment and relationships. Front. Hum. Neurosci. **9**, 266 (2015)
7. Bloch, C., Vogeley, K., Georgescu, A.L., Falter-Wagner, C.M.: INTRApersonal synchrony as constituent of INTERpersonal synchrony and its relevance for autism spectrum disorder. Front. Robot. AI **73** (2019)
8. Boot, N., Baas, M., Van Gaal, S., Cools, R., De Dreu, C.K.W.: Creative cognition and dopaminergic modulation of fronto-striatal networks: integrative review and research agenda. Neurosci. Biobehav. Rev. **78**, 13–23 (2017)
9. Bowlby, J.: Attachment. Basic Books (2008)
10. Chandra, N., Barkai, E.: A non-synaptic mechanism of complex learning: modulation of intrinsic neuronal excitability. Neurobiol. Learn. Mem. **154**, 30–36 (2018)
11. Coan, J.A.: Towards a neuroscience of attachment. In: Cassidy, J., Shaver, P.R. (eds.) Handbook of Attachment: Theory, Research, and Clinical Applications, 3rd edn., pp. 242–269. Guilford, New York, NY (2016)
12. Cronin, T.J., Pepping, C.A., O'donovan, A.: Attachment to friends and psychosocial well-being: the role of emotion regulation. Clin. Psychol. **22**(2), 158–167 (2018). https://doi.org/10.1111/cp.12159
13. Debanne, D., Inglebert, Y., Russier, M.: Plasticity of intrinsic neuronal excitability. Curr. Opin. Neurobiol. **54**, 73–82 (2019)
14. Fairhurst, M.T., Janata, P., Keller, P.E.: Being and feeling in sync with an adaptive virtual partner: brain mechanisms underlying dynamic cooperativity. Cerebr. Cort. **23**(11), 2592–2600 (2013). https://doi.org/10.1093/cercor/bhs243
15. Feeney, J.A.: Transfer of attachment from parents to romantic partners: effects of individual and relationship variables. J. Fam. Stud. **10**(2), 220–238 (2004). https://doi.org/10.5172/jfs.327.10.2.220; https://www.tandfonline.com/doi/abs/10.5172/jfs.327.10.2.220
16. Feldman, R.: Parent–infant synchrony biological foundations and developmental outcomes. Curr. Dir. Psychol. Sci. **16**, 340–345 (2007). https://doi.org/10.1111/j.1467-8721.2007.00532.x
17. Fonagy, P.: Attachment Theory and Psychoanalysis. Random House, New York (2001)
18. Fraley, R.C., Hudson, N.W.: The development of attachment styles. In: Personality Development Across the Lifespan, pp. 275–292. Elsevier Inc. (2017). https://doi.org/10.1016/B978-0-12-804674-6.00017-X
19. Fraley, R.C., Hudson, N.W., Heffernan, M.E., Segal, N.: Are adult attachment styles categorical or dimensional? A taxometric analysis of general and relationship-specific attachment orientations. J. Pers. Soc. Psychol. **109**(2), 354–368 (2015). https://doi-org.vu-nl.idm.oclc.org/10.1037/pspp0000027
20. Gagliardi, M.: Human attachment as a multi-dimensional control system: a computational implementation. Front. Psychol. **13**, 844012 (2022)
21. Hebb, D.O.: The Organization of Behavior: A Neuropsychological Theory. Wiley, New York (1949)
22. Heinze, J.E., Cook, S.H., Wood, E.P., Dumadag, A.C., Zimmerman, M.A.: Friendship attachment style moderates the effect of adolescent exposure to violence on emerging adult depression and anxiety trajectories. J. Youth Adoles. **47**, 177–193 (2018). https://doi-org.vu-nl.idm.oclc.org/10.1007/s10964-017-0729-x
23. Hendrikse, S.C.F., Kluiver, S., Treur, J., Wilderjans, T.F., Dikker, S., Koole, S.L.: How virtual agents can learn to synchronize: an adaptive joint decision-making model of psychotherapy. Cogn. Syst. Res. **79**, 138–155 (2023a)
24. Hendrikse, S.C.F., Treur, J., Koole, S.L.: Modeling emerging interpersonal synchrony and its related adaptive short-term affiliation and long-term bonding: a second-order multi-adaptive neural agent model. Int. J. Neural Syst. **33**(7) (2023b). https://doi.org/10.1142/S0129065723500387
25. Hendrikse, S.C.F., Treur, J., Koole, S.L.: Relationship-specific and relationship-independent behavioural adaptivity in affiliation and bonding: a multi-adaptive dynamical systems approach. Cogn. Syst. Res. **84**, 101182 (2024)

26. Hendrikse, S.C.F., Treur, J., Koole, S.L. (eds.): New Analysis and Modeling Directions in Social Interaction Science: Emergent Multimodal Interpersonal Synchrony, Affiliation, and Bonding. Springer Nature (2025a)
27. Hendrikse, S.C.F., Treur, J., Koole, S.L.: Introduction to the role of adaptivity in social interaction science. In: Hendrikse, S.C.F., Treur, J., Koole, S.L. (eds.) New Analysis and Modeling Directions in Social Interaction Science: Emergent Multimodal Interpersonal Synchrony, Affiliation, and Bonding, Ch 1, this volume. Springer Nature (2025b)
28. Hendrikse, S.C.F., Treur, J., Koole, S.L.: The adaptive interplay of interpersonal synchrony, short-term affiliation and long-term bonding: a multi-timescale multi-adaptive agent model. In: Hendrikse, S.C.F., Treur, J., Koole, S.L. (eds.) New Analysis and Modeling Directions in Social Interaction Science: Emergent Multimodal Interpersonal Synchrony, Affiliation, and Bonding, Ch 12, this volume. Springer Nature (2025c)
29. Hendrikse, S.C.F., Treur, J., Wilderjans, T.F., Dikker, S., Koole, S.L.: On the same wavelengths: emergence of multiple synchronies among multiple agents. In: Proceedings of the 22nd International Workshop on Multi-Agent-Based Simulation, MABS'21. Lecture Notes in AI, vol. 13128, pp. 57–71. Springer Nature (2022a)
30. Hendrikse, S.C.F., Treur, J., Wilderjans, T.F., Dikker, S., Koole, S.L.: On becoming in sync with yourself and others: an adaptive agent model for how persons connect by detecting intra- and interpersonal synchrony. Human-Centric Intelligent Systems, vol. 3, pp. 123–146. Also presented as: In Sync with Yourself and with Others: Detection of Intra- and Interpersonal Synchrony Within an Adaptive Agent Model. In: Face2face: Advancing the Science of Social Interaction. Royal Society, London (2023c). https://www.researchgate.net/publication/358964043
31. Hendrikse, S.C.F., Treur, J., Wilderjans, T.F., Dikker, S., Koole, S.L.: On the interplay of interpersonal synchrony, short-term affiliation and long-term bonding: a second-order multi-adaptive neural agent model. In: Maglogiannis et al. (eds.) Proceedings of the 18th International Conference on Artificial Intelligence Applications and Innovations, AIAI'22. Advances in Information and Communication Technology, vol. 646, pp. 37–57. Springer Nature (2022b)
32. Hendrikse, S.C.F., Treur, J., Wilderjans, T.F., Dikker, S., Koole, S.L.: The emergence of multimodal interpersonal synchrony during social interaction: a network model. In: Hendrikse, S.C.F., Treur, J., Koole, S.L. (eds.) New Analysis and Modeling Directions in Social Interaction Science: Emergent Multimodal Interpersonal Synchrony, Affiliation, and Bonding, Ch 3, this volume. Springer Nature (2025a)
33. Hendrikse, S.C.F., Treur, J., Wilderjans, T.F., Dikker, S., Koole, S.L.: Growing connectedness through synchrony: the role of subjective analysis of intrapersonal synchrony and interpersonal synchrony within an adaptive agent model. In: Hendrikse, S.C.F., Treur, J., Koole, S.L. (eds.) New Analysis and Modeling Directions in Social Interaction Science: Emergent Multimodal Interpersonal Synchrony, Affiliation, and Bonding, Ch 11, this volume. Springer Nature (2025b)
34. Hermans, A.M.M., Muhammad, S., Treur, J.: An adaptive network model of attachment theory. In: Proceedings of the 21th International Conference on Computational Science, ICCS'21. Lecture Notes in Computer Science, vol. 12744, pp. 462–475. Springer Nature Publishers (2021)
35. Hermans, A.M.M., Muhammad, S., Treur, J.: You feel so familiar, you feel so different: a controlled adaptive network model for attachment patterns as adaptive mental models. In: Treur, J., Van Ments, L. (eds.) Mental Models and Their Dynamics, Adaptation and Control: A Self-Modeling Network Modeling Approach, Ch 12, pp. 321–346. Springer Nature (2022)
36. Hesslow, G.: Conscious thought as simulation of behaviour and perception. Trends Cogn. Sci. **6**, 242–247 (2002)
37. Hove, M.J., Risen, J.L.: It's all in the timing: interpersonal synchrony increases affiliation. Soc. Cogn. **27**(6), 949–960 (2009)
38. Iacoboni, M.: Mirroring People: The New Science of How We Connect with Others. Farrar, Straus and Giroux, NY (2008)
39. Johnson, S.M.: Attachment Theory in Practice: Emotionally Focused Therapy with Individuals, Couples, and Families. Guilford, New York, NY (2019)

40. Kirschner, S., Tomasello, M.: Joint music making promotes prosocial behavior in 4-year-old children. Evol. Hum. Behav. **31**, 354–364 (2010). https://doi.org/10.1016/j.evolhumbehav.2010.04.004
41. Koole, S.L., Tschacher, W.: Synchrony in psychotherapy: a review and an integrative framework for the therapeutic alliance. Front. Psychol. **7**, 862 (2016)
42. Koole, S.L., Tschacher, W., Butler, E., Dikker, S., Wilderjans, T.F.: In sync with your shrink. In: Forgas, J.P., Crano, W.D., Fiedler, K. (eds.) Applications of Social Psychology, pp. 161–184. Taylor and Francis, Milton Park (2020)
43. Laws, H.B., Sayer, A.G., Pietromonaco, P.R., Powers, S.I.: Longitudinal changes in spouses' HPA responses: convergence in cortisol patterns during the early years of marriage. Health Psychol. **34**(11), 1076 (2015)
44. Lisman, J., Cooper, K., Sehgal, M., Silva, A.J.: Memory formation depends on both synapse-specific modifications of synaptic strength and cell-specific increases in excitability. Nat. Neurosci. **2018**(21), 309–314 (2018)
45. Marmarosh, C., Markin, R., Spiegel, E.: Attachment in group psychotherapy. The American Psychological Association, Washington, D.C. (2013)
46. Maurer, R.E., Tindall, J.H.: Effect of postural congruence on client's perception of counselor empathy. J. Counsel. Psychol. **30**, 158 (1983). https://doi.org/10.1037/0022-0167.30.2.158
47. Palumbo, R.V., Marraccini, M.E., Weyandt, L.L., Wilder-Smith, O., McGee, H.A., Liu, S., Goodwin, M.S.: Interpersonal autonomic physiology: a systematic review of the literature. Pers. Soc. Psychol. Rev. **21**(2), 99–141 (2017)
48. Port, R.F., Van Gelder, T.: Mind as Motion: Explorations in the Dynamics of Cognition. MIT Press, Cambridge, MA (1995)
49. Prince, K., Brown, S.: Neural correlates of partnered interaction as revealed by cross-domain ALE meta-analysis. Psychol. Neurosci. (2022). https://doi.org/10.1037/pne0000282
50. Ramseyer, F., Tschacher, W.: Nonverbal synchrony in psychotherapy: coordinated body movement reflects relationship quality and outcome. J. Consult. Clin. Psychol. **79**, 284–295 (2011). https://doi.org/10.1037/a0023419a
51. Robinson, B.L., Harper, N.S., McAlpine, D.: Meta-adaptation in the auditory midbrain under cortical influence. Nat. Commun. **7**, e13442 (2016)
52. Salter Ainsworth, M.D.: Infancy in Uganda. Johns Hopkins, Baltimore (1967)
53. Salter Ainsworth, M.D., Blehar, M., Waters, E., Wall, S.: Patterns of Attachment. Erlbaum, Hillsdale, NJ (1978)
54. Salter Ainsworth, M.D., Bowlby, J.: Child Care and the Growth of Love. Penguin Books, London (1965)
55. Salter Ainsworth, M.D., Bowlby, J.: An ethological approach to personality development. Am. Psychol. **46**(4), 333–341 (1991)
56. Samsonovich, A.V.: Socially emotional brain-inspired cognitive architecture framework for artificial intelligence. Cogn. Syst. Res. **60**, 57–76 (2020)
57. Schurger, A., Uithol, S.: Nowhere and everywhere: the causal origin of voluntary action. Rev. Philos. Psychol. **6**, 761–778 (2015). https://doi.org/10.1007/s13164-014-0223-2
58. Shatz, C.J.: The developing brain. Sci. Am. **267**, 60–67 (1992)
59. Sharpley, C.F., Halat, J., Rabinowicz, T., Weiland, B., Stafford, J.: Standard posture, postural mirroring and client-perceived rapport. Couns. Psychol. Q. **14**, 267–280 (2001). https://doi.org/10.1080/09515070110088843
60. Stanton, P.K.: LTD, LTP, and the sliding threshold for long-term synaptic plasticity. Hippocampus, **6**(1), 35–42 (1996). https://doi.org/10.1002/(sici)1098-1063(1996)6:1<35::aid-hipo7>3.0.co;2-6
61. Tarr, B., Launay, J., Dunbar, R.I.M.: Silent disco: dancing in synchrony leads to elevated pain thresholds and social closeness. Evol. Hum. Behav. **37**(5), 343–349 (2016)
62. Treur, J.: Temporal factorisation: a unifying principle for dynamics of the world and of mental states. Cogn. Syst. Res. **8**(2), 57–74 (2007)
63. Treur, J.: Temporal factorisation: realisation of mediating state properties for dynamics. Cogn. Syst. Res. **8**(2), 75–88 (2007)

64. Treur, J.: Network-Oriented Modeling: Addressing Complexity of Cognitive, Affective and Social Interactions. Springer Nature (2016)
65. Treur, J.: Network-Oriented Modeling for Adaptive Networks: Designing Higher-Order Adaptive Biological, Mental and Social Network Models. Springer Nature (2020a)
66. Treur, J.: Modeling multi-order adaptive processes by self-modeling networks. In: Tallón-Ballesteros, A.J., Chen, C.-H. (eds.) Proceedings of the 2nd International Conference on Machine Learning and Intelligent Systems, MLIS'20. Frontiers in Artificial Intelligence and Applications, vol. 332, pp. 206–217. IOS Press (2020b)
67. Treur, J.: Modeling the emergence of informational content by adaptive networks for temporal factorisation and criterial causation. Cogn. Syst. Res. **68**, 34–52 (2021)
68. Treur, J.: On the dynamics and adaptivity of mental processes: relating adaptive dynamical systems and self-modeling network models by mathematical analysis. Cogn. Syst. Res. **70**, 93–100 (2021)
69. Trout, D.L., Rosenfeld, H.M.: The effect of postural lean and body congruence on the judgment of psychotherapeutic rapport. J. Nonverbal Behav. **4**, 176–190 (1980)
70. Valdesolo, P., DeSteno, D.: Synchrony and the social tuning of compassion. Emotion **11**, 262 (2011). https://doi.org/10.1037/a0021302
71. Valdesolo, P., Ouyang, J., DeSteno, D.: The rhythm of joint action: synchrony promotes cooperative ability. J. Exp. Soc. Psychol. **46**(4), 693–695 (2010)
72. Van Gelder, T., Port, R.: It's about time: an overview of the dynamical approach to cognition. In: Port, R.F., van Gelder, T. (eds.) Mind as Motion: Explorations in the Dynamics of Cognition, pp. 1–43. MIT Press, Cambridge, MA (1995)
73. Welch, R.D., Houser, M.E.: Extending the four-category model of adult attachment: an interpersonal model of friendship attachment. J. Soc. Pers. Relat. **27**(3), 351–366 (2010). https://doi.org/10.1177/0265407509349632; https://journals.sagepub.com/doi/abs/10.1177/0265407509349632?journalCode=spra
74. White, L., Kungl, M., Vrticka, P.: Charting the social neuroscience of human attachment (SoNeAt). Attach. Hum. Dev. **25**(1), 1–18 (2023)
75. Williams, A.H., O'Leary, T., Marder, E.: Homeostatic regulation of neuronal excitability. Scholarpedia **8**, 1656 (2013)
76. Wiltermuth, S.S., Heath, C.: Synchrony and cooperation. Psychol. Sci. **20**(1), 1–5 (2009)
77. Zhang, A., Li, X., Gao, Y., Niu, Y.: Event-driven intrinsic plasticity for spiking convolutional neural networks. IEEE Trans. Neural Netw. Learn. Syst. (2021). https://doi.org/10.1109/tnnls.2021.3084955

Further Computational Analysis of Social Interaction Adaptivity Induced by Subjective Interpersonal Synchrony

In this part, some more refined analyses for adaptivity in social interaction are contributed. In "The Role of Timing in Synchrony Detection and Social Interaction Adaptivity: Including Time Lags in a Second-Order Adaptive Agent Model", the role of time lags for subjective synchrony analysis is computationally explored in more detail. In contrast to objective synchrony detection by an external (third-party) observer, subjective synchrony detection relies solely on information that is perceived by each of the synchronizing persons. Simultaneous actions of the two persons in principle cannot be sensed instantaneously by one of the two persons but will involve different time lags due to the different pathways followed. In this chapter, we explore the role of time lags in different types of subjective synchrony detection and its involvement in synchrony-induced behavioural adaptivity. Furthermore, in "The Effect of Transition Dynamics in Synchrony on Social Interaction Adaptivity: a Multi-Adaptive Network Model" it is explored computationally how not only subjective synchrony detection itself leads to behavioural adaptivity but also detected transitions or interruptions in synchrony can contribute to such an effect. This chapter introduces a second-order adaptive network model that explains how persons' affiliation might benefit from transitions in synchrony in addition to mean levels of synchrony. Finally, in "The Emergence of Group Synchrony and Adaptivity for Group Bonding: a Multi-Adaptive Network Model" the interplay between group synchrony and behavioural adaptivity for group bonding is studied. By simulating multimodal interactions within a group of four agents, the study uncovers patterns in group cohesion in the sense of emerging multimodal group synchrony and related group bonding. The obtained insights offer an understanding of adaptive group interaction dynamics.

The Role of Timing in Synchrony Detection and Social Interaction Adaptivity: Including Time Lags in a Second-Order Adaptive Agent Model

Sophie C. F. Hendrikse, Jan Treur, Tom F. Wilderjans, Suzanne Dikker, and Sander L. Koole

Abstract Interpersonal synchrony usually means that people mutually adapt their behavior to each other over time. Such behavioral adaptivity is assumed to be driven by some form of subjective internal synchrony detection. In contrast to objective synchrony detection by an external (third-party) observer, subjective synchrony detection relies solely on information that is perceived by each of the synchronizing persons. Simultaneous actions of the two persons in principle cannot be sensed instantaneously by one of the two persons, but will involve time lags. These time lags reflect the time differences between the pathways for a person's sensing of the own actions and for the sensing of the actions of the other person. In the computational agent

S. C. F. Hendrikse
Department of Methodology and Statistics, Tilburg School of Social and Behavioral Sciences, Tilburg University, Tilburg, The Netherlands
e-mail: S.C.F.Hendrikse@tilburguniversity.edu

J. Treur (✉)
Social AI Group, Department of Computer Science, Vrije Universiteit Amsterdam, Amsterdam, The Netherlands
e-mail: j.treur@vu.nl

T. F. Wilderjans
Methodology and Statistics Research Unit, Institute of Psychology, Faculty of Social and Behavioural Sciences, Leiden University, Leiden, The Netherlands

Research Group of Quantitative Psychology and Individual Differences, Faculty of Psychology and Educational Sciences, Katholieke Universiteit (KU) Leuven, Leuven, Belgium

T. F. Wilderjans
e-mail: t.f.wilderjans@leidenuniv.nl

S. Dikker
NYU – Max Planck Center for Language, Music and Emotion, New York University, New York, NY, USA

S. L. Koole
Amsterdam Emotion Regulation Lab, Department of Clinical Psychology, Vrije Universiteit Amsterdam, Amsterdam, The Netherlands
e-mail: s.l.koole@vu.nl

S. C. F. Hendrikse et al. (eds.), *New Analysis and Modeling Directions in Social Interaction Science*, Studies in Systems, Decision and Control 614,
https://doi.org/10.1007/978-3-031-99968-0_14

model described in this chapter, we explore the role of time lags in different types of subjective synchrony detection and its involvement in behavioral adaptivity. Multiple simulation experiments show expected types of patterns of subjective time-lagged synchrony detection and related behavioral adaptivity.

Keywords Time lags · Subjective synchrony detection · Computational agent model

1 Introduction

When two persons become synchronized in their behavior, they tend to experience a number of relation benefits such as more closeness, concentration, coordination, cooperation, affiliation, alliance, connection, and bonding [2, 23, 24, 27, 28, 31–33, 38, 41, 48–50]. For more background on this, see this volume [16], Chap. "Introduction to the Role of Adaptivity in Social Interaction Science" [17]. In order for the above-mentioned relational benefits to emerge, it stands to reason that people are capable of detecting synchrony based on the (subjective) information that is available via sensing of their own actions and sensing the other person's actions (e.g., [9]). The pathways for sensing own actions are based on internal mechanisms whereas the pathways for sensing actions of the other person partly involve mechanisms used in the external world and for sensing. Therefore, it has to be taken into account that simultaneous actions of the two persons in principle cannot be sensed instantaneously, but will involve time lags. Any mechanism for subjective synchrony detection therefore has to incorporate this time lag.

The computational adaptive neural agent model described in the current chapter introduces two computational mechanisms for subjective synchrony detection using time lags and in addition covers how these different types of synchrony detection lead to behavioral adaptivity concerning the interaction with the other person. The model addresses intrapersonal synchrony and interpersonal synchrony for three different modalities: movement, affective and verbal responses. Simulation experiments were based on an experimental scenario, in which different stimuli for each of the two agents alternate over time independently and the same holds for periods in which communication is allowed. This work extends our previous work included in this [16] Part IV, Chaps. "Growing Connectedness Through Synchrony: The Role of Subjective Analysis of Intrapersonal Synchrony and Interpersonal Synchrony Within an Adaptive Agent Model"–"A Multi-timescale Multi-adaptive Dynamical System Approach for Relationship-Specific and Relationship-Independent Adaptivity in Affiliation and Bonding" [18, 19] where time differences in the pathways toward the detector states from the own actions and sensing the actions of the other were assumed to be non-existent. In the current chapter, we describe simulation experiments in which we varied the computations and sizes of these time lags.

The design of the adaptive neural agent model was based on the self-modeling network modeling approach introduced in [43, 44]; this is briefly described in Sect. 3.

The simulation experiments were conducted using the available dedicated software environment; some are described in Sect. 5. In Sect. 2 background literature is discussed and in Sect. 4 the designed adaptive neural agent model is described in more detail. Finally, Sect. 6 contains a discussion and conclusion.

2 Background Perspectives

There is an extensive literature in the behavioral sciences describing how synchronies between two persons can be detected from an objective external third-person (observer) point of view; e.g., [3, 4, 36]. A basic assumption here is that an external observer has objective information available on the (movement, affective, verbal) actions of the two persons. For the sake of brevity, this can be termed objective synchrony detection. In contrast, subjective synchrony detection occurs when synchrony is detected by each of the two persons in the considered dyad. These persons usually do not have access to objective information about their joint actions. Subjective synchrony detection has a distinct profile of characteristics. The following characteristics have been addressed in the model:

Asynchrony of pathways: fast sensing of self, slow sensing of other. For someone to notice or experience synchrony for themselves, several dedicated mechanisms and pathways have to be used. More specifically, this subjective form of synchrony detection essentially depends on subjective information acquired by a person on actions (e.g., movements, affective reactions, verbal responses) by the self and actions by the other. Therefore, it is crucial to consider the pathways that enable a person to sense own and other's actions. Roughly speaking, the pathways leading to subjective synchrony detection within a person may be sketched as follows:

mutual interaction →
patterns of own actions and patterns of actions of the other →
internal sensing and external sensing →
patterns of sensed own actions and patterns of sensed actions of the other →
patterns of synchrony for sensed own actions and sensed actions of the other →
subjective synchrony detector states

Even in this simplified sequence of events, it should be noted that the internal and external sensing processes occur in parallel, but differ in timing: generally, sensing self is faster than sensing the other; see also, e.g., [34, 40, 42]. This makes that the generated subjective information from both sides does not refer to the same time point; due to the differently paced parallel pathways, a time lag is created between when the two types of information become available for processing in the brain. In the literature addressing objective synchrony detection from an external observer viewpoint, a number of approaches have been developed to handle time lags; e.g.,

[36]. These methods have been used as inspiration for the way in which, for the subjective synchrony detection introduced here, time lags have been handled based on sliding (rolling) time windows.

Asymmetry of time: only past, no future

Subjective synchrony detection is also different from objective synchrony detection in a second way. In subjective synchrony detection, people only have information from the past available. People do not have information available from the future, given that sensing does not address future states. In contrast, researchers who analyze objective synchrony detection within already completed time series, in many cases future time points are freely used as well, for example, to determine a global mean level of a signal over time. Regarding subjective synchrony detection, this can only be done for the past. Even if these computations are only done locally by considering some type of sliding (or rolling) window, for the subjective case such a window will necessarily have a focus only on values from the past, as values from the future are simply not available yet.

Anormality of signals: non-normalized levels

A third issue that distinguishes subjective synchrony detection from objective synchrony detection is that it is plausible, at least for humans, that the level of detected synchrony depends both on timing and strength of the signals. This is in contrast to some approaches for objective synchrony detection where the absolute strength of signals is changed to a more relative strength by applying a normalizing factor to the signals during the detection process.

From subjective synchrony detection to behavioral adaptivity

After subjective synchrony detection occurs, a subsequent issue is how this affects the interaction. For example, Prince and Brown [32] suggested that synchrony induces what they call partnered interaction, which is: '…real-time behavioral adaptivity between the partners, as studied in a large literature devoted to joint action and joint agency (e.g., [2, 10, 25, 26, 30])'. This perspective is also adopted here. Two types of behavioral adaptivity are addressed:

- short term affiliation, modeled based on changed thresholds for intrinsic excitabilities of states (nonsynaptic plasticity)
- long-term bonding, modeled based on the changed strengths of connections (synaptic plasticity).

In the neuroscientific literature, such as Chandra and Barkai [6], the distinction between synaptic and nonsynaptic (intrinsic) adaptation is discussed in some detail. A classical notion of synaptic plasticity is Hebbian learning; e.g., [12, 39]. Nonsynaptic adaptation of intrinsic excitability of (neural) states has been addressed in more detail recently; e.g., [6, 8, 51]. The latter form of adaptation has been related, for example, to how deviant dopamine levels during sleep can result in dreams that might use more associations due to easier excitable neurons; e.g., [5]. A difference in pace is assumed here as well between the two adaptation processes that happen in parallel. The adaptation of connections is assumed to take place at a slow pace and the adaptation of intrinsic excitabilities at a fast pace. Finally, plasticity is often highly context-dependent according to what is called metaplasticity; e.g., [1, 35]. To enable context-sensitive control of plasticity, second-order adaptation has been included, which makes the model more realistic.

3 The Self-modeling Network Modeling Approach Used

In this section, the network-oriented modeling approach used as described in [43, 44] is briefly introduced. This modeling approach allows to model in a network-oriented manner any adaptive dynamical systems model, as shown in [15, 45], see also this volume [16], Chap. 12. As discussed above, in this case the challenge is to model an interplay of a number of dynamic and adaptive processes, which typically requires a complex adaptive dynamical system.

Following [43, 44], a temporal-causal network model is characterized by (here X and Y denote nodes of the network, also called states):

- **Connectivity characteristics** Connections from a state X to a state Y and their weights $\boldsymbol{\omega}_{X,Y}$
- **Aggregation characteristics** For any state Y, some combination function $\mathbf{c}_Y(..)$ defines the aggregation that is applied to the impacts $\boldsymbol{\omega}_{X,Y}X(t)$ on Y from its incoming connections from states X
- **Timing characteristics** Each state Y has a speed factor $\boldsymbol{\eta}_Y$ defining how fast it changes for a given causal impact.

The following difference equations that are used for simulation, incorporate these network characteristics $\boldsymbol{\omega}_{X,Y}$, $\mathbf{c}_Y(..)$ and $\boldsymbol{\eta}_Y$ in a standard, canonical numerical format:

$$Y(t+\Delta t) = Y(t) + \boldsymbol{\eta}_Y\big[\mathbf{c}_Y\big(\boldsymbol{\omega}_{X_1,Y}X_1(t), \ldots, \boldsymbol{\omega}_{X_k,Y}X_k(t)\big) - Y(t)\big]\Delta t \qquad (1)$$

for any state Y and where X_1 to X_k are the states from which Y receives its incoming connections. Within the software environment described in [43], Chap. 9, a large number of currently around 60 useful basic combination functions are included in a combination function library. The combination functions that are applied in the model introduced here can be found in Table 1. Here the W indicates a sliding window; for further explanation, see Sect. 4.

The two combination functions in the second and third row address different approaches for synchrony detection based on sliding windows for time lags; they were specifically developed and included in the library for the research reported here. They will be explained in Sect. 4.

By using a *self-modeling network* (also called a *reified* network), a similar network-oriented conceptualization can also be applied to *adaptive* networks; see [43, 44]. This works through the addition of new states to the network (called *self-model states*) which represent (adaptive) network characteristics. In the graphical

Table 1 The combination functions used in the introduced neural agent model

Function	Notation	Formula	Parameters	Used for
Advanced logistic sum	$\mathbf{alogistic}_{\sigma,\tau}(V_1, \ldots, V_k)$	$\left[\frac{1}{1+e^{-\sigma(V_1+\cdots+V_k-\tau)}} - \frac{1}{1+e^{\sigma\tau})}\right](1+e^{-\sigma\tau})$	Steepness $\boldsymbol{\sigma}$ Excitability threshold $\boldsymbol{\tau}$	$X_4 - X_5$, $X_{10} - X_{16}$, $X_{24} - X_{26}$, $X_{31} - X_{38}$, $X_{45} - X_{47}$, $X_{54} - X_{59}$, $X_{63} - X_{71}$, $X_{75} - X_{93}$
Complemental lagged difference	$\mathbf{compdifflag}_{\iota\delta,\lambda}(W_1, W_2)$	0 if $W_1(\boldsymbol{\iota\delta}, \lambda) = W_2(\boldsymbol{\iota\delta}, 1) = 0$ $1 - \frac{\lvert W_2(\boldsymbol{\iota\delta},1)-W_1(\iota\delta,\lambda)\rvert}{\max(W_1(\iota\delta,\lambda),W_2(\boldsymbol{\iota\delta},1))}$ else	$\boldsymbol{\iota\delta}$ state identifier $\boldsymbol{\lambda}$ time lag	$X_{18} - X_{23}$, $X_{39} - X_{44}$ (synchrony detectors)
Complemental one-sided lagged average difference	$\mathbf{compdifflag1av}_{\iota\delta,\lambda}(W_1, W_2)$	$1 - \text{av}(\frac{\lvert W_2(\boldsymbol{\iota\delta},1)-W_1(\iota\delta,\boldsymbol{\nu})\rvert}{\max(W_1(\iota\delta,\boldsymbol{\nu}),W_2(\boldsymbol{\iota\delta},1))}$ for $1 \leq \boldsymbol{\nu} \leq \boldsymbol{\lambda}$ (average)	$\boldsymbol{\iota\delta}$ state identifier $\boldsymbol{\lambda}$ max time lag	$X_{18} - X_{23}$, $X_{39} - X_{44}$ (synchrony detectors)
Stepmod	$\mathbf{stepmod}_{\rho,\delta}(V)$	0 if $0 \leq$ time t mod $\boldsymbol{\rho} \leq \delta$ 1 else	Repetition $\boldsymbol{\rho}$ Step time $\boldsymbol{\delta}$	X_2 (stimulus person A) $X_{60} - X_{62}$, $X_{72} - X_{74}$ (communication channels)
Stepmodopp	$\mathbf{stepmodopp}_{\rho,\delta}(V)$	0 if $\boldsymbol{\delta} \leq$ time t mod $\boldsymbol{\rho} \leq \rho$ 1 else	Repetition $\boldsymbol{\rho}$ Step time $\boldsymbol{\delta}$	X_1 (stimulus person B)
Euclidean	$\mathbf{eucl}_{\boldsymbol{n},\lambda}(V_1, \ldots, V_k)$	$\sqrt[n]{\frac{V_1^n+\cdots+V_k^{\boldsymbol{n}}}{\boldsymbol{\lambda}}}$	Order $\boldsymbol{n}$ Scaling factor $\boldsymbol{\lambda}$	$X_6 - X_9$, $X_{27} - X_{30}$ (sensing) $X_{48} - X_{53}$ (communication)

3D-format as shown in Sect. 4, such additional states are depicted at a next (higher) level (called *self-model level* or *reification level*), where the original network is at the *base level*.

As an example, the weight $\boldsymbol{\omega}_{X,Y}$ of a connection from state X to state Y can be represented (at a next/higher self-model level) by a self-model state named $\mathbf{W}_{X,Y}$. Similarly, all other network characteristics from $\boldsymbol{\omega}_{X,Y}$, $\mathbf{c}_Y(..)$ and $\boldsymbol{\eta}_Y$ can be made adaptive by including self-model states for them. As another example, an adaptive excitability threshold $\boldsymbol{\tau}_Y$ (as parameter for a logistic combination function) for state Y can be represented by a self-model state named $\mathbf{T}_Y$ and an adaptive speed factor $\boldsymbol{\eta}_Y$ can be represented by a self-model state named $\mathbf{H}_Y$. As the outcome of self-modeling is also a temporal-causal network model itself, as has been shown in [43], Chap. 10, this self-modeling network construction can easily be applied iteratively to obtain multiple orders of self-models (first-order, second-order, …). The second-order self-model level can be used to make the adaptation speed context-sensitive as addressed by metaplasticity literature such as [1, 35]. For instance, the metaplasticity principle 'Adaptation accelerates with increasing stimulus exposure' formulated by Robinson et al. [35] has been modeled by using second-order self-model states for the introduced model, as will be discussed in Sect. 4. Such a second-order self-model may include a second-order self-model state $\mathbf{H}_{\mathbf{W}_{X,Y}}$ representing the speed factor $\boldsymbol{\eta}_{\mathbf{W}_{X,Y}}$ for the dynamics of first-order self-model state $\mathbf{W}_{X,Y}$, which in turn represents the adaptation of connection weight $\boldsymbol{\omega}_{X,Y}$. Similarly, a second-order self-model may include a second-order self-model state $\mathbf{H}_{\mathbf{T}_Y}$ representing the speed factor $\boldsymbol{\eta}_{\boldsymbol{T}_Y}$ for the dynamics of first-order self-model state $\mathbf{T}_Y$, which in turn represents the adaptation of excitability threshold $\boldsymbol{\tau}_Y$ for Y.

For more details of the modeling approach used, see [43, 44] or this volume [16], Chap. "Modeling Coevolution of Dynamics, Adaptivity and Control of Adaptivity by Self-modeling Networks" [46].

4 The Adaptive Agent Model

In this section, the introduced adaptive agent model is explained in some detail. The controlled adaptive agent design uses a self-modeling network architecture of three levels as discussed in Sect. 3: a base level, a first-order self-model level, and a second-order self-model level. Here the (middle) first-order self-model level models how connections of the base level are adapted over time, and the (upper) second-order self-model level models the control over the adaptation. In the Appendix in Sect. 7, detailed explanations of all states and a complete specification of the model by role matrices can be found. Figure 1 shows a graphical overview of the base level of the agent model (agents are indicated by the big boxes).

Sensing, sensory processing, conscious emotions, preparing and acting In Fig. 1, the boxes indicate agents, where states involved in sensing can be found on the left-hand side, and states involved in execution or expression of actions (move, exp_affect, talk) on the right-hand side. Within a box the agent's internal mental

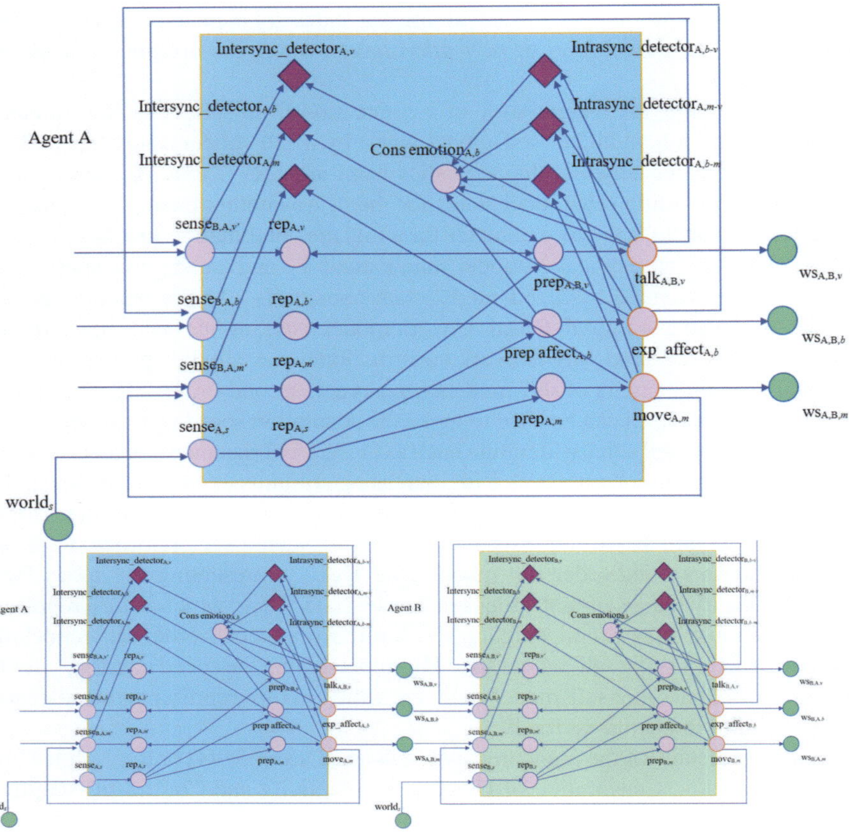

Fig. 1 Base level of (1) the introduced adaptive neural agent model (upper picture) with three modalities and (in dark pink) six synchrony detector states for intrapersonal synchrony and interpersonal synchrony and (2) how the agents interact (lower picture) according to the three modalities

states are depicted: sensory representation states (rep), preparation states (prep) for each of the three modalities (movement *m*, expression of affect *b*, and verbal action *v*), and a conscious emotion state. The representation states for the modalities have outgoing connections to the corresponding preparation states and incoming (prediction) connections back to model internal mental simulation [7, 22]. World states are modeled for dynamic stimuli sensed by the agents and the world situation's suitability for enabling interaction is modeled similarly. Finally, two context states model the (default) conditions to maintain excitability thresholds.

The synchrony detector states and their incoming pathways Six synchrony detector states (depicted in Fig. 1 by the darker pink diamond shapes) are introduced, three *intrapersonal synchrony detector states* for the three pairs of the three modalities: movement-emotion (*m-b*), movement-verbal (*m-v*), emotion-verbal (*b-v*).

These synchrony detector states have incoming connections from the two execution states for the modalities they address. Following Grandjean et al. [11], the conscious emotion state is triggered by incoming connections from the preparation state for affective response *b* together with the three intrapersonal synchrony detector states. In addition, the conscious emotion state has an incoming connection from the verbal action execution state (for noticing the emotion in the verbal utterance) and an outgoing connection to the preparation of the verbal action (for emotion integration in the verbal action preparation).

In addition, three *interpersonal synchrony detector states* are included for the three modalities *m*, *b*, and *v*. Each of them has two incoming connections: from the sensing state (used for representing the action of the other person) and the execution state (representing the own action) of the modality addressed.

As discussed conceptually in Sect. 2, acquiring information about the actions from the other person follows a longer and therefore slower pathway (via the external world and sensors) than for the own actions. In Figs. 1 and 2 it can be seen that the pathways for sensing the own actions and for sensing the actions of the other person are indeed of different lengths.

The pathways for sensing own actions are the one-connection pathways from execution states directly to detector states: the upward arrows from states move, exp_affect and talk in Fig. 1. More specifically, for internal sensing for intrapersonal synchrony detectors within an agent *A* just the following one-step internal pathways are used.

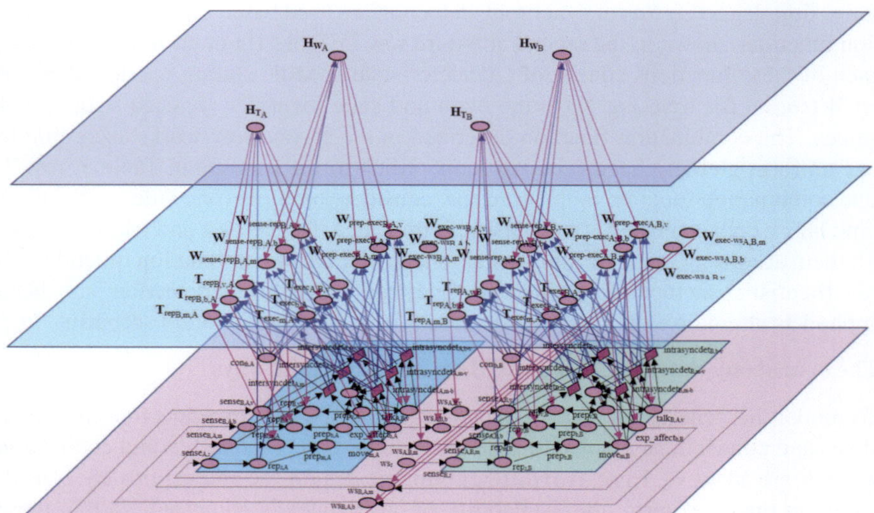

Fig. 2 Overview of the overall second-order adaptive network model

$move_{m,A}$	→	$intrasyncdet_{A,b\text{-}m}$	$move_{m,A}$	→	$intrasyncdet_{A,m\text{-}v}$
$exp\_affect_{b,A}$	→	$intrasyncdet_{A,b\text{-}v}$	$exp\_affect_{b,A}$	→	$intrasyncdet_{A,b\text{-}m}$
$talk_{A,B,v}$	→	$intrasyncdet_{A,m\text{-}v}$	$talk_{A,B,v}$	→	$intrasyncdet_{A,b\text{-}v}$

Moreover, for internal sensing for interpersonal synchrony detectors the following one-step internal pathways are used.

$move_{m,A}$	→	$intersyncdet_{B,A,m}$
$exp\_affect_{b,A}$	→	$intersyncdet_{B,A,b}$
$talk_{A,B,v}$	→	$intersyncdet_{B,A,v}$

In contrast, the pathways for external sensing of the actions from the other person are three-connection pathways, first from move, exp_affect and talk to world states, next from world states to sensing states, and finally from sensing states to detector states.

$move_{m,B}$	→	$ws_{m,B,}$	→	$sense_{B,m,A}$	→	$intersyncdet_{B,A,m}$
$exp\_affect_{b,B}$	→	$ws_{b,B,A}$	→	$sense_{B,b,A}$	→	$intersyncdet_{B,A,b}$
$talk_{B,A,v}$	→	$ws_{v,B,A}$	→	$sense_{B,v,A}$	→	$intersyncdet_{B,A,v}$

To cope with the time lags caused by these differences in pathways, for the detector states time-lagged forms of synchrony detection are used based on the two combination functions shown in the second and third row in Table 1. For each detector state, each of these functions consist of a detector-state-specific sliding window denoted by $W(\iota\delta, \lambda)$ for time lag λ (to the past) and state identifier $\iota\delta$, each with $\lambda + 1$ values. The combination function described in the second row works according to one uniform globally fixed time lag λ according to the formula in Table 1, row 2. The combination function in the third row considers multiple (one-sided to the past) time lags ν between 0 and some maximal time lag λ for the past and takes averages for them according to the formula in Table 1, row 3. These combination functions for two (heuristic) methods for subjective time-lagged synchrony detection have been applied as alternative options and compared for different simulation scenarios.

The synchrony detector states and their outgoing pathways

To model the behavioral adaptivity induced by the synchrony detection, for some states and connections involved in the interaction, their excitability and connection weights are adaptive. Here, two different time scales for the adaptations are considered. On the short-term, the excitability of such states is enhanced, so that these states become more responsive or sensitive. On the long-term, the weights of such connections are made stronger so that propagation between states is strengthened (a form of a more endurable bonding). The more synchrony is detected, the lower the excitability thresholds will become (short-term adaptation) and weights for these connections will slowly become stronger (long-term adaptation); each type of all these adaptations contributes in its own way (and time scale). This makes the sensed signals more accessible to the brain and expressions of modalities to the other agent stronger. These forms of adaptivity were modeled using the notion of self-modeling of the network model:

- first-order self-model **T**-states $\mathbf{T}_Y$ for short-term adaptation of the adaptive base excitability thresholds $\boldsymbol{\tau}_Y$, for the internal representation states and execution states Y for the three considered modalities (movement, affective response and verbal action)
- first-order self-model **W**-states $\mathbf{W}_{X,Y}$ for adaptation of the adaptive connection weights $\boldsymbol{\omega}_{X,Y}$ for internal connections from sense states to representation states and from preparation states to execution states
- second-order self-model $\mathbf{H}_\mathbf{T}$-states to control the adaptation of the adaptive excitability thresholds $\boldsymbol{\tau}_Y$ and second-order self-model $\mathbf{H}_\mathbf{W}$-states to control the adaptation of the adaptive base connection weights $\boldsymbol{\omega}_{X,Y}$; this follows the second-order adaptation (or metaplasticity) principle 'Adaptation accelerates with stimulus exposure' [35].

5 Simulation Results

To investigate the effect of time lags on the synchrony detection and consequently on the induced behavioral adaptivity, several simulation experiments during 1440 time units (2880 iterations) have been conducted for a variety of lags, both for synchrony detection based on a single time lag (using the combination function in row 2 of Table 1) and for synchrony detection based on an average over multiple time lags (using the combination function in row 3 of Table 1). Experiments were done with lag sizes of 2, 3 and 5, and a lag of 0 as a base simulation to compare the results of the simulations with the different lag sizes against. These different lag sizes have been chosen in such a way that the estimated highest value of synchrony detection based on the lengths of the different pathways is approximately in the middle (a lag size of 2.5). All simulations use two different stimuli for the two agents in a repetitive way: for agent A value 1 for time 0–120 and value 0 for time 120–180, for agent B value 0 for time 0–60 and value 1 for time 60–180, and so on. Moreover, repetitive intervals of 30 time units were used where the environment enables or does not enable (verbal and nonverbal) communication: 0–30 not enabling communication, 30–60 enabling communication, and so on.

To get an idea of the general behavior of the agents, the results of all the states of one example of these simulations (for the specific settings of the network characteristics specified in the Appendix) is depicted by the three graphs in Fig. 3. Note that first a period of relatively complex and variable patterns occurred (Fig. 3), whereas after a while more ordered forms of behavior emerged, presumingly due to the synchrony-induced long-term behavioral adaptivity. In the latter period it is seen that interpersonal synchrony is systematically high in the communication enabling intervals. The two forms of (short-term and long-term) synchrony-induced behavioral adaptivity are illustrated in the lower graph of Fig. 3: the **W**-states (representing connection weights) for long-term adaptivity start at 0.4 and meander upward to around 0.8 and the **T**-states (representing excitability thresholds) for short-term adaptivity start at 0.7 and fluctuate with the communication enabling periods: low, down to 0.2, when communication is enabled (and therefore high interpersonal synchrony is detected), and higher, around 0.4, when communication is not enabled. The same patterns of results, including the apparently two kinds of episodes (more chaotic patterns followed by ordered patterns), were obtained in all simulation patterns. Therefore, the simulation results are evaluated over the first and second half of the time interval within a simulation.

An overview of the outcomes of the simulation experiments is shown in Tables 2 (averages over time period 0–720) and 3 (averages over time period 721–1440). In each of these tables the upper part summarizes the results for the single lag detection and the lower part for the averages over multiple lags in the past. As can be seen, these averages do differ for different lags but the differences are not dramatic.

The yellow highlighted numbers indicate the highest values per category. As an example, in Table 2, in the first column of agent B for movement the base number for lag 0 is 0.522 whereas for multiple lags up to 2 (or up to 5) the number is 0.565. The difference between these numbers is 0.043, which makes a relative difference (with respect to the base number 0.522) of 8%. The other differences vary but in general are smaller. Roughly, it can be seen that especially for the first phase most of these highest numbers occur for lags of 2 or 3 time units.

6 Discussion

Interpersonal synchrony relates to behavioral adaptivity in the interaction and relationship between the synchronized persons; e.g., [2, 23, 27, 28, 31, 32, 41, 49, 50]. To explain the relation between synchrony and behavioral adaptivity, in Hendrikse et al. [21] subjective synchrony detector states were introduced and the pathways from these detector states to the behavioral adaptivity were explored in more detail; see also this volume [16], Chap. 11. However, the pathways toward the detector states from the own actions and sensing the actions of the other were assumed to be simple. As discussed in the current chapter, acquiring information about the actions from the other person follows slower pathways (via the external world and sensors) than for the own actions. Therefore, two combination functions for two (heuristic) methods for subjective synchrony detection with time lags have been applied as alternative options and compared for different simulation scenarios. The first function works

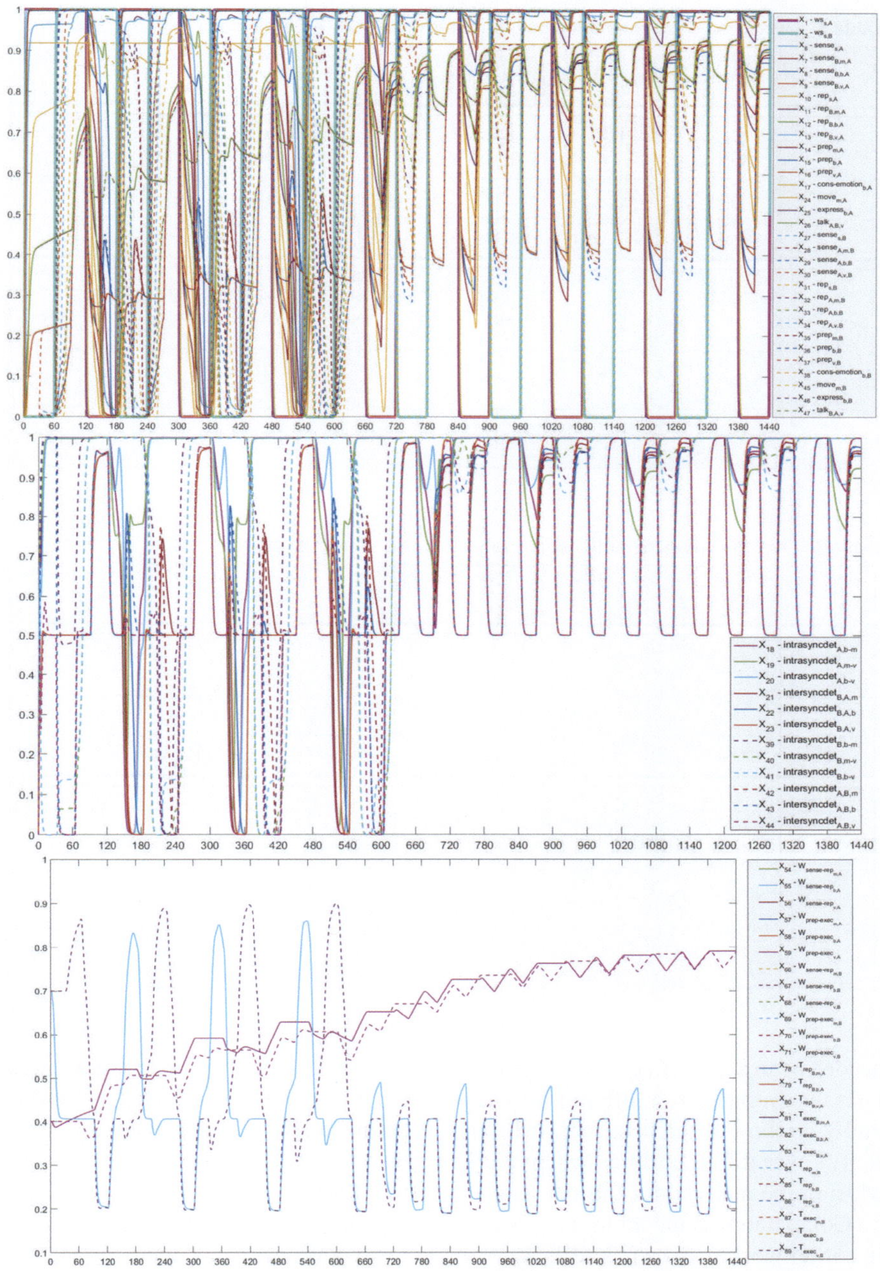

Fig. 3 Different views at one example simulation for averaged multiple time lags up to 2 time units. Upper graph: base states except the synchrony detectors. Middle graph: time-lagged synchrony detectors. Lower graph: the behavioral adaptivity **T**-states (short term adaptation) and **W**-states (long-term adaptation)

Table 2 Overview of the simulation outcomes for the first 0–720 time units

	Intersync movement A	Intersync emotion A	Intersync verbal A	Intersync Average A	W-states A	Intersync movement B	Intersync emotion B	Intersync verbal B	Intersync Average B	W-states B
Lag 0	0.541	0.565	0.538	0.548	0.539	0.522	0.511	0.528	0.520	0.523
Fixed lag: 2 time units	0.539	0.577	0.546	0.554	0.569	0.526	0.507	0.530	0.521	0.535
Fixed lag: 3 time units	0.535	0.580	0.544	0.553	0.578	0.528	0.506	0.480	0.505	0.548
Fixed lag: 5 time units	0.503	0.574	0.518	0.532	0.578	0.526	0.498	0.493	0.506	0.548
Lag 0	0.541	0.565	0.538	0.548	0.539	0.522	0.511	0.528	0.520	0.523
Multiple lags: up to 2 time units	0.541	0.571	0.543	0.552	0.551	0.565	0.549	0.552	0.555	0.527
Multiple lags: up to 3 time units	0.540	0.573	0.544	0.552	0.559	0.564	0.547	0.506	0.539	0.531
Multiple lags: up to 5 time units	0.536	0.576	0.542	0.551	0.570	0.565	0.545	0.537	0.549	0.539

Table 3 Overview of the simulation outcomes for time units 721–1440

	Intersync movement A	Intersync emotion A	Intersync verbal A	Intersync Average A	W-states A	Intersync movement B	Intersync emotion B	Intersync verbal B	Intersync Average B	W-states B
Lag 0	0.741	0.740	0.747	0.743	0.743	0.738	0.741	0.747	0.742	0.740
Fixed lag: 2 time units	0.736	0.736	0.746	0.739	0.745	0.735	0.736	0.746	0.739	0.740
Fixed lag: 3 time units	0.731	0.733	0.743	0.736	0.745	0.732	0.733	0.743	0.736	0.742
Fixed lag: 5 time units	0.694	0.705	0.703	0.701	0.723	0.711	0.696	0.683	0.697	0.712
Lag 0	0.741	0.740	0.747	0.743	0.743	0.738	0.741	0.747	0.742	0.740
Multiple lags: up to 2 time units	0.739	0.738	0.747	0.742	0.743	0.737	0.739	0.747	0.741	0.738
Multiple lags: up to 3 time units	0.737	0.737	0.746	0.740	0.743	0.736	0.737	0.746	0.740	0.739
Multiple lags: up to 5 time units	0.733	0.734	0.743	0.737	0.744	0.733	0.734	0.743	0.737	0.739

according to a uniform globally fixed time lag. The other combination function calculates the averages over different time lags. Simulation results, as depicted in Tables 2 and 3, do show differences for synchrony detection (and the induced behavioral adaptivity) over different time lags applied but these differences are relatively modest. Most of the material for this chapter comes from [21].

Computational modeling of synchrony between agents was already investigated in earlier work such as [13, 20]. However, in the models described there, no internal detection of synchrony is incorporated. Moreover, in [20] no behavioral adaptivity was covered, whereas in [13] another type of adaptivity was modeled, namely of internal responding connections from representations to preparations. As already mentioned above, in [15, 21] also internal synchrony detector states were used, in that case with the focus on the induced behavioral adaptivity, thereby neglecting the differences in timing of the incoming signals to these detector states by implicitly assuming the sensing processes being synchronous, see also this volume [16] and, for example, Chaps. "The Adaptive Interplay of Interpersonal Synchrony, Short-Term Affiliation and Long-Term Bonding: A Multi-timescale Multi-adaptive Agent Model" and "A Multi-timescale Multi-adaptive Dynamical System Approach for Relationship-Specific and Relationship-Independent Adaptivity in Affiliation and Bonding". In contrast, the current chapter explored different ways in which asynchronous sensing processes for subjective synchrony detection can be addressed computationally by considering time lags. Thus, a more realistic human-like adaptive agent model was obtained for subjectively detected synchrony. The model can provide a basis to develop adaptive virtual agents that are able to concentrate and bond with each other and in the further future with humans by behavioral adaptivity. Furthermore, future research could consider the notion of social presence [47] by dynamically detecting the time differences between the actions of an agent itself and another agent. Our model implies that time differences might slightly affect the simultaneously experienced presence of the two agents, because higher interpersonal synchrony scores have been reached when a time lag was taken into account.

Appendix

Explanation of all states of the model

Table 4 provides explanations for all of its base states and Table 5 for explanations of all self-model states.

Specification by role matrices

In Tables 6, 7, 8, 9 and 10, the full specification of the adaptive network model by role matrices is shown. Each role matrix has 93 rows for all states X_1 to X_{93} of the model.

The connectivity characteristics are specified by role matrices **mb** and **mcw** shown in Tables 6 and 7. Role matrix **mb** lists for each state the states (at the same or lower level) from which the state gets its incoming connections, while in role matrix **mcw** the connection weights are listed for these connections.

Nonadaptive connection weights are indicated in **mcw** (in Table 7) by a number (in a green shaded cell), but adaptive connection weights are indicated by a reference to the (self-model) **W**-state representing the adaptive value (in a peach-red shaded cell). This can be seen for states X_7 to X_9 (with self-model **W**-states X_{63} to X_{65}), states X_{11} to X_{13} (with self-model **W**-states X_{54} to X_{56}), X_{24} to X_{26} (with self-model **W**-states X_{57} to X_{59}), X_{28} to X_{30} (with self-model **W**-states X_{75} to X_{77}), X_{32} to X_{34} (with self-model **W**-states X_{66} to X_{68}), and X_{45} to X_{53} (with self-model **W**-states X_{69} to X_{71}, X_{60} to X_{62}, and X_{72} to X_{74}).

Table 4 Base states of the computational agent model

State	Name	Explanation
X_1	$ws_{s,A}$	World state for stimulus *s* for *A*
X_2	$ws_{s,B}$	World state for stimulus *s* for *B*
X_3	$ws_{s,AB}$	World state for stimulus *s* both for *A* and *B*
X_4	$con_{th,A}$	Context state for excitability threshold for *A*
X_5	$con_{th,B}$	Context state for excitability threshold for *B*
X_6	$sense_{s,A}$	Sensor state of *A* for stimulus *s* for *A*
X_7	$sense_{B,m,A}$	Sensor state of *A* for movement *m* of *B*
X_8	$sense_{B,b,A}$	Sensor state of *A* for expressed affective response *b* of *B*
X_9	$sense_{B,v,A}$	Sensor state of *A* for verbal action *v* of *B*
X_{10}	$rep_{s,A}$	Sensory representation state of *A* for stimulus *s* for *A*
X_{11}	$rep_{B,m,A}$	Sensory representation state of *A* for movement *m* of *B*
X_{12}	$rep_{B,b,A}$	Sensory representation state of *A* for expressed affective response *b* of *B*
X_{13}	$rep_{B,v,A}$	Sensory representation state of *A* for verbal action *v* of *B*
X_{14}	$prep_{m,A}$	Preparation state for movement *m* of *A*
X_{15}	$prep_{b,A}$	Preparation state for affective response *b* of *A*
X_{16}	$prep_{v,A}$	Preparation state for verbal action *v* of *A*
X_{17}	$cons\_emotion_{b,A}$	Conscious emotion state for *b* of *A*
X_{18}	$intrasyncdet_{A,\text{b-m}}$	Intrapersonal synchrony detection of *A* for executing *b* and *m* by *A*
X_{19}	$intrasyncdet_{A,\text{m-v}}$	Intrapersonal synchrony detection of *A* for executing *m* and *v* by *A*
X_{20}	$intrasyncdet_{A,\text{b-v}}$	Intrapersonal synchrony detection of *A* for executing *b* and *v* by *A*
X_{21}	$intersyncdet_{B,A,\text{m}}$	Interpersonal synchrony detection of *A* for executing *m* by *B* and *A*
X_{22}	$intersyncdet_{B,A,\text{b}}$	Interpersonal synchrony detection of *A* for executing *b* by *B* and *A*
X_{23}	$intersyncdet_{B,A,\text{v}}$	Interpersonal synchrony detection of *A* for executing *v* by *B* and *A*
X_{24}	$move_{m,A}$	Executing movement *m* by *A*
X_{25}	$exp\_affect_{b,A}$	Executing expression of *b* by *A*
X_{26}	$talk_{A,B,v}$	Executing verbal action *v* by *A*
X_{27}	$sense_{s,B}$	Sensor state of *B* for stimulus *s* for *B*
X_{28}	$sense_{A,m,B}$	Sensor state of *B* for movement *m* of *A*
X_{29}	$sense_{A,b,B}$	Sensor state of *B* for expressed affective response *b* of *A*
X_{30}	$sense_{A,v,B}$	Sensor state of *B* for verbal action *v* of *A*
X_{31}	$rep_{s,B}$	Sensory representation state of *B* for stimulus *s* for *B*
X_{32}	$rep_{A,m,B}$	Sensory representation state of *B* for movement *m* of *A*
X_{33}	$rep_{A,b,B}$	Sensory representation state of *B* for expressed affective response *b* of *A*
X_{34}	$rep_{A,v,B}$	Sensory representation state of *B* for verbal action *v* of *A*
X_{35}	$prep_{m,B}$	Preparation state for movement *m* of *B*
X_{36}	$prep_{b,B}$	Preparation state for affective response *b* of *B*
X_{37}	$prep_{v,B}$	Preparation state for verbal action *v* of *B*
X_{38}	$cons\_emotion_{b,B}$	Conscious emotion state for *b* of *B*
X_{39}	$intrasyncdet_{B,\text{b-m}}$	Intrapersonal synchrony detection of *B* for executing *b* and *m* by *B*
X_{40}	$intrasyncdet_{B,\text{m-v}}$	Intrapersonal synchrony detection of *B* for executing *m* and *v* by *B*
X_{41}	$intrasyncdet_{B,\text{b-v}}$	Intrapersonal synchrony detection of *B* for executing *b* and *v* by *B*
X_{42}	$intersyncdet_{A,B,\text{m}}$	Interpersonal synchrony detection of *B* for executing *m* by *A* and *B*
X_{43}	$intersyncdet_{A,\text{B},\text{b}}$	Interpersonal synchrony detection of *B* for executing *b* by *A* and *B*
X_{44}	$intersyncdet_{A,\text{B},\text{v}}$	Interpersonal synchrony detection of *B* for executing *v* by *A* and *B*
X_{45}	$move_{m,B}$	Executing movement *m* by *B*
X_{46}	$exp\_affect_{b,B}$	Executing expression of *b* by *B*
X_{47}	$talk_{B,A,v}$	Executing verbal action *v* by *B*
X_{48}	$ws_{m,A,B}$	World state for transmitting movement *m* of *A* to *B*
X_{49}	$ws_{b,A,B}$	World state for transmitting affective response *b* of *A* to *B*
X_{50}	$ws_{v,A,B}$	World state for transmitting verbal action *v* of *A* to *B*
X_{51}	$ws_{m,B,A}$	World state for transmitting movement *m* of *B* to *A*
X_{52}	$ws_{b,B,A}$	World state for transmitting affective response *b* of *B* to *A*
X_{53}	$ws_{v,B,A}$	World state for transmitting verbal action *v* of *B* to *A*

Table 5 First-order self-model **T**-states and **W**-states for excitability thresholds and connection weights and second-order self-model $\mathbf{H_T}$-states and $\mathbf{H_W}$-states for the adaptation speed of the **T**-states and **W**-states of the computational network model

State	Name	Explanation
X_{54}	$\mathbf{W}_{\text{sense-rep\_}m,A}$	First-order self-model state for the weight of *A*'s internal connection from sensing to representing movement *m*
X_{55}	$\mathbf{W}_{\text{sense-rep\_}b,A}$	First-order self-model state for the weight of *A*'s internal connection from sensing to representing affective response *b*
X_{56}	$\mathbf{W}_{\text{sense-rep\_}v,A}$	First-order self-model state for the weight of *A*'s internal connection from sensing to representing verbal action *v*
X_{57}	$\mathbf{W}_{\text{prep-exec\_}m,A}$	First-order self-model state for the weight of *A*'s internal connection from preparing to executing movement *m*
X_{58}	$\mathbf{W}_{\text{prep-exec\_}b,A}$	First-order self-model state for the weight of *A*'s internal connection from preparing to expressing affective response *b*
X_{59}	$\mathbf{W}_{\text{prep-exec\_}v,A}$	First-order self-model state for the weight of *A*'s internal connection from preparing to executing verbal action *v*
X_{60}	$\mathbf{W}_{\text{exec-ws\_}m,A,B}$	First-order self-model state for the weight of *A*'s external connection from executing to world state for movement *m* for *B* (enabling interaction from *A* to *B*)
X_{61}	$\mathbf{W}_{\text{exec-ws\_}b,A,B}$	First-order self-model state for the weight of *A*'s external connection from executing to world state for affective response *b* for *B* (enabling interaction from *A* to *B*)
X_{62}	$\mathbf{W}_{\text{exec-ws\_}v,A,B}$	First-order self-model state for the weight of *A*'s external connection from executing to world state for verbal action *v* for *B* (enabling interaction from *A* to *B*)
X_{63}	$\mathbf{W}_{\text{ws-sense\_}m,B,A}$	First-order self-model state for the weight of *A*'s external connection from world state to *A*'s sensing of movement *m* from *B*
X_{64}	$\mathbf{W}_{\text{ws-sense\_}b,B,A}$	First-order self-model state for the weight of *A*'s external connection from world state to *A*'s sensing of affective response *b* from *B*
X_{65}	$\mathbf{W}_{\text{ws-sense\_}v,B,A}$	First-order self-model state for the weight of *A*'s external connection from world state to *A*'s sensing of verbal action *v* from *B*
X_{66}	$\mathbf{W}_{\text{sense-rep\_}m,B}$	First-order self-model state for the weight of *B*'s internal connection from sensing to representing movement *m*
X_{67}	$\mathbf{W}_{\text{sense-rep\_}b,B}$	First-order self-model state for the weight of *B*'s internal connection from sensing to representing affective response *b*
X_{68}	$\mathbf{W}_{\text{sense-rep\_}v,B}$	First-order self-model state for the weight of *B*'s internal connection from sensing to representing verbal action *v*
X_{69}	$\mathbf{W}_{\text{prep-exec\_}m,B}$	First-order self-model state for the weight of *B*'s internal connection from preparing to executing movement *m*
X_{70}	$\mathbf{W}_{\text{prep-exec\_}b,B}$	First-order self-model state for the weight of *B*'s internal connection from preparing to expressing affective response *b*
X_{71}	$\mathbf{W}_{\text{prep-exec\_}v,B}$	First-order self-model state for the weight of *B*'s internal connection from preparing to executing verbal action *v*

(continued)

Table 5 (continued)

X_{72}	$\mathbf{W}_{\text{exec-ws\_}m,B,A}$	First-order self-model state for the weight of *B*'s external connection from executing to world state for movement *m* for *A* (enabling interaction from *B* to *A*)
X_{73}	$\mathbf{W}_{\text{exec-ws\_}b,B,A}$	First-order self-model state for the weight of *B*'s external connection from executing to world state for affective response *b* for *A* (enabling interaction from *B* to *A*)
X_{74}	$\mathbf{W}_{\text{exec-ws\_}v,B,A}$	First-order self-model state for the weight of *B*'s external connection from executing to world state for verbal action *v* for *A* (enabling interaction from *B* to *A*)
X_{75}	$\mathbf{W}_{\text{ws-sense\_}m,A,B}$	First-order self-model state for the weight of *B*'s external connection from world state to *B*'s sensing of movement *m* from *A*
X_{76}	$\mathbf{W}_{\text{ws-sense\_}b,A,B}$	First-order self-model state for the weight of *B*'s external connection from world state to *B*'s sensing of affective response *b* from *A*
X_{77}	$\mathbf{W}_{\text{ws-sense\_}v,A,B}$	First-order self-model state for the weight of *B*'s external connection from world state to *B*'s sensing of verbal action *v* from *A*
X_{78}	$\mathbf{T}_{\text{rep\_}m,A}$	First-order self-model state for the excitability threshold of *A*'s sensory representation state $\text{rep}_{m,\text{A}}$ for movement *m*
X_{79}	$\mathbf{T}_{\text{rep\_}b,A}$	First-order self-model state for the excitability threshold of *A*'s sensory representation state $\text{rep}_{b,\text{A}}$ for affective response *b*
X_{80}	$\mathbf{T}_{\text{rep\_}v,A}$	First-order self-model state for the excitability threshold of *A*'s sensory representation state $\text{rep}_{v,\text{A}}$ for verbal response *v*
X_{81}	$\mathbf{T}_{\text{exec\_}m,A}$	First-order self-model state for the excitability threshold of *A*'s execution state state $\text{move}_{m,\text{A}}$ for movement *m*
X_{82}	$\mathbf{T}_{\text{exec\_}b,A}$	First-order self-model state for the excitability threshold of *A*'s execution state state $\text{exp\_affect}_{b,\text{A}}$ for affective response *b*
X_{83}	$\mathbf{T}_{\text{exec\_}v,A}$	First-order self-model state for the excitability threshold of *A*'s execution state state $\text{talk}_{v,\text{A}}$ for verbal response *v*
X_{84}	$\mathbf{T}_{\text{rep\_}m,B}$	First-order self-model state for the excitability threshold of B's sensory representation state $\text{rep}_{m,B}$ for movement *m*
X_{85}	$\mathbf{T}_{\text{rep\_}b,B}$	First-order self-model state for the excitability threshold of *B*'s sensory representation state $\text{rep}_{b,\text{B}}$ for affective response *b*
X_{86}	$\mathbf{T}_{\text{rep\_}v,B}$	First-order self-model state for the excitability threshold of *B*'s sensory representation state $\text{rep}_{v,\text{B}}$ for verbal response *v*
X_{87}	$\mathbf{T}_{\text{exec\_}m,B}$	First-order self-model state for the excitability threshold of *B*'s execution state state $\text{move}_{m,\text{B}}$ for movement *m*
X_{88}	$\mathbf{T}_{\text{exec\_}b,B}$	First-order self-model state for the excitability threshold of *B*'s execution state state $\text{exp\_affect}_{b,\text{B}}$ for affective response *b*
X_{89}	$\mathbf{T}_{\text{exec\_}v,B}$	First-order self-model state for the excitability threshold of *B*'s execution state state $\text{talk}_{v,\text{B}}$ for verbal response *v*
X_{90}	$\mathbf{H}_{\mathbf{W}_A}$	Second-order self-model state for the speed factor of the first-order self-model **W**-states for *A*
X_{91}	$\mathbf{H}_{\mathbf{W}_B}$	Second-order self-model state for the speed factor of the first-order self-model **W**-states for *B*
X_{92}	$\mathbf{H}_{\mathbf{T}_A}$	Second-order self-model state for the speed factor of the first-order self-model **T**-states for *A*
X_{93}	$\mathbf{H}_{\mathbf{T}_B}$	Second-order self-model state for the speed factor of the first-order self-model **T**-states for *B*

Table 6 Role matrix **mb** for base connectivity

mb **base connectivity**	1	2	3	4	5	6	7
X_1 ws$_{s,A}$	X_1						
X_2 ws$_{s,B}$	X_2						
X_3 ws$_{s,AB}$	X_3						
X_4 con$_{th,A}$	X_4						
X_5 con$_{th,B}$	X_5						
X_6 sense$_{s,A}$	X_3	X_1					
X_7 sense$_{B,m,A}$	X_{24}	X_{51}					
X_8 sense$_{B,b,A}$	X_{25}	X_{52}					
X_9 sense$_{B,v,A}$	X_{26}	X_{53}					
X_{10} rep$_{s,A}$	X_6						
X_{11} rep$_{B,m,A}$	X_7	X_{14}					
X_{12} rep$_{B,b,A}$	X_8	X_{15}					
X_{13} rep$_{B,v,A}$	X_9	X_{16}					
X_{14} prep$_{m,A}$	X_{10}	X_{11}					
X_{15} prep$_{b,A}$	X_{10}	X_{12}					
X_{16} prep$_{v,A}$	X_{10}	X_{13}	X_{17}				
X_{17} cons_emotion$_{b,A}$	X_{15}	X_{26}	X_{18}	X_{19}	X_{20}		
X_{18} intrasyncdet$_{A,\text{b-m}}$	X_{24}	X_{25}					
X_{19} intrasyncdet$_{A,\text{m-v}}$	X_{24}	X_{26}					
X_{20} intrasyncdet$_{A,\text{b-v}}$	X_{25}	X_{26}					
X_{21} intersyncdet$_{B,A,\text{m}}$	X_{24}	X_7					
X_{22} intersyncdet$_{B,A,\text{b}}$	X_{25}	X_8					
X_{23} intersyncdet$_{B,A,\text{v}}$	X_{26}	X_9					
X_{24} move$_{m,A}$	X_{14}						
X_{25} exp_affect$_{b,A}$	X_{15}						
X_{26} talk$_{A,B,v}$	X_{16}						
X_{27} sense$_{s,B}$	X_3	X_2					
X_{28} sense$_{A,m,B}$	X_{45}	X_{48}					
X_{29} sense$_{A,b,B}$	X_{46}	X_{49}					
X_{30} sense$_{A,v,B}$	X_{47}	X_{50}					
X_{31} rep$_{s,B}$	X_{27}						
X_{32} rep$_{A,m,B}$	X_{28}	X_{35}					
X_{33} rep$_{A,b,B}$	X_{29}	X_{36}					
X_{34} rep$_{A,v,B}$	X_{30}	X_{37}					
X_{35} prep$_{m,B}$	X_{31}	X_{32}					
X_{36} prep$_{b,B}$	X_{31}	X_{33}					
X_{37} prep$_{v,B}$	X_{31}	X_{34}	X_{38}				
X_{38} cons_emotion$_{b,B}$	X_{36}	X_{47}	X_{39}	X_{40}	X_{41}		
X_{39} intrasyncdet$_{B,\text{b-m}}$	X_{45}	X_{46}					
X_{40} intrasyncdet$_{B,\text{m-v}}$	X_{45}	X_{47}					
X_{41} intrasyncdet$_{B,\text{b-v}}$	X_{46}	X_{47}					
X_{42} intersyncdet$_{A,B,\text{m}}$	X_{45}	X_{28}					
X_{43} intersyncdet$_{A,\text{B,b}}$	X_{46}	X_{29}					
X_{44} intersyncdet$_{A,\text{B,v}}$	X_{47}	X_{30}					
X_{45} move$_{m,B}$	X_{35}						
X_{46} exp_affect$_{b,B}$	X_{36}						
X_{47} talk$_{B,A,v}$	X_{37}						
X_{48} ws$_{m,A,B}$	X_{24}						
X_{49} ws$_{b,A,B}$	X_{25}						
X_{50} ws$_{v,A,B}$	X_{26}						
X_{51} ws$_{m,B,A}$	X_{45}						
X_{52} ws$_{b,B,A}$	X_{46}						
X_{53} ws$_{v,B,A}$	X_{47}						

(continued)

Table 6 (continued)

X_{54} $\mathbf{W}_{\text{sense-rep\_}m,A}$	X_{18}	X_{19}	X_{20}	X_{21}	X_{22}	X_{23}		
X_{55} $\mathbf{W}_{\text{sense-rep\_}b,A}$	X_{18}	X_{19}	X_{20}	X_{21}	X_{22}	X_{23}		
X_{56} $\mathbf{W}_{\text{sense-rep\_}v,A}$	X_{18}	X_{19}	X_{20}	X_{21}	X_{22}	X_{23}		
X_{57} $\mathbf{W}_{\text{prep-exec\_}m,A}$	X_{18}	X_{19}	X_{20}	X_{21}	X_{22}	X_{23}		
X_{58} $\mathbf{W}_{\text{prep-exec\_}b,A}$	X_{18}	X_{19}	X_{20}	X_{21}	X_{22}	X_{23}		
X_{59} $\mathbf{W}_{\text{prep-exec\_}v,A}$	X_{18}	X_{19}	X_{20}	X_{21}	X_{22}	X_{23}		
X_{60} $\mathbf{W}_{\text{exec-ws\_}m,A,B}$	X_{60}							
X_{61} $\mathbf{W}_{\text{exec-ws\_}b,A,B}$	X_{61}							
X_{62} $\mathbf{W}_{\text{exec-ws\_}v,A,B}$	X_{62}							
X_{63} $\mathbf{W}_{\text{ws-sense\_}m,B,A}$	X_{63}							
X_{64} $\mathbf{W}_{\text{ws-sense\_}b,B,A}$	X_{64}							
X_{65} $\mathbf{W}_{\text{ws-sense\_}v,B,A}$	X_{65}							
X_{66} $\mathbf{W}_{\text{sense-rep\_}m,B}$	X_{39}	X_{40}	X_{41}	X_{42}	X_{43}	X_{44}		
X_{67} $\mathbf{W}_{\text{sense-rep\_}b,B}$	X_{39}	X_{40}	X_{41}	X_{42}	X_{43}	X_{44}		
X_{68} $\mathbf{W}_{\text{sense-rep\_}v,B}$	X_{39}	X_{40}	X_{41}	X_{42}	X_{43}	X_{44}		
X_{69} $\mathbf{W}_{\text{prep-exec\_}m,B}$	X_{39}	X_{40}	X_{41}	X_{42}	X_{43}	X_{44}		
X_{70} $\mathbf{W}_{\text{prep-exec\_}b,B}$	X_{39}	X_{40}	X_{41}	X_{42}	X_{43}	X_{44}		
X_{71} $\mathbf{W}_{\text{prep-exec\_}v,B}$	X_{39}	X_{40}	X_{41}	X_{42}	X_{43}	X_{44}		
X_{72} $\mathbf{W}_{\text{exec-ws\_}m,B,A}$	X_{72}							
X_{73} $\mathbf{W}_{\text{exec-ws\_}b,B,A}$	X_{73}							
X_{74} $\mathbf{W}_{\text{exec-ws\_}v,B,A}$	X_{74}							
X_{75} $\mathbf{W}_{\text{ws-sense\_}m,A,B}$	X_{75}							
X_{76} $\mathbf{W}_{\text{ws-sense\_}b,A,B}$	X_{76}							
X_{77} $\mathbf{W}_{\text{ws-sense\_}v,A,B}$	X_{77}							
X_{78} $\mathbf{T}_{\text{rep\_}m,A}$	X_{18}	X_{19}	X_{20}	X_{21}	X_{22}	X_{23}		X_4
X_{79} $\mathbf{T}_{\text{rep\_}b,A}$	X_{18}	X_{19}	X_{20}	X_{21}	X_{22}	X_{23}		X_4
X_{80} $\mathbf{T}_{\text{rep\_}v,A}$	X_{18}	X_{19}	X_{20}	X_{21}	X_{22}	X_{23}		X_4
X_{81} $\mathbf{T}_{\text{exec\_}m,A}$	X_{18}	X_{19}	X_{20}	X_{21}	X_{22}	X_{23}		X_4
X_{82} $\mathbf{T}_{\text{exec\_}b,A}$	X_{18}	X_{19}	X_{20}	X_{21}	X_{22}	X_{23}		X_4
X_{83} $\mathbf{T}_{\text{exec\_}v,A}$	X_{18}	X_{19}	X_{20}	X_{21}	X_{22}	X_{23}		X_4
X_{84} $\mathbf{T}_{\text{rep\_}m,B}$	X_{39}	X_{40}	X_{41}	X_{42}	X_{43}	X_{44}		X_5
X_{85} $\mathbf{T}_{\text{rep\_}b,B}$	X_{39}	X_{40}	X_{41}	X_{42}	X_{43}	X_{44}		X_5
X_{86} $\mathbf{T}_{\text{rep\_}v,B}$	X_{39}	X_{40}	X_{41}	X_{42}	X_{43}	X_{44}		X_5
X_{87} $\mathbf{T}_{\text{exec\_}m,B}$	X_{39}	X_{40}	X_{41}	X_{42}	X_{43}	X_{44}		X_5
X_{88} $\mathbf{T}_{\text{exec\_}b,B}$	X_{39}	X_{40}	X_{41}	X_{42}	X_{43}	X_{44}		X_5
X_{89} $\mathbf{T}_{\text{exec\_}v,B}$	X_{39}	X_{40}	X_{41}	X_{42}	X_{43}	X_{44}		X_5
X_{90} $\mathbf{H}_{\mathbf{W}_A}$	X_{10}							
X_{91} $\mathbf{H}_{\mathbf{W}_B}$	X_{31}							
X_{92} $\mathbf{H}_{\mathbf{T}_A}$	X_{10}							
X_{93} $\mathbf{H}_{\mathbf{T}_B}$	X_{31}							

The network characteristics for aggregation are defined by the selection of combination functions from the library and values for their parameters. In role matrix **mcfw** it is specified by weights which state uses which combination function; see Table 8.

In role matrix **mcfp** (see Table 9) it is indicated what the parameter values are for the chosen combination functions. A number of them are adaptive: their adaptive excitability thresholds are represented by self-model **T**-states. These concern for agent A states X_{11} to X_{13} (with excitability threshold self-model **T**-states X_{78} to X_{80}) and X_{24} to X_{26} (with self-model **T**-states X_{81} to X_{83}), and for agent B states X_{32} to X_{34} (with excitability threshold self-model **T**-states X_{84} to X_{86}) and X_{35} to X_{47} (with self-model **T**-states X_{87} to X_{89}).

Table 7 Role matrix **mcw** for connection weights

mcw	connection weights	1	2	3	4	5	6	7	8
X_1	$ws_{s,A}$	1							
X_2	$ws_{s,B}$	1							
X_3	$ws_{s,AB}$	1							
X_4	$con_{th,A}$	1							
X_5	$con_{th,B}$	1							
X_6	$sense_{s,A}$	1	1						
X_7	$sense_{B,m,A}$	1	X_{63}						
X_8	$sense_{B,b,A}$	1	X_{64}						
X_9	$sense_{B,v,A}$	1	X_{65}						
X_{10}	$rep_{s,A}$	1							
X_{11}	$rep_{B,m,A}$	X_{54}	1						
X_{12}	$rep_{B,b,A}$	X_{55}	1						
X_{13}	$rep_{B,v,A}$	X_{56}	1						
X_{14}	$prep_{m,A}$	1	1						
X_{15}	$prep_{b,A}$	1	1						
X_{16}	$prep_{v,A}$	1	1	1					
X_{17}	$cons\_emotion_{b,A}$	1	1	1	1	1			
X_{18}	$intrasyncdet_{A,\text{b-m}}$	1	1						
X_{19}	$intrasyncdet_{A,\text{m-v}}$	1	1						
X_{20}	$intrasyncdet_{A,\text{b-v}}$	1	1						
X_{21}	$intersyncdet_{B,A,\text{m}}$	1	1						
X_{22}	$intersyncdet_{B,A,\text{b}}$	1	1						
X_{23}	$intersyncdet_{B,A,\text{v}}$	1	1						
X_{24}	$move_{m,A}$	X_{57}							
X_{25}	$exp\_affect_{b,A}$	X_{58}							
X_{26}	$talk_{A,B,v}$	X_{59}							
X_{27}	$sense_{s,B}$	1	1						
X_{28}	$sense_{A,m,B}$	1	X_{75}						
X_{29}	$sense_{A,b,B}$	1	X_{76}						
X_{30}	$sense_{A,v,B}$	1	X_{77}						
X_{31}	$rep_{s,B}$	1							
X_{32}	$rep_{A,m,B}$	X_{66}	1						
X_{33}	$rep_{A,b,B}$	X_{67}	1						
X_{34}	$rep_{A,v,B}$	X_{68}	1						
X_{35}	$prep_{m,B}$	1	1						
X_{36}	$prep_{b,B}$	1	1						
X_{37}	$prep_{v,B}$	1	1	1					
X_{38}	$cons\_emotion_{b,B}$	1	1	1	1	1			
X_{39}	$intrasyncdet_{B,\text{b-m}}$	1	1						
X_{40}	$intrasyncdet_{B,\text{m-v}}$	1	1						
X_{41}	$intrasyncdet_{B,\text{b-v}}$	1	1						
X_{42}	$intersyncdet_{A,B,\text{m}}$	1	1						
X_{43}	$intersyncdet_{A,\text{B},\text{b}}$	1	1						
X_{44}	$intersyncdet_{A,\text{B},\text{v}}$	1	1						
X_{45}	$move_{m,B}$	X_{69}							
X_{46}	$exp\_affect_{b,B}$	X_{70}							
X_{47}	$talk_{B,A,v}$	X_{71}							

(continued)

Table 7 (continued)

X_{48}	$\text{ws}_{m,A,B}$	X_{60}							
X_{49}	$\text{ws}_{b,A,B}$	X_{61}							
X_{50}	$\text{ws}_{v,A,B}$	X_{62}							
X_{51}	$\text{ws}_{m,B,A}$	X_{72}							
X_{52}	$\text{ws}_{b,B,A}$	X_{73}							
X_{53}	$\text{ws}_{v,B,A}$	X_{74}							
X_{54}	$\mathbf{W}_{\text{sense-rep\_}m,A}$	1	1	1	1	1	1		
X_{55}	$\mathbf{W}_{\text{sense-rep\_}b,A}$	1	1	1	1	1	1		
X_{56}	$\mathbf{W}_{\text{sense-rep\_}v,A}$	1	1	1	1	1	1		
X_{57}	$\mathbf{W}_{\text{prep-exec\_}m,A}$	1	1	1	1	1	1		
X_{58}	$\mathbf{W}_{\text{prep-exec\_}b,A}$	1	1	1	1	1	1		
X_{59}	$\mathbf{W}_{\text{prep-exec\_}v,A}$	1	1	1	1	1	1		
X_{60}	$\mathbf{W}_{\text{exec-ws\_}m,A,B}$	1							
X_{61}	$\mathbf{W}_{\text{exec-ws\_}b,A,B}$	1							
X_{62}	$\mathbf{W}_{\text{exec-ws\_}v,A,B}$	1							
X_{63}	$\mathbf{W}_{\text{ws-sense\_}m,B,A}$	1							
X_{64}	$\mathbf{W}_{\text{ws-sense\_}b,B,A}$	1							
X_{65}	$\mathbf{W}_{\text{ws-sense\_}v,B,A}$	1							
X_{66}	$\mathbf{W}_{\text{sense-rep\_}m,B}$	1	1	1	1	1	1		
X_{67}	$\mathbf{W}_{\text{sense-rep\_}b,B}$	1	1	1	1	1	1		
X_{68}	$\mathbf{W}_{\text{sense-rep\_}v,B}$	1	1	1	1	1	1		
X_{69}	$\mathbf{W}_{\text{prep-exec\_}m,B}$	1	1	1	1	1	1		
X_{70}	$\mathbf{W}_{\text{prep-exec\_}b,B}$	1	1	1	1	1	1		
X_{71}	$\mathbf{W}_{\text{prep-exec\_}v,B}$	1	1	1	1	1	1		
X_{72}	$\mathbf{W}_{\text{exec-ws\_}m,B,A}$	1							
X_{73}	$\mathbf{W}_{\text{exec-ws\_}b,B,A}$	1							
X_{74}	$\mathbf{W}_{\text{exec-ws\_}v,B,A}$	1							
X_{75}	$\mathbf{W}_{\text{ws-sense\_}m,A,B}$	1							
X_{76}	$\mathbf{W}_{\text{ws-sense\_}b,A,B}$	1							
X_{77}	$\mathbf{W}_{\text{ws-sense\_}v,A,B}$	1							
X_{78}	$\mathbf{T}_{\text{rep\_}m,A}$	-0.12	-0.12	-0.12	-0.12	-0.12	-0.12		1
X_{79}	$\mathbf{T}_{\text{rep\_}b,A}$	-0.12	-0.12	-0.12	-0.12	-0.12	-0.12		1
X_{80}	$\mathbf{T}_{\text{rep\_}v,A}$	-0.12	-0.12	-0.12	-0.12	-0.12	-0.12		1
X_{81}	$\mathbf{T}_{\text{exec\_}m,A}$	-0.12	-0.12	-0.12	-0.12	-0.12	-0.12		1
X_{82}	$\mathbf{T}_{\text{exec\_}b,A}$	-0.12	-0.12	-0.12	-0.12	-0.12	-0.12		1
X_{83}	$\mathbf{T}_{\text{exec\_}v,A}$	-0.12	-0.12	-0.12	-0.12	-0.12	-0.12		1
X_{84}	$\mathbf{T}_{\text{rep\_}m,B}$	-0.12	-0.12	-0.12	-0.12	-0.12	-0.12		1
X_{85}	$\mathbf{T}_{\text{rep\_}b,B}$	-0.12	-0.12	-0.12	-0.12	-0.12	-0.12		1
X_{86}	$\mathbf{T}_{\text{rep\_}v,B}$	-0.12	-0.12	-0.12	-0.12	-0.12	-0.12		1
X_{87}	$\mathbf{T}_{\text{exec\_}m,B}$	-0.12	-0.12	-0.12	-0.12	-0.12	-0.12		1
X_{88}	$\mathbf{T}_{\text{exec\_}b,B}$	-0.12	-0.12	-0.12	-0.12	-0.12	-0.12		1
X_{89}	$\mathbf{T}_{\text{exec\_}v,B}$	-0.12	-0.12	-0.12	-0.12	-0.12	-0.12		1
X_{90}	$\mathbf{H}\mathbf{w}_A$	0.04							
X_{91}	$\mathbf{H}\mathbf{w}_B$	0.04							
X_{92}	$\mathbf{H}\mathbf{T}_A$	1							
X_{93}	$\mathbf{H}\mathbf{T}_B$	1							

In Table 10, the role matrix **ms** for speed factors is shown, which lists all speed factors. Next to it, the list of initial values can be found. Also, for **ms** some entries are adaptive. For agent A the speed factors of **W**-states X_{54} to X_{65} are represented by (second-order) self-model $\mathbf{H_W}$-state X_{90} and for agent B the speed factors of **W**-states X_{66} to X_{77} are represented by (second-order) self-model $\mathbf{H_W}$-state X_{91}.

Table 8 Role matrix **mcfw** for combination function weights

mcfw combination function weights		**1 alo-gistic**	**2 comp-diff**	**3 stepmod**	**4 stepmodopp**	**5 eucl**
X_1	ws$_{s,A}$				1	
X_2	ws$_{s,B}$			1		
X_3	ws$_{s,AB}$			1		
X_4	con$_{th,A}$	1				
X_5	con$_{th,B}$	1				
X_6	sense$_{s,A}$					1
X_7	sense$_{B,m,A}$					1
X_8	sense$_{B,b,A}$					1
X_9	sense$_{B,v,A}$					1
X_{10}	rep$_{s,A}$	1				
X_{11}	rep$_{B,m,A}$	1				
X_{12}	rep$_{B,b,A}$	1				
X_{13}	rep$_{B,v,A}$	1				
X_{14}	prep$_{m,A}$	1				
X_{15}	prep$_{b,A}$	1				
X_{16}	prep$_{v,A}$	1				
X_{17}	cons_emotion$_{b,A}$	1				
X_{18}	intrasyncdet$_{A,\text{b-m}}$		1			
X_{19}	intrasyncdet$_{A,\text{m-v}}$		1			
X_{20}	intrasyncdet$_{A,\text{b-v}}$		1			
X_{21}	intersyncdet$_{B,A,\text{m}}$		1			
X_{22}	intersyncdet$_{B,A,\text{b}}$		1			
X_{23}	intersyncdet$_{B,A,\text{v}}$		1			
X_{24}	move$_{m,A}$	1				
X_{25}	exp_affect$_{b,A}$	1				
X_{26}	talk$_{A,B,v}$	1				
X_{27}	sense$_{s,B}$					1
X_{28}	sense$_{A,m,B}$					1
X_{29}	sense$_{A,b,B}$					1
X_{30}	sense$_{A,v,B}$					1
X_{31}	rep$_{s,B}$	1				
X_{32}	rep$_{A,m,B}$	1				
X_{33}	rep$_{A,b,B}$	1				
X_{34}	rep$_{A,v,B}$	1				
X_{35}	prep$_{m,B}$	1				
X_{36}	prep$_{b,B}$	1				
X_{37}	prep$_{v,B}$	1				
X_{38}	cons_emotion$_{b,B}$	1				
X_{39}	intrasyncdet$_{B,\text{b-m}}$		1			
X_{40}	intrasyncdet$_{B,\text{m-v}}$		1			
X_{41}	intrasyncdet$_{B,\text{b-v}}$		1			
X_{42}	intersyncdet$_{A,B,\text{m}}$		1			
X_{43}	intersyncdet$_{A,\text{B},\text{b}}$		1			
X_{44}	intersyncdet$_{A,\text{B},\text{v}}$		1			

(continued)

Table 8 (continued)

X_{45}	$move_{m,B}$	1				
X_{46}	$exp\_affect_{b,B}$	1				
X_{47}	$talk_{B,A,v}$	1				
X_{48}	$ws_{m,A,B}$					1
X_{49}	$ws_{b,A,B}$					1
X_{50}	$ws_{v,A,B}$					1
X_{51}	$ws_{m,B,A}$					1
X_{52}	$ws_{b,B,A}$					1
X_{53}	$ws_{v,B,A}$					1
X_{54}	$\mathbf{W}_{sense\text{-}rep\_m,A}$	1				
X_{55}	$\mathbf{W}_{sense\text{-}rep\_b,A}$	1				
X_{56}	$\mathbf{W}_{sense\text{-}rep\_v,A}$	1				
X_{57}	$\mathbf{W}_{prep\text{-}exec\_m,A}$	1				
X_{58}	$\mathbf{W}_{prep\text{-}exec\_b,A}$	1				
X_{59}	$\mathbf{W}_{prep\text{-}exec\_v,A}$	1				
X_{60}	$\mathbf{W}_{exec\text{-}ws\_m,A,B}$			1		
X_{61}	$\mathbf{W}_{exec\text{-}ws\_b,A,B}$			1		
X_{62}	$\mathbf{W}_{exec\text{-}ws\_v,A,B}$			1		
X_{63}	$\mathbf{W}_{ws\text{-}sense\_m,B,A}$	1				
X_{64}	$\mathbf{W}_{ws\text{-}sense\_b,B,A}$	1				
X_{65}	$\mathbf{W}_{ws\text{-}sense\_v,B,A}$	1				
X_{66}	$\mathbf{W}_{sense\text{-}rep\_m,B}$	1				
X_{67}	$\mathbf{W}_{sense\text{-}rep\_b,B}$	1				
X_{68}	$\mathbf{W}_{sense\text{-}rep\_v,B}$	1				
X_{69}	$\mathbf{W}_{prep\text{-}exec\_m,B}$	1				
X_{70}	$\mathbf{W}_{prep\text{-}exec\_b,B}$	1				
X_{71}	$\mathbf{W}_{prep\text{-}exec\_v,B}$	1				
X_{72}	$\mathbf{W}_{exec\text{-}ws\_m,B,A}$			1		
X_{73}	$\mathbf{W}_{exec\text{-}ws\_b,B,A}$			1		
X_{74}	$\mathbf{W}_{exec\text{-}ws\_v,B,A}$			1		
X_{75}	$\mathbf{W}_{ws\text{-}sense\_m,A,B}$	1				
X_{76}	$\mathbf{W}_{ws\text{-}sense\_b,A,B}$	1				
X_{77}	$\mathbf{W}_{ws\text{-}sense\_v,A,B}$	1				
X_{78}	$\mathbf{T}_{rep\_m,A}$	1				
X_{79}	$\mathbf{T}_{rep\_b,A}$	1				
X_{80}	$\mathbf{T}_{rep\_v,A}$	1				
X_{81}	$\mathbf{T}_{exec\_m,A}$	1				
X_{82}	$\mathbf{T}_{exec\_b,A}$	1				
X_{83}	$\mathbf{T}_{exec\_v,A}$	1				
X_{84}	$\mathbf{T}_{rep\_m,B}$	1				
X_{85}	$\mathbf{T}_{rep\_b,B}$	1				
X_{86}	$\mathbf{T}_{rep\_v,B}$	1				
X_{87}	$\mathbf{T}_{exec\_m,B}$	1				
X_{88}	$\mathbf{T}_{exec\_b,B}$	1				
X_{89}	$\mathbf{T}_{exec\_v,B}$	1				
X_{90}	$\mathbf{H}_{\mathbf{W}_A}$	1				
X_{91}	$\mathbf{H}_{\mathbf{W}_B}$	1				
X_{92}	$\mathbf{H}_{\mathbf{T}_A}$	1				
X_{93}	$\mathbf{H}_{\mathbf{T}_B}$	1				

Table 9 Role matrix **mcfp** for combination function parameters

mcfp combination function parameters		**1 alogistic**		**2 compdiff***		**3 stepmod**		**4 stepmodopp**		**5 eucl**	
X_1	ws$_{s,A}$							180	120		
X_2	ws$_{s,B}$					240	60				
X_3	ws$_{s,AB}$										
X_4	con$_{th,A}$	5	0.5								
X_5	con$_{th,B}$	5	0.5								
X_6	sense$_{s,A}$									1	1
X_7	sense$_{B,m,A}$									1	2
X_8	sense$_{B,b,A}$									1	2
X_9	sense$_{B,v,A}$									1	2
X_{10}	rep$_{s,A}$	5	0.5								
X_{11}	rep$_{B,m,A}$	5	X_{78}								
X_{12}	rep$_{B,b,A}$	5	X_{79}								
X_{13}	rep$_{B,v,A}$	5	X_{80}								
X_{14}	prep$_{m,A}$	5	0.8								
X_{15}	prep$_{b,A}$	5	0.7								
X_{16}	prep$_{v,A}$	5	0.75								
X_{17}	cons_emotion$_{b,A}$	5	3.25								
X_{18}	intrasyncdet$_{A,\text{b-m}}$			2	1						
X_{19}	intrasyncdet$_{A,\text{m-v}}$			3	1						
X_{20}	intrasyncdet$_{A,\text{b-v}}$			4	1						
X_{21}	intersyncdet$_{B,A,\text{m}}$			5	5						
X_{22}	intersyncdet$_{B,A,\text{b}}$			6	5						
X_{23}	intersyncdet$_{B,A,\text{v}}$			7	5						
X_{24}	move$_{m,A}$	5	X_{81}								
X_{25}	exp_affect$_{b,A}$	5	X_{82}								
X_{26}	talk$_{A,B,v}$	5	X_{83}								
X_{27}	sense$_{s,B}$									1	1
X_{28}	sense$_{A,m,B}$									1	2
X_{29}	sense$_{A,b,B}$									1	2
X_{30}	sense$_{A,v,B}$									1	2
X_{31}	rep$_{s,B}$	5	0.5								
X_{32}	rep$_{A,m,B}$	5	X_{84}								
X_{33}	rep$_{A,b,B}$	5	X_{85}								
X_{34}	rep$_{A,v,B}$	5	X_{86}								
X_{35}	prep$_{m,B}$	5	0.6								
X_{36}	prep$_{b,B}$	5	0.75								
X_{37}	prep$_{v,B}$	5	1.2								
X_{38}	cons_emotion$_{b,B}$	5	3.25								
X_{39}	intrasyncdet$_{B,\text{b-m}}$			8	1						
X_{40}	intrasyncdet$_{B,\text{m-v}}$			9	1						
X_{41}	intrasyncdet$_{B,\text{b-v}}$			10	1						
X_{42}	intersyncdet$_{A,B,\text{m}}$			11	5						
X_{43}	intersyncdet$_{A,\text{B,b}}$			12	5						
X_{44}	intersyncdet$_{A,\text{B,v}}$			13	5						
X_{45}	move$_{m,B}$	5	X_{87}								
X_{46}	exp_affect$_{b,B}$	5	X_{88}								
X_{47}	talk$_{B,A,v}$	5	X_{89}								
X_{48}	ws$_{m,A,B}$									1	1
X_{49}	ws$_{b,A,B}$									1	1
X_{50}	ws$_{v,A,B}$									1	1
X_{51}	ws$_{m,B,A}$									1	1
X_{52}	ws$_{b,B,A}$									1	1
X_{53}	ws$_{v,B,A}$									1	1

(continued)

Table 9 (continued)

X_{54}	$\mathbf{W}_{\text{sense-rep\_}m,A}$	5	4.5		
X_{55}	$\mathbf{W}_{\text{sense-rep\_}b,A}$	5	4.5		
X_{56}	$\mathbf{W}_{\text{sense-rep\_}v,A}$	5	4.5		
X_{57}	$\mathbf{W}_{\text{prep-exec\_}m,A}$	5	4.5		
X_{58}	$\mathbf{W}_{\text{prep-exec\_}b,A}$	5	4.5		
X_{59}	$\mathbf{W}_{\text{prep-exec\_}v,A}$	5	4.5		
X_{60}	$\mathbf{W}_{\text{exec-ws\_}m,A,B}$			60	30
X_{61}	$\mathbf{W}_{\text{exec-ws\_}b,A,B}$			60	30
X_{62}	$\mathbf{W}_{\text{exec-ws\_}v,A,B}$			60	30
X_{63}	$\mathbf{W}_{\text{ws-sense\_}m,B,A}$	5	4.5		
X_{64}	$\mathbf{W}_{\text{ws-sense\_}b,B,A}$	5	4.5		
X_{65}	$\mathbf{W}_{\text{ws-sense\_}v,B,A}$	5	4.5		
X_{66}	$\mathbf{W}_{\text{sense-rep\_}m,B}$	5	4.5		
X_{67}	$\mathbf{W}_{\text{sense-rep\_}b,B}$	5	4.5		
X_{68}	$\mathbf{W}_{\text{sense-rep\_}v,B}$	5	4.5		
X_{69}	$\mathbf{W}_{\text{prep-exec\_}m,B}$	5	4.5		
X_{70}	$\mathbf{W}_{\text{prep-exec\_}b,B}$	5	4.5		
X_{71}	$\mathbf{W}_{\text{prep-exec\_}v,B}$	5	4.5		
X_{72}	$\mathbf{W}_{\text{exec-ws\_}m,B,A}$			60	30
X_{73}	$\mathbf{W}_{\text{exec-ws\_}b,B,A}$			60	30
X_{74}	$\mathbf{W}_{\text{exec-ws\_}v,B,A}$			60	30
X_{75}	$\mathbf{W}_{\text{ws-sense\_}m,A,B}$	5	4.5		
X_{76}	$\mathbf{W}_{\text{ws-sense\_}b,A,B}$	5	4.5		
X_{77}	$\mathbf{W}_{\text{ws-sense\_}v,A,B}$	5	4.5		
X_{78}	$\mathbf{T}_{\text{rep\_}m,A}$	5	0.5		
X_{79}	$\mathbf{T}_{\text{rep\_}b,A}$	5	0.5		
X_{80}	$\mathbf{T}_{\text{rep\_}v,A}$	5	0.5		
X_{81}	$\mathbf{T}_{\text{exec\_}m,A}$	5	0.5		
X_{82}	$\mathbf{T}_{\text{exec\_}b,A}$	5	0.5		
X_{83}	$\mathbf{T}_{\text{exec\_}v,A}$	5	0.5		
X_{84}	$\mathbf{T}_{\text{rep\_}m,B}$	5	0.5		
X_{85}	$\mathbf{T}_{\text{rep\_}b,B}$	5	0.5		
X_{86}	$\mathbf{T}_{\text{rep\_}v,B}$	5	0.5		
X_{87}	$\mathbf{T}_{\text{exec\_}m,B}$	5	0.5		
X_{88}	$\mathbf{T}_{\text{exec\_}b,B}$	5	0.5		
X_{89}	$\mathbf{T}_{\text{exec\_}v,B}$	5	0.5		
X_{90}	$\mathbf{H}_{\mathbf{W}_A}$	5	0.7		
X_{91}	$\mathbf{H}_{\mathbf{W}_B}$	5	0.7		
X_{92}	$\mathbf{H}_{\mathbf{T}_A}$	5	0.7		
X_{93}	$\mathbf{H}_{\mathbf{T}_B}$	5	0.7		

Moreover, for agent A the speed factors for **T**-states X_{78} to X_{83} are represented by (second-order) self-model $\mathbf{H_T}$-state X_{92} and for agent B the speed factors of **T**-states X_{84} to X_{89} are represented by (second-order) self-model $\mathbf{H_T}$-state X_{93}.

Table 10 Role matrix **ms** for speed factors and **iv** for initial values

ms speed factors		1	iv initial values	1
X_1	$ws_{s,A}$	1		1
X_2	$ws_{s,B}$	1		0
X_3	$ws_{s,AB}$	1		0
X_4	$con_{th,A}$	0		1
X_5	$con_{th,B}$	0		1
X_6	$sense_{s,A}$	1		0
X_7	$sense_{B,m,A}$	1		0
X_8	$sense_{B,b,A}$	1		0
X_9	$sense_{B,v,A}$	1		0
X_{10}	$rep_{s,A}$	1		0
X_{11}	$rep_{B,m,A}$	1		0
X_{12}	$rep_{B,b,A}$	1		0
X_{13}	$rep_{B,v,A}$	1		0
X_{14}	$prep_{m,A}$	1		0
X_{15}	$prep_{b,A}$	1		0
X_{16}	$prep_{v,A}$	1		0
X_{17}	$cons\_emotion_{b,A}$	1		0
X_{18}	$intrasyncdet_{A,\text{b-m}}$	0.5		0
X_{19}	$intrasyncdet_{A,\text{m-v}}$	0.5		0
X_{20}	$intrasyncdet_{A,\text{b-v}}$	0.5		0
X_{21}	$intersyncdet_{B,A,\text{m}}$	0.5		0
X_{22}	$intersyncdet_{B,A,\text{b}}$	0.5		0
X_{23}	$intersyncdet_{B,A,\text{v}}$	0.5		0
X_{24}	$move_{m,A}$	1		0
X_{25}	$exp\_affect_{b,A}$	1		0
X_{26}	$talk_{A,B,v}$	1		0
X_{27}	$sense_{s,B}$	1		0
X_{28}	$sense_{A,m,B}$	1		0
X_{29}	$sense_{A,b,B}$	1		0
X_{30}	$sense_{A,v,B}$	1		0
X_{31}	$rep_{s,B}$	1		0
X_{32}	$rep_{A,m,B}$	1		0
X_{33}	$rep_{A,b,B}$	1		0
X_{34}	$rep_{A,v,B}$	1		0
X_{35}	$prep_{m,B}$	1		0
X_{36}	$prep_{b,B}$	1		0
X_{37}	$prep_{v,B}$	1		0
X_{38}	$cons\_emotion_{b,B}$	1		0
X_{39}	$intrasyncdet_{B,\text{b-m}}$	0.5		0
X_{40}	$intrasyncdet_{B,\text{m-v}}$	0.5		0
X_{41}	$intrasyncdet_{B,\text{b-v}}$	0.5		0
X_{42}	$intersyncdet_{A,B,\text{m}}$	0.5		0
X_{43}	$intersyncdet_{A,\text{B},\text{b}}$	0.5		0
X_{44}	$intersyncdet_{A,\text{B},\text{v}}$	0.5		0
X_{45}	$move_{m,B}$	1		0
X_{46}	$exp\_affect_{b,B}$	1		0
X_{47}	$talk_{B,A,v}$	1		0
X_{48}	$ws_{m,A,B}$	1		0
X_{49}	$ws_{b,A,B}$	1		0
X_{50}	$ws_{v,A,B}$	1		0

(continued)

Table 10 (continued)

X_{51}	$WS_{m,B,A}$	1		0
X_{52}	$WS_{b,B,A}$	1		0
X_{53}	$WS_{v,B,A}$	1		0
X_{54}	$\mathbf{W}_{sense\text{-}rep\_m,A}$	X_{90}		0.4
X_{55}	$\mathbf{W}_{sense\text{-}rep\_b,A}$	X_{90}		0.4
X_{56}	$\mathbf{W}_{sense\text{-}rep\_v,A}$	X_{90}		0.4
X_{57}	$\mathbf{W}_{prep\text{-}exec\_m,A}$	X_{90}		0.4
X_{58}	$\mathbf{W}_{prep\text{-}exec\_b,A}$	X_{90}		0.4
X_{59}	$\mathbf{W}_{prep\text{-}exec\_v,A}$	X_{90}		0.4
X_{60}	$\mathbf{W}_{exec\text{-}ws\_m,A,B}$	1		0
X_{61}	$\mathbf{W}_{exec\text{-}ws\_b,A,B}$	1		0
X_{62}	$\mathbf{W}_{exec\text{-}ws\_v,A,B}$	1		0
X_{63}	$\mathbf{W}_{ws\text{-}sense\_m,B,A}$	0		1
X_{64}	$\mathbf{W}_{ws\text{-}sense\_b,B,A}$	0		1
X_{65}	$\mathbf{W}_{ws\text{-}sense\_v,B,A}$	0		1
X_{66}	$\mathbf{W}_{sense\text{-}rep\_m,B}$	X_{91}		0.4
X_{67}	$\mathbf{W}_{sense\text{-}rep\_b,B}$	X_{91}		0.4
X_{68}	$\mathbf{W}_{sense\text{-}rep\_v,B}$	X_{91}		0.4
X_{69}	$\mathbf{W}_{prep\text{-}exec\_m,B}$	X_{91}		0.4
X_{70}	$\mathbf{W}_{prep\text{-}exec\_b,B}$	X_{91}		0.4
X_{71}	$\mathbf{W}_{prep\text{-}exec\_v,B}$	X_{91}		0.4
X_{72}	$\mathbf{W}_{exec\text{-}ws\_m,B,A}$	1		0
X_{73}	$\mathbf{W}_{exec\text{-}ws\_b,B,A}$	1		0
X_{74}	$\mathbf{W}_{exec\text{-}ws\_v,B,A}$	1		0
X_{75}	$\mathbf{W}_{ws\text{-}sense\_m,A,B}$	0		1
X_{76}	$\mathbf{W}_{ws\text{-}sense\_b,A,B}$	0		1
X_{77}	$\mathbf{W}_{ws\text{-}sense\_v,A,B}$	0		1
X_{78}	$\mathbf{T}_{rep\_m,A}$	X_{92}		0.7
X_{79}	$\mathbf{T}_{rep\_b,A}$	X_{92}		0.7
X_{80}	$\mathbf{T}_{rep\_v,A}$	X_{92}		0.7
X_{81}	$\mathbf{T}_{exec\_m,A}$	X_{92}		0.7
X_{82}	$\mathbf{T}_{exec\_b,A}$	X_{92}		0.7
X_{83}	$\mathbf{T}_{exec\_v,A}$	X_{92}		0.7
X_{84}	$\mathbf{T}_{rep\_m,B}$	X_{93}		0.7
X_{85}	$\mathbf{T}_{rep\_b,B}$	X_{93}		0.7
X_{86}	$\mathbf{T}_{rep\_v,B}$	X_{93}		0.7
X_{87}	$\mathbf{T}_{exec\_m,B}$	X_{93}		0.7
X_{88}	$\mathbf{T}_{exec\_b,B}$	X_{93}		0.7
X_{89}	$\mathbf{T}_{exec\_v,B}$	X_{93}		0.7
X_{90}	$\mathbf{H}\mathbf{w}_{A}$	1		0
X_{91}	$\mathbf{H}\mathbf{w}_{B}$	1		0
X_{92}	$\mathbf{H}\mathbf{T}_{A}$	1		0
X_{93}	$\mathbf{H}\mathbf{T}_{B}$	1		0

References

1. Abraham, W.C., Bear, M.F.: Metaplasticity: the plasticity of synaptic plasticity. Trends Neurosci. **19**(4), 126–130 (1996)
2. Accetto, M., Treur, J., Villa, V.: An adaptive cognitive-social model for mirroring and social bonding during synchronous joint action. Procedia Computer Science 145, 3–12 (2018)
3. Altmann, U.: Investigation of movement synchrony using windowed cross-lagged regression. In: Esposito, A., Vinciarelli, A., Vicsi, K., Pelachaud, C., Nijholt, A. (eds.) Analysis of Verbal

and Nonverbal Communication and Enactment. The Processing Issues. LNCS, vol. 6800, pp. 335–345. Springer, Heidelberg (2011). https://doi.org/10.1007/978-3-642-25775-9_31
4. Behrens, F., Moulder, R.G., Boker, S.M., Kret, M.E.: Quantifying physiological synchrony through windowed cross-correlation analysis: statistical and theoretical considerations. BioRxiv (2020)
5. Boot, N., Baas, M., Van Gaal, S., Cools, R., De Dreu, C.K.W.: Creative cognition and dopaminergic modulation of fronto-striatal networks: integrative review and research agenda. Neurosci. Biobehav. Rev. **78**, 13–23 (2017)
6. Chandra, N., Barkai, E.: A non-synaptic mechanism of complex learning: modulation of intrinsic neuronal excitability. Neurobiol. Learn. Mem. **154**, 30–36 (2018)
7. Damasio, A.R.: The Feeling of What Happens: Body and Emotion in the Making of Consciousness. Houghton Mifflin Harcourt (1999)
8. Debanne, D., Inglebert, Y., Russier, M.: Plasticity of intrinsic neuronal excitability. Curr. Opin. Neurobiol. **54**, 73–82 (2019)
9. Dhamala, M., Assisi, C.G., Jirsa, V.K., Steinberg, F.L., Kelso, J.A.S.: Multisensory integration for timing engages different brain networks. Neuroimage **34**, 764–773 (2007)
10. Fairhurst, M.T., Janata, P., Keller, P.E.: Being and feeling in sync with an adaptive virtual partner: brain mechanisms underlying dynamic cooperativity. Cereb. Cortex **23**(11), 2592–2600 (2013). https://doi.org/10.1093/cercor/bhs243
11. Grandjean, D., Sander, D., Scherer, K.R.: Conscious emotional experience emerges as a function of multilevel, appraisal-driven response synchronization. Conscious. Cogn. **17**(2), 484–495 (2008)
12. Hebb, D.O.: The Organization of Behavior: A Neuropsychological Theory. Wiley, New York (1949)
13. Hendrikse, S.C.F., Kluiver, S., Treur, J., Wilderjans, T.F., Dikker, S., Koole, S.L.: How virtual agents can learn to synchronize: an adaptive joint decision-making model of psychotherapy. Cogn. Syst. Res. **79**, 138–155 (2022)
14. Hendrikse, S.C.F., Kluiver, S., Treur, J., Wilderjans, T.F., Dikker, S., Koole, S.L. (2025). Learning to synchronize: adaptive joint decision-making by virtual agents. In: Hendrikse, S.C.F., Treur, J., Koole, S.L. (eds.) New Analysis and Modeling Directions in Social Interaction Science: Emergent Multimodal Interpersonal Synchrony, Affiliation, and Bonding, Chap. 5. Springer Nature (2025)
15. Hendrikse, S.C.F., Treur, J., Koole, S.L.: Modeling emerging interpersonal synchrony and its related adaptive short-term affliation and long-term bonding: a second-order multi-adaptive neural agent model. Inter. J. Neural Syst **33**(7), 2350038 (2023)
16. Hendrikse, S.C.F., Treur, J., Koole, S.L. (eds.) New Analysis and Modeling Directions in Social Interaction Science: Emergent Multimodal Interpersonal Synchrony, Affiliation, and Bonding. Springer Nature (2025a)
17. Hendrikse, S.C.F., Treur, J., Koole, S.L.: Introduction to the role of adaptivity in social interaction science. In: Hendrikse, S.C.F., Treur, J., Koole, S.L. (eds.) New Analysis and Modeling Directions in Social Interaction Science: Emergent Multimodal Interpersonal Synchrony, Affiliation, and Bonding, Chap. 1. Springer Nature (2025b)
18. Hendrikse, S.C.F., Treur, J., Koole, S.L.: The adaptive interplay of interpersonal synchrony, short-term affiliation and long-term bonding: a multi-timescale multi-adaptive agent model. In: Hendrikse, S.C.F., Treur, J., Koole, S.L. (eds.), New Analysis and Modeling Directions in Social Interaction Science: Emergent Multimodal Interpersonal Synchrony, Affiliation, and Bonding, Chap. 12. Springer Nature (2025c)
19. Hendrikse, S.C.F., Treur, J., Koole, S.L.: A multi-timescale multi-adaptive dynamical system approach for relationship-specific and relationship-independent adaptivity in affiliation and bonding. In: Hendrikse, S.C.F., Treur, J., Koole, S.L. (eds.) New Analysis and Modeling Directions in Social Interaction Science: Emergent Multimodal Interpersonal Synchrony, Affiliation, and Bonding, Chap. 13. Springer Nature (2025d)
20. Hendrikse, S.C.F., Treur, J., Wilderjans, T.F., Dikker, S., Koole, S.L.: On the same wavelengths: emergence of multiple synchronies among multiple agents. In: Van Dam, K.H., Verstaevel, N.

(eds.) Proceedings MABS 2021. LNAI, vol. 13128, pp. 57–71. Springer, Cham (2022b) https://doi.org/10.1007/978-3-030-94548-0_5
21. Hendrikse, S.C.F., Treur, J., Wilderjans, T.F., Dikker, S., Koole, S.L.: Becoming attuned to each other over time: a computational neural agent model for the role of time lags in subjective synchrony detection and related behavioral adaptivity. In: Proceedings of the 15th International Conference on Brain Informatics, BI'22. Lecture Notes in AI, vol. 13406, pp. 369–383. Springer Nature (2022c)
22. Hesslow, G.: Conscious thought as simulation of behaviour and perception. Trends Cogn. Sci. **6**, 242–247 (2002)
23. Hove, M.J., Risen, J.L.: It's all in the timing: interpersonal synchrony increases affiliation. Soc. Cogn. **27**(6), 949–960 (2009)
24. Hu, Y., Cheng, X., Pan, Y., Hu, Y.: The intrapersonal and interpersonal consequences of interpersonal synchrony. Acta Psychol. **224**, e103513 (2022)
25. Izawa, J., Rane, T., Donchin, O., Shadmehr, R.: Motor adaptation as a process of reoptimization. J. Neurosci. Offic. J. Soc. Neurosci. **28**(11), 2883–2891 (2008). https://doi.org/10.1523/JNEUROSCI.5359-07.2008
26. Keller, P.E., Novembre, G., Hove, M.J.: Rhythm in joint action: psychological and neurophysiological mechanisms for real-time interpersonal coordination. Philos. Trans. R. Soc. Lond. Ser. B. Biol. Sci. **369**(1658), Article 20130394 (2014)
27. Kirschner, S., Tomasello, M.: Joint music making promotes prosocial behavior in 4-year-old children. Evol. Hum. Behav. **31**, 354–364 (2010)
28. Koole, S.L., Tschacher, W.: Synchrony in psychotherapy: a review and an integrative framework for the therapeutic alliance. Front. Psychol. **7**, 862 (2016)
29. Koole, S.L., Tschacher, W., Butler, E., Dikker, S., Wilderjans, T.F.: In sync with your shrink. In: Forgas, J.P., Crano, W.D., Fiedler, K. (eds.) Applications of Social Psychology, pp. 161–184. Taylor and Francis, Milton Park (2020)
30. Pacherie, E.: The phenomenology of joint action: self-agency versus joint-agency. In: Seeman, A. (ed.) Joint Attention: New Developments, pp. 343–389. MIT Press (2012)
31. Palumbo, R.V., et al.: Interpersonal autonomic physiology: a systematic review of the literature. Pers. Soc. Psychol. Rev. **21**(2), 99–141 (2017)
32. Prince, K., Brown, S.: Neural correlates of partnered interaction as revealed by cross-domain ALE meta-analysis. Psychol. Neurosci. **15**(1), 1–13 (2022)
33. Ramseyer, F., Tschacher, W.: Nonverbal synchrony in psychotherapy: coordinated body movement reflects relationship quality and outcome. J. Consult. Clin. Psychol. **79**, 284–295 (2011)
34. Rayner, K., Smith, T.J., Malcolm, G.L., Henderson, J.M.: Eye movements and visual encoding during scene perception. Psychol. Sci. **20**(1), 6–10 (2009)
35. Robinson, B.L., Harper, N.S., McAlpine, D.: Meta-adaptation in the auditory midbrain under cortical influence. Nat. Commun. **7**, e13442 (2016)
36. Schoenherr, D., et al.: Identification ofmovement synchrony: validation ofwindowed cross-lagged correlation and -regression with peak-picking algorithm. PLoS ONE **14**(2), e0211494 (2019). https://doi.org/10.1371/journal.pone.0211494
37. Schoenherr, D., et al.: Quantification of nonverbal synchrony using linear time series analysis methods: lack of convergent validity and evidence for facets of synchrony. Behav. Res. Methods **51**(1), 361–383 (2018). https://doi.org/10.3758/s13428-018-1139-z
38. Sharpley, C.F., Halat, J., Rabinowicz, T., Weiland, B., Stafford, J.: Standard posture, postural mirroring and client-perceived rapport. Couns. Psychol. Q. **14**, 267–280 (2001). https://doi.org/10.1080/09515070110088843
39. Shatz, C.J.: The developing brain. Sci. Am. **267**, 60–67 (1992)
40. Shelton, J., Kumar, G.: Comparison between auditory and visual simple reaction times. Neurosci. Med. **1**(1), 30–32 (2010)
41. Tarr, B., Launay, J., Dunbar, R.I.M.: Silent disco: dancing in synchrony leads to elevated pain thresholds and social closeness. Evol. Hum. Beh. **37**(5), 343–349 (2016)

42. Thompson, J.J., Blair, M.R., Henrey, A.J.: Over the hill at 24: persistent age-related cognitive-motor decline in reaction times in an ecologically valid video game task begins in early adulthood. PLoS ONE **9**(4), e94215 (2014)
43. Treur, J.: Network-oriented modeling for adaptive networks: designing higher-order adaptive biological, mental and social network models. Springer Nature, Cham (2020a). https://doi.org/10.1007/978-3-030-31445-3
44. Treur, J.: Modeling multi-order adaptive processes by self-modeling networks. In: Tallón-Ballesteros, A.J., Chen, C.-H. (eds.) Proceedings of the 2nd International Conference on Machine Learning and Intelligent Systems, MLIS 2020. Frontiers in Artificial Intelligence and Applications, vol. 332, pp. 206–217. IOS Press (2020b)
45. Treur, J.: On the dynamics and adaptivity of mental processes: relating adaptive dynamical systems and self-modeling network models by mathematical analysis. Cogn. Syst. Res. **70**, 93–100 (2021)
46. Treur, J.: Modeling coevolution of dynamics, adaptivity and control of adaptivity by self-modeling networks. In: Hendrikse, S.C.F., Treur, J., Koole, S.L. (eds.) New Analysis and Modeling Directions in Social Interaction Science: Emergent Multimodal Interpersonal Synchrony, Affiliation, and Bonding, Chapter 2. Springer Nature (2025)
47. Tschacher, W., Ramseyer, F., Koole, S.L.: Sharing the now in the social present: duration of nonverbal synchrony is linked with personality. J. Pers. **86**(2), 129–138 (2018)
48. Valdesolo, P., DeSteno, D.: Synchrony and the social tuning of compassion. Emotion **11**, 262 (2011). https://doi.org/10.1037/a0021302
49. Valdesolo, P., Ouyang, J., DeSteno, D.: The rhythm of joint action: synchrony promotes cooperative ability. J. Exp. Soc. Psychol. **46**(4), 693–695 (2010)
50. Wiltermuth, S.S., Heath, C.: Synchrony and cooperation. Psychol. Sci. **20**(1), 1–5 (2009)
51. Zhang, A., Li, X., Gao, Y., Niu, Y.: Event-driven intrinsic plasticity for spiking convolutional neural networks. IEEE Trans. Neural Netw. Learn. Syst. (2021). https://doi.org/10.1109/tnnls.2021.3084955

The Effect of Transition Dynamics in Synchrony on Social Interaction Adaptivity: A Multi-adaptive Network Model

Sophie C. F. Hendrikse, Jan Treur, Tom F. Wilderjans, Suzanne Dikker, and Sander L. Koole

Abstract Interpersonal synchrony is associated with better interpersonal affiliation. No matter how well-affiliated people are, interruptions or transitions in synchrony rebound to occur. One might intuitively expect that transitions in synchrony negatively affect affiliation or liking. Empirical evidence, however, suggests that time periods with interruptions in synchrony may favor affiliation or liking even more than time periods without interruptions in synchrony. This chapter introduces a controlled adaptive network model to explain how persons' affiliation might benefit from transitions in synchrony over and above mean levels of synchrony. The adaptive network model was evaluated in a series of simulation experiments for two persons with a

S. C. F. Hendrikse
Department of Methodology and Statistics, Tilburg School of Social and Behavioral Sciences, Tilburg University, Tilburg, The Netherlands
e-mail: S.C.F.Hendrikse@tilburguniversity.edu

S. L. Koole
Amsterdam Emotion Regulation Lab, Department of Clinical Psychology, Vrije Universiteit Amsterdam, Amsterdam, The Netherlands
e-mail: s.l.koole@vu.nl

J. Treur (✉)
Social AI Group, Department of Computer Science, Vrije Universiteit Amsterdam, Amsterdam, The Netherlands
e-mail: j.treur@vu.nl

T. F. Wilderjans
Methodology and Statistics Research Unit, Faculty of Social and Behavioural Sciences, Institute of Psychology, Leiden University, Leiden, The Netherlands

Research Group of Quantitative Psychology and Individual Differences, Faculty of Psychology and Educational Sciences, Katholieke Universiteit (KU) Leuven, Leuven, Belgium

T. F. Wilderjans
e-mail: t.f.wilderjans@leidenuniv.nl

S. Dikker
NYU – Max Planck Center for Language, Music and Emotion, New York University, New York, NY, USA

S. C. F. Hendrikse et al. (eds.), *New Analysis and Modeling Directions in Social Interaction Science*, Studies in Systems, Decision and Control 614,
https://doi.org/10.1007/978-3-031-99968-0_15

setup in which a number of scenarios were encountered in different (time) episodes. Our controlled adaptive network model may serve as a foundation for more realistic virtual agents with regard to synchrony transitions and their role in affiliation.

Keywords Controlled adaptive network model · Interpersonal synchrony · Synchrony transitions · Liking · Affiliation

1 Introduction

When two or more individuals are interacting, their behavior tends to become mutually coordinated in time, or synchronized. Such interpersonal synchrony has been found to lead to behavioral adaptivity by enhancing, for example, closeness, concentration, cooperation, affiliation, alliance, connection, or bonding; e.g., [2, 22, 24, 31, 35, 40, 44]. However, perhaps counterintuitively, interruptions in interpersonal synchrony may positively affect behavioral adaptivity as well [11]. Furthermore, interpersonal liking is highest when a balance between interpersonal synchrony and complexity/unpredictability during a mirror game is obtained [32]. In this chapter, we demonstrate how a controlled adaptive network model can computationally capture how behavioral adaptivity is induced not only by synchrony, but also by transitions of synchrony.

The controlled adaptive network model is based on an adaptive interplay of a number of mechanisms drawn from cognitive, behavioral, and affective neuroscience. For example, a neural basis for behavioral adaptivity in the form of adaptive affiliation can be found in recent work on the (nonsynaptic, intrinsic) adaptive excitability of (neural) states; e.g., [4, 7, 43, 45]. The extent of adaptation that a person requires may vary from situation to situation, which is called metaplasticity (e.g., [1, 33]). This controls the plasticity in a context-sensitive manner. The resulting network model yields a controlled adaptive network model.

The adaptive network model was evaluated in a series of simulation experiments for two persons with a setup in which a number of scenarios were encountered in different (time) episodes. The simulations included not only episodes with a common stimulus for the two persons, but also episodes with different stimuli for the persons. Moreover, to analyze the role of communication (interaction), circumstances were also included for episodes when communication was enabled by the environment and episodes when communication was not enabled.

2 Background Literature

Interpersonal synchrony usually leads to behavioral adaptivity in the form of mutual adaptation of interactive behavior; e.g., [2, 9, 22–25, 28, 30, 35, 41, 42, 44]. For example, therapists were rated more favorably and as more empathic when, they were instructed to make their movements more synchronized with the client [26, 34, 39]. Ramseyer and Tschacher [31] found that initial movement synchrony between client and therapist was predictive of the client's experience of the quality of the therapeutic alliance at the end of each session. Moreover, Koole and Tschacher [24] reviewed converging evidence that movement synchrony has a positive effect on the working alliance between patient and therapist. Interpersonal synchrony in face-to-face interactions has been found to promote interpersonal alliance [10, 44]. For more background on synchrony, see this volume [13], Chapter "Introduction to the Role of Adaptivity in Social Interaction Science" [14].

However, it seems that more interpersonal synchrony is not always better for behavioral outcomes, such as good (working) relationships. A moderate range instead of a very low or very high behavioral synchrony between children and their parent is related to children who are more securely attached to their parent [3]. Another study [29] found a comparable pattern regarding the success of a therapy and movement synchrony. Patients who improved during therapy had a medium level of movement synchrony at the beginning of therapy, whereas patients who did not improve and consensually terminated their therapy had the highest level of movement synchrony at the beginning of therapy. Other findings indicating that high levels of synchrony do not always link to the best behavioral outcomes by Deres-Cohen et al. [8], suggest that therapists increase their levels of movement synchrony towards their patient to keep a strong alliance when difficulties occur.

Also, a more nuanced view of the link between interpersonal synchrony and behavioral adaptivity is found in several experimental paradigms with the mirror game, a task in which participants have the goal to move as synchronously as possible. On real world experimental data with the mirror game, Dahan et al. [5] have fitted multiple mathematical models. The models that only included moving in synchrony fit the real data worse than an alternative model that included a tendency to withdraw from synchrony (i.e., moving in and out of synchrony). As another example with the mirror game, Ravreby et al. [32] found that a combination of movement synchrony and interruptions by events with more complexity explains liking almost two times better than movement synchrony by itself does. Thus, in addition to synchrony, maintaining interest may be essential for bonding.

Partly inspired by the few available empirical studies, some conceptual studies have been published on the importance of both interpersonal synchrony and transitions in interpersonal synchrony with regard to other outcomes like affiliation. García and Di Paolo [11] propose that transitions in interpersonal synchrony play an important role, instead of only interpersonal synchrony itself, in the relation between movement synchrony and phenomena, such as therapeutic alliance. Similarly, Mayo and Gordon [27] claim that, to have an adaptive interpersonal system within a social context, people have two simultaneous tendencies: (1) to achieve interpersonal synchrony and (2) to switch in and out of interpersonal synchrony. The above empirical and conceptual literature on how interrupted synchrony can lead to stronger affiliation than 'just' synchrony was a main inspiration for the development of the controlled adaptive network model introduced here.

3 Self-Modelling Network Modelling

The presented controlled adaptive network model is based on network-oriented modelling. Following [36, 37], a temporal-causal network model uses nodes X and Y, also called states, with values $X(t)$ and $Y(t)$ over time t and the following characteristics:

- *Connectivity characteristics*

Connections from a state X to a state Y and their weights $\boldsymbol{\omega}_{X,Y}$

- *Aggregation characteristics*

For any state Y, some combination function $\mathbf{c}_Y(..)$ defines the aggregation that is applied to the impacts $\boldsymbol{\omega}_{X,Y}X(t)$ on Y from its incoming connections from states X

- *Timing characteristics*

Each state Y has a speed factor $\boldsymbol{\eta}_Y$ defining how fast it changes for given causal impact

The following difference (or related differential) equations that are used for simulation purposes and also for analysis of temporal-causal networks, incorporate these network characteristics $\boldsymbol{\omega}_{X,Y}$, $\mathbf{c}_Y(..)$ and $\boldsymbol{\eta}_Y$ in a standard numerical format:

$$Y(t + \Delta t) = Y(t) + \boldsymbol{\eta}_Y\left[\mathbf{c}_Y\left(\boldsymbol{\omega}_{X_1,Y}X_1(t), \ldots, \boldsymbol{\omega}_{X_k,Y}X_k(t)\right) - Y(t)\right]\Delta t \quad (1)$$

for any state Y and where X_1 to X_k are the states from which Y gets its incoming connections. Within the software environment described in [36, Chap. 9], a large number of currently around 70 useful basic combination functions are included in a combination function library. The above concepts make it possible to design network models and their dynamics in a declarative manner, based on mathematically defined functions and relations. The examples of combination functions that were applied in the model introduced here can be found in the Appendix, in Table 9. In Table 1, the new transition detection functions are shown. Here, W is a representation of the sliding windows.

Realistic network models are usually adaptive: often not only their states but also some of their network characteristics change over time. By using a *self-modelling network* (also called a *reified* network), a similar network-oriented conceptualization can also be applied to adaptive networks to obtain a declarative description using

Table 1 The new transition detection combination functions used in the introduced network model

	Notation	Formula	Parameters
Average transition	$\mathbf{transdetavabs}_{\boldsymbol{\iota\delta},\boldsymbol{\sigma\omega}}$ (W)	$\lvert \text{mean}_{1 \leq \mathbf{v} \leq \boldsymbol{\sigma\omega}}(W(\boldsymbol{\iota\delta}, \mathbf{v})) - \text{mean}_{\boldsymbol{\sigma\omega} + 1 \leq \mathbf{v} \leq 2\boldsymbol{\sigma\omega}}(W(\boldsymbol{\iota\delta}, \mathbf{v}))\rvert$	$\boldsymbol{\iota\delta}$ state identifier $\boldsymbol{\sigma\omega}$ sliding window size
Maxmin transition	$\mathbf{transdetmaxminabs}_{\boldsymbol{\iota\delta},\boldsymbol{\sigma\omega}}$ (W)	$\lvert \max_{1 \leq \mathbf{v} \leq \boldsymbol{\sigma\omega}}(W(\boldsymbol{\iota\delta}, \mathbf{v})) - \min_{1 \leq \mathbf{v} \leq \boldsymbol{\sigma\omega}}(W(\boldsymbol{\iota\delta}, \mathbf{v}))\rvert$	$\boldsymbol{\iota\delta}$ state identifier $\boldsymbol{\sigma\omega}$ sliding window size
Standard deviation transition	$\mathbf{transdetstdev}_{\boldsymbol{\iota\delta},\boldsymbol{\sigma\omega}}$ (W)	$2\surd[\text{mean}_{1 \leq \mathbf{v} \leq \boldsymbol{\sigma\omega}}(W(\boldsymbol{\iota\delta}, \mathbf{v}) - \text{mean}(W))^2]$	$\boldsymbol{\iota\delta}$ state identifier $\boldsymbol{\sigma\omega}$ sliding window size

mathematically defined functions and relations for them as well; see [36, 37]. This works through the addition of new states to the network (called *self-model states*) which represent (adaptive) network characteristics. In the graphical 3D-format as shown in Sect. 4, such additional states are depicted at a next level (called *self-model level* or *reification level*), where the original network is at the *base level*.

For instance, the weight $\boldsymbol{\omega}_{X,Y}$ of a connection from state X to state Y can be represented (at a next self-model level) by a self-model state named $\mathbf{W}_{X,Y}$. Similarly, all other network characteristics from $\boldsymbol{\omega}_{X,Y}$, $\mathbf{c}_Y(..)$ and $\boldsymbol{\eta}_Y$ can be made adaptive by including self-model states for them. For example, an adaptive excitability threshold $\boldsymbol{\tau}_Y$ (e.g., [4, 7, 45]) for a logistic combination function for state Y can be represented by a self-model state named $\mathbf{T}_Y$ and an adaptive speed factor $\boldsymbol{\eta}_Y$ can be represented by a self-model state named $\mathbf{H}_Y$.

As the outcome of such a process of network reification is also a temporal-causal network model itself, as has been shown in [36], Ch 10, this self-modelling network construction can easily be applied iteratively to obtain multiple orders of self-models at multiple (first-order, second-order, …) self-model levels. For example, a second-order self-model may include a second-order self-model state $\mathbf{H}_{\mathbf{T}_Y}$ representing the speed factor $\boldsymbol{\eta}_{\mathbf{T}_Y}$ for the dynamics of first-order self-model state $\mathbf{T}_Y$ which in turn represents the adaptative excitability threshold $\boldsymbol{\tau}_Y$ for Y. In the current chapter, this multi-level self-modelling network perspective is applied to obtain a second-order adaptive network architecture addressing controlled adaptation induced by detected synchrony and detected synchrony transitions. The control level is used to make the adaptation speed context-sensitive as addressed by the metaplasticity literature such as [1, 33]: the metaplasticity principle 'Adaptation accelerates with increasing stimulus exposure' formulated by Robinson et al. [33] can easily be modelled by using second-order self-model states; this is discussed in Sect. 4.

For more details of the network modeling approach used, see [36, 37] or this volume [13], Chapter "Modeling Coevolution of Dynamics, Adaptivity and Control of Adaptivity by Self-Modeling Networks" [38].

4 The Adaptive Network Model

In this section, we introduce an adaptive neural agent model covering the detection of interpersonal synchrony and transitions of interpersonal synchrony and its related behavioral adaptivity. It applies a self-modelling network architecture of three levels: a base level, a first-order self-model level, and a second-order self-model level. The (middle) first-order self-model level models how excitability of states at the base level are adapted over time, and the (upper) second-order self-model level models the control over this adaptation.

4.1 Base Level

Figure 1 shows a graphic overview of the base level of the person model (persons are indicated by the big boxes) and the Appendix section provides explanations for all of its states. Each person uses six states for the interaction with the other person: three states (indicated by sense) for sensing the other person on the left-hand side of each box, and three states for execution or expression of actions (move, exp_affect, talk) on the right-hand side. Within a box the person's internal mental states can be found, outside the boxes are the world states. Note that the ws-states are sensed by one agent and determined by another agent, while the state world$_s$ represents an external world state that is independent from the agents. Each person also senses its own actions, modelled by the arrows from right to left outside the box.

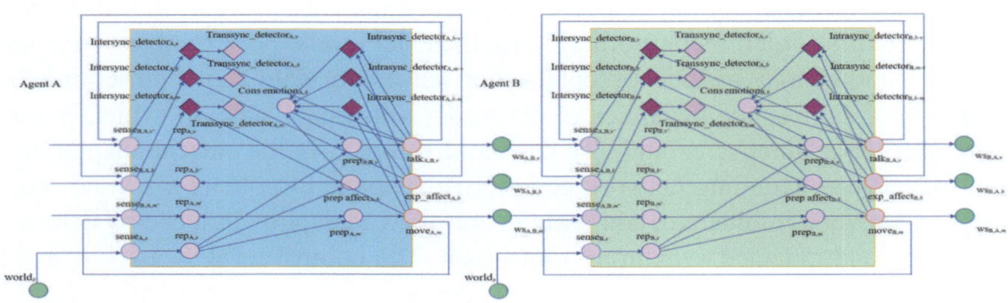

Fig. 1 Base level of the introduced adaptive network model with three modalities, six synchrony detection states for intrapersonal and interpersonal synchrony (dark pink diamonds) and three states for interpersonal synchrony transition detection (light pink diamonds)

The internal mental states cover sensory representation states (rep) and preparation states (prep) for each of the three modalities: movement *m*, expression of affect *b*, and verbal action *v*. In addition, each person has a conscious emotion state for affective response *b* (cons_emotion). Each of the mentioned states is depicted in Fig. 1 by a light pink circle shape. For each modality, the corresponding representation state has an outgoing response connection to the related preparation state and it has an incoming (prediction) connection back from that preparation state to model internal mental simulation [6, 21].

Six synchrony detector states are depicted in Figs. 1 and 2 by the darker pink diamond shapes: (1) three of them for intrapersonal synchrony for the three pairs of the three modalities movement-emotion (*m-b*), movement-verbal action (*m-v*), emotion-verbal action (*b-v*), and (2) the other three for interpersonal synchrony for each other three modalities [15]. In addition, we introduce three interpersonal synchrony transition detector states (the light purple diamonds) that are able to note transitions in the synchrony.

Following [32], we assume that both the detected interpersonal synchrony and the detected transitions of interpersonal synchrony contribute to behavioral adaptivity of each person: detected synchrony and synchrony transition lead to becoming more sensitive to sensing a person and to expressing to that person (a form of homeostatic regulation). This is modelled through self-model **T**-states for adaptive excitability thresholds [4, 7, 45].

Thus, more synchrony and synchrony transition detection will enhance excitability for the representation and execution states. More sensitive states for representations will lead to having better images of the modalities of the other person, which will

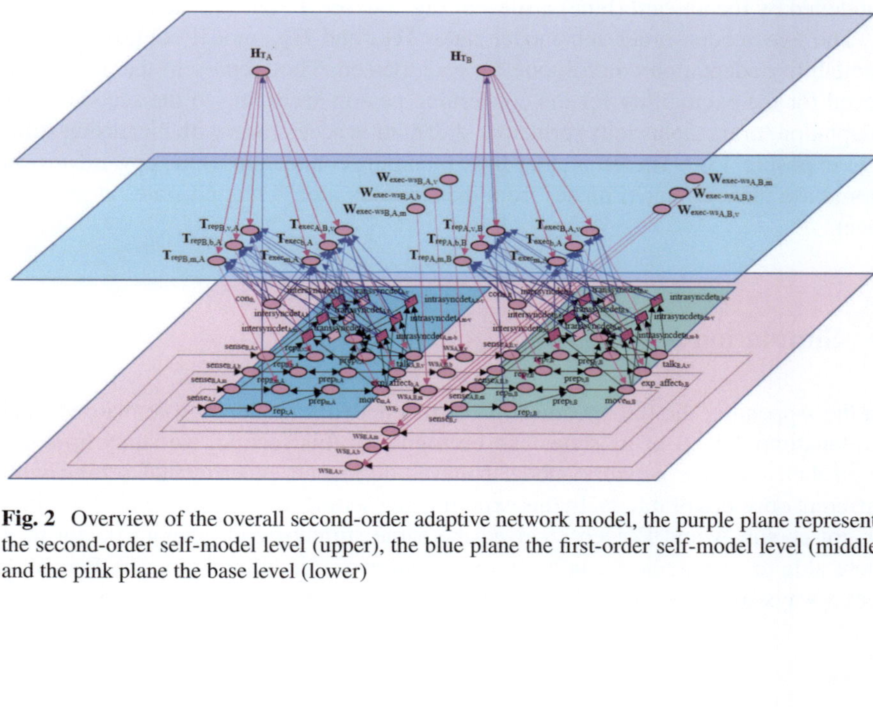

Fig. 2 Overview of the overall second-order adaptive network model, the purple plane represents the second-order self-model level (upper), the blue plane the first-order self-model level (middle) and the pink plane the base level (lower)

make the sensed signals better available and accessible for the brain. More sensitive states for execution will lead to better expressed own modalities, so that the other person will sense them better. In Sect. 4.2, we discuss in more detail how we modelled this behavioral adaptivity and its control.

At the base level some world states (world_s) are modelled for stimuli s that are sensed by the persons. The world's suitability for enabling communication between the two persons is modelled by **W**-states. Two general context states are included to model the conditions to maintain excitability thresholds represented by **T**-states.

4.2 Modeling Controlled Adaptation

Following what has been described in Sects. 3 and 4.1, the behavioral adaptivity is modelled by first-order self-model states $\mathbf{T}_Y$ for adaptation of the excitability thresholds $\boldsymbol{\tau}_Y$ for the internal representation states Y and execution states Y, covering the three considered modalities (movement, affective response and verbal action). In addition, second-order self-model $\mathbf{H}_\mathbf{T}$-states are used for context-sensitive control of the adaptation of these adaptive excitability thresholds $\boldsymbol{\tau}_Y$ for the internal representation states and execution states Y.

Figure 2 shows a 3D picture of the overall design of the model; here, the first-order self-model states are in the middle (blue) plane and the second-order self-model states in the upper (purple) plane. By changing the activation values of the **T**-states, the corresponding excitability thresholds change accordingly [4, 7, 45]. This change occurs due to the influences from the detected synchronies and synchrony transitions, modelled by the upward (blue) arrows in Fig. 2 to the **T**-states in the middle plane.

The two second-order self-model states $\mathbf{H}_{\mathbf{T}_\mathrm{A}}$ and $\mathbf{H}_{\mathbf{T}_\mathrm{B}}$ model context-sensitive excitability adaptation control, one for each person. They represent the adaptation speed for the excitability for the concerning person according to the second-order adaptation (or metaplasticity) principle 'Adaptation accelerates with increasing stimulus exposure' [33]. Therefore, they have incoming connections (blue upward arrows from base plane to upper plane) from the stimulus representation states at the base level.

5 Simulation Results

In the Appendix, the full specification of this network model by role matrices and explanations for all of its states can be found. In this section, we study how our controlled adaptive network model behaves during an experimental setting with different episodes (Table 2). In this experimental setting, it was manipulated whether or not person A and B each received the same stimulus and whether or not they were able to communicate, in such a way that each condition happened. Person A was always stimulated for 120 time units, followed by 60 time units of no stimulus

Table 2 Stimuli and communication for the first 4 episodes

Type	Stimulus person A, B	Communication enabled	Duration	Time intervals
Episode 1	Different	No	30	0–30
Episode 2	Common	Yes	30	30–60
Episode 3	Different	No	30	60–90
Episode 4	Common	Yes	30	90–120

(180 time units in total), and thereafter this process was repeated. Regarding person B, the first 60 time units no stimulus was present, followed by 180 time units with the stimulus (i.e., 240 time units in total), and this (non-)stimulus episode of 240 time units was thereafter repeated. The communication enabling conditions in the environment are indicated by the self-model states $\mathbf{W}_{\text{exec-ws}_{x,\text{A,B}}}$ (from A to B) and the states $\mathbf{W}_{\text{exec-ws}_{x,\text{B,A}}}$ (from B to A). They are activated to activation value 1 from time 30 to time 60 and then repeated every 60 time units; when they are not activated, they have activation value 0 (from time 0 to time 30 and then repeated every 60 time units). Limit cycles in the results and repetition of conditions occur after a certain amount of time. We will depict the simulations until 360 time units were reached with a step size (Δt) of 0.5, meaning that our shown simulation run contains 720 computational steps. In Figs. 3 and 4 examples of simulation outcomes are shown. All initial state values can be consulted in the Appendix.

5.1 *Comparing Time Intervals with Transitions and Time Intervals Without Transitions*

Since the focus of this chapter is the role of intrapersonal synchrony, interpersonal synchrony and interpersonal synchrony transitions in relation to affiliation, we highlight the results of these processes in the controlled adaptive network model. The results of the other states in the model can be consulted in the Appendix section. Within a simulation, in different time periods different phenomena can be observed. First of all, in the period 0–30 mainly startup phenomena can be observed. After that, the following more specific patterns can be observed.

- If a person does not get a stimulus and communication is not enabled then this results in a low level of detected subjective intrapersonal synchrony for that person; e.g., for person A in Fig. 3a, time period 120–150; when communication is enabled with just one person having a stimulus, then there is increasing intrapersonal synchrony for both persons
- Large values for interpersonal synchrony occur in periods of enabled communication, even when a person has no stimulus; e.g., Fig. 3b, time 30–60
- When the stimulus is absent for a person, then the **H**-state (not depicted) for that person becomes 0 and therefore the related **T**-states of that person show flat lines for: e.g., Fig. 3d, time 120–180 for person A

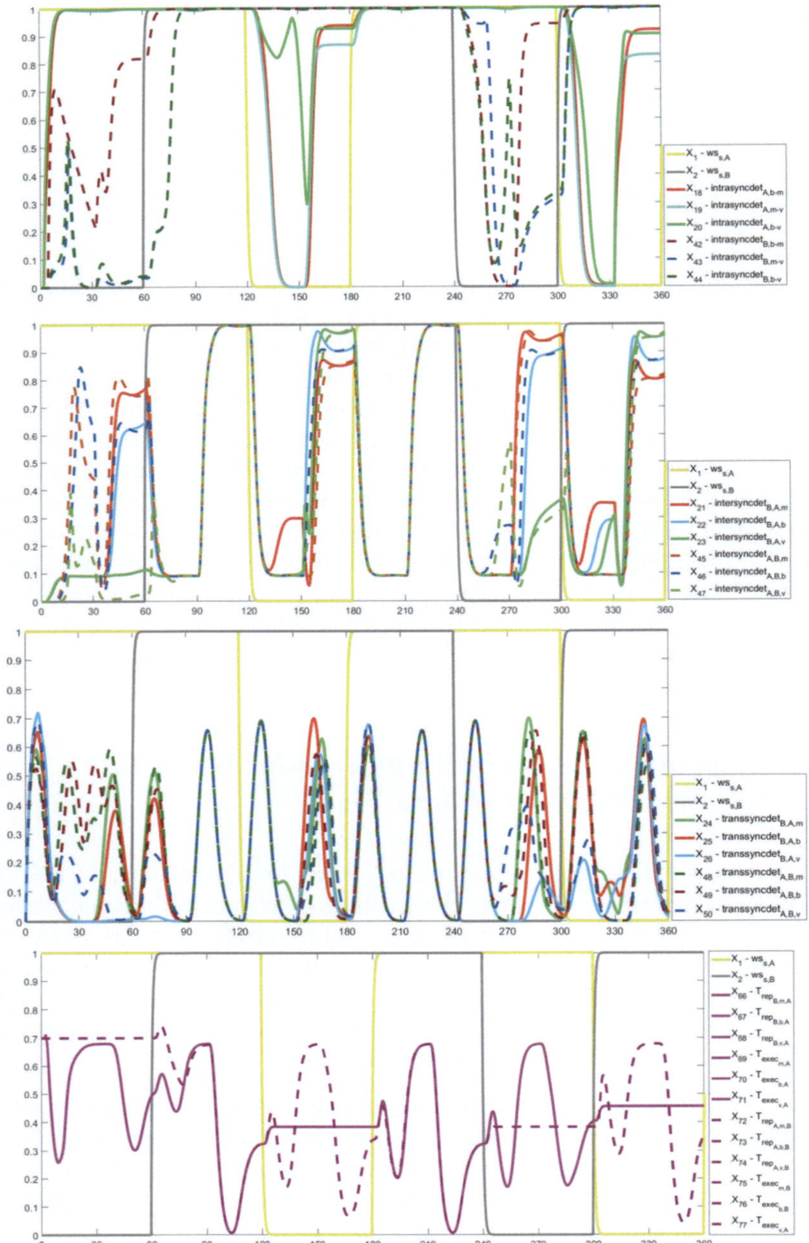

Fig. 3 The trajectory over the stimuli episodes of **a** the intrapersonal synchrony detector states (first), **b** the interpersonal synchrony detector states (second), **c** the interpersonal synchrony transition detector states (third), **d** the affiliation states when both interpersonal synchrony and interpersonal synchrony transitions are connected with the affiliation (fourth). The horizontal axis represents time and the vertical axis represents the activation levels (from 0 to 1)

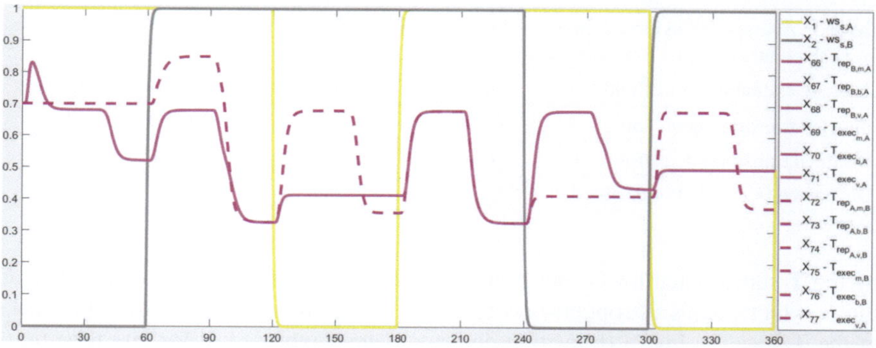

Fig. 4 The trajectory over the stimuli episodes of the affiliation states when only interpersonal synchrony (and not interpersonal synchrony transitions) is connected with the affiliation. The horizontal axis represents time and the vertical axis represents the activation levels (from 0 to 1)

- When no communication is enabled, then sensing becomes 0 and when the stimulus is absent also actions become 0 as well; therefore, interpersonal synchrony increases; e.g., Fig. 3b, time 240–270 (person B); note that this does not really hold at time 120–150 (only a small effect of movement only of person A)
- When detected interpersonal synchrony goes from high to low, then a peak in the transition follows results in lower **T**-states due to the transitions and not due to the interpersonal synchrony; e.g., Fig. 3b–d, time 180–210

  Oppositely, when detected interpersonal synchrony goes from low to high, then a peak in the transition follows resulting in lower **T**-states due to both the transitions and the interpersonal synchrony; e.g., Fig. 3b–d, time 90–120
- The lowest values for the **T**-states are reached when both interpersonal synchrony and transition are high; e.g., Fig. 3b–d, time 90–120, time 210–240

As shown in Fig. 3, for each person the activation of intrapersonal synchrony detection seems to rely heavily on the stimulus the person itself receives. Moreover, in time intervals in which transitions occur the **T**-states reach lower values than in time intervals where only synchrony occurs; this means stronger behavioral adaptivity in the former type of time intervals. So, overall, the results are in accordance with findings reported in literature such as [32].

5.2 *Comparing Continuous Synchrony Without Transitions with Synchrony with Transitions*

In Fig. 4 the simulation results of the **T**-states (regarding other simulation results, see Appendix) are shown in case that the connections from the transition detectors to the **T**-states are disabled. For both persons, the excitability thresholds represented

Table 3 Average behavioral adaptivity through excitability thresholds (represented by **T**-states) for enabled and disabled transition detection over 1080 time units

Average excitability threshold	Person A	Person B	Person A and B
Enabled transition detection	0.394	0.400	0.397
Disabled transition detection	0.507	0.526	0.517
Difference % enabled of disabled	22	24	23

by the **T**-states reach lower values in the case with enabled transition detectors in Fig. 3, which means a stronger short-term behavioral adaptivity. Moreover, the peaks of the **T**-states in Fig. 3 (transition detector states enabled) last for less time units than those in Fig. 4, likely due to the effect of the enabled transition detector states that change at a higher frequency.

To quantify the precise role of the transition detector states, Table 3 displays the specific averages of the adaptive excitability thresholds over 1080 time units for both simulations in Figs. 3 and 4. It turns out that the transitions make that the excitability thresholds are 20–25% lower, which is substantial: it makes activation much easier. This once more is in accordance with findings reported in the literature such as [32].

6 Discussion

Within psychotherapy sessions, more interpersonal synchrony usually leads to a better therapeutical affiliation. However, in practice also switching in and out of interpersonal synchrony often occurs, which breaks synchrony for some period of time. Given that in general interpersonal synchrony positively affect affiliation, a reasonable expectation would be that transitions in synchrony negatively influence affiliation. In contrast, it has been put forward [11, 27, 32] that such transitions also positively affect affiliation and do this even to such an extent that time periods with interpersonal synchrony interrupted by transitions may positively affect affiliation or liking more than time periods with synchrony without such transitions. This chapter has introduced a controlled adaptive network model that addresses this effect. The material from this chapter is based on Hendrikse et al. [20].

Computational modelling of interpersonal synchrony was already addressed in earlier work such as [12, 17–19], see also this volume [13], for example, Chapters "Growing Connectedness Through Synchrony: The Role of Subjective Analysis of Intrapersonal Synchrony and Interpersonal Synchrony within an Adaptive Agent Model" to "A Multi-timescale Multi-adaptive Dynamical System Approach for Relationship-Specific and Relationship-Independent Adaptivity in Affiliation and Bonding" [15, 16]. However, in the models described in [12, 17], no (subjective) internal detection of synchrony takes place. Furthermore, in [17] no adaptivity was covered, whereas in [12] another type of adaptivity was incorporated, namely of internal responding connections from representation states to preparation states.

Moreover, [18, 19] are publications on other computational models where subjective synchrony detection is addressed in relation to affiliation. A difference is that in these publications no subjective detection of synchrony transitions was addressed, which is the novelty introduced by the current chapter. The controlled adaptive network model introduced in the current chapter has adopted part of the model of Hendrikse et al. [18] as point of departure but has extended it by detector states for interpersonal synchrony transitions and their effect on behavioral adaptivity. In this way, the introduced controlled adaptive network model was obtained for the way in which detected interpersonal synchrony transitions also lead to different types of behavioral adaptivity concerning affiliation between two persons. The model can provide a basis to develop adaptive virtual agents that are able to concentrate on each other by short-term behavioral adaptivity in a human-like manner not only during periods of interpersonal synchrony but also during periods in which from time-to-time transitions of the interpersonal synchrony occur.

Appendix

Specification by Role Matrices

In Tables 4, 5, 6, 7 and 8, the full specification of the adaptive network model by role matrices is shown. Each role matrix has 79 rows for all states X_1 to X_{79} of the model.

The connectivity characteristics are specified by role matrices **mb** and **mcw** shown in Tables 4 and 5. Role matrix **mb** lists for each state the states (at the same or lower level) from which the state gets its incoming connections, while in role matrix **mcw** the connection weights are listed for these connections.

Nonadaptive connection weights are indicated in **mcw** (in Table 5) by a number (in a green shaded cell), but adaptive connection weights are indicated by a reference to the (self-model) **W**-state representing the adaptive value (in a peach-red shaded cell). This can be seen for states X_{54} to X_{59} (with self-model **W**-states X_{60} to X_{65}).

The network characteristics for aggregation are defined by the selection of combination functions from the library and values for their parameters. In role matrix **mcfw** it is specified by weights which state uses which combination function; see Table 6.

In role matrix **mcfp** (see Table 7) it is indicated what the parameter values are for the chosen combination functions. A number of them are adaptive: their adaptive excitability thresholds are represented by self-model **T**-states. These concern for person A states X_{11} to X_{13} (with excitability threshold self-model **T**-states X_{66} to X_{68}) and X_{27} to X_{29} (with self-model **T**-states X_{69} to X_{71}), and for person B states X_{35} to X_{37} (with excitability threshold self-model **T**-states X_{72} to X_{74}) and X_{51} to X_{53} (with self-model **T**-states X_{75} to X_{77}).

In Table 8, the role matrix **ms** for speed factors is shown, which lists all speed factors. Next to it, the list of initial values can be found. Also for **ms** some entries

Table 4 Role matrix **mb** for base connectivity

mb	**base connectivity**	1	2	3	4	5	6	7	8	9	10
X_1	$ws_{s,A}$	X_1									
X_2	$ws_{s,B}$	X_2									
X_3	$ws_{s,AB}$	X_3									
X_4	$con_{th,A}$	X_4									
X_5	$con_{th,B}$	X_5									
X_6	$sense_{s,A}$	X_3	X_1								
X_7	$sense_{B,m,A}$	X_{27}	X_{57}								
X_8	$sense_{B,b,A}$	X_{28}	X_{58}								
X_9	$sense_{B,v,A}$	X_{29}	X_{59}								
X_{10}	$rep_{s,A}$	X_6									
X_{11}	$rep_{B,m,A}$	X_7	X_{14}								
X_{12}	$rep_{B,b,A}$	X_8	X_{15}								
X_{13}	$rep_{B,v,A}$	X_9	X_{16}								
X_{14}	$prep_{m,A}$	X_{10}	X_{11}								
X_{15}	$prep_{b,A}$	X_{10}	X_{12}								
X_{16}	$prep_{v,A}$	X_{10}	X_{13}	X_{17}							
X_{17}	$cons\_emotion_{b,A}$	X_{15}	X_{29}	X_{18}	X_{19}	X_{20}					
X_{18}	$intrasyncdet_{A,\text{b-m}}$	X_{27}	X_{28}								
X_{19}	$intrasyncdet_{A,\text{m-v}}$	X_{27}	X_{29}								
X_{20}	$intrasyncdet_{A,\text{b-v}}$	X_{28}	X_{29}								
X_{21}	$intersyncdet_{B,A,\text{m}}$	X_{27}	X_7								
X_{22}	$intersyncdet_{B,A,\text{b}}$	X_{28}	X_8								
X_{23}	$intersyncdet_{B,A,\text{v}}$	X_{29}	X_9								
X_{24}	$transsyncdet_{B,A,\text{m}}$	X_{21}									
X_{25}	$transsyncdet_{B,A,\text{b}}$	X_{22}									
X_{26}	$transsyncdet_{B,A,\text{v}}$	X_{23}									
X_{27}	$move_{m,A}$	X_{14}									
X_{28}	$exp\_affect_{b,A}$	X_{15}									
X_{29}	$talk_{A,B,v}$	X_{16}									
X_{30}	$sense_{s,B}$	X_3	X_2								
X_{31}	$sense_{A,m,B}$	X_{51}	X_{54}								
X_{32}	$sense_{A,b,B}$	X_{52}	X_{55}								
X_{33}	$sense_{A,v,B}$	X_{53}	X_{56}								
X_{34}	$rep_{s,B}$	X_{30}									
X_{35}	$rep_{A,m,B}$	X_{31}	X_{38}								
X_{36}	$rep_{A,b,B}$	X_{32}	X_{39}								
X_{37}	$rep_{A,v,B}$	X_{33}	X_{40}								
X_{38}	$prep_{m,B}$	X_{34}	X_{35}								
X_{39}	$prep_{b,B}$	X_{34}	X_{36}								
X_{40}	$prep_{v,B}$	X_{34}	X_{37}	X_{41}							
X_{41}	$cons\_emotion_{b,B}$	X_{39}	X_{53}	X_{42}	X_{43}	X_{44}					
X_{42}	$intrasyncdet_{B,\text{b-m}}$	X_{51}	X_{52}								
X_{43}	$intrasyncdet_{B,\text{m-v}}$	X_{51}	X_{53}								
X_{44}	$intrasyncdet_{B,\text{b-v}}$	X_{52}	X_{53}								
X_{45}	$intersyncdet_{A,B,\text{m}}$	X_{51}	X_{31}								
X_{46}	$intersyncdet_{A,\text{B,b}}$	X_{52}	X_{32}								
X_{47}	$intersyncdet_{A,\text{B,v}}$	X_{53}	X_{33}								

(continued)

Table 4 (continued)

X_{48}	$\text{transsyncdet}_{A,B,\text{m}}$	X_{45}									
X_{49}	$\text{transsyncdet}_{A,B,\text{b}}$	X_{46}									
X_{50}	$\text{transsyncdet}_{A,B,\text{v}}$	X_{47}									
X_{51}	$\text{move}_{m,B}$	X_{38}									
X_{52}	$\text{exp\_affect}_{b,B}$	X_{39}									
X_{53}	$\text{talk}_{B,A,v}$	X_{40}									
X_{54}	$\text{ws}_{m,A,B}$	X_{27}									
X_{55}	$\text{ws}_{b,A,B}$	X_{28}									
X_{56}	$\text{ws}_{v,A,B}$	X_{29}									
X_{57}	$\text{ws}_{m,B,A}$	X_{51}									
X_{58}	$\text{ws}_{b,B,A}$	X_{52}									
X_{59}	$\text{ws}_{v,B,A}$	X_{53}									
X_{60}	$\mathbf{W}_{\text{exec-ws\_}m,A,B}$	X_{60}									
X_{61}	$\mathbf{W}_{\text{exec-ws\_}b,A,B}$	X_{61}									
X_{62}	$\mathbf{W}_{\text{exec-ws\_}v,A,B}$	X_{62}									
X_{63}	$\mathbf{W}_{\text{exec-ws\_}m,B,A}$	X_{63}									
X_{64}	$\mathbf{W}_{\text{exec-ws\_}b,B,A}$	X_{64}									
X_{65}	$\mathbf{W}_{\text{exec-ws\_}v,B,A}$	X_{65}									
X_{66}	$\mathbf{T}_{\text{rep\_}m,A}$	X_4	X_{18}	X_{19}	X_{20}	X_{21}	X_{22}	X_{23}	X_{24}	X_{25}	X_{26}
X_{67}	$\mathbf{T}_{\text{rep\_}b,A}$	X_4	X_{18}	X_{19}	X_{20}	X_{21}	X_{22}	X_{23}	X_{24}	X_{25}	X_{26}
X_{68}	$\mathbf{T}_{\text{rep\_}v,A}$	X_4	X_{18}	X_{19}	X_{20}	X_{21}	X_{22}	X_{23}	X_{24}	X_{25}	X_{26}
X_{69}	$\mathbf{T}_{\text{exec\_}m,A}$	X_4	X_{18}	X_{19}	X_{20}	X_{21}	X_{22}	X_{23}	X_{24}	X_{25}	X_{26}
X_{70}	$\mathbf{T}_{\text{exec\_}b,A}$	X_4	X_{18}	X_{19}	X_{20}	X_{21}	X_{22}	X_{23}	X_{24}	X_{25}	X_{26}
X_{71}	$\mathbf{T}_{\text{exec\_}v,A}$	X_4	X_{18}	X_{19}	X_{20}	X_{21}	X_{22}	X_{23}	X_{24}	X_{25}	X_{26}
X_{72}	$\mathbf{T}_{\text{rep\_}m,B}$	X_5	X_{42}	X_{43}	X_{44}	X_{45}	X_{46}	X_{47}	X_{48}	X_{49}	X_{50}
X_{73}	$\mathbf{T}_{\text{rep\_}b,B}$	X_5	X_{42}	X_{43}	X_{44}	X_{45}	X_{46}	X_{47}	X_{48}	X_{49}	X_{50}
X_{74}	$\mathbf{T}_{\text{rep\_}v,B}$	X_5	X_{42}	X_{43}	X_{44}	X_{45}	X_{46}	X_{47}	X_{48}	X_{49}	X_{50}
X_{75}	$\mathbf{T}_{\text{exec\_}m,B}$	X_5	X_{42}	X_{43}	X_{44}	X_{45}	X_{46}	X_{47}	X_{48}	X_{49}	X_{50}
X_{76}	$\mathbf{T}_{\text{exec\_}b,B}$	X_5	X_{42}	X_{43}	X_{44}	X_{45}	X_{46}	X_{47}	X_{48}	X_{49}	X_{50}
X_{77}	$\mathbf{T}_{\text{exec\_}v,B}$	X_5	X_{42}	X_{43}	X_{44}	X_{45}	X_{46}	X_{47}	X_{48}	X_{49}	X_{50}
X_{78}	$\mathbf{H}_{\mathbf{T}_A}$	X_{10}									
X_{79}	$\mathbf{H}_{\mathbf{T}_B}$	X_{34}									

are adaptive. For person A the speed factors of **T**-states X_{66} to X_{71} are represented by (second-order) self-model $\mathbf{H_T}$-state X_{78} and for person B the speed factors of **T**-states X_{72} to X_{77} are represented by (second-order) self-model $\mathbf{H_T}$-state X_{79}.

Additional Results and Complete Model Description

See Fig. 5 and Tables 9, 10 and 11.

Table 5 Role matrix **mcw** for connection weights

mcw	**connection weights**	1	2	3	4	5	6	7	8	9	10
X_1	$ws_{s,A}$	1									
X_2	$ws_{s,B}$	1									
X_3	$ws_{s,AB}$	1									
X_4	$con_{th,A}$	1									
X_5	$con_{th,B}$	1									
X_6	$sense_{s,A}$	1	1								
X_7	$sense_{B,m,A}$	1	1								
X_8	$sense_{B,b,A}$	1	1								
X_9	$sense_{B,v,A}$	1	1								
X_{10}	$rep_{s,A}$	1									
X_{11}	$rep_{B,m,A}$	1	1								
X_{12}	$rep_{B,b,A}$	1	1								
X_{13}	$rep_{B,v,A}$	1	1								
X_{14}	$prep_{m,A}$	1	1								
X_{15}	$prep_{b,A}$	1	1								
X_{16}	$prep_{v,A}$	1	1	1							
X_{17}	$cons\_emotion_{b,A}$	1	1	1	1	1					
X_{18}	$intrasyncdet_{A,\text{b-m}}$	1	1								
X_{19}	$intrasyncdet_{A,\text{m-v}}$	1	1								
X_{20}	$intrasyncdet_{A,\text{b-v}}$	1	1								
X_{21}	$intersyncdet_{B,A,\text{m}}$	1	1								
X_{22}	$intersyncdet_{B,A,\text{b}}$	1	1								
X_{23}	$intersyncdet_{B,A,\text{v}}$	1	1								
X_{24}	$transsyncdet_{B,A,\text{m}}$	1									
X_{25}	$transsyncdet_{B,A,\text{b}}$	1									
X_{26}	$transsyncdet_{B,A,\text{v}}$	1									
X_{27}	$move_{m,A}$	1									
X_{28}	$exp\_affect_{b,A}$	1									
X_{29}	$talk_{A,B,v}$	1									
X_{30}	$sense_{s,B}$	1	1								
X_{31}	$sense_{A,m,B}$	1	1								
X_{32}	$sense_{A,b,B}$	1	1								
X_{33}	$sense_{A,v,B}$	1	1								
X_{34}	$rep_{s,B}$	1									
X_{35}	$rep_{A,m,B}$	1	1								
X_{36}	$rep_{A,b,B}$	1	1								
X_{37}	$rep_{A,v,B}$	1	1								
X_{38}	$prep_{m,B}$	1	1								
X_{39}	$prep_{b,B}$	1	1								
X_{40}	$prep_{v,B}$	1	1	1							

(continued)

Table 5 (continued)

X_{41}	cons_emotion$_{b,B}$	1	1	1	1	1					
X_{42}	intrasyncdet$_{B,\text{b-m}}$	1	1								
X_{43}	intrasyncdet$_{B,\text{m-v}}$	1	1								
X_{44}	intrasyncdet$_{B,\text{b-v}}$	1	1								
X_{45}	intersyncdet$_{A,B,\text{m}}$	1	1								
X_{46}	intersyncdet$_{A,\text{B},\text{b}}$	1	1								
X_{47}	intersyncdet$_{A,\text{B},\text{v}}$	1	1								
X_{48}	transsyncdet$_{A,B,\text{m}}$	1									
X_{49}	transsyncdet$_{A,B,\text{b}}$	1									
X_{50}	transsyncdet$_{A,B,\text{v}}$	1									
X_{51}	move$_{m,B}$	1									
X_{52}	exp_affect$_{b,B}$	1									
X_{53}	talk$_{B,A,v}$	1									
X_{54}	ws$_{m,A,B}$	X_{60}									
X_{55}	ws$_{b,A,B}$	X_{61}									
X_{56}	ws$_{v,A,B}$	X_{62}									
X_{57}	ws$_{m,B,A}$	X_{63}									
X_{58}	ws$_{b,B,A}$	X_{64}									
X_{59}	ws$_{v,B,A}$	X_{65}									
X_{60}	$\mathbf{W}_{\text{exec-ws\_}m,A,B}$	1									
X_{61}	$\mathbf{W}_{\text{exec-ws\_}b,A,B}$	1									
X_{62}	$\mathbf{W}_{\text{exec-ws\_}v,A,B}$	1									
X_{63}	$\mathbf{W}_{\text{exec-ws\_}m,B,A}$	1									
X_{64}	$\mathbf{W}_{\text{exec-ws\_}b,B,A}$	1									
X_{65}	$\mathbf{W}_{\text{exec-ws\_}v,B,A}$	1									
X_{66}	$\mathbf{T}_{\text{rep\_}m,A}$	1	-0.1	-0.1	-0.1	-0.1	-0.1	-0.1	-0.2	-0.2	-0.2
X_{67}	$\mathbf{T}_{\text{rep\_}b,A}$	1	-0.1	-0.1	-0.1	-0.1	-0.1	-0.1	-0.2	-0.2	-0.2
X_{68}	$\mathbf{T}_{\text{rep\_}v,A}$	1	-0.1	-0.1	-0.1	-0.1	-0.1	-0.1	-0.2	-0.2	-0.2
X_{69}	$\mathbf{T}_{\text{exec\_}m,A}$	1	-0.1	-0.1	-0.1	-0.1	-0.1	-0.1	-0.2	-0.2	-0.2
X_{70}	$\mathbf{T}_{\text{exec\_}b,A}$	1	-0.1	-0.1	-0.1	-0.1	-0.1	-0.1	-0.2	-0.2	-0.2
X_{71}	$\mathbf{T}_{\text{exec\_}v,A}$	1	-0.1	-0.1	-0.1	-0.1	-0.1	-0.1	-0.2	-0.2	-0.2
X_{72}	$\mathbf{T}_{\text{rep\_}m,B}$	1	-0.1	-0.1	-0.1	-0.1	-0.1	-0.1	-0.2	-0.2	-0.2
X_{73}	$\mathbf{T}_{\text{rep\_}b,B}$	1	-0.1	-0.1	-0.1	-0.1	-0.1	-0.1	-0.2	-0.2	-0.2
X_{74}	$\mathbf{T}_{\text{rep\_}v,B}$	1	-0.1	-0.1	-0.1	-0.1	-0.1	-0.1	-0.2	-0.2	-0.2
X_{75}	$\mathbf{T}_{\text{exec\_}m,B}$	1	-0.1	-0.1	-0.1	-0.1	-0.1	-0.1	-0.2	-0.2	-0.2
X_{76}	$\mathbf{T}_{\text{exec\_}b,B}$	1	-0.1	-0.1	-0.1	-0.1	-0.1	-0.1	-0.2	-0.2	-0.2
X_{77}	$\mathbf{T}_{\text{exec\_}v,B}$	1	-0.1	-0.1	-0.1	-0.1	-0.1	-0.1	-0.2	-0.2	-0.2
X_{78}	$\mathbf{H}_{\mathbf{T}_A}$	1									
X_{79}	$\mathbf{H}_{\mathbf{T}_B}$	1									

Table 6 Role matrix **mcfw** for combination function weights

mcfw combination function weights	1 alogistic	2 syncdet	3 stepmod	4 stepmo-dopp	5 eucl	6 transdet
X_1 $ws_{s,A}$				1		
X_2 $ws_{s,B}$			1			
X_3 $ws_{s,AB}$			1			
X_4 $con_{th,A}$	1					
X_5 $con_{th,B}$	1					
X_6 $sense_{s,A}$					1	
X_7 $sense_{B,m,A}$					1	
X_8 $sense_{B,b,A}$					1	
X_9 $sense_{B,v,A}$					1	
X_{10} $rep_{s,A}$	1					
X_{11} $rep_{B,m,A}$	1					
X_{12} $rep_{B,b,A}$	1					
X_{13} $rep_{B,v,A}$	1					
X_{14} $prep_{m,A}$	1					
X_{15} $prep_{b,A}$	1					
X_{16} $prep_{v,A}$	1					
X_{17} $cons\_emotion_{b,A}$	1					
X_{18} $intrasyncdet_{A,b\text{-}m}$		1				
X_{19} $intrasyncdet_{A,m\text{-}v}$		1				
X_{20} $intrasyncdet_{A,b\text{-}v}$		1				
X_{21} $intersyncdet_{B,A,m}$		1				
X_{22} $intersyncdet_{B,A,b}$		1				
X_{23} $intersyncdet_{B,A,v}$		1				
X_{24} $transsyncdet_{B,A,m}$						1
X_{25} $transsyncdet_{B,A,b}$						1
X_{26} $transsyncdet_{B,A,v}$						1
X_{27} $move_{m,A}$	1					
X_{28} $exp\_affect_{b,A}$	1					
X_{29} $talk_{A,B,v}$	1					
X_{30} $sense_{s,B}$					1	
X_{31} $sense_{A,m,B}$					1	
X_{32} $sense_{A,b,B}$					1	
X_{33} $sense_{A,v,B}$					1	
X_{34} $rep_{s,B}$	1					
X_{35} $rep_{A,m,B}$	1					
X_{36} $rep_{A,b,B}$	1					
X_{37} $rep_{A,v,B}$	1					

(continued)

Table 6 (continued)

X_{38} prep$_{m,B}$	1					
X_{39} prep$_{b,B}$	1					
X_{40} prep$_{v,B}$	1					
X_{41} cons_emotion$_{b,B}$	1					
X_{42} intrasyncdet$_{B,\text{b-m}}$		1				
X_{43} intrasyncdet$_{B,\text{m-v}}$		1				
X_{44} intrasyncdet$_{B,\text{b-v}}$		1				
X_{45} intersyncdet$_{A,B,\text{m}}$		1				
X_{46} intersyncdet$_{A,\text{B,b}}$		1				
X_{47} intersyncdet$_{A,\text{B,v}}$		1				
X_{48} transsyncdet$_{A,B,\text{m}}$						1
X_{49} transsyncdet$_{A,B,\text{b}}$						1
X_{50} transsyncdet$_{A,B,\text{v}}$						1
X_{51} move$_{m,B}$	1					
X_{52} exp_affect$_{b,B}$	1					
X_{53} talk$_{B,A,v}$	1					
X_{54} ws$_{m,A,B}$					1	
X_{55} ws$_{b,A,B}$					1	
X_{56} ws$_{v,A,B}$					1	
X_{57} ws$_{m,B,A}$					1	
X_{58} ws$_{b,B,A}$					1	
X_{59} ws$_{v,B,A}$					1	
X_{60} $\mathbf{W}_{\text{exec-ws}\_m,A,B}$			1			
X_{61} $\mathbf{W}_{\text{exec-ws}\_b,A,B}$			1			
X_{62} $\mathbf{W}_{\text{exec-ws}\_v,A,B}$			1			
X_{63} $\mathbf{W}_{\text{exec-ws}\_m,B,A}$			1			
X_{64} $\mathbf{W}_{\text{exec-ws}\_b,B,A}$			1			
X_{65} $\mathbf{W}_{\text{exec-ws}\_v,B,A}$			1			
X_{66} $\mathbf{T}_{\text{rep}\_m,A}$	1					
X_{67} $\mathbf{T}_{\text{rep}\_b,A}$	1					
X_{68} $\mathbf{T}_{\text{rep}\_v,A}$	1					
X_{69} $\mathbf{T}_{\text{exec}\_m,A}$	1					
X_{70} $\mathbf{T}_{\text{exec}\_b,A}$	1					
X_{71} $\mathbf{T}_{\text{exec}\_v,A}$	1					
X_{72} $\mathbf{T}_{\text{rep}\_m,B}$	1					
X_{73} $\mathbf{T}_{\text{rep}\_b,B}$	1					
X_{74} $\mathbf{T}_{\text{rep}\_v,B}$	1					
X_{75} $\mathbf{T}_{\text{exec}\_m,B}$	1					
X_{76} $\mathbf{T}_{\text{exec}\_b,B}$	1					
X_{77} $\mathbf{T}_{\text{exec}\_v,B}$	1					
X_{78} $\mathbf{H}_{\mathbf{T}_A}$	1					
X_{79} $\mathbf{H}_{\mathbf{T}_B}$	1					

Table 7 Role matrix **mcfp** for combination function parameters

mcfp combination function parameters		1 alogistic σ	τ	2 syncdet ιδ	σω	3 stepmod ρ	δ	4 stepmodopp ρ	δ	5 eucl *n*	λ	6 transdet ιδ	σω
X_{1}	$\mathrm{ws}_{s,A}$							180	120				
X_{2}	$\mathrm{ws}_{s,B}$					240	60						
X_{3}	$\mathrm{ws}_{s,AB}$												
X_{4}	$\mathrm{con}_{th,A}$	5	0.5										
X_{5}	$\mathrm{con}_{th,B}$	5	0.5										
X_{6}	$\mathrm{sense}_{s,A}$									1	1		
X_{7}	$\mathrm{sense}_{B,m,A}$									1	2		
X_{8}	$\mathrm{sense}_{B,b,A}$									1	2		
X_{9}	$\mathrm{sense}_{B,v,A}$									1	2		
X_{10}	$\mathrm{rep}_{s,A}$	5	0.5										
X_{11}	$\mathrm{rep}_{B,m,A}$	5	X_{66}										
X_{12}	$\mathrm{rep}_{B,b,A}$	5	X_{67}										
X_{13}	$\mathrm{rep}_{B,v,A}$	5	X_{68}										
X_{14}	$\mathrm{prep}_{m,A}$	5	0.8										
X_{15}	$\mathrm{prep}_{b,A}$	5	0.7										
X_{16}	$\mathrm{prep}_{v,A}$	5	0.75										
X_{17}	$\mathrm{cons\_emotion}_{b,A}$	5	3.25										
X_{18}	$\mathrm{intrasyncdet}_{A,\mathrm{b\text{-}m}}$			2	1								
X_{19}	$\mathrm{intrasyncdet}_{A,\mathrm{m\text{-}v}}$			3	1								
X_{20}	$\mathrm{intrasyncdet}_{A,\mathrm{b\text{-}v}}$			4	1								
X_{21}	$\mathrm{intersyncdet}_{B,A,\mathrm{m}}$			5	5								
X_{22}	$\mathrm{intersyncdet}_{B,A,\mathrm{b}}$			6	5								
X_{23}	$\mathrm{intersyncdet}_{B,A,\mathrm{v}}$			7	5								
X_{24}	$\mathrm{transsyncdet}_{B,A,\mathrm{m}}$											14	25
X_{25}	$\mathrm{transsyncdet}_{B,A,\mathrm{b}}$											15	25
X_{26}	$\mathrm{transsyncdet}_{B,A,\mathrm{v}}$											16	25
X_{27}	$\mathrm{move}_{m,A}$	5	X_{69}										
X_{28}	$\mathrm{exp\_affect}_{b,A}$	5	X_{70}										
X_{29}	$\mathrm{talk}_{A,B,v}$	5	X_{71}										
X_{30}	$\mathrm{sense}_{s,B}$									1	1		
X_{31}	$\mathrm{sense}_{A,m,B}$									1	2		
X_{32}	$\mathrm{sense}_{A,b,B}$									1	2		
X_{33}	$\mathrm{sense}_{A,v,B}$									1	2		
X_{34}	$\mathrm{rep}_{s,B}$	5	0.5										
X_{35}	$\mathrm{rep}_{A,m,B}$	5	X_{72}										
X_{36}	$\mathrm{rep}_{A,b,B}$	5	X_{73}										
X_{37}	$\mathrm{rep}_{A,v,B}$	5	X_{74}										
X_{38}	$\mathrm{prep}_{m,B}$	5	0.6										
X_{39}	$\mathrm{prep}_{b,B}$	5	0.75										
X_{40}	$\mathrm{prep}_{v,B}$	5	1.2										
X_{41}	$\mathrm{cons\_emotion}_{b,B}$	5	3.25										
X_{42}	$\mathrm{intrasyncdet}_{B,\mathrm{b\text{-}m}}$			8	1								
X_{43}	$\mathrm{intrasyncdet}_{B,\mathrm{m\text{-}v}}$			9	1								
X_{44}	$\mathrm{intrasyncdet}_{B,\mathrm{b\text{-}v}}$			10	1								
X_{45}	$\mathrm{intersyncdet}_{A,B,\mathrm{m}}$			11	5								
X_{46}	$\mathrm{intersyncdet}_{A,\mathrm{B},\mathrm{b}}$			12	5								
X_{47}	$\mathrm{intersyncdet}_{A,\mathrm{B},\mathrm{v}}$			13	5								
X_{48}	$\mathrm{transsyncdet}_{A,B,\mathrm{m}}$											17	25
X_{49}	$\mathrm{transsyncdet}_{A,B,\mathrm{b}}$											18	25
X_{50}	$\mathrm{transsyncdet}_{A,B,\mathrm{v}}$											19	25
X_{51}	$\mathrm{move}_{m,B}$	5	X_{75}										
X_{52}	$\mathrm{exp\_affect}_{b,B}$	5	X_{76}										
X_{53}	$\mathrm{talk}_{B,A,v}$	5	X_{77}										
X_{54}	$\mathrm{ws}_{m,A,B}$									1	1		
X_{55}	$\mathrm{ws}_{b,A,B}$									1	1		
X_{56}	$\mathrm{ws}_{v,A,B}$									1	1		
X_{57}	$\mathrm{ws}_{m,B,A}$									1	1		
X_{58}	$\mathrm{ws}_{b,B,A}$									1	1		
X_{59}	$\mathrm{ws}_{v,B,A}$									1	1		
X_{60}	$\mathbf{W}_{\mathrm{exec\text{-}ws}\_m,A,B}$					60	30						
X_{61}	$\mathbf{W}_{\mathrm{exec\text{-}ws}\_b,A,B}$					60	30						
X_{62}	$\mathbf{W}_{\mathrm{exec\text{-}ws}\_v,A,B}$					60	30						
X_{63}	$\mathbf{W}_{\mathrm{exec\text{-}ws}\_m,B,A}$					60	30						
X_{64}	$\mathbf{W}_{\mathrm{exec\text{-}ws}\_b,B,A}$					60	30						
X_{65}	$\mathbf{W}_{\mathrm{exec\text{-}ws}\_v,B,A}$					60	30						
X_{66}	$\mathbf{T}_{\mathrm{rep}\_m,A}$	5	0.5										
X_{67}	$\mathbf{T}_{\mathrm{rep}\_b,A}$	5	0.5										
X_{68}	$\mathbf{T}_{\mathrm{rep}\_v,A}$	5	0.5										
X_{69}	$\mathbf{T}_{\mathrm{exec}\_m,A}$	5	0.5										
X_{70}	$\mathbf{T}_{\mathrm{exec}\_b,A}$	5	0.5										
X_{71}	$\mathbf{T}_{\mathrm{exec}\_v,A}$	5	0.5										
X_{72}	$\mathbf{T}_{\mathrm{rep}\_m,B}$	5	0.5										
X_{73}	$\mathbf{T}_{\mathrm{rep}\_b,B}$	5	0.5										
X_{74}	$\mathbf{T}_{\mathrm{rep}\_v,B}$	5	0.5										
X_{75}	$\mathbf{T}_{\mathrm{exec}\_m,B}$	5	0.5										
X_{76}	$\mathbf{T}_{\mathrm{exec}\_b,B}$	5	0.5										
X_{77}	$\mathbf{T}_{\mathrm{exec}\_v,B}$	5	0.5										
X_{78}	$\mathbf{H}_{\mathbf{T}_A}$	5	0.5										
X_{79}	$\mathbf{H}_{\mathbf{T}_B}$	5	0.5										

Table 8 Role matrix **ms** for speed factors and **iv** for initial values

ms	speed factors	1	iv initial values	1
X_1	$ws_{s,A}$	1		1
X_2	$ws_{s,B}$	1		0
X_3	$ws_{s,AB}$	1		0
X_4	$con_{th,A}$	0		1
X_5	$con_{th,B}$	0		1
X_6	$sense_{s,A}$	1		0
X_7	$sense_{B,m,A}$	1		0
X_8	$sense_{B,b,A}$	1		0
X_9	$sense_{B,v,A}$	1		0
X_{10}	$rep_{s,A}$	1		0
X_{11}	$rep_{B,m,A}$	1		0
X_{12}	$rep_{B,b,A}$	1		0
X_{13}	$rep_{B,v,A}$	1		0
X_{14}	$prep_{m,A}$	1		0
X_{15}	$prep_{b,A}$	1		0
X_{16}	$prep_{v,A}$	1		0
X_{17}	$cons\_emotion_{b,A}$	1		0
X_{18}	$intrasyncdet_{A,\text{b-m}}$	0.5		0
X_{19}	$intrasyncdet_{A,\text{m-v}}$	0.5		0
X_{20}	$intrasyncdet_{A,\text{b-v}}$	0.5		0
X_{21}	$intersyncdet_{B,A,\text{m}}$	0.5		0
X_{22}	$intersyncdet_{B,A,\text{b}}$	0.5		0
X_{23}	$intersyncdet_{B,A,\text{v}}$	0.5		0
X_{24}	$transsyncdet_{B,A,\text{m}}$	0.5		0
X_{25}	$transsyncdet_{B,A,\text{b}}$	0.5		0
X_{26}	$transsyncdet_{B,A,\text{v}}$	0.5		0
X_{27}	$move_{m,A}$	1		0
X_{28}	$exp\_affect_{b,A}$	1		0
X_{29}	$talk_{A,B,v}$	1		0
X_{30}	$sense_{s,B}$	1		0
X_{31}	$sense_{A,m,B}$	1		0
X_{32}	$sense_{A,b,B}$	1		0
X_{33}	$sense_{A,v,B}$	1		0
X_{34}	$rep_{s,B}$	1		0
X_{35}	$rep_{A,m,B}$	1		0
X_{36}	$rep_{A,b,B}$	1		0
X_{37}	$rep_{A,v,B}$	1		0
X_{38}	$prep_{m,B}$	1		0
X_{39}	$prep_{b,B}$	1		0
X_{40}	$prep_{v,B}$	1		0

(continued)

Table 8 (continued)

X_{41}	cons_emotion$_{b,B}$	1		0
X_{42}	intrasyncdet$_{B,\text{b-m}}$	0.5		0
X_{43}	intrasyncdet$_{B,\text{m-v}}$	0.5		0
X_{44}	intrasyncdet$_{B,\text{b-v}}$	0.5		0
X_{45}	intersyncdet$_{A,B,\text{m}}$	0.5		0
X_{46}	intersyncdet$_{A,\text{B,b}}$	0.5		0
X_{47}	intersyncdet$_{A,\text{B,v}}$	0.5		0
X_{48}	transsyncdet$_{A,B,\text{m}}$	0.5		0
X_{49}	transsyncdet$_{A,B,\text{b}}$	0.5		0
X_{50}	transsyncdet$_{A,B,\text{v}}$	0.5		0
X_{51}	move$_{m,B}$	1		0
X_{52}	exp_affect$_{b,B}$	1		0
X_{53}	talk$_{B,A,v}$	1		0
X_{54}	ws$_{m,A,B}$	1		0
X_{55}	ws$_{b,A,B}$	1		0
X_{56}	ws$_{v,A,B}$	1		0
X_{57}	ws$_{m,B,A}$	1		0
X_{58}	ws$_{b,B,A}$	1		0
X_{59}	ws$_{v,B,A}$	1		0
X_{60}	$\mathbf{W}_{\text{exec-ws}\_m,A,B}$	1		0.4
X_{61}	$\mathbf{W}_{\text{exec-ws}\_b,A,B}$	1		0.4
X_{62}	$\mathbf{W}_{\text{exec-ws}\_v,A,B}$	1		0.4
X_{63}	$\mathbf{W}_{\text{exec-ws}\_m,B,A}$	1		0.4
X_{64}	$\mathbf{W}_{\text{exec-ws}\_b,B,A}$	1		0.4
X_{65}	$\mathbf{W}_{\text{exec-ws}\_v,B,A}$	1		0.4
X_{66}	$\mathbf{T}_{\text{rep}\_m,A}$	X_{78}		0.7
X_{67}	$\mathbf{T}_{\text{rep}\_b,A}$	X_{78}		0.7
X_{68}	$\mathbf{T}_{\text{rep}\_v,A}$	X_{78}		0.7
X_{69}	$\mathbf{T}_{\text{exec}\_m,A}$	X_{78}		0.7
X_{70}	$\mathbf{T}_{\text{exec}\_b,A}$	X_{78}		0.7
X_{71}	$\mathbf{T}_{\text{exec}\_v,A}$	X_{78}		0.7
X_{72}	$\mathbf{T}_{\text{rep}\_m,B}$	X_{79}		0.7
X_{73}	$\mathbf{T}_{\text{rep}\_b,B}$	X_{79}		0.7
X_{74}	$\mathbf{T}_{\text{rep}\_v,B}$	X_{79}		0.7
X_{75}	$\mathbf{T}_{\text{exec}\_m,B}$	X_{79}		0.7
X_{76}	$\mathbf{T}_{\text{exec}\_b,B}$	X_{79}		0.7
X_{77}	$\mathbf{T}_{\text{exec}\_v,B}$	X_{79}		0.7
X_{78}	$\mathbf{H}_{\mathbf{T}_A}$	1		0
X_{79}	$\mathbf{H}_{\mathbf{T}_B}$	1		0

Table 9 The combination functions used in the introduced network model

	Notation	Formula	Parameters
Advanced logistic sum	$\mathbf{alogistic}_{\sigma,\tau}(V_1, \ldots, V_k)$	$\left[\frac{1}{1+e^{-\sigma(V_1+\cdots+V_k-\tau)}} - \frac{1}{1+e^{\sigma\tau}}\right](1 + e^{-\sigma\tau})$	Steepness $\boldsymbol{\sigma}$ Excitability threshold $\boldsymbol{\tau}$
Complemental difference	$\mathbf{compdiff}(V_1, V_2)$	0 if $V_1 = V_2 = 0$ $1 - \frac{\lvert V_1 - V_2\rvert}{\max(V_1, V_2)}$ else	–
Complemental one-sided lagged average difference	$\mathbf{compdifflag1av}_{\iota\delta,\lambda}(W_1, W_2)$	$1 - \mathrm{av}\left(\frac{\lvert W_2(\iota\delta,1) - W_1(\iota\delta,\nu)\rvert}{\max(W_1(\iota\delta,\nu), W_2(\iota\delta,1))}\right)$ for $1 \leq \boldsymbol{\nu} \leq \boldsymbol{\lambda}$ (average)	$\boldsymbol{\iota\delta}$ state identifier $\boldsymbol{\lambda}$ max time lag
Average transition	$\mathbf{transdetavabs}_{\iota\delta,\sigma\omega}(W)$	$\lvert \mathrm{mean}_{1 \leq \nu \leq \sigma\omega}(W(\iota\delta, \nu)) - \mathrm{mean}_{\sigma\omega+1 \leq \nu \leq 2\sigma\omega}(W(\iota\delta, \nu))\rvert$	$\boldsymbol{\iota\delta}$ state identifier $\boldsymbol{\sigma\omega}$ sliding window size
Maxmin transition	$\mathbf{transdetmaxminabs}_{\iota\delta,\sigma\omega}(W)$	$\lvert \max_{1 \leq \nu \leq \sigma\omega}(W(\iota\delta, \nu)) - \min_{1 \leq \nu \leq \sigma\omega}(W(\iota\delta, \nu))\rvert$	$\boldsymbol{\iota\delta}$ state identifier $\boldsymbol{\sigma\omega}$ sliding window size
Standard deviation transition	$\mathbf{transdetstdev}_{\iota\delta,\sigma\omega}(W)$	$2\sqrt{[\mathrm{mean}_{1\leq\nu\leq\sigma\omega}(W(\iota\delta, \nu) - \mathrm{mean}(W))^2]}$	$\boldsymbol{\iota\delta}$ state identifier $\boldsymbol{\sigma\omega}$ sliding window size
Stepmod	$\mathbf{stepmod}_{\rho,\delta}(V)$	0 if $0 \leq$ time t mod $\boldsymbol{\rho} \leq \boldsymbol{\delta}$ 1 else	Repitition $\boldsymbol{\rho}$ Step time $\boldsymbol{\delta}$
Stepmodopp	$\mathbf{stepmodopp}_{\rho,\delta}(V)$	1 if $0 \leq$ time t mod $\boldsymbol{\rho} \leq \boldsymbol{\delta}$ 0 else	Repitition $\boldsymbol{\rho}$ Step time $\boldsymbol{\delta}$
Euclidean	$\mathbf{eucl}_{n,\lambda}(V_1, \ldots, V_k)$	$\sqrt[n]{\frac{V_1^n + \ldots + V_k^n}{\lambda}}$	Order $\boldsymbol{n}$ Scaling factor $\boldsymbol{\lambda}$

Table 10 Base states of the computational network model

state	Name	Explanation
X_1	$ws_{s,A}$	World state for stimulus *s* for *A*
X_2	$ws_{s,B}$	World state for stimulus *s* for *B*
X_3	$ws_{s,AB}$	World state for stimulus *s* both for *A* and *B*
X_4	$con_{th,A}$	Context state for excitability threshold for *A*
X_5	$con_{th,B}$	Context state for excitability threshold for *B*
X_6	$sense_{s,A}$	Sensor state of *A* for stimulus *s* for *A*
X_7	$sense_{B,m,A}$	Sensor state of *A* for movement *m* of *B*
X_8	$sense_{B,b,A}$	Sensor state of *A* for expressed affective response *b* of *B*
X_9	$sense_{B,v,A}$	Sensor state of *A* for verbal action *v* of *B*
X_{10}	$rep_{s,A}$	Sensory representation state of *A* for stimulus *s* for *A*
X_{11}	$rep_{B,m,A}$	Sensory representation state of *A* for movement *m* of *B*
X_{12}	$rep_{B,b,A}$	Sensory representation state of *A* for expressed affective response *b* of *B*
X_{13}	$rep_{B,v,A}$	Sensory representation state of *A* for verbal action *v* of *B*
X_{14}	$prep_{m,A}$	Preparation state for movement *m* of *A*
X_{15}	$prep_{b,A}$	Preparation state for affective response *b* of *A*
X_{16}	$prep_{v,A}$	Preparation state for verbal action *v* of *A*
X_{17}	$cons\_emotion_{b,A}$	Conscious emotion state for *b* of *A*
X_{18}	$intrasyncdet_{A,\text{b-m}}$	Intrapersonal synchrony detection of *A* for executing *b* and *m* by *A*
X_{19}	$intrasyncdet_{A,\text{m-v}}$	Intrapersonal synchrony detection of *A* for executing *m* and *v* by *A*
X_{20}	$intrasyncdet_{A,\text{b-v}}$	Intrapersonal synchrony detection of *A* for executing *b* and *v* by *A*
X_{21}	$intersyncdet_{B,A,\text{m}}$	Interpersonal synchrony detection of *A* for executing *m* by *B* and *A*
X_{22}	$intersyncdet_{B,A,\text{b}}$	Interpersonal synchrony detection of *A* for executing *b* by *B* and *A*
X_{23}	$intersyncdet_{B,A,\text{v}}$	Interpersonal synchrony detection of *A* for executing *v* by *B* and *A*
X_{24}	$transsyncdet_{B,A,\text{m}}$	Synchrony transition detection of *A* for executing *m* by *B* and *A*
X_{25}	$transsyncdet_{B,A,\text{b}}$	Synchrony transition detection of *A* for executing *b* by *B* and *A*
X_{26}	$transsyncdet_{B,A,\text{v}}$	Synchrony transition detection of *A* for executing *v* by *B* and *A*
X_{27}	$move_{m,A}$	Executing movement *m* by *A*
X_{28}	$exp\_affect_{b,A}$	Executing expression of *b* by *A*
X_{29}	$talk_{A,B,v}$	Executing verbal action *v* by *A*
X_{30}	$sense_{s,B}$	Sensor state of *B* for stimulus *s* for *B*
X_{31}	$sense_{A,m,B}$	Sensor state of *B* for movement *m* of *A*
X_{32}	$sense_{A,b,B}$	Sensor state of *B* for expressed affective response *b* of *A*
X_{33}	$sense_{A,v,B}$	Sensor state of *B* for verbal action *v* of *A*
X_{34}	$rep_{s,B}$	Sensory representation state of *B* for stimulus *s* for *B*
X_{35}	$rep_{A,m,B}$	Sensory representation state of *B* for movement *m* of *A*
X_{36}	$rep_{A,b,B}$	Sensory representation state of *B* for expressed affective response *b* of *A*
X_{37}	$rep_{A,v,B}$	Sensory representation state of *B* for verbal action *v* of *A*
X_{38}	$prep_{m,B}$	Preparation state for movement *m* of *B*
X_{39}	$prep_{b,B}$	Preparation state for affective response *b* of *B*
X_{40}	$prep_{v,B}$	Preparation state for verbal action *v* of *B*
X_{41}	$cons\_emotion_{b,B}$	Conscious emotion state for *b* of *B*
X_{42}	$intrasyncdet_{B,\text{b-m}}$	Intrapersonal synchrony detection of *B* for executing *b* and *m* by *B*
X_{43}	$intrasyncdet_{B,\text{m-v}}$	Intrapersonal synchrony detection of *B* for executing *m* and *v* by *B*

(continued)

Table 10 (continued)

X_{44}	intrasyncdet$_{B,b\text{-}v}$	Intrapersonal synchrony detection of *B* for executing *b* and *v* by *B*
X_{45}	intersyncdet$_{A,B,m}$	Interpersonal synchrony detection of *B* for executing *m* by *A* and *B*
X_{46}	intersyncdet$_{A,B,b}$	Interpersonal synchrony detection of *B* for executing *b* by *A* and *B*
X_{47}	intersyncdet$_{A,B,v}$	Interpersonal synchrony detection of *B* for executing *v* by *A* and *B*
X_{48}	transsyncdet$_{A,B,m}$	Synchrony transition detection of *B* for executing *m* by *A* and *B*
X_{49}	transsyncdet$_{A,B,b}$	Synchrony transition detection of *B* for executing *b* by *A* and *B*
X_{50}	transsyncdet$_{A,B,v}$	Synchrony transition detection of *B* for executing *v* by *A* and *B*
X_{51}	move$_{m,B}$	Executing movement *m* by *B*
X_{52}	exp_affect$_{b,B}$	Executing expression of *b* by *B*
X_{53}	talk$_{B,A,v}$	Executing verbal action *v* by *B*
X_{54}	ws$_{m,A,B}$	World state for transmitting movement *m* of *A* to *B*
X_{55}	ws$_{b,A,B}$	World state for transmitting affective response *b* of *A* to *B*
X_{56}	ws$_{v,A,B}$	World state for transmitting verbal action *v* of *A* to *B*
X_{57}	ws$_{m,B,A}$	World state for transmitting movement *m* of *B* to *A*
X_{58}	ws$_{b,B,A}$	World state for transmitting affective response *b* of *B* to *A*
X_{59}	ws$_{v,B,A}$	World state for transmitting verbal action *v* of *B* to *A*

Table 11 First-order self-model **T**-states for excitability thresholds and second-order self-model $\mathbf{H_T}$-states for the adaptation speed of the **T**-states of the computational network model

state	name	Explanation
X_{66}	$\mathbf{T}_{rep\_m,A}$	First-order self-model state for the excitability threshold of *A*'s sensory representation state rep$_{m,A}$ for movement *m*
X_{67}	$\mathbf{T}_{rep\_b,A}$	First-order self-model state for the excitability threshold of *A*'s sensory representation state rep$_{b,A}$ for affective response *b*
X_{68}	$\mathbf{T}_{rep\_v,A}$	First-order self-model state for the excitability threshold of *A*'s sensory representation state rep$_{v,A}$ for verbal response *v*
X_{69}	$\mathbf{T}_{exec\_m,A}$	First-order self-model state for the excitability threshold of *A*'s execution state state move$_{m,A}$ for movement *m*
X_{70}	$\mathbf{T}_{exec\_b,A}$	First-order self-model state for the excitability threshold of *A*'s execution state state exp_affect$_{b,A}$ for affective response *b*
X_{71}	$\mathbf{T}_{exec\_v,A}$	First-order self-model state for the excitability threshold of *A*'s execution state state talk$_{v,A}$ for verbal response *v*
X_{72}	$\mathbf{T}_{rep\_m,B}$	First-order self-model state for the excitability threshold of B's sensory representation state rep$_{m,B}$ for movement *m*
X_{73}	$\mathbf{T}_{rep\_b,B}$	First-order self-model state for the excitability threshold of *B*'s sensory representation state rep$_{b,B}$ for affective response *b*
X_{74}	$\mathbf{T}_{rep\_v,B}$	First-order self-model state for the excitability threshold of *B*'s sensory representation state rep$_{v,B}$ for verbal response *v*
X_{75}	$\mathbf{T}_{exec\_m,B}$	First-order self-model state for the excitability threshold of *B*'s execution state state move$_{m,B}$ for movement *m*
X_{76}	$\mathbf{T}_{exec\_b,B}$	First-order self-model state for the excitability threshold of *B*'s execution state state exp_affect$_{b,B}$ for affective response *b*
X_{77}	$\mathbf{T}_{exec\_v,B}$	First-order self-model state for the excitability threshold of *B*'s execution state state talk$_{v,B}$ for verbal response *v*
X_{78}	$\mathbf{H}_{\mathbf{T}_A}$	Second-order self-model state for the speed factor of the first-order self-model **T**-states for *A*
X_{79}	$\mathbf{H}_{\mathbf{T}_B}$	Second-order self-model state for the speed factor of the first-order self-model **T**-states for *B*

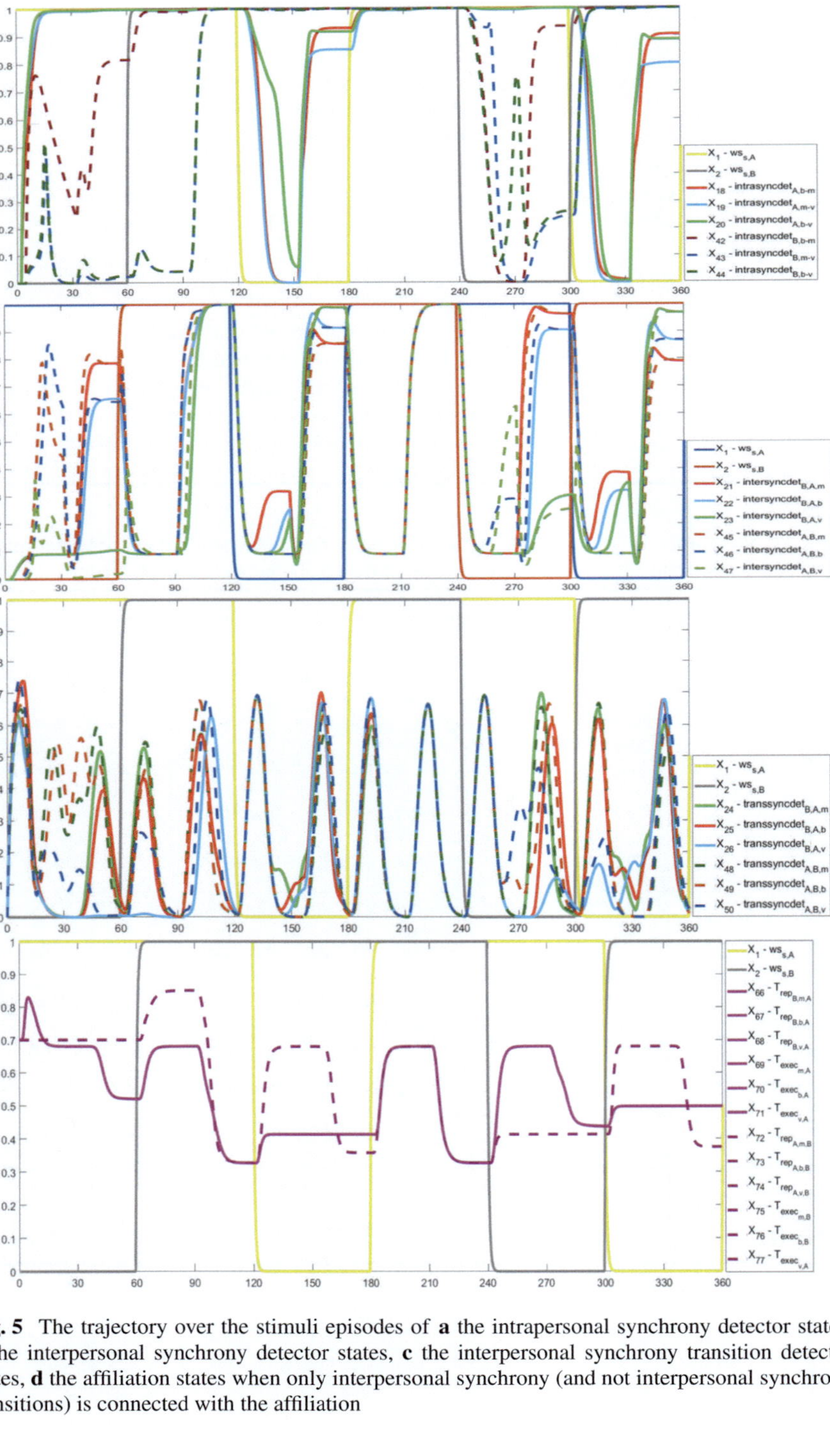

Fig. 5 The trajectory over the stimuli episodes of **a** the intrapersonal synchrony detector states, **b** the interpersonal synchrony detector states, **c** the interpersonal synchrony transition detector states, **d** the affiliation states when only interpersonal synchrony (and not interpersonal synchrony transitions) is connected with the affiliation

References

1. Abraham, W.C., Bear, M.F.: Metaplasticity: the plasticity of synaptic plasticity. Trends Neurosci. **19**(4), 126–130 (1996)
2. Accetto, M., Treur, J., Villa, V.: An adaptive cognitive-social model for mirroring and social bonding during synchronous joint action. Procedia Comput. Sci. **145**, 3–12 (2018). https://doi.org/10.1016/j.procs.2018.11.002
3. Beebe, B., Steele, M.: How does microanalysis of mother–infant communication inform maternal sensitivity and infant attachment? Attach. Hum. Dev. **15**(5–6), 583–602 (2013). https://doi.org/10.1080/14616734.2013.841050
4. Chandra, N., Barkai, E.: A non-synaptic mechanism of complex learning: modulation of intrinsic neuronal excitability. Neurobiol. Learn. Mem. **154**, 30–36 (2018)
5. Dahan, A., Noy, L., Hart, Y., Mayo, A., Alon, U.: Exit from synchrony in joint improvised motion. PLoS One **11**(10), e0160747 (2016). https://doi.org/10.1371/journal.pone.0160747
6. Damasio, A.R.: The Feeling of What Happens: Body and Emotion in the Making of Consciousness. Houghton Mifflin Harcourt (1999)
7. Debanne, D., Inglebert, Y., Russier, M.: Plasticity of intrinsic neuronal excitability. Curr. Opin. Neurobiol. **54**, 73–82 (2019)
8. Deres-Cohen, K., Dolev-Amit, T., Peysachov, G., Ramseyer, F.T., Zilcha-Mano, S.: Nonverbal synchrony as a marker of alliance ruptures. Psychotherapy **58**(4), 499–509 (2021)
9. Fairhurst, M.T., Janata, P., Keller, P.E.: Being and feeling in sync with an adaptive virtual partner: brain mechanisms underlying dynamic cooperativity. Cereb. Cortex. **23**(11), 2592–2600 (2013). https://doi.org/10.1093/cercor/bhs243
10. Feldman, R.: Parent–infant synchrony biological foundations and developmental outcomes. Curr. Dir. Psychol. Sci. **16**, 340–345 (2007). https://doi.org/10.1111/j.1467-8721.2007.00532.x
11. García, E., Di Paolo, E.A.: Embodied coordination and psychotherapeutic outcome: beyond direct mappings. Front. Psychol. 1257 (2018)
12. Hendrikse, S.C.F., Kluiver, S., Treur, J., Wilderjans, T.F., Dikker, S., Koole, S.L.: How virtual agents can learn to synchronize: an adaptive joint decision-making model of psychotherapy. Cogn. Syst. Res. 79, 138-155 (2022a)
13. Hendrikse, S.C.F., Treur, J., Koole, S.L. (eds.): New Analysis and Modeling Directions in Social Interaction Science: Emergent Multimodal Interpersonal Synchrony, Affiliation, and Bonding. Springer Nature (2025a)
14. Hendrikse, S.C.F., Treur, J., Koole, S.L.: Introduction to the role of adaptivity in social interaction science. In: Hendrikse, S.C.F., Treur, J., Koole, S.L. (eds.) New Analysis and Modeling Directions in Social Interaction Science: Emergent Multimodal Interpersonal Synchrony, Affiliation, and Bonding, Ch 1. Springer Nature (2025b)
15. Hendrikse, S.C.F., Treur, J., Koole, S.L.: The adaptive interplay of interpersonal synchrony, short-term affiliation and long-term bonding: a multi-timescale multi-adaptive agent model. In: Hendrikse, S.C.F., Treur, J., Koole, S.L. (eds.) New Analysis and Modeling Directions in Social Interaction Science: Emergent Multimodal Interpersonal Synchrony, Affiliation, and Bonding, Ch 12. Springer Nature (2025c)
16. Hendrikse, S.C.F., Treur, J., Koole, S.L.: A multi-timescale multi-adaptive dynamical system approach for relationship-specific and relationship-independent adaptivity in affiliation and bonding. In: Hendrikse, S.C.F., Treur, J., Koole, S.L. (eds.) New Analysis and Modeling Directions in Social Interaction Science: Emergent Multimodal Interpersonal Synchrony, Affiliation, and Bonding, Ch 13. Springer Nature (2025d)
17. Hendrikse, S.C.F., Treur, J., Wilderjans, T.F., Dikker, S., Koole, S.L.: On the same wavelengths: emergence of multiple synchronies among multiple agents. In: Proceedings of the 22nd International Workshop on Multi-Agent-Based Simulation, MABS'21. Lecture Notes in AI, vol. 13128, pp. 57–71. Springer Nature (2022b)
18. Hendrikse, S.C.F., Treur, J., Wilderjans, T.F., Dikker, S., Koole, S.L.: On the interplay of interpersonal synchrony, short-term affiliation and long-term bonding: a second-order multi-adaptive neural agent model. In: Maglogiannis, et al. (eds.) Proceedings of the 18th International

Conference on Artificial Intelligence Applications and Innovations, AIAI'22. Advances in Information and Communication Technology, vol. 646, pp. 37–57. Springer Nature (2022d)
19. Hendrikse, S.C.F., Treur, J., Wilderjans, T.F., Dikker, S., Koole, S.L.: Becoming attuned to each other over time: a computational neural agent model for the role of time lags in subjective synchrony detection and related behavioral adaptivity. In: Proceedings of the 15th International Conference on Brain Informatics, BI'22. Lecture Notes in AI, vol 13406, pp 369–383. Springer Nature (2022e)
20. Hendrikse, S.C.F., Treur, J., Wilderjans, T.F., Dikker, S., Koole, S.L.: Switching in and out of sync: a controlled adaptive network model of transition dynamics in the effects of interpersonal synchrony on affiliation. In: Proceedings of the 11th International Conference on Complex Networks and their Applications. Studies in Computational Intelligence, vol. 1078, pp. 81–95. Springer Nature Switzerland, Cham (2022f)
21. Hesslow, G.: Conscious thought as simulation of behaviour and perception. Trends Cogn. Sci. **6**, 242–247 (2002)
22. Hove, M.J., Risen, J.L.: It's all in the timing: interpersonal synchrony increases affiliation. Soc. Cogn. **27**(6), 949–960 (2009)
23. Kirschner, S., Tomasello, M.: Joint music making promotes prosocial behavior in 4-year-old children. Evol. Hum. Behav. **31**, 354–364 (2010). https://doi.org/10.1016/j.evolhumbehav.2010.04.004
24. Koole, S.L., Tschacher, W.: Synchrony in psychotherapy: a review and an integrative framework for the therapeutic alliance. Front. Psychol. **7**, 862 (2016)
25. Koole, S.L., Tschacher, W., Butler, E., Dikker, S., Wilderjans, T.F.: In sync with your shrink. In: Forgas, J.P., Crano, W.D., Fiedler, K. (eds.) Applications of Social Psychology, pp. 161–184. Taylor and Francis, Milton Park (2020)
26. Maurer, R.E., Tindall, J.H.: Effect of postural congruence on client's perception of counselor empathy. J. Counsel. Psychol. **30**, 158 (1983). https://doi.org/10.1037/0022-0167.30.2.158
27. Mayo, O., Gordon, I.: In and out of synchrony—behavioral and physiological dynamics of dyadic interpersonal coordination. Psychophysiology **57**, e13574 (2020)
28. Palumbo, R.V., Marraccini, M.E., Weyandt, L.L., Wilder-Smith, O., McGee, H.A., Liu, S., Goodwin, M.S.: Interpersonal autonomic physiology: a systematic review of the literature. Pers. Soc. Psychol. Rev. **21**(2), 99–141 (2017)
29. Paulick, J., Deisenhofer, A.-K., Ramseyer, F., Tschacher, W., Boyle, K., Rubel, J., Lutz, W.: Nonverbal synchrony: a new approach to better understand psychotherapeutic processes and drop-out. J. Psychother. Integr. **28**(3), 367–384 (2018). https://doi.org/10.1037/int0000099
30. Prince, K., Brown, S.: Neural correlates of partnered interaction as revealed by cross-domain ALE meta-analysis. Psychol. Neurosci. **15**(1), 1–13 (2022). https://doi.org/10.1037/pne0000282
31. Ramseyer, F., Tschacher, W.: Nonverbal synchrony in psychotherapy: coordinated body movement reflects relationship quality and outcome. J. Consult. Clin. Psychol. **79**, 284–295 (2011). https://doi.org/10.1037/a0023419a
32. Ravreby, I., Shilat, Y., Yeshurun, Y.: Liking as a balance between synchronization, complexity and novelty. Sci. Rep. **12**(1), 3181 (2022). https://doi.org/10.1038/s41598-022-06610-z
33. Robinson, B.L., Harper, N.S., McAlpine, D.: Meta-adaptation in the auditory midbrain under cortical influence. Nat. Commun. **7**, e13442 (2016)
34. Sharpley, C.F., Halat, J., Rabinowicz, T., Weiland, B., Stafford, J.: Standard posture, postural mirroring and client-perceived rapport. Couns. Psychol. Q. **14**, 267–280 (2001). https://doi.org/10.1080/09515070110088843
35. Tarr, B., Launay, J., Dunbar, R.I.M.: Silent disco: dancing in synchrony leads to elevated pain thresholds and social closeness. Evol. Hum. Behav. **37**(5), 343–349 (2016)
36. Treur, J.: Network-Oriented Modeling for Adaptive Networks: Designing Higher-Order Adaptive Biological, Mental and Social Network Models. Springer Nature (2020a)
37. Treur, J.: Modeling multi-order adaptive processes by self-modeling networks. In: Tallón-Ballesteros, A.J., Chen, C.-H. (eds.) Proceedings of the 2nd International Conference on Machine Learning and Intelligent Systems, MLIS'20. Frontiers in Artificial Intelligence and Applications, vol. 332, pp. 206–217. IOS Press (2020b)

38. Treur, J.: Modeling coevolution of dynamics, adaptivity and control of adaptivity by self-modeling networks. In: Hendrikse, S.C.F., Treur, J., Koole, S.L. (eds.) New Analysis and Modeling Directions in Social Interaction Science: Emergent Multimodal Interpersonal Synchrony, Affiliation, and Bonding, Ch 2. Springer Nature (2025)
39. Trout, D.L., Rosenfeld, H.M.: The effect of postural lean and body congruence on the judgment of psychotherapeutic rapport. J. Nonverbal Behav. **4**, 176–190 (1980)
40. Tschacher, W., Ramseyer, F., Koole, S.L.: Sharing the now in the social present: duration of nonverbal synchrony is linked with personality. J. Pers. **86**(2), 129–138 (2018)
41. Valdesolo, P., DeSteno, D.: Synchrony and the social tuning of compassion. Emotion **11**, 262 (2011). https://doi.org/10.1037/a0021302
42. Valdesolo, P., Ouyang, J., DeSteno, D.: The rhythm of joint action: synchrony promotes cooperative ability. J. Exp. Soc. Psychol. **46**(4), 693–695 (2010)
43. Williams, A.H., O'Leary, T., Marder, E.: Homeostatic regulation of neuronal excitability. Scholarpedia **8**, 1656 (2013)
44. Wiltermuth, S.S., Heath, C.: Synchrony and cooperation. Psychol. Sci. **20**(1), 1–5 (2009)
45. Zhang, A., Li, X., Gao, Y., Niu, Y.: Event-driven intrinsic plasticity for spiking convolutional neural networks. IEEE Trans. Neural Netw. Learn. Syst. (2021). https://doi.org/10.1109/tnnls.2021.3084955

The Emergence of Group Synchrony and Adaptivity for Group Bonding: A Multi-Adaptive Network Model

Francesco Mattera, Sophie C. F. Hendrikse, and Jan Treur

Abstract The research reported here analyses the relationship between group synchrony and group bonding through a novel adaptive computational dynamical system model. By simulating multimodal interactions within a group of four agents, the study uncovers patterns in group cohesion in the sense of emerging multimodal group synchrony and related group bonding. Findings include patterns for emerging group synchrony (logarithmical) and group bonding (logistic). The obtained insights offer an understanding of group interaction dynamics. Future research may consider larger groups and more variations of synchrony detection functions to widen the obtained findings.

Keywords Group dynamics · Adaptive network model · Interpersonal · Group synchrony · Group bonding

1 Introduction

Interpersonal and group relationships frequently start from a sense of synchrony, which may arise in many different ways, and is an essential and crucial part of the way we bond with others. One important form of synchrony is called multimodal interpersonal synchrony [4]. This occurs when two or more people spontaneously attune their speech, feelings and movement across time [18, 31]. While previous results have examined this synchrony in pairs of agents computationally, we want to

F. Mattera · J. Treur (✉)
Social AI Group, Department of Computer Science, Vrije Universiteit Amsterdam, Amsterdam, The Netherlands
e-mail: j.treur@vu.nl

S. C. F. Hendrikse
Department of Methodology and Statistics, Tilburg School of Social and Behavioral Sciences, Tilburg University, Tilburg, The Netherlands
e-mail: S.C.F.Hendrikse@tilburguniversity.edu

S. C. F. Hendrikse et al. (eds.), *New Analysis and Modeling Directions in Social Interaction Science*, Studies in Systems, Decision and Control 614,
https://doi.org/10.1007/978-3-031-99968-0_16

expand our knowledge on the role it plays in dynamics of larger groups: dynamics that involve many more lines of interaction between agents [1].

In [14] the concept of subjective synchrony detection state was introduced and used to show how emerging multimodal synchrony can lead to behavioral adaptivity concerning bonding, see also this volume, [10], Chap. 11 [15]. Similarly, for bonding by homophily, in [35] the concept of subjective representation states for homophily was introduced and related to bonding, applied in [2, 23, 20], see also this volume [10], Chap. 7 [36] and Chaps. 8–10 [3, 24, 25]. In all these cases only dyads were considered.

In line with the above, this chapter proposes the idea of subjective group synchrony detection states in addition to those for dyads only. These states co-exists with the dyad synchrony detection states and are related to behavioural adaptivity in the form of group bonding. The scenario for this study focuses on four agents and their multimodel interactions within the group for the three modalities verbal action, movement and expressed affective response.

To analyse this computationally, we developed a novel agent-based adaptive dynamical system model. The latter is a temporal-causal adaptive network, including synchrony and bonding states for both pairs of agents and the whole group.

By this, we aim to contribute to the growing field of research on interpersonal synchrony and its relation to behavioral adaptivity underlying bonding. The analysis by the model and the model's results may be applied to the psychological field, for example, aiming for synchrony to enhance bonding in group formation and group therapy [26, 29].

2 Background Research

Behavioral adaptivity underlying bonding, is often linked to the concept of synchrony, which plays a key role in communication and how relationships work [38]. Many studies, especially in the field of psychotherapy, have shown that therapists who consciously synchronize their movements with their clients have a better perspective because they are felt to show insight and understanding [29] and lead to better affiliation or bonding between them [22]. For more background about this, see this volume [10], Chap. 1 [11]. As previous studies on this matter showed how interpersonal synchrony between two agents can predict their cohesion (bonding), this study aims at answering the same question for a group of agents instead. Note that the focus is not on straightforward application of existing methods applied to all dyads within a group and aggregating them, but on whole group phenomena, following literature in this area such as [30, 32, 33, 39, 41]. However, next to this whole group perspective, the dyadic perspective on synchrony will be covered as well in the model for the purpose of comparison.

In order for us to draw insights from the model, some main assumptions need to be made beforehand. The first assumption is that all agents unconsciously and subjectively detect group synchrony [27]. By that, we mean that all agents are able to perform and detect synchrony patterns, having hypothetical internal states for holding information about it. These detection states will be differentiated to address both pair-wise interpersonal synchrony and group-wise synchrony. The latter can be analysed in an empirical context by applying multivariate synchrony detection methods [16, 21, 17] to measured data for movements, expressed emotions or verbal utterances of all group members (not to the dyads), for example. The second main assumption is that bonding between agents happens via (behavioral) adaptivity of connections involved in their interaction. This can be analysed in an empirical context based on observing the quality of coordination, cooperation or affiliation as a group, for example as reported via questionnaires, e.g., [5, 17].

Note that we are building the novel computational model to have pathways between interactions within the group, emerging synchrony patterns, synchrony detection states and the adaptive connections related to their bonding. Overall, the analysed pathways roughly form a circular structure as follows:

$$\begin{aligned}&\text{group interaction} \rightarrow \text{synchrony patterns}\\&\rightarrow \text{detected synchrony} \rightarrow \text{adaptation for bonding}\\&\rightarrow \text{group interaction} \rightarrow \ldots.\end{aligned}$$

This means that there is not one unidirectional form of causality from one to the other but an adaptive dynamical system describing how both synchrony and bonding can emerge and strengthen via their mutual dynamic and adaptive influences. We believe that studying synchronization within groups and their related behavioural adaptivity from an adaptive dynamical system perspective will provide a better understanding of the dynamics and adaptivity of human groups and add new perspectives to the field.

3 Network Representations for Adaptive Dynamical Systems

In this section, we discuss the modeling approach for adaptive dynamical systems by their self-modeling temporal-causal network representation as described in [34, 8], see also this volume [10], Chaps. 2 and 12. This approach has been used for the adaptive dynamical system model introduced here. Temporal-causal networks models describe a number of states that can change their value overtime, like in our simulations. These types of models can be represented in a conceptual representation, as in Figs. 3 and 4 (a labeled graph), or in Fig. 2 (a matrix). The network characteristics specifying these models are listed here:

Connectivity Characteristics: These define how different states X and Y of the model are connected by some weight $\boldsymbol{\omega}_{X,Y}$.

Aggregation Characteristics: These define how each state applies a combination function $\mathbf{c}_Y$ on every state Y to aggregate its incoming impacts.

Timing Characteristics: These define for each state Y a speed factor $\boldsymbol{\eta}_Y$ controlling how fast it changes overtime.

Based on these network characteristics, the following difference equation is used to compute from a current time t the impact of all states with incoming connections on state Y at later time $t + \Delta t$:

$$Y(t + \Delta t) = Y(t) + \boldsymbol{\eta}_Y\left[\mathbf{c}_Y\left(\boldsymbol{\omega}_{X_1,Y}X_1(t), \ldots, \boldsymbol{\omega}_{X_k,Y}X_k(t)\right) - Y(t)\right]\Delta t \tag{1}$$

Here, X_i are the states from which state Y gets incoming connections.

To model adaptive networks, the self-modeling principle (also called reification) is used. This means that for any adaptive network characteristic a self-model state is added to the network that represents it. For example, for an adaptive connection weight $\boldsymbol{\omega}_{X,Y}$, a self-model state is $\mathbf{W}_{X,Y}$ added that represents this weight and for an adaptive speed factor $\boldsymbol{\eta}_Y$, a self-model state $\mathbf{H}_Y$ is added representing it; see [34]. Then within Eq. (1) the values of these self-model states are used for the network characteristics, for example:

$$Y(t + \Delta t) = Y(t) + \mathbf{H}_Y(t)\left[c_Y\left(W_{X_1,Y}(t)X_1(t), \ldots, W_{X_k,,Y}(t)X_k(t)\right) - Y(t)\right]\Delta t \tag{2}$$

This can also be done iteratively. For example, the speed of a self-model state $Z = \mathbf{W}_{X,Y}$ can be made adaptive by adding a self-model state $\mathbf{H}_Z = \mathbf{H}_{\mathbf{W}_{X,Y}}$. Thus an adaptive learning rate can be introduced, that can make learning context-sensitive. Such self-model states $\mathbf{H}_{\mathbf{W}_{X,Y}}$ are called second-order self-model states whereas self-model states $\mathbf{W}_{X,Y}$ are called first-order self-model states. This can be used to model first-order and second-order adaptation of networks.

For the adaptive network model introduced in this chapter, the bonding both between pairs of agents and for a group, we made it adaptive [40, 43, 34] see also this volume, [10], Chap. 2 [37]. This means that 2 self-modeling levels were added on top of the base level network model, as shown in Fig. 1. Starting from the bottom, we have a base level network holding the connections between agents and their synchrony states. The first-order self-model level has states for group bonding and bonding in pairs. Moreover, for each agent, the second-order self-model level holds a single state that controls the adaptation speed of the bonding states below it. In Fig. 1, the downward arrows represent the effectuation of the different orders of adaptiveness.

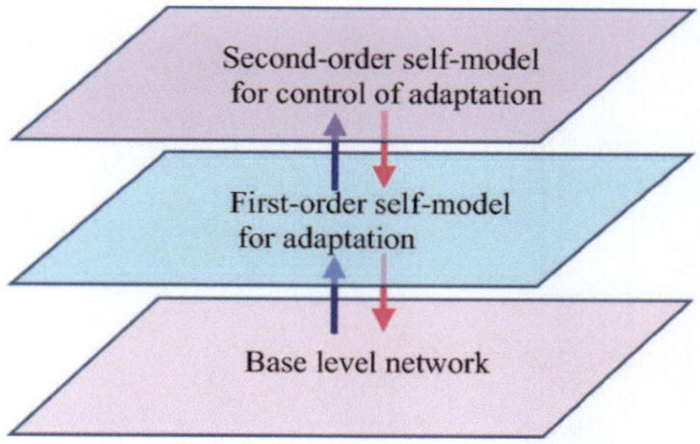

Fig. 1 Three-level network architecture used for the introduced network model for group synchrony and group bonding to obtain a second-order adaptive network

State	Name	Explanation	Level
X1	world_s,A	world state for stimulus s for A	base
X2	senseA,s	sensor state of A for stimulus s	base
X3	repA,s	sensory representation state of A for stimulus s	base
X4	senseB,A,m	sensor state of A for movement m of B	base
X5	senseB,A,b	sensor state of A for expressed affective response b of B	base
X6	senseB,A,v	sensor state of A for verbal action v of B	base
X7	senseC,A,m	sensor state of A for movement m of C	base
X8	senseC,A,b	sensor state of A for expressed affective response b of C	base
X9	senseC,A,v	sensor state of A for verbal action v of C	base
X10	senseD,A,m	sensor state of A for movement m of D	base
X11	senseD,A,b	sensor state of A for expressed affective response b of D	base
X12	senseD,A,v	sensor state of A for verbal action v of D	base
X13	repB,A,m	sensory representation state of A for movement m of B	base
X14	repB,A,b	sensory representation state of A for expressed affective response b of B	base
X15	repB,A,v	sensory representation state of A for verbal action v of B	base
X16	repC,A,m	sensory representation state of A for movement m of C	base
X17	repC,A,b	sensory representation state of A for expressed affective response b of C	base
X18	repC,A,v	sensory representation state of A for verbal action v of C	base
X19	repD,A,m	sensory representation state of A for movement m of D	base
X20	repD,A,b	sensory representation state of A for expressed affective response b of D	base
X21	repD,A,v	sensory representation state of A for verbal action v of D	base
X22	inter_group,v	intrapersonal group synchrony detection for executing v	base
X23	inter_group,b	intrapersonal group synchrony detection for executing b	base
X24	inter_group,m	intrapersonal group synchrony detection for executing m	base
X25	interAC,v	interpersonal synchrony detection of A for executing v by A and C	base
X26	interAC,b	interpersonal synchrony detection of A for executing b by A and C	base
X27	interAC,m	interpersonal synchrony detection of A for executing m by A and C	base
X28	interAB,v	interpersonal synchrony detection of A for executing v by A and B	base
X29	interAB,b	interpersonal synchrony detection of A for executing b by A and B	base
X30	interAB,m	interpersonal synchrony detection of A for executing m by A and B	base
X31	interAD,v	interpersonal synchrony detection of A for executing v by A and D	base
X32	interAD,b	interpersonal synchrony detection of A for executing b by A and D	base
X33	interAD,m	interpersonal synchrony detection of A for executing m by A and D	base
X34	prepA,v	preparation state for verbal action v of A	base
X35	prep_affectA,b	preparation state for expressed affective response b of A	base
X36	prepA,m	preparation state for movement m of A	base
X37	talkA,v	executing verbal action v by A	base
X38	exp_affectA,b	executing expression b by A	base
X39	moveA,m	executing movement m by A	base
X40	WBCD,A	group bonding	1st
X41	WB,A	bonding B & A	1st
X42	WC,A	bonding C & A	1st
X43	WD,A	bonding D & A	1st
X44	HW,A	self-model state W for A	2nd

Fig. 2 Explanation of all states for agent A

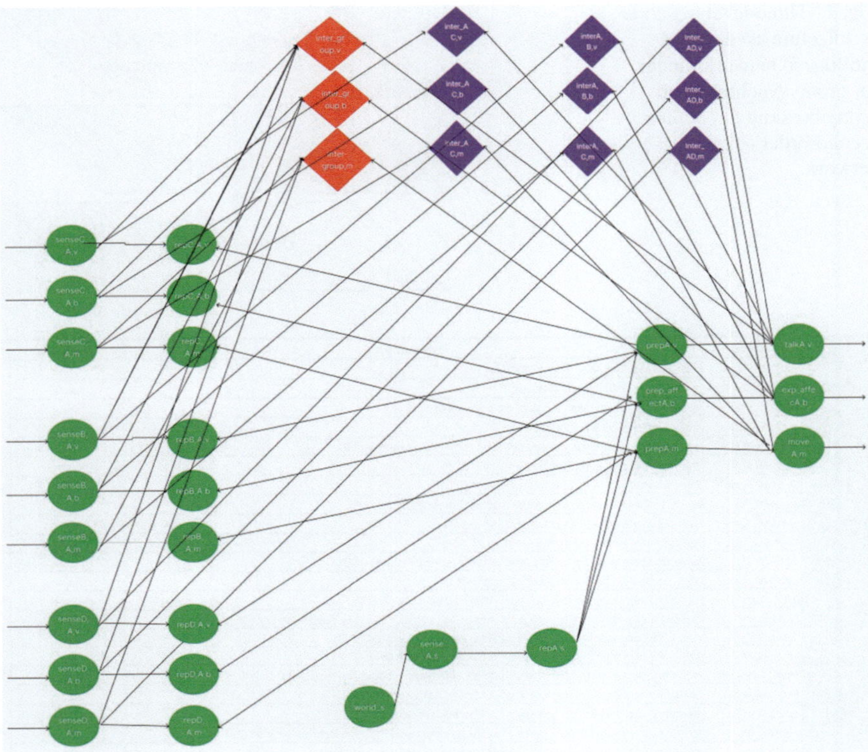

Fig. 3 The base level of the network architecture for agent A

4 A Network Model for Group Synchrony and Group Bonding

In this study, we focus on a group of four agents with (1) their emerging group synchronization and (2) their group bonding related to it [19]. To accomplish this, we have included for each agent a novel type of state for group synchrony detection to our base model that represents that the agent detects synchrony of all agents within the whole group. We designed the model in a way that each agent is able to sense and process three types of modalities: verbal action, movement and expressed affective response.

The design of the adaptive agent model uses a network architecture consisting of three levels, as explained earlier in Sect. 3, see Fig. 1. These levels include a base level, a first-order self-modeling level, and a second-order self-modeling level, with states as explained in Fig. 2: for each agent 39 states at the base level, 4 at the first-order and 1 at the second-order adaptation level. Since our study involves 4 agents designed in the same way, all with the same number and type of states, we will focus

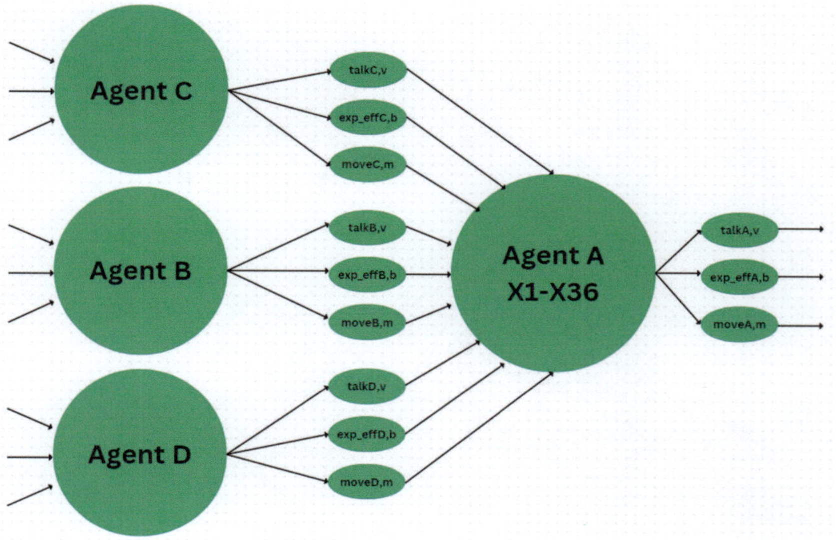

Fig. 4 How the different agents interact

only on how Agent A is structured and how it interacts with the other agents of the experiment. The base level structure of Agent A is shown in Fig. 3. In Fig. 4 it is shown how the different agents interact via three modalities.

To facilitate understanding, we refer to Figs. 3 and 4 to showcase the connections between base-level states in agent A. Figure 3 shows interpersonal group synchrony detection states and their connections as red diamonds with four incoming connections each for all four agents. In the same figure, purple diamonds represent interpersonal dyadic synchrony between pairs of agents, only having 2 inputs for each state for the two agents of the pair. Finally, the green circles represent sensor states (X_4 to X_{12}), sensory representation states (X_{13} to X_{21}), preparation states (X_{34} to X_{36}) and executing states (X_{37} to X_{39}). An understanding of how agents interact between each other at the base level can be found in Fig. 4, each agent has 9 incoming and 3 outgoing connections:

Incoming connections: for each other agent (B, C, D), agent A has sensor states for verbal action (v), movement (m) and expressed affective response (b).

Outgoing connections: agent A executes verbal action (v), movement (m) and expressed affective response (b).

The first-order self-modeling level, located in the middle, is responsible for holding the information of how different agents bond modeled by **W**-states X_{40} to X_{43}. These W-states adaptively change connection weights of the base level (between states X_4 to X_{12}). More specifically, Figs. 5 and 6 show how base states from X_{13} and X_{21} (for sensory representations) and from X_{37} to X_{39} (for action executions) have the weights of their incoming connections regulated by the first-order adaptation

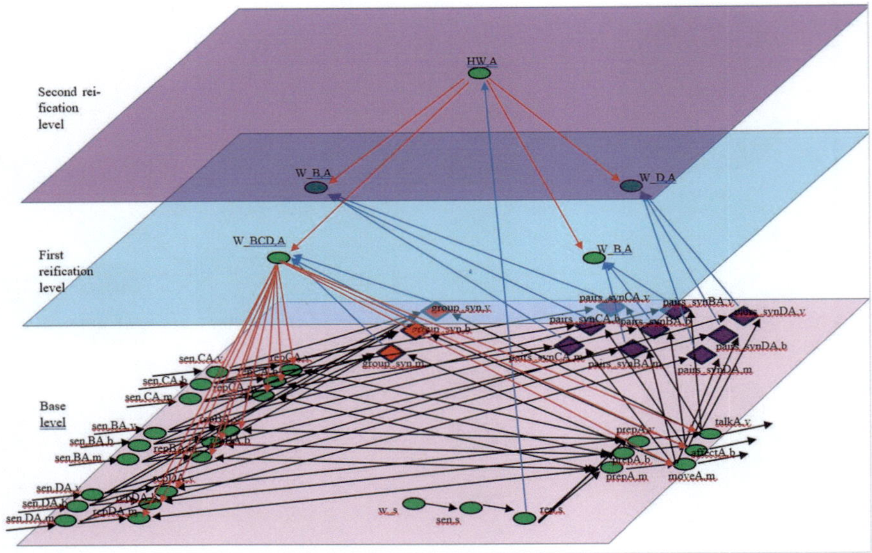

Fig. 5 3D network visualization using detection of group synchrony

level. In this setup, the first-order adaptation level receives incoming connections from the base level (in particular from the synchrony detection states), as explained in Sect. 3. The top level of the network, on the other hand, holds a single $\mathbf{H_W}$-state per agent (X_{44} for Agent A) that is responsible for controlling the adaptation process of the the bonding **W**-states present in the level just below it.

Put simply, the adaptive network model uses a system that has different levels to detect emerging synchrony and predict group bonding. The middle level focuses on understanding and representing how bonding occurs based on synchrony detection, thereby in turn changing connection weights of the base level. Finally, the top level is responsible for controlling the adaption speed of the middle level **W**-states.

For the way in which the middle level **W-**states affect the connection weights at the base level, two options have been considered in a comparative manner: the **W**-states $\mathbf{W}_{B,A}$, $\mathbf{W}_{C,A}$, $\mathbf{W}_{D,A}$ based on dyadic synchrony detection or the **W**-states $\mathbf{W}_{BCD,A}$ based on group synchrony detection. These two options are depicted respectively in Fig. 5 and 6.

5 Simulation Results

We've conducted experiments based on a model that simulates the emerging synchrony that occurs among a group of four friends, and how it can predict their group cohesion and bonding. We've completed approximately 20 iterations of this

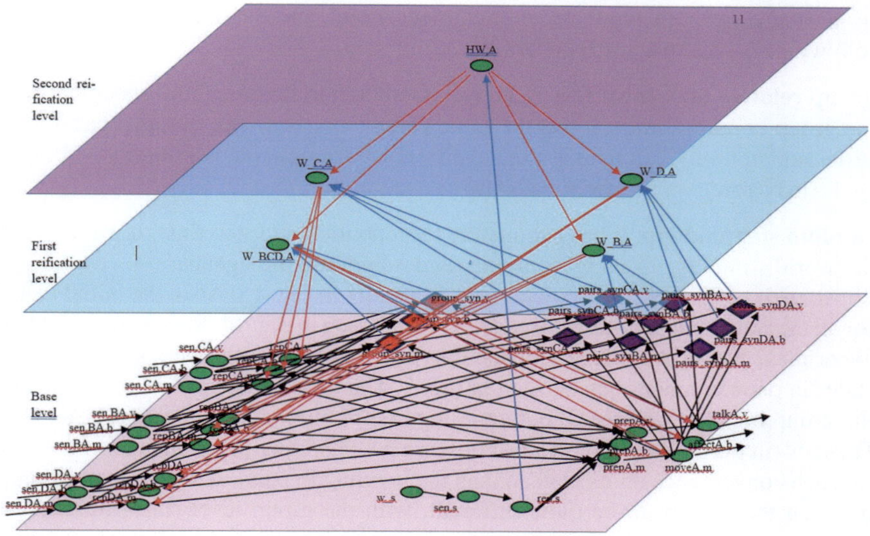

Fig. 6 3D network visualization using detection of synchrony in pairs

model with our four simulated individuals. Each simulation ran for 180 units of time, having each unit divided into 0.5 subunits. This means that each simulation incorporated 360 individual calculation steps. All 20 simulations followed the same framework regarding the type and length of the random elements involved: a 40 unit period with no stimulus influences acting on the four individuals, succeeded by an equivalent 40 unit period with influences. No matter the circumstances, the simulated individuals were always able to sense each other's verbal action, movement and expressed affective response. We will dive into 2 representative simulations that showed common patterns, frequently seen in the 20 iterations. In the Appendix Sect. 7 the full specification can be found by role matrices Figs. 11, 12, 13, 14 and 15.

Base Level Regulated by Synchrony Detection for Pairs

As discussed before in Sect. 3, the states are connected between each other as specified in Fig. 11 in Sect. 7. Most of these connections all have unit weights, as shown in Fig. 13 in Sect. 7. For the ones that are adaptive, this first simulation used the interpersonal synchrony states to regulate those connections. A full 3D visualization of the model architecture can be seen in Fig. 6. The aggregration characteristics are specified by 4 different combination functions:

alogistic: It has parameters for steepness and excitability threshold. It was used for all states aside the synchrony ones and the world stimulus.

swsm: It has parameters for identification state number and length of (the sliding) window. It was applied on all synchrony states between pairs of agents, working by normalizing both input series (achieving nser1 and nser2), to then apply the following: 1—*mean*(|*nser*2—*nser*1|).

group relative max-min: This is a novel combination function built specifically for the group bonding states. It has the same parameters as **swsm**, but as it has 4 input series and not only 2, it uses a relative difference formula on their rows:

1–(max(row)—min(row))/max(row).

random stepmodopp: It has parameters for repetition and step time. It was used for the world stimulus, it acts by applying random noise to all agents.

All their parameters are shown in the Appendix in Fig. 15. Also, the initial values for all the states and their respective speed can be seen in Fig. 14. Results for the whole experiment having base level connections adapted by the interpersonal synchrony states in pairs of agents can be seen in Fig. 7. The lower graph shows the curves for the group bonding (in green), bonding in pairs (in blue) and the second-order level **H**-states (in purple). From it, we can observe how bonding in group takes more time to reach convergence on its plateau than the bonding in pairs. Also, the curve shapes between the 2 seem to be quite different, with the group level states following a logistic curve and the states of pairs of agents follow more of a cubic curve. Clearly, the ways in which humans bond between each other is fundamentally different, when being in group. Finally, in Fig. 8 we have a comparison between group synchrony states (in blue) and the group bonding (in green). From it, we can determine that it is possible to predict the bonding from group synchrony, since their curves are of similar shape. Note that the plateaus in the curves relate to the absence of the repetitive stimulus that is used to trigger episodes of group activity.

Base Level Regulated by Group Synchrony States

For this setup, the matrix **mcw** shown in Fig. 13 is changed. This time, we use the group synchrony states of each agent to control the base level connections. In agent A, for example, the matrix shown in 9 would have X_{40} for all the adaptive states, instead of X_{41}, X_{42} or X_{43}. Aside this, all other parameters and combination functions were kept the same. A full 3D visualization of the model architecture can be seen in Fig. 5. The simulation timing and the curve colors are kept consistent with each other.

Lower graph: Group bonding (green), bonding in pairs (blue) and **H**-states (purple).

Having the adaptive base level connections regulated by the group synchrony instead of synchrony in pairs seem to show some different results. In Fig. 9. Upper graph, we have a full graph of the simulation results. In Fig. 9, lower graph we see how the curve representing group bonding, bonding in pairs or the $\mathbf{H_W}$ states change. Bonding in pairs seem to be more fluctuant overtime, reaching convergence in about the same timing as in Sect. 5.1. The group bonding follows a kind of logistic curve (same as before) but reaches lower plateau values between 80 and 120 time points. We have a comparison of synchrony and bonding in this setup as well (Fig. 10). Even when using different synchrony detection functions influencing the adaptive connections, it is possible to predict the group bonding from group synchrony.

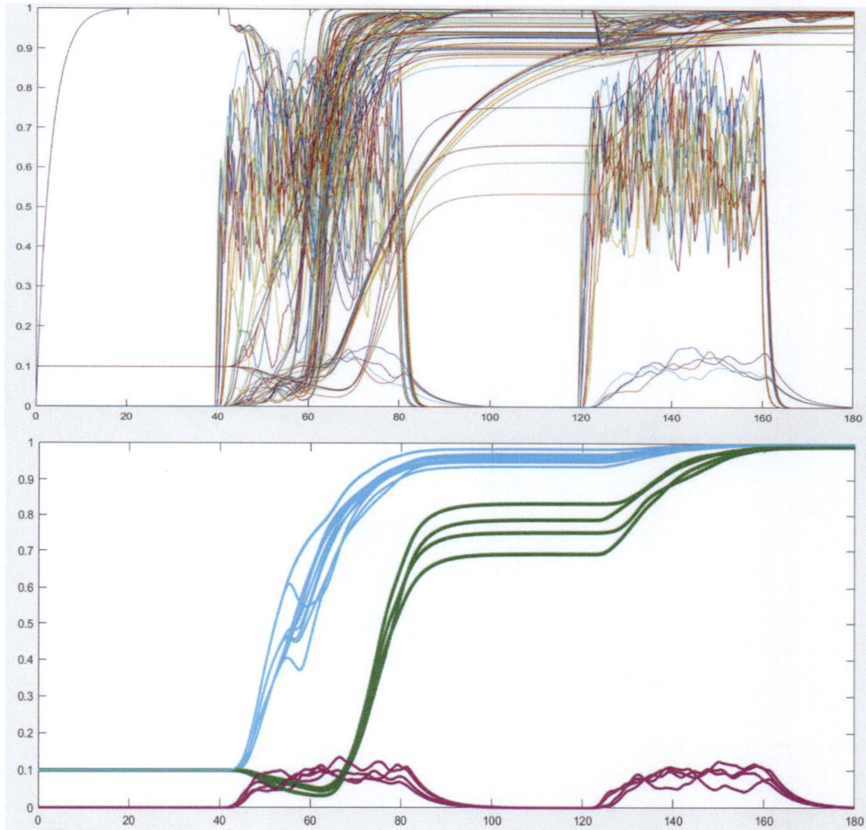

Fig. 7 Upper graph: Full graph of the experiment. Lower graph: group bonding (green), bonding in pairs (blue) and H-states for bonding speed (purple)

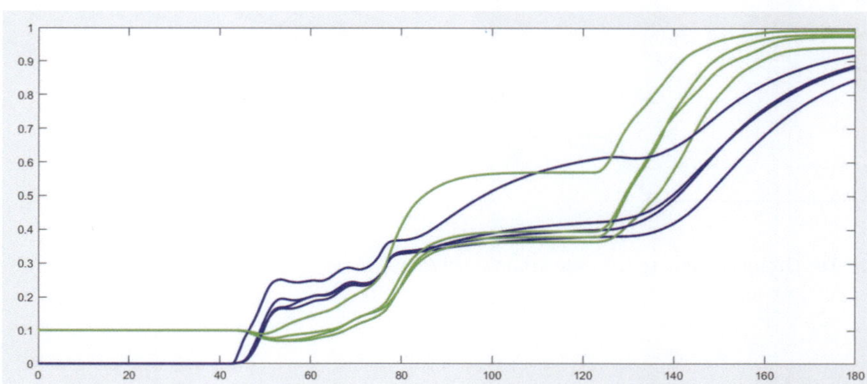

Fig. 8 Group bonding (green) and group synchrony (blue)

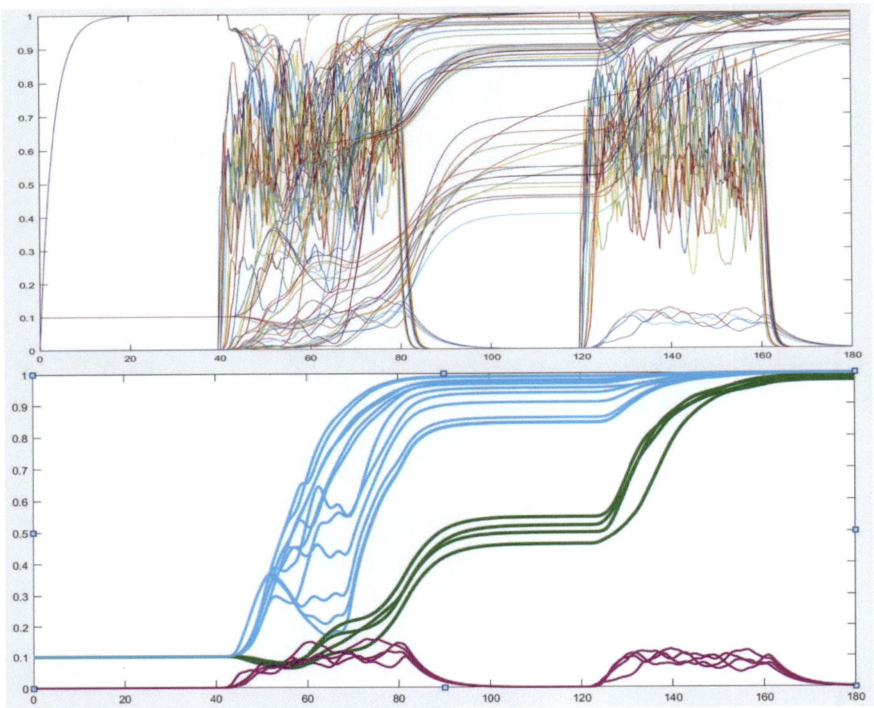

Fig. 9 Upper graph: Full graph of the experiment

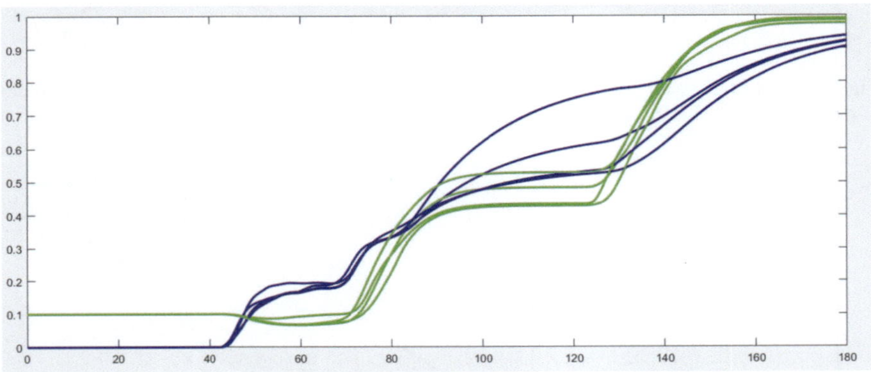

Fig. 10 Group bonding (green) and group synchrony (blue)

6 Discussion

This section aims to interpret the findings from our simulations and identify potential directions for future research. Our findings indicate how we can predict behavioural adaptivity manifested as group bonding from interpersonal group synchrony. We have attempted to represent and understand this complex phenomenon using a second-order adaptive network model. Most of the material for this chapter comes from [20].

Interpretation of Results

In our simulations, we observed a clear relationship between group synchrony and group bonding. We found that as group synchrony increased, the level of bonding within the group also increased. This connection was demonstrated in two scenarios: one with the base level connections regulated by the synchrony in pairs, and one where group synchrony was used.

In the scenario with synchrony in pairs, we observed that bonding reached high values (around 0.8) as the synchrony within the group increased (between time points 80 and 120). The increase in group bonding was smooth and followed a logistic curve. On the other hand, the bonding in pairs had a less smooth nature.

In the scenario with group synchrony, the dynamics of bonding were slightly different. While the relationship (and their curve shapes) between synchrony and bonding remained consistent, bonding within the group reached a lower plateau value between 80 and 120 time points. This indicates that adapting the base model with different synchrony states will impact the final group bonding.

In both scenarios, it has been proven possible to predict group bonding from group synchrony.

In previous work, in [6] it was analysed how persons can synchronize in the context of adaptive joint decision making, see also this volume [10], Chap. 5 [7]. In [14] it was analysed how subjective detection of synchrony can play a causal role for bonding, see also this volume [10], Chap. 11 [15]. Finally, in [8] the distinction between short-term affiliation and long-term bonding and their relation to synchronisation was introduced, and relationship-specific vs relationship-independent adaptivity in [9], see also this volume [10], Chaps. 12 and 13 [12, 13]. These papers focus on interpersonal synchrony in dyads and do not consider group synchrony and group bonding like the current chapter does.

7 Future Research

This work has presented a novel computational model that has shown promise for understanding interpersonal group synchrony and its relation to group bonding. However, there are many directions for future research that could improve our knowledge on it.

Firstly, it would be valuable to investigate how varying the size of the group influences the dynamics of synchrony and bonding. In this study, we focused on a group of four agents. Future work could explore different group sizes and configurations to see if the observed patterns remain consistent, confirming this chapter's simulations findings.

Also, different types of combination functions could be used on this model. Specifically, the group relative max–min function designed for the group bonding states' values could be improved.

Finally, an interesting next step would be to get empirical data over time and compare them to the simulation data. For example, a video recording of a joint dance activity may be analysed by a synchrony analysis method [16] and compared to reported affiliation.

8 Appendix: Specification by Role Matrices

See Figs. 11, 12, 13, 14 and 15.

mb		1	2	3	4
X1	world_s,A	X1			
X2	senseA,s	X1			
X3	repA,s	X2			
X4	senseB,A,m	X83			
X5	senseB,A,b	X82			
X6	senseB,A,v	X81			
X7	senseC,A,m	X127			
X8	senseC,A,b	X126			
X9	senseC,A,v	X125			
X10	senseD,A,m	X171			
X11	senseD,A,b	X170			
X12	senseD,A,v	X169			
X13	repB,A,m	X4	X36		
X14	repB,A,b	X5	X35		
X15	repB,A,v	X6	X34		
X16	repC,A,m	X7	X36		
X17	repC,A,b	X8	X35		
X18	repC,A,v	X9	X34		
X19	repD,A,m	X10	X36		
X20	repD,A,b	X11	X35		
X21	repD,A,v	X12	X34		
X22	inter_group,v	X9	X6	X12	X37
X23	inter_group,b	X8	X5	X11	X38
X24	inter_group,m	X7	X4	X10	X39
X25	interAC,v	X9	X37		
X26	interAC,b	X8	X38		
X27	interAC,m	X7	X39		
X28	interAB,v	X6	X37		
X29	interAB,b	X5	X38		
X30	interAB,m	X4	X39		
X31	interAD,v	X12	X37		
X32	interAD,b	X11	X38		
X33	interAD,m	X10	X39		
X34	prepA,v	X15	X18	X21	X3
X35	prep_affectA,b	X14	X17	X20	X3
X36	prepA,m	X13	X16	X19	X3
X37	talkA,v	X34			
X38	exp_affectA,b	X35			
X39	moveA,m	X36			
X40	WBCD,A	X22	X23	X24	
X41	WB,A	X28	X29	X30	
X42	WC,A	X25	X26	X27	
X43	WD,A	X31	X32	X33	
X44	HW,A	X3			

Fig. 11 Role Matrix **mb**

mcfw		alogistic	rho	randstepmodopp	compdiff
X1	world_s,A			1	
X2	senseA,s	1			
X3	repA,s	1			
X4	senseB,A,m	1			
X5	senseB,A,b	1			
X6	senseB,A,v	1			
X7	senseC,A,m	1			
X8	senseC,A,b	1			
X9	senseC,A,v	1			
X10	senseD,A,m	1			
X11	senseD,A,b	1			
X12	senseD,A,v	1			
X13	repB,A,m	1			
X14	repB,A,b	1			
X15	repB,A,v	1			
X16	repC,A,m	1			
X17	repC,A,b	1			
X18	repC,A,v	1			
X19	repD,A,m	1			
X20	repD,A,b	1			
X21	repD,A,v	1			
X22	inter_group,v		1		
X23	inter_group,b		1		
X24	inter_group,m		1		
X25	interAC,v				1
X26	interAC,b				1
X27	interAC,m				1
X28	interAB,v				1
X29	interAB,b				1
X30	interAB,m				1
X31	interAD,v				1
X32	interAD,b				1
X33	interAD,m				1
X34	prepA,v	1			
X35	prep_affectA,b	1			
X36	prepA,m	1			
X37	talkA,v	1			
X38	exp_affectA,b	1			
X39	moveA,m	1			
X40	WBCD,A	1			
X41	WB,A	1			
X42	WC,A	1			
X43	WD,A	1			
X44	HW,A	1			

Fig. 12 Role Matrix **mcfw**

mcw		1	2	3	4
X1	world_s,A	1			
X2	senseA,s	1			
X3	repA,s	1			
X4	senseB,A,m	1			
X5	senseB,A,b	1			
X6	senseB,A,v	1			
X7	senseC,A,m	1			
X8	senseC,A,b	1			
X9	senseC,A,v	1			
X10	senseD,A,m	1			
X11	senseD,A,b	1			
X12	senseD,A,v	1			
X13	repB,A,m	**X41**	1		
X14	repB,A,b	**X41**	1		
X15	repB,A,v	**X41**	1		
X16	repC,A,m	**X42**	1		
X17	repC,A,b	**X42**	1		
X18	repC,A,v	**X42**	1		
X19	repD,A,m	**X43**	1		
X20	repD,A,b	**X43**	1		
X21	repD,A,v	**X43**	1		
X22	inter_group,v	1	1	1	1
X23	inter_group,b	1	1	1	1
X24	inter_group,m	1	1	1	1
X25	interAC,v	1	1		
X26	interAC,b	1	1		
X27	interAC,m	1	1		
X28	interAB,v	1	1		
X29	interAB,b	1	1		
X30	interAB,m	1	1		
X31	interAD,v	1	1		
X32	interAD,b	1	1		
X33	interAD,m	1	1		
X34	prepA,v	1	1	1	1
X35	prep_affectA,b	1	1	1	1
X36	prepA,m	1	1	1	1
X37	talkA,v	**X40**			
X38	exp_affectA,b	**X40**			
X39	moveA,m	**X40**			
X40	WBCD,A	1	1	1	
X41	WB,A	1	1	1	
X42	WC,A	1	1	1	
X43	WD,A	1	1	1	
X44	HW,A	1			

Fig. 13 Role Matrix **mcw**

Fig. 14 Role Matrix **ms**

ms		speed	(i.v.)
X1	world_s,A	1	0
X2	senseA,s	1	0
X3	repA,s	1	0
X4	senseB,A,m	1	0
X5	senseB,A,b	1	0
X6	senseB,A,v	1	0
X7	senseC,A,m	1	0
X8	senseC,A,b	1	0
X9	senseC,A,v	1	0
X10	senseD,A,m	1	0
X11	senseD,A,b	1	0
X12	senseD,A,v	1	0
X13	repB,A,m	1	0
X14	repB,A,b	1	0
X15	repB,A,v	1	0
X16	repC,A,m	1	0
X17	repC,A,b	1	0
X18	repC,A,v	1	0
X19	repD,A,m	1	0
X20	repD,A,b	1	0
X21	repD,A,v	1	0
X22	inter_group,v	0.1	0
X23	inter_group,b	0.1	0
X24	inter_group,m	0.1	0
X25	interAC,v	0.5	0
X26	interAC,b	0.5	0
X27	interAC,m	0.5	0
X28	interAB,v	0.5	0
X29	interAB,b	0.5	0
X30	interAB,m	0.5	0
X31	interAD,v	0.5	0
X32	interAD,b	0.5	0
X33	interAD,m	0.5	0
X34	prepA,v	1	0
X35	prep_affectA,b	1	0
X36	prepA,m	1	0
X37	talkA,v	1	0
X38	exp_affectA,b	1	0
X39	moveA,m	1	0
X40	WBCD,A	**X44**	0.1
X41	WB,A	**x44**	0.1
X42	WC,A	**X44**	0.1
X43	WD,A	**X44**	0.1
X44	HW,A	0.2	0

mcfp		alogistic		group relative max-min		randstepmodopp		swsm	
X1	world_s,A					80	40		
X2	senseA,s	5	0.3						
X3	repA,s	5	0.7						
X4	senseB,A,m	5	0.3						
X5	senseB,A,b	5	0.3						
X6	senseB,A,v	5	0.3						
X7	senseC,A,m	5	0.3						
X8	senseC,A,b	5	0.3						
X9	senseC,A,v	5	0.3						
X10	senseD,A,m	5	0.3						
X11	senseD,A,b	5	0.3						
X12	senseD,A,v	5	0.3						
X13	repB,A,m	5	0.9						
X14	repB,A,b	5	0.9						
X15	repB,A,v	5	0.9						
X16	repC,A,m	5	0.9						
X17	repC,A,b	5	0.9						
X18	repC,A,v	5	0.9						
X19	repD,A,m	5	0.9						
X20	repD,A,b	5	0.9						
X21	repD,A,v	5	0.9						
X22	inter_group,v			2	2				
X23	inter_group,b			3	2				
X24	inter_group,m			4	2				
X25	interAC,v							14	10
X26	interAC,b							15	10
X27	interAC,m							16	10
X28	interAB,v							17	10
X29	interAB,b							18	10
X30	interAB,m							19	10
X31	interAD,v							20	10
X32	interAD,b							21	10
X33	interAD,m							22	10
X34	prepA,v	5	0.8						
X35	prep_affectA,b	5	0.8						
X36	prepA,m	5	0.8						
X37	talkA,v	5	0.5						
X38	exp_affectA,b	5	0.5						
X39	moveA,m	5	0.5						
X40	WBCD,A	5	1.5						
X41	WB,A	5	1.5						
X42	WC,A	5	1.5						
X43	WD,A	5	1.5						
X44	HW,A	5	1						

Fig. 15 Role Matrix **mcfp**

References

1. Abbas, H.A., Shaheen, S.I., Amin, M.H.: Organization of multi-agent systems: an overview (2015). arXiv:1506.09032
2. Bössenecker, G., Mreijen, P., Treur, J.: A second-order adaptive network model for connecting social media platforms and users through bonding by faked homophily. Procedia Comput. Sci. **213**, 32–40 (2022)
3. Bössenecker, G., Mreijen, P., Treur, J: Connecting social media platforms and users through bonding by faked homophily: a second-order adaptive network model. In: Hendrikse, S.C.F.,

Treur, J., Koole, S.L. (eds.), New Analysis and Modeling Directions in Social Interaction Science: Emergent Multimodal Interpersonal Synchrony, Affiliation, and Bonding, Ch 9, this volume. Springer Nature (2025)

4. Delaherche, E., Chetouani, M., Mahdhaoui, A., Saint-Georges, C., Viaux, S., Cohen, D.: Interpersonal synchrony: a survey of evaluation methods across disciplines. IEEE Trans. Affect. Comput. **3**(3), 349–365 (2012)
5. dos Santos, N.R., Figueiredo, C., Pais, L.: Development and validation of the organisational cooperation questionnaire. Eur. Rev. Appl. Psychol. **70**(4), 100555 (2020)
6. Hendrikse, S.C.F., Kluiver, S., Treur, J., Wilderjans, T.F., Dikker, S., Koole, S.L.: How Virtual Agents Can Learn to Synchronize: an Adaptive Joint Decision-Making Model of Psychotherapy. Cogn. Syst. Res. **79**, 138–155 (2023)
7. Hendrikse, S.C.F., Kluiver, S., Treur, J., Wilderjans, T.F., Dikker, S., Koole, S.L.: Learning to synchronize: adaptive joint decision-making by virtual agents. In: Hendrikse, S.C.F., Treur, J., Koole, S.L. (eds.), New Analysis and Modeling Directions in Social Interaction Science: Emergent Multimodal Interpersonal Synchrony, Affiliation, and Bonding, Ch 5, this volume. Springer Nature (2025)
8. Hendrikse, S.C.F., Treur, J., Koole, S.L.: Modeling emerging interpersonal synchrony and its related adaptive short-term affliation and long-term bonding: a second-order multi-adaptive neural agent model. Int. J. Neural Syst. **33**(7), 2350038 (2023)
9. Hendrikse, S. C., Treur, J., Koole, S. L.: Relationship-specific and relationship-independent behavioural adaptivity in affiliation and bonding: a multi-adaptive dynamical systems approach. Cogn. Syst. Res. **84**, 101182 (2024)
10. Hendrikse, S.C.F., Treur, J., Koole, S.L, eds.: New Analysis and Modeling Directions in Social Interaction Science: Emergent Multimodal Interpersonal Synchrony, Affiliation, and Bonding. Springer Nature (this volume) (2025a)
11. Hendrikse, S.C.F., Treur, J., Koole, S.L.: Introduction to the role of adaptivity in social interaction science. In: Hendrikse, S.C.F., Treur, J., Koole, S.L. (eds.), New Analysis and Modeling Directions in Social Interaction Science: Emergent Multimodal Interpersonal Synchrony, Affiliation, and Bonding, Ch 1, this volume. Springer Nature (2025b)
12. Hendrikse, S.C.F., Treur, J., Koole, S.L.: The adaptive interplay of interpersonal synchrony, short-term affiliation and long-term bonding: a multi-timescale multi-adaptive agent model. In: Hendrikse, S.C.F., Treur, J., Koole, S.L. (eds.), New Analysis and Modeling Directions in Social Interaction Science: Emergent Multimodal Interpersonal Synchrony, Affiliation, and Bonding, Ch 12, this volume. Springer Nature (2025c)
13. Hendrikse, S.C.F., Treur, J., Koole, S.L.: A multi-timescale multi-adaptive dynamical system approach for relationship-specific and relationship-independent adaptivity in affiliation and bonding. In: Hendrikse, S.C.F., Treur, J., Koole, S.L. (eds.), New Analysis and Modeling Directions in Social Interaction Science: Emergent Multimodal Interpersonal Synchrony, Affiliation, and Bonding, Ch 13, this volume. Springer Nature (2025d)
14. Hendrikse, S.C.F., Treur, J., Wilderjans, T.F., Dikker, S., Koole, S.L.: On becoming in sync with yourself and others: an adaptive agent model for how persons connect by detecting intra-and interpersonal synchrony. Hum.-Centric Intell. Syst. **3**, 123–146 (2023)
15. Hendrikse, S.C.F., Treur, J., Wilderjans, T.F., Dikker, S., Koole, S.L.: Growing connectedness through synchrony: the role of subjective analysis of intrapersonal synchrony and interpersonal synchrony within an adaptive agent model. In: Hendrikse, S.C.F., Treur, J., Koole, S.L. (eds.), New Analysis and Modeling Directions in Social Interaction Science: Emergent Multimodal Interpersonal Synchrony, Affiliation, and Bonding, Ch 11, this volume. Springer Nature (2025)
16. Hudson, D., Wiltshire, T.J., Atzmueller, M.: MultiSyncPy: a python package for assessing multivariate coordination dynamics. Behav. Res. Methods **55**(2), 932–962 (2023)
17. León-del-Barco, B., Mendo-Lázaro, S., Felipe-Castaño, E., Fajardo-Bullón, F., Iglesias-Gallego, D.: Measuring responsibility and cooperation in learning teams in the university setting: validation of a questionnaire. Front. Psychol. **9**, 326 (2018)
18. Louwerse, M.M., Dale, R., Bard, E.G., Jeuniaux, P.: Behavior matching in multimodal communication is synchronized. Cogn. Sci. **36**(8), 1404–1426 (2012)

19. Mataric, M.J.: Designing and understanding adaptive group behavior. Adapt. Behav. **4**(1), 51–80 (1995)
20. Mattera, F., Hendrikse, S.C.F., Treur, J.: An adaptive network model for the emergence of group synchrony and behavioral adaptivity for group bonding. In: Proceeding of the 12th International Conference on Complex Networks and Their Applications. Studies in Computational Intelligence, vol. 1144, pp. 53–66. Cham: Springer Nature (2024)
21. Meier, D., Tschacher, W.: Beyond dyadic coupling: the method of multivariate Surrogate Synchrony (mv-SUSY). Entropy **23**(11), 1385 (2021)
22. Miles, L.K., Lumsden, J., Flannigan, N., Allsop, J.S., Marie, D.: Coordination matters: Interpersonal synchrony influences collaborative problem-solving. Psychology (2017)
23. Mukeriia, Y., Treur, J., Hendrikse, S.C.F.: An adaptive network model for learning and bonding during a varying in rhythm synchronous joint action. In: Proceeding of the 12th International Conference on Complex Networks and Their Applications. Studies in Computational Intelligence, vol. 1144, pp. 41–52. Cham: Springer Nature Switzerland (2023)
24. Mukeriia, Y., Treur, J., Hendrikse, S.C.F.: The interplay of bonding based on subjective homophily and learning during a varied rhythmic action: an adaptive network model. In: Hendrikse, S.C.F., Treur, J., Koole, S.L. (eds.), New Analysis and Modeling Directions in Social Interaction Science: Emergent Multimodal Interpersonal Synchrony, Affiliation, and Bonding, Ch 8, this volume. Springer Nature (2025a)
25. Mukeriia, Y., Treur, J., Hendrikse, S.C.F.: Human learning, attunement and social bonding based on adaptive homophily: a multi-adaptive network model. In: Hendrikse, S.C.F., Treur, J., Koole, S.L. (eds.), New Analysis and Modeling Directions in Social Interaction Science: Emergent Multimodal Interpersonal Synchrony, Affiliation, and Bonding, Ch 10, this volume. Springer Nature (2025b)
26. Ramseyer, F., Tschacher, W.: Nonverbal synchrony of head-and body-movement inpsychotherapy: different signals have different associations with outcome. Front. Psychol. **5**, 979 (2014)
27. Richardson, D., Dale, R., Shockley, K.: Synchrony and swing in conversation: Coordination, temporal dynamics, and communication. In: Embodied communication in humans and machines, pp. 75–94 (2008)
28. Richardson, M.J., Garcia, R.L., Frank, T.D., Gergor, M., Marsh, K.L.: Measuring group synchrony: a cluster-phase method for analyzing multivariate movement time-series. Front. Physiol. **3**, 405 (2012)
29. Rothgangel, A.S., Braun, S.M., Beurskens, A.J., Seitz, R.J., Wade, D.T.: The clinical aspects of mirror therapy in rehabilitation: a systematic review of the literature. Int. J. Rehabil. Res. **34**(1), 1–13 (2011)
30. Rothenberg-Elder, K.: Transitional rituals for group initiation and group bonding. In: Farewell and new beginning: The Psychosocial Effects of Religiously Traditional Rites of Passage, pp. 205–225. Wiesbaden: Springer Fachmedien Wiesbaden (2023)
31. Rotondo, J.L., Boker, S.M.: Behavioral synchronization in human conversational interaction. Mirror Neurons Evol. Brain Lan-Guage **42**, 151–162 (2002)
32. Spoor, J.R., Kelly, J.R.: The evolutionary significance of affect in groups: Communication and group bonding. Group Process. Intergroup Relat. **7**(4), 398–412 (2004)
33. Tarr, B., Dunbar, R.I.: The evolutionary role of dance: group bonding but not prosocial altruism. Evol. Behav. Sci. (2023). https://doi.org/10.1037/ebs0000330
34. Treur, J.: Network-oriented modeling for adaptive networks: designing higher-order adaptive biological, mental and social network models. Springer Nature Publishing, Cham, Switzerland (2020)
35. Treur, J.: Controlled Social Network Adaptation:Subjective Elements in an Objective Social World. In: Yang, XS., Sherratt, S., Dey, N., Joshi, A. (eds) Proceedings of Sixth International Congress on Information and Communication Technology, ICICT'21. Lecture Notes in Networks and Systems, vol. 235, pp. 263–274. Springer Nature (2022)
36. Treur, J.: A Second-Order Adaptive Network Model for Bonding Based on Subjectively Perceived Homophily. In: Hendrikse, S.C.F., Treur, J., Koole, S.L. (eds.), New Analysis

and Modeling Directions in Social Interaction Science: Emergent Multimodal Interpersonal Synchrony, Affiliation, and Bonding, Ch 7. Springer Nature (2025)

37. Treur, J.: Modeling coevolution of dynamics, adaptivity and control of adaptivity by self-modeling networks. In: Hendrikse, S.C.F., Treur, J., Koole, S.L. (eds.), New Analysis and Modeling Directions in Social Interaction Science: Emergent Multimodal Interpersonal Synchrony, Affiliation, and Bonding, Ch 2, this volume. Springer Nature (2025)
38 Tschacher, W., Rees, G.M., Ramseyer, F.: Nonverbal synchrony and affect indyadic interactions. Front. Psychol. **5**, 1323 (2014)
39. von Zimmermann, J., Vicary, S., Sperling, M., Orgs, G., Richardson, D.C.: The choreography of group affiliation. Top. Cogn. Sci. **10**(1), 80–94 (2018)
40. Wayne, G.D.: Self-modeling neural systems. Columbia University (2013)
41. Wood, C., Caldwell-Harris, C., Stopa, A.: The rhythms of discontent: synchrony impedes performance and group functioning in an interdependent coordination task. J. Cogn. Cult. **18**(1–2), 154–179 (2018)
42. Yu, B., Venkatraman, M., Singh, M.P.: An adaptive social network for informationaccess: theoretical and experimental results. Appl. Artif. Intell. **17**(1), 21–38 (2003)
43. Yu, W., Chen, G., Cao, M., Kurths, J.: Second-order consensus for multiagent systems with directed topologies and nonlinear dynamics. IEEE Trans. Syst., Man, Cybern., Part B (Cybern.) **40**(3), 881–891 (2009)

Computational Analysis of Real-World Scenarios for Synchrony-Induced Social Interaction Adaptivity

In this part, the methods and models developed in the previous parts are applied to analyse three real-world cases, two of them concerning events with elderly where synchrony-induced adaptivity occurs. With the aging population increasing, there is a crucial need to focus on the health and social interaction of older adults. "Joint Action and Memory Recall for Elderly by Detecting Interpersonal Synchrony: A Higher-Order Adaptive Network Model" computationally recreates, simulates and subsequently inspects a real-life scenario in which positive impacts from a religious meeting for elderly were experienced. In order to do so, a first-order adaptive agent model was designed. The model exploits concepts from the previous parts such as emerging multimodel interpersonal synchrony and synchrony-induced adaptivity of interaction behaviour. "Interpersonal Emotion Regulation via Multimodal Interaction with a Bonding Bot: An Adaptive Network Model" explores joint action and adaptive memory recall among elderly induced by detecting subjective interpersonal synchrony. This chapter presents a computational analysis of how to enhance positive adaptive effects of social interactions among older individuals by applying a second-order adaptive network model. The research examines the concept of interpersonal synchrony and its impact on joint action, memory, and emotional well-being in elderly populations. The study demonstrates the potential benefits for music in memory recall for older adults with cognitive decline, highlighting the importance of social interaction and emotional resonance. In "Interpersonal Emotion Regulation via Multimodal Interaction with a Bonding Bot: An Adaptive Network Model", it is shown how the developed computational approach can be applied to model adaptivity in interpersonal emotion regulation between a human and a bot. This chapter provides and analyses a model and simulated data about human-bot interaction to analyse to what extent mapping human-like mental processes and behaviors on their companion bots has valuable effects. The analysis results found here provide a promising perspective for interpersonal emotion regulation using such human-like bonding bots.

Reliving Memory and Religious Experience Through Multimodal Synchrony and Adaptation: An Adaptive Network Model

Damien Vervoordeldonk, Jan Treur, Sophie C. F. Hendrikse, and Peter H. M. P. Roelofsma

Abstract During a group worship session, through intersensory facilitation participants can experience interpersonal synchrony. In addition it may occur that seemingly forgotten memories are suddenly relived and religious experiences come up. This chapter addresses the question whether a higher-order adaptive network model can enhance our understanding of such an interplay of interpersonal synchrony and the mentioned phenomena. In the developed adaptive dynamical system model, feeling of togetherness and oneness was modelled making use of synchrony detection states. The felt presence of God was modelled through a mental God-model. The improvements in cognition perceived were modelled by introducing adaptivity of certain connections between cognitive states. The subsequent simulation provided a closer look into the inner workings of the interplay of the mechanisms involved. It was found that synchronization suddenly intensifies after some time of mutual adaptation resulting in improvements in cognition of the cognitive impaired participant.

Keywords Interplay of synchrony · Adaptation · Reliving memory · Multimodal interaction · Religious experience · Adaptive network model

---

D. Vervoordeldonk · J. Treur (✉)
Social AI Group, Department of Computer Science, Vrije Universiteit Amsterdam, Amsterdam, The Netherlands
e-mail: j.treur@vu.nl

S. C. F. Hendrikse
Department of Methodology and Statistics, Tilburg School of Social and Behavioral Sciences, Tilburg University, Tilburg, The Netherlands
e-mail: S.C.F.Hendrikse@tilburguniversity.edu

P. H. M. P. Roelofsma
Risk Management and Cybersecurity Group, Centre of Expertise Cybersecurity, The Hague University of Applied Sciences, The Hague, The Netherlands
e-mail: P.H.M.P.Roelofsma@hhs.nl

S. C. F. Hendrikse et al. (eds.), *New Analysis and Modeling Directions in Social Interaction Science*, Studies in Systems, Decision and Control 614,
https://doi.org/10.1007/978-3-031-99968-0_17

1 Introduction

This chapter addresses computational analysis of religious experiences during a worship session. Religion can be a controversial topic to discuss. Depending on the person asked, religion can seem to have many wonderful or awful aspects and impacts. Regardless the view, emotion often takes the upper hand over objectivity. This chapter aims to contribute to a more objective look on the impact religion can have. This is done by computationally modeling, simulating and subsequently analysing a real-life scenario in which strong impacts were experienced.

Important concepts from social psychology taken into account are the emergence of multimodal synchrony during social interaction, subjective detection of such synchrony by the person, and different forms of adaptation induced by such detected synchrony [27, 24], see also this volume [23], Chaps. 11–13 [28, 25, 26]. Moreover, to integrate a dynamical and adaptive mental God representation, earlier work such as [54, 55] was used as inspiration.

The ultimate aim behind this research is to obtain a base model that can be used as a basis for the development of an embodied virtual agent that can guide groups in joint events like the one analysed here.

In order to address the above, a dynamical system modeling approach was used to design an adaptive agent represented as network model. This modeling approach can be used to model real-world processes involving human beings and the representation of mental mechanisms inspired by neurological processes [44–46]. To incorporate adaptivity, the introduced adaptive agent model also uses the basics of the self-modeling network modeling approach as found in particular in [45, 46], see also this volume see also this volume [23], Chap. 2.

The chapter can especially be of interest for multidisciplinary researchers from AI and social psychology, as well as from theological sciences, and for developers of virtual agents or bots to support humans.

The explanation of the scenario at hand with the resulting research question is given in Sect. 2. Then some background knowledge and literature is provided in Sect. 3. Next, the network-oriented approach employed is shortly explained in Sect. 4. Section 5 explains the introduced model which is then followed by Sect. 6 describing the specifics of the simulation. Finally, an overview is given by providing a conclusion and discussion in Sect. 7. The appendix contains additional information about the values, connections and formulas used in the model and simulation as well as the entire scenario description.

2 Type of Scenario Addressed

An example of religion having a positive impact on peoples' life can be seen in the real-life scenario of a Christian group worship session. This scenario can be summarized as follows. An elderly couple with cognitive impairments participates in a group worship session led by a worship leader. During this session various activities are undertaken that are highly comparable with the past religious experiences of the participants. The worship leader introduces these activities and other facets of the session. For example, singing, music, lyrics/text, pictures explaining the text and the breaking of bread together whilst singing. In essence, the session could be regarded as one long prayer. The participants are constantly encouraged and instructed by the worship leader. During the session, the participants get more into the different facets of the session by synchronizing their actions with each other including the worship leader. This results in improvements in their cognitive abilities that can be observed. Moreover, at the end of the session the elderly couple report having had a religious experience during the worship session that brought back memories, a feeling of togetherness and other positive feelings with the breaking of bread being the final catalyst. Essentially, it concerns a group worship session with triggers of music making, pictures and breaking of bread for reliving memory and religious experiencing through intersensory facilitation. Therefore, the following *research question* can be formulated:

> Can a higher-order adaptive network model enhance our understanding of the interplay of interpersonal synchrony and brain plasticity during a group worship session for reliving memory and religious experiences through intersensory facilitation by computationally recreating the scenario?

3 Background Knowledge

The addressed scenario can be broken down into three main concepts:

- The feeling of togetherness and oneness
- Exercising some belief in God
- An observed improvement in cognition

These three trends eventually result in the emotional and religious breakthrough. Exploring possibilities on how to model these main ideas computationally using the network-oriented approach will be based on the literature discussed below.

3.1 The Feeling of Togetherness and Oneness

As described in the scenario, the worship leader and participants as well as the participants together form a connection resulting in a feeling of togetherness. Such connection between persons often originates from experiencing a form of synchrony with each other where their behavior tends to become mutually coordinated in time. This synchrony is shown to foster enhancements in the functioning of the relation between the synchronized persons. These enhgancements cover inducing greater levels of closeness, coordination, bonding and concentration [1, 30, 33, 41]. Furthermore, according to various researchers [18, 57] synchrony in face-to-face interactions turns out to play a key role in interpersonal bonding or in other words the feeling of togetherness.

Synchrony can be manifested in various forms. Intrapersonal synchrony happens when a person synchronizes different modalities with each other, such as movement, emotion and audio. This type of synchrony has the possibility of leading towards some conscious emotions [20]. Interpersonal synchrony is when different persons synchronize corresponding modalities. Intrapersonal and interpersonal synchrony are frequently studied independently, despite the possibility of their interconnection and association with other outcomes [11].

In two papers [27, 22] this interconnection and association with other outcomes does play a key role, see also this volume [23] Chaps. 11 and 13. Therefore these contributions gives rise to a great tool to model the observed feeling of togetherness of the scenario. The main innovations brought forward in [27], based on the aforementioned literature and additional literary argumentation, are the built in intra- and interpersonal synchrony detection states. These states can be used to model the observed feeling of togetherness which can be seen as the subject noticing forms of synchronization. Within this paper three substantiated assumptions were made. Firstly, it is assumed that individuals unconsciously detect synchrony in some manner and form stronger connections based on that perception. Secondly, connecting can be seen as a form of adaption of behavior (instead of a form of mentalizing) as put forward by [35] in their metareview. Thirdly, the detected intrapersonal synchrony that underlies a conscious emotion can have a causal impact on the individual's responses and their associated behavioral adaptivity. Naturally these assumptions carry over.

3.2 Exercising Some Belief in God

Two distinct approaches exist when studying the various aspects of religion:

- The evolutionary approach that aims to explain the origin of religion and its different facets from the evolutionary standpoint [2, 15],
- The neurotheologist approach which has its focus on investigating the neurological basis of religion and religious behavior by examining the neurological processes [6, 13, 17, 19, 31].

Although the idea to use a network-oriented modelling approach seems to mainly focus on the latter, it's not entirely the case. Since the two perspectives are not intrinsically contradictory and as scientific and philosophical advancements from both sides are made, various aspects from the two fields combined would yield the most promising look on religion. In [31, 32] a framework that does this can be found which proposes an integrative cognitive neuroscience approach to comprehend the cognitive and neural underpinnings of religions. [31] and other researchers have identified three dimensions that form an individual's religious belief:

- God's perceived involvement
- God's perceived emotion
- The religious knowledge of the individual

These dimensions could be considered as nodes in a network emphasizing their interconnectedness and neural underpinnings.

Other literature shows that evidence exists of God being considered a real person by religious individuals praying [39]. It was shown that praying activates brain regions responsible for active interpersonal interactions which enables persons to create an internal representation of God. This notion of God being represented internally as a person is also brought forward in [37, 38]. As the worship session in the scenario can be seen as one long prayer, such presence of God is mentally represented at all times within all individuals participating. The question remains how such mental God representation should be created.

In line with the idea of God as a real significant person, the notion of three dimensions which form an individual's religious belief and the possibility of this being represented as nodes in a network, the following idea was developed in [55]. There the aforementioned literature in addition to additional sources formed the literary foundation of an extensive God model created. This chapter studied the influence of religion on human empathy and disempathy. As a result of an extensive literature study done on the processes related to religion and human behavior, a design of a temporal-causal network model was proposed that captures the process on how religion influences human behavior. Furthermore, it was discussed how the religion-related external input that an individual receives is processed and activates and strengthens an internal personal mental God-model together with the resulting influences of this God-model on the individual's behavior and emotions. The extensive mental God-model created in that chapter as a result of that discussion forms the basis of a smaller mental God-model used in the introduced model in the current paper.

This mental God-model consists of an emotional and cognitive part as brought forward by [31, 32, 37]. In addition, the personal belief of the person (that displays how religious a person is) is also represented in a third state. This makes it that the three main dimensions, God's perceived involvement, God's perceived emotion, and the possessed religious knowledge of the individual are represented. In [34] as well as [42] it is described how a person is able to develop an empathic understanding of others through mirroring and internal simulation mechanisms. These mechanisms influence the individuals' beliefs and actions of that person. Such mechanisms can also be applied to the mental God-model in which God is represented like a real person. This results in the person mirroring the internally represented beliefs, actions and emotions of the perceived God. Because of these mechanisms, the mental Godmodel of the person influences their own beliefs, actions and emotions in a similar way as to how another individual would.

3.3 Cognition and Plasticity

Cognition usually concerns the mental action or process of acquiring knowledge and understanding through thought, experience and the senses. It is something that is at play at all times during the session. Both with partaking in actions as with sensing the actions. As a result, due to plasticity improvements in cognition can be inferred from seeing improvements in various aspects of the session such as the executing and sensing of actions by the cognitive impaired participants. In the empirically founded literature it has been found that emerging interpersonal synchrony often leads to forms of plasticity (behavioural adaptivity) for social interaction. Examples of this can be found in [30, 33, 41, 57]. For more background on this, see this volume [23], Chap. 1. Based on such empirically founded literature it has been analysed in this volume Chaps. 11, 12 and 13 [28, 25, 26] how from an adaptive dynamical system perspective such complex adaptive processes can be modeled computationally. This has formed an important source of inspiration for the approach contributed by the current chapter.

3.4 All Aspects Combined

By combining all topics discussed above, a higher-order adaptive network model has been introduced that can be used to generate simulations that closely resemble the events of the scenario. The synchrony detection states model the feeling of togetherness, the internal mental God-model models the strong felt presence of God and its effects and the improvement in cognition can be a consequence of that. By this model, the research question can be investigated. A closer look to the makeup of the three parts and how they are combined will be given in the next two sections.

Note that the notion of the mental God-model has not been specifically tailored towards one certain God but rather to the personal God of the individual. Furthermore, as will be seen in the next section, the specific activities and triggers described in the scenario are not modelled as such. This allows for the more general and main idea of the past religious experiences of the participants being similar to the introduced activities and triggers. Therefore, it can be said that the introduced model is not religion-specific and a translation to multiple ('mainly Abrahamic') religions is possible.

4 The Self-Modeling Network Modeling Approach Used

This section provides a concise overview of the network-oriented modeling approach employed. Drawing from [45, 46], a temporal-causal network model can be defined by the ensuing attributes (with X and Y representing network nodes, also referred to as states).

Connectivity characteristics

Connections from state X to a state Y and their weights $\boldsymbol{\omega}_{X,Y}$.

Aggregation characteristics

For every state Y, an aggregation function $\mathbf{c}_Y(..)$ is used to combine the impacts $\boldsymbol{\omega}_{X,Y}X(t)$ on Y that stem from its incoming connections from states X.

Timing characteristics

Every state Y uses a speed factor η_Y to determine how fast the state changes for the given causal impact.

The difference and differential equations employed for both simulation and analysis purposes of the temporal-causal networks incorporate the aforementioned network characteristics $\boldsymbol{\omega}_{X,Y}$, $\mathbf{c}_Y(..)$ and $\boldsymbol{\eta}_Y$ in a standerd numerical format for every state Y with incoming connections from states X_1 to X_k:

$$Y(t + \Delta t) = Y(t) + \boldsymbol{\eta}_Y\left[\mathbf{c}_Y\left(\boldsymbol{\omega}_{X_1,Y}X_1(t), \ldots, \boldsymbol{\omega}_{X_k,Y}X_k(t)\right) - Y(t)\right]\Delta t \qquad (1)$$

Notice that the Eq. (1) bears resemblance to the structure of recurrent neural networks.

In Eq. (1), at the current time t, we have a state value $Y(t)$. This represents the current value of state Y. The equation determines how this state $Y(t)$ will change in the future, specifically at time $t + \Delta t$. To calculate this change, the concept of a speed factor is introduced, denoted as η_Y, which characterizes how rapidly state Y responds to impacts from its environment. A higher $\boldsymbol{\eta}_Y$ implies quicker change. In addition to the speed factor, the influences from other states are considered, denoted as X_1 to X_k, which are interconnected toward state Y. These influences are modulated by weighted connections with weights $\boldsymbol{\omega}_{X_1,Y}$ to $\boldsymbol{\omega}_{X_k,Y}$. These weights modulate the strength of the impact that each state X has on Y by multiplying each state value $X_i(t)$ by $\boldsymbol{\omega}_{X_i,Y}$, thus obtaining impact $\boldsymbol{\omega}_{X_i,Y}X_i(t)$. Importantly, the current state values of these influencing states at time t, which are X_1 *(t)* and X_k *(t)*, need to be known at this point. These states represent where X_1 and X_k are at the present moment.

The next step involves a combination function $\mathbf{c}_Y(\ldots)$, which takes into account all the incoming influences from states X_1 to X_k. This function aggregates the impacts $\boldsymbol{\omega}_{X_i,Y}X_i(t)$ and determines how they collectively affect the change in state Y. Finally, the aggregated impact is multiplied by a small time interval Δt and incorporates this to obtain the future state value $Y(t+\Delta t)$ following state $Y(t)$: this calculation results in the prediction of this future state value $Y(t+\Delta t)$. These calculations were computed by the network engine implemented within the dedicated software environment in MATLAB and described in detail in [45, 46]. The software framework detailed in [45] Chapter 9 offers a comprehensive combination function library comprising of around 60 useful basic combination functions. These concepts enable the design of network models and their dynamics in a declarative fashion, relying on mathematically defined functions and relations. The model introduced uses in particular the combination functions described in Table 1.

Typically, realistic networks show adaptivity where not only their states change over time, but also certain characteristics of it. Through the utilization of a self-modeling network, also referred to as a reified network, a similar network-oriented approach can be extended to adaptive networks. This allows for a declarative description utilizing mathematically defined functions and relations as seen in [45, 46]. This process involves the incorporation of new states into the network which represent the adaptive network characteristics. These new states are referred to as self-model states and are visually depicted at a higher level known as the self-model level or reification level, while the original network resides at the base level.

All network characteristics from $\boldsymbol{\omega}_{X,Y}$, $\mathbf{c}_Y(..)$ and $\boldsymbol{\eta}_Y$ have the possibility of being made adaptive by introducing such self-model states. Two instances would be the weight $\boldsymbol{\omega}_{X,Y}$ of the connection from state X to Y with the introduction of the self-model state named $\mathbf{W}_{X,Y}$ (henceforth referred to as $\mathbf{W}$-states) and the adaptive speed factor $\boldsymbol{\eta}_Y$ introduced with the selfmodel state named $\mathbf{H}_{\mathbf{W}_{X,Y}}$. By employing this concept, adaptation processes can be modeled in a declarative manner by specifying the appropriate combination functions from the library to be applied to the self-model states. This allows for adaptivity to be represented in a declarative format distinct from procedural descriptions such as learning algorithms commonly found in the literature on learning. As a result, the modeling of adaptivity becomes more intuitive and transparent. When using self-model states for adaptive network characteristics, for example for the connection weights, this is incorporated in Eq. (1) as follows:

$$Y(t+\Delta t) = Y(t) + \boldsymbol{\eta}_Y[\boldsymbol{c}_Y(W_{X_1,Y}(t)X_1(t), \ldots, \mathbf{W}_{X_k,Y}(t)X_k(t)) - Y(t)]\Delta t \quad (2)$$

Since it has been shown in [45] Chap. 10 that the outcome of this procedure of network reification is a temporal-causal network as well, the self-modeling network assembly can simply be applied repeatedly in order to obtain multiple orders of self-models. An example of this would be a second-order self-model which has a second-order self-model state $\mathbf{H}_{\mathbf{W}_{X,Y}}$ that represents the speed factor $\boldsymbol{\eta}_{\mathbf{W}_{X,Y}}$ of the first-order self-model state $\mathbf{W}_{X,\,Y}$ that represents the adaptation of the connection weight $\boldsymbol{\omega}_{X,Y.}$ Note that in the model introduced the network is limited to the first reification level and as such, this exact example will not be illustrated. For more details of this modeling approach, see this volume, Chap. 2 [23].

Also adaptivity of order higher than two has been modeled in such a network-oriented manner, up to order five. This enables to model processes with a tower of control levels as considered in biology and philosophy; see, for example, [9, 7, 48, 49]. In [10, 40] this has been used to model how bad environmental circumstances such as a bad childhood can lead to lifelong mental disorders via five adaptation

Table 1 Combination functions used

	Notation	Formula	Parameters	Used for states
Advanced logistic sum	$\text{alogistic}_{\boldsymbol{\sigma},\boldsymbol{\tau}}(V_1,\ldots,V_k)$	$\left[\frac{1}{1+e^{-\sigma(V_1+\ldots+V_k-\tau)}} - \frac{1}{1+e^{\sigma\tau})}\right](1+e^{-\sigma\tau})$	**σ** steepness **τ** threshold	X10-12, X16-18, X20-26, X36-38, X42-44, X46-52 (Base states); X54-59, X66-72, X79 (**W**-states)
Identity	$\text{id}(V_1,\ldots,V_k)$	V_1	No parameter	X1-9, X13-15, X19, X27-35, X39-41, X45
Hebbian learning	$\text{hebb}_{\boldsymbol{\mu}}(V_1,V_2,\text{W})$	$V_1V_2(1\text{-}W) + \boldsymbol{\mu}W$	**μ** persistence factor	X60-65, X73-78 (Preconditioning **W**-states)
Step once	$\text{steponce}_{\boldsymbol{\alpha},\boldsymbol{\beta}}(V_1,\ldots,V_k)$	1 if $\boldsymbol{\alpha} \leq t \leq \boldsymbol{\beta}$ else 0 (time t)	**α** start time **β** end time	X53 (Context state)

levels embodied by epigenetic factors, gene expression, mRNA production, protein production, brain physiology, respectively.

This modeling approach could be regarded as a generalized form of neural networks as the availability of the vast library of combination functions allows for any activation function that is used to be followed up with a related combination function [47]. For more details, see also this volume [23] Chap. 2 [50]. In addition, in [47, 21] it has been demonstrated in a broader context that any (smooth) dynamical system can be modeled as a (temporal-causal) network model and that any (smooth) adaptive dynamical system can be represented as a self-modeling network model, see also Chap. 12 [25] of this volume. This renders the modeling approach highly versatile an applicable across a wide range of scenarios. This has also been confirmed by recent applications in other disciplines, for example, for internal mental models in cognitive science [51], for organisational learning within management science [16], for social interaction and bonding in social psychology [22], and for shared mentals and organisational learning for teams in health sciences [36].

The model introduced will be created utilizing the discussed modeling approach to create a first-order adaptive network model using among others synchrony detection states and a mental God-model at the base level and W-states at the first reification level. This model is in turn used in a simulation in order to further discuss the research question.

5 The Introduced Adaptive Agent Model

This section introduces in more detail the adaptive agent model used. As mentioned in the previous section, the model was designed based on a self-modeling network architecture with two levels, the base level and first reification level. The feeling of togetherness and presence of God will be modelled at the base level. The improvements in cognition will be modelled with the help of **W**-states introducing adaptivity at the first reification level. Despite the fact of the worship session concerning an elderly cognitive impaired couple, for the sake of simplicity only one cognitive impaired participant is modelled. Thus, two agents are modelled, the worship leader and the participant. The synchronization between participants can still be modelled as the worship leader can be regarded as a participant as well. The worship leader takes an important role as a guide throughout the session guiding the cognitive impaired participant. The cognitive impaired participant also plays an important role as the participant has to show the improvements in cognition experienced during the session. By modeling these two persons as agents the most important aspects of the interactions at play with respect to the feeling of togetherness, presence of God and improvements in cognition can be modelled and researched regardless. Both levels with their states and choices made for each agent will be explained alongside with supporting graphical overviews. At the end, an overview with the main differences between the two agents and the choices made for the simulation are given.

5.1 Base Level

An overview of the agents' base states can be seen in Figs. 1, 2 and 3 with Table 2 explaining the base states.

Different Modalities

The internal activities and external actions taking place in the session are generalized to three modalities. That of affection (smiling, making eye contact, laughing, etc.), movement (dancing, moving on the beat, tapping your foot or hand, etc.) and audio (talking, singing, humming, etc.). Externally, these modalities are represented by world states. World states are represented outside the box and the mental states of an agent inside the box. These states can be sensed and subsequently represented internally by the agents. These internal representations of the three modalities influence preparations for the same three modalities that the agent eventually executes subsequently, thus updating or creating new world states that can be sensed by the other agent. This creates a loop. The related connections are all one to one for the respective modalities except for the representation—preparation connections due to the fact that different modalities sensed help towards preparing for different modalities executed. For example, when hearing music, it not only helps towards singing along, it also helps towards moving along with the beat and maybe feeling some affection. This is done by having cross connections where each modality has its representation state connected to each preparation state.

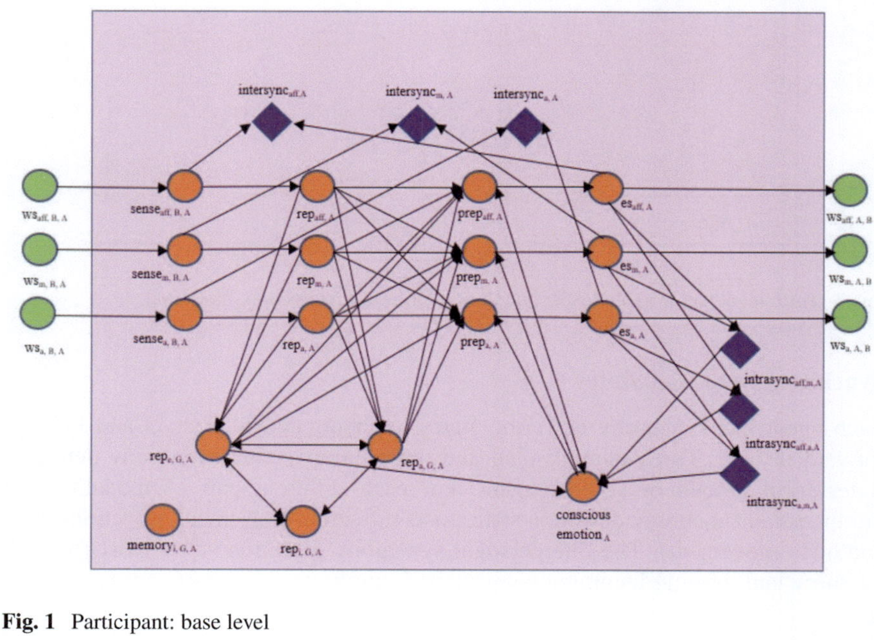

Fig. 1 Participant: base level

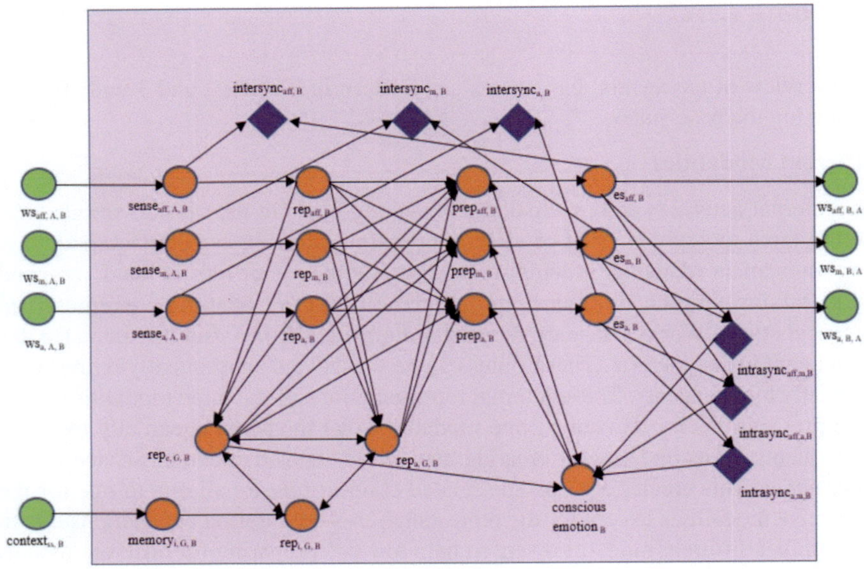

Fig. 2 Worship leader: base level

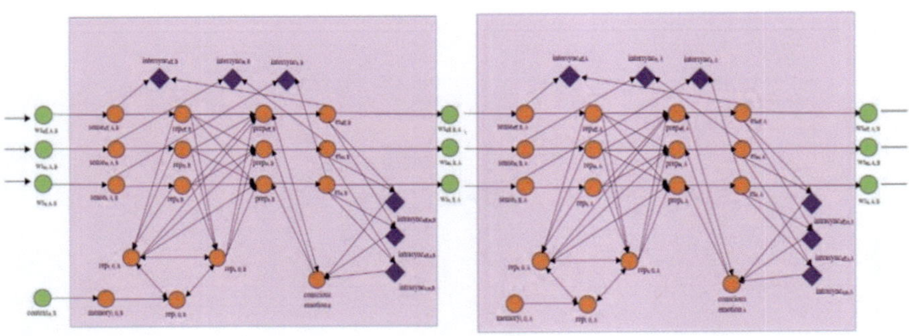

Fig. 3 Participant and worship leader and their multimodal interaction: base level

Synchrony Detection States

Each agent has synchrony detection states as mentioned in Sect. 3 and brought forward in [27]. Three interpersonal and three intrapersonal synchrony detection states form a total of six per agent. For each different pair of modalities an intrapersonal synchrony detection state exists (affection-movement, affection-audio and audiomovement). The interpersonal synchrony detection states exist for each modality and have an incoming connection from the sense and execution state for

Table 2 All states explained

State nr	State name	Explanation
X_1	$ws_{aff, B, A}$	World state for affection (aff) for person A.
X_2	$ws_{m, B, A}$	World state for movement (m) for person A.
X_3	$ws_{a, B, A}$	World state for audio (a) for person A.
X_4	$sense_{aff, B, A}$	Sensor state of person A for affection expressed by person B.
X_5	$sense_{m, B, A}$	Sensor state of person A for movement expressed by person B.
X_6	$sense_{a, B, A}$	Sensor state of person A for audio expressed by person B.
X_7	$rep_{aff, A}$	Sensory representation state of person A for affection.
X_8	$rep_{m, A}$	Sensory representation state of person A for movement.
X_9	$rep_{a, A}$	Sensory representation state of person A for audio.
X_{10}	$prep_{aff, A}$	Preparation state of person A for affection action.
X_{11}	$prep_{m, A}$	Preparation state of person A for movement action.
X_{12}	$prep_{a, A}$	Preparation state of person A for audio action.
X_{13}	$es_{aff, A}$	Execution state of person A for affection action.
X_{14}	$es_{m, A}$	Execution state of person A for movement action.
X_{15}	$es_{a, A}$	Execution state of person A for audio action.
X_{16}	$rep_{e, G, A}$	Sensory representation state of person A for God's emotion.
X_{17}	$rep_{a, G, A}$	Sensory representation state of person A for God's action.
X_{18}	$rep_{i, G, A}$	Sensory representation state of person A's image of God.
X_{19}	$memory_{i, G, A}$	Memory state of person A for God image.
X_{20}	conscious $emotion_A$	Conscious emotion state of person A.
X_{21}	$intersync_{aff,A}$	Inter synchrony detection state of person A for affection.
X_{22}	$intersync_{m, A}$	Inter synchrony detection state of person A for movement.
X_{23}	$intersync_{a, A}$	Inter synchrony detection state of person A for audio.
X_{24}	$intrasync_{aff,m,A}$	Intra synchrony detection state of person A for affection and movement.
X_{25}	$intrasync_{aff,a,A}$	Intra synchrony detection state of person A for affection and audio.
X_{26}	$intrasync_{a,m,A}$	Intra synchrony detection state of person A for audio and movement.
X_{27}	$ws_{aff, A, B}$	World state for affection (aff) for person B.
X_{28}	$ws_{m, A, B}$	World state for movement (m) for person B.
X_{29}	$ws_{a, A, B}$	World state for audio (a) for person B.
X_{30}	$sense_{aff, A, B}$	Sensor state of person B for affection expressed by person A.
X_{31}	$sense_{m, A, B}$	Sensor state of person B for movement expressed by person A.
X_{32}	$sense_{a, A, B}$	Sensor state of person B for audio expressed by person A.
X_{33}	$rep_{aff, B}$	Sensory representation state of person B for affection.
X_{34}	$rep_{m, B}$	Sensory representation state of person B for movement.
X_{35}	$rep_{a, B}$	Sensory representation state of person B for audio.
X_{36}	$prep_{aff, B}$	Preparation state of person B for affection action.
X_{37}	$prep_{m, B}$	Preparation state of person B for movement action.
X_{38}	$prep_{a, B}$	Preparation state of person B for audio action.
X_{39}	$es_{aff, B}$	Execution state of person B for affection action.
X_{40}	$es_{m, B}$	Execution state of person B for movement action.
X_{41}	$es_{a, B}$	Execution state of person B for audio action.
X_{42}	$rep_{e, G, B}$	Sensory representation state of person B for God's emotion.
X_{43}	$rep_{a, G, B}$	Sensory representation state of person B for God's action.
X_{44}	$rep_{i, G, B}$	Sensory representation state of person B's image of God.
X_{45}	$memory_{i, G, B}$	Memory state of person B for the God image.
X_{46}	conscious $emotion_B$	Conscious emotion state of person B.
X_{47}	$intersync_{aff,B}$	Inter synchrony detection state of person B for affection.
X_{48}	$intersync_{m, B}$	Inter synchrony detection state of person B for movement.
X_{49}	$intersync_{a, B}$	Inter synchrony detection state of person B for audio.
X_{50}	$intrasync_{aff,m,B}$	Intra synchrony detection state of person B for affection and movement.
X_{51}	$intrasync_{aff,a,B}$	Intra synchrony detection state of person B for affection and audio.
X_{52}	$intrasync_{a,m,B}$	Intra synchrony detection state of person B for audio and movement.
X_{53}	$context_{ss, B}$	Context state 'start session' for person B.

(continued)

Table 2 (continued)

X_{54}	$\mathbf{W}_{senseaff-repaff,A}$	First-order self-model state for the weight of the sense to rep state of person A for the affection action.
X_{55}	$\mathbf{W}_{sensem-repm,A}$	First-order self-model state for the weight of the sense to rep state of person A for the movement action.
X_{56}	$\mathbf{W}_{sensea-repa,A}$	First-order self-model state for the weight of the sense to rep state of person A for the audio action.
X_{57}	$\mathbf{W}_{prepaff-esaff,A}$	First-order self-model state for the weight of the prep to es state of person A for the affection action.
X_{58}	$\mathbf{W}_{prepm-esm,A}$	First-order self-model state for the weight of the prep to es state of person A for the movement action.
X_{59}	$\mathbf{W}_{prepa-esa,A}$	First-order self-model state for the weight of the prep to es state of person A for the audio action.
X_{60}	$\mathbf{W}_{repaff-repe,G,A}$	First-order self-model state for the weight of the rep to rep state of person A's God emotion (with respect to affection).
X_{61}	$\mathbf{W}_{repm-repe,G,A}$	First-order self-model state for the weight of the rep to rep state of person A's God emotion (with respect to movement).
X_{62}	$\mathbf{W}_{repa-repe,G,A}$	First-order self-model state for the weight of the rep to rep state of person A's God emotion (with respect to audio).
X_{63}	$\mathbf{W}_{repaff-repa,G,A}$	First-order self-model state for the weight of the rep to rep state of person A's God action (with respect to affection).
X_{64}	$\mathbf{W}_{repm-repa,G,A}$	First-order self-model state for the weight of the rep to rep state of person A's God action (with respect to movement).
X_{65}	$\mathbf{W}_{repa-repa,G,A}$	First-order self-model state for the weight of the rep to rep state of person A's God action (with respect to audio).
X_{66}	$\mathbf{W}_{memory-repi,G,A}$	First-order self-model state for the weight of the memory to rep state of person A's image of God.
X_{67}	$\mathbf{W}_{senseaff-repaff,B}$	First-order self-model state for the weight of the sense to rep state of person B for the affection action.
X_{68}	$\mathbf{W}_{sensem-repm,B}$	First-order self-model state for the weight of the sense to rep state of person B for the movement action.
X_{69}	$\mathbf{W}_{sensea-repa,B}$	First-order self-model state for the weight of the sense to rep state of person B for the audio action.
X_{70}	$\mathbf{W}_{prepaff-esaff,B}$	First-order self-model state for the weight of the prep to es state of person B for the affection action.
X_{71}	$\mathbf{W}_{prepm-esm,B}$	First-order self-model state for the weight of the prep to es state of person B for the movement action.
X_{72}	$\mathbf{W}_{prepa-esa,B}$	First-order self-model state for the weight of the prep to es state of person B for the audio action.
X_{73}	$\mathbf{W}_{repaff-repe,G,B}$	First-order self-model state for the weight of the rep to rep state of person B's God emotion (with respect to affection).
X_{74}	$\mathbf{W}_{repm-repe,G,B}$	First-order self-model state for the weight of the rep to rep state of person B's God emotion (with respect to movement).
X_{75}	$\mathbf{W}_{repa-repe,G,B}$	First-order self-model state for the weight of the rep to rep state of person B's God emotion (with respect to audio).
X_{76}	$\mathbf{W}_{repaff-repa,G,B}$	First-order self-model state for the weight of the rep to rep state of person B's God action (with respect to affection).
X_{77}	$\mathbf{W}_{repm-repa,G,B}$	First-order self-model state for the weight of the rep to rep state of person B's God action (with respect to movement).
X_{78}	$\mathbf{W}_{repa-repa,G,B}$	First-order self-model state for the weight of the rep to rep state of person B's God action (with respect to audio).
X_{79}	$\mathbf{W}_{memory-repi,G,B}$	First-order self-model state for the weight of the memory to rep state of person B's image of God.

that respective modality. These synchronization indicators play an important role for the **W**-states on the reification level.

Mental God-Model

The mental God-model of the agents exists of four states covering the three main dimensions as discussed in Sect. 2. The God image representation state shows the current level of religiousness of the individual. The God memory state shows the level of religiousness the person has built up so far in their life and can access when not cognitively impaired. The God emotion representation state shows the level of God's emotion felt by the individual which can be accessed and conveyed through actions. The God action representation state shows the level of God's actions felt by the individual which can be conveyed through actions. These states mutually affect each other except for the God memory state that only affects the God image representation state. The God emotion and action representation states have incoming connections from the representation states of the modalities as the emotions and actions of God conveyed by the other person can be detected. Vice versa, the God action representation state has effect on the preparation states so that the action of God can be conveyed. The God emotion representation state has effect on the conscious emotion state as will be discussed next. This mental God model starts of the session after being activated by a context state eventually influencing the preparations states of the worship leader.

Conscious Emotion

One particular important state of the agents is the conscious emotion state. This represents the overall emotional state during the session and the emotional investment of the person into the session. It can subsequently show an emotional and religious breakthrough or high persons may experience. This state has incoming connections from the affection preparation state, intrapersonal synchrony detection states and God emotion representation state as all these states help the person either get more into the session or experience more emotions. In turn, as the state reflects the investment into the session, the state has outgoing connections to each modality's preparation state.

5.2 *First-Order Self-Modeling Level*

An overview of the agents first-order self-modeling level states can be seen in Figs. 4, 5 and 6 with Table 1 explaining the **W**-states.

Synchronization, Plasticity and Cognition

Behavioural adaptivity helps with all facets of sensing and executing actions. Therefore, the sense—representation and preparation–execution connections of each modality are adaptive. The connections have weights represented by their **W**-states which will start relatively low. In addition, the God memory–God image representation connection is adaptive and uses the same sort of **W**-state. These **W**-states have each as input the six synchrony detection states. This means that as synchronization takes place during the session, the above-described connections strengthen. Therefore, as synchronization takes effect, an improvement in cognition as discussed in Sect. 3 can be measured by better translating the modalities and the 'hidden' religiousness of the participant. This way the improvement in cognition can be indirectly inferred.

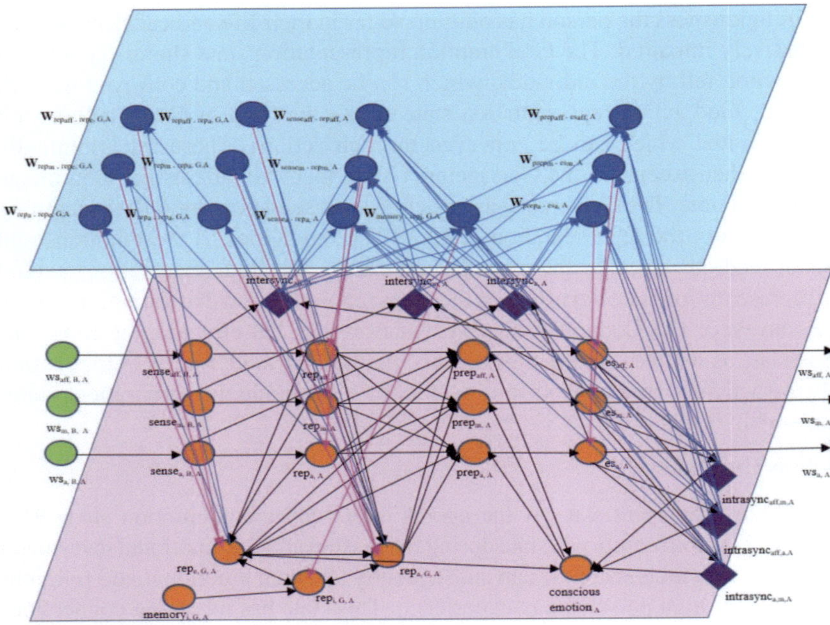

Fig. 4 Participant: first-order self-model level

Sensory Preconditioning

All other **W**-states are related to the connection between the representation state of each modality and the God emotion and action representation state. These **W**-states model the sensory preconditioning taking place. This preconditioning is the linking of the observed modalities to the mental God-model [8, 53]. By preconditioning these connections are learned or strengthened over time.

5.3 Simulation Choices and Agent Differences

The worship leader has one additional context state compared to the participant. This state influences the God memory state and can simulate different scenarios where the worship leader activates his God memory states at different stages. This allows for more possibilities with modeling scenarios but in this simulation it only indicates the start of the session. As a result of this, the God Image representation of the worship leader is modelled as a constant value due to the fact it is constantly activated throughout the session given there are no interruptions in the scenario.

As the God memory state reflects the build-up level of religiousness of a person it also allows for multiple modelling possibilities. In this case, the worship leader

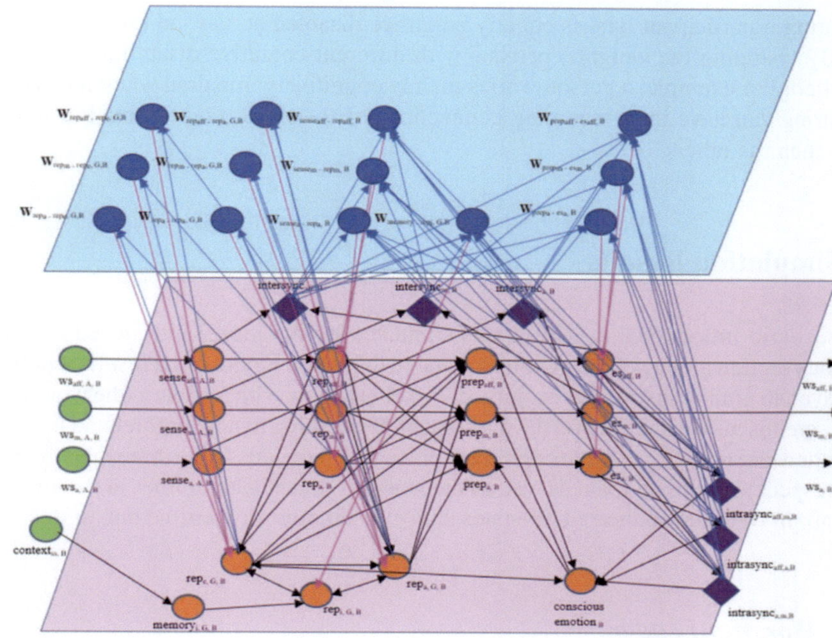

Fig. 5 Worship leader: first-order self-model level

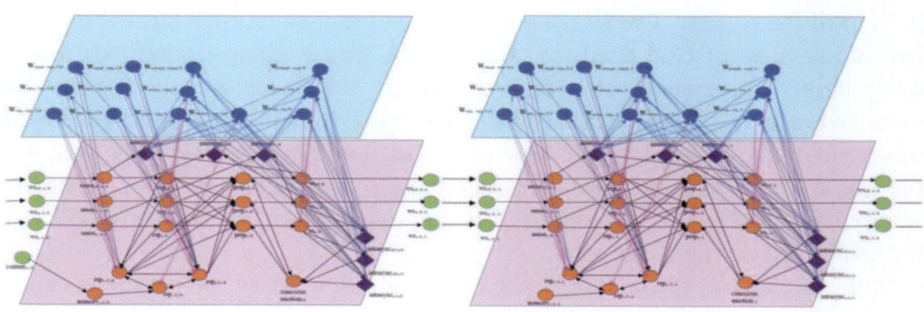

Fig. 6 Participant and worship leader in interaction: first-order self-model level

is very religious making this state the max value. However, another simulation may address a less religious worship leader by lowering the value of this state. The same goes for the participant. This simulation has a participant who has been religiously active more than average their entire life with the used value reflecting that. In the case of the participant the state can act as a sort of hidden religiousness as this religiousness will be uncovered as adaptivity occurs during the session.

Another way to create different simulations is by changing the weights of the cross connections. In this simulation the cross connections have the same weight indicating a participant who is equally cognitive disabled across the three modalities. By changing the weights, persons with different cognitive disabilities can be modelled. For example, a person who is mainly cognitively impaired when it comes to hearing can have their incoming connections of their audio representation state lower than the others.

6 Simulation Results

The scenario introduced in Sect. 2 was simulated by the model introduced in the previous section along with the choices made in that same section. The results will be displayed in graphs and further explained. First the results with respect to the worship leader are discussed and second the participant. Something to note beforehand is that since there are no differences in cognitive abilities with regards to the three modalities with respect to the participant as well as the worship leader, a lot of lines in the graph overlap. In turn, the connected **W**-states also overlap most of the time due to this.

6.1 Worship Leader

The session begins with the worship leader who kicks everything off by activating the own mental God-model. First, the *context state* $\text{context}_{\text{ss, B}}$ is activated to indicate the start of the session. The worship leader knows that the session has to begin and that the message of God needs to be conveyed. The worship leader does this by accessing the *God memory* state $\text{memory}_{\text{i,G,B}}$ and subsequently activating the *God image* state $\text{rep}_{\text{i, G, B}}$. In the graph the *God memory* stste $\text{memory}_{\text{i, G, B}}$ is constant as it is a constant state as mentioned in the previous section. From this the *God image* state $\text{rep}_{\text{i, G, B}}$, a *God action* representation state $\text{rep}_{\text{a,G,B}}$ and *God emotion* representation state $\text{rep}_{\text{e,G,B}}$ is activated in order to fully convey the emotions and actions of God that can be felt during the session.

With the mindset of the worship leader in the right place, preparations can be made for the actions that will be taken during the session. This is shown by the rise of the *preparation states* $\text{prep}_{\text{aff, B}}$, $\text{prep}_{\text{m, B}}$, $\text{prep}_{\text{a, B}}$ for the three modalities. Because the worship leader knows what actions are going to take place, as many similar sessions have come before, the states can rise to a high level soon. Note that this does not mean these preparations will lead to execution as this will depend on things. With the preparations states a final increase is seen. Each participant (or group of participants) is different and may require a different final nudge or trigger, this case being the breaking of bread, in order to experience the emotional and or religious climax. This means that the worship leader cannot prepare for this too far in advance explaining the final increase (Fig. 7).

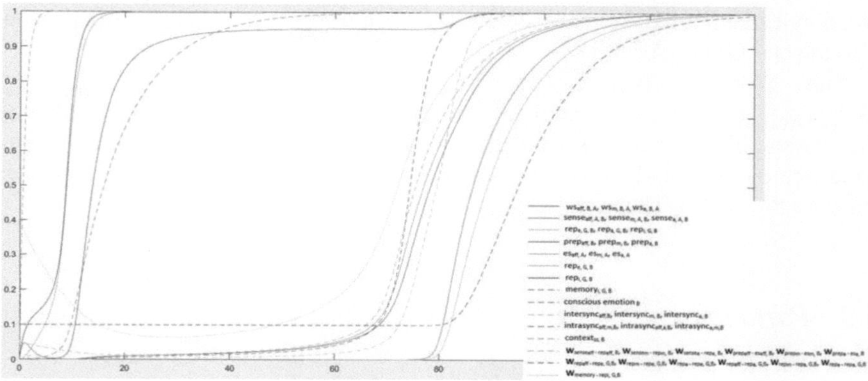

Fig. 7 Worship leader in the simulation

With the preparations for the actions ready, the worship leader starts executing these actions. This is shown by the *execution states* $es_{aff,\ B}$, $es_{m,\ B}$, $es_{a,\ B}$ and subsequent *world states* $ws_{aff,\ B,}$ A, $ws_{m,\ B,\ A}$, $ws_{a,\ B,\ A}$ activated for the participant. These states rise very slowly at the beginning as the worship leader takes the lead introducing the participant to the different actions and no improvement in the participant's cognition has taken place yet. The worship leader is waiting for the participant to pick up on these actions before increasing the intensity. Eventually these actions are picked up by the participant and reflected. This means that the worship leader gets feedback by sensing the actions of the participant. This is shown by the *sense* states $sense_{aff,}$ A, B, $sense_{m,\ A,\ B}$, $sense_{a,\ A,\ B}$ which start to rise in a similar fashion after the *execution* states of the participant $es_{aff,\ A}$, $es_{m,\ A}$, $es_{a,\ A}$ and are subsequently followed by the *representation* states rep, repa, G, B, repi, G, B of each modality.

Because synchronization means that the participant is ready for more difficult actions and that the participant is getting more into the actions, the worship leader can really start to increase the intensity of the actions when the participant starts to synchronize with the worship leader. This can be seen in the graph by the sudden rise in the states described in the paragraph above. This rise is explained by looking at the *synchrony detection states* $intersync_{aff,B}$, $intersync_{m,\ B}$, $intersync_{a,\ B}$, $intrasync_{aff,m,B}$, $intrasync_{aff,a,B}$, $intrasync_{a,m,B}$ and the **W**-*states* $\mathbf{W}_{sense_{aff}-rep_{aff,B}}$, $\mathbf{W}_{sense_{m}-rep_{m,B}}$, $\mathbf{W}_{sense_{a}-rep_{a,B}}$, $\mathbf{W}_{prep_{aff}-es_{aff,B}}$, $\mathbf{W}_{prep_{m}-es_{m,B}}$, $\mathbf{W}_{prep_{a}-es_{a,B}}$ influenced by them. When the synchrony detection states start to rise rapidly after some time has passed, the **W**-states having incoming connections from these synchrony detection states also start to rise. This allows for the adaptive connection weights to increase explaining this sudden rise. This event shows the improvements in cognition as a result of the synchronization that has taken place.

The *conscious emotion* state conscious $emotion_B$ of the worship leader displays the emotional state. That of the worship leader rises relatively early since many sessions have gone before and there are no emotional surprises left. Moreover, the worship leader has to take the lead not only when it comes to the physical aspect but also the emotional aspect.

The **W**-*states* $\mathbf{W}_{\text{rep}_{\text{aff}}-\text{rep}_{\text{e,G,B}}}$, $\mathbf{W}_{\text{rep}_{\text{m}}-\text{rep}_{\text{e,G,B}}}$, $\mathbf{W}_{\text{rep}_{\text{a}}-\text{rep}_{\text{e,G,B}}}$, $\mathbf{W}_{\text{rep}_{\text{aff}}-\text{rep}_{\text{a,G,B}}}$, $\mathbf{W}_{\text{rep}_{\text{m}}-\text{rep}_{\text{a,G,B}}}$, $\mathbf{W}_{\text{rep}_{\text{a}}-\text{rep}_{\text{a,G,B}}}$ showing the preconditioning rise against the end of the session. This is because it takes some time before the actions represented can be attributed to the individual's God model as well as that the representation states start out low and rise later on.

6.2 *Participant*

Overall, the states and graph of the participant follow the same pattern and explanation as the worship leader but with a slight timing difference as the participant reacts to the demonstrations and instructions of the worship leader. The point of synchronization and the resulting improvements in cognition can also be found in this graph as the participant is the source of this. However, there are two main differences: the mental God model and the conscious emotion.

Unlike the worship leader, the *God image* $\text{rep}_{\text{i, G, A}}$, *action* $\text{rep}_{\text{a, G, A}}$ and *emotion* $\text{rep}_{\text{e, G, A}}$ states rise around the same time as all other states when synchronization takes place as they all rely on the cognitive improvements. This in addition to the mental God image being activated from the 'hidden' God memory of the participant, makes the mental God model of the participant spring into action once synchronization starts to take place. Since the 'hidden' God memory has a certain level, the God image activated can only rise to this certain height as seen in the graph (Fig. 8).

The *conscious emotion* state conscious $\text{emotion}_{\text{A}}$ of the participant also rises later than that of the worship leader. That is because the participant is not mentally equipped enough to predict what is to come and cannot feel the emotions in anticipation of what is to come. The emotional state rises together with the increase in

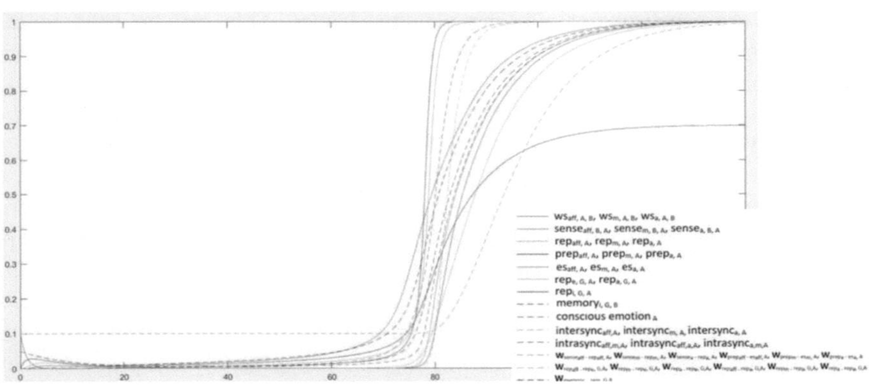

Fig. 8 Participant in the simulation

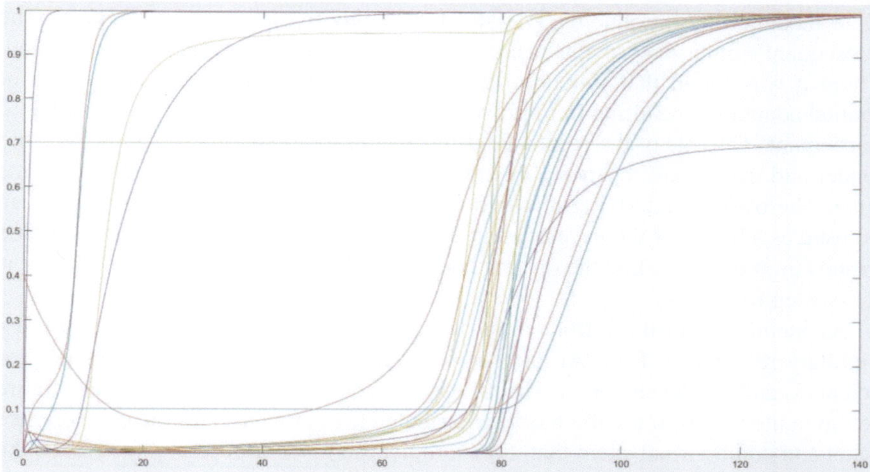

Fig. 9 Worship leader and participant in the simulation

cognition. That is because the more the participants gets into the session by executing the actions better and activating his mental God model, the closer he gets to the religious and emotional climax felt at the end. This climax is one of the last events to happen which is reflected by the conscious emotion state being one of the last states to peak (Fig. 9).

Altogether, the graph shows the sudden synchronization that takes place after some time resulting in the increase of cognition. This increase in cognition can be concluded from the increase in the weights of the adaptive connections allowing for the modalities sensed to be better represented, the God-model to be activated better and the prepared actions to be executed better. All this eventually results in an emotional and religious peak that the participant experiences.

7 Conclusion and Discussion

The material for this chapter came from [56]. The chapter aimed to provide a more objective look on religion and its positive impact by attempting to answer the research question as formulated in Sect. 2 on the basis of a reallife scenario. Namely, can a higher-order adaptive network model enhance our understanding of the interplay of interpersonal synchrony during a group worship session for reliving memory and religious experiences through intersensory facilitation by computationally recreating the scenario? This was done by introducing relevant literature that addressed how to model the three main themes found in the scenario. The feeling of togetherness and oneness was modelled by synchrony detection states. The felt presence of God was modelled by the mental God-model. The improvements in cognition perceived were modelled by introducing **W**-states for adaptivity of certain connections. The subsequent simulation provided a closer look into the inner workings of the session. There it was found that synchronization suddenly intensifies after some time of mutual adaptation resulting in improvements in cognition of the cognitive impaired participant. The role of the mental God-model and differences between the worship leader and participant regarding this mental model was also addressed at the same time. The overall underlying research question is to obtain a base model that can be used as a basis for the development of an embodied virtual agent that can guide groups in joint events like the one analysed here. The contribution in this chapter is a first step for this.

As mentioned in the introduction, the chapter can be of interest for multidisciplinary researchers from AI and social psychology, as well as from theological sciences, and for developers of virtual agents or bots to support humans. For future research the idea is to use the model contributed here as a basis for the development of an embodied virtual agent that, for example, can guide groups in joint events like the one analysed here.

The computational model contributed here has had inspiration from the models described in [27, 21, 22] see also this volume [23], Chaps. 11–13 [25, 28, 26]. Important differences are that in the current model (1) the integration of an adaptive mental God model has been addressed as well based on sensory preconditioning, and (2) the plasticity for cognition related to social interaction has been addressed for a wider scope also covering memory elements, (3) the model is focusing on a joint session concerning religious experiences.

Note that it is the research subject that is the difference with other papers using the same modeling approach such as the ones mentioned above and, for example, [4, 5, 12, 14]. This can be compared to the extensive use of the methodological tool of differential (or difference) equations for formalisation and simulation dynamics of different phenomena in practically all scientific disciplines. These papers have contributed to scientific progress because the (same) methodological tool enables to address these phenomena. Similarly, the scientific literature that use a given (same) method for statistical analysis for different phenomena has enabled much progress in empirical research. In all these cases, it is the research subject to which the method is applied that makes the difference, not the general method used.

When the model was introduced in Sect. 4, certain simulation choices were made. These choices had to be made because the makeup of the model allows for different simulations being done. This means that there are still a lot of interesting further possibilities for future simulations. As mentioned before, different cognitively impaired persons can be modeled by changing the weights of the cross connections between the representation–preparation states. Furthermore, the level of religiousness of the worship leader and participant can be changed by addressing the values of their God memory states. The context state can also be activated to simulate different time scales or interruptions during the session. A comparison could be made between the worship leader and the participant and a therapist and client. In the context of psychotherapy, synchronization of movements between the therapist and patient could lead to a stronger feeling of sharing the present moment [52] and with multiple sessions this may strengthen the therapeutic bond resulting in a more effective cooperation [33]. This could be researched in a religious context with the use of this model or some variation where the same worship leader is used multiple times. Subsequently, the differences between the same worship leader multiple times and different leaders can be researched.

Another important note is that the model and simulation are not religion-specific. The reason for this is that the notion of the mental God-model has not been specifically tailored towards one certain God but rather to the personal God of the individual and that the specific activities and triggers described are not modelled as such. This allows for the more general and main idea of the past religious experiences of the participants being similar to the introduced activities and triggers. Therefore, it can be said that the introduced model is not religion-specific and a translation to multiple (mainly Abrahamic) religions is possible. Therefore, this model and variations on the model could be used to simulate different scenarios and analogies from different religions.

As final note, the model addresses a real-life scenario that actually has been reported. Verification or validation of the model took place by reproducing this scenario by the simulation. However, it more scenarios and more detailed data will be available, in future research further verification and validation can be performed. The scenario used happens to convey positive elements of religion but it could as well convey negative elements. As mentioned before, the aim was to provide a more objective look on religion. This means that the objective look given does not make any claims towards the nature of religion but rather the innerworkings of related mechanisms that play a role in the scenarios and observed events. Similar models could also be used to give more insight into the negative impacts of religion. This is another possibility to explore.

8 Appendix: Specification by Role Matrices and Scenario Description

Role Matrices

mb	base connectivity	1	2	3	4	5	6
X_{1}	$ws_{aff, B, A}$	X_{39}					
X_{2}	$ws_{m, B, A}$	X_{40}					
X_{3}	$ws_{a, B, A}$	X_{41}					
X_{4}	$sense_{aff, B, A}$	X_{1}					
X_{5}	$sense_{m, B, A}$	X_{2}					
X_{6}	$sense_{a, B, A}$	X_{3}					
X_{7}	$rep_{aff, A}$	X_{4}					
X_{8}	$rep_{m, A}$	X_{5}					
X_{9}	$rep_{a, A}$	X_{6}					
X_{10}	$prep_{aff, A}$	X_{7}	X_{8}	X_{9}	X_{16}	X_{17}	X_{20}
X_{11}	$prep_{m, A}$	X_{7}	X_{8}	X_{9}	X_{16}	X_{17}	X_{20}
X_{12}	$prep_{a, A}$	X_{7}	X_{8}	X_{9}	X_{16}	X_{17}	X_{20}
X_{13}	$es_{aff, A}$	X_{10}					
X_{14}	$es_{m, A}$	X_{11}					
X_{15}	$es_{a, A}$	X_{12}					
X_{16}	$rep_{e, G, A}$	X_{18}	X_{17}	X_{7}	X_{8}	X_{9}	
X_{17}	$rep_{a, G, A}$	X_{18}	X_{16}	X_{7}	X_{8}	X_{9}	
X_{18}	$rep_{i, G, A}$	X_{19}					
X_{19}	$memory_{i, G, A}$	X_{19}					
X_{20}	conscious emotion $_{A}$	X_{16}	X_{10}	X_{24}	X_{25}	X_{26}	
X_{21}	$intersync_{aff,A}$	X_{4}	X_{13}				
X_{22}	$intersync_{m, A}$	X_{5}	X_{14}				
X_{23}	$intersync_{a, A}$	X_{6}	X_{15}				
X_{24}	$intrasync_{aff,m,A}$	X_{13}	X_{14}				
X_{25}	$intrasync_{aff,a,A}$	X_{13}	X_{15}				
X_{26}	$intrasync_{a,m,A}$	X_{15}	X_{14}				
X_{27}	$ws_{aff, A, B}$	X_{13}					
X_{28}	$ws_{m, A, B}$	X_{14}					
X_{29}	$ws_{a, A, B}$	X_{15}					
X_{30}	$sense_{aff, A, B}$	X_{27}					
X_{31}	$sense_{m, A, B}$	X_{28}					
X_{32}	$sense_{a, A, B}$	X_{29}					
X_{33}	$rep_{aff, B}$	X_{30}					
X_{34}	$rep_{m, B}$	X_{31}					
X_{35}	$rep_{a, B}$	X_{32}					
X_{36}	$prep_{aff, B}$	X_{33}	X_{34}	X_{35}	X_{42}	X_{43}	X_{46}
X_{37}	$prep_{m, B}$	X_{33}	X_{34}	X_{35}	X_{42}	X_{43}	X_{46}
X_{38}	$prep_{a, B}$	X_{33}	X_{34}	X_{35}	X_{42}	X_{43}	X_{46}
X_{39}	$es_{aff, B}$	X_{36}					
X_{40}	$es_{m, B}$	X_{37}					
X_{41}	$es_{a, B}$	X_{38}					
X_{42}	$rep_{e, G, B}$	X_{44}	X_{43}	X_{33}	X_{34}	X_{35}	
X_{43}	$rep_{a, G, B}$	X_{44}	X_{42}	X_{33}	X_{34}	X_{35}	
X_{44}	$rep_{i, G, B}$	X_{45}	X_{43}	X_{42}			

X_{45}	$\text{memory}_{i,G,B}$	X_{53}					
X_{46}	$\text{conscious emotion}_{B}$	X_{42}	X_{36}	X_{50}	X_{51}	X_{52}	
X_{47}	$\text{intersync}_{aff,B}$	X_{30}	X_{39}				
X_{48}	$\text{intersync}_{m,B}$	X_{31}	X_{40}				
X_{49}	$\text{intersync}_{a,B}$	X_{32}	X_{41}				
X_{50}	$\text{intrasync}_{aff,m,B}$	X_{39}	X_{40}				
X_{51}	$\text{intrasync}_{aff,a,B}$	X_{39}	X_{41}				
X_{52}	$\text{intrasync}_{a,m,B}$	X_{41}	X_{40}				
X_{53}	$\text{context}_{ss,B}$						
X_{54}	$\mathbf{W}_{sense\text{-}rep\ aff,A}$	X_{21}	X_{22}	X_{23}	X_{24}	X_{25}	X_{26}
X_{55}	$\mathbf{W}_{sense\text{-}rep\ m,A}$	X_{21}	X_{22}	X_{23}	X_{24}	X_{25}	X_{26}
X_{56}	$\mathbf{W}_{sense\text{-}rep\ a,A}$	X_{21}	X_{22}	X_{23}	X_{24}	X_{25}	X_{26}
X_{57}	$\mathbf{W}_{prep\text{-}es\ aff,A}$	X_{21}	X_{22}	X_{23}	X_{24}	X_{25}	X_{26}
X_{58}	$\mathbf{W}_{prep\text{-}es\ m,A}$	X_{21}	X_{22}	X_{23}	X_{24}	X_{25}	X_{26}
X_{59}	$\mathbf{W}_{prep\text{-}es\ a,A}$	X_{21}	X_{22}	X_{23}	X_{24}	X_{25}	X_{26}
X_{60}	$\mathbf{W}_{rep\text{-}rep(e),aff,G,A}$	X_{16}	X_{7}	X_{60}			
X_{61}	$\mathbf{W}_{rep\text{-}rep(e),m,G,A}$	X_{16}	X_{8}	X_{61}			
X_{62}	$\mathbf{W}_{rep\text{-}rep(e),a,G,A}$	X_{16}	X_{9}	X_{62}			
X_{63}	$\mathbf{W}_{rep\text{-}rep(a),aff,G,A}$	X_{17}	X_{7}	X_{63}			
X_{64}	$\mathbf{W}_{rep\text{-}rep(a),m,G,A}$	X_{17}	X_{8}	X_{64}			
X_{65}	$\mathbf{W}_{rep\text{-}rep(a),a,G,A}$	X_{17}	X_{9}	X_{65}			
X_{66}	$\mathbf{W}_{memory\text{-}rep(i),G,A}$	X_{21}	X_{22}	X_{23}	X_{24}	X_{25}	X_{26}
X_{67}	$\mathbf{W}_{sense\text{-}rep\ aff,B}$	X_{47}	X_{48}	X_{49}	X_{50}	X_{51}	X_{52}
X_{68}	$\mathbf{W}_{sense\text{-}rep\ m,B}$	X_{47}	X_{48}	X_{49}	X_{50}	X_{51}	X_{52}
X_{69}	$\mathbf{W}_{sense\text{-}rep\ a,B}$	X_{47}	X_{48}	X_{49}	X_{50}	X_{51}	X_{52}
X_{70}	$\mathbf{W}_{prep\text{-}es\ aff,B}$	X_{47}	X_{48}	X_{49}	X_{50}	X_{51}	X_{52}
X_{71}	$\mathbf{W}_{prep\text{-}es\ m,B}$	X_{47}	X_{48}	X_{49}	X_{50}	X_{51}	X_{52}
X_{72}	$\mathbf{W}_{prep\text{-}es\ a,B}$	X_{47}	X_{48}	X_{49}	X_{50}	X_{51}	X_{52}
X_{73}	$\mathbf{W}_{rep\text{-}rep(e),aff,G,B}$	X_{42}	X_{33}	X_{73}			
X_{74}	$\mathbf{W}_{rep\text{-}rep(e),m,G,B}$	X_{42}	X_{34}	X_{74}			
X_{75}	$\mathbf{W}_{rep\text{-}rep(e),a,G,B}$	X_{42}	X_{35}	X_{75}			
X_{76}	$\mathbf{W}_{rep\text{-}rep(a),aff,G,B}$	X_{43}	X_{33}	X_{76}			
X_{77}	$\mathbf{W}_{rep\text{-}rep(a),m,G,B}$	X_{43}	X_{34}	X_{77}			
X_{78}	$\mathbf{W}_{rep\text{-}rep(a),a,G,B}$	X_{43}	X_{35}	X_{78}			
X_{79}	$\mathbf{W}_{memory\text{-}rep\ i,G,B}$	X_{47}	X_{48}	X_{49}	X_{50}	X_{51}	X_{52}

mcw	**base connectivity**	1	2	3	4	5	6
X_{1}	$\text{ws}_{aff,B,A}$	1					
X_{2}	$\text{ws}_{m,B,A}$	1					
X_{3}	$\text{ws}_{a,B,A}$	1					
X_{4}	$\text{sense}_{aff,B,A}$	1					
X_{5}	$\text{sense}_{m,B,A}$	1					
X_{6}	$\text{sense}_{a,B,A}$	1					
X_{7}	$\text{rep}_{aff,A}$	X_{54}					
X_{8}	$\text{rep}_{m,A}$	X_{55}					
X_{9}	$\text{rep}_{a,A}$	X_{56}					
X_{10}	$\text{prep}_{aff,A}$	1	1	1	1	1	1
X_{11}	$\text{prep}_{m,A}$	1	1	1	1	1	1
X_{12}	$\text{prep}_{a,A}$	1	1	1	1	1	1
X_{13}	$\text{es}_{aff,A}$	X_{57}					
X_{14}	$\text{es}_{m,A}$	X_{58}					
X_{15}	$\text{es}_{a,A}$	X_{59}					
X_{16}	$\text{rep}_{e,G,A}$	1	1	X_{60}	X_{61}	X_{62}	
X_{17}	$\text{rep}_{a,G,A}$	1	1	X_{63}	X_{64}	X_{65}	
X_{18}	$\text{rep}_{i,G,A}$	X_{66}					
X_{19}	$\text{memory}_{i,G,A}$	1					
X_{20}	$\text{conscious emotion}_{A}$	1	1	1	1	1	
X_{21}	$\text{intersync}_{aff,A}$	1	1				

X_{22}	$\text{intersync}_{m, A}$	1	1				
X_{23}	$\text{intersync}_{a, A}$	1	1				
X_{24}	$\text{intrasync}_{aff,m,A}$	1	1				
X_{25}	$\text{intrasync}_{aff,a,A}$	1	1				
X_{26}	$\text{intrasync}_{a,m,A}$	1	1				
X_{27}	$\text{ws}_{aff, A, B}$	1					
X_{28}	$\text{ws}_{m, A, B}$	1					
X_{29}	$\text{ws}_{a, A, B}$	1					
X_{30}	$\text{sense}_{aff, A, B}$	1					
X_{31}	$\text{sense}_{m, A, B}$	1					
X_{32}	$\text{sense}_{a, A, B}$	1					
X_{33}	$\text{rep}_{aff, B}$	X_{67}					
X_{34}	$\text{rep}_{m, B}$	X_{68}					
X_{35}	$\text{rep}_{a, B}$	X_{69}					
X_{36}	$\text{prep}_{aff, B}$	1	1	1	1	1	1
X_{37}	$\text{prep}_{m, B}$	1	1	1	1	1	1
X_{38}	$\text{prep}_{a, B}$	1	1	1	1	1	1
X_{39}	$\text{es}_{aff, B}$	X_{70}					
X_{40}	$\text{es}_{m, B}$	X_{71}					
X_{41}	$\text{es}_{a, B}$	X_{72}					
X_{42}	$\text{rep}_{e, G, B}$	1	1	X_{73}	X_{74}	X_{75}	
X_{43}	$\text{rep}_{a, G, B}$	1	1	X_{76}	X_{77}	X_{78}	
X_{44}	$\text{rep}_{i, G, B}$	X_{79}	0.9	0.9			
X_{45}	$\text{memory}_{i, G, B}$	1					
X_{46}	$\text{conscious emotion}_{B}$	1	1	1	1	1	
X_{47}	$\text{intersync}_{aff,B}$	1	1				
X_{48}	$\text{intersync}_{m, B}$	1	1				
X_{49}	$\text{intersync}_{a, B}$	1	1				
X_{50}	$\text{intrasync}_{aff,m,B}$	1	1				
X_{51}	$\text{intrasync}_{aff,a,B}$	1	1				
X_{52}	$\text{intrasync}_{a,m,B}$	1	1				
X_{53}	$\text{context}_{ss, B}$						
X_{54}	$\mathbf{W}_{\text{sense-rep aff, A}}$	1	1	1	1	1	1
X_{55}	$\mathbf{W}_{\text{sense-rep m, A}}$	1	1	1	1	1	1
X_{56}	$\mathbf{W}_{\text{sense-rep a, A}}$	1	1	1	1	1	1
X_{57}	$\mathbf{W}_{\text{prep-es aff, A}}$	1	1	1	1	1	1
X_{58}	$\mathbf{W}_{\text{prep-es m, A}}$	1	1	1	1	1	1
X_{59}	$\mathbf{W}_{\text{prep-es a, A}}$	1	1	1	1	1	1
X_{60}	$\mathbf{W}_{\text{rep-rep(e), aff, G,A}}$	1	1	1			
X_{61}	$\mathbf{W}_{\text{rep-rep(e), m, G,A}}$	1	1	1			
X_{62}	$\mathbf{W}_{\text{rep-rep(e), a, G,A}}$	1	1	1			
X_{63}	$\mathbf{W}_{\text{rep-rep(a), aff, G,A}}$	1	1	1			
X_{64}	$\mathbf{W}_{\text{rep-rep(a), m, G,A}}$	1	1	1			
X_{65}	$\mathbf{W}_{\text{rep-rep(a), a, G,A}}$	1	1	1			
X_{66}	$\mathbf{W}_{\text{memory- rep(i), G, A}}$	1	1	1	1	1	1
X_{67}	$\mathbf{W}_{\text{sense-rep aff ,B}}$	1	1	1	1	1	1
X_{68}	$\mathbf{W}_{\text{sense-rep m, B}}$	1	1	1	1	1	1
X_{69}	$\mathbf{W}_{\text{sense-rep a, B}}$	1	1	1	1	1	1
X_{70}	$\mathbf{W}_{\text{prep-es aff, B}}$	1	1	1	1	1	1
X_{71}	$\mathbf{W}_{\text{prep-es m, B}}$	1	1	1	1	1	1
X_{72}	$\mathbf{W}_{\text{prep-es a, B}}$	1	1	1	1	1	1
X_{73}	$\mathbf{W}_{\text{rep-rep(e), aff, G,B}}$	1	1	1			
X_{74}	$\mathbf{W}_{\text{rep-rep(e), m, G,B}}$	1	1	1			
X_{75}	$\mathbf{W}_{\text{rep-rep(e), a, G,B}}$	1	1	1			
X_{76}	$\mathbf{W}_{\text{rep-rep(a), aff, G,B}}$	1	1	1			
X_{77}	$\mathbf{W}_{\text{rep-rep(a), m, G,B}}$	1	1	1			
X_{78}	$\mathbf{W}_{\text{rep-rep(a), a, G,B}}$	1	1	1			
X_{79}	$\mathbf{W}_{\text{memory- rep i, G, B}}$	1	1	1	1	1	1

ms speed factors		1	iv initial values	1
X_1	$ws_{aff, B, A}$	0.8		0
X_2	$ws_{m, B, A}$	0.8		0
X_3	$ws_{a, B, A}$	0.8		0
X_4	$sense_{aff, B, A}$	1		0
X_5	$sense_{m, B, A}$	1		0
X_6	$sense_{a, B, A}$	1		0
X_7	$rep_{aff, A}$	1		0
X_8	$rep_{m, A}$	1		0
X_9	$rep_{a, A}$	1		0
X_{10}	$prep_{aff, A}$	1		0
X_{11}	$prep_{m, A}$	1		0
X_{12}	$prep_{a, A}$	1		0
X_{13}	$es_{aff, A}$	1		0
X_{14}	$es_{m, A}$	1		0
X_{15}	$es_{a, A}$	1		0
X_{16}	$rep_{e, G, A}$	1		0
X_{17}	$rep_{a, G, A}$	1		0
X_{18}	$rep_{i, G, A}$	1		0
X_{19}	$memory_{i, G, A}$	0		0.7
X_{20}	conscious emotion $_A$	0.1		0
X_{21}	$intersync_{aff,A}$	0.5		0
X_{22}	$intersync_{m, A}$	0.5		0
X_{23}	$intersync_{a, A}$	0.5		0
X_{24}	$intrasync_{aff,m,A}$	0.5		0
X_{25}	$intrasync_{aff,a,A}$	0.5		0
X_{26}	$intrasync_{a,m,A}$	0.5		0
X_{27}	$ws_{aff, A, B}$	0.8		0.1
X_{28}	$ws_{m, A, B}$	0.8		0.1
X_{29}	$ws_{a, A, B}$	0.8		0.1
X_{30}	$sense_{aff, A, B}$	1		0
X_{31}	$sense_{m, A, B}$	1		0
X_{32}	$sense_{a, A, B}$	1		0
X_{33}	$rep_{aff, B}$	1		0
X_{34}	$rep_{m, B}$	1		0
X_{35}	$rep_{a, B}$	1		0
X_{36}	$prep_{aff, B}$	0.5		0
X_{37}	$prep_{m, B}$	0.5		0
X_{38}	$prep_{a, B}$	0.5		0
X_{39}	$es_{aff, B}$	1		0
X_{40}	$es_{m, B}$	1		0
X_{41}	$es_{a, B}$	1		0
X_{42}	$rep_{e, G, B}$	0.5		0
X_{43}	$rep_{a, G, B}$	0.5		0
X_{44}	$rep_{i, G, B}$	0.7		0
X_{45}	$memory_{i, G, B}$	0		1
X_{46}	conscious emotion $_B$	0.1		0
X_{47}	$intersync_{aff,B}$	0.5		0
X_{48}	$intersync_{m, B}$	0.5		0
X_{49}	$intersync_{a, B}$	0.5		0
X_{50}	$intrasync_{aff,m,B}$	0.5		0
X_{51}	$intrasync_{aff,a,B}$	0.5		0
X_{52}	$intrasync_{a,m,B}$	0.5		0
X_{53}	$context_{ss, B}$	1		0

X_{54}	$\mathbf{W}_{\text{sense-rep aff, A}}$	0.1		0.05
X_{55}	$\mathbf{W}_{\text{sense-rep m, A}}$	0.1		0.05
X_{56}	$\mathbf{W}_{\text{sense-rep a, A}}$	0.1		0.05
X_{57}	$\mathbf{W}_{\text{prep-es aff, A}}$	0.1		0.05
X_{58}	$\mathbf{W}_{\text{prep-es m, A}}$	0.1		0.05
X_{59}	$\mathbf{W}_{\text{prep-es a, A}}$	0.1		0.05
X_{60}	$\mathbf{W}_{\text{rep-rep(e), aff, G,A}}$	0.1		0.1
X_{61}	$\mathbf{W}_{\text{rep-rep(e), m, G,A}}$	0.1		0.1
X_{62}	$\mathbf{W}_{\text{rep-rep(e), a, G,A}}$	0.1		0.1
X_{63}	$\mathbf{W}_{\text{rep-rep(a), aff, G,A}}$	0.1		0.1
X_{64}	$\mathbf{W}_{\text{rep-rep(a), m, G,A}}$	0.1		0.1
X_{65}	$\mathbf{W}_{\text{rep-rep(a), a, G,A}}$	0.1		0.1
X_{66}	$\mathbf{W}_{\text{memory- rep(i), G, A}}$	0.1		0.05
X_{67}	$\mathbf{W}_{\text{sense-rep aff ,B}}$	0.1		0.05
X_{68}	$\mathbf{W}_{\text{sense-rep m, B}}$	0.1		0.05
X_{69}	$\mathbf{W}_{\text{sense-rep a, B}}$	0.1		0.05
X_{70}	$\mathbf{W}_{\text{prep-es aff, B}}$	0.1		0.05
X_{71}	$\mathbf{W}_{\text{prep-es m, B}}$	0.1		0.05
X_{72}	$\mathbf{W}_{\text{prep-es a, B}}$	0.1		0.05
X_{73}	$\mathbf{W}_{\text{rep-rep(e), aff, G,B}}$	0.1		0.1
X_{74}	$\mathbf{W}_{\text{rep-rep(e), m, G,B}}$	0.1		0.1
X_{75}	$\mathbf{W}_{\text{rep-rep(e), a, G,B}}$	0.1		0.1
X_{76}	$\mathbf{W}_{\text{rep-rep(a), aff, G,B}}$	0.1		0.1
X_{77}	$\mathbf{W}_{\text{rep-rep(a), m, G,B}}$	0.1		0.1
X_{78}	$\mathbf{W}_{\text{rep-rep(a), a, G,B}}$	0.1		0.1
X_{79}	$\mathbf{W}_{\text{memory- rep i, G, B}}$	0.1		0.4

mcfw combination function weights		**1 id**	**2 alogistic**	**3 hebbian**	**4 steponce**
X_{1}	$ws_{\text{aff, B, A}}$	1			
X_{2}	$ws_{\text{m, B, A}}$	1			
X_{3}	$ws_{\text{a, B, A}}$	1			
X_{4}	$sense_{\text{aff, B, A}}$	1			
X_{5}	$sense_{\text{m, B, A}}$	1			
X_{6}	$sense_{\text{a, B, A}}$	1			
X_{7}	$rep_{\text{aff, A}}$	1			
X_{8}	$rep_{\text{m, A}}$	1			
X_{9}	$rep_{\text{a, A}}$	1			
X_{10}	$prep_{\text{aff, A}}$		1		
X_{11}	$prep_{\text{m, A}}$		1		
X_{12}	$prep_{\text{a, A}}$		1		
X_{13}	$es_{\text{aff, A}}$	1			
X_{14}	$es_{\text{m, A}}$	1			
X_{15}	$es_{\text{a, A}}$	1			
X_{16}	$rep_{\text{e, G, A}}$		1		
X_{17}	$rep_{\text{a, G, A}}$		1		
X_{18}	$rep_{\text{i, G, A}}$		1		
X_{19}	$memory_{\text{i, G, A}}$	1			
X_{20}	conscious emotion $_{\text{A}}$		1		
X_{21}	$intersync_{\text{aff,A}}$		1		
X_{22}	$intersync_{\text{m, A}}$		1		
X_{23}	$intersync_{\text{a, A}}$		1		
X_{24}	$intrasync_{\text{aff,m,A}}$		1		
X_{25}	$intrasync_{\text{aff,a,A}}$		1		
X_{26}	$intrasync_{\text{a,m,A}}$		1		
X_{27}	$ws_{\text{aff, A, B}}$	1			
X_{28}	$ws_{\text{m, A, B}}$	1			
X_{29}	$ws_{\text{a, A, B}}$	1			
X_{30}	$sense_{\text{aff, A, B}}$	1			
X_{31}	$sense_{\text{m, A, B}}$	1			
X_{32}	$sense_{\text{a, A, B}}$	1			
X_{33}	$rep_{\text{aff, B}}$	1			
X_{34}	$rep_{\text{m, B}}$	1			
X_{35}	$rep_{\text{a, B}}$	1			
X_{36}	$prep_{\text{aff, B}}$		1		

X_{37}	$\mathrm{prep}_{m, B}$		1		
X_{38}	$\mathrm{prep}_{a, B}$		1		
X_{39}	$\mathrm{es}_{aff, B}$	1			
X_{40}	$\mathrm{es}_{m, B}$	1			
X_{41}	$\mathrm{es}_{a, B}$	1			
X_{42}	$\mathrm{rep}_{e, G, B}$		1		
X_{43}	$\mathrm{rep}_{a, G, B}$		1		
X_{44}	$\mathrm{rep}_{i, G, B}$		1		
X_{45}	$\mathrm{memory}_{i, G, B}$	1			
X_{46}	conscious emotion $_{B}$		1		
X_{47}	$\mathrm{intersync}_{aff,B}$		1		
X_{48}	$\mathrm{intersync}_{m, B}$		1		
X_{49}	$\mathrm{intersync}_{a, B}$		1		
X_{50}	$\mathrm{intrasync}_{aff,m,B}$		1		
X_{51}	$\mathrm{intrasync}_{aff,a,B}$		1		
X_{52}	$\mathrm{intrasync}_{a,m,B}$		1		
X_{53}	$\mathrm{context}_{ss, B}$				1
X_{54}	$\mathbf{W}_{sense\text{-}rep\ aff, A}$		1		
X_{55}	$\mathbf{W}_{sense\text{-}rep\ m, A}$		1		
X_{56}	$\mathbf{W}_{sense\text{-}rep\ a, A}$		1		
X_{57}	$\mathbf{W}_{prep\text{-}es\ aff, A}$		1		
X_{58}	$\mathbf{W}_{prep\text{-}es\ m, A}$		1		
X_{59}	$\mathbf{W}_{prep\text{-}es\ a, A}$		1		
X_{60}	$\mathbf{W}_{rep\text{-}rep(e), aff, G,A}$			1	
X_{61}	$\mathbf{W}_{rep\text{-}rep(e), m, G,A}$			1	
X_{62}	$\mathbf{W}_{rep\text{-}rep(e), a, G,A}$			1	
X_{63}	$\mathbf{W}_{rep\text{-}rep(a), aff, G,A}$			1	
X_{64}	$\mathbf{W}_{rep\text{-}rep(a), m, G,A}$			1	
X_{65}	$\mathbf{W}_{rep\text{-}rep(a), a, G,A}$			1	
X_{66}	$\mathbf{W}_{memory\text{-}\ rep(i), G, A}$		1		
X_{67}	$\mathbf{W}_{sense\text{-}rep\ aff ,B}$		1		
X_{68}	$\mathbf{W}_{sense\text{-}rep\ m, B}$		1		
X_{69}	$\mathbf{W}_{sense\text{-}rep\ a, B}$		1		
X_{70}	$\mathbf{W}_{prep\text{-}es\ aff, B}$		1		
X_{71}	$\mathbf{W}_{prep\text{-}es\ m, B}$		1		
X_{72}	$\mathbf{W}_{prep\text{-}es\ a, B}$		1		
X_{73}	$\mathbf{W}_{rep\text{-}rep(e), aff, G,B}$			1	
X_{74}	$\mathbf{W}_{rep\text{-}rep(e), m, G,B}$			1	
X_{75}	$\mathbf{W}_{rep\text{-}rep(e), a, G,B}$			1	
X_{76}	$\mathbf{W}_{rep\text{-}rep(a), aff, G,B}$			1	
X_{77}	$\mathbf{W}_{rep\text{-}rep(a), m, G,B}$			1	
X_{78}	$\mathbf{W}_{rep\text{-}rep(a), a, G,B}$			1	
X_{79}	$\mathbf{W}_{memory\text{-}\ rep\ i, G, B}$		1		

mcfp combination function parameters		1 id	2 alogistic		3 hebbian	4 steponce	
X_1	$ws_{aff, B, A}$						
X_2	$ws_{m, B, A}$						
X_3	$ws_{a, B, A}$						
X_4	$sense_{aff, B, A}$						
X_5	$sense_{m, B, A}$						
X_6	$sense_{a, B, A}$						
X_7	$rep_{aff, A}$						
X_8	$rep_{m, A}$						
X_9	$rep_{a, A}$						
X_{10}	$prep_{aff, A}$		5	0.6			
X_{11}	$prep_{m, A}$		5	0.6			
X_{12}	$prep_{a, A}$		5	0.6			
X_{13}	$es_{aff, A}$						
X_{14}	$es_{m, A}$						
X_{15}	$es_{a, A}$						
X_{16}	$rep_{e, G, A}$		5	0.5			
X_{17}	$rep_{a, G, A}$		5	0.5			
X_{18}	$rep_{i, G, A}$						
X_{19}	$memory_{i, G, A}$						
X_{20}	conscious emotion $_A$		5	0.5			
X_{21}	$intersync_{aff,A}$		5	0.5			
X_{22}	$intersync_{m, A}$		5	0.5			
X_{23}	$intersync_{a, A}$		5	0.5			
X_{24}	$intrasync_{aff,m,A}$		5	0.5			
X_{25}	$intrasync_{aff,a,A}$		5	0.5			
X_{26}	$intrasync_{a,m,A}$		5	0.5			
X_{27}	$ws_{aff, A, B}$						
X_{28}	$ws_{m, A, B}$						
X_{29}	$ws_{a, A, B}$						
X_{30}	$sense_{aff, A, B}$						
X_{31}	$sense_{m, A, B}$						
X_{32}	$sense_{a, A, B}$						
X_{33}	$rep_{aff, B}$						
X_{34}	$rep_{m, B}$						
X_{35}	$rep_{a, B}$						
X_{36}	$prep_{aff, B}$		3	2			
X_{37}	$prep_{m, B}$		3	2			
X_{38}	$prep_{a, B}$		3	2			
X_{39}	$es_{aff, B}$						
X_{40}	$es_{m, B}$						
X_{41}	$es_{a, B}$						
X_{42}	$rep_{e, G, B}$		5	0.5			
X_{43}	$rep_{a, G, B}$		5	0.5			
X_{44}	$rep_{i, G, B}$		5	0.7			
X_{45}	$memory_{i, G, B}$						
X_{46}	conscious emotion $_B$		5	0.5			
X_{47}	$intersync_{aff,B}$		5	0.5			
X_{48}	$intersync_{m, B}$		5	0.5			
X_{49}	$intersync_{a, B}$		5	0.5			
X_{50}	$intrasync_{aff,m,B}$		5	0.5			
X_{51}	$intrasync_{aff,a,B}$		5	0.5			
X_{52}	$intrasync_{a,m,B}$		5	0.5			
X_{53}	$context_{ss, B}$					0	160

X_{54}	$\mathbf{W}_{\text{sense-rep aff, A}}$	5	0.5	
X_{55}	$\mathbf{W}_{\text{sense-rep m, A}}$	5	0.5	
X_{56}	$\mathbf{W}_{\text{sense-rep a, A}}$	5	0.5	
X_{57}	$\mathbf{W}_{\text{prep-es aff, A}}$	5	0.5	
X_{58}	$\mathbf{W}_{\text{prep-es m, A}}$	5	0.5	
X_{59}	$\mathbf{W}_{\text{prep-es a, A}}$	5	0.5	
X_{60}	$\mathbf{W}_{\text{rep-rep(e), aff, G,A}}$			1
X_{61}	$\mathbf{W}_{\text{rep-rep(e), m, G,A}}$			1
X_{62}	$\mathbf{W}_{\text{rep-rep(e), a, G,A}}$			1
X_{63}	$\mathbf{W}_{\text{rep-rep(a), aff, G,A}}$			1
X_{64}	$\mathbf{W}_{\text{rep-rep(a), m, G,A}}$			1
X_{65}	$\mathbf{W}_{\text{rep-rep(a), a, G,A}}$			1
X_{66}	$\mathbf{W}_{\text{memory- rep(i), G, A}}$	5	0.2	
X_{67}	$\mathbf{W}_{\text{sense-rep aff ,B}}$	5	0.5	
X_{68}	$\mathbf{W}_{\text{sense-rep m, B}}$	5	0.5	
X_{69}	$\mathbf{W}_{\text{sense-rep a, B}}$	5	0.5	
X_{70}	$\mathbf{W}_{\text{prep-es aff, B}}$	5	0.5	
X_{71}	$\mathbf{W}_{\text{prep-es m, B}}$	5	0.5	
X_{72}	$\mathbf{W}_{\text{prep-es a, B}}$	5	0.5	
X_{73}	$\mathbf{W}_{\text{rep-rep(e), aff, G,B}}$			1
X_{74}	$\mathbf{W}_{\text{rep-rep(e), m, G,B}}$			1
X_{75}	$\mathbf{W}_{\text{rep-rep(e), a, G,B}}$			1
X_{76}	$\mathbf{W}_{\text{rep-rep(a), aff, G,B}}$			1
X_{77}	$\mathbf{W}_{\text{rep-rep(a), m, G,B}}$			1
X_{78}	$\mathbf{W}_{\text{rep-rep(a), a, G,B}}$			1
X_{79}	$\mathbf{W}_{\text{memory- rep i, G, B}}$	5	0.5	

Group Dynamics Scenario Description

The scenario concerns a group worship session with triggers of music making, pictures and breaking of bread for reliving memory and religious experiencing through intersensory facilitation.

1. Linda and Hans are a married couple. They live in a nursery home. The are 93 and 95 years old respectively. They are suffering from cognitive impairment and are vulnerable elderly with dementia. They used to go to worship every Sunday in their former live. However, they are unable to read very well anymore and have difficulty with using the liturgy booklet of the Sunday service. Also they have difficulty with experiencing what a specific activity is about.
2. They have indicated that they wanted to join the worship meeting today's service of breaking of bread with other elderly on the Sunday morning in the nursery home community room.
3. During the breaking of bread a song is sung together and the music is made together in the group with the elderly. The elderly all have received and hold a small musical instrument in their hand.
4. Hans has a cymbals and Linda has a drum.
5. The worship leader stimulates them to use the instruments:

‘You can all hold your instrument now and make a joyfull noise’, Ok?

6. The worship leader demonstrates and takes first the cymbal and demonstrates how to make a sound with it.

7. He also demonstrates this for the hand drum, the tambourine, the sambal balls and some bells for the other eldery.
8. Then he introduce rythmns on the tambourine. Slowly, a 'ONE, two, three four'; 'ONE two, three, four', 'ONE, two, three, four rythmm' is heard. The worship band of piano, drums and 2 guitars slowly join in.
9. He keeps on regularly to stimulate the eldery to 'keep on using the instruments'.
10. Then the piano introduces the melody line of the worship song 'How great though art' this a song that all elderly know from their youth.
11. While the music and rythmn continues the worship leaders says.

Let us sing the following while we share in the breaking of bread together. All elderly used to break bread regularly in their earlier lives.

12. The first lines of the song text is portrayed at a large screen in front. All elders can read the text.

The music and rythmn goes on while he reads aloud.

'O Lord my God, When I, in awesome wonder,

Consider all the worlds Thy Hands have made;

I see the stars, I hear the rolling thunder,

Thy power throughout the universe displayed

How your wisdom guides the threads of life,

And How All Beings Are Fed at Your Table'

Then Sings My Soul My Savior God to Thee How Great Thou Art, How Great Thou Art.

13. Simultanously while reading the words, at the background of the screen text pictures are deplayed at the large screen, behind the words. The pictures demonstrate visually the content of words of the e.g. pictures of nature, stars, following the content of the words.
14. Then the worship leader says: Let us sing the whole song together.
15. He also says: 'While we sing meanwhile bread and wine will be handed out to each of you individually. You can join in eating and drinking it together
16. Then they sing together the song.

O Lord my God, When I, in awesome wonder,
Consider all the worlds Thy Hands have made;
I see the stars, I hear the rolling thunder,
Thy power throughout the universe displayed.
Chorus:

Then Sings My Soul, My Saviour God, to Thee, How Great Thou Art, How Great Thou Art.

Then Sings My Soul, My Saviour God, to T

When through the woods and forest glades I wander, And hear the birds sing sweetly in the trees. When I look down, from lofty mountain grandeur And see the brook, and feel the gentle breeze.

Then sings my soul, My Saviour God, to Thee, How great Thou art, How great Thou art.

Then sings my soul, My Saviour God, to Thee, How great Thou art, How great Thou art!

And when I think, that God, His Son not sparing;

Sent Him to die, I scarce can take it in;

That on a Cross, my burdens gladly bearing, He bled and died to take away my sin.

Then sings my soul, My Saviour God, to Thee, How great Thou art, How great Thou art. Then sings my soul, My Saviour God, to Thee, How great Thou art, How great Thou art!

When Christ shall come, with shout of acclamation, And take me home, what joy shall fill my heart.

Then I shall bow, in humble adoration,

And then proclaim: "My God, how great Thou art!".

Then sings my soul, My Saviour God, to Thee, How great Thou art, How great Thou art.

Then sings my soul, My Saviour God, to Thee, How great Thou art, How great Thou art!

The song/music continues / is repeated till all eldery are finished with eating a piece of bread and drink a bit of wine.

17. Hans and Linda report to both have had a religious experience during this episode of singing together with breaking of bread.
18. Hans sayed about the event: 'Everything came back again suddenly and I felt so alive. My faith came back to me. I experience the presences of God. Something so great, I was so small and He is so great and I felt so much love from Him. I cried silently for increasing joy.'
19. Linda says, 'I felt such a longing to go home during the song. So many memories came back. So strange it was as if it was now again and new also again. I had such a longing to go home and I felt that I was not alone again. I felt all the people in the room are my family. I felt so much peace and patience more and more. When can we have this again?'

References

1. Accetto, M., Treur, J., Villa, V.: An adaptive cognitive-social model for mirroring and social bonding during synchronous joint action. Procedia Comput. Sci.**145**, 3–12 (2018)
2. Atran, S.: In Gods we trust: the evolutionary landscape of religion. Oxford University Press, New York (2002)
3. Augustinus, F. Hendrikse, S.C.F., Treur, J.: Psychopathic genes: a multi-level adaptive dynamic modelling approach. In: Proceeding of the 8th Computational Methods in Systems and Software Conference, CoMeSySo'24. Lecture Notes in Networks and Systems, vol. 1471, pp. 477–498. Springer Nature (2025)

4. Banerjee, S., Chitty, C., Lee, Q., Treur, J.: An adaptive network model of the role of the microbiome-gut-brain axis in insomnia. Cogn. Syst. Res. **78**, 106–121 (2023)
5. Bashkirova, A., Compagner, A., Henningsen, D.M., Treur, J.: An adaptive modelling approach to employee burnout in the context of the big five personality traits. Cogn. Syst. Res. **79**, 109–125 (2023)
6. Beauregard, M., Paquette, V.: Neural correlates of a mystical experience in Carmelite nuns. Neurosci. Lett. **405**(3), 186–190 (2006)
7. Bechtel, W., Bich, L.: Organisms need mechanisms; mechanisms need organisms. New Mechanism. Explan. Emerg. Reduct. **35**, 85–108 (2024)
8. Benbassat, J.: Role modeling in medical education: the importance of a reflective imitation. Acad. Med. **89**(4), 550–554 (2014)
9. Bich, L., Bechtel, W.: Control mechanisms: explaining the integration and versatility of biological organisms. Adapt. Behav. **30**(5), 389–407 (2022)
10. Blei, Z.A., Hendrikse, S.C.F., Treur, J.: Modeling Epigenetic modification due to childhood abuse and its relation to ASPD: a fifth-order adaptive network model. In: Proceeding of 15th Biologically Inspired Cognitive Architectures for AI Conference, BICA*AI'24. Studies in Computational Intelligence, vol. 477, pp. 59–70. Springer Nature (2024)
11. Bloch, C., Vogeley, K., Georgescu, A.L., Falter-Wagner, C.M.: INTRApersonal synchrony as constituent of INTERpersonal synchrony and its relevance for autism spectrum disorder. Frontiers in Robotics and AI **6**, 73 (2019)
12. Bouma, D., Canbaloğlu, G., Treur, J., Wiewiora, A.: Adaptive network modeling of the influence of leadership and communication on learning within an organization. Cogn. Syst. Res. **79**, 55–70 (2023)
13. Boyer, P.: Religious thought and behaviour as by-products of brain function. Trends Cogn. Sci. **7**(3), 119–124 (2003)
14. Budding, D., Doornkamp, S., Treur, J.: A second-order adaptive mental network model relating dreaming to creativity. Cogn. Syst. Res. **80**, 37–49 (2023)
15. Bulbulia, J.: The cognitive and evolutionary psychology of religion. Biol. Philos. **19**(5), 655–686 (2004)
16. Canbaloğlu, G., Treur, J., & Wiewiora, A. (eds.): Computational Modeling of Multilevel Organisational Learning and Its Control Using Self-Modeling Network Models. Springer Nature (2023)
17. Cooke, P., Elcoro, M.: Neurotheology: neuroscience of the soul. J. Young Investigators **25**(3), 1–6 (2013)
18. Feldman, R.: Parent–infant synchrony biological foundations and developmental outcomes. Curr. Dir. Psychol. Sci. **16**, 340–345 (2007). https://doi.org/10.1111/j.14678721.2007.00532.x
19. Grafman, J., Cristofori, I., Zhong, W., Bulbulia, J.: The neural basis of religious cognition. Curr. Dir. Psychol. Sci. **29**, 126–133 (2020)
20. Grandjean, D., Sander, D., Scherer, K.R.: Conscious emotional experience emerges as a function of multilevel, appraisal-driven response synchronization. Conscious. Cogn. **17**(2), 484–495 (2008)
21. Hendrikse, S.C.F., Treur, J., Koole, S.L.: Modeling emerging interpersonal synchrony and its related adaptive short-term affiliation and long-term bonding: a second-order multi-adaptive neural agent model. Int. J. Neural Syst. **33**(7), 2350038 (2023). https://doi.org/10.1142/S0129065723500387
22. Hendrikse, S.C.F., Treur, J., Koole, S.L.: Relationship-specific and relationship-independent behavioural adaptivity in affiliation and bonding: a multi-adaptive dynamical systems approach. Cogn. Syst. Res. **84**, 101182 (2024)
23. Hendrikse, S.C.F., Treur, J., Koole, S.L. (eds.): New analysis and modeling directions in social interaction science: emergent multimodal interpersonal synchrony, affiliation, and bonding. Springer Nature (2025a)
24. Hendrikse, S.C.F., Treur, J., Koole, S.L.: Introduction to the role of adaptivity in social interaction science. In: Hendrikse, S.C.F., Treur, J., Koole, S.L. (eds.): New Analysis and Modeling Directions in Social Interaction Science: Emergent Multimodal Interpersonal Synchrony, Affiliation, and Bonding, Ch 1. Springer Nature (2025b)

25. Hendrikse, S.C.F., Treur, J., Koole, S.L.: The adaptive interplay of interpersonal synchrony, short-term affiliation and long-term bonding: a multi-timescale multi-adaptive agent model. In: Hendrikse, S.C.F., Treur, J., Koole, S.L. (eds.): New Analysis and Modeling Directions in Social Interaction Science: Emergent Multimodal Interpersonal Synchrony, Affiliation, and Bonding, Ch 12. Springer Nature (2025c)
26. Hendrikse, S.C.F., Treur, J., Koole, S.L.: A multi-timescale multi-adaptive dynamical system approach for relationship-specific and relationship-independent adaptivity in affiliation and bonding. In: Hendrikse, S.C.F., Treur, J., Koole, S.L. (eds.): New Analysis and Modeling Directions in Social Interaction Science: Emergent Multimodal Interpersonal Synchrony, Affiliation, and Bonding, Ch 13. Springer Nature (2025d)
27. Hendrikse, S.C.F., Treur, J., Wilderjans, T.F., Dikker, S., Koole, S.L.: On becoming in sync with yourself and others: an adaptive agent model for how persons connect by detecting intrapersonal and interpersonal synchrony. Hum-Cent Intell Syst **3**, 123–146 (2023)
28. Hendrikse, S.C.F., Treur, J., Wilderjans, T.F., Dikker, S., Koole, S.L.: Growing connectedness through synchrony: the role of subjective analysis of intrapersonal synchrony and interpersonal synchrony within an adaptive agent model. In: Hendrikse, S.C.F., Treur, J., Koole, S.L. (eds.), New Analysis and Modeling Directions in Social Interaction Science: Emergent Multimodal Interpersonal Synchrony, Affiliation, and Bonding, Ch 11. Springer Nature (2025)
29. Hoffmans, C., Hendrikse, S.C.F., Treur, J., Roelofsma, P.H.M.P.: Modeling the adaptive interplay of epigenetics and insecurity in social media in the development of anorexia nervosa. In: Proceeding of 15th biologically inspired cognitive architectures for AI conference, BICA*AI'24. Studies in Computational Intelligence, vol. 477, pp. 134–145. Springer Nature (2024)
30. Hove, M.J., Risen, J.L.: It's all in the timing: Interpersonal synchrony increases affiliation. Soc. Cogn. **27**(6), 949–960 (2009)
31. Kapogiannis, D., Barbey, A.K., Su, M., Zamboni, G., Krueger, F., Grafman, J.: Cognitive and neural foundations of religious belief. Proc. Natl. Acad. Sci. **106**(12), 4876–4881 (2009)
32. Kapogiannis, D., Deshpande, G., Krueger, F., Thornburg, M.P., Grafman, J.H.: Brain networks shaping religious belief. Brain Connectivity **4**(1), 70–79 (2014)
33. Koole, S.L., Tschacher, W.: Synchrony in psychotherapy: a review and an integrative framework for the therapeutic alliance. Front. Psychol. **7**, 862 (2016)
34. Memon, Z.A., Treur, J.: An agent model for cognitive and affective empathic understanding of other agents. In: Transactions on Computational Collective Intelligence VI, pp. 56–83. Springer, Berlin, Heidelberg (2012)
35. Prince, K., Brown, S.: Neural correlates of partnered interaction as revealed by cross-domain ALE meta-analysis. Psychol. Neurosci (2022). https://doi.org/10.1037/pne0000282
36. Roelofsma, P.M.H.P., Jabeen, F., Taal, H.R., Treur, J. eds.: Using shared mental models and organisational learning to support safety and security through cyberspace: a computational analysis approach. Springer Nature (2025)
37. Schaap-Jonker, H., Sizoo, B., van Schothorst-van Roekel, J., Corveleyn, J.: Autism spectrum disorders and the image of God as a core aspect of religiousness. Int. J. Psychol. Relig. **23**(2), 145–160 (2013)
38. Schaap-Jonker, H., Corveleyn, J.M.: Mentalizing and religion. Arch. Psychol. Relig. **36**(3), 303–322 (2014)
39. Schjoedt, U., Stodkilde-Jorgensen, H., Geerts, A.W., Roepstorff, A.: Highly religious participants recruit areas of social cognition in personal prayer. SocCog Affect Neurosci. **4**, 199–207 (2009)
40. Schlöder, D., Hendrikse, S.C.F., Treur, J.: Higher-order adaptive dynamical system modeling of the role of epigenetics in substance use disorder. In: Proceeding of 15th Biologically Inspired Cognitive Architectures for AI Conference, BICA*AI'24. Studies in Computational Intelligence, vol. 477, pp. 358–368. Springer Nature (2024)
41. Tarr, B., Launay, J., Dunbar, R.I.M.: Silent disco: dancing in synchrony leads to elevated pain thresholds and social closeness. Evol. Hum. Behav. **37**(5), 343–349 (2016)

42. Treur, J.: Modelling joint decision making processes involving emotion-related valuing and empathic understanding. In: Agents in Principle, Agents in Practice. In: Proceedings of PRIMA'11. Lecture Notes in AI, vol. 7047, pp. 410–423. Springer, Berlin, Heidelberg (2011a)
43. Treur, J.: From mirroring to the emergence of shared understanding and collective power. In: Computational Collective Intelligence. Technologies and Applications, Lecture Notes in Artifical Intelligence, vol. 6922, pp. 1–16. Springer, Berlin, Heidelberg (2011b)
44. Treur, J.: Network-oriented modeling: addressing complexity of cognitive. Springer International Publishers, Affective and Social Interactions (2016)
45. Treur, J.: Network-oriented modeling for adaptive networks: designing higher-order adaptive biological, mental and social network models. Springer Nature (2020a)
46. Treur, J.: Modeling multi-order adaptive processes by self-modeling networks. In: Antonio, J., Tallón-B., Chi-Hua Chen (eds.): In: Proceeding of the 2nd Internation-al Conference on Machine Learning and Intelligent Systems, MLIS'20. Frontiers in Artificial Intelligence and Applications, vol. 332, pp. 206–217. IOS Press (2020b)
47. Treur, J.: On the dynamics and adaptivity of mental processes: relating adaptive dynamical systems and self-modeling network models by mathematical analysis. Cogn. Syst. Res. **70**, 93–100 (2021)
48. Treur, J.: On structure, dynamics, and adaptivity for biological and mental processes: a higher-order adaptive dynamical system modeling perspective. In: Proceedings of the 46th Annual Meeting of the Cognitive Science Society–Dynamics of Cognition, CogSci'24 (2024a)
49. Treur, J.: Using multilevel temporal factorisation to analyse structure and dynamics for higher-order adaptive and evolutionary processes. In: Proceeding of the 16th International Conference on Computational Collective Intelligence, ICCCI'24. Lecture Notes in AI, vol 14811, pp 378–392. Springer Nature (2024b)
50. Treur, J.: Modeling coevolution of dynamics, adaptivity and control of adaptivity by self-modeling networks. In: Hendrikse, S.C.F., Treur, J., Koole, S.L. (eds.): New Analysis and Modeling Directions in Social Interaction Science: Emergent Multimodal Interpersonal Synchrony, Affiliation, and Bonding, Ch 2. Springer Nature (2025)
51. Treur, J., Van Ments, L. eds.: Mental models and their dynamics, adaptation, and control: a self-modeling network modeling approach. Cham: Springer Nature (2022)
52. Tschacher, W., Ramseyer, F., Koole, S.L.: Sharing the now in the social present: duration of nonverbal synchrony is linked with personality. J. Pers. **86**(2), 129–138 (2018)
53. Van Gog, T., Paas, F., Marcus, N., Ayres, P., Sweller, J.: The mirror neuron system and observational learning: implications for the effectiveness of dynamic visualizations. Educ. Psychol. Rev. **21**(1), 21–30 (2009)
54. Van Ments, L., Roelofsma, P.H.M.P., Treur, J.: Modeling the effect of religion on human empathy based on an adaptive temporal-causal network model. Comput. Soc. Netw. **5**(1) (2018)
55. Van Ments, L., Roelofsma, P.H.M.P., Treur, J.: How empathic is your God: and Adaptive network model for formation and use of a mental God-model and its effect on human empathy. In Treur, J., Van Ments, L. (eds.): Mental Models and Their Dynamics, Adaptation and Control, Ch. 11. Springer (2022)
56. Vervoordeldonk, D., Treur, J., Hendrikse, S.C.F., Roelofsma, P.H.M.P.: An adaptive network model on the interplay of synchrony and adaptation for reliving memory and religious experience through multimodal interaction. In: Proceedings of the 8th International Conference on Computational Methods in Systems and Software, CoMeSySo'24. Lecture Notes in Networks and Systems, vol. 1490, pp 59–93. Springer Nature (2025)
57. Wiltermuth, S.S., Heath, C.: Synchrony and cooperation. Psychol. Sci. **20**(1), 1–5 (2009)

Joint Action and Memory Recall for Elderly by Detecting Interpersonal Synchrony: A Higher-Order Adaptive Network Model

Yijie Xu, Sophie C. F. Hendrikse, Jan Treur, and Peter H. M. P. Roelofsma

Abstract This paper explores the potential of adaptive network modeling for joint action and memory recall among elderly through detecting interpersonal synchrony. With the aging population increasing, there is a crucial need to focus on the health and social interaction of older adults. Based on research of the significance of social interaction and memory use for the elderly, as well as the role of interpersonal synchrony in joint action, this paper aims to analyse computationally how to enhance positive effects of social interactions among older individuals by applying an adaptive network model. The research examines the concept of interpersonal synchrony and its impact on joint action, memory, and emotional well-being in elderly populations. Through simulation experiments and analysis, the study demonstrates the potential benefits for music in memory recall for older adults with cognitive decline, highlighting the importance of social interaction and emotional resonance. This study offers a valuable contribution to understanding and improving social interactions and memory recall among the elderly.

Keywords Adaptive network · Memory recall · Elderly · Interpersonal synchrony

Y. Xu · J. Treur (✉)
Social AI Group, Department of Computer Science, Vrije Universiteit Amsterdam, Amsterdam, Netherlands
e-mail: j.treur@vu.nl

S. C. F. Hendrikse
Department of Methodology and Statistics, Tilburg School of Social and Behavioral Sciences, Tilburg University, Tilburg, Netherlands
e-mail: S.C.F.Hendrikse@tilburguniversity.edu

P. H. M. P. Roelofsma
Risk Management and Cybersecurity Group, Centre of Expertise Cybersecurity, The Hague University of Applied Sciences, The Hague, Netherlands
e-mail: P.H.M.P.Roelofsma@hhs.nl

S. C. F. Hendrikse et al. (eds.), *New Analysis and Modeling Directions in Social Interaction Science*, Studies in Systems, Decision and Control 614,
https://doi.org/10.1007/978-3-031-99968-0_18

1 Introduction

Today, with the ongoing intensification of the aging population, the significance of focusing on the health and social engagement of the elderly has become increasingly important. The World Health Organization released a fact sheet in 2022 Ageing and health [2] underscoring the impact of global aging, highlighting the imperative to focus on personalized healthcare strategies for the elderly population to boost their health outcomes and overall well-being. The National Institute on Aging (NIA) in the United States had also proposed the social interaction of elderly individuals not only directly influences their psychological and emotional health but also plays a significant role in their physiological aspects Social isolation, loneliness in older people pose health risks [50]. NIA emphasizes that a lack of social activity is associated with an increased susceptibility of the elderly to chronic diseases and cognitive decline. Memory loss is a significant sign of cognitive decline, also it is the first symptom reported by patients with Alzheimer's disease [34]. Dr. Collie has performed an experiment about the memory decline in a healthy elderly group, and the study found that in a group of 101 neurologically healthy older adults, about 34.7 percent showed memory decline over a 2-year period [11].

Social interaction plays a pivotal role in the lives of the elderly, it has a positive impact to their psychological status, potentially reliving anxiety, depression to a stable emotional state; see The Importance of Social Interaction for Seniors [52]. As for the interpersonal synchrony, this refers to the temporal coordination and behavioral adaptation occurring between individuals during social interactions. This synchronization plays a crucial role not only in early development and language learning but also in fostering social connections [12]. Currently, there is relatively little research on interpersonal synchrony among older adults. The major research direction focuses on its impact on social interaction, emotional resonance [68], and physical and mental health. For instance, the GiraffPlus System helping independent living old people at home. It can monitor the home activities, physical health of elderly and promote the social interaction [14].

The research reported here explores the *research question* whether and how through adaptive network modeling detected interpersonal synchrony and induced adaptivity can support and improve joint actions among the elderly with improved memory as a consequence.

2 Background Information

2.1 Interpersonal Synchrony

Interpersonal synchrony has been studied in various fields, including social psychology, linguistics, and marketing [39, 42, 43]. The definition of interpersonal synchrony primarily involves three aspects: timing matching of behaviors, content comparability of movements, and dynamic interaction between individuals [32]. In essence, interpersonal synchrony refers to instances when two or more individuals perform actions almost simultaneously, with overlapping or consistent movements within the overall timeframe, executed in identical, similar, or connected forms. This coordination is measured through factors such as temporal, emotional, and behavioral synchrony [44]. Cacioppo et al. [9] have discussed how the mirror neuron system promotes interpersonal synchrony The results of the fMRI image demonstrated that the brain areas related to social cognition, embodied cognition, self–other expansion, and action observation had a correlation of interpersonal synchrony. These findings have positive impact for understanding interpersonal synchrony and social coordination [16]. For more background of synchrony, see this volume [26, 27], Chapter 1.

Interpersonal synchrony plays a crucial role for different age groups. For young children, it would help foster early prosocial behavior by empathy, cooperation, and altruism in infants. In adolescents and adults, it's seen as vital for fostering social connections, coordinating behaviors, and enhancing emotional resonance [64, 69]. As for the elderly, [70] use functional Near-Infrared Spectroscopy (fNIRS) technology to investigate brain synchrony among elderly couples during a competitive task. The research showed elderly couples exhibited a significant increase in inter-brain synchrony during competitive interactions comparing to unfamiliar cross-sex pairs. This result indicates that long-term relationships have a positive impact on interpersonal interactions at both behavioral and neural levels in the elderly and also emphasizes the importance of intimate relationships among older adults [70]. Additionally, research concerning a mirror game in old age showed that it improved attention and the ability to detect spoken words in noise. Participants mimic each other's movements in a mirrored way, creating a synchronous interaction. The mirror game may be an effective intervention to enhance mood and attention functions in the elderly [36].

2.2 *Joint Action*

Joint actions are prevalent in various scenarios, spanning from basic interactions like a handshake to complex coordination within a group, such as a choir. Joint actions involve the coordination of both verbal and non-verbal interactions among individuals [65]. A joint action for cognitive and neural mechanisms is not only a physical collaboration but also a representation of social interaction. Sebanz et al. [46] explore how joint attention, action observation, task sharing, action coordination and agency contribute allow individuals to cooperate with each other without verbal communication, only by the goals of joint action to anticipate each other's actions. Marsh et al. [40] investigate how joint action and interpersonal coordination build social connections through a dynamic and ecological perspective. It also proves that joint action is not only about simple physical movements match, but includes the interaction among cognitive, emotional, and social aspects. Through such interactions, individuals can naturally connect with others without directed instructions [40]. The experiment performed by [66] has examined the higher the complexity joint action, the lower of the interpersonal synchrony. Their results suggest that task context significantly affects the role of the synchrony in joint action. However, existing research rarely mention studies on the elderly. For instance, Brownell et al. and Warneken et al. discuss children's collaborative performance in problem-solving tasks and social games. Also, young-aged children simply cooperate together with adults [7, 8, 41, 67]. People of different age stages have different behavior patterns [37], cognitive functions [19], and social interactions [20] when engaging in joint actions. Especially for the elderly, the levels of social interaction were a stronger predictor of cognitive decline than age [20], this could influence the way and efficiency when they participate in joint actions.

2.3 Memory

Memory refers to the psychological process by which individuals acquire, store, and recall information [51]. It can be broadly categorized into several types, including working memory, short-term memory, and long-term memory. Memory is also classified into explicit or implicit forms. The explicit memory associated with the hippocampus and medial temporal lobe system in the brain includes the conscious recall of past experiences. And the implicit memory operates without conscious awareness and involves basic associative learning and memory processes, it controlled by cerebellum and amygdala [38, 53]. With the increase of age, most elderly would suffer memory decline [21], including the reduction of working memory [15], long-term memory and the memory speed and accuracy [18]. In 2022, a study was reported on the impact of social engagement patterns on memory among Chinese elderly. The research found that elderly individuals who often participate recreational activities showed better memory compared to those with lower levels of engagement [23]. Another study found that maintaining positive social relationships may help delay the decline in memory and cognitive function. This research highlighted that superaged individuals tend to have more positive relationships compared to others [13]. Research indicates that active social engagement is beneficial for the memory and cognitive function of elderly individuals. To keep a healthy condition, elderly people need social activities.

2.4 Related Modeling Work

The area of computational modeling of synchronisation during social interaction, internal detection of synchrony within a mental process, and related adaptivity of interaction behaviour has not received much attention in the literature yet. As an exception, this topic has been addressed using adaptive network modeling by

Hendrikse et al. [25, 30], see also this volume [26], Chapters 11–13. Here, [30] introduces the notion of multimodal synchrony detectors based on noticing the own actions and the actions of another person and comparing them as part of mental processes and how that has an adaptive effect on interaction behaviour. Furthermore, (Hendrikse et al. 2022) provides an analysis of time differences in noticing the own actions and noticing another person's actions. Moreover, [25] in the same context short-term adaptivity of the interaction behaviour (affiliation) is distinguished from long-term adaptivity (bonding), so that together with the direct responding processes three different time scales are considered. In the topic addressed in the current paper some elements from this literature are used as part of the designed model, but new elements needed for the topic here have been added, such as other forms of adaptivity and addressing memory.

2.5 Scenarios: Joint Action Through Interpersonal Synchrony in Elderly Care

The following real-world scenario is addressed. In a nursing home in the Netherlands, two over 80 years old gentlemen, Hans and Tom, share a silent existence side by side in wheelchairs, their cognitive faculties dimmed by dementia. During a group singing activity in the community room, they remain quiet, each holding a rhythm instrument that had accompanied the last song.

As the music leader strums his guitar softly, Tom with eyes closed, begins to utter unrecognized sound. The music leader, attuned to Tom's attempts, plays chords that harmonize with the sounds. This prompts Tom to add a melody and wave his hands, and the leader follows and encourages Tom. Moved by the music, Hans recognizes the melody of'Panis Angelicus'. Tears come as memories return. He starts to sing in his tenor voice, even trying to stand up and move. While Tom struggles with the words, Hans remembers them, they look at each other together, they complete the song, and their voices echo in the quiet room.

This scenario illustrates a moment of interpersonal synchrony, a spontaneous alignment of rhythm and melody between two individuals and improved memory recall.

3 The Self-Modeling Network Modeling Approach Used

The network-oriented modeling approach is briefly introduced in this section. A temporal- causal network model is characterized as follows: here *X* and *Y* represent nodes of the network, also referred to states [56–58], see also this volume [26], Chap. 2 [59].

- **Connectivity characteristics**
  Connections from a state X to a state Y and their weights $\boldsymbol{\omega}_{X,Y}$
- **Aggregation characteristics**
  For any state Y, some combination function $\mathbf{c}_Y(..)$ defines the aggregation that is applied to the impacts $\boldsymbol{\omega}_{X,Y}\,X(t)$ on Y from its incoming connections from states X.
- **Timing characteristics**
  Each state Y has a speed factor $\boldsymbol{\eta}_Y$ defining how fast it changes for given causal impact.

The standard difference equations used for simulation and mathematical analysis incorporate the above three types of network characteristics:

$$Y(t+\Delta t) = Y(t) + \boldsymbol{\eta}_Y\left[\mathbf{c}_Y\left(\boldsymbol{\omega}_{X_1,Y}X_1(t), \ldots, \boldsymbol{\omega}_{X_k,Y}X_k(t)\right) - Y(t)\right]\Delta t \qquad (1)$$

Here, Y is any state Y and X_1 to X_k are the states from which Y gets its incoming connections. Equation (1) is automatically generated by the network characteristics within the software environment [57, 58]; around 70 combination function are included in the library.

The two combination functions that are applied in this model can be seen in Table 1. The combination functions may contain some parameters, the advanced logistic sum function **alogistic** has excitability threshold $\boldsymbol{\tau}$ and the steepness $\boldsymbol{\sigma}$ as parameters, but the complemental difference function **compdiff** has no parameters.

An adaptive network can be defined as a network where connections between nodes and/or other network structure characteristics change adaptively in response to the states of network or other influences [22, 27, 58]. Almost every realistic network is adaptive, for instance, the simulations of neural networks in biology, the modeling of emotions and cognitive processes in psychology, and analysis of social behavior networks in social sciences [22]. The adaptive network modeling approach based on the self-modeling (or reification) principle [56, 58] was used here, see Sect. 3 and this volume [26, 59], Chapter 2. It employs a modeling method that enables specification of the network's dynamics and adaptivity by declarative elements. The network characteristics of such a temporal-causal network for connectivity (connection weights), aggregation (combination functions), and timing (speed factors) are represented by declarative mathematical objects. The applied difference Eq. (1) expressed in terms of those factors are declarative temporal specifications which determine the behaviors

Table 1 Combination functions used in the adaptive network model

Function	Formula	Parameters
$\text{alogistic}_{\boldsymbol{\sigma},\boldsymbol{\tau}}(V_1,\ldots,V_k)$	$\left[\frac{1}{1+e^{-\sigma(V_1+\ldots+V_k-\tau)}} - \frac{1}{1+e^{\sigma\tau}}\right)](1+e^{-\sigma\tau})$	Steepness $\boldsymbol{\sigma}$ Excitability threshold $\boldsymbol{\tau}$
compdiff $(V_1,, V_2)$	1-$\lvert V_1$-$V_2\rvert$/max (V_1, V_2) if $V_1 > 0$ or $V_2 > 0$ else 0	None

of the model, when given them initial values. With this, complex dynamic patterns can be modeled easily [57, 58].

By using the self-modeling principle to obtain an adaptive network, additional states (called self-model states or reification states) are added to the network that represent the network's own characteristics, for example, connectivity self-model states $\mathbf{W}_{X,Y}$ (also called **W**-states) representing adaptive connection weights $\boldsymbol{\omega}_{X,Y}$, aggregation self-model states $\mathbf{T}_Y$ (also called **T**-states) representing adaptive excitability thresholds $\boldsymbol{\tau}_Y$ of the combination function **alogistic** or timing self-model states $\mathbf{H}_Y$ (also called **H**-states) representing adaptive speed factors $\boldsymbol{\eta}_Y$. When such network reification is applied to a temporal-causal network, it results in a temporal-causal network again. Therefore, this self-modeling network construction can be easily applied iteratively to obtain (first-order, second-order, …) self-model states of multiple orders, modeled at multiple levels. For example, adaptive speed factors $\boldsymbol{\eta}_{\mathbf{W}_{X,Y}}$ and $\boldsymbol{\eta}_{\mathbf{T}_Y}$ of (first-order) self-model states $\mathbf{W}_{X,Y}$ and $\mathbf{T}_Y$ can be represented by (second-order) self-model states $\mathbf{H}_{\mathbf{W}_{X,Y}}$ and $\mathbf{H}_{\mathbf{T}_Y}$ (also called $\mathbf{H}_\mathbf{W}$-states and $\mathbf{H}_\mathbf{T}$-states). In this study a second-order adaptive network is applied to address the self-adaptation that may happen during detection of synchrony. This second-order network model is based on adaptation of connection weights that model synapses, wherein synaptic plasticity refers to the changes in the structure and function of synapses, typically influenced by factors such as learning or memory [3, 24]. The first-order adaptation principle, based on Bainian or Hebbian learning [3, 24], describes the fundamental process of synaptic adaptation, and the second-order adaptation regulates the process of first-order adaptation (amplify or suppress), and can create metaplasticity [1] based on the principle that more stimulus exposure leads to faster adaptation [45].

Also adaptivity of order higher than two has been modeled in such a network-oriented manner, up to order five. This enables to model processes with a tower of control levels as considered in biology and philosophy; see, for example, [4, 5, 62, 63]. For example, in [6, 17, 35, 47] this has been used to model how a bad childhood can lead to lifelong mental disorders via five adaptation levels embodied by epigenetic factors, gene expression, mRNA production, protein production, brain physiology, respectively.

A dedicated MATLAB-based software environment is available which uses a table format called role matrices to specify the above network characteristics $\boldsymbol{\omega}$, **c**, $\boldsymbol{\eta}$ in a systematic way, and supports the execution of network models within MATLAB. The environment is enriched with a combination function library and a computational network engine, enabling the adaptation of network structures through well-defined role matrices [55].

The self-modeling network concept allows for the design of complex multi-level network models with sophisticated adaptive dynamics. By combining the base network and all adaptation levels in one overall unified network structure, simulation can be done in a unified way as well. This is supported by a software environment that includes a unified network engine that simulates the canonical difference Eq. (1) for all states at all self-modeling levels; for more details, see [57, 58]. For example, if all connection weights $\boldsymbol{\omega}$ for the base level are represented by first-order self-model **W**-states, the for each base level state Y difference Eq. (1) is simulated while substituting for each connection weight $\boldsymbol{\omega}_{X_i,Y}$ the activation value $\mathbf{W}_{X_i,Y}(t)$ of the corresponding **W**-state $\mathbf{W}_{X_i,Y}$:

$$Y(t + \Delta t) = Y(t) + \boldsymbol{\eta}_Y\big[\mathbf{c}_Y\big(\mathbf{W}_{X_1,Y}(t)X_1(t), \ldots, \mathbf{W}_{X_k,Y}(t)X_k(t)\big) - Y(t)\big]\Delta t \quad (2)$$

This works similarly for all adaptation levels and all network characteristics.

It has been found by mathematical analysis that any smooth dynamical system has a canonical representation as a network and any smooth (multi-order) adaptive dynamical system has a canonical representation as a (multi-order) self-modeling network, as is shown in detail in [25, 61]. Therefore, compared to adaptive dynamical systems in general, the network-oriented modeling approach used does not introduce any fundamental limitations concerning what it can model. This has also been confirmed by recent applications in other disciplines, for example, for internal mental models in cognitive science [60], for organisational learning within management science [10]

4 The Introduced Adaptive Network Model

The proposed network model is structured as a multilevel adaptive social network model. It covers 3 levels, the base level, first-order adaptation level (for learning and bonding) and second-order adaptation level (for control over the learning and bonding). The base level represents the fundamental connections and states in the network. The first-order adaptation level models the principles for adaptation of the base connections. The second-order adaptation level controls and adjusts the adaptation principles from the first-order adaptation level, allowing the network to self-adjust at a higher level, such as modifying the speed of adaptation of the base connection weights.

4.1 Base Level

The base level of the network model is shown in Fig. 1. The base level contains two blocks, the left orange box indicates the agent A which represents Hans, the right green box represents Tom as agent B. Table 2 describes the details of each of the base states.

The base level of the network model describes the connections between internal mental states of two agents. The internal states can be categorized as sense, representation, preparation, and execution states (emotion, singing, and movement) from left to right. In addition, the model also includes three interpersonal synchrony detector states for the different modalities (the small square shapes), each of the detectors has two incoming connections (from a sense and execution state). Both agents have the same structure but with different incoming connections from world states which

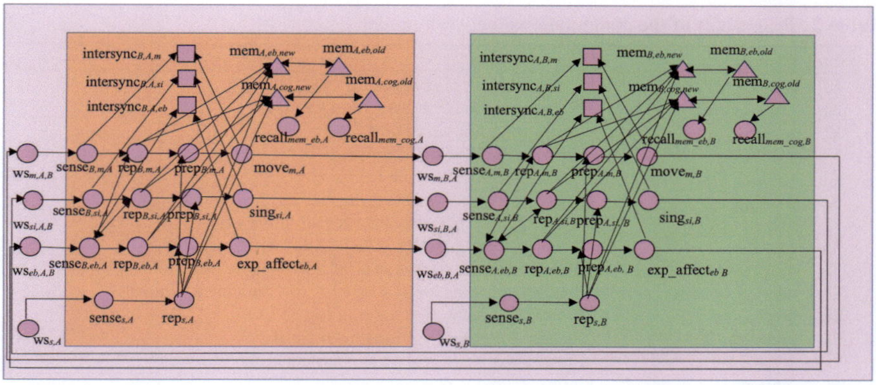

Fig. 1 Base level of the network model. Left block (Agent A—Hans), right block (Agent B—Tom). Each block includes the connections among states, the interpersonal synchrony detector states, the memory states, and the connection to the other agent through three modalities (movement, singing, and emotion)

were transmitted by the others' modality outputs. For three modalities, *eb* represents the level of emotion, *si* for level of singing, and *m* for level of movement. Besides, the memory states include the emotion and cognitive aspects, it is divided into new memory and old memory, here old memory can be awakened through new stimuli or information. This process involves the mechanism of memory reconsolidation. When a memory is retrieved, the new information can be integrated with the old memory, and then the memory is consolidated again and restored in the brain [48]. Outside of those blocks for the agents are the world states, which include the world stimulus states and the world states from another agent's execution states. Based on [25, 30], the adaptive connections between two agents include both external and internal ones, those connections adaptively change by synchrony, when more synchrony is detected, these connections will become stronger. Meanwhile each type of connection contributes in its own way to the interaction between agents. For example, a stronger executing connection will result in stronger expressions and actions towards the other [25, 30].

4.2 First and Second Reification Level

Similar to Hendrikse et al. [25, 30], it was assumed that the processes at the base level are first- and second-order adaptive concerning their connectivity and sensitivity in responding. To address this, as discussed in Sect. 3, first-order self-model **W**-states and **T**-states have been used to model first-order adaptivity of connectivity and sensitivity, respectively. Moreover, as also indicated in Sect. 3, second-order self-model $\mathbf{H_W}$-states and $\mathbf{H_T}$-states have been used to model adaptive adaptation speed of the first-order **W**-states and **T**-states. The first- and second-order adaptation levels are

Table 2 Base states of the computational network model

state	name	Explanation
X_1	$ws_{s,A}$	World state for stimulus *s* for agent *A*
X_2	$ws_{s,B}$	World state for stimulus *s* for agent *B*
X_3	$sense_{s,A}$	Sensor state for stimulus *s* for *A*
X_4	$sense_{B,eb,A}$	Sensor state of *A* for emotion *eb* for *B*
X_5	$sense_{B,si,A}$	Sensor state of *A* for sing *si* for *B*
X_6	$sense_{B,m,A}$	Sensor state of *A* for movement *m* for *B*
X_7	$rep_{s,A}$	Sensory representation state for stimulus *s* for *A*
X_8	$rep_{B,eb,A}$	Sensory representation state of *A* for emotion *eb* of *B*
X_9	$rep_{B,si,A}$	Sensory representation state of *A* for sing *si* for *B*
X_{10}	$rep_{B,m,A}$	Sensory representation state of A for movement *m* for *B*
X_{11}	$prep_{B,eb,A}$	Preparation state of *A* for emotion *eb* of *B*
X_{12}	$prep_{B,si,A}$	Preparation state of *A* for sing *si* for *B*
X_{13}	$prep_{B,m,A}$	Preparation state of *A* for movement *m* for *B*
X_{14}	$intersync_{B,A,eb}$	Interpersonal synchrony detection of *A* for executing *eb* by *B* and *A*
X_{15}	$intersync_{B,A,si}$	Interpersonal synchrony detection of *A* for executing *si* by *B* and *A*
X_{16}	$intersync_{B,A,m}$	Interpersonal synchrony detection of *A* for executing *m* by *B* and *A*
X_{17}	$mem_{A,eb,new}$	New memory state of *A* for emotion aspect
X_{18}	$mem_{A,eb,old}$	Old memory state of *A* for emotion aspect
X_{19}	$mem_{A,cog,new}$	New memory state of *A* for cognitive aspect
X_{20}	$mem_{A,cog,old}$	Old memory state of *A* for cognitive aspect
X_{21}	$recall_{mem\_eb,A}$	Memory recall state of *A* for emotion
X_{22}	$recall_{mem\_cog,A}$	Memory recall state of *A* for cognitive
X_{23}	$exp\_affect_{eb,A}$	Executing expression affect *eb* of *A*
X_{24}	$sing_{si,A}$	Executing sing *si* of *A*
X_{25}	$move_{m,A}$	Executing movement *m* of *A*
X_{26}	$sense_{s,B}$	Sensor state for stimulus *s* for *B*
X_{27}	$sense_{A,eb,B}$	Sensor state of *B* for emotion *eb* for *A*
X_{28}	$sense_{A,si,B}$	Sensor state of *B* for sing *si* for *A*
X_{29}	$sense_{A,m,B}$	Sensor state of *B* for movement *m* for *A*
X_{30}	$rep_{s,B}$	Sensory representation state for stimulus *s* for *B*
X_{31}	$rep_{A,eb,B}$	Sensory representation state of *B* for emotion *eb* of *A*
X_{32}	$rep_{A,si,B}$	Sensory representation state of *B* for sing *si* for *A*
X_{33}	$rep_{A,m,B}$	Sensory representation state of *B* for movement *m* for *A*
X_{34}	$prep_{A,eb,B}$	Preparation state of *B* for emotion *eb* of *A*
X_{35}	$prep_{A,si,B}$	Preparation state of *B* for sing *si* for *A*
X_{36}	$prep_{A,m,B}$	Preparation state of *B* for movement *m* for *A*
X_{37}	$intersync_{A,B,eb}$	Interpersonal synchrony detection of *B* for executing *eb* by *A* and *B*
X_{38}	$intersync_{A,B,si}$	Interpersonal synchrony detection of *B* for executing *si* by *A* and *B*
X_{39}	$intersync_{A,B,m}$	Interpersonal synchrony detection of *B* for executing *m* by *A* and *B*
X_{40}	$mem_{B,eb,new}$	New memory state of *B* for emotion aspect
X_{41}	$mem_{B,eb,old}$	Old memory state of *B* for emotion aspect
X_{42}	$mem_{B,cog,new}$	New memory state of *B* for cognitive aspect
X_{43}	$mem_{B,cog,old}$	Old memory state of *B* for cognitive aspect
X_{44}	$recall_{mem\_eb,B}$	Memory recall state of *B* for emotion
X_{45}	$recall_{mem\_cog,B}$	Memory recall state of *B* for cognitive
X_{46}	$exp\_affect_{eb,B}$	Executing expression affect *eb* of *B*
X_{47}	$sing_{si,B}$	Executing sing *si* of *B*
X_{48}	$move_{m,B}$	Executing movement *m* of *B*
X_{49}	$ws_{eb,A,B}$	World state for transmitting emotion *eb* of *B to A*
X_{50}	$ws_{si,A,B}$	World state for transmitting sing *si* of *B to A*
X_{51}	$ws_{m,A,B}$	World state for transmitting movement *m* of *B to A*
X_{52}	$ws_{eb,B,A}$	World state for transmitting emotion *eb* of *A to B*
X_{53}	$ws_{si,B,A}$	World state for transmitting sing *si* of *A to B*
X_{54}	$ws_{m,B,A}$	World state for transmitting movement *m* of *A to B*

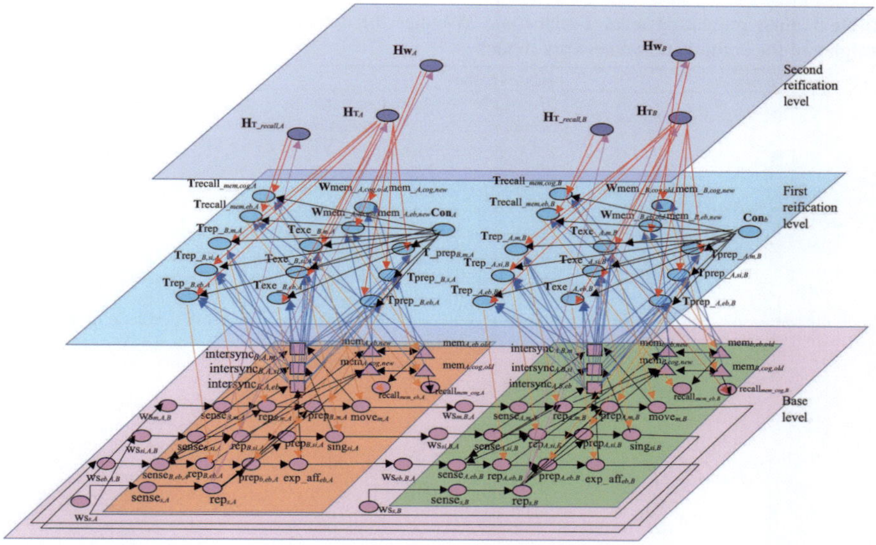

Fig. 2 Overview of the adaptive network model including base level, 1st and 2nd reification level

shown in Fig. 2. The middle blue plane describes the first-order adaptation level by self-model states for weights of some of the base connections (**W-**states), some of the excitability thresholds (by **T-**states) and context states for the excitability thresholds (**Con-**states). The purple top plane includes **H-**states ($\mathbf{H_T}$-states and $\mathbf{H_W}$-states) for the speed factor characteristic of the **T**-states and **W**-states. The corresponding explanations of each state for both reification levels are illustrated in Tables 3 and 4.

The **W-**states represent the weights of the adaptive base connections, specifically those for the memory. The adaptation principle used for them is that these weights strengthen upon stimulus exposure at the base level, which may be considered a variant of connectivity learning such as Bainian learning [3], p. 117, or Hebbian learning [24]. The **T**-states in the middle plane represent the adaptive excitability thresholds for representation, execution, preparation states and memory recall states. The **T**-states change based on the synchrony detection (the more synchrony is detected, the lower the thresholds become). To let these first-order adaptation principles work, the upward arrows from the base plane provide input to the **T**-states and **W**-states. The incoming states are from the three synchrony detection states indicated in base level as square symbols in both agents. The two context states maintain the excitability thresholds. The downward arrows indicate how the self-model states for adaptive connection weights and thresholds (**W**-states and **T**-states) affect the base level processes.

For the second-order adaptation level, there are six **H**-states presented to control the adaptation for the first reification level. $\mathbf{H}_{\mathbf{W}_A}$ and $\mathbf{H}_{\mathbf{W}_B}$ are for the weight adaptation for each agent, $\mathbf{H}_{\mathbf{T}_A}$ and $\mathbf{H}_{\mathbf{T}_{recall,B}}$ are for the excitability adaptation, $\mathbf{H}_{\mathbf{T}_{recall,B}}$ and $\mathbf{H}_{\mathbf{T}_{recall,B}}$ are for the excitability adaptation of memory recall for both agents.

Table 3 First reification level **T**-states and **W**-states for excitability thresholds and connection weights of the computational network model

state name	Explanation
X_{55} $\mathbf{T}_{rep\_B,eb,A}$	First-order self-model state for the excitability threshold of *A*'s sensory representation state $rep_{B,eb,A}$ for emotion *eb*
X_{56} $\mathbf{T}_{rep\_B,si,A}$	First-order self-model state for the excitability threshold of *A*'s sensory representation state $rep_{B,si,A}$ for sing *si*
X_{57} $\mathbf{T}_{rep\_B,m,A}$	First-order self-model state for the excitability threshold of *A*'s sensory representation state $rep_{B,m,A}$ for movement *m*
X_{58} $\mathbf{T}_{exe\_B,eb,A}$	First-order self-model state for the excitability threshold of *A*'s execution state $exp\_affect_{eb,A}$ for emotion *eb*
X_{59} $\mathbf{T}_{exe\_B,si,A}$	First-order self-model state for the excitability threshold of *A*'s execution state $sing_{si,A}$ for sing *s*
X_{60} $\mathbf{T}_{exe\_B,m,A}$	First-order self-model state for the excitability threshold of *A*'s execution state $move_{m,A}$ for movement *m*
X_{61} $\mathbf{T}_{prep\_B,eb,A}$	First-order self-model state for the excitability threshold of *A*'s preparation state $prep_{B,eb,A}$ for emotion *eb*
X_{62} $\mathbf{T}_{prep\_B,si,A}$	First-order self-model state for the excitability threshold of *A*'s preparation state $prep_{B,si,A}$ for sing *si*
X_{63} $\mathbf{T}_{prep\_B,m,A}$	First-order self-model state for the excitability threshold of *A*'s preparation state $prep_{B,m,A}$for movement *m*
X_{64} $\mathbf{T}_{recall\_mem,eb,A}$	First-order self-model state for the excitability threshold of *A*'s memory recall state $recall_{mem\ eb,A}$ for old memory emotion *eb* state and new memory emotion *eb* state
X_{65} $\mathbf{T}_{recall\_mem,cog,A}$	First-order self-model state for the excitability threshold of *A*'s memory recall state $recall_{mem\ cog,A}$ for old memory cognitive *cog* state and new memory cognitive *cog* state
X_{66} $\mathbf{W}_{mem\_A,eb,old,mem\_A,eb,new}$	First-order self-model state for the weight of *A*'s old memory emotion *eb* state to *A*'s new memory emotion *eb* state
X_{67} $\mathbf{W}_{mem\_A,cog,old,mem\_A,cog,new}$	First-order self-model state for the weight of *A*'s old memory cognitive *cog* state to *A*'s new memory cognitive *cog* state
X_{68} $\mathbf{T}_{rep\_A,eb,B}$	First-order self-model state for the excitability threshold of *B*'s sensory representation state $rep_{A,eb,B}$ for emotion *eb*
X_{69} $\mathbf{T}_{rep\_A,si,B}$	First-order self-model state for the excitability threshold of *B*'s sensory representation state $rep_{A,si,B}$ for sing *si*
X_{70} $\mathbf{T}_{rep\_A,m,B}$	First-order self-model state for the excitability threshold of *B*'s sensory representation state $rep_{A,m,B}$ for movement *m*
X_{71} $\mathbf{T}_{exe\_A,eb,B}$	First-order self-model state for the excitability threshold of *B*'s execution state $exp_affect_{eb,B}$for emotion *eb*
X_{72} $\mathbf{T}_{exe\_A,si,B}$	First-order self-model state for the excitability threshold of *B*'s execution state $sing_{si,B}$ for sing *si*
X_{73} $\mathbf{T}_{exe\_A,m,B}$	First-order self-model state for the excitability threshold of *B*'s execution state $move_{m,B}$ for movement *m*
X_{74} $\mathbf{T}_{prep\_A,eb,B}$	First-order self-model state for the excitability threshold of *B*'s preparation state $prep_{A,eb,B}$ for emotion *eb*
X_{75} $\mathbf{T}_{prep\_A,si,B}$	First-order self-model state for the excitability threshold of *B*'s preparation state $prep_{A,si,B}$ for sing *si*
X_{76} $\mathbf{T}_{prep\_A,m,B}$	First-order self-model state for the excitability threshold of *B*'s preparation state $prep_{A,m,B}$ for movement *m*
X_{77} $\mathbf{T}_{recall\_mem,eb\_B}$	First-order self-model state for the excitability threshold of *B*'s memory recall state $recall_{mem\ eb,B}$ for old memory emotion *eb* state and new memory emotion *eb* state
X_{78} $\mathbf{T}_{recall\_mem,cog\_B}$	First-order self-model state for the excitability threshold of *B*'s memory recall state $recall_{mem\ cog,B}$ for old memory cognitive *cog* state and new memory cognitive *cog* state
X_{79} $\mathbf{W}_{mem\_B,eb,old,mem\_B,eb,new}$	First-order self-model state for the weight of *B*'s old memory emotion *eb* state to *B*'s new memory emotion *eb* state
X_{80} $\mathbf{W}_{mem\_B,cog,old,mem\_B,cog,new}$	First-order self-model state for the weight of *B*'s old memory cognitive *cog* state to *B*'s new memory cognitive *cog* state
X_{81} $\mathbf{Con}_A$	Context state for excitablity threshold for *A*
X_{82} $\mathbf{Con}_B$	Context state for excitablity threshold for *B*

Table 4 Second-order adaptation level $\mathbf{H_W}$ – and $\mathbf{H_T}$ – states for the adaptation speed of the **T**-states and **W**-states of the computational network model

X_{83} $\mathbf{H_{WA}}$	Second-order self-model state for the speed factor of the first-order self-model **W**-states for *A*
X_{84} $\mathbf{H_{WB}}$	Second-order self-model state for the speed factor of the first-order self-model **W**-states for *B*
X_{85} $\mathbf{H_{TA}}$	Second-order self-model state for the speed factor of the first-order self-model **T**-states for *A*
X_{86} $\mathbf{H_{TB}}$	Second-order self-model state for the speed factor of the first-order self-model **T**-states for *B*
X_{87} $\mathbf{H}_{\mathbf{T}\_recall,A}$	Second-order self-model state for the speed factor of the first-order self-model **T-recall** states for *A*
X_{88} $\mathbf{H}_{\mathbf{T}\_recall,B}$	Second-order self-model state for the speed factor of the first-order self-model **T-recall** states for *B*

All of these second-order self-model **H**-states are activated upon stimulus exposure from the base level, following the metaplasticity principle 'Adaptation accelerates with increasing stimulus exposure' formulated by Robinson et al. [45] (see the long upward arrows from the pink base plane to the second-order **H**-states in the purple upper plane in Fig. 2).

5 Simulation Results

A few simulation experiments have been designed and performed based on the scenarios. Through analysis of these scenarios, it was discovered that music facilitated emotional release for the two elderly agents, shown in their excited standing and tears. They recalled memories, including from initial utterances to singing lyrics. Through background research, it was found that music indeed aided Alzheimer's patients in enhancing memory for relevant linguistic information [48], but it doesn't have a significant impact on healthy older adults [49]. However, since the cognitive faculties of the agents in this scenario are dimmed by dementia without specifying Alzheimer's disease, the effect of music on memory in elderly individuals with cognitive decline could be a direction for further research. Secondly, research on synchrony in elderly individuals is crucial. Despite extensive studies on synchrony in babies and adults [9, 54], research on synchrony in the elderly is relatively scarce.

Results of an example simulation are shown in Figs. 3 and 4. The total time of the simulation is 40, and the step size Δt is 0.5. The whole simulation duration is 80 computational steps. The remaining section will present in some detail the simulation results, including the base level of the two agents, as well as the 1st and 2nd order adaptation levels. It will also cover the simulation on interpersonal synchrony between the two agents, and the memory and memory recall status.

Figures 3 and 4 describe the simulation results of the base level states of agent A and agent B. From both of the graphs, it can be seen that around 30 to 35 time units, the simulation tends to stabilize. Dashed line, dotted line and diamond line represent the

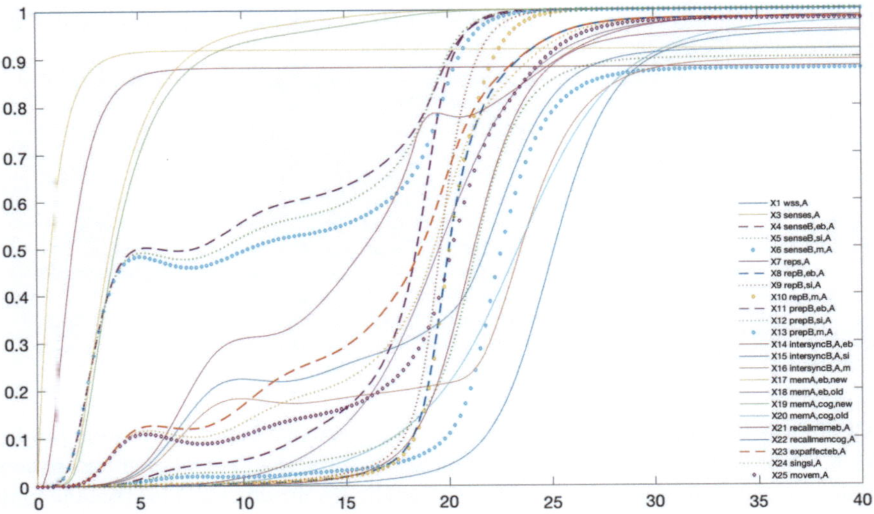

Fig. 3 Simulation activation level for Base level of Agent A

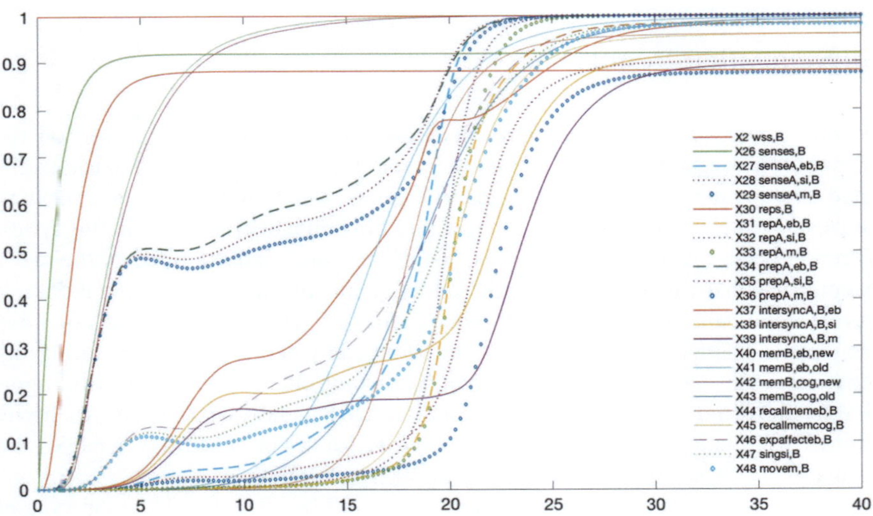

Fig. 4 Simulation activation level for Base level of Agent B

sense, representation, preparation, and execution states of three different modalities, emotion, singing and movement. Through the plot's visualization, it was observed that the trends of the two agents were generally similar, but there were a bit differences in the timing. It can be seen from the plots that the first activation occurred in the world sense and world representation state. Subsequently, new memory and other sense states were rapidly activated, completing the simulation approximately between the 5th and 20th time units, and then getting flat. Meanwhile, it is evident that the activation of emotion precedes singing, followed by singing preceding movement.

Figure 5 shows the first and second reification level for both agents. The triangle line represents agent A, the square line is for agent B. As can be seen, the weight of new and old memory for both agents are increased (X_{66}, X_{67}, X_{79}, X_{80}). From the 5th time unit to the 35th time unit. For the excitability threshold factors, all of them start with the initial value 0.7, and had been activated at around the 2nd time unit, it increased sharply and dropped gradually in around 5 time units. The excitability threshold of the three modalities overlapped during the representation, preparation, and execution status, but there are slight differences among the different modalities. Regarding the excitability threshold of memory recall, it exhibits a similar trend but is smoother, with its value stabilizing around 0.35. The dashed and solid lines represent the speed factors control (adaptive learning rates) of Agent A and B at the second reification level. $\mathbf{H_W}$-states represent the speed factor for weights, $\mathbf{H_T}$-states show the excitability threshold speed factor control. The four $\mathbf{H_T}$ states (X_{85}, X_{86}, X_{87}, X_{88}) overlap with each other and rise from the start of the simulation, reaching a constant value of 0.7 at the 10th time unit. The $\mathbf{H_W}$-states of Agent A and Agent B also occur at the same time, with a gradual upward trend. And the value is around 0.2 at about the 7th time unit.

Figure 6 illustrates the results of Agent A and Agent B during simulation regarding memory-related factors, including emotions and cognitive aspects, along with the corresponding memory recall. The solid line represents Agent A and the dashed line represents Agent B. It can be observed that new memories are initially activated and

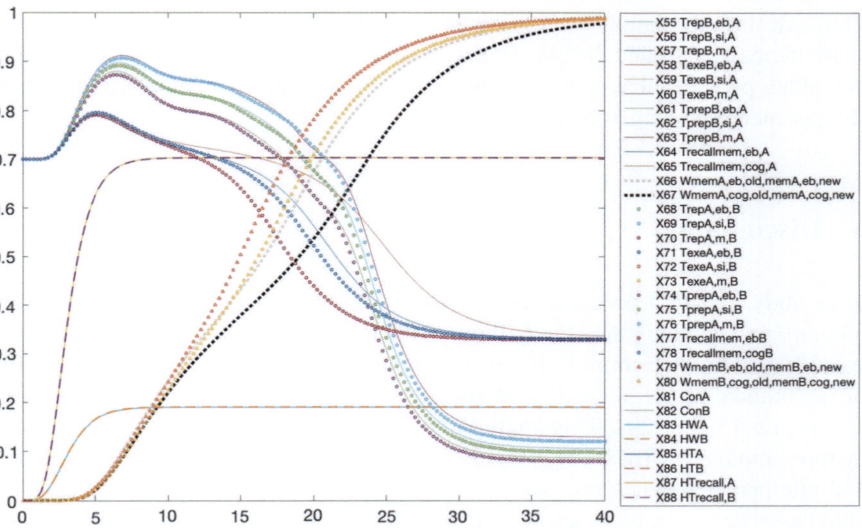

Fig. 5 Activation levels for 1st and 2nd order adaptation level for both agents

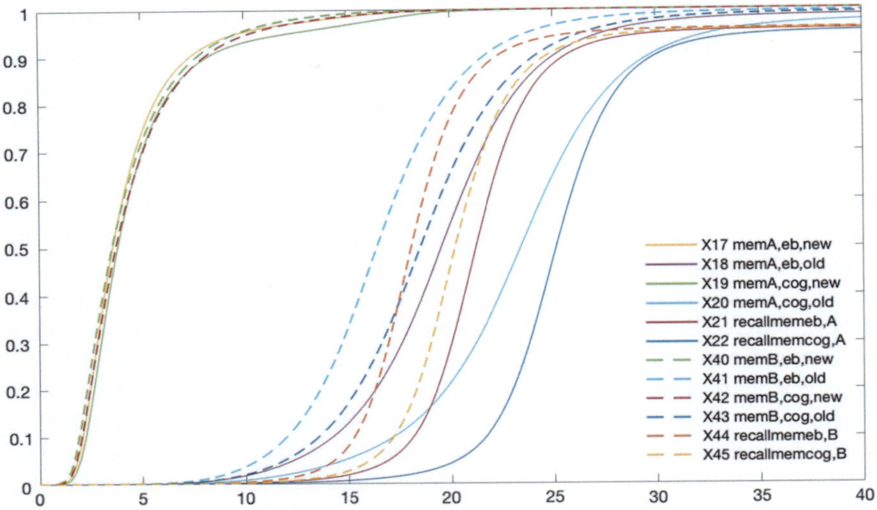

Fig. 6 The simulation result for memory related states for both agents

reach the peak value at the 15th time unit. The old memories of Agent B are activated earlier than Agent A. Furthermore, it is noted that emotional memories are activated prior to cognitive memories.

Figure 7 describes the simulation of detection of interpersonal synchrony for the two agents, where solid line represents agent A and dashed line represents agent B. It is observed that the synchrony of singing, and movement exhibits a similar trend. They are initially rapidly increasing and followed by a gradual rise, then another sharp increase at around the 20th time unit, and finally remain stable. The synchrony in singing precedes that of movement. As for the synchrony in emotion, it shows a sharper increase compared to the others.

6 Discussion

This study explored computational analysis methods to support joint action and memory recall among the elderly through adaptive network modeling and interpersonal synchrony detection. It makes use of previous research on adaptive modeling of multimodal social interaction in Hendrikse et al. [25, 30], see also this volume, Chaps 12–13 [28, 29]. This study simulated interactions among two old adults in three modalities (emotion, singing and movement) and analyzed the memory and interpersonal synchrony status by a multi-order adaptive network model. The simulation results indicate that music assists in memory recall for the elderly, especially for those with cognitive decline. Additionally, the study analyzes the interpersonal synchrony and illustrates the changing of singing, movement, and emotional

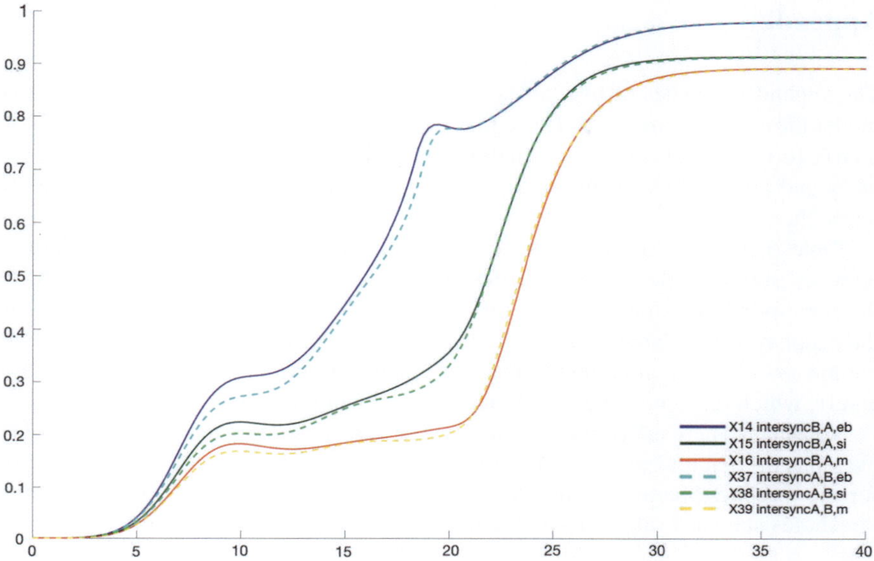

Fig. 7 The simulation result for interpersonal synchrony detection states for both agents

synchrony between two agents. These results further show the feasibility of this analysis approach.

However, the study also has limitations. The simulation results are based on some assumptions which may be too idealized. For example, potentially influential factors such as health conditions and medication have been left out of consideration. Meanwhile, as the selection and setting of parameters in the model may affect the accuracy and reliability of the results, it may need further optimization or validation. Furthermore, the study only focuses on interpersonal synchrony and memory recall, without fully consideration of the impact of other factors such as health status, therapy influence, or other factors. Therefore, future research could explore the interactions between different factors and establish a more comprehensive model. In conclusion, this study provides a new perspective and method for understanding and improving social interactions and memory recall among the elderly, but further research is needed to validate its effectiveness and feasibility. Another limitation is that the model and its simulation was based on one case study. Their was no systematic experiment going on from which multiple cases could be addressed and analysed in a comparative manner.

Appendix

The appendix consists of five tables with the specification of the adaptive network model through role matrices. These matrices are readable by MATLAB software in matrix form and utilized for simulation. These data not only validate the reproductivity and authenticity of this study but also serve as valuable reference for future research.

Table 5 displays the role matrices **mb**, illustrating the connectivity relations between states. Table 6 presents **mcw** which indicates the connection weight of the corresponding connection. Meanwhile, the pink shaded cells in Table 6 represent the adaptive value. Tables 7 and 8 are the role matrices **mcfp** for the weight and the parameters of used combination functions. Table 9 describes the role matrices of **ms** and **iv**, which represents the speed factors and the initial values.

Note that all network characteristics have been set according to a default range. For example, the default value 1 was used for all nonadaptive positive connection weights, for the steepness parameter the default values 5 or 10 were used, for excitability thresholds standard values 0.4–0.5, and so on.

Table 5 Role matrix **mb** for base connectivity

mb base connectivity	1	2	3	4	5
X_{1} $ws_{s,A}$	X_{1}				
X_{2} $ws_{s,B}$	X_{2}				
X_{3} $sense_{s,A}$	X_{1}				
X_{4} $sense_{B,eb,A}$	X_{9}	X_{10}	X_{49}		
X_{5} $sense_{B,si,A}$	X_{50}				
X_{6} $sense_{B,m,A}$	X_{51}				
X_{7} $rep_{s,A}$	X_{3}				
X_{8} $rep_{B,eb,A}$	X_{4}				
X_{9} $rep_{B,si,A}$	X_{4}	X_{5}			
X_{10} $rep_{B,m,A}$	X_{4}	X_{6}			
X_{11} $prep_{B,eb,A}$	X_{7}	X_{8}			
X_{12} $prep_{B,si,A}$	X_{7}	X_{9}			
X_{13} $prep_{B,m,A}$	X_{7}	X_{10}			
X_{14} $intersync_{B,A,eb}$	X_{4}	X_{23}			
X_{15} $intersync_{B,A,si}$	X_{5}	X_{24}			
X_{16} $intersync_{B,A,m}$	X_{6}	X_{25}			
X_{17} $mem_{A,eb,new}$	X_{7}	X_{8}	X_{9}	X_{10}	X_{18}
X_{18} $mem_{A,eb,old}$	X_{17}				
X_{19} $mem_{A,cog,new}$	X_{7}	X_{8}	X_{9}	X_{10}	X_{20}
X_{20} $mem_{A,cog,old}$	X_{19}				
X_{21} $recall_{mem\ eb,A}$	X_{18}				
X_{22} $recall_{mem\ cog,A}$	X_{20}				
X_{23} $exp\_affect_{eb,A}$	X_{11}				
X_{24} $sing_{si,A}$	X_{12}				
X_{25} $move_{m,A}$	X_{13}				
X_{26} $sense_{s,B}$	X_{2}				
X_{27} $sense_{A,eb,B}$	X_{32}	X_{33}	X_{52}		
X_{28} $sense_{A,si,B}$	X_{53}				
X_{29} $sense_{A,m,B}$	X_{54}				
X_{30} $rep_{s,B}$	X_{26}				
X_{31} $rep_{A,eb,B}$	X_{27}				
X_{32} $rep_{A,si,B}$	X_{27}	X_{28}			
X_{33} $rep_{A,m,B}$	X_{27}	X_{29}			
X_{34} $prep_{A,eb,B}$	X_{30}	X_{31}			
X_{35} $prep_{A,si,B}$	X_{30}	X_{32}			
X_{36} $prep_{A,m,B}$	X_{30}	X_{33}			
X_{37} $intersync_{A,B,eb}$	X_{27}	X_{46}			
X_{38} $intersync_{A,B,si}$	X_{28}	X_{47}			
X_{39} $intersync_{A,B,m}$	X_{29}	X_{48}			
X_{40} $mem_{B,eb,new}$	X_{30}	X_{31}	X_{32}	X_{33}	X_{41}
X_{41} $mem_{B,eb,old}$	X_{40}				
X_{42} $mem_{B,cog,new}$	X_{30}	X_{31}	X_{32}	X_{33}	X_{43}
X_{43} $mem_{B,cog,old}$	X_{42}				
X_{44} $recall_{mem\ eb,B}$	X_{41}				
X_{45} $recall_{mem\ cog,B}$	X_{43}				
X_{46} $exp\_affect_{eb,B}$	X_{34}				
X_{47} $sing_{si,B}$	X_{35}				
X_{48} $move_{m,B}$	X_{36}				
X_{49} $ws_{eb,A,B}$	X_{46}				
X_{50} $ws_{si,A,B}$	X_{47}				
X_{51} $ws_{m,A,B}$	X_{48}				
X_{52} $ws_{eb,B,A}$	X_{23}				
X_{53} $ws_{si,B,A}$	X_{24}				
X_{54} $ws_{m,B,A}$	X_{25}				
X_{55} $\mathbf{T}_{rep\ B,eb,A}$	X_{14}	X_{15}	X_{16}	X_{81}	
X_{56} $\mathbf{T}_{rep\ B,si,A}$	X_{14}	X_{15}	X_{16}	X_{81}	
X_{57} $\mathbf{T}_{rep\ B,m,A}$	X_{14}	X_{15}	X_{16}	X_{81}	
X_{58} $\mathbf{T}_{exe\ B,eb,A}$	X_{14}	X_{15}	X_{16}	X_{81}	
X_{59} $\mathbf{T}_{exe\ B,si,A}$	X_{14}	X_{15}	X_{16}	X_{81}	
X_{60} $\mathbf{T}_{exe\ B,m,A}$	X_{14}	X_{15}	X_{16}	X_{81}	
X_{61} $\mathbf{T}_{prep\ B,eb,A}$	X_{14}	X_{15}	X_{16}	X_{81}	
X_{62} $\mathbf{T}_{prep\ B,si,A}$	X_{14}	X_{15}	X_{16}	X_{81}	
X_{63} $\mathbf{T}_{prep\ B,m,A}$	X_{14}	X_{15}	X_{16}	X_{81}	
X_{64} $\mathbf{T}_{recall\ mem,eb,A}$	X_{17}	X_{18}	X_{81}		
X_{65} $\mathbf{T}_{recall\ mem,cog,A}$	X_{19}	X_{20}	X_{81}		
X_{66} $\mathbf{W}_{mem\ A,eb,old,mem\ A,eb,new}$	X_{17}	X_{18}			
X_{67} $\mathbf{W}_{mem\ A,cog,old,mem\ A,cog,new}$	X_{19}	X_{20}			
X_{68} $\mathbf{T}_{rep\ A,eb,B}$	X_{37}	X_{38}	X_{39}	X_{82}	
X_{69} $\mathbf{T}_{rep\ A,si,B}$	X_{37}	X_{38}	X_{39}	X_{82}	
X_{70} $\mathbf{T}_{rep\ A,m,B}$	X_{37}	X_{38}	X_{39}	X_{82}	
X_{71} $\mathbf{T}_{exe\ A,eb,B}$	X_{37}	X_{38}	X_{39}	X_{82}	
X_{72} $\mathbf{T}_{exe\ A,si,B}$	X_{37}	X_{38}	X_{39}	X_{82}	
X_{73} $\mathbf{T}_{exe\ A,m,B}$	X_{37}	X_{38}	X_{39}	X_{82}	
X_{74} $\mathbf{T}_{prep\ A,eb,B}$	X_{37}	X_{38}	X_{39}	X_{82}	
X_{75} $\mathbf{T}_{prep\ A,si,B}$	X_{37}	X_{38}	X_{39}	X_{82}	
X_{76} $\mathbf{T}_{prep\ A,m,B}$	X_{37}	X_{38}	X_{39}	X_{82}	
X_{77} $\mathbf{T}_{recall\ mem,eb\ B}$	X_{40}	X_{41}	X_{82}		
X_{78} $\mathbf{T}_{recall\ mem,cog\ B}$	X_{42}	X_{43}	X_{82}		
X_{79} $\mathbf{W}_{mem\ A,eb,old,mem\ A,eb,new}$	X_{40}	X_{41}			
X_{80} $\mathbf{W}_{mem\ A,cog,old,mem\ A,cog,new}$	X_{42}	X_{43}			
X_{81} $\mathbf{Con}_{A}$	X_{81}				
X_{82} $\mathbf{Con}_{B}$	X_{82}				
X_{83} $\mathbf{H}_{\mathbf{W_A}}$	X_{7}				
X_{84} $\mathbf{H}_{\mathbf{W_B}}$	X_{30}				
X_{85} $\mathbf{H}_{\mathbf{T_A}}$	X_{7}				
X_{86} $\mathbf{H}_{\mathbf{T_B}}$	X_{30}				
X_{87} $\mathbf{H}_{\mathbf{T}\ recall,A}$	X_{7}				
X_{88} $\mathbf{H}_{\mathbf{T}\ recall,B}$	X_{30}				

Table 6 Role matrix **mcw** for connection weights

mcw	connection weights	1	2	3	4	5
X_1	$ws_{s,A}$	1				
X_2	$ws_{s,B}$	1				
X_3	$sense_{s,A}$	1				
X_4	$sense_{B,eb,A}$	1	1	1		
X_5	$sense_{B,si,A}$	1				
X_6	$sense_{B,m,A}$	1				
X_7	$rep_{s,A}$	1				
X_8	$rep_{B,eb,A}$	1				
X_9	$rep_{B,si,A}$	1	1			
X_{10}	$rep_{B,m,A}$	1	1			
X_{11}	$prep_{B,eb,A}$	1	1			
X_{12}	$prep_{B,si,A}$	1	1			
X_{13}	$prep_{B,m,A}$	1	1			
X_{14}	$intersync_{B,A,eb}$	1	1			
X_{15}	$intersync_{B,A,si}$	1	1			
X_{16}	$intersync_{B,A,m}$	1	1			
X_{17}	$mem_{A,eb,new}$	1	1	1	1	1
X_{18}	$mem_{A,eb,old}$	X_{66}				
X_{19}	$mem_{A,cog,new}$	1	1	1	1	1
X_{20}	$mem_{A,cog,old}$	X_{67}				
X_{21}	$recall_{mem\_eb,A}$	1				
X_{22}	$recall_{mem\_cog,A}$	1				
X_{23}	$exp\_affect_{eb,A}$	1				
X_{24}	$sing_{si,A}$	1				
X_{25}	$move_{m,A}$	1				
X_{26}	$sense_{s,B}$	1				
X_{27}	$sense_{A,eb,B}$	1	1	1		
X_{28}	$sense_{A,si,B}$	1				
X_{29}	$sense_{A,m,B}$	1				
X_{30}	$rep_{s,B}$	1				
X_{31}	$rep_{A,eb,B}$	1				
X_{32}	$rep_{A,si,B}$	1	1			
X_{33}	$rep_{A,m,B}$	1	1			
X_{34}	$prep_{A,eb,B}$	1	1			
X_{35}	$prep_{A,si,B}$	1	1			
X_{36}	$prep_{A,m,B}$	1	1			
X_{37}	$intersync_{A,B,eb}$	1	1			
X_{38}	$intersync_{A,B,si}$	1	1			
X_{39}	$intersync_{A,B,m}$	1	1			
X_{40}	$mem_{B,eb,new}$	1	1	1	1	1
X_{41}	$mem_{B,eb,old}$	X_{79}				
X_{42}	$mem_{B,cog,new}$	1	1	1	1	1
X_{43}	$mem_{B,cog,old}$	X_{80}				
X_{44}	$recall_{mem\_eb,B}$	1				
X_{45}	$recall_{mem\_cog,B}$	1				
X_{46}	$exp\_affect_{eb,B}$	1				
X_{47}	$sing_{si,B}$	1				
X_{48}	$move_{m,B}$	1				
X_{49}	$ws_{eb,A,B}$	1				
X_{50}	$ws_{si,A,B}$	1				
X_{51}	$ws_{m,A,B}$	1				
X_{52}	$ws_{eb,B,A}$	1				
X_{53}	$ws_{si,B,A}$	1				
X_{54}	$ws_{m,B,A}$	1				
X_{55}	$\mathbf{T}_{rep\_B,eb,A}$	-0.3	-0.3	-0.3	1	
X_{56}	$\mathbf{T}_{rep\_B,si,A}$	-0.3	-0.3	-0.3	1	
X_{57}	$\mathbf{T}_{rep\_B,m,A}$	-0.3	-0.3	-0.3	1	
X_{58}	$\mathbf{T}_{exe\_B,eb,A}$	-0.3	-0.3	-0.3	1	
X_{59}	$\mathbf{T}_{exe\_B,si,A}$	-0.3	-0.3	-0.3	1	
X_{60}	$\mathbf{T}_{exe\_B,m,A}$	-0.3	-0.3	-0.3	1	
X_{61}	$\mathbf{T}_{prep\_B,eb,A}$	-0.3	-0.3	-0.3	1	
X_{62}	$\mathbf{T}_{prep\_B,si,A}$	-0.3	-0.3	-0.3	1	
X_{63}	$\mathbf{T}_{prep\_B,m,A}$	-0.3	-0.3	-0.3	1	
X_{64}	$\mathbf{T}_{recall\_mem,eb,A}$	-0.3	-0.3	1		
X_{65}	$\mathbf{T}_{recall\_mem,cog,A}$	-0.3	-0.3	1		
X_{66}	$\mathbf{W}_{mem\_A,eb,old,mem\_A,eb,new}$	1	1			
X_{67}	$\mathbf{W}_{mem\_A,cog,old,mem\_A,cog,new}$	1	1			
X_{68}	$\mathbf{T}_{rep\_A,eb,B}$	-0.3	-0.3	-0.3	1	
X_{69}	$\mathbf{T}_{rep\_A,si,B}$	-0.3	-0.3	-0.3	1	
X_{70}	$\mathbf{T}_{rep\_A,m,B}$	-0.3	-0.3	-0.3	1	
X_{71}	$\mathbf{T}_{exe\_A,eb,B}$	-0.3	-0.3	-0.3	1	
X_{72}	$\mathbf{T}_{exe\_A,si,B}$	-0.3	-0.3	-0.3	1	
X_{73}	$\mathbf{T}_{exe\_A,m,B}$	-0.3	-0.3	-0.3	1	
X_{74}	$\mathbf{T}_{prep\_A,eb,B}$	-0.3	-0.3	-0.3	1	
X_{75}	$\mathbf{T}_{prep\_A,si,B}$	-0.3	-0.3	-0.3	1	
X_{76}	$\mathbf{T}_{prep\_A,m,B}$	-0.3	-0.3	-0.3	1	
X_{77}	$\mathbf{T}_{recall\_mem,eb\_B}$	-0.3	-0.3	1		
X_{78}	$\mathbf{T}_{recall\_mem,cog\_B}$	-0.3	-0.3	1		
X_{79}	$\mathbf{W}_{mem\_A,eb,old,mem\_A,eb,new}$	1	1			
X_{80}	$\mathbf{W}_{mem\_A,cog,old,mem\_A,cog,new}$	1	1			
X_{81}	$\mathbf{Con}_A$	1				
X_{82}	$\mathbf{Con}_B$	1				
X_{83}	$\mathbf{H}_{\mathbf{W_A}}$	0.5				
X_{84}	$\mathbf{H}_{\mathbf{W_B}}$	0.5				
X_{85}	$\mathbf{H}_{\mathbf{T_A}}$	1				
X_{86}	$\mathbf{H}_{\mathbf{T_B}}$	1				
X_{87}	$\mathbf{H}_{\mathbf{T}\_recall,A}$	1				
X_{88}	$\mathbf{H}_{\mathbf{T}\_recall,B}$	1				

Table 7 Role matrix **mcfw** for combination function weights

mcfw combination function weights	**1 alogistic**	**2 compdiff**
X_{1} $\mathbf{ws}_{s,A}$	1	
X_{2} $\mathbf{ws}_{s,B}$	1	
X_{3} $\mathbf{sense}_{s,A}$	1	
X_{4} $\mathbf{sense}_{B,eb,A}$	1	
X_{5} $\mathbf{sense}_{B,si,A}$	1	
X_{6} $\mathbf{sense}_{B,m,A}$	1	
X_{7} $\mathbf{rep}_{s,A}$	1	
X_{8} $\mathbf{rep}_{B,eb,A}$	1	
X_{9} $\mathbf{rep}_{B,si,A}$	1	
X_{10} $\mathbf{rep}_{B,m,A}$	1	
X_{11} $\mathbf{prep}_{B,eb,A}$	1	
X_{12} $\mathbf{prep}_{B,si,A}$	1	
X_{13} $\mathbf{prep}_{B,m,A}$	1	
X_{14} $\mathbf{intersync}_{B,A,eb}$		1
X_{15} $\mathbf{intersync}_{B,A,si}$		1
X_{16} $\mathbf{intersync}_{B,A,m}$		1
X_{17} $\mathbf{mem}_{A,eb,new}$	1	
X_{18} $\mathbf{mem}_{A,eb,old}$	1	
X_{19} $\mathbf{mem}_{A,cog,new}$	1	
X_{20} $\mathbf{mem}_{A,cog,old}$	1	
X_{21} $\mathbf{recall}_{mem\_eb,A}$	1	
X_{22} $\mathbf{recall}_{mem\_cog,A}$	1	
X_{23} $\mathbf{exp\_affect}_{eb,A}$	1	
X_{24} $\mathbf{sing}_{si,A}$	1	
X_{25} $\mathbf{move}_{m,A}$	1	
X_{26} $\mathbf{sense}_{s,B}$	1	
X_{27} $\mathbf{sense}_{A,eb,B}$	1	
X_{28} $\mathbf{sense}_{A,si,B}$	1	
X_{29} $\mathbf{sense}_{A,m,B}$	1	
X_{30} $\mathbf{rep}_{s,B}$	1	
X_{31} $\mathbf{rep}_{A,eb,B}$	1	
X_{32} $\mathbf{rep}_{A,si,B}$	1	
X_{33} $\mathbf{rep}_{A,m,B}$	1	
X_{34} $\mathbf{prep}_{A,eb,B}$	1	
X_{35} $\mathbf{prep}_{A,si,B}$	1	
X_{36} $\mathbf{prep}_{A,m,B}$	1	
X_{37} $\mathbf{intersync}_{A,B,eb}$		1
X_{38} $\mathbf{intersync}_{A,B,si}$		1
X_{39} $\mathbf{intersync}_{A,B,m}$		1
X_{40} $\mathbf{mem}_{B,eb,new}$	1	
X_{41} $\mathbf{mem}_{B,eb,old}$	1	
X_{42} $\mathbf{mem}_{B,cog,new}$	1	
X_{43} $\mathbf{mem}_{B,cog,old}$	1	
X_{44} $\mathbf{recall}_{mem\_eb,B}$	1	
X_{45} $\mathbf{recall}_{mem\_cog,B}$	1	
X_{46} $\mathbf{exp\_affect}_{eb,B}$	1	
X_{47} $\mathbf{sing}_{si,B}$	1	
X_{48} $\mathbf{move}_{m,B}$	1	
X_{49} $\mathbf{ws}_{eb,A,B}$	1	
X_{50} $\mathbf{ws}_{si,A,B}$	1	
X_{51} $\mathbf{ws}_{m,A,B}$	1	
X_{52} $\mathbf{ws}_{eb,B,A}$	1	
X_{53} $\mathbf{ws}_{si,B,A}$	1	
X_{54} $\mathbf{ws}_{m,B,A}$	1	
X_{55} $\mathbf{T}_{rep\_B,eb,A}$	1	
X_{56} $\mathbf{T}_{rep\_B,si,A}$	1	
X_{57} $\mathbf{T}_{rep\_B,m,A}$	1	
X_{58} $\mathbf{T}_{exe\_B,eb,A}$	1	
X_{59} $\mathbf{T}_{exe\_B,si,A}$	1	
X_{60} $\mathbf{T}_{exe\_B,m,A}$	1	
X_{61} $\mathbf{T}_{prep\_B,eb,A}$	1	
X_{62} $\mathbf{T}_{prep\_B,si,A}$	1	
X_{63} $\mathbf{T}_{prep\_B,m,A}$	1	
X_{64} $\mathbf{T}_{recall\_mem,eb,A}$	1	
X_{65} $\mathbf{T}_{recall\_mem,cog,A}$	1	
X_{66} $\mathbf{W}_{mem\_A,eb,old,mem\_A,eb,new}$	1	
X_{67} $\mathbf{W}_{mem\_A,cog,old,mem\_A,cog,new}$	1	
X_{68} $\mathbf{T}_{rep\_A,eb,B}$	1	
X_{69} $\mathbf{T}_{rep\_A,si,B}$	1	
X_{70} $\mathbf{T}_{rep\_A,m,B}$	1	
X_{71} $\mathbf{T}_{exe\_A,eb,B}$	1	
X_{72} $\mathbf{T}_{exe\_A,si,B}$	1	
X_{73} $\mathbf{T}_{exe\_A,m,B}$	1	
X_{74} $\mathbf{T}_{prep\_A,eb,B}$	1	
X_{75} $\mathbf{T}_{prep\_A,si,B}$	1	
X_{76} $\mathbf{T}_{prep\_A,m,B}$	1	
X_{77} $\mathbf{T}_{recall\_mem,eb\_B}$	1	
X_{78} $\mathbf{T}_{recall\_mem,cog\_B}$	1	
X_{79} $\mathbf{W}_{mem\_A,eb,old,mem\_A,eb,new}$	1	
X_{80} $\mathbf{W}_{mem\_A,cog,old,mem\_A,cog,new}$	1	
X_{81} $\mathbf{Con}_{A}$	1	
X_{82} $\mathbf{Con}_{B}$	1	
X_{83} $\mathbf{H}_{\mathbf{W_A}}$	1	
X_{84} $\mathbf{H}_{\mathbf{W_B}}$	1	
X_{85} $\mathbf{H}_{\mathbf{T_A}}$	1	
X_{86} $\mathbf{H}_{\mathbf{T_B}}$	1	
X_{87} $\mathbf{H}_{\mathbf{T}\_recall,A}$	1	
X_{88} $\mathbf{H}_{\mathbf{T}\_recall,B}$	1	

Table 8 Role matrix **mcfp** for combination function parameters

mcfp combination function parameters	1 alogistic		2 compdiff
X_1 $ws_{s,A}$	5	0.5	
X_2 $ws_{s,B}$	5	0.5	
X_3 $sense_{s,A}$	5	0.5	
X_4 $sense_{B,eb,A}$	5	0.4	
X_5 $sense_{B,si,A}$	5	0.45	
X_6 $sense_{B,m,A}$	5	0.5	
X_7 $rep_{s,A}$	5	0.5	
X_8 $rep_{B,eb,A}$	5	X_{55}	
X_9 $rep_{B,si,A}$	5	X_{56}	
X_{10} $rep_{B,m,A}$	5	X_{57}	
X_{11} $prep_{B,eb,A}$	5	X_{61}	
X_{12} $prep_{B,si,A}$	5	X_{62}	
X_{13} $prep_{B,m,A}$	5	X_{63}	
X_{14} $intersync_{B,A,eb}$			
X_{15} $intersync_{B,A,si}$			
X_{16} $intersync_{B,A,m}$			
X_{17} $mem_{A,eb,new}$	10	0.55	
X_{18} $mem_{A,eb,old}$	10	0.55	
X_{19} $mem_{A,cog,new}$	10	0.6	
X_{20} $mem_{A,cog,old}$	10	0.6	
X_{21} $recall_{mem\_eb,A}$	5	X_{64}	
X_{22} $recall_{mem\_cog,A}$	5	X_{65}	
X_{23} $exp\_affect_{eb,A}$	5	X_{58}	
X_{24} $sing_{si,A}$	5	X_{59}	
X_{25} $move_{m,A}$	5	X_{60}	
X_{26} $sense_{s,B}$	5	0.5	
X_{27} $sense_{A,eb,B}$	5	0.4	
X_{28} $sense_{A,si,B}$	5	0.45	
X_{29} $sense_{A,m,B}$	5	0.5	
X_{30} $rep_{s,B}$	5	0.5	
X_{31} $rep_{A,eb,B}$	5	X_{68}	
X_{32} $rep_{A,si,B}$	5	X_{69}	
X_{33} $rep_{A,m,B}$	5	X_{70}	
X_{34} $prep_{A,eb,B}$	5	X_{74}	
X_{35} $prep_{A,si,B}$	5	X_{75}	
X_{36} $prep_{A,m,B}$	5	X_{76}	
X_{37} $intersync_{A,B,eb}$			
X_{38} $intersync_{A,B,si}$			
X_{39} $intersync_{A,B,m}$			
X_{40} $mem_{B,eb,new}$	10	0.45	
X_{41} $mem_{B,eb,old}$	10	0.45	
X_{42} $mem_{B,cog,new}$	10	0.5	
X_{43} $mem_{B,cog,old}$	10	0.5	
X_{44} $recall_{mem\_eb,B}$	5	X_{77}	
X_{45} $recall_{mem\_cog,B}$	5	X_{78}	
X_{46} $exp\_affect_{eb,B}$	5	X_{71}	
X_{47} $sing_{si,B}$	5	X_{72}	
X_{48} $move_{m,B}$	5	X_{73}	
X_{49} $ws_{eb,A,B}$	5	0.5	
X_{50} $ws_{si,A,B}$	5	0.5	
X_{51} $ws_{m,A,B}$	5	0.5	
X_{52} $ws_{eb,B,A}$	5	0.5	
X_{53} $ws_{si,B,A}$	5	0.5	
X_{54} $ws_{m,B,A}$	5	0.5	
X_{55} $\mathbf{T}_{rep\_B,eb,A}$	5	0.5	
X_{56} $\mathbf{T}_{rep\_B,si,A}$	5	0.45	
X_{57} $\mathbf{T}_{rep\_B,m,A}$	5	0.4	
X_{58} $\mathbf{T}_{exe\_B,eb,A}$	5	0.5	
X_{59} $\mathbf{T}_{exe\_B,si,A}$	5	0.45	
X_{60} $\mathbf{T}_{exe\_B,m,A}$	5	0.4	
X_{61} $\mathbf{T}_{prep\_B,eb,A}$	5	0.5	
X_{62} $\mathbf{T}_{prep\_B,si,A}$	5	0.45	
X_{63} $\mathbf{T}_{prep\_B,m,A}$	5	0.4	
X_{64} $\mathbf{T}_{recall\_mem,eb,A}$	5	0.5	
X_{65} $\mathbf{T}_{recall\_mem,cog,A}$	5	0.5	
X_{66} $\mathbf{W}_{mem\_A,eb,old,mem\_A,eb,new}$	5	1	
X_{67} $\mathbf{W}_{mem\_A,cog,old,mem\_A,cog,new}$	5	1	
X_{68} $\mathbf{T}_{rep\_A,eb,B}$	5	0.52	
X_{69} $\mathbf{T}_{rep\_A,si,B}$	5	0.47	
X_{70} $\mathbf{T}_{rep\_A,m,B}$	5	0.42	
X_{71} $\mathbf{T}_{exe\_A,eb,B}$	5	0.52	
X_{72} $\mathbf{T}_{exe\_A,si,B}$	5	0.47	
X_{73} $\mathbf{T}_{exe\_A,m,B}$	5	0.42	
X_{74} $\mathbf{T}_{prep\_A,eb,B}$	5	0.52	
X_{75} $\mathbf{T}_{prep\_A,si,B}$	5	0.47	
X_{76} $\mathbf{T}_{prep\_A,m,B}$	5	0.42	
X_{77} $\mathbf{T}_{recall\_mem,eb\_B}$	5	0.5	
X_{78} $\mathbf{T}_{recall\_mem,cog\_B}$	5	0.5	
X_{79} $\mathbf{W}_{mem\_A,eb,old,mem\_A,eb,new}$	5	1	
X_{80} $\mathbf{W}_{mem\_A,cog,old,mem\_A,cog,new}$	5	1	
X_{81} $\mathbf{Con}_A$	5	3.25	
X_{82} $\mathbf{Con}_B$	5	3.25	
X_{83} $\mathbf{H}_{\mathbf{W_A}}$	5	0.7	
X_{84} $\mathbf{H}_{\mathbf{W_B}}$	5	0.7	
X_{85} $\mathbf{H}_{\mathbf{T_A}}$	5	0.7	
X_{86} $\mathbf{H}_{\mathbf{T_B}}$	5	0.7	
X_{87} $\mathbf{H}_{\mathbf{T}\_recall,A}$	5	0.7	
X_{88} $\mathbf{H}_{\mathbf{T}\_recall,B}$	5	0.7	

Table 9 Role matrix **ms** for speed factors and **iv** for initial values

ms speed factors	1	iv initial values	1
X_{1} $\mathrm{ws}_{s,A}$	0		1
X_{2} $\mathrm{ws}_{s,B}$	0		1
X_{3} $\mathrm{sense}_{s,A}$	1		0
X_{4} $\mathrm{sense}_{B,eb,A}$	1		0
X_{5} $\mathrm{sense}_{B,si,A}$	1		0
X_{6} $\mathrm{sense}_{B,m,A}$	1		0
X_{7} $\mathrm{rep}_{s,A}$	1		0
X_{8} $\mathrm{rep}_{B,eb,A}$	1		0
X_{9} $\mathrm{rep}_{B,si,A}$	1		0
X_{10} $\mathrm{rep}_{B,m,A}$	1		0
X_{11} $\mathrm{prep}_{B,eb,A}$	1		0
X_{12} $\mathrm{prep}_{B,si,A}$	1		0
X_{13} $\mathrm{prep}_{B,m,A}$	1		0
X_{14} $\mathrm{intersync}_{B,A,eb}$	0.5		0
X_{15} $\mathrm{intersync}_{B,A,si}$	0.5		0
X_{16} $\mathrm{intersync}_{B,A,m}$	0.5		0
X_{17} $\mathrm{mem}_{A,eb,new}$	0.5		0
X_{18} $\mathrm{mem}_{A,eb,old}$	0.5		0
X_{19} $\mathrm{mem}_{A,cog,new}$	0.5		0
X_{20} $\mathrm{mem}_{A,cog,old}$	0.5		0
X_{21} $\mathrm{recall}_{mem\_eb,A}$	1		0
X_{22} $\mathrm{recall}_{mem\_cog,A}$	1		0
X_{23} $\mathrm{exp\_affect}_{eb,A}$	1		0
X_{24} $\mathrm{sing}_{si,A}$	1		0
X_{25} $\mathrm{move}_{m,A}$	1		0
X_{26} $\mathrm{sense}_{s,B}$	1		0
X_{27} $\mathrm{sense}_{A,eb,B}$	1		0
X_{28} $\mathrm{sense}_{A,si,B}$	1		0
X_{29} $\mathrm{sense}_{A,m,B}$	1		0
X_{30} $\mathrm{rep}_{s,B}$	1		0
X_{31} $\mathrm{rep}_{A,eb,B}$	1		0
X_{32} $\mathrm{rep}_{A,si,B}$	1		0
X_{33} $\mathrm{rep}_{A,m,B}$	1		0
X_{34} $\mathrm{prep}_{A,eb,B}$	1		0
X_{35} $\mathrm{prep}_{A,si,B}$	1		0
X_{36} $\mathrm{prep}_{A,m,B}$	1		0
X_{37} $\mathrm{intersync}_{A,B,eb}$	0.5		0
X_{38} $\mathrm{intersync}_{A,B,si}$	0.5		0
X_{39} $\mathrm{intersync}_{A,B,m}$	0.5		0
X_{40} $\mathrm{mem}_{B,eb,new}$	0.5		0
X_{41} $\mathrm{mem}_{B,eb,old}$	0.5		0
X_{42} $\mathrm{mem}_{B,cog,new}$	0.5		0
X_{43} $\mathrm{mem}_{B,cog,old}$	0.5		0
X_{44} $\mathrm{recall}_{mem\_eb,B}$	1		0
X_{45} $\mathrm{recall}_{mem\_cog,B}$	1		0
X_{46} $\mathrm{exp\_affect}_{eb,B}$	1		0
X_{47} $\mathrm{sing}_{si,B}$	1		0
X_{48} $\mathrm{move}_{m,B}$	1		0
X_{49} $\mathrm{ws}_{eb,A,B}$	1		0
X_{50} $\mathrm{ws}_{si,A,B}$	1		0
X_{51} $\mathrm{ws}_{m,A,B}$	1		0
X_{52} $\mathrm{ws}_{eb,B,A}$	1		0
X_{53} $\mathrm{ws}_{si,B,A}$	1		0
X_{54} $\mathrm{ws}_{m,B,A}$	1		0
X_{55} $\mathbf{T}_{rep\_B,eb,A}$	X_{90}		0.7
X_{56} $\mathbf{T}_{rep\_B,si,A}$	X_{85}		0.7
X_{57} $\mathbf{T}_{rep\_B,m,A}$	X_{85}		0.7
X_{58} $\mathbf{T}_{exe\_B,eb,A}$	X_{85}		0.7
X_{59} $\mathbf{T}_{exe\_B,si,A}$	X_{85}		0.7
X_{60} $\mathbf{T}_{exe\_B,m,A}$	X_{85}		0.7
X_{61} $\mathbf{T}_{prep\_B,eb,A}$	X_{85}		0.7
X_{62} $\mathbf{T}_{prep\_B,si,A}$	X_{85}		0.7
X_{63} $\mathbf{T}_{prep\_B,m,A}$	X_{85}		0.7
X_{64} $\mathbf{T}_{recall\_mem,eb,A}$	X_{87}		0.7
X_{65} $\mathbf{T}_{recall\_mem,cog,A}$	X_{87}		0.7
X_{66} $\mathbf{W}_{mem\_A,eb,old,mem\_A,eb,new}$	X_{83}		0
X_{67} $\mathbf{W}_{mem\_A,cog,old,mem\_A,cog,new}$	X_{83}		0
X_{68} $\mathbf{T}_{rep\_A,eb,B}$	X_{86}		0.7
X_{69} $\mathbf{T}_{rep\_A,si,B}$	X_{86}		0.7
X_{70} $\mathbf{T}_{rep\_A,m,B}$	X_{86}		0.7
X_{71} $\mathbf{T}_{exe\_A,eb,B}$	X_{86}		0.7
X_{72} $\mathbf{T}_{exe\_A,si,B}$	X_{86}		0.7
X_{73} $\mathbf{T}_{exe\_A,m,B}$	X_{86}		0.7
X_{74} $\mathbf{T}_{prep\_A,eb,B}$	X_{86}		0.7
X_{75} $\mathbf{T}_{prep\_A,si,B}$	X_{86}		0.7
X_{76} $\mathbf{T}_{prep\_A,m,B}$	X_{86}		0.7
X_{77} $\mathbf{T}_{recall\_mem,eb\_B}$	X_{88}		0.7
X_{78} $\mathbf{T}_{recall\_mem,cog\_B}$	X_{88}		0.7
X_{79} $\mathbf{W}_{mem\_A,eb,old,mem\_A,eb,new}$	X_{84}		0
X_{80} $\mathbf{W}_{mem\_A,cog,old,mem\_A,cog,new}$	X_{84}		0
X_{81} $\mathbf{Con}_{A}$	0		1
X_{82} $\mathbf{Con}_{B}$	0		1
X_{83} $\mathbf{H}_{\mathbf{W_A}}$	1		0
X_{84} $\mathbf{H}_{\mathbf{W_B}}$	1		0
X_{85} $\mathbf{H}_{\mathbf{T_A}}$	1		0
X_{86} $\mathbf{H}_{\mathbf{T_B}}$	1		0
X_{87} $\mathbf{H}_{\mathbf{T}\_recall,A}$	1		0
X_{88} $\mathbf{H}_{\mathbf{T}\_recall,B}$	1		0

References

1. Abraham, W.C., Bear, M.F.: Metaplasticity: the plasticity of synaptic plasticity. Trends Neurosci. **19**(4), 126–130 (1996)
2. Ageing and health: https://www.who.int/news-room/fact-sheets/detail/ageing-and-health. Last accessed 16 Nov 2023
3. Bain, A.: Mind and body: the theories of their relation. HS King & Company 1873 Farnborough Eng. Gregg International (1873)
4. Bechtel, W., Bich, L.: Organisms need mechanisms; mechanisms need organisms. In: New Mechanism: Explanation, Emergence and Reduction vol. 35, pp. 85–108 (2024)
5. Bich, L., Bechtel, W.: Control mechanisms: explaining the integration and versatility of biological organisms. Adapt. Behav. **30**(5), 389–407 (2022)

6. Bei, Z.A., Hendrikse, S.C.F., Treur, J.: Modeling epigenetic modification due to childhood abuse and its relation to ASPD: a fifth-order adaptive network model. In: Samsonovich, A.V., Liu, T. (eds), Proceedings of 15th Biologically Inspired Cognitive Architectures for AI Conference, BICA*AI'24. Studies in Computational Intelligence, vol 477, pp 59–70. Springer Nature (2024)
7. Brownell, C.A., Carriger, M.S.: Changes in cooperation and self-other differentiation during the second year. Child Dev. **61**(4), 1164–1174 (1990)
8. Brownell, C.A., Ramani, G.B., Zerwas, S.L.: Becoming a social partner with peers: cooperation and social understanding in one-and two-year-olds. Child Dev. **77**(4), 803–821 (2006)
9. Cacioppo, S., Zhou, H., Monteleone, G., Majka, E.A., Quinn, K.A., Ball, A.B., Norman, G.J., Semin, G.R., Cacioppo, J.T.: You are in sync with me: neural correlates of interpersonal synchrony with a partner. Neuroscience **277**, 842–858 (2014)
10. Canbaloğlu, G., Treur, J., Wiewiora, A. (eds.): Computational modeling of multilevel organisational learning and its control using self-modeling network models. Springer Nature (2023)
11. Collie, A., Maruff, P., Shafiq-Antonacci, R., Smith, M., Hallup, M., Schofield, P.R., Currie, J.: Memory decline in healthy older people: implications for identifying mild cognitive impairment. Neurology **56**(11), 1533–1538 (2001)
12. Chetouani, M., Delaherche, E., Dumas, G., Cohen, D.: 15 Interpersonal synchrony: from social perception to social interaction, pp. 202–212. Social Signal Processing Part II—Machine Analysis of Social Signals, Cambridge University Press (2017)
13. Cook Maher, A., Kielb, S., Loyer, E., Connelley, M., Rademaker, A., Mesulam, M.M., Weintraub, S., McAdams, D., Logan, R., Rogalski, E.: Psychological well-being in elderly adults with extraordinary episodic memory. PLoS ONE **12**(10), e0186413 (2017)
14. Coradeschi, S., Cesta, A., Cortellessa, G., Coraci, L., Galindo, C., Gonzalez, J., Ötslund, B.: GiraffPlus: a system for monitoring activities and physiological parameters and promoting social interaction for elderly. In: Human-computer systems interaction: backgrounds and applications vol. 3, pp. 261–271 (2014)
15. Cavallini, E., Cornoldi, C., Vecchi, T.: The effects of age and professional expertise on working memory performance. Appl Cogn Psychol: Off J Soc Appl Res Mem Cogn **23**(3), 382–395 (2009)
16. Cirelli, L.K.: How interpersonal synchrony facilitates early prosocial behavior. Curr. Opin. Psychol. **20**, 35–39 (2018)
17. David, F., Kalibala, G., Pichon, B., Treur, J.: A network model for modulating sensory processing sensitivity in autism spectrum disorder: epigenetics, adaptivity, and other factors. Cogn. Syst. Res. **87**, 101240 (2024)
18. Davis, H.P., Small, S.A., Stern, Y., Mayeux, R., Feldstein, S.N., Keller, F.R.: Acquisition, recall, and forgetting of verbal information in long-term memory by young, middle-aged, and elderly individuals. Cortex **39**(4–5), 1063–1091 (2003)
19. Deary, I.J., Corley, J., Gow, A.J., Harris, S.E., Houlihan, L.M., Marioni, R.E., Penke, L., Rafnsson, S.B., Starr, J.M.: Age-associated cognitive decline. Br. Med. Bull. **92**(1), 135–152 (2009)
20. Dolen, L.S., Bearison, D.J.: Social interaction and social cognition in aging: a contextual analysis. Hum. Dev. **25**(6), 430–442 (1982)
21. Gagnon, M., Dartigues, J.F., Mazaux, J.M., Dequae, L., Letenneur, L., Giroire, J.M., Barberger-Gateau, P.: Self-reported memory complaints and memory performance in elderly French community residents: results of the PAQUID Research Program. Neuroepidemiology **13**(4), 145–154 (1994)
22. Gross, T., Sayama, H.: Adaptive networks. Springer, Berlin Heidelberg (2009)
23. Han, H., Hengyuan, Z., Yonggang, T.: Patterns of social participation and impacts on memory among the older people. Front. Public Health **10**, 963215 (2022)
24. Hebb, D.O.: The organization of behavior: a neuropsychological theory (1949)
25. Hendrikse, S.C.F., Treur, J., Koole, S.L.: Modeling emerging interpersonal synchrony and its related adaptive short-term affliation and long-term bonding: a second-order multi-adaptive neural agent model. Inter. J. Neural Syst **33**(7), 2350038 (2023)

26. Hendrikse, S.C.F., Treur, J., Koole, S.L. (eds.) (2025a). New analysis and modeling directions in social interaction science: emergent multimodal interpersonal synchrony, affiliation, and bonding. Springer Nature (this volume)
27. Hendrikse, S.C.F., Treur, J., Koole, S.L.: Introduction to the role of adaptivity in social interaction science. In: Hendrikse, S.C.F., Treur, J., Koole, S.L. (eds.) New analysis and modeling directions in social interaction science: emergent multimodal interpersonal synchrony, affiliation, and bonding, Ch 1, this volume. Springer Nature (2025b)
28. Hendrikse, S.C.F., Treur, J., Koole, S.L.: The Adaptive interplay of interpersonal synchrony, short-term affiliation and long-term bonding: a multi-timescale multi-adaptive agent model. In: Hendrikse, S.C.F., Treur, J., Koole, S.L. (eds.) New analysis and modeling directions in social interaction science: emergent multimodal interpersonal synchrony, affiliation, and bonding, Ch 12, this volume. Springer Nature (2025c)
29. Hendrikse, S.C.F., Treur, J., Koole, S.L.: A multi-timescale multi-adaptive dynamical system approach for relationship-specific and relationship-independent adaptivity in affiliation and bonding. In: Hendrikse, S.C.F., Treur, J., Koole, S.L. (eds.) New analysis and modeling directions in social interaction science: emergent multimodal interpersonal synchrony, affiliation, and bonding, Ch 13, this volume. Springer Nature (2025d)
30. Hendrikse, S.C.F., Treur, J., Wilderjans, T.F., Dikker, S., Koole, S.L.: On becoming in sync with yourself and others: an adaptive agent model for how persons connect by detecting intrapersonal and interpersonal synchrony. Hum-Centric Intell Syst **3**(2), 123–146 (2023)
31. Hendrikse, S.C.F., Treur, J., Wilderjans, T.F., Dikker, S., Koole, S.L.: Growing connectedness through synchrony: the role of subjective analysis of intrapersonal synchrony and interpersonal synchrony within an adaptive agent model. In: Hendrikse, S.C.F., Treur, J., Koole, S.L. (eds.) New analysis and modeling directions in social interaction science: emergent multimodal interpersonal synchrony, affiliation, and bonding, Ch 11, this volume. Springer Nature (2025)
32. Hu, Y., Cheng, X., Pan, Y., Hu, Y.: The intrapersonal and interpersonal consequences of interpersonal synchrony. Acta Physiol **224**, 103513 (2022)
33. Inda, M.C., Muravieva, E.V., Alberini, C.M.: Memory retrieval and the passage of time: from reconsolidation and strengthening to extinction. J. Neurosci. **31**(5), 1635–1643 (2011)
34. Jahn, H.: Memory loss in Alzheimer's disease. Dialogues Clin. Neurosci. **15**(4) (2013)
35. Kathusing, S., Samhan, N., Treur, J.: Higher-order adaptive dynamical system modeling of the role of epigenetics in anxiety disorders. Cogn. Syst. Res. **83**, 101177 (2024)
36. Keisari, S., Feniger-Schaal, R., Palgi, Y., Golland, Y., Gesser-Edelsburg, A., Ben-David, B.: Synchrony in old age: playing the mirror game improves cognitive performance. Clin. Gerontol. **45**(2), 312–326 (2022)
37. Liang, W., Shediac-Rizkallah, M.C., Celentano, D.D., Rohde, C.: A population-based study of age and gender differences in patterns of health-related behaviors. Am. J. Prev. Med. **17**(1), 8–17 (1999)
38. Loftus, E.F.: Memory. Rowman & Littlefield Publishers (1988)
39. McFarland, D.H.: Respiratory markers of conversational interaction. J. Speech Lang. Hear. Res. **44**(1), 128–143 (2001)
40. Marsh, K.L., Richardson, M.J., Schmidt, R.C.: Social connection through joint action and interpersonal coordination. Top. Cogn. Sci. **1**(2), 320–339 (2009)
41. Meyer, M., Bekkering, H., Paulus, M., Hunnius, S.: Joint action coordination in 2½-and 3-year-old children. Front. Hum. Neurosci. **4**, 220 (2010)
42. Polansky, N., Lippitt, R., Redl, F.: An investigation of behavioral contagion in groups. Hum. Relat. **3**(4), 319–348 (1950)
43. Ramanathan, S., McGill, A.L.: Consuming with others: Social influences on moment-to-moment and retrospective evaluations of an experience. J. Consum. Res. **34**(4), 506–524 (2007)
44. Rennung, M., Goritz, A.S.: Prosocial consequences of interpersonal synchrony. Z. Für Psychol. **224**(3), 168–189 (2016)
45. Robinson, B.L., Harper, N.S., McAlpine, D.: Meta-adaptation in the auditory midbrain under cortical influence. Nat. Commun. **7**, e13442 (2016)

46. Sebanz, N., Bekkering, H., Knoblich, G.: Joint action: bodies and minds moving together. Trends Cogn. Sci. **10**(2), 70–76 (2006)
47. Schlöder, D., Hendrikse, S.C.F., Treur, J.: Higher-order adaptive dynamical system modeling of the role of epigenetics in substance use disorder. In: Samsonovich, A.V., Liu, T. (eds), Proceedings of 15th Biologically Inspired Cognitive Architectures for AI Conference, BICA*AI'24. Studies in Computational Intelligence, vol 477, pp 358–368. Springer (2024)
48. Simmons-Stern, N.R., Budson, A.E., Ally, B.A.: Music as a memory enhancer in patients with Alzheimer's disease. Neuropsychologia **48**(10), 3164–3167 (2010)
49. Simmons-Stern, N.R., Deason, R.G., Brandler, B.J., Frustace, B.S., O'Connor, M.K., Ally, B.A., Budson, A.E.: Music-based memory enhancement in Alzheimer's disease: promise and limitations. Neuropsychologia **50**(14), 3295–3303 (2012)
50. Social isolation, loneliness in older people pose health risks: https://www.nia.nih.gov/news/social-isolation-loneliness-older-people-pose-health-risks. Last accessed 20 Nov 2023
51. Squire, L.R.: Memory and brain systems: 1969–2009. J. Neurosci. **29**(41), 12711–12716 (2009)
52. The Importance of Social Interaction for Seniors: https://thelegacyoffarmington.com/the-importance-of-social-interaction-for-seniors/. Last accessed 19 Nov 2023
53. Thompson, R.F., Kim, J.J.: Memory systems in the brain and localization of a memory. Proc. Natl. Acad. Sci. **93**(24), 13438–13444 (1996)
54. Trainor, L.J., Cirelli, L.: Rhythm and interpersonal synchrony in early social development. Ann. N. Y. Acad. Sci. **1337**(1), 45–52 (2015)
55. Treur, J.: With a little help: a modeling environment for self-modeling network models. In: Treur, J., Van Ments, L. (eds.), Mental models and their dynamics, adaptation, and control: a self-modeling network modeling approach, Ch 17. Springer Nature, Cham (2022)
56. Treur, J.: Adaptive networks at the crossroad of artificial intelligence and formal, biological, medical, and social sciences. In: Rezaei, N. (eds.) Integrated science: science without borders. Integrated science series, vol. 1, pp. 335–375. Springer Nature, Cham (2021)
57. Treur, J.: Network-oriented modeling for adaptive networks: designing higher-order adaptive biological, mental and social network models. Springer Nature (2020a)
58. Treur, J.: Modeling multi-order adaptive processes by self-modeling networks. In: Antonio J. Tallon-Ballesteros, Chen, C.-H. (eds.) Proceedings of the 2nd Internation-al Conference on Machine Learning and Intelligent Systems, MLIS'20. Frontiers in Artificial Intelligence and Applications, vol. 332, pp. 206 –217. IOS Press (2020b)
59. Treur, J.: Modeling coevolution of dynamics, adaptivity and control of adaptivity by self-modeling networks. In: Hendrikse, S.C.F., Treur, J., Koole, S.L. (eds.) New analysis and modeling directions in social interaction science: emergent multimodal interpersonal synchrony, affiliation, and bonding, Ch 2, this volume. Springer Nature (2025)
60. Treur, J., Van Ments, L. (eds.): Mental models and their dynamics, adaptation, and control: a self-modeling network modeling approach. Springer Nature, Cham (2022)
61. Treur, J.: On the dynamics and adaptivity of mental processes: relating adaptive dynamical systems and self-modeling network models by mathematical analysis. Cogn. Syst. Res. **70**, 93–100 (2021)
62. Treur, J.: On structure, dynamics, and adaptivity for biological and mental processes: a higher-order adaptive dynamical system modeling perspective. In: Proceedings of the 46th Annual Meeting of the Cognitive Science Society—Dynamics of Cognition, CogSci'24 https://escholarship.org/uc/item/9911192q (2024a)
63. Treur, J.: Using multilevel temporal factorisation to analyse structure and dynamics for higher-order adaptive and evolutionary processes. In: Nguyen, N.T., et al. (eds.) Proceedings of the 16th International Conference on Computational Collective Intelligence, ICCCI'24. Lecture Notes in AI, vol. 14811, pp. 378–392. Springer Nature (2024b)
64. Valdesolo, P., Ouyang, J., DeSteno, D.: The rhythm of joint action: synchrony promotes cooperative ability. J. Exp. Soc. Psychol. **46**(4), 693–695 (2010)
65. van der Wel, R.P., Becchio, C., Curioni, A., Wolf, T.: Understanding joint action: current theoretical and empirical approaches. Acta Physiol **215**, 103285 (2021)

66. Wallot, S., Mitkidis, P., McGraw, J.J., Roepstorff, A.: Beyond synchrony: joint action in a complex production task reveals beneficial effects of decreased interpersonal synchrony. PLoS ONE **11**(12), e0168306 (2016)
67. Warneken, F., Chen, F., Tomasello, M.: Cooperative activities in young children and chimpanzees. Child Dev. **77**(3), 640–663 (2006)
68. Wells, J.L., Haase, C.M., Rothwell, E.S., Naugle, K.G., Otero, M.C., Brown, C.L., Fredrickson, B.L.: Positivity resonance in long-term married couples: Multimodal characteristics and consequences for health and longevity. J. Pers. Soc. Psychol. **123**(5) 983 (2022)
69. Wiltermuth, S.S., Heath, C.: Synchrony and cooperation. Psychol. Sci. **20**(1), 1–5 (2009)
70. Zhang, Q., Liu, Z., Qian, H., Hu, Y., Gao, X.: Interpersonal competition in elderly couples: a functional near-infrared spectroscopy hyperscanning study. Brain Sci. **13**(4), 600 (2023)

Interpersonal Emotion Regulation via Multimodal Interaction with a Bonding Bot: An Adaptive Network Model

Edgar Eler, Jan Treur, Sophie C. F. Hendrikse, Tara Donker, and Sander L. Koole

Abstract This chapter provides and analyses an adaptive network model and simulated data about interaction of a human with a bonding bot to help guide designers in assessing to what extent mapping human-like mental processes and behaviors on their companion bots has valuable effects. The analysis results found here provide a promising perspective for interpersonal emotion regulation using such human-like bonding bots. The bot's multimodal interactions helped to regulate the human's emotions by effects of emerging synchrony via synchrony-induced bonding on social contagion even under less favorable circumstances such as poor human emotion regulation capabilities.

Keywords Interpersonal emotion regulation · Multimodal interaction · Bonding · Adaptive network model · Human-bot interaction

---

E. Eler · J. Treur (✉)
Social AI Group, Department of Computer Science, Vrije Universiteit Amsterdam, Amsterdam, The Netherlands
e-mail: j.treur@vu.nl

E. Eler
e-mail: e.eler@vu.nl

E. Eler · T. Donker · S. L. Koole
Department of Clinical Psychology, Vrije Universiteit Amsterdam, Amsterdam, The Netherlands
e-mail: t.donker@vu.nl

S. L. Koole
e-mail: s.l.koole@vu.nl

S. C. F. Hendrikse
Department of Methodology and Statistics, Tilburg School of Social and Behavioral Sciences, Tilburg University, Tilburg, The Netherlands
e-mail: S.C.F.Hendrikse@tilburguniversity.edu

S. C. F. Hendrikse et al. (eds.), *New Analysis and Modeling Directions in Social Interaction Science*, Studies in Systems, Decision and Control 614,
https://doi.org/10.1007/978-3-031-99968-0_19

1 Introduction

Since the 2010s, mental health problems have increased substantially with each younger generation in every measured country around the world [6, 20, 46]. For instance, in Australia, reported mental and behavioral disorders increased from 9.6% of Australians aged 15 years old and over in 2001 to 21.4% in 2020/2021: Australian Bureau of Statistics [3], cited by [6]. The reasons for the rise of mental health problems among adolescents and young adults are not completely understood. Nevertheless, it seems clear that the treatment of adolescents and young adults will place great demands on mental health systems, systems that are currently already overburdened. There is thus a pressing need for innovative solutions in mental health services.

One such solution may be the use of artificial agents in mental health services. In recent years, mental health services are increasingly being offered online [4, 8, 14]. However, it appears that only treatments tend to be more effective when they are guided by a therapist [2]. If this guidance could be delivered by an artificial agent, this would make online psychotherapy more scalable, thereby greatly expanding the reach of mental health services.

From research on human psychotherapists, it is known that their effectiveness depends in large part on their ability to form a beneficial working relationship with clients [12]. It thus stands to reason that social-emotional functions would be of considerable utility for artificial agents operating in the mental health domain. Such social-emotional functions are the main focus of the present contribution.

Over the years, strategies to provide personal companionship through AI have been developed, mostly focused on providing specific assistance after voice command inputs [5]. Recent advances in natural language processing (NLP), particularly the emergence of large language models (LLM), have led to numerous efforts to develop personal companions through chatbots created on top of those models [40]. Despite their advanced, near-human semantics comprehension, they still lack embodiment, which is a key aspect of human social-emotional functions [47].

To create embodied bots that bond with humans, it's important to consider modeling an abstraction of the dynamic processes underlying the human mind into the bot's synthetic mind. The work from [13] proposed an abstract model in which a bot had multimodal interaction with a distressed human and supported regulating their emotions through co-regulation [7]. The bot and the human were so-called 'mapped': mirroring their internal processes, and the simulations showed improvements in the human's emotions. This mapping type can prevent uncanny aspects of the bot's reactions that could deprecate an ongoing bonding process in this interpersonal interaction.

Emotions and feelings are claimed to be involved in a cyclic process between the brain and the body, in which a felt emotion is a result of a combination of, amongst other components, the body sensors, the preparation for the expressed emotion, and the expressed emotion itself in the body [43, Ch. 3]. To improve the representation of the bot's synthetic mind, it would be useful to enhance its embodiment by modeling multiple modalities for its communication interfaces with humans and embedding this multimodal network into the cyclic process that builds up the bot's emotions.

Like the human mind, real-world dynamic networks exhibit adaptation. This chapter aims to expand previous work described in [13, 15] by designing a higher-order adaptive network model based on the framework proposed by [44] to simulate both bonding and interpersonal emotion regulation process in multimodal human-bot interaction. The results found by these simulations are promising, showing some level of interpersonal regulation in all experiments. Future work can be developed with additional scenarios and experimental approaches that could apply to the same or an extended model.

2 Interpersonal Emotion Regulation

Emotion regulation refers to how individuals can manage their emotions [19]. When these same individuals try to find help from others to manage their emotions, they are engaging in interpersonal emotion regulation [48]. Social interactions with family, friends, or therapists can be seen as coping mechanisms, and, depending on the circumstances, the mere presence of others can help regulate emotions [7].

Prior research has documented multiple types of interpersonal synchrony [32] and related adaptivity, see also this volume [24, 25] Chapter 1. We focus on the following types in this chapter: movement synchrony, linguistic style matching, acoustic matching, and facial expression synchrony. Each of them has been linked to positive skills from social interaction (and thereby relevant for emotion regulation) in the literature.

Movement Synchrony. Movement synchrony, the coordination of bodily movements between individuals in social interactions, has emerged as a prominent area of investigation across various disciplines, including psychology and neuroscience. Research has revealed the multifaceted role of movement synchrony in shaping interpersonal dynamics and social outcomes. For instance, studies [33] have demonstrated that synchronized movements enhance rapport, cooperation, and affiliation between individuals, fostering a sense of connectedness and mutual understanding. Moreover, studying group behavior has highlighted the significance of movement synchrony in facilitating collective action and cohesion, as evidenced by studies in dance, sports, and music performance (e.g., [30, 37]). Neuroscientific studies (e.g., [31, 36]) have further elucidated some neural mechanisms underlying movement synchrony, implicating brain regions associated with motor coordination, empathy, and social reward processing.

Linguistic Style Matching. Linguistic style matching, the convergence of linguistic patterns between individuals during communication, plays a role in various aspects of social interaction. For instance, studies [29, 42] have demonstrated that individuals tend to adapt their language style, including word choice, syntax, and discourse markers, to align with that of their conversation partners, facilitating rapport and interpersonal connection. Furthermore, linguistic alignment promotes cooperation and cohesion within teams and dyads (e.g., [11, 18]). Neurocognitive studies (e.g., [9, 49]) have also shed light on the neural mechanisms underlying linguistic style matching, implicating brain regions associated with language processing, social cognition, and empathy.

Acoustic Matching. Acoustic matching, the process by which individuals synchronize their vocal characteristics during communication, is crucial in various interpersonal contexts, including turn-taking, emotional expression, and social bonding. [38] has shown that speakers tend to unconsciously adjust their speech characteristics, such as pitch and rhythm, to match those of their conversational partners, facilitating smoother communication and rapport. Moreover, studies into emotional contagion and empathy have underscored the importance of acoustic matching in conveying and understanding affective states during interactions [34]. Additionally, neuroscientific inquiries (e.g., [1, 17]) have provided insights into the neural mechanisms underlying acoustic matching, highlighting the involvement of brain regions associated with social cognition, auditory processing, and emotion regulation.

Facial Expression Synchrony. Facial expression synchrony, the phenomenon wherein individuals mirror each other's facial expressions during social interactions, is also related to multiple social dynamics. For example, individuals naturally synchronize their facial expressions with those of their interaction partners, leading to increased rapport, empathy, and social bonding [28, 39]. Neuroscientific studies (e.g., [35, 41]) have further unraveled the neural mechanisms underlying facial expression synchrony, implicating brain regions associated with empathy, social cognition, and mirror neuron systems.

Emotion Regulation. In the model introduced in the current chapter, emotion generation is conceptualized in a network-oriented manner by an emotion-feeling cycle that involves an as-if body loop for internal simulation of the body's responses to a stimulus and the felt emotion [10, 43, Ch. 3]. For a constant environment, this cyclic process leads to an equilibrium state for the emotional response (preparation) and feeling. Note that the emotion considered here is considered a negative emotion for which regulation is desirable. Therefore, in addition individual emotion regulation is incorporated, involving a control state within the network that (by incoming monitoring connections) is activated when a too high level of emotion and feeling occurs. This activation in turn takes care for individual emotion regulation, which is based on suppression here, for example, see [43], Sect. 3.3.

3 Adaptive Network Modeling

The model proposed by this chapter follows the adaptive network modeling approach [43, 44], by which a temporal-causal network model is represented by *nodes* Y with activation value $Y(t)$ at time t; they are also called *states*. States can be connected and affect each other, via the network structure characterized by:

- *connectivity characteristics*: connections from a state X to a state Y and their *weights* $\boldsymbol{\omega}_{X,Y}$

Table 1 Combination functions from the library used here

Name	Notation	Formula	Parameters
Euclidean	$\mathbf{eucl}_{n,\lambda}(V_1, \ldots, V_k)$	$\sqrt[n]{\frac{V_1^n+\cdots+V_k^n}{\lambda}}$	Order n Scaling factor λ
Advanced logistic sum	$\mathbf{alogistic}_{\sigma,\tau}(V_1, \ldots, V_k)$	$[\frac{1}{1+e^{-\sigma(V_1+\cdots+V_k-\tau)}} - \frac{1}{1+e^{\sigma\tau}}]\,(1 + e^{-\sigma\tau})$	Steepness $\sigma > 0$ Threshold τ
Complemental difference	$\mathbf{compdiff}(V_1, V_2)$	0 if $V_1 = V_2 = 0$ $1 - \frac{\lvert V_1-V_2\rvert}{\max(V_1,V_2)}$ else	–
Random stepmod	$\mathbf{randstepmod}_{\rho,\delta}(V)$	0 if $0 \le$ time t mod $\rho \le \delta$ aV + (1-a) rand(1,1) else	Repitition ρ Step time δ
Random stepmodopp	$\mathbf{randstepmodopp}_{\rho,\delta}(V)$	0 if $\delta \le$ time t mod $\rho \le \rho$ aV + (1-a) rand(1,1) else	Repitition ρ Step time δ

- *aggregation characteristics*: for any state Y, some *combination function* $\mathbf{c}_Y(..)$ defines aggregation that is applied to the *single impacts* $\boldsymbol{\omega}_{X_i,Y}X_i(t)$ on Y from its incoming connections from states $X_1, \ldots, X_k$
- timing characteristics: each state Y has a *speed factor* $\boldsymbol{\eta}_Y$ defining how fast it changes upon a given impact

The dynamics of such networks are described by the following canonical difference (or differential) equation that incorporates the network characteristics $\boldsymbol{\omega}_{X,Y}$, $\mathbf{c}_Y(..)$, $\boldsymbol{\eta}_Y$:

$$Y(t + \Delta t) = Y(t) + \boldsymbol{\eta}_Y[\mathbf{c}_Y(\boldsymbol{\omega}_{X_1,Y}X_1(t), \ldots, \boldsymbol{\omega}_{X_k,Y}X_k(t)) - Y(t)]\Delta t \tag{1}$$

for any state Y and where $X_1 \ldots X_k$ are the states from which Y gets its incoming connections. The Eq. (1) is useful for simulation purposes and also for analysis of properties of the emerging behaviour of such network models. The overall combination function $\mathbf{c}_Y(..)$ for state Y is taken as the weighted average of some of the available basic combination functions $\mathbf{c}_j(..)$ by specified weights $\boldsymbol{\gamma}_{j,Y}$, and parameters $\boldsymbol{\pi}_{1,j,Y}$, $\boldsymbol{\pi}_{2,j,Y}$ of $\mathbf{c}_j(..)$, for Y:

$$\mathbf{c}_Y(V_1, \ldots, V_k) = \frac{\boldsymbol{\gamma}_{1,Y}\mathrm{c}_1(V_1, \ldots, V_k) + \cdots + \boldsymbol{\gamma}_{m,Y}\mathrm{c}_m(V_1, \ldots, V_k)}{\boldsymbol{\gamma}_{1,Y} + \cdots + \boldsymbol{\gamma}_{m,Y}} \tag{2}$$

The Eqs. (1), (2) are part the available dedicated software environment; see [44], Ch 9. This software environment is freely downloadable from URL: https://www.researchgate.net/publication/368775720.

For the combination functions used here, see Table 1. For more details of the modeling approach, see [44] or this volume [24], Chapter 2 [45].

4 The Adaptive Network Model for Human-Bot Interaction

The multimodal network model designed for this chapter aims at the analysis of social interaction between a human and a bot and in particular how interpersonal emotion regulation can enhance individual emotion regulation. It works as an adaptive dynamical system that simulates the inner mental processes of the human and the bot's synthetic mental processes where via sensor and actuator states human-bot interaction takes place.

4.1 Proposed Scenario

The base scenario addressed in this chapter involves a distressed human interacting with a companion bot that helps them regulate their emotions, among others, by creating bonding. To model this scenario, the two agents, human and bot, have the same network architecture, with slight variations in certain characteristics for the human to cover higher distress levels. These variations will be described in Sect. 5.

As illustrated by Fig. 1, the adaptive network model has three levels: one base level and two adaptation levels. This base level is partially inspired by the work of [22], most prominently using interpersonal synchrony detectors; see also ([24], this volume), Chapter 12 and 13 [26, 27]. In this model, the ellipses represent the states: pink (base level), blue (first adaptation level), and purple (second adaptation level). The diamonds on the base level indicate states that represent interpersonal synchrony detectors. The black links represent the same-level connections between the states, with the dashed lines representing suppression or monitoring effects. Links with two arrows represent a bi-directional connection between the states. The red downward links represent the adaptivity of a particular characteristic on the impacted state, like connection weight, excitability threshold, or speed factor. The blue upward links represent the input the adaptation states receive from the base-level states. The role matrices containing the detailed specification for this network can be found in the Appendix.

4.2 Base Level

The model's base level covers a mental network to simulate the mind/body processes of each agent (human and bot) and a social network to simulate the interactions between those two agents, as illustrated by Fig. 2. Those networks have the same structure for both agents. The mental networks include a set of sensor states (*sense_* prefix) that compose the multimodal communication from the other agent. The modalities modeled for each agent are verbal, tone of voice, world (context that the agent brings to the interaction), facial expression, and movement. Each sensor

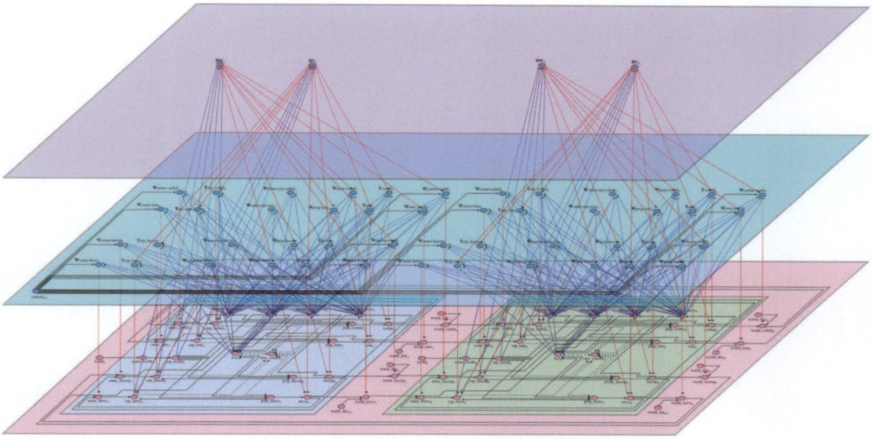

Fig. 1 Overall adaptive network model with one base level (pink plane), containing the human (blue area) and bot (green area), and two adaptation levels (cyan and purple planes)

state connects to a corresponding representation state (*rep_* prefix) that in turn have connections to respective preparation states (*prep_* prefix). The modalities involved the interaction with the other agent (verbal, tone, facial expression, and movement) also have their related preparation and execution (no prefix) states. The preparation states do not only receive influences from their respective representation states, but as well from the world stimulus representation state and the feeling state. Moreover, they receive suppression influences from the individual emotion regulation control state. The execution states only receive influences from their respective preparation states.

The feeling state (*feel*) receives (positive) influences from all preparation states and suppression influences from the emotion regulation control state; keep in mind that we consider a negative emotion and feeling here. The emotion regulation control state receives positive influences from the feeling state, all preparation states and the representation states for verbal, tone, facial expression, and movement.

The social interaction involves world states affected by each agent. For each execution state, a respective world state (*world_* prefix) is modeled. To simulate potential noise in the interface between the agents, disturbance states (*world_dist* prefix) were also modeled. Thus, the specific world states receive influences from their respective execution and disturbance states.

4.3 Adaptation Levels

The model's two adaptation levels are illustrated in Fig. 3. Here, the upward links from the base level were hidden for better visibility. The first-order adaptation level

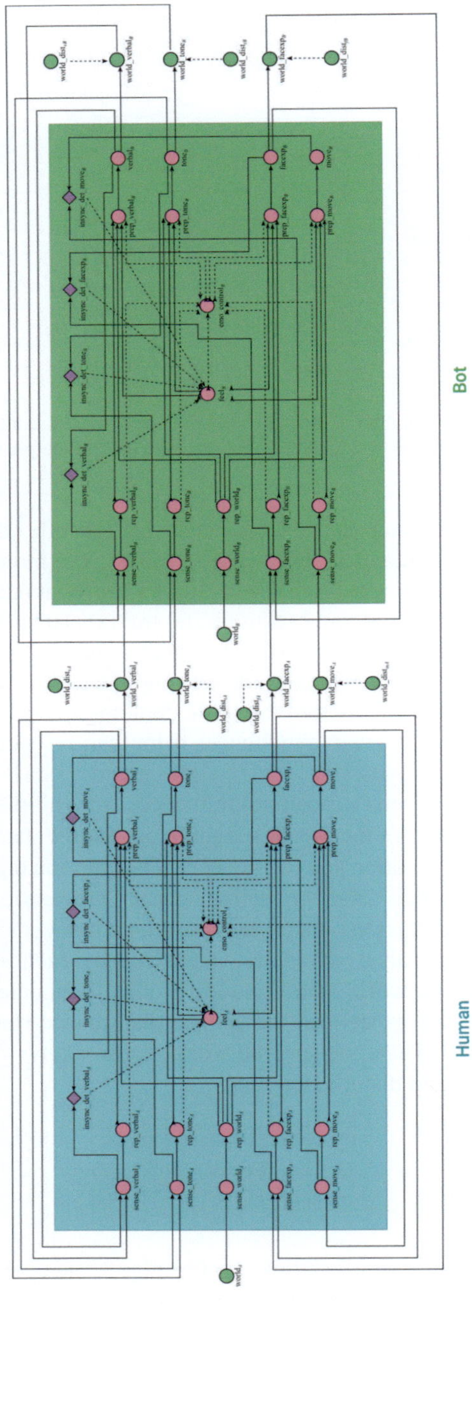

Fig. 2 A 2D representation of the base level, with the human in the blue area and the bot in the green area

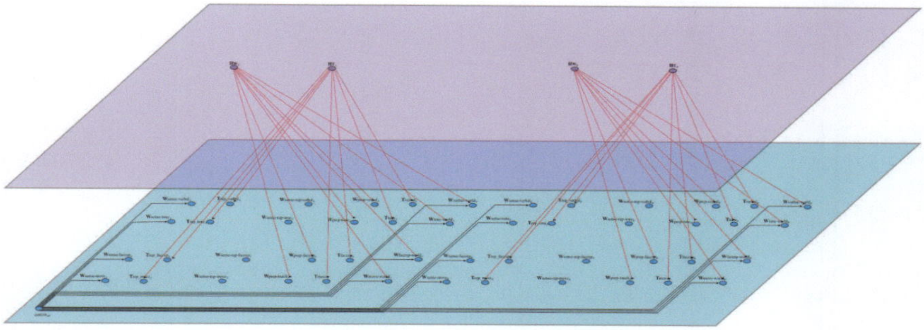

Fig. 3 The two adaptation levels of the proposed model, with the cyan plane representing the first-order adaptation level and the purple plane representing the second-order adaptation level

(cyan plane) via the self-modeling principle (also called reification) [44] models the adaptation of characteristics from the base-level network, with first-order self-model **W**-states $\mathbf{W}_{X,Y}$ representing the adaptive weights $\boldsymbol{\omega}_{X,Y}$ of base level connections $X \rightarrow Y$ and first-order self-model **T**-states $\mathbf{T}_Y$ representing the adaptive excitability thresholds $\boldsymbol{\tau}_Y$ (of the advanced logistic combination functions) in base level states Y. This first level also includes a *context* state, which represents how the presence of both agents in the same interaction impacts the adaptation of the sensor states and the modal world states. The second-order adaptation level (purple plane) includes second-order self-model $\mathbf{H}_\mathbf{W}$-states $\mathbf{H}_{\mathbf{W}_A}$, $\mathbf{H}_{\mathbf{W}_B}$ and $\mathbf{H}_\mathbf{T}$-states $\mathbf{H}_{\mathbf{T}_A}$, $\mathbf{H}_{\mathbf{T}_B}$ which represent the speed factors $\boldsymbol{\eta}_{\mathbf{W}_{X,Y}}$ and $\boldsymbol{\eta}_{\mathbf{T}_Y}$ for self-model **W**-states and self-model **T**-states from the first adaptation level of the two agents.

5 Simulation Results

The scenario described in Sect. 4.1 has been analyzed following a 2 × 2 design, as described in Table 2. In this analysis, two independent characteristics of the human network will be considered: the effect of Individual Emotion Regulation Control and the effect of detected Interpersonal Synchrony (In-Sync). The objective of this analysis is to estimate to what extent a bot can help a human regulate their emotions even with the human's lower levels of individual emotion regulation control and/or effects from In-Sync.

Table 2 2 × 2 design for the simulation experiments

	Regular individual emotion control (RC)	Low individual emotion control (LC)
Regular In-sync effect (RI)	RC x RI (Sect. 5.1)	LC x RI (Sect. 5.2)
Low In-sync effect (LI)	RC × LI (Sect. 5.3)	LC × LI (Sect. 5.4)

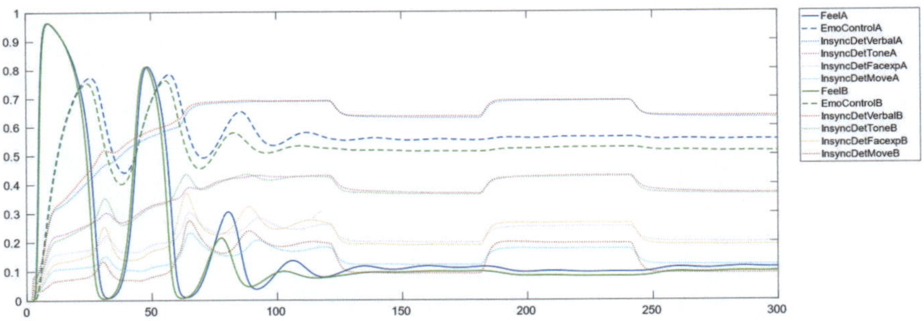

Fig. 4 Results for the simulation of the human with regular individual emotion control and regular In-Sync effects

Low individual emotion regulation control will be simulated by reducing the influences that reach the human's emotion regulation control state ($emo\_control_A$). Low In-Sync effects will be simulated by reducing the suppression effect that the In-Sync detectors ($insync\_det\_\{modal\}_A$) apply to the feeling state of the human ($feel_A$). The network characteristic to customize the effects for each analysis is the connection weight of the incoming connections at $emo\_control_A$ (all links) and $feel_A$ (In-Sync links). Note that synchronization detection still takes place and still has its effects on adaptation.

The simulations analyze the effects in interaction sessions of 300 time units (assumed seconds here). In the simulation results (Figs. 4, 5, 6 and 7), the solid curves represent the feeling states, the dashed curves represent the individual emotion regulation control states, and the dotted curves represent the In-Sync states. The stroke color for the curves for the feeling states and the control states is set to blue for the human and green for the bot. More detailed results, considering all states, can be found in the Appendix.

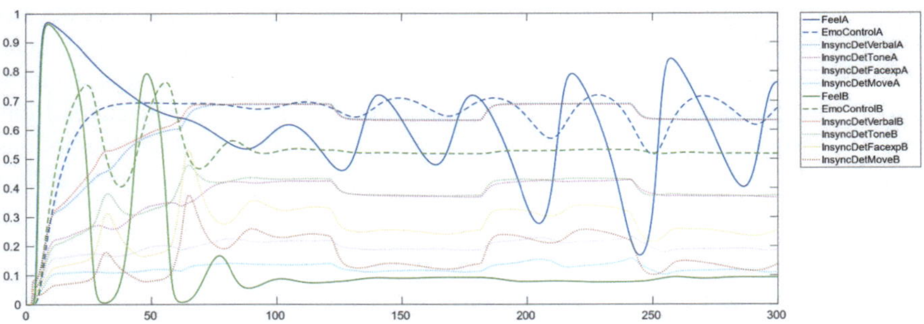

Fig. 5 Results for the simulation of the human with low individual emotion control and regular In-Sync effects

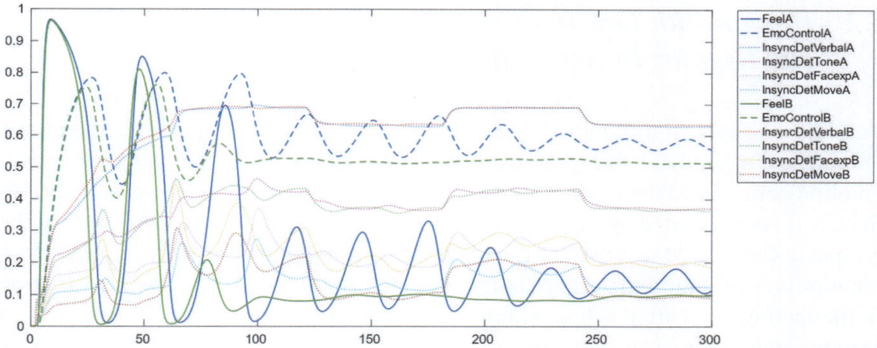

Fig. 6 Results for the simulation of the human with regular individual emotion control and low In-Sync effects

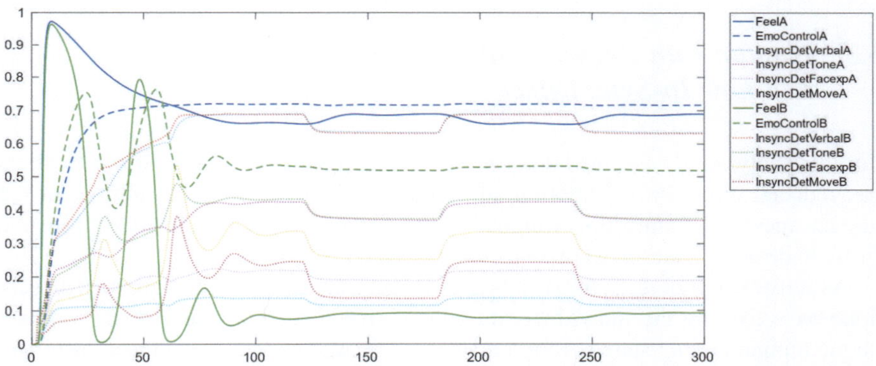

Fig. 7 Results for the simulation of the human with low individual emotion control and low In-Sync effects

5.1 Human with Regular Individual Emotion Control and Regular In-Sync Effects

In this simulation, the human and the bot are configured with the same values for the individual emotion control and the In-Sync effects. As illustrated by Fig. 4, the results show a significant impact on the feeling states, with strong fluctuations in the first 60 s, mostly due to the sensitive activation of the individual emotion control states. From 100 s, the feeling gradually stabilizes at a low level, around 0.1, thus demonstrating successful emotion regulation.

5.2 Human with Low Individual Emotion Control and Regular In-Sync Effect

For this analysis, it's assumed that the human has issues regulating their own emotions. To simulate that, the incoming connections to the human's individual emotion control state were reduced by 20% compared to the simulation described in Sect. 5.1. As illustrated by Fig. 5, there is a stable reduction in the feeling state of the human in the first 90 s, mostly impacted by the combination of the lower individual emotion control state and the regular In-Sync effects, followed by a stronger variation in the feeling state after 100 s. Although this demonstrates a successful short-term emotion regulation, over time, the feeling state of the human becomes more chaotic around level 0.5 or 0.55 as now it is more dependent on external stimuli which are not constant here.

5.3 Human with Regular Individual Emotion Control and Low In-Sync Effect

For this analysis, it's assumed that for the human agent synchronization with others has reduced effect on its emotion and feeling. To simulate that condition, in the human agent the connection weights of the outgoing suppression links from the In-Sync detector states to the feeling state are reduced by 50%.

As illustrated by Fig. 6, the results show successful emotion regulation, primarily done by activating the individual emotion control state. This has a deeper impact on the human feeling state before the 100 s of simulation and maintains a sustained variation of around 0.1 after 200 s.

5.4 Human with Low Individual Emotion Control and Low In-Sync Effects

For this analysis, it is assumed that both the human's capacity to control their own emotions and the effects of synchronization with others are reduced. As done in Sects. 5.2 and 5.3, the low individual emotion control is achieved, respectively, by reducing the connection weights of the incoming links to the emotion control state by 20% and the low effect of synchronization by reducing the connection weights of the outgoing suppression links from the In-Sync detector states to the feeling state by 50%. Both changes are done only in the human network.

As illustrated in Fig. 7, the results show a slow decrease in the human feeling state up to 100 s, then a stable, high level (around 0.65) for the feeling state curve, slightly impacted by the In-Sync variations, with negligible reaction from the individual emotion control state.

6 Discussion

This chapter examined to what extent human-like mental processes and behaviors can be mapped onto their companion bots to enable (1) a form of synchrony-induced bonding between human and bot and (2) interpersonal emotion regulation via the bot's individual emotion regulation capability and social contagion supported by the bonding. The analysis of our simulated data provides a framework for interpersonal emotion regulation using human-like bonding bots. The bonding bot's interactions through its multimodal interfaces helped to regulate the human's emotions by interpersonal synchronization and bonding effects even under less favorable situations. This chapter is mostly based on material from [16].

The proposed adaptive network model was designed in line with the computational architecture [21] that was developed through earlier case studies on synchrony-induced bonding such as reported in [22, 23], see also ([24], this volume) Chapters 12, 13 and 20.

Across all four analyses of the conducted simulation experiments, some level of emotion regulation was found. In all of them, there was a combined effect of the individual (intrapersonal) emotion regulation (*emo_control* states) and the interpersonal emotion regulation via social contagion influenced by the extent of bonding, although with different intensities. The results can help designers make proper decisions about the design and characteristics of their agents toward more human-like synthetic mental processes for bonding bots.

Future work can establish which modalities are more effective for interpersonal synchrony and, thus, for regulating human emotions. It could also analyze whether a stable or a low-variant curve is more realistic for representing the bonding bot's human-like emotion regulation. Ultimately, we hope that this work will pave the way for artificial agents with advanced social-emotional functions, of the kinds that may be used, for instance, in guiding online psychotherapy [2].

Appendix

Role Matrices

Shared Role Matrices

mb	base connectivity	1	2	3	4	5	6	7	8	9
X_1	world_a	X_1								
X_2	sense_world_a	X_1								
X_3	sense_verbal_a	X_{18}	X_{30}							
X_4	sense_tone_a	X_{19}	X_{31}							
X_5	sense_facexp_a	X_{20}	X_{32}							
X_6	sense_move_a	X_{21}	X_{33}							
X_7	rep_world_a	X_2								
X_8	rep_verbal_a	X_3	X_{14}							
X_9	rep_tone_a	X_4	X_{15}							
X_{10}	rep_facexp_a	X_5	X_{16}							
X_{11}	rep_move_a	X_6	X_{17}							
X_{12}	feel_a	X_{13}	X_{14}	X_{15}	X_{16}	X_{17}	X_{22}	X_{23}	X_{24}	X_{25}
X_{13}	control_a	[illegible]	[illegible]	[illegible]	X_{11}	X_{12}	[illegible]	[illegible]	[illegible]	[illegible]
X_{14}	prep_verbal_a	X_7	X_8	X_{12}	X_{13}					
X_{15}	prep_tone_a	X_7	X_9	X_{12}	X_{13}					
X_{16}	prep_facexp_a	X_7	X_{10}	X_{12}	X_{13}					
X_{17}	prep_move_a	X_7	X_{11}	X_{12}	X_{13}					
X_{18}	verbal_a	X_{14}								
X_{19}	tone_a	X_{15}								
X_{20}	facexp_a	X_{16}								
X_{21}	move_a	X_{17}								
X_{22}	insync_det_verbal_a	X_3	X_{18}							
X_{23}	insync_det_tone_a	X_4	X_{19}							
X_{24}	insync_det_facexp_a	X_5	X_{20}							
X_{25}	insync_det_move_a	X_6	X_{21}							
X_{26}	world_verbal_a	X_{18}	X_{22}							
X_{27}	world_tone_a	X_{19}	X_{23}							
X_{28}	world_facexp_a	X_{20}	X_{24}							
X_{29}	world_move_a	X_{21}	X_{25}							
X_{30}	world_dist_va	X_{26}								
X_{31}	world_dist_ta	X_{27}								
X_{32}	world_dist_fa	X_{28}								
X_{33}	world_dist_ma	X_{29}								
X_{34}	world_b	X_{34}								
X_{35}	sense_world_b	X_{34}								
X_{36}	sense_verbal_b	X_{51}	X_{63}							
X_{37}	sense_tone_b	X_{52}	X_{64}							
X_{38}	sense_facexp_b	X_{53}	X_{65}							
X_{39}	sense_move_b	X_{54}	X_{66}							
X_{40}	rep_world_b	X_{35}								
X_{41}	rep_verbal_b	X_{36}	X_{47}							
X_{42}	rep_tone_b	X_{37}	X_{48}							
X_{43}	rep_facexp_b	X_{38}	X_{49}							
X_{44}	rep_move_b	X_{39}	X_{50}							
X_{45}	feel_b	X_{46}	X_{47}	X_{48}	X_{49}	X_{50}	X_{55}	X_{56}	X_{57}	X_{58}
X_{46}	control_b	[illegible]	[illegible]	[illegible]	X_{44}	X_{45}	[illegible]	[illegible]	[illegible]	[illegible]
X_{47}	prep_verbal_b	X_{40}	X_{41}	X_{45}	X_{46}					
X_{48}	prep_tone_b	X_{40}	X_{42}	X_{45}	X_{46}					
X_{49}	prep_facexp_b	X_{40}	X_{43}	X_{45}	X_{46}					
X_{50}	prep_move_b	X_{40}	X_{44}	X_{45}	X_{46}					
X_{51}	verbal_b	X_{47}								
X_{52}	tone_b	X_{48}								
X_{53}	facexp_b	X_{49}								
X_{54}	move_b	X_{50}								
X_{55}	insync_det_verbal_b	X_{36}	X_{51}							
X_{56}	insync_det_tone_b	X_{37}	X_{52}							
X_{57}	insync_det_facexp_b	X_{38}	X_{53}							
X_{58}	insync_det_move_b	X_{39}	X_{54}							
X_{59}	world_verbal_b	X_{51}	X_{55}							
X_{60}	world_tone_b	X_{52}	X_{56}							
X_{61}	world_facexp_b	X_{53}	X_{57}							
X_{62}	world_move_b	X_{54}	X_{58}							
X_{63}	world_dist_vb	X_{59}								
X_{64}	world_dist_tb	X_{60}								
X_{65}	world_dist_fb	X_{61}								
X_{66}	world_dist_mb	X_{62}								
X_{67}	w_sense_verbal_a	X_{22}	X_{23}	X_{24}	X_{25}	X_{115}				
X_{68}	w_sense_tone_a	X_{22}	X_{23}	X_{24}	X_{25}	X_{115}				
X_{69}	w_sense_facexp_a	X_{22}	X_{23}	X_{24}	X_{25}	X_{115}				
X_{70}	w_sense_move_a	X_{22}	X_{23}	X_{24}	X_{25}	X_{115}				
X_{71}	t_rep_verbal_a	X_{12}	X_{22}	X_{23}	X_{24}	X_{25}				
X_{72}	t_rep_tone_a	X_{12}	X_{22}	X_{23}	X_{24}	X_{25}				
X_{73}	t_rep_facexp_a	X_{12}	X_{22}	X_{23}	X_{24}	X_{25}				
X_{74}	t_rep_move_a	X_{12}	X_{22}	X_{23}	X_{24}	X_{25}				
X_{75}	w_sense_rep_verbal_a	X_{22}	X_{23}	X_{24}	X_{25}					
X_{76}	w_sense_rep_tone_a	X_{22}	X_{23}	X_{24}	X_{25}					
X_{77}	w_sense_rep_facexp_a	X_{22}	X_{23}	X_{24}	X_{25}					
X_{78}	w_sense_rep_move_a	X_{22}	X_{23}	X_{24}	X_{25}					
X_{79}	w_prep_verbal_a	X_{22}	X_{23}	X_{24}	X_{25}					
X_{80}	w_prep_tone_a	X_{22}	X_{23}	X_{24}	X_{25}					
X_{81}	w_prep_facexp_a	X_{22}	X_{23}	X_{24}	X_{25}					
X_{82}	w_prep_move_a	X_{22}	X_{23}	X_{24}	X_{25}					
X_{83}	t_verbal_a	X_{12}	X_{22}	X_{23}	X_{24}	X_{25}				
X_{84}	t_tone_a	X_{12}	X_{22}	X_{23}	X_{24}	X_{25}				
X_{85}	t_facexp_a	X_{12}	X_{22}	X_{23}	X_{24}	X_{25}				
X_{86}	t_move_a	X_{12}	X_{22}	X_{23}	X_{24}	X_{25}				
X_{87}	w_verbal_world_a	X_{22}	X_{23}	X_{24}	X_{25}	X_{115}				
X_{88}	w_tone_world_a	X_{22}	X_{23}	X_{24}	X_{25}	X_{115}				
X_{89}	w_facexp_world_a	X_{22}	X_{23}	X_{24}	X_{25}	X_{115}				
X_{90}	w_move_world_a	X_{22}	X_{23}	X_{24}	X_{25}	X_{115}				
X_{91}	w_sense_verbal_b	X_{55}	X_{56}	X_{57}	X_{58}	X_{115}				
X_{92}	w_sense_tone_b	X_{55}	X_{56}	X_{57}	X_{58}	X_{115}				
X_{93}	w_sense_facexp_b	X_{55}	X_{56}	X_{57}	X_{58}	X_{115}				
X_{94}	w_sense_move_b	X_{55}	X_{56}	X_{57}	X_{58}	X_{115}				
X_{95}	t_rep_verbal_b	X_{45}	X_{55}	X_{56}	X_{57}	X_{58}				
X_{96}	t_rep_tone_b	X_{45}	X_{55}	X_{56}	X_{57}	X_{58}				
X_{97}	t_rep_facexp_b	X_{45}	X_{55}	X_{56}	X_{57}	X_{58}				
X_{98}	t_rep_move_b	X_{45}	X_{55}	X_{56}	X_{57}	X_{58}				
X_{99}	w_sense_rep_verbal_b	X_{55}	X_{56}	X_{57}	X_{58}					
X_{100}	w_sense_rep_tone_b	X_{55}	X_{56}	X_{57}	X_{58}					
X_{101}	w_sense_rep_facexp_b	X_{55}	X_{56}	X_{57}	X_{58}					
X_{102}	w_sense_rep_move_b	X_{55}	X_{56}	X_{57}	X_{58}					
X_{103}	w_prep_verbal_b	X_{55}	X_{56}	X_{57}	X_{58}					
X_{104}	w_prep_tone_b	X_{55}	X_{56}	X_{57}	X_{58}					
X_{105}	w_prep_facexp_b	X_{55}	X_{56}	X_{57}	X_{58}					
X_{106}	w_prep_move_b	X_{55}	X_{56}	X_{57}	X_{58}					
X_{107}	t_verbal_b	X_{45}	X_{55}	X_{56}	X_{57}	X_{58}				
X_{108}	t_tone_b	X_{45}	X_{55}	X_{56}	X_{57}	X_{58}				
X_{109}	t_facexp_b	X_{45}	X_{55}	X_{56}	X_{57}	X_{58}				
X_{110}	t_move_b	X_{45}	X_{55}	X_{56}	X_{57}	X_{58}				
X_{111}	w_verbal_world_b	X_{55}	X_{56}	X_{57}	X_{58}	X_{115}				
X_{112}	w_tone_world_b	X_{55}	X_{56}	X_{57}	X_{58}	X_{115}				
X_{113}	w_facexp_world_b	X_{55}	X_{56}	X_{57}	X_{58}	X_{115}				
X_{114}	w_move_world_b	X_{55}	X_{56}	X_{57}	X_{58}	X_{115}				
X_{115}	context_ab	X_{111}								
X_{116}	hw_a	X_7	X_8	X_9	X_{10}	X_{11}				
X_{117}	ht_a	X_7	X_8	X_9	X_{10}	X_{11}				
X_{118}	hw_b	X_{40}	X_{41}	X_{42}	X_{43}	X_{44}				
X_{119}	ht_b	X_{40}	X_{41}	X_{42}	X_{43}	X_{44}				

msv	speed factors	1
X_1	world_a	0
X_2	sense_world_a	1
X_3	sense_verbal_a	1
X_4	sense_tone_a	1
X_5	sense_facexp_a	1
X_6	sense_move_a	1
X_7	rep_world_a	1
X_8	rep_verbal_a	1
X_9	rep_tone_a	1
X_{10}	rep_facexp_a	1
X_{11}	rep_move_a	1
X_{12}	feel_a	1
X_{13}	control_a	0.1
X_{14}	prep_verbal_a	1
X_{15}	prep_tone_a	1
X_{16}	prep_facexp_a	1
X_{17}	prep_move_a	1
X_{18}	verbal_a	1
X_{19}	tone_a	1
X_{20}	facexp_a	1
X_{21}	move_a	1
X_{22}	insync_det_verbal_a	0.5
X_{23}	insync_det_tone_a	0.5
X_{24}	insync_det_facexp_a	0.5
X_{25}	insync_det_move_a	0.5
X_{26}	world_verbal_a	1
X_{27}	world_tone_a	1
X_{28}	world_facexp_a	1
X_{29}	world_move_a	1
X_{30}	world_dist_va	1
X_{31}	world_dist_ta	1
X_{32}	world_dist_fa	1
X_{33}	world_dist_ma	1
X_{34}	world_b	0
X_{35}	sense_world_b	1
X_{36}	sense_verbal_b	1
X_{37}	sense_tone_b	1
X_{38}	sense_facexp_b	1
X_{39}	sense_move_b	1
X_{40}	rep_world_b	1
X_{41}	rep_verbal_b	1
X_{42}	rep_tone_b	1
X_{43}	rep_facexp_b	1
X_{44}	rep_move_b	1
X_{45}	feel_b	1
X_{46}	control_b	0.1
X_{47}	prep_verbal_b	1
X_{48}	prep_tone_b	1
X_{49}	prep_facexp_b	1
X_{50}	prep_move_b	1
X_{51}	verbal_b	1
X_{52}	tone_b	1
X_{53}	facexp_b	1
X_{54}	move_b	1
X_{55}	insync_det_verbal_b	0.5
X_{56}	insync_det_tone_b	0.5
X_{57}	insync_det_facexp_b	0.5
X_{58}	insync_det_move_b	0.5
X_{59}	world_verbal_b	1
X_{60}	world_tone_b	1
X_{61}	world_facexp_b	1
X_{62}	world_move_b	1
X_{63}	world_dist_vb	1
X_{64}	world_dist_tb	1
X_{65}	world_dist_fb	1
X_{66}	world_dist_mb	1
X_{67}	w_sense_verbal_a	0
X_{68}	w_sense_tone_a	0
X_{69}	w_sense_facexp_a	0
X_{70}	w_sense_move_a	0
X_{71}	t_rep_verbal_a	X_{117}
X_{72}	t_rep_tone_a	X_{117}
X_{73}	t_rep_facexp_a	X_{117}
X_{74}	t_rep_move_a	X_{117}
X_{75}	w_sense_rep_verbal_a	X_{116}
X_{76}	w_sense_rep_tone_a	X_{116}
X_{77}	w_sense_rep_facexp_a	X_{116}
X_{78}	w_sense_rep_move_a	X_{116}
X_{79}	w_prep_verbal_a	X_{116}
X_{80}	w_prep_tone_a	X_{116}
X_{81}	w_prep_facexp_a	X_{116}
X_{82}	w_prep_move_a	X_{116}
X_{83}	t_verbal_a	X_{117}
X_{84}	t_tone_a	X_{117}
X_{85}	t_facexp_a	X_{117}
X_{86}	t_move_a	X_{117}
X_{87}	w_verbal_world_a	X_{116}
X_{88}	w_tone_world_a	X_{116}
X_{89}	w_facexp_world_a	X_{116}
X_{90}	w_move_world_a	X_{116}
X_{91}	w_sense_verbal_b	0
X_{92}	w_sense_tone_b	0
X_{93}	w_sense_facexp_b	0
X_{94}	w_sense_move_b	0
X_{95}	t_rep_verbal_b	X_{119}
X_{96}	t_rep_tone_b	X_{119}
X_{97}	t_rep_facexp_b	X_{119}
X_{98}	t_rep_move_b	X_{119}
X_{99}	w_sense_rep_verbal_b	X_{118}
X_{100}	w_sense_rep_tone_b	X_{118}
X_{101}	w_sense_rep_facexp_b	X_{118}
X_{102}	w_sense_rep_move_b	X_{118}
X_{103}	w_prep_verbal_b	X_{118}
X_{104}	w_prep_tone_b	X_{118}
X_{105}	w_prep_facexp_b	X_{118}
X_{106}	w_prep_move_b	X_{118}
X_{107}	t_verbal_b	X_{119}
X_{108}	t_tone_b	X_{119}
X_{109}	t_facexp_b	X_{119}
X_{110}	t_move_b	X_{119}
X_{111}	w_verbal_world_b	X_{118}
X_{112}	w_tone_world_b	X_{118}
X_{113}	w_facexp_world_b	X_{118}
X_{114}	w_move_world_b	X_{118}
X_{115}	context_ab	0
X_{116}	hw_a	1
X_{117}	ht_a	1
X_{118}	hw_b	1
X_{119}	ht_b	1

mcfwv	combination function weights	1 eucl	2 alogistic	3 compdiff	4 randstepmod	5 randstepmodopp
X_1	world_a					1
X_2	sense_world_a	1				
X_3	sense_verbal_a		1			
X_4	sense_tone_a		1			
X_5	sense_facexp_a		1			
X_6	sense_move_a		1			
X_7	rep_world_a	1				
X_8	rep_verbal_a		1			
X_9	rep_tone_a		1			
X_{10}	rep_facexp_a		1			
X_{11}	rep_move_a		1			
X_{12}	feel_a		1			
X_{13}	control_a		1			
X_{14}	prep_verbal_a		1			
X_{15}	prep_tone_a		1			
X_{16}	prep_facexp_a		1			
X_{17}	prep_move_a		1			
X_{18}	verbal_a		1			
X_{19}	tone_a		1			
X_{20}	facexp_a		1			
X_{21}	move_a		1			
X_{22}	insync_det_verbal_a			1		
X_{23}	insync_det_tone_a			1		
X_{24}	insync_det_facexp_a			1		
X_{25}	insync_det_move_a			1		
X_{26}	world_verbal_a		1			
X_{27}	world_tone_a		1			
X_{28}	world_facexp_a		1			
X_{29}	world_move_a		1			
X_{30}	world_dist_va				1	
X_{31}	world_dist_ta				1	
X_{32}	world_dist_fa				1	
X_{33}	world_dist_ma				1	
X_{34}	world_b					1
X_{35}	sense_world_b	1				
X_{36}	sense_verbal_b		1			
X_{37}	sense_tone_b		1			
X_{38}	sense_facexp_b		1			
X_{39}	sense_move_b		1			
X_{40}	rep_world_b	1				
X_{41}	rep_verbal_b		1			
X_{42}	rep_tone_b		1			
X_{43}	rep_facexp_b		1			
X_{44}	rep_move_b		1			
X_{45}	feel_b		1			
X_{46}	control_b		1			
X_{47}	prep_verbal_b		1			
X_{48}	prep_tone_b		1			
X_{49}	prep_facexp_b		1			
X_{50}	prep_move_b		1			
X_{51}	verbal_b		1			
X_{52}	tone_b		1			
X_{53}	facexp_b		1			
X_{54}	move_b		1			
X_{55}	insync_det_verbal_b			1		
X_{56}	insync_det_tone_b			1		
X_{57}	insync_det_facexp_b			1		
X_{58}	insync_det_move_b			1		
X_{59}	world_verbal_b		1			
X_{60}	world_tone_b		1			
X_{61}	world_facexp_b		1			
X_{62}	world_move_b		1			
X_{63}	world_dist_vb				1	
X_{64}	world_dist_tb				1	
X_{65}	world_dist_fb				1	
X_{66}	world_dist_mb				1	
X_{67}	w_sense_verbal_a		1			
X_{68}	w_sense_tone_a		1			
X_{69}	w_sense_facexp_a		1			
X_{70}	w_sense_move_a		1			
X_{71}	t_rep_verbal_a		1			
X_{72}	t_rep_tone_a		1			
X_{73}	t_rep_facexp_a		1			
X_{74}	t_rep_move_a		1			
X_{75}	w_sense_rep_verbal_a		1			
X_{76}	w_sense_rep_tone_a		1			
X_{77}	w_sense_rep_facexp_a		1			
X_{78}	w_sense_rep_move_a		1			
X_{79}	w_prep_verbal_a		1			
X_{80}	w_prep_tone_a		1			
X_{81}	w_prep_facexp_a		1			
X_{82}	w_prep_move_a		1			
X_{83}	t_verbal_a		1			
X_{84}	t_tone_a		1			
X_{85}	t_facexp_a		1			
X_{86}	t_move_a		1			
X_{87}	w_verbal_world_a		1			
X_{88}	w_tone_world_a		1			
X_{89}	w_facexp_world_a		1			
X_{90}	w_move_world_a		1			
X_{91}	w_sense_verbal_b		1			
X_{92}	w_sense_tone_b		1			
X_{93}	w_sense_facexp_b		1			
X_{94}	w_sense_move_b		1			
X_{95}	t_rep_verbal_b		1			
X_{96}	t_rep_tone_b		1			
X_{97}	t_rep_facexp_b		1			
X_{98}	t_rep_move_b		1			
X_{99}	w_sense_rep_verbal_b		1			
X_{100}	w_sense_rep_tone_b		1			
X_{101}	w_sense_rep_facexp_b		1			
X_{102}	w_sense_rep_move_b		1			
X_{103}	w_prep_verbal_b		1			
X_{104}	w_prep_tone_b		1			
X_{105}	w_prep_facexp_b		1			
X_{106}	w_prep_move_b		1			
X_{107}	t_verbal_b		1			
X_{108}	t_tone_b		1			
X_{109}	t_facexp_b		1			
X_{110}	t_move_b		1			
X_{111}	w_verbal_world_b		1			
X_{112}	w_tone_world_b		1			
X_{113}	w_facexp_world_b		1			
X_{114}	w_move_world_b		1			
X_{115}	context_ab	1				
X_{116}	hw_a		1			
X_{117}	ht_a		1			
X_{118}	hw_b		1			
X_{119}	ht_b		1			

mcfpv	combination function parameters	1		2		3	4		5	
	function	eucl		alogistic		compdiff	randstepmod		randstepmodopp	
	parameter	η	λ	σ	τ		ρ	δ	ρ	δ
X_1	world_a								120	60
X_2	sense_world_a	1	1							
X_3	sense_verbal_a			2.5	0.5					
X_4	sense_tone_a			2	0.6					
X_5	sense_facexp_a			1.5	0.7					
X_6	sense_move_a			1	0.8					
X_7	rep_world_a	1	1							
X_8	rep_verbal_a			2.5	X_{71}					
X_9	rep_tone_a			2	X_{72}					
X_{10}	rep_facexp_a			1.5	X_{73}					
X_{11}	rep_move_a			1	X_{74}					
X_{12}	feel_a			3	2.5					
X_{13}	control_a			10	0.4					
X_{14}	prep_verbal_a			5	0.6					
X_{15}	prep_tone_a			4.5	0.7					
X_{16}	prep_facexp_a			4	0.8					
X_{17}	prep_move_a			3.5	0.9					
X_{18}	verbal_a			2.5	X_{83}					
X_{19}	tone_a			2	X_{84}					
X_{20}	facexp_a			1.5	X_{85}					
X_{21}	move_a			1	X_{86}					
X_{22}	insync_det_verbal_a									
X_{23}	insync_det_tone_a									
X_{24}	insync_det_facexp_a									
X_{25}	insync_det_move_a									
X_{26}	world_verbal_a			2.5	0.5					
X_{27}	world_tone_a			2	0.6					
X_{28}	world_facexp_a			1.5	0.7					
X_{29}	world_move_a			1	0.8					
X_{30}	world_dist_va						120	60		
X_{31}	world_dist_ta						120	60		
X_{32}	world_dist_fa						120	60		
X_{33}	world_dist_ma						120	60		
X_{34}	world_b								120	60
X_{35}	sense_world_b	1	1							
X_{36}	sense_verbal_b			2.5	0.5					
X_{37}	sense_tone_b			2	0.6					
X_{38}	sense_facexp_b			1.5	0.7					
X_{39}	sense_move_b			1	0.8					
X_{40}	rep_world_b	1	1							
X_{41}	rep_verbal_b			2.5	X_{95}					
X_{42}	rep_tone_b			2	X_{96}					
X_{43}	rep_facexp_b			1.5	X_{97}					
X_{44}	rep_move_b			1	X_{98}					
X_{45}	feel_b			3	2.5					
X_{46}	control_b			10	0.4					
X_{47}	prep_verbal_b			5	0.6					
X_{48}	prep_tone_b			4.5	0.7					
X_{49}	prep_facexp_b			4	0.8					
X_{50}	prep_move_b			3.5	0.9					
X_{51}	verbal_b			2.5	X_{107}					
X_{52}	tone_b			2	X_{108}					
X_{53}	facexp_b			1.5	X_{109}					
X_{54}	move_b			1	X_{110}					
X_{55}	insync_det_verbal_b									
X_{56}	insync_det_tone_b									
X_{57}	insync_det_facexp_b									
X_{58}	insync_det_move_b									
X_{59}	world_verbal_b			2.5	0.5					
X_{60}	world_tone_b			2	0.6					
X_{61}	world_facexp_b			1.5	0.7					
X_{62}	world_move_b			1	0.8					
X_{63}	world_dist_vb						120	60		
X_{64}	world_dist_tb						120	60		
X_{65}	world_dist_fb						120	60		
X_{66}	world_dist_mb						120	60		
X_{67}	w_sense_verbal_a			5	0.75					
X_{68}	w_sense_tone_a			4	0.75					
X_{69}	w_sense_facexp_a			3	0.75					
X_{70}	w_sense_move_a			2	0.75					
X_{71}	t_rep_verbal_a			8	1.3					
X_{72}	t_rep_tone_a			7	1.2					
X_{73}	t_rep_facexp_a			6	1.1					
X_{74}	t_rep_move_a			5	1					
X_{75}	w_sense_rep_verbal_a			5	0.75					
X_{76}	w_sense_rep_tone_a			4	0.75					
X_{77}	w_sense_rep_facexp_a			3	0.75					
X_{78}	w_sense_rep_move_a			2	0.75					
X_{79}	w_prep_verbal_a			5	0.75					
X_{80}	w_prep_tone_a			4	0.75					
X_{81}	w_prep_facexp_a			3	0.75					
X_{82}	w_prep_move_a			2	0.75					
X_{83}	t_verbal_a			8	1.3					
X_{84}	t_tone_a			7	1.2					
X_{85}	t_facexp_a			6	1.1					
X_{86}	t_move_a			5	1					
X_{87}	w_verbal_world_a			5	0.75					
X_{88}	w_tone_world_a			4	0.75					
X_{89}	w_facexp_world_a			3	0.75					
X_{90}	w_move_world_a			2	0.75					
X_{91}	w_sense_verbal_b			5	0.75					
X_{92}	w_sense_tone_b			4	0.75					
X_{93}	w_sense_facexp_b			3	0.75					
X_{94}	w_sense_move_b			2	0.75					
X_{95}	t_rep_verbal_b			8	1.3					
X_{96}	t_rep_tone_b			7	1.2					
X_{97}	t_rep_facexp_b			6	1.1					
X_{98}	t_rep_move_b			5	1					
X_{99}	w_sense_rep_verbal_b			5	0.75					
X_{100}	w_sense_rep_tone_b			4	0.75					
X_{101}	w_sense_rep_facexp_b			3	0.75					
X_{102}	w_sense_rep_move_b			2	0.75					
X_{103}	w_prep_verbal_b			5	0.75					
X_{104}	w_prep_tone_b			4	0.75					
X_{105}	w_prep_facexp_b			3	0.75					
X_{106}	w_prep_move_b			2	0.75					
X_{107}	t_verbal_b			8	1.3					
X_{108}	t_tone_b			7	1.2					
X_{109}	t_facexp_b			6	1.1					
X_{110}	t_move_b			5	1					
X_{111}	w_verbal_world_b			5	0.75					
X_{112}	w_tone_world_b			4	0.75					
X_{113}	w_facexp_world_b			3	0.75					
X_{114}	w_move_world_b			2	0.75					
X_{115}	context_ab	1	1							
X_{116}	hw_a			5	0.7					
X_{117}	ht_a			5	0.7					
X_{118}	hw_b			5	0.7					
X_{119}	ht_b			5	0.7					

iv	initial values	1
X_1	world_a	1
X_2	sense_world_a	0
X_3	sense_verbal_a	0
X_4	sense_tone_a	0
X_5	sense_facexp_a	0
X_6	sense_move_a	0
X_7	rep_world_a	0
X_8	rep_verbal_a	0
X_9	rep_tone_a	0
X_{10}	rep_facexp_a	0
X_{11}	rep_move_a	0
X_{12}	feel_a	0
X_{13}	control_a	0
X_{14}	prep_verbal_a	0
X_{15}	prep_tone_a	0
X_{16}	prep_facexp_a	0
X_{17}	prep_move_a	0
X_{18}	verbal_a	0
X_{19}	tone_a	0
X_{20}	facexp_a	0
X_{21}	move_a	0
X_{22}	insync_det_verbal_a	0
X_{23}	insync_det_tone_a	0
X_{24}	insync_det_facexp_a	0
X_{25}	insync_det_move_a	0
X_{26}	world_verbal_a	0
X_{27}	world_tone_a	0
X_{28}	world_facexp_a	0
X_{29}	world_move_a	0
X_{30}	world_dist_va	0.05
X_{31}	world_dist_ta	0.05
X_{32}	world_dist_fa	0.05
X_{33}	world_dist_ma	0.05
X_{34}	world_b	1
X_{35}	sense_world_b	0
X_{36}	sense_verbal_b	0
X_{37}	sense_tone_b	0
X_{38}	sense_facexp_b	0
X_{39}	sense_move_b	0
X_{40}	rep_world_b	0
X_{41}	rep_verbal_b	0
X_{42}	rep_tone_b	0
X_{43}	rep_facexp_b	0
X_{44}	rep_move_b	0
X_{45}	feel_b	0
X_{46}	control_b	0
X_{47}	prep_verbal_b	0
X_{48}	prep_tone_b	0
X_{49}	prep_facexp_b	0
X_{50}	prep_move_b	0
X_{51}	verbal_b	0
X_{52}	tone_b	0
X_{53}	facexp_b	0
X_{54}	move_b	0
X_{55}	insync_det_verbal_b	0
X_{56}	insync_det_tone_b	0
X_{57}	insync_det_facexp_b	0
X_{58}	insync_det_move_b	0
X_{59}	world_verbal_b	0
X_{60}	world_tone_b	0
X_{61}	world_facexp_b	0
X_{62}	world_move_b	0
X_{63}	world_dist_vb	0.05
X_{64}	world_dist_tb	0.05
X_{65}	world_dist_fb	0.05
X_{66}	world_dist_mb	0.05
X_{67}	w_sense_verbal_a	0.2
X_{68}	w_sense_tone_a	0.2
X_{69}	w_sense_facexp_a	0.2
X_{70}	w_sense_move_a	0.2
X_{71}	t_rep_verbal_a	0.7
X_{72}	t_rep_tone_a	0.7
X_{73}	t_rep_facexp_a	0.7
X_{74}	t_rep_move_a	0.7
X_{75}	w_sense_rep_verbal_a	0.2
X_{76}	w_sense_rep_tone_a	0.2
X_{77}	w_sense_rep_facexp_a	0.2
X_{78}	w_sense_rep_move_a	0.2
X_{79}	w_prep_verbal_a	0.2
X_{80}	w_prep_tone_a	0.2
X_{81}	w_prep_facexp_a	0.2
X_{82}	w_prep_move_a	0.2
X_{83}	t_verbal_a	0.7
X_{84}	t_tone_a	0.7
X_{85}	t_facexp_a	0.7
X_{86}	t_move_a	0.7
X_{87}	w_verbal_world_a	0.2
X_{88}	w_tone_world_a	0.2
X_{89}	w_facexp_world_a	0.2
X_{90}	w_move_world_a	0.2
X_{91}	w_sense_verbal_b	0.2
X_{92}	w_sense_tone_b	0.2
X_{93}	w_sense_facexp_b	0.2
X_{94}	w_sense_move_b	0.2
X_{95}	t_rep_verbal_b	0.7
X_{96}	t_rep_tone_b	0.7
X_{97}	t_rep_facexp_b	0.7
X_{98}	t_rep_move_b	0.7
X_{99}	w_sense_rep_verbal_b	0.2
X_{100}	w_sense_rep_tone_b	0.2
X_{101}	w_sense_rep_facexp_b	0.2
X_{102}	w_sense_rep_move_b	0.2
X_{103}	w_prep_verbal_b	0.2
X_{104}	w_prep_tone_b	0.2
X_{105}	w_prep_facexp_b	0.2
X_{106}	w_prep_move_b	0.2
X_{107}	t_verbal_b	0.7
X_{108}	t_tone_b	0.7
X_{109}	t_facexp_b	0.7
X_{110}	t_move_b	0.7
X_{111}	w_verbal_world_b	0.2
X_{112}	w_tone_world_b	0.2
X_{113}	w_facexp_world_b	0.2
X_{114}	w_move_world_b	0.2
X_{115}	context_ab	0.5
X_{116}	hw_a	0
X_{117}	ht_a	0
X_{118}	hw_b	0
X_{119}	ht_b	0

Custom Role Matrices

Human with Regular Individual Emotion Control and Regular In-Sync Effects

mcwv	connection weights	1	2	3	4	5	6	7	8	9
X_1	world_a	1								
X_2	sense_world_a	1								
X_3	sense_verbal_a	0.5	X_{67}							
X_4	sense_tone_a	0.45	X_{68}							
X_5	sense_facexp_a	0.4	X_{69}							
X_6	sense_move_a	0.35	X_{70}							
X_7	rep_world_a	1								
X_8	rep_verbal_a	X_{75}	0.5							
X_9	rep_tone_a	X_{76}	0.5							
X_{10}	rep_facexp_a	X_{77}	0.5							
X_{11}	rep_move_a	X_{78}	0.5							
X_{12}	feel_a	-1	1	1	1	1	-0.1	-0.1	-0.1	-0.1
X_{13}	control_a	0.1	0.1	0.1	0.1	0.1	0.1	0.1	0.1	0.1
X_{14}	prep_verbal_a	1	1	1	-1					
X_{15}	prep_tone_a	1	1	1	-1					
X_{16}	prep_facexp_a	1	1	1	-1					
X_{17}	prep_move_a	1	1	1	-1					
X_{18}	verbal_a	X_{79}								
X_{19}	tone_a	X_{80}								
X_{20}	facexp_a	X_{81}								
X_{21}	move_a	X_{82}								
X_{22}	insync_det_verbal_a	1	1							
X_{23}	insync_det_tone_a	1	1							
X_{24}	insync_det_facexp_a	1	1							
X_{25}	insync_det_move_a	1	1							
X_{26}	world_verbal_a	X_{87}	0.2							
X_{27}	world_tone_a	X_{88}	0.2							
X_{28}	world_facexp_a	X_{89}	0.2							
X_{29}	world_move_a	X_{90}	0.2							
X_{30}	world_dist_va	1								
X_{31}	world_dist_ta	1								
X_{32}	world_dist_fa	1								
X_{33}	world_dist_ma	1								
X_{34}	world_b	1								
X_{35}	sense_world_b	0.95								
X_{36}	sense_verbal_b	0.5	X_{91}							
X_{37}	sense_tone_b	0.45	X_{92}							
X_{38}	sense_facexp_b	0.4	X_{93}							
X_{39}	sense_move_b	0.35	X_{94}							
X_{40}	rep_world_b	1								
X_{41}	rep_verbal_b	X_{99}	0.5							
X_{42}	rep_tone_b	X_{100}	0.5							
X_{43}	rep_facexp_b	X_{101}	0.5							
X_{44}	rep_move_b	X_{102}	0.5							
X_{45}	feel_b	-1	1	1	1	1	-0.1	-0.1	-0.1	-0.1
X_{46}	control_b	0.1	0.1	0.1	0.1	0.1	0.1	0.1	0.1	0.1
X_{47}	prep_verbal_b	1	1	1	-1					
X_{48}	prep_tone_b	1	1	1	-1					
X_{49}	prep_facexp_b	1	1	1	-1					
X_{50}	prep_move_b	1	1	1	-1					
X_{51}	verbal_b	X_{103}								
X_{52}	tone_b	X_{104}								
X_{53}	facexp_b	X_{105}								
X_{54}	move_b	X_{106}								
X_{55}	insync_det_verbal_b	1	1							
X_{56}	insync_det_tone_b	1	1							
X_{57}	insync_det_facexp_b	1	1							
X_{58}	insync_det_move_b	1	1							
X_{59}	world_verbal_b	X_{111}	0.2							
X_{60}	world_tone_b	X_{112}	0.2							
X_{61}	world_facexp_b	X_{113}	0.2							
X_{62}	world_move_b	X_{114}	0.2							
X_{63}	world_dist_vb	1								
X_{64}	world_dist_tb	1								
X_{65}	world_dist_fb	1								
X_{66}	world_dist_mb	1								
X_{67}	w_sense_verbal_a	1	0.9	0.8	0.8	0.7				
X_{68}	w_sense_tone_a	0.9	1	0.8	0.8	0.7				
X_{69}	w_sense_facexp_a	0.8	0.8	1	0.9	0.7				
X_{70}	w_sense_move_a	0.8	0.8	0.9	1	0.7				
X_{71}	t_rep_verbal_a	1	-0.12	-0.06	-0.03	-0.03				
X_{72}	t_rep_tone_a	1	-0.06	-0.12	-0.03	-0.03				
X_{73}	t_rep_facexp_a	1	-0.03	-0.03	-0.12	-0.06				
X_{74}	t_rep_move_a	1	-0.03	-0.03	-0.06	-0.12				
X_{75}	w_sense_rep_verbal_a	1	0.9	0.8	0.8					
X_{76}	w_sense_rep_tone_a	0.9	1	0.8	0.8					
X_{77}	w_sense_rep_facexp_a	0.8	0.8	1	0.9					
X_{78}	w_sense_rep_move_a	0.8	0.8	0.9	1					
X_{79}	w_prep_verbal_a	1	0.9	0.8	0.8					
X_{80}	w_prep_tone_a	0.9	1	0.8	0.8					
X_{81}	w_prep_facexp_a	0.8	0.8	1	0.9					
X_{82}	w_prep_move_a	0.8	0.8	0.9	1					
X_{83}	t_verbal_a	1	-0.12	-0.06	-0.03	-0.03				
X_{84}	t_tone_a	1	-0.06	-0.12	-0.03	-0.03				
X_{85}	t_facexp_a	1	-0.03	-0.03	-0.12	-0.06				
X_{86}	t_move_a	1	-0.03	-0.03	-0.06	-0.12				
X_{87}	w_verbal_world_a	1	0.9	0.8	0.8	0.7				
X_{88}	w_tone_world_a	0.9	1	0.8	0.8	0.7				
X_{89}	w_facexp_world_a	0.8	0.8	1	0.9	0.7				
X_{90}	w_move_world_a	0.8	0.8	0.9	1	0.7				
X_{91}	w_sense_verbal_b	1	0.9	0.8	0.8	0.7				
X_{92}	w_sense_tone_b	0.9	1	0.8	0.8	0.7				
X_{93}	w_sense_facexp_b	0.8	0.8	1	0.9	0.7				
X_{94}	w_sense_move_b	0.8	0.8	0.9	1	0.7				
X_{95}	t_rep_verbal_b	1	-0.12	-0.06	-0.03	-0.03				
X_{96}	t_rep_tone_b	1	-0.06	-0.12	-0.03	-0.03				
X_{97}	t_rep_facexp_b	1	-0.03	-0.03	-0.12	-0.06				
X_{98}	t_rep_move_b	1	-0.03	-0.03	-0.06	-0.12				
X_{99}	w_sense_rep_verbal_b	1	0.9	0.8	0.8					
X_{100}	w_sense_rep_tone_b	0.9	1	0.8	0.8					
X_{101}	w_sense_rep_facexp_b	0.8	0.8	1	0.9					
X_{102}	w_sense_rep_move_b	0.8	0.8	0.9	1					
X_{103}	w_prep_verbal_b	1	0.9	0.8	0.8					
X_{104}	w_prep_tone_b	0.9	1	0.8	0.8					
X_{105}	w_prep_facexp_b	0.8	0.8	1	0.9					
X_{106}	w_prep_move_b	0.8	0.8	0.9	1					
X_{107}	t_verbal_b	1	-0.12	-0.06	-0.03	-0.03				
X_{108}	t_tone_b	1	-0.06	-0.12	-0.03	-0.03				
X_{109}	t_facexp_b	1	-0.03	-0.03	-0.12	-0.06				
X_{110}	t_move_b	1	-0.03	-0.03	-0.06	-0.12				
X_{111}	w_verbal_world_b	1	0.9	0.8	0.8	0.7				
X_{112}	w_tone_world_b	0.9	1	0.8	0.8	0.7				
X_{113}	w_facexp_world_b	0.8	0.8	1	0.9	0.7				
X_{114}	w_move_world_b	0.8	0.8	0.9	1	0.7				
X_{115}	context_ab	1								
X_{116}	hw_a	0.1	0.1	0.1	0.1	0.1				
X_{117}	ht_a	1	1	1	1	1				
X_{118}	hw_b	0.1	0.1	0.1	0.1	0.1				
X_{119}	ht_b	1	1	1	1	1				

Human with Low Individual Emotion Control and Regular In-Sync Effects

mcwv	connection weights	1	2	3	4	5	6	7	8	9
X_{1}	world_a	1								
X_{2}	sense_world_a	1								
X_{3}	sense_verbal_a	0.5	X_{67}							
X_{4}	sense_tone_a	0.45	X_{68}							
X_{5}	sense_facexp_a	0.4	X_{69}							
X_{6}	sense_move_a	0.35	X_{70}							
X_{7}	rep_world_a	1								
X_{8}	rep_verbal_a	X_{75}	0.5							
X_{9}	rep_tone_a	X_{76}	0.5							
X_{10}	rep_facexp_a	X_{77}	0.5							
X_{11}	rep_move_a	X_{78}	0.5							
X_{12}	feel_a	-1	1	1	1	1	-0.1	-0.1	-0.1	-0.1
X_{13}	control_a	0.08	0.08	0.08	0.08	0.08	0.08	0.08	0.08	0.08
X_{14}	prep_verbal_a	1	1	1	-1					
X_{15}	prep_tone_a	1	1	1	-1					
X_{16}	prep_facexp_a	1	1	1	-1					
X_{17}	prep_move_a	1	1	1	-1					
X_{18}	verbal_a	X_{79}								
X_{19}	tone_a	X_{80}								
X_{20}	facexp_a	X_{81}								
X_{21}	move_a	X_{82}								
X_{22}	insync_det_verbal_a	1	1							
X_{23}	insync_det_tone_a	1	1							
X_{24}	insync_det_facexp_a	1	1							
X_{25}	insync_det_move_a	1	1							
X_{26}	world_verbal_a	X_{87}	0.2							
X_{27}	world_tone_a	X_{88}	0.2							
X_{28}	world_facexp_a	X_{89}	0.2							
X_{29}	world_move_a	X_{90}	0.2							
X_{30}	world_dist_va	1								
X_{31}	world_dist_ta	1								
X_{32}	world_dist_fa	1								
X_{33}	world_dist_ma	1								
X_{34}	world_b	1								
X_{35}	sense_world_b	0.95								
X_{36}	sense_verbal_b	0.5	X_{91}							
X_{37}	sense_tone_b	0.45	X_{92}							
X_{38}	sense_facexp_b	0.4	X_{93}							
X_{39}	sense_move_b	0.35	X_{94}							
X_{40}	rep_world_b	1								
X_{41}	rep_verbal_b	X_{99}	0.5							
X_{42}	rep_tone_b	X_{100}	0.5							
X_{43}	rep_facexp_b	X_{101}	0.5							
X_{44}	rep_move_b	X_{102}	0.5							
X_{45}	feel_b	-1	1	1	1	1	-0.1	-0.1	-0.1	-0.1
X_{46}	control_b	0.1	0.1	0.1	0.1	0.1	0.1	0.1	0.1	0.1
X_{47}	prep_verbal_b	1	1	1	-1					
X_{48}	prep_tone_b	1	1	1	-1					
X_{49}	prep_facexp_b	1	1	1	-1					
X_{50}	prep_move_b	1	1	1	-1					
X_{51}	verbal_b	X_{103}								
X_{52}	tone_b	X_{104}								
X_{53}	facexp_b	X_{105}								
X_{54}	move_b	X_{106}								
X_{55}	insync_det_verbal_b	1	1							
X_{56}	insync_det_tone_b	1	1							
X_{57}	insync_det_facexp_b	1	1							
X_{58}	insync_det_move_b	1	1							
X_{59}	world_verbal_b	X_{111}	0.2							
X_{60}	world_tone_b	X_{112}	0.2							
X_{61}	world_facexp_b	X_{113}	0.2							
X_{62}	world_move_b	X_{114}	0.2							
X_{63}	world_dist_vb	1								
X_{64}	world_dist_tb	1								
X_{65}	world_dist_fb	1								
X_{66}	world_dist_mb	1								
X_{67}	w_sense_verbal_a	1	0.9	0.8	0.8	0.7				
X_{68}	w_sense_tone_a	0.9	1	0.8	0.8	0.7				
X_{69}	w_sense_facexp_a	0.8	0.8	1	0.9	0.7				
X_{70}	w_sense_move_a	0.8	0.8	0.9	1	0.7				
X_{71}	t_rep_verbal_a	1	-0.12	-0.06	-0.03	-0.03				
X_{72}	t_rep_tone_a	1	-0.06	-0.12	-0.03	-0.03				
X_{73}	t_rep_facexp_a	1	-0.03	-0.03	-0.12	-0.06				
X_{74}	t_rep_move_a	1	-0.03	-0.03	-0.06	-0.12				
X_{75}	w_sense_rep_verbal_a	1	0.9	0.8	0.8					
X_{76}	w_sense_rep_tone_a	0.9	1	0.8	0.8					
X_{77}	w_sense_rep_facexp_a	0.8	0.8	1	0.9					
X_{78}	w_sense_rep_move_a	0.8	0.8	0.9	1					
X_{79}	w_prep_verbal_a	1	0.9	0.8	0.8					
X_{80}	w_prep_tone_a	0.9	1	0.8	0.8					
X_{81}	w_prep_facexp_a	0.8	0.8	1	0.9					
X_{82}	w_prep_move_a	0.8	0.8	0.9	1					
X_{83}	t_verbal_a	1	-0.12	-0.06	-0.03	-0.03				
X_{84}	t_tone_a	1	-0.06	-0.12	-0.03	-0.03				
X_{85}	t_facexp_a	1	-0.03	-0.03	-0.12	-0.06				
X_{86}	t_move_a	1	-0.03	-0.03	-0.06	-0.12				
X_{87}	w_verbal_world_a	1	0.9	0.8	0.8	0.7				
X_{88}	w_tone_world_a	0.9	1	0.8	0.8	0.7				
X_{89}	w_facexp_world_a	0.8	0.8	1	0.9	0.7				
X_{90}	w_move_world_a	0.8	0.8	0.9	1	0.7				
X_{91}	w_sense_verbal_b	1	0.9	0.8	0.8	0.7				
X_{92}	w_sense_tone_b	0.9	1	0.8	0.8	0.7				
X_{93}	w_sense_facexp_b	0.8	0.8	1	0.9	0.7				
X_{94}	w_sense_move_b	0.8	0.8	0.9	1	0.7				
X_{95}	t_rep_verbal_b	1	-0.12	-0.06	-0.03	-0.03				
X_{96}	t_rep_tone_b	1	-0.06	-0.12	-0.03	-0.03				
X_{97}	t_rep_facexp_b	1	-0.03	-0.03	-0.12	-0.06				
X_{98}	t_rep_move_b	1	-0.03	-0.03	-0.06	-0.12				
X_{99}	w_sense_rep_verbal_b	1	0.9	0.8	0.8					
X_{100}	w_sense_rep_tone_b	0.9	1	0.8	0.8					
X_{101}	w_sense_rep_facexp_b	0.8	0.8	1	0.9					
X_{102}	w_sense_rep_move_b	0.8	0.8	0.9	1					
X_{103}	w_prep_verbal_b	1	0.9	0.8	0.8					
X_{104}	w_prep_tone_b	0.9	1	0.8	0.8					
X_{105}	w_prep_facexp_b	0.8	0.8	1	0.9					
X_{106}	w_prep_move_b	0.8	0.8	0.9	1					
X_{107}	t_verbal_b	1	-0.12	-0.06	-0.03	-0.03				
X_{108}	t_tone_b	1	-0.06	-0.12	-0.03	-0.03				
X_{109}	t_facexp_b	1	-0.03	-0.03	-0.12	-0.06				
X_{110}	t_move_b	1	-0.03	-0.03	-0.06	-0.12				
X_{111}	w_verbal_world_b	1	0.9	0.8	0.8	0.7				
X_{112}	w_tone_world_b	0.9	1	0.8	0.8	0.7				
X_{113}	w_facexp_world_b	0.8	0.8	1	0.9	0.7				
X_{114}	w_move_world_b	0.8	0.8	0.9	1	0.7				
X_{115}	context_ab	1								
X_{116}	hw_a	0.1	0.1	0.1	0.1	0.1				
X_{117}	ht_a	1	1	1	1	1				
X_{118}	hw_b	0.1	0.1	0.1	0.1	0.1				
X_{119}	ht_b	1	1	1	1	1				

Human with Regular Individual Emotion Control and Low In-Sync Effects

mcwv	connection weights	1	2	3	4	5	6	7	8	9
X_{1}	world_a	1								
X_{2}	sense_world_a	1								
X_{3}	sense_verbal_a	0.5	X_{67}							
X_{4}	sense_tone_a	0.45	X_{68}							
X_{5}	sense_facexp_a	0.4	X_{69}							
X_{6}	sense_move_a	0.35	X_{70}							
X_{7}	rep_world_a	1								
X_{8}	rep_verbal_a	X_{75}	0.5							
X_{9}	rep_tone_a	X_{76}	0.5							
X_{10}	rep_facexp_a	X_{77}	0.5							
X_{11}	rep_move_a	X_{78}	0.5							
X_{12}	feel_a	-1	1	1	1	1	-0.05	-0.05	-0.05	-0.05
X_{13}	control_a	0.1	0.1	0.1	0.1	0.1	0.1	0.1	0.1	0.1
X_{14}	prep_verbal_a	1	1	1	-1					
X_{15}	prep_tone_a	1	1	1	-1					
X_{16}	prep_facexp_a	1	1	1	-1					
X_{17}	prep_move_a	1	1	1	-1					
X_{18}	verbal_a	X_{79}								
X_{19}	tone_a	X_{80}								
X_{20}	facexp_a	X_{81}								
X_{21}	move_a	X_{82}								
X_{22}	insync_det_verbal_a	1	1							
X_{23}	insync_det_tone_a	1	1							
X_{24}	insync_det_facexp_a	1	1							
X_{25}	insync_det_move_a	1	1							
X_{26}	world_verbal_a	X_{87}	0.2							
X_{27}	world_tone_a	X_{88}	0.2							
X_{28}	world_facexp_a	X_{89}	0.2							
X_{29}	world_move_a	X_{90}	0.2							
X_{30}	world_dist_va	1								
X_{31}	world_dist_ta	1								
X_{32}	world_dist_fa	1								
X_{33}	world_dist_ma	1								
X_{34}	world_b	1								
X_{35}	sense_world_b	0.95								
X_{36}	sense_verbal_b	0.5	X_{91}							
X_{37}	sense_tone_b	0.45	X_{92}							
X_{38}	sense_facexp_b	0.4	X_{93}							
X_{39}	sense_move_b	0.35	X_{94}							
X_{40}	rep_world_b	1								
X_{41}	rep_verbal_b	X_{99}	0.5							
X_{42}	rep_tone_b	X_{100}	0.5							
X_{43}	rep_facexp_b	X_{101}	0.5							
X_{44}	rep_move_b	X_{102}	0.5							
X_{45}	feel_b	-1	1	1	1	1	-0.1	-0.1	-0.1	-0.1
X_{46}	control_b	0.1	0.1	0.1	0.1	0.1	0.1	0.1	0.1	0.1
X_{47}	prep_verbal_b	1	1	1	-1					
X_{48}	prep_tone_b	1	1	1	-1					
X_{49}	prep_facexp_b	1	1	1	-1					
X_{50}	prep_move_b	1	1	1	-1					
X_{51}	verbal_b	X_{103}								
X_{52}	tone_b	X_{104}								
X_{53}	facexp_b	X_{105}								
X_{54}	move_b	X_{106}								
X_{55}	insync_det_verbal_b	1	1							
X_{56}	insync_det_tone_b	1	1							
X_{57}	insync_det_facexp_b	1	1							
X_{58}	insync_det_move_b	1	1							
X_{59}	world_verbal_b	X_{111}	0.2							
X_{60}	world_tone_b	X_{112}	0.2							
X_{61}	world_facexp_b	X_{113}	0.2							
X_{62}	world_move_b	X_{114}	0.2							
X_{63}	world_dist_vb	1								
X_{64}	world_dist_tb	1								
X_{65}	world_dist_fb	1								
X_{66}	world_dist_mb	1								
X_{67}	w_sense_verbal_a	1	0.9	0.8	0.8	0.7				
X_{68}	w_sense_tone_a	0.9	1	0.8	0.8	0.7				
X_{69}	w_sense_facexp_a	0.8	0.8	1	0.9	0.7				
X_{70}	w_sense_move_a	0.8	0.8	0.9	1	0.7				
X_{71}	t_rep_verbal_a	1	-0.12	-0.06	-0.03	-0.03				
X_{72}	t_rep_tone_a	1	-0.06	-0.12	-0.03	-0.03				
X_{73}	t_rep_facexp_a	1	-0.03	-0.03	-0.12	-0.06				
X_{74}	t_rep_move_a	1	-0.03	-0.03	-0.06	-0.12				
X_{75}	w_sense_rep_verbal_a	1	0.9	0.8	0.8					
X_{76}	w_sense_rep_tone_a	0.9	1	0.8	0.8					
X_{77}	w_sense_rep_facexp_a	0.8	0.8	1	0.9					
X_{78}	w_sense_rep_move_a	0.8	0.8	0.9	1					
X_{79}	w_prep_verbal_a	1	0.9	0.8	0.8					
X_{80}	w_prep_tone_a	0.9	1	0.8	0.8					
X_{81}	w_prep_facexp_a	0.8	0.8	1	0.9					
X_{82}	w_prep_move_a	0.8	0.8	0.9	1					
X_{83}	t_verbal_a	1	-0.12	-0.06	-0.03	-0.03				
X_{84}	t_tone_a	1	-0.06	-0.12	-0.03	-0.03				
X_{85}	t_facexp_a	1	-0.03	-0.03	-0.12	-0.06				
X_{86}	t_move_a	1	-0.03	-0.03	-0.06	-0.12				
X_{87}	w_verbal_world_a	1	0.9	0.8	0.8	0.7				
X_{88}	w_tone_world_a	0.9	1	0.8	0.8	0.7				
X_{89}	w_facexp_world_a	0.8	0.8	1	0.9	0.7				
X_{90}	w_move_world_a	0.8	0.8	0.9	1	0.7				
X_{91}	w_sense_verbal_b	1	0.9	0.8	0.8	0.7				
X_{92}	w_sense_tone_b	0.9	1	0.8	0.8	0.7				
X_{93}	w_sense_facexp_b	0.8	0.8	1	0.9	0.7				
X_{94}	w_sense_move_b	0.8	0.8	0.9	1	0.7				
X_{95}	t_rep_verbal_b	1	-0.12	-0.06	-0.03	-0.03				
X_{96}	t_rep_tone_b	1	-0.06	-0.12	-0.03	-0.03				
X_{97}	t_rep_facexp_b	1	-0.03	-0.03	-0.12	-0.06				
X_{98}	t_rep_move_b	1	-0.03	-0.03	-0.06	-0.12				
X_{99}	w_sense_rep_verbal_b	1	0.9	0.8	0.8					
X_{100}	w_sense_rep_tone_b	0.9	1	0.8	0.8					
X_{101}	w_sense_rep_facexp_b	0.8	0.8	1	0.9					
X_{102}	w_sense_rep_move_b	0.8	0.8	0.9	1					
X_{103}	w_prep_verbal_b	1	0.9	0.8	0.8					
X_{104}	w_prep_tone_b	0.9	1	0.8	0.8					
X_{105}	w_prep_facexp_b	0.8	0.8	1	0.9					
X_{106}	w_prep_move_b	0.8	0.8	0.9	1					
X_{107}	t_verbal_b	1	-0.12	-0.06	-0.03	-0.03				
X_{108}	t_tone_b	1	-0.06	-0.12	-0.03	-0.03				
X_{109}	t_facexp_b	1	-0.03	-0.03	-0.12	-0.06				
X_{110}	t_move_b	1	-0.03	-0.03	-0.06	-0.12				
X_{111}	w_verbal_world_b	1	0.9	0.8	0.8	0.7				
X_{112}	w_tone_world_b	0.9	1	0.8	0.8	0.7				
X_{113}	w_facexp_world_b	0.8	0.8	1	0.9	0.7				
X_{114}	w_move_world_b	0.8	0.8	0.9	1	0.7				
X_{115}	context_ab	1								
X_{116}	hw_a	0.1	0.1	0.1	0.1	0.1				
X_{117}	ht_a	1	1	1	1	1				
X_{118}	hw_b	0.1	0.1	0.1	0.1	0.1				
X_{119}	ht_b	1	1	1	1	1				

Human with Low Individual Emotion Control and Low In-Sync Effects

mcwv	connection weights	1	2	3	4	5	6	7	8	9
X_{1}	world_a	1								
X_{2}	sense_world_a	1								
X_{3}	sense_verbal_a	0.5	X_{67}							
X_{4}	sense_tone_a	0.45	X_{68}							
X_{5}	sense_facexp_a	0.4	X_{69}							
X_{6}	sense_move_a	0.35	X_{70}							
X_{7}	rep_world_a	1								
X_{8}	rep_verbal_a	X_{75}	0.5							
X_{9}	rep_tone_a	X_{76}	0.5							
X_{10}	rep_facexp_a	X_{77}	0.5							
X_{11}	rep_move_a	X_{78}	0.5							
X_{12}	feel_a	-1	1	1	1	1	-0.05	-0.05	-0.05	-0.05
X_{13}	control_a	0.08	0.08	0.08	0.08	0.08	0.08	0.08	0.08	0.08
X_{14}	prep_verbal_a	1	1	1	-1					
X_{15}	prep_tone_a	1	1	1	-1					
X_{16}	prep_facexp_a	1	1	1	-1					
X_{17}	prep_move_a	1	1	1	-1					
X_{18}	verbal_a	X_{79}								
X_{19}	tone_a	X_{80}								
X_{20}	facexp_a	X_{81}								
X_{21}	move_a	X_{82}								
X_{22}	insync_det_verbal_a	1	1							
X_{23}	insync_det_tone_a	1	1							
X_{24}	insync_det_facexp_a	1	1							
X_{25}	insync_det_move_a	1	1							
X_{26}	world_verbal_a	X_{87}	0.2							
X_{27}	world_tone_a	X_{88}	0.2							
X_{28}	world_facexp_a	X_{89}	0.2							
X_{29}	world_move_a	X_{90}	0.2							
X_{30}	world_dist_va	1								
X_{31}	world_dist_ta	1								
X_{32}	world_dist_fa	1								
X_{33}	world_dist_ma	1								
X_{34}	world_b	1								
X_{35}	sense_world_b	0.95								
X_{36}	sense_verbal_b	0.5	X_{91}							
X_{37}	sense_tone_b	0.45	X_{92}							
X_{38}	sense_facexp_b	0.4	X_{93}							
X_{39}	sense_move_b	0.35	X_{94}							
X_{40}	rep_world_b	1								
X_{41}	rep_verbal_b	X_{99}	0.5							
X_{42}	rep_tone_b	X_{100}	0.5							
X_{43}	rep_facexp_b	X_{101}	0.5							
X_{44}	rep_move_b	X_{102}	0.5							
X_{45}	feel_b	-1	1	1	1	1	-0.1	-0.1	-0.1	-0.1
X_{46}	control_b	0.1	0.1	0.1	0.1	0.1	0.1	0.1	0.1	0.1
X_{47}	prep_verbal_b	1	1	1	-1					
X_{48}	prep_tone_b	1	1	1	-1					
X_{49}	prep_facexp_b	1	1	1	-1					
X_{50}	prep_move_b	1	1	1	-1					
X_{51}	verbal_b	X_{103}								
X_{52}	tone_b	X_{104}								
X_{53}	facexp_b	X_{105}								
X_{54}	move_b	X_{106}								
X_{55}	insync_det_verbal_b	1	1							
X_{56}	insync_det_tone_b	1	1							
X_{57}	insync_det_facexp_b	1	1							
X_{58}	insync_det_move_b	1	1							
X_{59}	world_verbal_b	X_{111}	0.2							
X_{60}	world_tone_b	X_{112}	0.2							
X_{61}	world_facexp_b	X_{113}	0.2							
X_{62}	world_move_b	X_{114}	0.2							
X_{63}	world_dist_vb	1								
X_{64}	world_dist_tb	1								
X_{65}	world_dist_fb	1								
X_{66}	world_dist_mb	1								
X_{67}	w_sense_verbal_a	1	0.9	0.8	0.8	0.7				
X_{68}	w_sense_tone_a	0.9	1	0.8	0.8	0.7				
X_{69}	w_sense_facexp_a	0.8	0.8	1	0.9	0.7				
X_{70}	w_sense_move_a	0.8	0.8	0.9	1	0.7				
X_{71}	t_rep_verbal_a	1	-0.12	-0.06	-0.03	-0.03				
X_{72}	t_rep_tone_a	1	-0.06	-0.12	-0.03	-0.03				
X_{73}	t_rep_facexp_a	1	-0.03	-0.03	-0.12	-0.06				
X_{74}	t_rep_move_a	1	-0.03	-0.03	-0.06	-0.12				
X_{75}	w_sense_rep_verbal_a	1	0.9	0.8	0.8					
X_{76}	w_sense_rep_tone_a	0.9	1	0.8	0.8					
X_{77}	w_sense_rep_facexp_a	0.8	0.8	1	0.9					
X_{78}	w_sense_rep_move_a	0.8	0.8	0.9	1					
X_{79}	w_prep_verbal_a	1	0.9	0.8	0.8					
X_{80}	w_prep_tone_a	0.9	1	0.8	0.8					
X_{81}	w_prep_facexp_a	0.8	0.8	1	0.9					
X_{82}	w_prep_move_a	0.8	0.8	0.9	1					
X_{83}	t_verbal_a	1	-0.12	-0.06	-0.03	-0.03				
X_{84}	t_tone_a	1	-0.06	-0.12	-0.03	-0.03				
X_{85}	t_facexp_a	1	-0.03	-0.03	-0.12	-0.06				
X_{86}	t_move_a	1	-0.03	-0.03	-0.06	-0.12				
X_{87}	w_verbal_world_a	1	0.9	0.8	0.8	0.7				
X_{88}	w_tone_world_a	0.9	1	0.8	0.8	0.7				
X_{89}	w_facexp_world_a	0.8	0.8	1	0.9	0.7				
X_{90}	w_move_world_a	0.8	0.8	0.9	1	0.7				
X_{91}	w_sense_verbal_b	1	0.9	0.8	0.8	0.7				
X_{92}	w_sense_tone_b	0.9	1	0.8	0.8	0.7				
X_{93}	w_sense_facexp_b	0.8	0.8	1	0.9	0.7				
X_{94}	w_sense_move_b	0.8	0.8	0.9	1	0.7				
X_{95}	t_rep_verbal_b	1	-0.12	-0.06	-0.03	-0.03				
X_{96}	t_rep_tone_b	1	-0.06	-0.12	-0.03	-0.03				
X_{97}	t_rep_facexp_b	1	-0.03	-0.03	-0.12	-0.06				
X_{98}	t_rep_move_b	1	-0.03	-0.03	-0.06	-0.12				
X_{99}	w_sense_rep_verbal_b	1	0.9	0.8	0.8					
X_{100}	w_sense_rep_tone_b	0.9	1	0.8	0.8					
X_{101}	w_sense_rep_facexp_b	0.8	0.8	1	0.9					
X_{102}	w_sense_rep_move_b	0.8	0.8	0.9	1					
X_{103}	w_prep_verbal_b	1	0.9	0.8	0.8					
X_{104}	w_prep_tone_b	0.9	1	0.8	0.8					
X_{105}	w_prep_facexp_b	0.8	0.8	1	0.9					
X_{106}	w_prep_move_b	0.8	0.8	0.9	1					
X_{107}	t_verbal_b	1	-0.12	-0.06	-0.03	-0.03				
X_{108}	t_tone_b	1	-0.06	-0.12	-0.03	-0.03				
X_{109}	t_facexp_b	1	-0.03	-0.03	-0.12	-0.06				
X_{110}	t_move_b	1	-0.03	-0.03	-0.06	-0.12				
X_{111}	w_verbal_world_b	1	0.9	0.8	0.8	0.7				
X_{112}	w_tone_world_b	0.9	1	0.8	0.8	0.7				
X_{113}	w_facexp_world_b	0.8	0.8	1	0.9	0.7				
X_{114}	w_move_world_b	0.8	0.8	0.9	1	0.7				
X_{115}	context_ab	1								
X_{116}	hw_a	0.1	0.1	0.1	0.1	0.1				
X_{117}	ht_a	1	1	1	1	1				
X_{118}	hw_b	0.1	0.1	0.1	0.1	0.1				
X_{119}	ht_b	1	1	1	1	1				

References

1. Abrams, D.A., Ryali, S., Chen, T., Chordia, P., Khouzam, A., Levitin, D.J., Menon, V.: Intersubject synchronization of brain responses during natural music listening. Eur. J. Neurosci. **37**(9), 1458–1469 (2013)
2. Andersson, G.: Internet-delivered psychological treatments. Annu. Rev. Clin. Psychol. **12**, 157–179 (2016)
3. Australian Bureau of Statistics (2020–2022): National Study of mental health and wellbeing. ABS. https://www.abs.gov.au/statistics/health/mental-health/national-study-mental-health-and-wellbeing/latest-release. Last accessed 01 June 2024
4. Barak, A., Hen, L., Boniel-Nissim, M., Shapira, N.A.: A comprehensive review and a meta-analysis of the effectiveness of internet-based psychotherapeutic interventions. J. Technol. Hum. Serv. **26**(2–4), 109–160 (2008)
5. Berdasco, A., López, G., Diaz I., Quesada, L., Guerrero, L. A.: User experience comparison of intelligent personal assistants: Alexa, Google assistant, Siri and Cortana. Proc. MDPI **31**(1), 51. https://doi.org/10.3390/proceedings2019031051 (2019)
6. Botha, F., Morris, R.W., Butterworth, P., Glozier, N.: Generational differences in mental health trends in the twenty-first century. Proc. Natl. Acad. Sci. **120**(49), e2303781120 (2023)
7. Butler, E.A., Randall, A.K.: Emotional coregulation in close relationships. Emot. Rev. **5**(2), 202–210 (2013)
8. Carlbring, P., Andersson, G., Cuijpers, P., Riper, H., Hedman-Lagerlöf, E.: Internet-based vs. face-to-face cognitive behavior therapy for psychiatric and somatic disorders: an updated systematic review and meta-analysis. Cogn. Behav. Ther. **47**(1), 1–18 (2018)
9. Chen, J., Gao, Y., Huang, X., Tian, Y.: Neural correlates of linguistic style matching in social interaction. Soc. Cogn. Affect. Neurosci. **15**(3), 333–343 (2020)
10. Damasio, A.R.: The feeling of what happens: body and emotion in the making of consciousness. Houghton Mifflin Harcourt (1999)
11. Danescu-Niculescu-Mizil, C., Sudhof, M., Jurafsky, D., Leskovec, J., Potts, C.: A computational approach to politeness with application to social factors. In: Proceedings of the 51st Annual Meeting of the Association for Computational Linguistics (Volume 1: Long Papers), pp. 250–259. arXiv preprint arXiv:1306.6078 (2012)
12. Del Re, A.C., Flückiger, C., Horvath, A.O., Wampold, B.E.: Examining therapist effects in the alliance–outcome relationship: a multilevel meta-analysis. J. Consult. Clin. Psychol. **89**(5), 371 (2021)
13. Dimopoulos, F., Eler, E., Timmers, M., Treur, J., Koole, S.L.: An agent-based network model for interpersonal emotion regulation in a human-bot interaction. In: Nguyen, N.T., et al., (eds.) Advances in computational collective intelligence, ICCCI 2023. Communications in computer and information science, vol. 1864, pp. 68–79. Springer, Cham (2023)
14. Donker, T., Bennett, K., Bennett, A., Mackinnon, A., van Straten, A., Cuijpers, P., Christensen, H., Griffiths, K.M.: Internet-delivered interpersonal psychotherapy versus internet-delivered cognitive behavioral therapy for adults with depressive symptoms: randomized controlled noninferiority trial. J. Med. Internet Res. **15**(5), e82 (2013)
15. Eler, E., Dimopoulos, F., Timmers, M., Treur, J., Koole, S.L.: An Adaptive network model for interpersonal emotion regulation in multimodal human-bot interaction. In: Hendrikse, S.C.F., Treur, J., Koole, S.L. (eds.) New analysis and modeling directions in social interaction science: emergent multimodal interpersonal synchrony, affiliation, and bonding, Ch. 4. Springer (2025)
16. Eler, E., Treur, J., Hendrikse, S. C. F., Donker, T., Koole, S. L.: An Adaptive network model for interpersonal emotion regulation in multimodal human-bot interaction. In: Nguyen, N.T., et al., (eds.) Computational collective intelligence. Proceedings of ICCCI 2024. Lecture Notes in Computer Science, vol. 14811, pp. 67–79. Springer, Cham (2024)
17. Ethofer, T., Anders, S., Erb, M., Droll, C., Royen, L., Saur, R., Reiterer, S., Grodd, W., Wildgruber, D.: Impact of voice on emotional judgment of faces: an event-related fMRI study. Hum. Brain Mapp. **27**(9), 707–714 (2006)

18. Gonçalves, P., Araújo, L., Coutinho, D., Santos, F.: Social interactions in online social networks: the case of Instagram. Front. Psychol. **2**, 139 (2011)
19. Gross, J.J.: The emerging field of emotion regulation: an integrative review. Rev. Gen. Psychol. **2**(3), 271–299 (1998)
20. Haidt, J.: The anxious generation: How the great rewiring of childhood is causing an epidemic of mental illness. Random House (2024)
21. Hendrikse, S.C.F., Treur, J.: Perspectives on social interaction science: from a multilevel multi-timescale computational architecture for synchrony- and homophily-induced attunement to designing bonding bots. In: Proc. of the 8th International Conference on Computational Methods in Systems and Software, CoMeSySo'24. Lecture Notes in Networks and Systems, vol. 1490. pp 518–545. Springer Nature (2025)
22. Hendrikse, S.C.F., Treur, J., Koole, S.L.: Modeling emerging interpersonal synchrony and its related adaptive short-term affiliation and long-term bonding: a second-order multi-adaptive neural agent model. Inter. J. Neural Syst. **33**(7), S038 (2023)
23. Hendrikse, S.C.F., Treur, J., Koole, S.L.: Relationship-specific and relationship-independent behavioural adaptivity in affiliation and bonding: a multi-adaptive dynamical systems approach. Cogn. Syst. Res. **84**, 101182 (2024)
24. Hendrikse, S.C.F., Treur, J., Koole, S.L. (eds.), New analysis and modeling directions in social interaction science: emergent multimodal interpersonal synchrony, affiliation, and bonding. Springer Nature (2025a)
25. Hendrikse, S.C.F., Treur, J., Koole, S.L.: Introduction to the role of adaptivity in social interaction science. In: Hendrikse, S.C.F., Treur, J., Koole, S.L. (eds.) New analysis and modeling directions in social interaction science: emergent multimodal interpersonal synchrony, affiliation, and bonding, Ch. 1. Springer (2025b)
26. Hendrikse, S.C.F., Treur, J., Koole, S.L.: The adaptive interplay of interpersonal synchrony, short-term affiliation and long-term bonding: a multi-timescale multi-adaptive agent model. In: Hendrikse, S.C.F., Treur, J., Koole, S.L. (eds.) New analysis and modeling directions in social interaction science: emergent multimodal interpersonal synchrony, affiliation, and bonding, Ch. 12. Springer Nature (2025c)
27. Hendrikse, S.C.F., Treur, J., Koole, S.L.: A multi-timescale multi-adaptive dynamical system approach for relationship-specific and relationship-independent adaptivity in affiliation and bonding. In: Hendrikse, S.C.F., Treur, J., Koole, S.L. (eds.) New analysis and modeling directions in social interaction science: emergent multimodal interpersonal synchrony, affiliation, and bonding, Ch. 13. Springer (2025d)
28. Hess, U., Fischer, A.: Emotional mimicry: why and when we mimic emotions. Soc. Pers. Psychol. Compass **7**(11), 1–17 (2013)
29. Ireland, M.E., Pennebaker, J.W.: Language style matching in writing: synchrony in essays, correspondence, and poetry. J. of Pers. and Soc. Psychology **99**(3), 549–571 (2010)
30. Keller, P.E., Novembre, G., Hove, M.J.: Rhythm in joint action: psychological and neurophysiological mechanisms for real-time interpersonal coordination. Philos. Trans. R. Soc. B: Biol. Sci. **369**(1658), 20130394 (2014)
31. Kokal, I., Engel, A., Kirschner, S., Keysers, C.: Synchronized drumming enhances activity in the caudate and facilitates prosocial commitment—if the rhythm comes easily. PLoS ONE **6**(11), e27272 (2011)
32. Koole, S.L., Tschacher, W.: Synchrony in psychotherapy: a review and an integrative framework for the therapeutic alliance. Front. Psychol. **7**, 862 (2016)
33. Launay, J., Dean, R.T., Bailes, F.: Synchronization can influence trust following virtual interaction. Exp. Psychol. **60**(1), 53–63 (2013)
34. Leongómez, J.D., Binter, J., Kubicová, L., Stolařová, P., Klapilová, K., Havlíček, J., Roberts, S.C.: Vocal modulation during courtship increases proceptivity even in naive listeners. Evol. Hum. Behav. **38**(2), 229–236 (2017)
35. Likowski, K.U., Mühlberger, A., Seibt, B., Pauli, P., Weyers, P.: Modulation of facial mimicry by attitudes. J. Exp. Soc. Psychol. **48**(1), 423–426 (2012)

36. Müller, V., Lindenberger, U., Düzel, S.: Intra- and inter-brain synchronization during musical improvisation on the guitar. PLoS ONE **14**(1), e0206854 (2019)
37. Néda, Z., Ravasz, E., Brechet, Y., Vicsek, T., Barabási, A.L.: The sound of many hands clapping. Nature **403**(6772), 849–850 (2000)
38. Pardo, J.S.: On phonetic convergence during conversational interaction. J. Acoust. Soc. Am. **133**(1), 588–594 (2013)
39. Paxton, A., Dale, R.: Argument disrupts interpersonal synchrony. Q. J. Exp. Psychol. **66**(11), 2092–2102 (2013)
40. Rivas, P., Liang, Z.: Marketing with ChatGPT: navigating the ethical terrain of GPT-based chatbot technology. AI **4**(2), 375–384 (2023)
41. Rymarczyk, K., Biele, C., Grabowska, A., Majczynski, H.: Empathy is related to differences in facial mimicry during positive and negative emotional processing. Front. Psychol. **11**, 1829 (2020)
42. Tausczik, Y.R., Pennebaker, J.W.: The psychological meaning of words: LIWC and computerized text analysis methods. J. Lang. Soc. Psychol. **29**(1), 24–54 (2010)
43. Treur, J.: Network-oriented modeling: addressing complexity of cognitive, affective and social interactions. Springer, Cham (2016)
44. Treur, J.: Network-Oriented modeling for adaptive networks: designing higher-order adaptive biological, mental and social network models. Springer Nature, Cham (2020)
45. Treur, J.: Modeling coevolution of dynamics, adaptivity and control of adaptivity by self-modeling networks. In: Hendrikse, S.C.F., Treur, J., Koole, S.L. (eds.) New analysis and modeling directions in social interaction science: emergent multimodal interpersonal synchrony, affiliation, and bonding, Ch. 2. Springer Nature (2025)
46. Twenge, J.M., Cooper, A.B., Joiner, T.E., Duffy, M.E., Binau, S.G.: Age, period, and cohort trends in mood disorder indicators and suicide-related outcomes in a nationally representative dataset, 2005–2017. J. Abnorm. Psychol. **128**(3), 185–199 (2019)
47. Willemse, C.J., Van Erp, J.B.F.: Social touch in human–robot interaction: robot-initiated touches can induce positive responses without extensive prior bonding. Int. J. Soc. Robot. **11**(2), 285–304 (2019)
48. Williams, W.C., Morelli, S.A., Ong, D.C., Zaki, J.: Interpersonal emotion regulation: implications for affiliation, perceived support, relationships, and well-being. J. Pers. Soc. Psychol. **115**(2), 224 (2018)
49. Yildirim, I., D'Mello, S.K., Kory, J.: Linguistic style matching and empathy in the therapeutic process. J. Lang. Soc. Psychol. **37**(1), 83–99 (2018)

Finalising

In this part, first in "From a Multilevel Multi-Timescale Computational Architecture for Synchrony-Induced and Homophily-Induced Attunement to Designing Bonding Bots" a unified computational architecture is presented that covers both homophily-induced and synchrony-induced social interaction adaptivity as addressed separately in earlier parts. In "Discussion and Concluding Remarks on Adaptivity in Social Interaction Science" the parts and chapters of this volume are reviewed and topics for further research suggested.

From a Multilevel Multi-Timescale Computational Architecture for Synchrony-Induced and Homophily-Induced Attunement to Designing Bonding Bots

Sophie C. F. Hendrikse and Jan Treur

Abstract In this chapter a generic computational architecture for social interaction adaptivity and attunement is presented that was developed based on unification of a number of specific case studies addressed earlier. This architecture describes a generic multilevel multi-timescale adaptive dynamical system and is formalised and represented as a multi-order self-modeling network. It is shown how this architecture covers different earlier case studies addressing social interaction adaptivity and attunement subsuming synchrony- or homophily-induced adaptivity. Moreover, it is shown how based on it a new application to the design of bonding bots is realised that supports interpersonal emotion regulation via adaptive human-bot interaction and attunement.

Keywords Social interaction · Homophily · Synchrony · Adaptivity · Similarity · Connectedness · Computational architecture · Multilevel · Multi-timescale · Bonding bot

---

S. C. F. Hendrikse
Department of Methodology, Tilburg School of Social and Behavioral Sciences, Tilburg University, Tilburg, The Netherlands
e-mail: S.C.F.Hendrikse@tilburguniversity.edu

J. Treur (✉)
Social AI Group, Department of Computer Science, Vrije Universiteit Amsterdam, Amsterdam, The Netherlands
e-mail: j.treur@vu.nl

S. C. F. Hendrikse et al. (eds.), *New Analysis and Modeling Directions in Social Interaction Science*, Studies in Systems, Decision and Control 614,
https://doi.org/10.1007/978-3-031-99968-0_20

1 Introduction: Social Interaction Adaptivity

A central focus in this chapter is on unification of the contributions in our research line on computational methods for social interaction science. The emphasis is on modeling and analysis of multimodal mental and social interaction dynamics adaptivity and attunement as occurs in affiliation and bonding from underlying causal mechanisms and pathways. The chapter will contribute a unified computational architecture that covers several case studies analysed in the past and was also evaluated based on a new analysis to design bonding bots that interact with humans in order to obtain interpersonal emotion regulation.

The processes where people mutually coordinate their behavior have been studied under various labels, including alignment [104], mimicry [23], interpersonal coordination (dynamics), coregulation [25], concordance [17], coupling [42], matching [12], and mirroring [44]. We use the term 'interpersonal synchrony' or simply 'synchrony', to refer to mutually coordinated behavioral patterns over time displayed by two or more agents [6, 9, 28, 30, 37, 77, 91, 103, 104]. Also in the area referred as 'bonding by homophily', it has been studied how during social interaction people can form bonds, e.g., McPherson et al. [83]. The term 'homophily' is used here: it concerns states or characteristics of two persons that are (approximately) the same at a given point in time, such as an emotion, opinion or a preference [83].

The computational architecture proposed in the current chapter provides a form of unification of the types of processes and phenomena mentioned. Throughout this work, we have sought to allow computational modeling and empirical science to mutually benefit from each other. More specifically, we used the causal relationships and pathways described in empirical literature to explain certain phenomena as the foundation for a programmatic series of computational adaptive dynamical system models. These were designed as higher-order adaptive network models based on the self-modeling or reification principle introduced in [118] and extended to multiple levels in [120, 121], see also [122, 123].

Such empirical literature usually has an emphasis on showing that these mechanisms and pathways are necessary for the phenomenon to be explained: if these mechanisms are (made) disfunctional, the phenomenon does not occur. But whether the presumed mechanisms on their own are sufficient to let the phenomenon emerge is another, much harder question to answer empirically due to the high extent of contextuality. Contributions from other mechanisms than the presumed ones that are left implicit but are present in the context cannot be excluded empirically in an easy manner as many of such mechanisms naturally occur in all real humans that act in experiments.

One of the major advantages of computational modeling is that it can test whether proposed correlations between such mechanisms and phenomena indeed can be confirmed computationally by in silico experiments (experiments through computer simulations), both in the necessary and sufficiency sense. This allows researchers to verify whether what leads to the emergence of certain observed phenomena are indeed the presumed causal relationships and not other mechanisms left implicit.

Moreover, in specific cases where there is doubt on the mechanisms involved it means that researchers can ask themselves whether any additional mechanisms that are tacitly assumed are needed explicitly and check this computationally. Furthermore, in cases where the empirical literature does not show all necessary specific mechanisms in detail, assumptions can be made about the missing details and based on that it can be verified computationally whether and to what extent they lead to the phenomena of interest. Consequently, in the research we have sought to achieve a mutual exchange between empirical science and computational science.

2 Coevolution of Similarity and Induced Connectedness

During social interaction different forms of 'becoming connected' may develop. This may concern shorter-term forms such as affiliation during a meeting or therapy session or a joint action that is undertaken. But also longer-term forms often develop, indicated, for example, by friendhips, lifelong relationships or bonding. In this chapter we will use the term 'connectedness' to subsume all such different types of becoming connected in one way or the other (Fig. 1).

There are two areas that particularly address how such connectedness may develop: homophily-induced connectedness and (interpersonal) synchrony-induced connectedness. The first area was developed mostly from sociology and social science (see Sect. 2.1), the second mostly from social and clinical psychology (see Sect. 2.2).

2.1 *Homophily-Induced Connectedness*

Already in early research in the area of social networks an important focus was on the relationship between connectedness between persons and the extent of similarity of them. For example, in McPherson et al. [83] it is put forward that the relationship between connectedness and similarity is mentioned by Aristotle by writing that people 'love those who are like themselves' [5], p. 1371) and Plato writing 'similarity begets friendship' [93], p. 837). Systematic observations of group formation and network ties by social scientists began in the 1920s and 1930s. For example, Bott [22], Wellman [140], Hubbard [70] reported that school children formed friendships and play groups at higher rates if they were similar on demographic characteristics

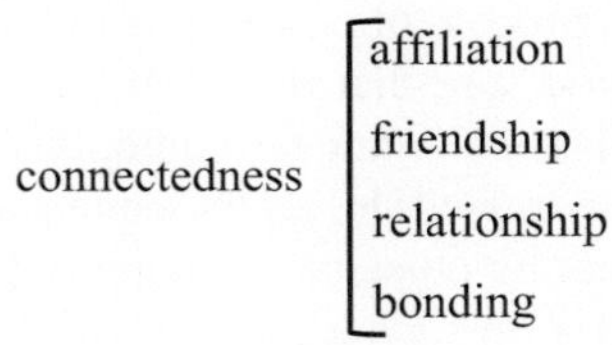

Fig. 1 Different forms of becoming connected: shorter-term and longer-term

[83]. An important contribution to the sociological literature is [73] about forming friendships. They use the term 'homophily' and to describe the process they also quoted the metaphoric expression.

> Birds of a feather flock together [78]

which has often been used in the literature since then [83]. For a further historical review, see Freeman [38]. Although in such literature often the emphasis may seem on the causal effect that the extent of homophily has on (strengthened) connectedness such as friendship, usually there is still a causal effect in the opposite direction as well. By social contagion, persons that have strong connectedness can strengthen their extent of homophily. Unfortunately, the above metaphor is not applicable to describe this opposite causal effect: it is very unlikely that feathers of birds can change due to flocking together. Nevertheless, for many cases a bidirectional causal mechanism is considered realistic and sometimes described by the term coevolution which means that over time:

- connectedness evolves via adaptation under influence of the extent of homophily
- homophily evolves via social contagion under influence of the extent of connectedness

Especially when homophily is considered for personal characteristics or states that can change over time, such as opinions, emotions, or preferences for types of movies or music, it is assumed that these can become attuned to each other due to social contagion. In turn, social contagion depends on strength of connectedness, so in this way you indeed get the above circular feedback loop.

In recent years this circular coevolution process in relation to homophily has been modeled from an adaptive dynamical system perspective [94], often in a network-oriented manner, see, for example, Blankendaal et al. [18], Luo et al. [81]; Hsieh et al. [69], King et al. (2024), Liu et al. [80], Sharpanskykh and Treur [106], Snijders et al. [110], Su et al. [111], Treur [119], Van den Beukel et al. [135]. Simulations based on such adaptive computational network models have demonstrated how such a circular coevolution process can explain, for example, the emergence of clusters or communities, or seggregation in society, e.g., [50, 73].

Perhaps because homophily-induced connectedness often has been addressed from a social science or sociology perspective, in computational models for coevolution as indicated in the previous paragraph usually no internal mental states are taken into account. In particular, the underlying criterion of homophily is usually based on objective states in the social world as considered from an external observer perspective. However, in reality, for one of the persons to notice a form of homophily, a process of perception and interpretation plays an important role. For example, are you sure that the other person really prefers the same type of music? Therefore, in some work on homophily-induced connectedness this subjective perspective is addressed. Here, the connectedness is induced by subjective homophily detection, not by objective homophily [84, 85, 125], see also Treur [129], Mukeriia et al. [86, 87]. In such a case also manipulation of homophily can be modelled, for example, by faking that you like the same type of movies as your friend does [20], see also Bössenecker et al. [21].

2.2 Synchrony-Induced Connectedness

In the analysis in Sect. 2.1, the term homophily can also be generalised so that also the notion interpersonal synchrony is covered. The concept 'synchrony-induced connectedness', has been explored extensively in literature such as [68, 112, 132–134, 143].

Within the area focusing on synchrony, research has documented how people spontaneously synchronize their activities in a variety of ways, including their movements [13, 28, 90], vocal pitch [102, 115, 146], linguistic expressions [32, 43, 105], their physiological responses [91] and neural responses [104]. These interpersonal dynamics play a role in many relationships over the lifetime. Already in utero, infants synchronize their heart beats with their mother's heart beats [71, 136]. After birth, children develop autonomous emotion regulation from physiological synchrony that they exhibit with their caretaker [36]. Moreover, when adults form close relationships, they are prone to synchronize their physiological responses [91]. Neural synchrony can also elevate within parent–child relationships [96], romantic relationships [75] or persons who play together guitar [89]. Strangers may even synchronize their physiological responses by merely watching the same event [76]. Interpersonal synchrony is thus a pervasive pattern in social life.

Importantly, interpersonal synchrony has been shown to have important psychological consequences concerning adaptation of the social interaction behaviour. For instance, it has been found that people bond more with each other after they were experimentally induced to synchronize their movements [112]. Furthemore, leading people to synchronize by singing or laughing together has been shown to stimulate more cooperative behavior [143]. Perhaps most impressively, experimentally inducing synchrony in brain-to-brain activity has been found to improve joint performance on cognitive tasks [92]. These and related studies demonstrate that interpersonal synchrony is not merely an epiphenomenon, but that it actually shapes interpersonal dynamics in an adaptive manner. Consequently, it seems important to learn more about the nature of interpersonal synchrony.

Given the above, like for homophily as discussed in Sect. 2.1, from a dynamical system perspective also for synchrony an overall circular pattern of coevolution over time has been found:

- connectedness evolves via adaptation under influence of the extent of synchrony
- synchrony evolves via social contagion under influence of the extent of connectedness

Recently, this type of coevolution processes have been analysed computationally in [54, 55, 59–62]; see also [56, 57, 63–66]. To achieve this, based on the temporal factorisation principle from [116, 117], the notion of subjective synchrony detection was postulated to indicate how a person can actually notice being in synchrony with someone else.

Adaptive self-modeling network models were very useful to address the complexity of this type of adaptive dynamical systems. This developed adaptive dynamical system modeling approach to synchrony-based connectedness has been successfully applied to model complex social phenomena such as adaptive interpersonal emotion regulation and adaptive collective phenomena involving elderly in [34, 137, 144], see also [35, 138, 145].

2.3 A Unified View on Coevolution: Similarity-Induced Connectedness

To unify the two types of coevolution discussed in Sects. 2.1 and 2.2, in this chapter, we will use the term similarity as generalisation subsuming different forms of both homophily and synchrony (Fig. 2).

The two co-evolution patterns indicated in Sects. 2.1 and 2.2 then can be unified as a more general circular coevolution pattern as shown in Fig. 3.

This perspective provides a conceptual basis for the unified computational view based on multilevel multi-timescale dynamical system modeling that is developed in the current chapter.

3 Conceptual Architecture for Social Interaction Adaptivity at Multiple Timescales

In Fig. 4 an informal global conceptual architecture is shown for adaptive (multimodal) social interaction for coevolution processes as depicted in Fig. 3. This is used as point of departure for the computational architecture proposed later on in this chapter. It considers base level mechanisms, multimodal interaction and

Fig. 2 Different forms of similarity subsuming homophily and synchrony

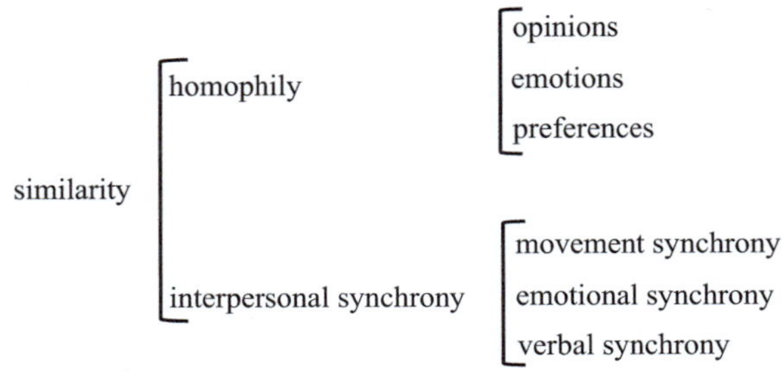

Fig. 3 Coevolution of similarity and connectedness as unified view

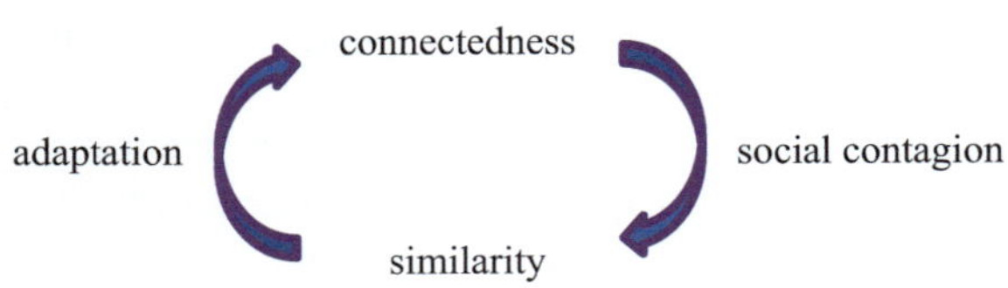

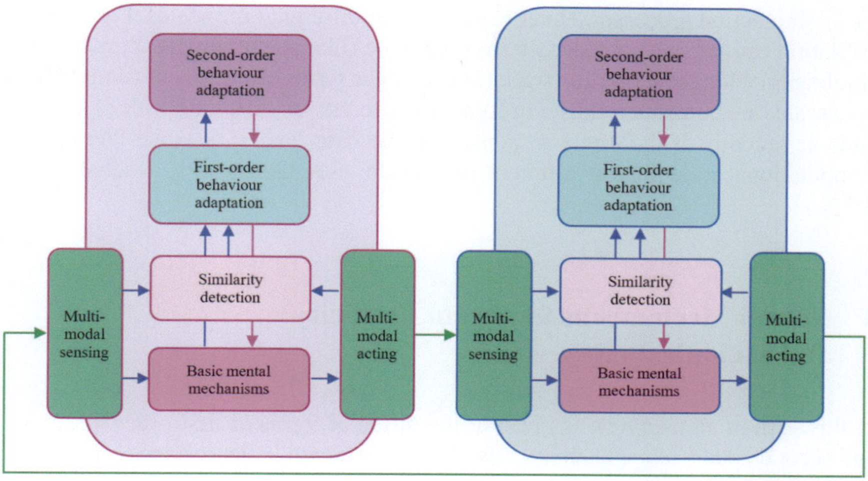

Fig. 4 Informal overall conceptual architecture for adaptive multimodal social interaction; generalised and refined version of Fig. 13 contributed by the current authors to [41]

different types of social interaction adaptivity. This conceptual architecture generalises and refines what is depicted in [45], Fig. 13, by this time subsuming both homophily-induced connectedness and synchrony-induced connectedness (using the more abstract term similarity for homophily and synchrony) and by distinguishing different levels of adaptivity, in particular, first-order and second-order adaptivity, see Fig. 4.

At the base level, basic mechanisms are covered (see the pink boxes in Fig. 4), for example, for representing based on (multimodal) sensory processing, for responding by preparing for actions, and for executing by performing prepared actions.

As part of the base level, also some form of (subjective) similarity detection is assumed by which each of the persons can subjectively notice similarity, for example, in terms of homophily or synchrony. At this base level, also the actual multimodal interaction between two agents takes place, indicated by the green boxes and green arrows in Fig. 4.

First-order adaptivity is indicated by the light-blue boxes in Fig. 4. It can cover, for example, plasticity of structures in the brain such as connectivity adaptation by Bainian or Hebbian learning [8, 49] or intrinsic excitability adaptation as described in Chandra and Barkai [39], Debanne et al. [30].

Second-order adaptivity is indicated by the purple boxes in Fig. 4. This provides context-sensitive control over first-order adaptivity, also called metaplasticity; see for example [1, 98, 109]. These forms of adaptivity and their interaction will be discussed in more detail in Sect. 4 and formalised in a proposed computational architecture in Sect. 6.

Note that in this chapter, we consider connectedness concepts such as affiliation and bonding in an operational manner from the perspective of changing mental and social interaction processes. Hereby we consider different time scales. For example, affiliation can happen on a short-term time scale (like one 45 min therapy session), bonding on a longer-term time scale (like a longer series of therapy sessions). These time scales are both much slower in comparison to the very fast time scale of milliseconds or seconds of the direct responses on the base level. For some literature on computational modeling at different time scales, see Geiser [40], Radhakrishnan [95].

4 Neural Mechanisms for Brain Plasticity and Metaplasticity

In this section an overview is given of the different forms of first-order (plasticity) and second-order (metaplasticity) adaptivity considered in neuroscience that form a foundation for the proposed computational architecture.

4.1 Neural Mechanisms for First-Order Adaptivity: Brain Plasticity

In this section a number of mechanisms for first-order adaptivity (plasticity) of the brain are discussed. For second-order adaptivity (metaplasticity), see Sect. 4.2. In this section, two examples of first-order adaptation principles are considered: Bainian [8] or Hebbian [49] learning for connectivity adaptation and intrinsic excitability modulation [31, 41].

Neural mechanisms for connectivity adaptation: synaptic plasticity

Some early work on connectivity adaptation can be found, for example, in contributions by Beale [10] and Bain [8] who as a first step have put forward the cyclic structure of networks ('circuits') in the brain:

> I venture to conclude that the typical anatomical arrangement of a nervous mechanism is not a cord with two ends—a point of origin and terminal extremity, but a cord without an end—a continuous circuit. Beale [10], p. 392

Bain [8] commented on this and points at the role of connection ('cell-crossing') points:

> He couples with this appearance the doctrine (maintained by him, although disputed by others) that the nerves terminate in loops, and consequently form an unbroken nervous circuit. He then suggests that the cell-crossing is the place where the inner bendings of a great many independent circuits come into close neighbourhood, and affect one another by a process of the nature of electrical induction. Any one of the circuits being active, or excited, would impart excitement to all that came near it in the same cell. Bain [8], p. 118–119

As one step further, also plasticity of brain networks was discussed in Bain [8]:

> Now to each impression, each sensation or thought, there corresponds physically a group or series of nerve-currents; when two impressions concur, or closely succeed one another, the nerve currents find some bridge or place of continuity, better or worse, according to the abundance of nerve matter available for the transition. In the cells or corpuscles where the currents meet and join, there is, in consequence of the meeting, a *strengthened connexion* or *diminished obstruction*—a preference track for that line over lines where no continuity has been established. Bain [8], p. 117, original italics

He continues in more detail in:

> This is merely a hypothetical rendering of the facts: yet it is a very probable rendering. In the nature and number of the nerve elements, and their mode of connexion, there is nothing hypothetical; and there is no departure from fact or strong probability, in assigning special and distinct tracks for the currents connected with each separate sensation, idea, emotion, or other conscious state. As to the precise mode of the plastic growth that unites separate impressions into trains and aggregates in the memory,—we know that the corpuscles or crossings are the points that must be operated upon; that a flow of healthy blood must co-operate to the effect; and that the process takes time. It is evidently a species of growth: but the precise molecular change effected in the lines of strengthened communication, or diminished obstruction, we can describe only as increasing the conducting tendency in those lines, as compared with the collateral openings where no such operation has taken place. Bain [8], pp. 117–118

> I can suppose that, at first, each one of the circuits would affect all others indiscriminately; but that, in consequence of two of them being independently made active at the same moment (which is the fact in acquisition), a strengthened connexion or diminished obstruction would arise between these two, by a change wrought in the intervening cell-substance; and that, afterwards, the induction from one of these circuits would not be indiscriminate, but select; being comparatively strong towards one, and weaker towards the rest. [8], p. 119

Much later, this perspective on connectivity adaptation has been presented as well by Hebb [49]. For example, in this form:

> When an axon of cell A is near enough to excite B and repeatedly or persistently takes part in firing it, some growth process or metabolic change takes place in one or both cells such that A's efficiency, as one of the cells firing B, is increased. Hebb [49], p. 62

This type of connectivity adaptation is often referred as Hebbian learning, but considering the above, might as well or even better be called Bainian learning. For more on this issue, see also Wilkes and Wade [141]. To indicate this type of adaptation, often the following summarised phrasing is used:

> What fires together, wires together Shatz [107], Keysers and Gazzola [69]

Neural mechanisms for intrinsic excitability adaptation: nonsynaptic plasticity

The brain network does not only show connectivity adaptation (synaptic plasticity). More recently, in the neuroscientific literature, a distinction was made between synaptic and nonsynaptic (or intrinsic) plasticity. Intrinsic nonsynaptic adaptation of (neural) states has been addressed in more detail more recently in literature such as [31, 41, 147]. For example:

> Learning-related cellular changes can be divided into two general groups: modifications that occur at synapses and modifications in the intrinsic properties of the neurons. While it is commonly agreed that changes in strength of connections between neurons in the relevant networks underlie memory storage, ample evidence suggests that modifications in intrinsic neuronal properties may also account for learning related behavioral changes. Chandra and Barkai, [41], p. 30

The following quote indicates that long-lasting modifications in excitability of neurons can evolve over time:

> Long-lasting modifications in intrinsic excitability are manifested in changes in the neuron's response to a given extrinsic current (generated by synaptic activity or applied via the recording electrode). Chandra and Barkai [41], p. 30

Intrinsic excitability adaptation has been related, for example, to homeostatic regulation [142] and also to how deviant dopamin levels during sleep make that dreams can make use of more associations due to easier excitable neurons, e.g., Budding et al. [24].

The interplay of connectivity adaptation and intrinsic excitability adaptation

Both synaptic and nonsynaptic forms of adaptation can work together, e.g., Alejandre-García et al. [2], Budding et al. [24], Lisman et al. [79], Hansel and Yuste [47], see for example:

> So, circuit plasticity could be due to the interplay of synaptic and intrinsic mechanisms and this balance could be different for short and long-term plasticity. In this view, ensembles could be quickly formed by fast intrinsic mechanisms but then be modified by slower synaptic plasticity, perhaps in relation with learning. Alejandre-García et al. [2], p. 18/25
>
> To explain how neuronal ensembles are generated, we propose an "iceberg" model, by which the increased neuronal excitability that results from repeated input stimulation makes subthreshold connections become suprathreshold, enhancing the postsynaptic effect of already existing synapses, and generating a neuronal ensemble. Hansel and Yuste [47], p. 4/10

Note that here it is assumed that intrinsic plasticity works on a faster timescale than synaptic plasticity. Based on this perspective, the model proposed in Budding et al. [24] explains how creativity can be based on dopamine-induced (DA) enhanced associativity during dreaming that is 'exported' to the awake life. More specifically, the model uses dopamine-induced enhanced intrinsic excitability by lowering socalled **T**-states for adaptive excitability thresholds (see also Sect. 5) as short-term adaptation during dreaming (only short-lived) which in turn boosts longer-term adaptation based on Hebbian learning that can cross the border from dreaming to being awake:

> The longer-term adaptation of connection weights by Hebbian learning creates the plasticity that is needed, as it has the potential to cross the border from sleeping to being awake (depending also on dream recall). This is in contrast to the DA-dependent **T**-state suppression which is a short-term adaptation that will not cross this border. Budding et al. [24], p. 42

In a different context, a similar perspective on short-term adaptation boosting long-term adaptation is used in Hendrikse et al. [54], see also Hendrikse et al. [56], Ch 12. Here, induced by detected emerging synchrony, also different time scales for short-term affiliation and long-term bonding are used where repeated episodes of nonsynaptic short-term adaptation (intrinsic excitability modulation) can boost the synaptic long-term adaptation (connectivity adaptation).

4.2 Neural Mechanisms for Second-Order Adaptivity: Metaplasticity

The two types of first-order adaptation principles discussed above refer to what in neuroscientific literature is called plasticity. However, this is usually not a constant feature. For an organism, in some circumstances it is better to learn (and change) fast, but in other circumstances, it is better to stay stable and let what has been learnt in the past persist: 'The Plasticity Versus Stability Conundrum' ([109], p. 773). Under which circumstances and to which extent such plasticity actually takes place is controlled by a form of socalled *metaplasticity*, e.g., [1, 39, 82, 98, 109]. Such control can address 'The Plasticity Versus Stability Conundrum' by only making plasticity happen in contextual circumstances when it is important for the person to change, and otherwise stabelize it. In literature as mentioned, various studies show how adaptation (for example, connectivity adaptation), is modulated depending on contextual circumstances by accelerating the adaptation process or decelerating or even blocking it. Among the reported factors affecting plasticity in such a way are stimulus exposure, activation, previous experiences, and stress. For example, in Robinson et al. [98] the following compact quote can be found summarizing that increasing stimulus exposure makes that the adaptation speed increases:

> Adaptation accelerates with increasing stimulus exposure Robinson et al. [98], p. 2)

Also, for learning of mental models by individuals or for learning within or of organisations such context-sensitive control of adaptation is a crucial concept, e.g., Treur and Van Ments [130], Canbaloğlu et al. [27]. For example, depending on the context, a manager can decide to organise organisational processes where individuals learn from other individuals. In Mukeriia 85, Treur [122], Ch 6 different variants of second-order adaptation to control bonding by adaptive homophily are addressed; see also this volume [56], Ch 8.

5 Modeling Adaptive Dynamical Systems by Self-Modeling Networks

The type of computational modeling considered here is based on a computational architecture of multilevel adaptive dynamical system models. Such an architecture can cover well the concept of multiple control levels for mechanisms in biological and mental processes as from a philosophical and biological perspective put forward, for example, in Bechtel and Bich [11], Bich and Bechtel [15, 16], Treur [126, 127]. In this chapter two of such adaptation or control levels are used: first-order and second-order. The second-order adaptation level exerts control over the first-order adaptation in a context-sensitive manner. For more details of underlying neural mechanisms addressing brain plasticity and metaplasticity, see Sect. 4.

To design multilevel adaptive dynamical systems for social interaction adaptivity, a network-oriented modeling format based on self-modeling networks can be useful. In Hendrikse et al. [54], Treur [124] it has been shown how any smooth multilevel adaptive dynamical system can be represented as a multilevel temporal-causal self-modeling network; see also this volume [56], Ch 11). Therefore, this network-oriented format does not introduce any fundamental limitation. A brief introduction to this modeling format is as follows.

In this approach *nodes* Y in a temporal-causal network have activation values $Y(t)$ that are dynamic over time t; therefore, they serve as state variables and will usually be called *states*. To specify these state dynamics, the states are considered to causally affect each other by the connections within the network. Following [122], a basic *network structure* is characterised by:

- **connectivity characteristics**
  Connections from a state X to a state Y and their *weights* $\boldsymbol{\omega}_{X,Y}$
- **aggregation characteristics**
  For any node Y, some *combination function* $\mathbf{c}_Y(..)$ defines aggregation that is applied to the *single impacts* $\mathbf{impact}_{X_i,Y}(t) = \boldsymbol{\omega}_{X_i,Y} X_i(t)$ on Y from its incoming connections from states $X_1, \ldots, X_k$:

$$\begin{aligned}\mathbf{aggimpact}_Y(t) &= \mathbf{c}_Y\left(\mathbf{impact}_{X_1,Y}(t), \ldots, \mathbf{impact}_{X_k,Y}(t)\right)\\ &= \mathbf{c}_Y(\boldsymbol{\omega}_{X_1,Y} X_1(t), \ldots, \boldsymbol{\omega}_{X_k,Y} X_k(t))\end{aligned} \tag{1}$$

- **timing characteristics**
  Each state Y has a *speed factor* $\boldsymbol{\eta}_Y$ defining how fast it changes upon given impact. This is an important feature enabling different time scales within one network model.

Here, the states X_i and Y have activation levels $X_i(t)$ and $Y(t)$ that vary (often within the [0, 1] interval) over time, described by real numbers t. Note that speed factors can be used to design network models where different states change on different time scales.

The dynamics of such networks are described by the following difference (or differential) equations that via (1) incorporate in a canonical manner the network characteristics $\boldsymbol{\omega}_{X,Y}$, $\mathbf{c}_Y(..)$, $\boldsymbol{\eta}_Y$:

$$Y(t + \Delta t) = Y(t) + \boldsymbol{\eta}_Y[\mathbf{aggimpact}_Y(t) - Y(t)]\Delta t \tag{2}$$

for any state Y.

The concept of self-modeling networks (or reified networks) was introduced in Treur [118] to model adaptation for a network by adding *self-model states* to it. This modeling approach was extended by introducting higher-order self-models in Treur [120, 121] and further extended by illustrations for case studies and mathematical analysis in Treur [122, 123], resulting in multilevel or higher-order *self-modeling networks*, also called *reified* networks.

The idea is that self-model states represent network characteristics by network states, e.g., a self-model state $\mathbf{W}_{X,Y}$ is used to represent connection weight $\boldsymbol{\omega}_{X,Y}$ or a self-model state $\mathbf{T}_Y$ is used to represent excitability threshold $\boldsymbol{\tau}_Y$. Such states are also called in general **W**-states or **T**-states. Then the impacts of the self-modeled characteristics $\boldsymbol{\omega}_{X,Y}$ or $\boldsymbol{\tau}_Y$ on a state Y at time t can be modelled as impacts from the state values $\mathbf{W}_{X,Y}(t)$ or $\mathbf{T}_Y(t)$ of such self-model states $\mathbf{W}_{X,Y}$ or $\mathbf{T}_Y$ on Y. In this way, the dynamics of these self-model states can model adaptation processes for connectivity learning and modulation of excitability as discussed in Sect. 4.1. This brings the impacts from the adaptive characteristics on a state Y in the standard form of a network model where via connections node activations affect activations of other nodes.

For example, if all connection weights $\boldsymbol{\omega}_{X_i,Y}$ in (1) are represented by self-model states $\mathbf{W}_{X_i,Y}$, then $\mathbf{impact}_{X_i,Y}(t) = \mathbf{W}_{X_i,Y}(t)X_i(t)$ for all i and therefore (1) becomes:

$$\begin{aligned}\mathbf{aggimpact}_Y(t) &= \mathbf{c}_Y\big(\mathbf{impact}_{X_1,Y}(t), \ldots, \mathbf{impact}_{X_k,Y}(t)\big)\\ &= \mathbf{c}_Y(\mathbf{W}_{X_1,Y}(t)X_1(t), \ldots, \mathbf{W}_{X_k,Y}(t)X_k(t))\end{aligned}$$

Within a self-modeling network model, such self-model states have their own network characteristics. For instance, **W**-states $\mathbf{W}_{X,Y}$ have their own speed factor $\boldsymbol{\eta}_{\mathbf{W}_{X,Y}}$, and their own connection weights $\boldsymbol{\omega}_{Z,\mathbf{W}_{X,Y}}$ of a connection from a state Z to $\mathbf{W}_{X,Y}$. As adaptation processes modeled by the dynamics of such self-model states in general take place on a much slower time scale than for the base states X and Y, the values of such speed factors $\boldsymbol{\eta}_{\mathbf{W}_{X,Y}}$ or $\boldsymbol{\eta}_{\mathbf{T}_Y}$ will be much lower than for those of these X and Y. This feature is then used to enable different time scales for base dynamics and for adaptivity.

As a base network model extended by self-model states is also a temporal-causal network model itself, as has been shown in detail [122], Ch 10), this self-modeling construction can easily be applied iteratively to include self-models of multiple self-modeling (or reification) levels. For example:

- a second-order self-model state $\mathbf{H}_{\mathbf{W}_{X,Y}}$ can represent the speed factor of first-order self-model state $\mathbf{W}_{X,Y}$ so that the adaptation speed of $\mathbf{W}_{X,Y}$ becomes adaptive
- a second-order self-model state $\mathbf{H}_{\mathbf{T}_Y}$ can represent the speed factor of first-order self-model state $\mathbf{T}_Y$ so that the adaptation speed of $\mathbf{T}_Y$ becomes adaptive
- a second-order self-model state $\mathbf{W}_{\mathbf{Z},\mathbf{W}_{X,Y}}$ can represent the weight $\boldsymbol{\omega}_{Z,\mathbf{W}_{X,Y}}$ of the connection from Z to first-order self-model state $\mathbf{W}_{X,Y}$ so that the adaptation speed of $\mathbf{W}_{X,Y}$ becomes adaptive

- a second-order self-model state $\mathbf{W}_{\mathbf{Z},\mathbf{T}_Y}$ can represent the weight $\boldsymbol{\omega}_{Z,\mathbf{T}_Y}$ of the connection from Z to first-order self-model state $\mathbf{T}_Y$ so that the adaptation speed of $\mathbf{T}_Y$ becomes adaptive

This can model higher-order adaptive dynamical systems by network models and has turned out quite useful to model metaplasticity in a unified form by a second-order self-model: context-sensitive control of adaptation, as discussed in Sect. 4.2. Such second-order self-model states can use yet other timescales by assigning proper values to their own speed factors $\boldsymbol{\eta}_{\mathbf{H}_{\mathbf{W}_{X,Y}}}$, $\boldsymbol{\eta}_{\mathbf{H}_{\mathbf{T}_Y}}$, $\boldsymbol{\eta}_{\mathbf{W}_{Z,\mathbf{W}_{X,Y}}}$, and $\boldsymbol{\eta}_{\mathbf{W}_{Z,\mathbf{T}_Y}}$. For more details of the modeling approach, see [122] or this volume, Chap. 2 [128].

In other, more biologically oriented research the role of environment-induced epigenetic changes for the development of mental disorders is addressed, such as Schizophrenia, ASPD, Binge eating, Insomnia, Psychopathy, SUD. In that case, a fifth-order control architecture turned out useful; e.g., [7, 19, 26, 48, 72, 99].

6 A Computational Architecture for Social Interaction Adaptivity

In this section, in addition to the informal conceptual architecture in Sect. 3, a formalised multilevel multi-timescale computational dynamical system architecture is discussed, represented via the self-modeling networks modeling approach briefly introduced in Sect. 5. This computational architecture refines and formalises the informal conceptual architecture from Fig. 4.

6.1 Base Level of the Computational Architecture

The architecture for the base level for three modalities (movement *m*, affect *b*, and verbal modality *v*) can be seen in Fig. 5 with states indicated by pink (within person) and green (external) circles and connections by blue arrows.

Here perception by sensory processing and representing takes place via the sense (for sensing) and rep (for representing) states and the connections between them. Moreover, the connections from rep states to prep (for preparation) states indicate responding, and the connections from prep states to talk, express and move state indicate executing. This refines and formalises what is in the lower pink boxes in Fig. 4. Moreover, subjective similarity detection states for the different modalities are included (pink diamond shapes). In this way, this picture refines and formalises what is in both the lower and upper pink boxes in Fig. 4.

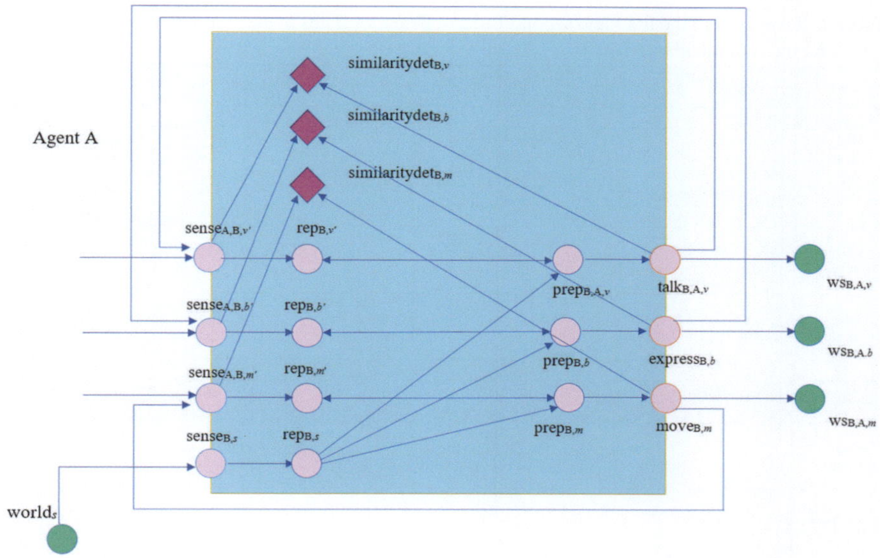

Fig. 5 Base level of the computational architecture with states and connectivity for basic mechanisms including subjective similarity detection states (pink diamond shapes)

6.2 *First-Order Adaptation Level of the Computational Architecture*

For the computational architecture, at the first-order adaptation level, the agents make use of one or two types of first-order self-model states, called **W**-states and **T**-states; see the blue area of Table 1:

- self-model **T**-states $\mathbf{T}_Y$ control adaptive excitability thresholds for the base level states Y indicated in their subscripts. This is a form of (nonsynaptic) adaptation at a relatively fast adaptation timescale, but slower than the timescales for the base level.
- self-model **W**-states $\mathbf{W}_{X,Y}$ control adaptive connection weights of connections $X \rightarrow Y$ for the base level states X,Y indicated in their subscripts. This is a form of (synaptic) adaptation at a relatively slow adaptation timescale.

In Table 1 more details can be found of these different types of first-order self-models states of the computational architecture for the different types of base states and base connections for representing, responding, executing and connecting. Within the computational architecture, the connections for these first-order self-model states are as follows; see also Table 2 for an overview of the connectivity of all self-model states:

Table 1 Overview of the first- and second-order self-model states in the computational architecture for social interaction adaptivity

	Representing		Responding		Executing		Connecting	
First-order adaptation								
Adaptive activation timing								
Ultra short term	**H**-states for speed of representing base states		**H**-states for speed of responding base states		**H**-states for speed of executing base states		**H**-states for speed of connecting base states	
Adaptive activation strength								
	connectivity	**excitability**	**connectivity**	**excitability**	**connectivity**	**excitability**	**connectivity**	**excitability**
Short term		**T**-states for excitability of representing base states		**T**-states for excitability of responding base states		**T**-states for excitability of executing base states		**T**-states for excitability of connecting base states
Long term	**W**-states for weights for representing base states		**W**-states for weights for responding base states		**W**-states for weights for executing base states		**W**-states for weights for connecting base states	
Second-order adaptation								
Adaptive adaptation timing								
	connectivity	**excitability**	**connectivity**	**excitability**	**connectivity**	**excitability**	**connectivity**	**excitability**
Short term		$\mathbf{H_T}$-states for excitability adaptation speed for		$\mathbf{H_T}$-states for excitability adaptation speed for		$\mathbf{H_T}$-states for excitability adaptation speed for		$\mathbf{H_T}$-states for excitability adaptation speed for
		representing base states		responding base states		executting base states		connceting base states
Long term	$\mathbf{H_W}$-states for weights adaptation speed for representing base states		$\mathbf{H_W}$-states for weights adaptation speed for responding base states		$\mathbf{H_W}$-states for weights adaptation speed for executing base states		$\mathbf{H_W}$-states for weights adaptation speed for connecting base states	
Adaptive adaptation strength								
	connectivity	**excitability**	**connectivity**	**excitability**	**connectivity**	**excitability**	**connectivity**	**excitability**
Short term		$\mathbf{W_T}$-states for excitability adaptation strength for representing base states		$\mathbf{W_T}$-states for excitability adaptation strength for responding base states		$\mathbf{W_T}$-states for excitability adaptation strength for executing base states		$\mathbf{W_T}$-states for excitability adaptation strength for connecting base states
Long term	$\mathbf{W_W}$-states for weights adaptation strength of representing base states		$\mathbf{W_W}$-states for weights adaptation strength of responding base states		$\mathbf{W_W}$-states for weights adaptation strength of executing base states		$\mathbf{W_W}$-states for weights adaptation strength of connecting base states	

Table 2 Concise overview of the connectivity for the self-model states by upward and downward connections in the computational architecture

Upward connections		Downward connections
For first-order self-model states		
Similarity detector base states →	**T**-state $\mathbf{T}_Y$	→ Base state Y
Similarity detector base states →	**W**-state $\mathbf{W}_{X,Y}$	→ Base state Y
For second-order self-model states		
Relevant context indicators from lower levels →	$\mathbf{H_T}$-state $\mathbf{H}_{\mathbf{T}_Y}$	→ **T**-state $\mathbf{T}_Y$
Relevant context indicators from lower levels →	$\mathbf{H_W}$-state $\mathbf{H}_{\mathbf{W}_{X,Y}}$	→ **W**-state $\mathbf{W}_{X,Y}$
Relevant context indicators from lower levels →	$\mathbf{W_T}$-state $\mathbf{W}_{Z,\mathbf{T}_Y}$	→ **T**-state $\mathbf{T}_Y$
Relevant context indicators from lower levels →	$\mathbf{W_W}$-state $\mathbf{W}_{Z,\mathbf{W}_{X,Y}}$	→ **W**-state $\mathbf{W}_{X,Y}$

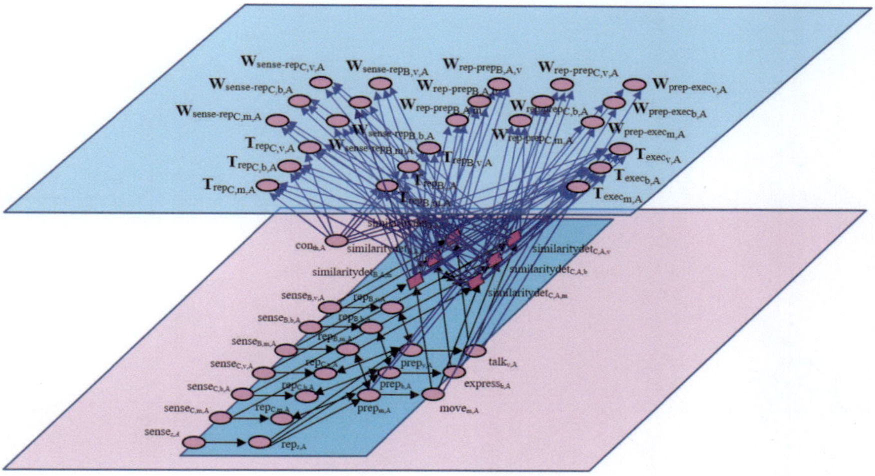

Fig. 6 A 3D visualisation of the lower two levels of the computational architecture; for the downward connections, see Fig. 7

- Upward connections from similarity detector base states to the self-model **T**-states and **W**-states to model the similarity-induced adaptation principle
- Downward connections for each self-model **T**-state $\mathbf{T}_Y$ to base state Y and from each self-model **W**-state to base state Y to effectuate the adaptivity

Figure 6 shows a 3D visualisation of the nodes and connections of the base level (pink plane) and first-order adaptivity level (blue plane) of the computational architecture. Here for the sake of transparency the downward connections have been left out. However, these downward connections are shown in Fig. 7. Note that this concerns agent A that has interaction with and senses two other agents B and C.

6.3 Second-Order Adaptation Level of the Computational Architecture

In Table 1 also the second-order self-model level (purple area and purple plane) is shown, with four types of second-order self-model states:

- $\mathbf{H_T}$-states $\mathbf{H}_{\mathbf{T}_Y}$ control the adaptation speed of **T**-state $\mathbf{T}_Y$
- $\mathbf{H_W}$-states $\mathbf{H}_{\mathbf{W}_{X,Y}}$ control the adaptation speed of **W**-state $\mathbf{W}_{X,Y}$
- $\mathbf{W_T}$-states $\mathbf{W}_{Z,\mathbf{T}_Y}$ control the adaptation strength of **T**-state $\mathbf{T}_Y$
- $\mathbf{W_W}$-states $\mathbf{W}_{Z,\mathbf{W}_{X,Y}}$ control the adaptation strength of **W**-state $\mathbf{W}_{X,Y}$

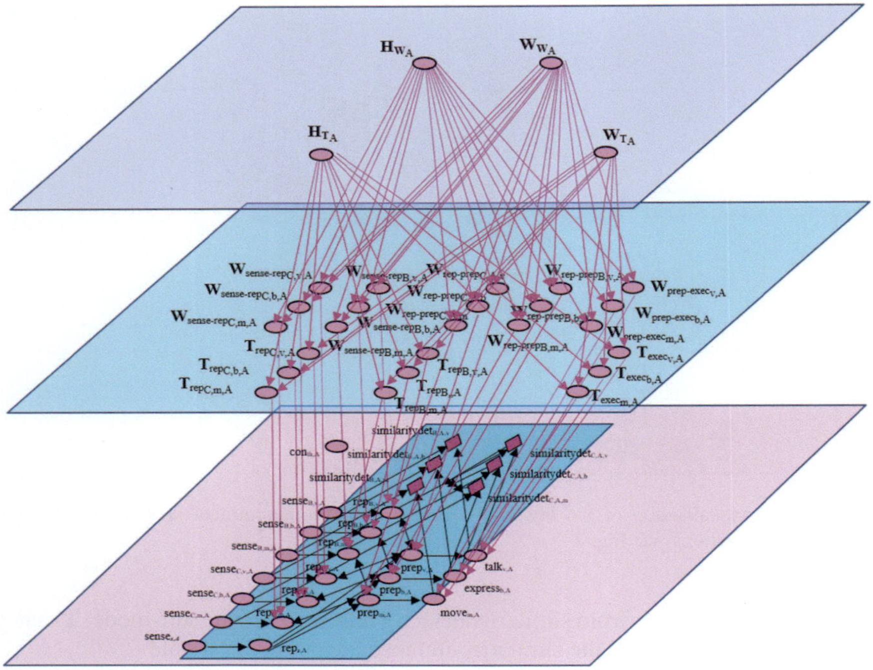

Fig. 7 A 3D visualisation of all three levels of the computational architecture with all nodes and their within-level and downward connections

In the purple area in Table 1 more details of the different second-order self-model states of the computational architecture as shown for the different types of base states and base connections for representing, responding, executing and connecting.

Within the computational architecture, the connections for these second-order self-model states are as follows; see also Figs. 6 and 7 and Table 2 for a visualisation and concise overview of the connectivity of all self-model states:

- To model context-sensitive control over the adaptivity, upward connections to the second-order self-model states are included from all base states and first-order self-model states that are considered indicators for the relevant context considered. For example, to model the second-order adaptation principle 'Adaptation accelerates with increasing stimulus exposure' from Chandra and Barkai (2016) by $\mathbf{H_T}$-states or $\mathbf{H_W}$-states, the (sensory) representation states from the base level can be used as such context indicators.
- To effectuate the adaptivity, downward connections from the second-order self-model states to the first-order self-model level are included:
  - for each second-order self-model $\mathbf{H_T}$-state $\mathbf{H}_{\mathbf{T}_Y}$ a downward connection to first-order self-model $\mathbf{T}$-state $\mathbf{T}_Y$

– for each second-order self-model $\mathbf{H_W}$-state $\mathbf{H}_{\mathbf{W}_{X,Y}}$ a downward connection to first-order self-model $\mathbf{W}$-state $\mathbf{W}_{X,Y}$
– for each second-order self-model $\mathbf{W_T}$-state $\mathbf{W}_{Z,\mathbf{T}_Y}$ a downward connection to first-order self-model $\mathbf{T}$-state $\mathbf{T}_Y$
– for each second-order self-model $\mathbf{W_W}$-state $\mathbf{W}_{Z,\mathbf{W}_{X,Y}}$ a downward connection to first-order self-model $\mathbf{W}$-state $\mathbf{W}_{X,Y}$.

Note that in Fig. 7, for the sake of transparancy at the second-order self-model level, a form of abstraction is applied. There is not a specific second-order self-model state for each specific first-order self-model state but one for all specific first-order self-model states of the same type, for example one $\mathbf{H_T}$-state for all $\mathbf{T}$-states of a given person A and one $\mathbf{H_W}$-state for all $\mathbf{W}$-states of person A, so no state $\mathbf{H}_{\mathbf{T}_Y}$ for each state $\mathbf{T}_Y$ and no state $\mathbf{H}_{\mathbf{W}_{X,Y}}$ for each state $\mathbf{T}_{X,Y}$.

7 The Computational Architecture Illustrated in Case Studies

Two case studies used as inspiration in particular for the computational architecture proposed here are briefly discussed in this section. Case study 1 addresses the distinction between and interplay of short-term affiliation and long-term bonding [54] and Case study 2 transference between different relationships [55]. In these two case studies different forms of synchrony-induced adaptivity are addressed:

- Some forms of adaptivity have a relatively fast timescale (e.g., within an episode of 45 min), modeled by self-model $\mathbf{T}$-states for excitability adaptation, and some have a slower timescale (e.g., over tens of episodes of 45 min), modeled by self-model $\mathbf{W}$-states for connectivity adaptation, in accordance with Sect. 4.1, last part.
- In Case study 1 in addition to (subjective) interpersonal synchrony detection, als intrapersonal synchrony detection is addressed and used for the adaptivity and for conscious emotion [46]
- In Case study 2, some forms of adaptivity are relationship-specific and some are relationship-independent so that transference takes place [3, 4, 29]

For examples of simulation experiments performed for these case studies, see Hendrikse et al. [55, 62]. In Fig. 8 the base level for Case study 1 is shown, with added states for intrapersonal synchrony detection and conscious emotion based on them [46].

In Fig. 9 all levels are shown with the within-level connections (black arrows) and the downward interlevel connections (pink arrows) from first- and second-order self-model states. These downward connections indicate the control effects of the different types of adaptation.

In accordance with the connectivity summarised in Table 2, at the first-order adaptation level $\mathbf{W}$-states $\mathbf{W}_{X,Y}$ control connection weights of connections $X \rightarrow$

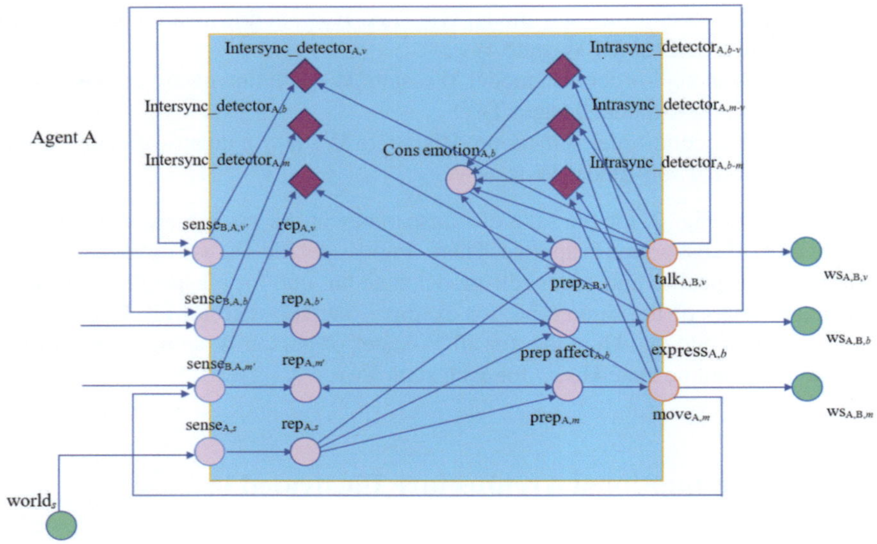

Fig. 8 Base level for Case study 1

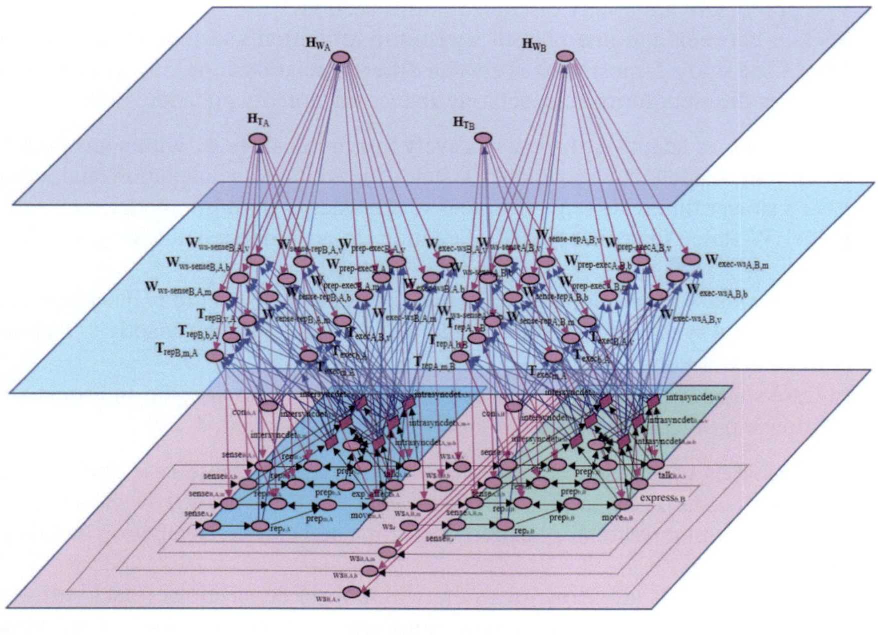

Fig. 9 A 3D visualisation of the computational architecture applied to the considered Case study 1, with excitability self-model **T**-states for fast adaptation timescale and connectivity self-model **W**-states for slow adaptation timescale. The downward connections indicate the control effects of the different types of adaptation

Y for base level states X,Y indicated in their subscripts, and **T**-states $\mathbf{T}_Y$ control excitability thresholds for base level states Y indicated in their subscripts. At the second-order adaptation level $\mathbf{H_W}$-states $\mathbf{H}_{\mathbf{W}_{X,Y}}$ control the adaptation speed of all **W**-states of agent X, $\mathbf{H_T}$-states $\mathbf{H}_{\mathbf{T}_X}$ control the adaptation speed of all **T**-states of agent X, $\mathbf{W_W}$-states $\mathbf{W}_{\mathbf{W}_{X,Y}}$ control the adaptation strength of all **W**-states of agent X, and $\mathbf{W_T}$-states $\mathbf{W}_{\mathbf{T}_W}$ control the adaptation strength of all **T**-states of agent X.

8 The Computational Architecture Applied for Bonding Bots

Recently, the described computational architecture has been applied to design bonding bots and how humans can interact and bond with them while achieving interpersonal emotion regulation [34], see Fig. 10.

To incorporate emotion regulation, in this case a feeling state (e.g., for a negative feeling) and an emotion control state to suppress the feeling state has been added. This control state gets incoming connections from the feeling and preparation states so that a form of monitoring occurs by which it gets activated when these activation levels are too high. If so, in turn the emotion control state suppresses these states by negative connections to them.

Weaker or stronger forms of emotion regulation can be modeled by weaker or stronger links to this emotion control state: if these links are too weak, the control state will not get activated when needed. If the human has weak emotion regulation, the emotion levels can stay high. However, if the bot has strong emotion regulation

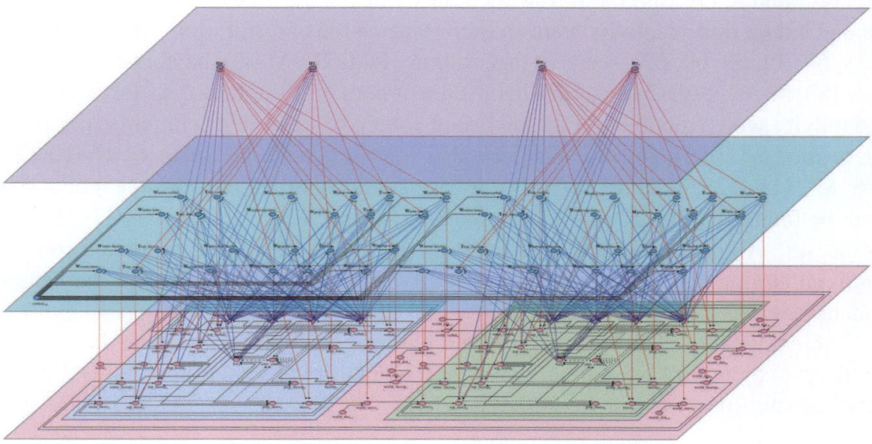

Fig. 10 The computational architecture applied to design bonding bots. Base level (pink plane), of the human (blue area) and the bot (green area), and first- and second-order adaptation level for both of them (blue and purple plane). Adopted from Eler et al. [34], Fig. 1

while the human has weak emotion regulation, then due to proper social contagion between the two based on mirroring (which also depends on how strong their bond is), interpersonal emotion regulation can emerge by which the human's emotion will still be downregulated. By the adaptivity built-in in the bonding bot according to the computation architecture, the bond can become stronger over time so that the interpersonal emotion regulation will become more and more effective. For more details, including example simulations, see Eler et al. [34].

9 Discussion

The generic computational architecture for social interaction adaptivity and attunement proposed in this chapter was developed based on unification of a number of specific case studies addressed earlier. The proposed architecture concerns a generic multilevel multi-timescale adaptive dynamical system. It is formalised and represented as a multi-order self-modeling network [121–123]. In the chapter it was discussed how this architecture covers earlier case studies addressing social interaction adaptivity and attunement subsuming synchrony- or homophily-induced adaptivity. It was also shown how based on it a new application to the design of bonding bots is realised that supports interpersonal emotion regulation via adaptive human-bot interaction and attunement. The material for this chapter comes from Hendrikse and Treur [53].

The computational architecture for adaptive social interaction presented in this chapter was based on a variety of case studies and related evaluations, some of which for synchrony-induced social interation adaptivity already have been discussed in Sects. 7 and 8. Overall, the following case studies in the area of synchrony-induced social interaction adaptivity were an important source of inspiration: Hendrikse et al. [51, 54, 55, 59, 60, 62], see also this volume, Parts II, IV–V, Chapters 3–5 and 11–16: [52, 58–61, 66]. Case studies in the area of homophily-induced social interaction adaptivity were equally well an important source of inspiration, in particular those described in [20, 84, 85, 113, 125], see also this volume, Part III, Chapters 6–10: [21, 86, 87, 114, 128, 129]. Finally, recent case studies that had a more evaluative role can be found in [34, 137, 144], see also this volume, Part VI, Chapters 17–19: [35, 138, 145]. All these case studies provide an indication of how the presented computational architecture has a wide applicability to various adaptive social interaction phenomena.

More background from biological and philosophical perspectives on the use of a multilevel dynamical system architecture with multiple adaptation or control levels at different timescales can be found in [11, 15, 16, 126, 127].

References

1. Abraham, W.C., Bear, M.F.: Metaplasticity: the plasticity of synaptic plasticity. Trends Neurosci. **19**(4), 126–130 (1996)
2. Alejandre-García, T., Kim, S., Pérez-Ortega, J., Yuste, R.: Intrinsic excitability mechanisms of neuronal ensemble formation. eLife **11**, e77470 (2022)
3. Andersen, S.M., Berk, M.S.: The social-cognitive model of transference: experiencing past relationships in the present. Curr. Dir. Psychol. Sci. **7**(4), 109–115 (1998)
4. Andersen, S.M., Przybylinski, E.: Experiments on transference in interpersonal relations: implications for treatment. Psychotherapy **49**(3), 370 (2012)
5. Aristotle: Rhetoric. Nichomachean ethics. In: Aristotle in 23 volumes. Rackman transl. Harvard Univ. Press, Cambridge (1934)
6. Atzil-Slonim, D., Soma, C.S., Zhang, X., Paz, A., Imel, Z.E.: Facilitating dyadic synchrony in psychotherapy sessions: systematic review and meta-analysis. Psychother. Res. **33**(7), 898–917 (2023)
7. Augustinus, F., Hendrikse, S.C.F., Treur, J.: Psychopathic genes: a multi-level adaptive dynamic modelling approach. In: Proceedings of the 8th Computational Methods in Systems and Software Conference, CoMeSySo'24. Lecture Notes in Networks and Systems, vol. 1471, pp 477–498. Springer Nature (2024)
8. Bain, A.: Mind and Body: The Theories of Their Relation. HS King & Company. Farnborough Eng. Gregg International (1873)
9. Basso, J.C., Satyal, M.K., Rugh, R.: Dance on the brain: enhancing intra-and interbrain synchrony. Front. Hum. Neurosci. **14**, 584312 (2021)
10. Beale, L.S.: XXII. Indications of the paths taken by the nerve-currents as they traverse the caudate nerve-cells of the spinal cord and encephalon. Proceed Royal Soc London (13), 386–392 (1864)
11. Bechtel, W., Bich, L.: Organisms need mechanisms; mechanisms need organisms. In: Cordovil, J.L., Santos, G., Vecchi, D. (eds.) New Mechanism: Explanation, Emergence and Reduction. History, Philosophy and Theory of the Life Sciences, vol. 35, pp. 85–108. Springer Nature (2024)
12. Bernieri, F.J., Rosenthal, R.: Interpersonal coordination: behavior matching and interactional synchrony. In: Feldman, R.S., Rimé, B. (eds.) Studies in Emotion & Social Interaction. Fundamentals of Nonverbal Behavior, pp. 401–432. Cambridge University Press (1991)
13. Bernieri, F.J.: Coordinated movement in human interaction: bynchrony, posture similarity, and rapport. Harvard University (1988)
14. Bevilacqua, A., Wilkinson, S.J., Dimelow, R., Murabito, E., Rehman, S., Nardelli, M., Westerhoff, H.V.: Vertical systems biology: from DNA to flux and back. In: Hetherington, A., Grierson, C. (eds.) Practical Systems Biology, pp. 79–106. Taylor & Francis (2008)
15. Bich, L., Bechtel, W.: Control mechanisms: explaining the integration and versatility of biological organisms. Adapt. Behav. **30**(5), 389–407 (2022)
16. Bich, L., Bechtel, W.: Organization needs organization: understanding integrated control in living organisms. Stud. Hist. Philos. Sci. **93**, 96–106 (2022)
17. Bilodeau-Houle, A., Morand-Beaulieu, S., Bouchard, V., Marin, M.F.: Parent–child physiological concordance predicts stronger observational fear learning in children with a less secure relationship with their parent. J. Exp. Child Psychol. **226**, 105553 (2023)
18. Blankendaal, R., Parinussa, S., Treur, J.: A temporal-causal modelling approach to integrated contagion and network change in social networks. In: Proceedings of the Twenty-second European Conference on AI, ECAI'16, pp. 1388–1396. IOS Press (2016)
19. Blei, Z.A., Hendrikse, S.C.F., Treur, J.: Modeling epigenetic modification due to childhood abuse and its relation to ASPD: a fifth-order adaptive network model. In: Proceedings of the 15th Biologically Inspired Cognitive Architectures for AI Conference, BICA*AI'24. Studies in Computational Intelligence, vol. 477, pp 59–70. Springer Nature (2024)

20. Bössenecker, G., Mreijen, P., Treur, J.: A second-order adaptive network model for connecting social media platforms and users through bonding by faked homophily. Proced. Comput. Sci. 213, 32–40 (2022). https://doi.org/10.1016/j.procs.2022.11.035
21. Bössenecker, G., Mreijen, P., Treur, J.: Connecting social media platforms and users through bonding by faked homophily: a second-order adaptive network model. In: Hendrikse, S.C.F., Treur, J., Koole, S.L. (eds.) New Analysis and Modeling Directions in Social Interaction Science: Emergent Multimodal Interpersonal Synchrony, Affiliation, and Bonding, Ch 9, this volume. Springer Nature (2025)
22. Bott, H.: Observation of play activities in a nursery school. Genet. Psychol. Monogr. **4**, 44–88 (1928)
23. Bourgeois, P., Hess, U.: The impact of social context on mimicry. Biol. Psychol. **77**(3), 343–352 (2008)
24. Budding, D., Doornkamp, S., Treur, J.: A second-order adaptive mental network model relating dreaming to creativity. Cogn. Syst. Res. **80**, 37–49 (2023)
25. Butler, E.A., Randall, A.K.: Emotional coregulation in close relationships. Emot. Rev. **5**(2), 202–210 (2013)
26. Călin, R.S., Hendrikse, S.C.F., Treur, J.: Environmental and Epigenetic Influences in Binge Eating: An Adaptive Network Modeling Approach. To appear (2024)
27. Canbaloğlu, G., Treur, J., Wiewiora, A. (eds.): Computational Modeling of Multilevel Organisational Learning and its Control Using Self-Modeling Network Models. Springer Nature (2023b)
28. Cirelli, L.K.: How interpersonal synchrony facilitates early prosocial behavior. Curr. Opin. Psychol. **20**, 35–39 (2018)
29. Chen, S., Andersen, S.M.: Relationships from the past in the present: significant other representations and transference in interpersonal life. In: Advances in experimental social psychology, vol. 31, pp. 123–190. Academic Press (1999)
30. Czeszumski, A., Hsin-Yi Liang, S., Dikker, S., König, P., Lee, C.-P., Koole, S. L., Kelsen, B.: Cooperative behavior evokes interbrain synchrony in the prefrontal and temporoparietal cortex: a systematic review and meta-analysis of fNIRS hyperscanning studies. eNeuro **9**(2) https://doi.org/10.1523/ENEURO.0268-21.2022 (2022)
31. Debanne, D., Inglebert, Y., Russier, M.: Plasticity of intrinsic neuronal excitability. Curr. Opin. Neurobiol. **54**, 73–82 (2019)
32. Doré, B.P., Morris, R.R.: Linguistic synchrony predicts the immediate and lasting impact of text-based emotional support. Psychol. Sci. **29**(10), 1716–1723 (2018)
33. Eler, E., Dimopoulos, F., Timmers, M., Treur, J., Koole, S.L.: An adaptive network model for interpersonal emotion regulation in multimodal human-bot interaction. In: Hendrikse, S.C.F., Treur, J., Koole, S.L. (eds.) New Analysis and Modeling Directions in Social Interaction Science: Emergent Multimodal Interpersonal Synchrony, Affiliation, and Bonding, Ch 4, this volume. Springer Nature (2025)
34. Eler, E., Treur, J., Hendrikse, S.C.F., Donker, T., Koole, S.L.: An adaptive network model for interpersonal emotion regulation in multimodal human-bot interaction. In: Nguyen, N.T., et al. (eds.) Computational Collective Intelligence. Proc. ICCCI 2024. Lecture Notes in Computer Science, vol. 14811, pp. 67–79. Springer Nature, Cham (2024)
35. Eler, E., Treur, J., Hendrikse, S.C.F., Donker, T., Koole, S.L.: An adaptive network model for interpersonal emotion regulation in multimodal human-bot interaction. In: Hendrikse, S.C.F., Treur, J., Koole, S.L. (eds.) New Analysis and Modeling Directions in Social Interaction Science: Emergent Multimodal Interpersonal Synchrony, Affiliation, and Bonding, Ch 19, this volume. Springer Nature (2025)
36. Feldman, R.: Parent–infant synchrony: biological foundations and developmental outcomes. Curr. Dir. Psychol. Sci. **16**(6), 340–345 (2007)
37. Feldman, R.: Parent-infant synchrony: a biobehavioral model of mutual influences in the formation of affiliative bonds. Monogr. Soc. Res. Child Dev. **77**(2), 42–51 (2012)
38. Freeman, L.: Some antecedents of social network analysis. Connections **19**, 39–42 (1996)
39. Garcia, R.: Stress, metaplasticity, and antidepressants. Curr. Mol. Med. **2**(7), 629–638 (2002)

40. Geiser, J.: Multicomponent and Multiscale Systems. Springer International Publishing (2016)
41. Chandra, N., Barkai, E.: A non-synaptic mechanism of complex learning: modulation of intrinsic neuronal excitability. Neurobiol. Learn. Mem. **154**, 30–36 (2018)
42. Grandjean, D., Sander, D., Scherer, K.R.: Conscious emotional experience emerges as a function of multilevel, appraisal-driven response synchronization. Conscious. Cogn. 17(2), 484–495 (2008)
43. Goode, J., Robinson, J.D.: L Linguistic synchrony in parasocial interaction. Commun. Stud. **64**(4), 453–466 (2013)
44. Gorman, J.C., Amazeen, P.G., Crites, M.J., Gipson, C.L.: Deviations from mirroring in interpersonal multifrequency coordination when visual information is occluded. Exp. Brain Res. **23**5(4), 1209–1221 (2017)
45. Górriz, J.M., Álvarez-Illán, I., Álvarez-Marquina, A., Arco, J.E., Atzmueller, M., Ballarini, F., Ferrández-Vicente, J.M.: Computational approaches to explainable artificial intelligence: advances in theory, applications and trends. Inf Fusion **100**, 101945 (2023)
46. Grandjean, D., Sander, D., Scherer, K.R.: Conscious emotional experience emerges as a function of multilevel, appraisal- driven response synchronization. Conscious. Cogn. **17**(2), 484–495 (2008)
47. Hansel, C., Yuste, R.: Neural ensembles: role of intrinsic excitability and its plasticity. Front. Cell. Neurosci. **18**, 1440588 (2024)
48. Hassouna, S., Hendrikse, S.C.F., Treur, J.: The Dynamics of Epigenetic Influence in Insomnia: A Higher-Order Adaptive Modeling Perspective. To appear (2024)
49. Hebb, D.O.: The Organization of Behavior **4**(60), 78–60 (1949)
50. Heijmans, P., van Stijn, J., Treur, J.: Modeling cultural segregation of the queer community through an adaptive social network model. In Fourth International Congress on Information and Communication Technology: ICICT 2019, London, vol. 2, pp. 233–248. Springer Singapore (2020)
51. Hendrikse, S.C.F., Kluiver, S., Treur, J., Wilderjans, T.F., Dikker, S., Koole, S.L.: How virtual agents can learn to synchronize: an adaptive joint decision-making model of psychotherapy. Cogn. Syst. Res. **79**, 138–155 (2022a). https://www.researchgate.net/publication/354403086
52. Hendrikse, S.C.F., Kluiver, S., Treur, J., Wilderjans, T.F., Dikker, S., Koole, S.L.: Learning to synchronize: adaptive joint decision-making by virtual agents. In: Hendrikse, S.C.F., Treur, J., Koole, S.L. (eds.) New Analysis and Modeling Directions in Social In-teraction Science: Emergent Multimodal Interpersonal Synchrony, Affiliation, and Bonding, Ch 5. Springer Nature (2025)
53. Hendrikse, S.C.F., Treur, J.: Perspectives on social interaction science: from a multilevel multi-timescale computational architecture for synchrony-induced and homophily-induced attunement to designing bonding bots. In: Proceedings of the 8th International Conference on Computational Methods in Systems and Software, CoMeSySo'24. Lecture Notes in Networks and Systems, vol. 1490, pp. 518–545. Springer Nature (2025)
54. Hendrikse, S.C.F., Treur, J., Koole, S.L.: Modeling emerging interpersonal synchrony and its related adaptive short-term affliation and long-term bonding: a second-order multi-adaptive neural agent model. Inter. J. Neural Syst **33**(7), 2350038 (2023)
55. Hendrikse, S.C.F., Treur, J., Koole, S.L.: Relationship-specific and relationship-independent behavioural adaptivity in affiliation and bonding: a multi-adaptive dynamical systems approach. Cogn. Syst. Res. **84**, 101182 (2024)
56. Hendrikse, S.C.F., Treur, J., Koole, S.L. (eds.): New Analysis and Modeling Directions in Social Interaction Science: Emergent Multimodal Interpersonal Synchrony, Affiliation, and Bonding. Springer Nature (2025a)
57. Hendrikse, S.C.F., Treur, J., Koole, S.L.: The adaptive interplay of interpersonal synchrony, short-term affiliation and long-term bonding: a multi-timescale multi-adaptive agent model. In: Hendrikse, S.C.F., Treur, J., Koole, S.L. (eds.) New Analysis and Modeling Directions in Social Interaction Science: Emergent Multimodal Interpersonal Synchrony, Affiliation, and Bonding, Ch 12. Springer Nature (2025b)

58. Hendrikse, S.C.F., Treur, J., Koole, S.L.: A multi-timescale multi-adaptive dynamical system approach for relationship-specific and relationship-independent adaptivity in affiliation and bonding. In: Hendrikse, S.C.F., Treur, J., Koole, S.L. (eds.) New Analysis and Modeling Directions in Social Interaction Science: Emergent Multimodal Interpersonal Synchrony, Affiliation, and Bonding, Ch 13. Springer Nature (2025c)
59. Hendrikse, S.C.F., Treur, J., Wilderjans, T.F., Dikker, S., Koole, S.L.: On the same wavelengths: emergence of multiple synchronies among multiple agents. In: Proceedings of the 22nd International Workshop on Multi-Agent-Based Simulation, MABS'21. Lecture Notes in AI, vol. 13128, pp. 57–71. Springer Nature (2022a)
60. Hendrikse, S.C.F., Treur, J., Wilderjans, T.F., Dikker, S., Koole, S.L.: Switching in and out of sync: a controlled adaptive network model of transition dynamics in the effects of interpersonal synchrony on affiliation. In: Proceedings of the 11th International Conference on Complex Networks and their Applications. Studies in Computational Intelligence, vol 1078, pp. 81–95. Springer Nature Switzerland, Cham (2022b)
61. Hendrikse, S.C.F., Treur, J., Wilderjans, T.F., Dikker, S., Koole, S.L.: Becoming attuned to each other over time: a computational neural agent model for the role of time lags in subjective synchrony detection and related behavioral adaptivity. In: Proceedings of the 15th International Conference on Brain Informatics, BI'22. Lecture Notes in AI, vol. 13406, pp. 369–383. Springer Nature (2022c)
62. Hendrikse, S.C.F., Treur, J., Wilderjans, T.F., Dikker, S., Koole, S.L.: On becoming in sync with yourself and others: an adaptive agent model for how persons connect by detecting intrapersonal and interpersonal synchrony. Hum. Centric Intell. Syst. **3**(2), 123–146 (2023)
63. Hendrikse, S.C.F., Treur, J., Wilderjans, T.F., Dikker, S., Koole, S.L.: The emergence of multimodal interpersonal synchrony during social interaction: a network model. In: Hendrikse, S.C.F., Treur, J., Koole, S.L. (eds.) New Analysis and Modeling Directions in Social Interaction Science: Emergent Multimodal Interpersonal Synchrony, Affiliation, and Bonding, Ch 3. Springer Nature (2025a)
64. Hendrikse, S.C.F., Treur, J., Wilderjans, T.F., Dikker, S., Koole, S.L.: Growing connectedness through synchrony: the role of subjective analysis of intrapersonal synchrony and interpersonal synchrony within an adaptive agent model. In: Hendrikse, S.C.F., Treur, J., Koole, S.L. (eds.) New Analysis and Modeling Directions in Social Interaction Science: Emergent Multimodal Interpersonal Synchrony, Affiliation, and Bonding, Ch 11. Springer Nature (2025b)
65. Hendrikse, S.C.F., Treur, J., Wilderjans, T.F., Dikker, S., Koole, S.L.: The role of timing in synchrony detection and social interaction adaptivity: including time lags in a second-order adaptive agent model. In: Hendrikse, S.C.F., Treur, J., Koole, S.L. (eds.) New Analysis and Modeling Directions in Social Interaction Science: Emergent Multimodal Interpersonal Synchrony, Affiliation, and Bonding, Ch 14. Springer Nature (2025c)
66. Hendrikse, S.C.F., Treur, J., Wilderjans, T.F., Dikker, S., Koole, S.L.: The effect of transition dynamics in synchrony on social interaction adaptivity: a multi-adaptive network model. In: Hendrikse, S.C.F., Treur, J., Koole, S.L. (eds.) New Analysis and Modeling Directions in Social Interaction Science: Emergent Multimodal Interpersonal Synchrony, Affiliation, and Bonding, Ch 15. Springer Nature (2025d)
67. Hesslow, G.: Conscious thought as simulation of behaviour and perception. Trends Cogn. Sci. **6**, 242–247 (2002)
68. Hove, M.J., Risen, J.L.: It's all in the timing: interpersonal synchrony increases affiliation. Soc. Cogn. **27**(6), 949–960 (2009)
69. Hsieh, C.S., König, M.D., Liu, X.: A structural model for the coevolution of networks and behavior. Rev. Econ. Stat. **104**(2), 355–367 (2022)
70. Hubbard R.M.: A method of studying spontaneous group formation. In: Thomas, D.S., (ed.) Some New Techniques for Studying Social Behavior, pp. 76–85. Child Dev. Monogr (1929)
71. Ivanov, P.C., Ma, Q.D., Bartsch, R.P.: Maternal–fetal heartbeat phase synchronization. Proc. Natl. Acad. Sci. **106**(33), 13641–13642 (2009)
72. Jaganjac, I., Hendrikse, S.C.F., Treur, J.: An adaptive dynamical system model for development of schizophrenia: epigenetics and false memories. Cogn. Syst. Res. 88, 101288 (2024)

73. Kappert, C., Rus, R., Treur, J.: On the emergence of segregation in society: network-oriented analysis of the effect of evolving friendships. In: Computational Collective Intelligence: 10th International Conference, ICCCI 2018, Bristol, UK, September 5–7, 2018, Proceedings, Part I 10, pp. 178–191. Springer International Publishing (2018)
74. Keysers, C., Gazzola, V.: Dissociating the ability and propensity for empathy. Trends Cogn. Sci. **18**(4), 163–166 (2014)
75. Kinreich, S., Djalovski, A., Kraus, L., Louzoun, Y., Feldman, R.: Brain-to-brain synchrony during naturalistic social interactions. Sci. Rep. **7**(1), 17060 (2017)
76. Konvalinka, I., Xygalatas, D., Bulbulia, J., Schjødt, U., Jegindø, E. M., Wallot, S., & Roepstorff, A.: Synchronized arousal between performers and related spectators in a fire-walking ritual. Proceed. Nat. Acad. Sci. **108**(20), 8514–8519 (2011)
77. Koole, S.L., Tschacher, W.: Synchrony in psychotherapy: a review and an integrative framework for the therapeutic alliance. Front. Psychol. **7**, 862 (2016)
78. Lazarsfeld, P.F., Merton, R.K.: Friendship as a social process: a substantive and methodological analysis. In: Berger, M. (ed.) Freedom and Control in Modern Society, pp. 18–66. Van Nostrand, New York (1954)
79. Lisman, J., Cooper, K., Sehgal, M., Silva, A.J.: Memory formation depends on both synapse-specific modifications of synaptic strength and cell-specific increases in excitability. Nat. Neurosci. **21**, 309–314 (2018)
80. Liu, J., Huang, S., Aden, N.M., Johnson, N.F., Song, C.: Emergence of polarization in coevolving networks. Phys. Rev. Lett. **130**(3), 037401 (2023)
81. Luo, S., Du, Y., Liu, P., Xuan, Z., Wang, Y.: A study on coevolutionary dynamics of knowledge diffusion and social network structure. Expert Syst. Appl. **42**(7), 3619–3633 (2015)
82. Magerl, W., Hansen, N., Treede, R.D., Klein, T.: The human pain system exhibits higher-order plasticity (metaplasticity). Neurobiol. Learn. Mem. **154**, 112–120 (2018)
83. McPherson, M., Smith-Lovin, L., Cook, J.M.: Birds of a feather: homophily in social networks. Annu. Rev. Sociol. **27**, 415–444 (2001)
84. Mukeriia, Y., Treur, J., Hendrikse, S.C.F.: An adaptive network model for learning and bonding during a varying in rhythm synchronous joint action. In: Proceedings of the 12th International Conference on Complex Networks and Their Applications. Studies in Computational Intelligence, vol. 1144, pp. 41–52. Springer Nature Switzerland, Cham (2023)
85. Mukeriia, Y., Treur, J., Hendrikse, S.C.F.: A multi-adaptive network model for human hebbian learning, synchronization and social bonding based on adaptive homophily. Cogn. Syst. Res. **84**, 101187 (2024)
86. Mukeriia, Y., Treur, J., Hendrikse, S.C.F.: The interplay of bonding based on subjective homophily and learning during a varied rhythmic action: an adaptive network model. In: Hendrikse, S.C.F., Treur, J., Koole, S.L. (eds.) New Analysis and Modeling Directions in Social Interaction Science: Emergent Multimodal Interpersonal Synchrony, Affiliation, and Bonding, Ch 8. Springer Nature (2025a)
87. Mukeriia, Y., Treur, J., Hendrikse, S.C.F.: Human learning, attunement and social bonding based on adaptive homophily: a multi-adaptive network model. In: Hendrikse, S.C.F., Treur, J., Koole, S.L. (eds.) New Analysis and Modeling Directions in Social Interaction Science: Emergent Multimodal Interpersonal Synchrony, Affiliation, and Bonding, Ch 10. Springer Nature (2025b)
88. Mulder, J., Leenders, R.T.A.: Modeling the evolution of interaction behavior in social networks: a dynamic relational event approach for real-time analysis. Chaos Solitons Fractals **119**, 73–85 (2019)
89. Müller, V., Sänger, J., Lindenberger, U.: Hyperbrain network properties of guitarists playing in quartet. Ann. N. Y. Acad. Sci. **1423**(1), 198–210 (2018)
90. Oullier, O., De Guzman, G.C., Jantzen, K.J., Lagarde, J., Scott Kelso, J. A.: Social coordination dynamics: Measuring human bonding. Soc. Neurosci. **3**(2), 178–192 (2008)
91. Palumbo, R.V., Marraccini, M.E., Weyandt, L.L., Wilder-Smith, O., McGee, H.A., Liu, S., Goodwin, M.S.: Interpersonal autonomic physiology: a systematic review of the literature. Pers. Soc. Psychol. Rev. **21**(2), 99–141 (2017)

92. Pan, Y., Novembre, G., Song, B., Zhu, Y., Hu, Y.: Dual brain stimulation enhances interpersonal learning through spontaneous movement synchrony. Soc. Cogn. Affect Neurosci. **16**(1–2), 210–221 (2021)
93. Plato.: Laws. Plato in Twelve Volumes, vol. 11. Bury translator. Harvard Univ. Press, Cambridge (1968)
94. Port, R. F., Van Gelder, T. (eds.): Mind as motion: Explorations in the dynamics of cognition. MIT press (1995)
95. Radhakrishnan, R.: A survey of multiscale modeling: foundations, historical milestones, current status, and future prospects. AIChE J. **67**(3), e17026 (2021)
96. Reindl, V., Gerloff, C., Scharke, W., Konrad, K.: Brain-to-brain synchrony in parent-child dyads and the relationship with emotion regulation revealed by fNIRS-based hyperscanning. Neuroimage **178**, 493–502 (2018)
97. Rizzolatti, G., Sinigaglia, C.: Mirrors in the Brain: How our Minds Share Actions and Emotions. Oxford University Press, Oxford (2008)
98. Robinson, B.L., Harper, N.S., McAlpine, D.: Meta-adaptation in the auditory midbrain under cortical influence. Nat. Commun. **7**(1), 13442 (2016)
99. Schlöder, D., Hendrikse, S.C.F., Treur, J.: Higher-order adaptive dynamical system modeling of the role of epigenetics in substance use disorder. In: Proceedings of the 15th Biologically Inspired Cognitive Architectures for AI Conference, BICA*AI'24. Studies in Computational Intelligence, vol 477, pp. 358–368. Springer Nature (2024)
100. Schoenherr, D., Paulick, J., Strauss, B.M., Deisenhofer, A.-K., Schwartz, B., Rubel, J.A., Lutz, W., Stangier, U., Altmann, U.: Identification of movement synchrony: validation of windowed cross-lagged correlation and -regression with peakpicking algorithm. PLoS ONE **14**, e0211494 (2019)
101. Schoenherr, D., Paulick, J., Worrack, S., Strauss, B.M., Rubel, J.A., Schwartz, B., Altmann, U.: Quantification of nonverbal synchrony using linear time series analysis methods: lack of convergent validity and evidence for facets of synchrony. Behav. Res. Methods 1–23 (2019b)
102. Schoenherr, D., Strauss, B., Stangier, U., Altmann, U.: The influence of vocal synchrony on outcome and attachment anxiety/avoidance in treatments of social anxiety disorder. Psychotherapy **58**(4), 510 (2021)
103. Semin, G.R.: Grounding communication: synchrony. In: Kruglanski, A.W., Higgins, E.T., (eds.) Social Psychology: Handbook of Basic Principles, pp. 630–649. The Guilford Press (2007)
104. Shamay-Tsoory, S.G., Saporta, N., Marton-Alper, I.Z., Gvirts, H.Z.: Herding brains: a core neural mechanism for social alignment. Trends Cogn. Sci. **23**(3), 174–186 (2019)
105. Shapira, N., Atzil-Slonim, D., Tuval-Mashiach, R., Shapira, O.: Measuring linguistic synchrony in psychotherapy. In: Proceedings of the Eighth Workshop on Computational Linguistics and Clinical Psychology, pp. 158–176 (2022)
106. Sharpanskykh, A., Treur, J.: Modelling and analysis of social contagion in dynamic networks. Neurocomputing **146**, 140–150 (2014)
107. Shatz, C.J.: The developing brain. Sci. Am. **267**(3), 60–67 (1992)
108. Shelton, J., Kumar, G.: Comparison between auditory and visual simple reaction times. Neurosci. Med. **1**(1), 30–32 (2010)
109. Sjöström, P.J., Rancz, E.A., Roth, A., Häusser, M.: Dendritic excitability and synaptic plasticity. Physiol. Rev. **88**(2), 769–840 (2008)
110. Snijders, T., Steglich, C., Schweinberger, M.: Modeling the coevolution of networks and behavior. In van Montfort, K., Oud, J., Satorra, A. (eds.) Longitudinal models in the behavioral and related sciences, pp. 41–71. Taylor and Francis (2017)
111. Su, J., Liu, B., Li, Q., Ma, H.: Coevolution of opinions and directed adaptive networks in a social group. J. Artif Soc. Soc. Simul. **17**(2) (2014)
112. Tarr, B., Launay, J., Dunbar, R.I.M.: Silent disco: dancing in synchrony leads to elevated pain thresholds and social closeness. Evol. Hum. Behav. **37**(5), 343–349 (2016)
113. Tichelaar, C., Treur, J.: On the adaptive interplay of mirroring and bonding by homophily in joint decision making: a secondorder adaptive network model. In: Proceedings of the

8th International Conference on Computer-Human Interaction Research and Applications, CHIRA'24. Communications in Computer and Information Science, vol. 2371, pp. 280–298. Springer Nature (2024)

114. Tichelaar, C., Treur, J.: Bonding by homophily and adaptive mirroring in joint decision making: a second-order adaptive network model. In: Hendrikse, S.C.F., Treur, J., Koole, S.L. (eds.) New Analysis and Modeling Directions in Social Interaction Science: Emergent Multimodal Interpersonal Synchrony, Affiliation, and Bonding, Ch 6. Springer Nature (this volume) (2025).
115. Tomprou, M., Kim, Y.J., Chikersal, P., Woolley, A.W., Dabbish, L.A.: Speaking out of turn: how video conferencing reduces vocal synchrony and collective intelligence. PLoS ONE **16**(3), e0247655 (2021)
116. Treur, J.: Temporal factorisation: a unifying principle for dynamics of the world and of mental states. Cogn. Syst. Res. **8**(2), 57–74 (2007)
117. Treur, J.: Temporal factorisation: realisation of mediating state properties for dynamics. Cogn. Syst. Res. **8**(2), 75–88 (2007)
118. Treur, J.: Network reification as a unified approach to represent network adaptation principles within a network. In: Theory and Practice of Natural Computing: Proceedings of the 7th International Conference TPNC'18. Lecture Notes in Computer Science, vol. 11324, pp. 344–358. Springer International Publishing (2018)
119. Treur, J.: Mathematical analysis of the emergence of communities based on coevolution of social contagion and bonding by homophily. Appl. Netw. Sci. **4**(1), 39 (2019)
120. Treur, J.: Multilevel network reification: representing higher-order adaptivity in a network. In: Aiello, L., Cherifi, C., Cherifi, H., Lambiotte, R., Lió, P., Rocha, L. (eds) Complex Networks and Their Applications VII: Volume 1 Proceedings The 7th International Conference on Complex Networks and Their Applications, COMPLEX NETWORKS'18. Studies in Computational Intelligence, vol. 812, pp. 635–651. Springer International Publishing (2019)
121. Treur, J.: Modeling higher-order adaptivity of a network by multilevel network reification. Netw. Sci. **8**, S110–S144 (2020)
122. Treur, J.: Network-Oriented Modeling for Adaptive Networks: Designing Higher-Order Adaptive Biological, Mental and Social Network Models. Springer Nature, Cham (2020a). https://www.researchgate.net/publication/334576216 (2020b)
123. Treur, J.: Modeling multi-order adaptive processes by self-modeling networks. In: Tallon-Ballesteros, A.J., Chen, C-H. (eds.) Machine Learning and Intelligent Systems: Proceedings of MLIS 2020. Frontiers in Artificial Intelligence and Applications, vol. 332, pp. 206–217. IOS Press (2020c)
124. Treur, J.: On the dynamics and adaptivity of mental processes: relating adaptive dynamical systems and self-modeling network models by mathematical analysis. Cogn. Syst. Res. **70**, 93–100 (2021)
125. Treur, J.: Controlled social network adaptation: subjective elements in an objective social world. In: Yang, X.S., Sherratt, S., Dey, N., Joshi, A. (eds) Proceedings of Sixth International Congress on Information and Communication Technology, ICICT'21. Lecture Notes in Networks and Systems, vol. 235, pp 263–274. Springer Nature (2022). https://doi.org/10.1007/978-981-16-2377-6_26
126. Treur, J.: On structure, dynamics, and adaptivity for biological and mental processes: a higher-order adaptive dynamical system modeling perspective. In: Proceedings of the 46th Annual Meeting of the Cognitive Science Society: Dynamics of Cognition, CogSci'24 (2024a)
127. Treur, J.: Using multilevel temporal factorisation to analyse structure and dynamics for higher-order adaptive and evolutionary processes. In: Nguyen, N.T., et al. (ed.) Computational Collective Intelligence. Proc. ICCCI 2024. Lecture Notes in Computer Science, vol. 14811, pp. 378–392. Springer Nature, Cham (2024b)
128. Treur, J.: Modeling coevolution of dynamics, adaptivity and control of adaptivity by self-modeling networks. In: Hendrikse, S.C.F., Treur, J., Koole, S.L. (eds.) New Analysis and Modeling Directions in Social Interaction Science: Emergent Multimodal Interpersonal Synchrony, Affiliation, and Bonding, Ch 2. Springer Nature (2025a)

129. Treur, J.: A second-order adaptive network model for bonding based on subjectively perceived homophily. In: Hendrikse, S.C.F., Treur, J., Koole, S.L. (eds.) New Analysis and Modeling Directions in Social Interaction Science: Emergent Multimodal Interpersonal Synchrony, Affiliation, and Bonding, Ch 7. Springer Nature (2025b)
130. Treur, J., van Ments, L. (eds.): Mental Models and their Dynamics, Adaptation and Control: a Self-Modeling Network Modeling Approach. Springer Nature Switzerland (2022)
131. Tschacher, W., Ramseyer, F., Koole, S.L.: Sharing the now in the social present: duration of nonverbal synchrony is linked with personality. J. Pers. **86**(2), 129–138 (2018)
132. Vacharkulksemsuk, T., Fredrickson, B.L.: Strangers in sync: achieving embodied rapport through shared movements. J. Exp. Soc. Psychol. **48**(1), 399–402 (2012)
133. Valdesolo, P., DeSteno, D.: Synchrony and the social tuning of compassion. Emotion **11**(2), 262–266 (2011)
134. Valdesolo, P., Ouyang, J., DeSteno, D.: The rhythm of joint action: synchrony promotes cooperative ability. J. Exp. Soc. Psychol. **46**(4), 693695 (2010)
135. van den Beukel, S., Goos, S.H., Treur, J.: An adaptive temporal-causal network model for social networks based on the homophily and more-becomes-more principle. Neurocomputing **338**, 361–371 (2019)
136. Van Leeuwen, P., Geue, D., Thiel, M., Cysarz, D., Lange, S., Romano, M.C., ... , Grönemeyer, D.H.: Influence of paced maternal breathing on fetal–maternal heart rate coordination.Proc. Natl. Acad. Sci. **106**(33), 13661–13666 (2009)
137. Vervoordeldonk, D., Treur, J., Hendrikse, S.C.F., Roelofsma, P.H.M.P.: An Adaptive Network Model of the Interplay of Synchrony and Adaptation for Reliving Memory and Religious Experience Through Multimodal Interaction. In: Silhavy, R., Silhavy, P. (eds) Proceedings CoMeSySo 2024. Lecture Notes in Networks and Systems, vol. 1490, pp 59–93. Cham. Springer Nature (2025)
138. Vervoordeldonk, D., Treur, J., Hendrikse, S.C.F., Roelofsma, P.H.M.P.: On the interplay of synchrony and adaptation for reliving memory and religious experience through multimodal interaction: An Adaptive Network Model. In: Hendrikse, S.C.F., Treur, J., Koole, S.L. (eds.) New Analysis and Modeling Directions in Social Interaction Science: Emergent Multimodal Interpersonal Synchrony, Affiliation, and Bonding, Ch 17. Springer Nature (2025)
139. Vieira, F., Leenders, R., Mulder, J.: A state-space relational event modeling approach for learning dynamic social interaction behavior. Methodological Innovations, 20597991241270299 (2024)
140. Wellman, B.: The school child's choice of companions. J. Educ. Res. **14**, 126–132 (1929)
141. Wilkes, A.L., Wade, N.J.: Bain on neural networks. Brain Cogn **33**(3), 295–305 (1997). https://doi.org/10.1006/brcg.1997.0869
142. Williams, A.H., O'Leary, T., Marder, E.: Homeostatic regulation of neuronal excitability. Scholarpedia **8**, 1656 (2013)
143. Wiltermuth, S.S., Heath, C.: Synchrony and cooperation. Psychol. Sci. **20**(1), 1–5 (2009)
144. Xu, Y., Treur, J., Hendrikse, S.C.F., Roelofsma, P.H.M.P.: Adaptive network modeling for joint action and memory recall for elderly by detecting interpersonal synchrony. Cogn. Syst. Res. **88**, 101280 (2024)
145. Xu, Y., Treur, J., Hendrikse, S.C.F., Roelofsma, P.H.M.P.: Modeling joint action and memory recall for elderly by detecting interpersonal synchrony: a higher-order adaptive network model. In: Hendrikse, S.C.F., Treur, J., Koole, S.L. (eds.) New Analysis and Modeling Directions in Social Interaction Science: Emergent Multimodal Interpersonal Synchrony, Affiliation, and Bonding, Ch 18. Springer Nature (2025)
146. Yokozuka, T., Miyamoto, H., Kasai, M., Miyake, Y., Nozawa, T.: The relationship between turn-taking, vocal pitch synchrony, and rapport in creative problem-solving communication. Speech Commun. **129**, 33–40 (2021)
147. Zhang, A., Li, X., Gao, Y., Niu, Y.: Event-Driven Intrinsic Plasticity for Spiking Convolutional Neural Networks. IEEE Transactions on Neural Networks and Learning Systems (2021). https://doi.org/10.1109/tnnls.2021.3084955

Discussion and Concluding Remarks on Adaptivity in Social Interaction Science

Sophie C. F. Hendrikse, Jan Treur, and Sander L. Koole

Abstract In this chapter an overview of the results achieved in this volume is presented and topics for further research in the field of social interaction science are discussed. It provides an overview of all different types of social interaction adaptivity addressed in the different chapters. Moreover, all parts and chapters are briefly reviewed. Finally, recommendations for further research are discussed.

Keywords Social interaction · Synchrony · Homophily · Adaptivity · Computational model

1 Introduction

A central focus in this volume [44] is on modeling and analysis of interpersonal interaction dynamics and adaptivity from underlying causal mechanisms and pathways. Throughout this work, we have sought to allow computational modeling and empirical science to mutually benefit from each other. More specifically, we used the causal relationships and pathways described in empirical literature to explain certain phenomena as the foundation for a programmatic series of computational adaptive dynamical system models. These were designed as higher-order adaptive network

S. C. F. Hendrikse
Department of Methodology and Statistics, Tilburg School of Social and Behavioral Sciences, Tilburg University, Tilburg, The Netherlands
e-mail: S.C.F.Hendrikse@tilburguniversity.edu

J. Treur (✉)
Social AI Group, Department of Computer Science, Vrije Universiteit Amsterdam, Amsterdam, The Netherlands
e-mail: j.treur@vu.nl

S. L. Koole
Department of Clinical Psychology, Vrije Universiteit Amsterdam, Amsterdam, The Netherlands
e-mail: s.l.koole@vu.nl

S. C. F. Hendrikse et al. (eds.), *New Analysis and Modeling Directions in Social Interaction Science*, Studies in Systems, Decision and Control 614,
https://doi.org/10.1007/978-3-031-99968-0_21

models based on the self-modeling or reification principle for networks introduced in [91] and further extended in [92, 93], see also [94, 95].

Such empirical literature usually has an emphasis on showing that the considered mechanisms are necessary for the phenomenon to be explained: if these mechanisms are (made) disfunctional, the phenomenon does not occur. But whether the presumed mechanisms on their own are sufficient to let the phenomenon emerge is another, much harder issue due to the high extent of contextuality. Contributions from other mechanisms than the presumed ones that are left implicit but are present in the context cannot be excluded empirically in an easy manner as many of such mechanisms naturally occur in all real humans that act in experiments.

One of the major advantages of computational modeling is that it can test whether proposed correlations between such mechanisms and phenomena indeed can be confirmed computationally by in silico experiments (experiments through computer simulations), both in the necessary and sufficiency sense. This allows researchers to verify whether what leads to the emergence of certain observed phenomena are indeed the presumed causal relationships and not other mechanisms left implicit. Moreover, in specific cases where there is doubt on the mechanisms involved it means that researchers can ask themselves whether any additional mechanisms that are tacitly assumed are needed explicitly and check this computationally. Furthermore, in cases where the empirical literature does not show the specific mechanisms in detail, assumptions can be made about these details and verified to what extent they lead to the phenomena of interest. Consequently, we have sought to achieve a mutual exchange between empirical science and computational science.

In this final chapter, we review the main conclusions that emerged in this volume [44]. First, in Sect. 2 we provide an overview of all types of adaptivity and underlying principles and concepts for all chapters from Chapters "The Emergence of Multimodal Interpersonal Synchrony During Social Interaction: A Network Model" to "Interpersonal Emotion Regulation via Multimodal Interaction with a Bonding Bot: An Adaptive Network Model" in Table 1. Next, in Sects. 3–7, we start by reviewing the main findings for the themes addressed in the different parts II to VI of the book. Next, in Sects. 8 and 9, we situate the contributed material in a broader intellectual context by considering implications and limitations of it, and future directions for the research.

2 The Different Types of Social Interaction Adaptivity Covered

In Table 1 an overview is presented of all types of adaptivity addressed in the chapters in this volume [44], Parts II to VI. For the basics, in the first row in the pink area of this table it is indicated that in all Parts and chapters social contagion is addressed. Moreover, it is indicated that in Part III, in Chapters "A Second-Order Adaptive Network Model for Bonding Based on Subjectively Perceived Homophily" to

Table 1 Overview of all types of adaptivity and the underlying principles and concepts addressed in Part II to VI, Chapters "The Emergence of Multimodal Interpersonal Synchrony During Social Interaction: A Network Model" to "Interpersonal Emotion Regulation via Multimodal Interaction with a Bonding Bot: An Adaptive Network Model" in this volume [44]

Part	II			III					IV			V			VI		
Level / **Chapter**	3	4	5	6	7	8	9	10	11	12	13	14	15	16	17	18	19
Basics	+	+	+	+	+	+	+	+	+	+	+	+	+	+	+	+	+
Social contagion	+	+	+	+	+	+	+	+	+	+	+	+	+	+	+	+	+
Subjective homophily detection					+	+	+	+									
Subjective synchrony detection									+	+	+	+	+	+	+	+	+
Subjective time-lagged synchrony detection												+					
Subjective synchrony transition detection													+				
Subjective group synchrony detection														+			
First-order adaptation			+	+	+	+	+	+	+	+	+	+	+	+	+	+	+
Connectedness: affiliation and bonding				+	+	+	+	+	+	+	+	+	+	+	+	+	+
Subjective homophily-induced					+	+	+	+									
Subjective synchrony-induced									+	+	+	+	+	+	+	+	+
Time-lagged subjective synchrony-induced												+					
Subjective synchrony-transition-induced													+				
Subjective group-synchrony-induced														+			
Bainian or Hebbian connectivity learning			+	+		+		+							+	+	
Responding learning			+	+		+		+									
Sensory preconditioning learning															+		
Memory updating learning																+	
Types of connectivity W-states			+	+	+	+	+	+	+	+	+	+		+	+	+	+
Representing links									+	+	+	+		+	+		+
Responding links			+	+		+		+			+						
Executing links									+	+	+	+		+	+		+
Sensory preconditiong links															+		
External connecting links				+	+	+	+	+									
Memory updating links																+	
Types of excitability T-states										+	+	+	+			+	+
Representing states thresholds										+	+	+	+			+	+
Responding states thresholds											+						
Executing states thresholds										+	+	+	+			+	+
Second-order adaptation				+	+	+		+	+	+	+	+	+	+		+	+
Adaptive adaptation speed				+	+	+		+	+	+	+	+	+	+		+	+
Activation-induced adaptation speed control				+	+	+		+	+	+	+	+	+	+		+	+
Adaptation speed H_W-states				+	+	+		+	+	+	+	+		+			+
Representing links									+	+	+	+		+			+
Responding links											+						
Executing links									+	+	+	+		+			+
External connecting links				+	+	+		+									
Memory updating links																+	
Adaptation speed H_T-states								+		+	+	+	+			+	+
Representing thresholds										+	+	+	+			+	+
Responding thresholds											+						
Executing thresholds										+	+	+	+			+	+
Adaptive adaptation connectivity											+						
Activation-induced adaptation strength control											+						
Adaptation connectivity W_W-states											+						
Representing links											+						
Responding links											+						
Executing links											+						
Adaptation excitability W_T-states											+						
Representing thresholds											+						
Responding thresholds											+						
Executing thresholds											+						
Adaptive homophily-induced bonding								+									
Context-sensitive homophily control								+									
Homophily tipping point H_{TP}-states								+									
External connecting links								+									

"Human Learning, Attunement and Social Bonding Based on Adaptive Homophily: A Multi-adaptive Network Model" the homophily-induced affiliation or bonding is based on subjective homophily detection and in Parts IV to VI in all chapters the synchrony-induced affiliation or bonding is based on subjective synchrony detection. In addition, it is shown that in Part V, Chapter "The Role of Timing in Synchrony Detection and Social Interaction Adaptivity: Including Time Lags in a Second-Order Adaptive Agent Model" also subjective time-lagged synchrony detection is addressed, in Chapter "The Effect of Transition Dynamics in Synchrony on Social Interaction Adaptivity: A Multi-adaptive Network Model" subjective synchrony transition detection, and in Chapter "The Emergence of Group Synchrony and Adaptivity for Group Bonding: A Multi-Adaptive Network Model" subjective group synchrony detection.

The blue area in Table 1 gives an overview of all forms of first-order adaptivity addressed in this volume. First, the types of adaptivity concerning connectedness are shown, in accordance with the different forms of subjective detection indicated in the pink area. Next, some cases of Bainian or Hebbian connectivity learning are indicated: responding learning for Chapter "Learning to Synchronize: Adaptive Joint Decision-Making by Virtual Agents" in Part II and Chapters "Bonding by Homophily and Adaptive Mirroring in Joint Decision Making: A Second-Order Adaptive Network Model", "The Interplay of Bonding Based on Subjective Homophily and Learning During a Varied Rhythmic Action: An Adaptive Network Model", and "Human Learning, Attunement and Social Bonding Based on Adaptive Homophily: A Multi-adaptive Network Model" in Part III, and in Part VI sensory preconditioning in Chapter "Reliving Memory and Religious Experience Through Multimodal Synchrony and Adaptation: An Adaptive Network Model" and memory updating learning.

In all these cases of adaptivity, the agents make use of one or two types of first-order self-model states, called **W**-states (for connectivity adaptation) and **T**-states (for excitability adaptation):

- self-model **T**-states $\mathbf{T}_Y$ control adaptive excitability thresholds for the base level states *Y* indicated in their subscripts. This is a form of (nonsynaptic) adaptation at a relatively fast adaptation timescale, but slower than the timescales for the base level.
- self-model **W**-states $\mathbf{W}_{X,Y}$ control adaptive connection weights of connections $X \rightarrow Y$ for the base level states *X*,*Y* indicated in their subscripts. This is a form of (synaptic) adaptation at a relatively slow adaptation timescale.

Bainian or Hebbian connectivity learning and homophily-induced adaptivity make use of **W**-states in Part III, these **W**-states concern connecting links that connect one agent to the other agent. In contrast, synchrony-induced adaptivity in Parts IV to VI makes use of either **W**-states or **T**-states or both. Moreover, these **W**-states can concern representing, executing, or responding links. Similarly, these **T**-states can concern representing states, executing states, or responding states thresholds. Here, representing states are the sensory representation states created by perception, responding states are the preparation states for actions and executing states are the states for performing the actions. The indicated links are the connections to these types of states.

The purple area in Table 1 gives an overview of all forms of second-order adaptivity addressed in this volume [44]. This level of adaptivity indicates forms of (context-sensitive) control over the first-order adaptation processes. Such control can address the adaptation speed or the adaptation strength or both and can be applied both to connectivity adaptation as self-modeled by first-order **W**-states and excitability adaptation a self-modeled by first-order **T**-states. At the second-order self-modeling level, such control is modeled by higher-order **H**-states (to control adaptation speed) or higher-order **W**-states (to control adaptation strength). Therefore, as shown in Table 1 four types of second-order self-model states are used in this volume:

- $\mathbf{H_T}$-states $\mathbf{H}_{\mathbf{T}_Y}$ to control the speed of first-order adaptation **T**-state $\mathbf{T}_Y$
- $\mathbf{H_W}$-states $\mathbf{H}_{\mathbf{W}_{X,Y}}$ to control the speed of first-order adaptation **W**-state $\mathbf{W}_{X,Y}$
- $\mathbf{W_T}$-states $\mathbf{W}_{Z,\mathbf{T}_Y}$ to control the strength of first-order adaptation **T**-state $\mathbf{T}_Y$
- $\mathbf{W_W}$-states $\mathbf{W}_{Z,\mathbf{W}_{X,Y}}$ to control the strength of first-order adaptation **W**-state $\mathbf{W}_{X,Y}$

The purple area in Table 1 shows where all these forms of second-order adaptivity have been addressed. In addition to these, the table also shows one other type of second-order adaptivity that has been addressed: adaptive homophily-induced bonding in Part III, Chapter "Human Learning, Attunement and Social Bonding Based on Adaptive Homophily: A Multi-adaptive Network Model", where the homophily detection tipping point is made adaptive in a context-sensitive manner.

3 Agent-Based Analysis of the Emergence of Multimodal Interpersonal Synchrony During Social Interaction

In part II it was explored how multimodal interpersonal synchrony can be simulated by interacting agents. One the one hand, this concerns the mechanisms that underlie the emergence of multimodal interpersonal synchrony among two agents taking into account internal cognitive and affective mental states and mechanisms from neuroscience. This was addressed in this volume, Chapter "The Emergence of Multimodal Interpersonal Synchrony During Social Interaction: A Network Model" [47]. Next, in this volume Chapter "Interpersonal Emotion Regulation Through Attunement of Emotions in Human-Bot Interaction: An Agent-Based Network Model" [30] it was explored how in social human-bot interaction coregulation for emotions occurs so that the human's emotions can be regulated due to the bot. Finally, it was analysed how such agents can learn to synchronize during their interaction, applied to a therapy session simulated by virtual agents based on avatars. The latter was addressed in this volume, Chapter "Learning to Synchronize: Adaptive Joint Decision-Making by Virtual Agents" [43].

The agent models that were developed in this part demonstrate that an agent-based dynamical systems modeling approach is capable of modeling the emergence of synchrony for different response modalities, such as movements, affective responses and verbal actions. Notably, synchrony emerged both at the intrapersonal and interpersonal levels. Intrapersonal synchrony refers to the synchronization of different response modalities within an agent. Interpersonal synchrony defines the synchronization of corresponding modalities between agents. These agent models serve as an overall agent-based approach to computational modeling. As part of this approach, we modeled general mechanisms from psychology and neuroscience, such as the use of prediction links and mirror links. Prediction links generate through internal simulation an internal sensory representation of the effect that a certain action is likely to have, i.e., mental predictions of actions are conducted before the actual actions are done [26, 27, 35, 38, 51]. Mirror neurons are activated due to links that connect a sensory representation of another agent's action or emotion to an own corresponding preparation state. These preparation states are activated like mirror neurons: both when an agent prepare for an own action or emotion and when an agent observe a corresponding action or emotion by another agent [53, 75].

3.1 Emergence of Multimodal Synchrony

The agent model developed in this volume Chapter "The Emergence of Multimodal Interpersonal Synchrony During Social Interaction: A Network Model" [47] shows that interpersonal synchrony patterns known from clinical studies and psychological theories emerge from such mechanisms for internal mental simulation and mirror neurons. This means that no other hidden mechanisms are needed for the emergence of interpersonal synchrony. Specifically, regarding synchrony patterns, the right order of response modalities of intrapersonal and interpersonal synchrony happened: Movement synchrony was followed by synchrony of conscious emotions followed by synchrony in linguistic responses [8, 55, 69, 102]. The different response modalities of interpersonal synchrony were interdependent: When the movements of the agents were disabled, synchrony in the other response modalities emerged much less readily. It was shown that the synchrony emerges from the interaction, not by a potential common stimulus between agents at some point in time that has a diminishing prolonged (stamping out/extinguishing) effect on the agent's (synchronizing) behavior. These findings are in line with embodied cognition theories [8, 55, 69, 102], which propose that elementary embodied processes provide the scaffolding for higher-order, sophisticated forms of cognition.

These findings support the rationale of the In-Sync model [63] with regard to the order (higher-order processes like language are preceded by lower-order processes like movement) and interdependence (when the movements were disabled, the other response modalities were weakened as well) of different response modalities of interpersonal synchrony.

The model developed in this volume Chapter "Interpersonal Emotion Regulation Through Attunement of Emotions in Human-Bot Interaction: An Agent-Based Network Model" [30] shows that a similar multimodal agent model can achieve emotion coregulation during social human-bot interaction (following [29]). Here a form of synchrony for emotion states can occur so that the emotion level for the human is downregulated in comparison to cases without interaction with a bot.

3.2 Learning to Synchronize

In this volume Chapter "Learning to Synchronize: Adaptive Joint Decision-Making by Virtual Agents" [43], we incorporated a form of adaptivity in an agent model with regard to interpersonal synchrony. This development of an adaptive agent model was based on causal mechanisms for synaptic plasticity [61, 83]. Synaptic plasticity regards adapting the strength of synapses over time, these synapses correspond to connections within network-oriented agent models. An important example of such synaptic plasticity is Hebbian learning, which in simplified form is often formulated as 'what fires together, wires together'. To facilitate plasticity in this agent model, Hebbian learning for connection weights was used [7, 42]. Specifically, Hebbian learning was applied to the mirroring connections from the sensory representation state of another agent's action to the preparation state for one's own action and from the sensory representation state of another agent's emotional body state (i.e., emotion) to the preparation state for one's own emotion. Here, only such internal stimulus–response like connections were made adaptive.

Plasticity changes over time. As it is beneficial for an organism to adapt the extent of plasticity depending on circumstances, there exists plasticity of the plasticity, also called metaplasticity [1, 36, 76, 85]. Such metaplasticity is a form of second-order adaptivity, whereby plasticity itself concerns first-order adaptivity. Besides plasticity (for first-order adaptivity) and metaplasticity (for second-order adaptivity), the adaptive agent model relied on the following general mechanisms from psychology and neuroscience: like in the other two chapters in this part, mirroring what can lead to imitation [53], emotion integration for affective understanding [19, 26, 27, 71], and self-other distinction for being able to distinguish between yourself and others [21, 25, 34, 53, 54, 60, 66].

This adaptive agent model was used to model an illustrative scenario of a psychotherapy session. Initially not much synchronisation between therapist and client emerged. However, during a number of therapy sessions, gradually adaptation between therapist and client occurred. This adaptation finally resulted in a breakthrough, in which the client synchronized more with the therapist. This interpersonal synchrony was expressed by positive facial expressions and leaning towards each other, along with more open body postures. It was further shown that whether or not a breakthrough in a therapy session occurs and how fast this happens, depends on the extent of adaptivity. The adaptive agent model was illustrated by visualizations through avatars (representing therapist and client) driven by the model. Such a scenario can be seen as a form of virtual role play and provides viewers with more insights in the dynamics of complex mental and social processes.

Our agent model and its simulations in this volume Chapter "Learning to Synchronize: Adaptive Joint Decision-Making by Virtual Agents" [43] also showed how interpersonal synchrony and its considered internal learning can provide a basis for joint decision-making processes. In such a joint decision process, a joint decision consists of three components: a common action, a shared positive emotion about the effect of this common action and a mutual empathic understanding about the action and emotion of each other. Since joint decision processes occur in almost all social interaction settings, our adaptive agent model offers a framework for understanding many contexts where virtual agents and persons need to reach a joint decision through synchronization of multiple response modalities.

4 Adaptive Agent-Based Simulation of Social Interaction and the Emergence of Homophily and Related Bonding

In Part III the focus was on a specific type of social interaction adaptivity, developed mainly from social science and sociology contexts: homophily-induced bonding.

4.1 Integrating Homophily-Induced Adaptivity and Hebbian Learning

In this volume Chapter "Bonding by Homophily and Adaptive Mirroring in Joint Decision Making: A Second-Order Adaptive Network Model" [88] the interplay of social contagion, Hebbian learning of mirroring and bonding is analysed in the context of joint decision making. An adaptive network model was designed, and scenarios were explored using this computational model. The results show that Hebbian learning of mirroring connections plays an important role in decision making, and that bonding can be necessary to make a joint decision, but also that conversely joint decisions strengthen the bonding and the mirroring connections.

4.2 Subjective Homophily-Induced Adaptivity

In this volume Chapter "A Second-Order Adaptive Network Model for Bonding Based on Subjectively Perceived Homophily" [98], the role of subjective elements and control in social network adaptation is analyzed computationally. In particular, it is analyzed: (1) how the coevolution of social contagion and bonding by homophily may be controlled by the persons involved, and (2) how subjective representation states (e.g., what they know or believe) can play a role in this coevolution and its control. To address this, a second-order adaptive social network model was designed in which for persons their bonding depends on their subjective representation states about themselves and about each other and in relation to this they do have a form of control over the coevolution process. Moreover, social contagion depends on their subjective representation states about their connections.

In line with Chapter "A Second-Order Adaptive Network Model for Bonding Based on Subjectively Perceived Homophily", in this volume Chapter "The Interplay of Bonding Based on Subjective Homophily and Learning During a Varied Rhythmic Action: An Adaptive Network Model" [67] an adaptive network model in the context of joint action and social bonding based on subjective homophily detection is presented. Exploration of mechanisms for mental processes and social interaction during joint rhythmic action are presented, inspired by Accetto et al. [2], but here specifically focusing on adaptation by bonding based on subjectively perceived homophily and Hebbian learning, and adaptation speed controlled by second-order adaptation. The paper provides a comprehensive explanation of these concepts and their role in controlled adaptation within illustrative scenarios.

In this volume Chapter "Connecting Social Media Platforms and Users Through Bonding by Faked Homophily: A Second-Order Adaptive Network Model" [18] the concept of subjective homophily detection has also been used to model processes where one person tries to mislead the other person concerning subjective homophily detection (following [17]). This easily happens, for example, when someone fakes interest in certain types of music or movies preferred by another person, just to increase chances on bonding. This idea of faked homophily is applied to obtain an adaptive network model for presenting a user with content they like on a social media platform. The platform's method is using a so-called fake state for its own preference. Simulation results have been performed with different scenarios for different starting values of the user and the platform. In all scenarios the platform's method can move towards the state of the user in order to achieve faked homophily leading to increased bonding. After the user indeed connects to the platform more strongly it becomes more easy for the platform to move the user's preferences in a desired direction by social contagion so that the platform can offer more adequate content more efficiently.

4.3 Context-Sensitive Adaptation by Second-Order Homophily-Induced Social Interaction Adaptivity

This volume Chapter "Human Learning, Attunement and Social Bonding Based on Adaptive Homophily: A Multi-adaptive Network Model" [68] provides an analysis of ways in which context-sensitivity can be introduced to homophily-induced bonding, so that the bonding process is faster or stronger in one context and slower or weaker in another type of context. To this end, a multi-adaptive agent model was designed integrating multiple adaptation mechanisms, specifically focusing on five types of such adaptation mechanisms. Two of them address first-order adaptation by learning of responding on others and first-order adaptation by bonding with others based on homophily. Three other adaptation mechanisms addressed are second-order adaptation of the speed of both Hebbian learning and bonding by homophily (making the speed of adaptation context-sensitive), and second-order adaptation of the homophily tipping point (making the strength of bonding context-sensitive). This provides a comprehensive explanation of these concepts and their role in context-sensitive, controlled adaptation, illustrated by the diverse contextual scenarios addressed in the chapter.

5 Subjective Synchrony Detection and Social Interaction Adaptivity

Part IV focused on modeling how adaptive agents can be able to subjectively detect interpersonal synchrony with other agents for different modalities and how they subsequently can adapt the interaction behavior based on this subjective interpersonal synchrony detection. This has been addressed first in this volume Chapter "Growing Connectedness Through Synchrony: The Role of Subjective Analysis of Intrapersonal Synchrony and Interpersonal Synchrony Within an Adaptive Agent Model" [48] introducing subjective synchrony detection for different modalities and its related adaptivity. This volume Chapter "The Adaptive Interplay of Interpersonal Synchrony, Short-Term Affiliation and Long-Term Bonding: A Multi-timescale Multi-adaptive Agent Model" [45] introduces different time scales for the social interaction adaptivity: short-term affiliation and long-term bonding. Finally, in this volume Part IV Chapter "A Multi-timescale Multi-adaptive Dynamical System Approach for Relationship-Specific and Relationship-Independent Adaptivity in Affiliation and Bonding" [46] it is analysed how behavioural adaptivity can partly be relationship-specific and partly relationship-independent.

Thus, in this part we established that an agent-based adaptive dynamical systems perspective with synchrony-induced adaptivity can be successfully used to model the emergence of interpersonal synchrony from an overall multi-agent perspective. In addition, the designed agent-based adaptive dynamical systems make effectively use of subjective detection of synchrony within each agent, from its own subjective perspective. It is this subjective synchrony detection that induces the adaptivity of the agent's interaction behavior.

5.1 *Subjective Detection of Multimodal Intra- and Interpersonal Synchrony Within an Adaptive Agent Model*

Just like the adaptive agent model described in Chapter "Learning to Synchronize: Adaptive Joint Decision-Making by Virtual Agents" [43], the adaptive agent model in this volume Chapter "Growing Connectedness Through Synchrony: The Role of Subjective Analysis of Intrapersonal Synchrony and Interpersonal Synchrony Within an Adaptive Agent Model" [48] relies on assumptions regarding the basic mechanisms such as mirroring links, prediction links, plasticity and metaplasticity of internal processes. We further made the assumption that agents can subjectively detect their intrapersonal and interpersonal synchrony for different modalities. Studies have shown that connections between persons occur by adapting their interaction behavior to each other [2, 32, 56, 58, 70, 72] and that interpersonal synchrony is linked to these forms of adaptivity. For example, sensing each other becomes more sensitive and actions towards each other become stronger in their expression. From this observation, intrapersonal and interpersonal synchrony detection have been assumed and built in in agents as a general mechanism.

According to the above, the patterns of actual interpersonal synchrony between interacting agents over time lead to patterns of behavioral adaptivity expressed as changing connections between persons. Adopting the temporal factorization principle from [89, 90], a mediating (mental) state can be postulated that is entailed by these patterns of synchrony and in turn entails the patterns of interaction behaviour adaptivity. Such a mediating state is what we call an interpersonal synchrony detection state. Postulating subjective interpersonal synchrony detection states is also based on the assumption or belief that when interpersonal synchrony at multiple response modalities is causally related to behavioral change (e.g., [33, 52, 103]), people should somehow (even unconsciously) be able to note such interpersonal synchrony. To allow virtual agents to synchronize with and accordingly adapt to other virtual agents or persons, such subjective interpersonal synchrony detector states provide an effective mechanism.

In this volume Chapter "Growing Connectedness Through Synchrony: The Role of Subjective Analysis of Intrapersonal Synchrony and Interpersonal Synchrony Within an Adaptive Agent Model" [48], we modeled adaptivity of interpersonal dynamics by adaptive connection weights of four types of connections between states within the adaptive agent model. Specifically, they are (observing) connections from world states to sensing states, (effectuating) connections from execution states to world states, (representing) connections from sensing to representation states for each of the three modalities and (executing) connections from preparation to execution states for each of the three modalities. If the weights of, for example, the executing connections become higher, agents will become more expressive in their actions, which makes it easier for other agents to sense the stronger expressed behavior of this agent. The first two types of connections are external, and the last two types of connections are internal to the agent, but interaction-specific in contrast to stimulus–response like internal connections (from representation to preparation states). This modeling of adaptivity of interaction behavior expands intrapersonal behavioral adaptivity within interpersonal dynamics.

In the simulations, with regard to the within-agent dynamics, conscious emotions emerged from the detection of high levels of intrapersonal synchronization of the three response modality combinations: movement-affect, affect-language and movement-language. These within-agent dynamics of synchronization were in line with the emotion literature [39]. Regarding the between-agent dynamics, in the majority of simulations, agents developed a stronger affiliation or connection during interpersonal synchronization [52, 62, 63, 72, 86, 100, 103]. This interpersonal synchrony detection between agents was higher during common stimulus intervals than during their interaction intervals without common stimulus, suggesting that the mutual effect of interaction is amplified by a shared contextual environment to which both agents respond.

In sum, the assumption of subjective interpersonal synchrony detection among agents lead to within-agent and between-agent dynamics in accordance with the literature. The agents for now relied on one synchrony detection function that can be classified as a variant of signal matching. However, because the agent models are modular, other synchrony detection functions—including statistical ones—can be easily incorporated. The subjective synchrony detection in turn positively affects interaction behaviour adaptivity which was modeled through newly introduced computational mechanisms as well.

In this volume Chapter "Interpersonal Emotion Regulation via Multimodal Interaction with a Bonding Bot: An Adaptive Network Model" [31] subjective multimodal synchrony detection was applied in an agent model for human-bot interaction addressing interpersonal emotion regulation. The analysis results found here provide a promising perspective for interpersonal emotion regulation using such human-like bots. The bot's multimodal interactions helped to regulate the human's emotions by effects of emerging synchrony during the interaction even under less favorable situations such as poor individual emotion regulation capabilities.

5.2 Short-Term and Long-Term Social Interaction Adaptivity in Multi-adaptive Agent Models

Adaptations in the short-term and in the long-term social interaction behaviour have so far been mostly reported separately in the literature. It thus appears that behavioral adaptivity occurs both on shorter and longer time scales, but how these time scales are interrelated has not been systematically considered. To fill this void in the literature, we developed a multi-adaptive agent model in this volume Chapter "The Adaptive Interplay of Interpersonal Synchrony, Short-Term Affiliation and Long-Term Bonding: A Multi-timescale Multi-adaptive Agent Model" [45]. We included mechanisms of both non-synaptic and synaptic plasticity from neuroscience to address respectively short-term and long-term behavioral adaptivity in this multi-adaptive agent model. This distinction and integration grounded in neuroscience principles contributes a new conceptual and computational perspective.

This multi-adaptive agent model demonstrated in simulation experiments that subjective interpersonal synchrony detection can contribute simultaneously to behavioral adaptivity in the short-term and in the long-term. The simulation experiments provide evidence that the mechanisms from neuroscience that were used for the two types of behavioral adaptivity were sufficient to explain how people can move from moment-to-moment interpersonal synchrony to short-term affiliation and, after repeated interactions, long-term bonding. Our simulations thus help to understand how brief moments in interpersonal interactions become organized into more complex interaction experiences and from there give rise to more enduring relationship qualities.

This finding implies that mechanisms known from neuroscience provide support for the theories from conceptual and empirical literatures that address how interaction behavior is affected in an adaptive manner by being in synchrony. Moreover, by explicitly distinguishing between short-term and long-term behavioral adaptivity, such as affiliation and bonding, our modeling approach provides mathematical and computational formalization of theories from empirical literature addressing the interplay of (nonsynaptic) short-term intrinsic adaptivity of excitability and (synaptic) long-term adaptivity of connectivity, such as [3, 40, 64]. See also this volume Chapter "Modeling Coevolution of Dynamics, Adaptivity and Control of Adaptivity by Self-modeling Networks".

In addition, the generality of the chosen modeling approach has been analyzed. In particular, we established that the chosen network-oriented modeling approach can model any adaptive dynamical systems. More precisely, we demonstrated mathematically how any smooth adaptive dynamical system (of any order of adaptivity) has a canonical representation by a self-modeling network. Therefore, the choice for self-modeling networks as modeling format does not introduce any fundamental limitations in modeling adaptive dynamical systems.

5.3 Learning Relationship-Specific and Relationship-Independent Social Interaction Behavior in Adaptive Agent Models

Within social interactions, relationship-specific and relationship-independent adaptation can be distinguished. The aim of this volume Chapter "A Multi-timescale Multi-adaptive Dynamical System Approach for Relationship-Specific and Relationship-Independent Adaptivity in Affiliation and Bonding" [46] was to examine such patterns of transferences and differentiation. So far, our analysis of interpersonal interaction dynamics was based on the behavior of fixed pairs of agents. In [46], we extended the analysis of interaction behavior to include multiple agents with multiple mutual interaction episodes. In that context, a main question is how adapted behaviors learned by a single agent within one dyad can also play a role within other dyads. In other words: How can transference over relationships happen and how can relationship-specific and relationship-independent (adaptivity of) social interaction behavior be distinguished? Such transference is a central focus in classical psychoanalysis and modern social-cognitive empirical research has shown the existence of transference [4, 5, 23, 59]. Moreover, in the attachment theory, it has been described that such transference from one relationship to another relationship does often occur, meaning that adaptations that happened within one relationship can also have their effects in other relationships [20, 77–80]. When transference is modeled within adaptive agent models, this might provide useful insights for empirical researchers and therapists.

For the developed adaptive agent model, we assumed that the speed and strength of the short-term and long-term behavioral adaptivity in specific relationships are generally adaptive in a relationship-independent manner. Furthermore, we assumed that also the strength of the connections towards the execution states are adaptive in a relationship-independent manner.

In simulation experiments, pairs of agents were randomly selected for interaction from a group of four agents. We demonstrated in these simulation experiments that the assumptions about neural underpinnings suffice for agents' ability to transfer adaptations in interaction behavior developed in one relationship to another relationship. These results can inspire future empirical research to study in more detail to what extent such transference of interaction behavior happen, what the underlying (neural) mechanisms are and how transference can be incorporated in therapeutical applications of virtual agents [59].

6 Further Computational Analysis of Social Interaction Adaptivity Induced by Subjective Interpersonal Synchrony Detection

In Part V, some more refined analyses are contributed. In this volume Chapter "The Role of Timing in Synchrony Detection and Social Interaction Adaptivity: Including Time Lags in a Second-Order Adaptive Agent Model" [49], the role of time lags for subjective synchrony analysis is computationally explored in more detail. Furthermore, in this volume Chapter "The Effect of Transition Dynamics in Synchrony on Social Interaction Adaptivity: A Multi-adaptive Network Model" [50] it is explored computationally how not only subjective synchrony detection itself leads to behavioural adaptivity but also transitions or interruptions in synchrony can contribute to such an effect. Finally, in this volume Part V Chapter "The Emergence of Group Synchrony and Adaptivity for Group Bonding: A Multi-adaptive Network Model" [65] the interplay between group synchrony and behavioural adaptivity for group bonding is studied.

6.1 Comparing Synchrony Detection Methods with Different Time Lags Within Adaptive Agent Models

In this volume, Chapter "The Role of Timing in Synchrony Detection and Social Interaction Adaptivity: Including Time Lags in a Second-Order Adaptive Agent Model" [49] we integrated and compared different synchrony detection methods in adaptive agent models. For the already used synchrony detection function relating to signal matching, we compared three different types of functions (no time lag, a uniform fixed time lag and a maximum time lag in which the synchrony score is an average over all time lags until the maximum time lag). All time lags were onesided toward the past because we assumed that our agents have no knowledge about future time points.

Moreover, sensing the action of the other agent follows a pathway different from the pathway of sensing one's own action; due to this, in general it goes faster to sense oneself than someone else [74, 84, 87]. For these reasons, the study of time lags makes sense. Both the resulting interpersonal synchrony scores and how the scores of behavioral adaptivity of the two agents evolve accordingly were evaluated.

As it turns out, there are differences in the synchrony detection scores and the induced behavioral adaptivity over the different time lags used, although these differences are relatively small. This adaptive agent model suggests that time differences might slightly impact the experienced notion of social presence [99] during interaction, both theoretically between persons and practically between agent(s) and/or persons. Namely, higher interpersonal synchrony scores can be obtained when proper time lags are involved during synchrony computations. This implies that future research and social interaction theories need to consider how time differences in actions and sensing of them between persons/agents can be found and how these time differences affect the social presence.

6.2 Disruptions and Transitions of Interpersonal Synchrony

In addition to detected synchrony itself, detected disruptions or transitions in synchrony can also have an effect on the adaptation of interaction behaviour. Our aim in this volume Chapter "The Effect of Transition Dynamics in Synchrony on Social Interaction Adaptivity: A Multi-adaptive Network Model" [50] was to examine these transition dynamics. From the literature, we know that transitions in interpersonal synchrony—over and above average levels of interpersonal synchrony—are related to stronger affiliation between persons [37, 73]. In this Chapter "The Effect of Transition Dynamics in Synchrony on Social Interaction Adaptivity: A Multi-adaptive Network Model", we studied underlying mechanisms for such transitions. The link between transitions in synchrony and stronger affiliation led to the assumption that persons should also be able to detect transitions in interpersonal synchrony over and above their detection of interpersonal synchrony itself. Such subjective interpersonal synchrony transition detector states were added as extra states to each agent within the adaptive agent model. Affiliation was modeled through non-synaptic plasticity by adaptive excitability. Agents had a stronger affiliation with each other when they were able to detect transitions in synchrony on top of the detection of interpersonal synchrony itself, as was demonstrated by simulation experiments. This suggests that not only for computational research, but also for empirical research towards the effect of synchrony transitions on affiliation, underlying mechanisms for the notion of synchrony transition detection can shed important light on changes in relationship dynamics.

6.3 Group Synchrony and Group Bonding

In the computational models and simulation experiments described above, social interaction between agents was always in dyads. But attunement or synchronisation can also occur for larger groups of persons. An interesting question is how this may lead to a form of group bonding. In this volume Chapter "The Emergence of Group Synchrony and Adaptivity for Group Bonding: a Multi-Adaptive Network Model" [65] this was explored computationally by generalising the synchrony-induced social interaction adaptivity principles for dyads to the case of more than two persons. Thus a novel adaptive computational dynamical system model was designed for the relationship between group synchrony and group bonding. By simulating multimodal interactions within a group of four agents, the study indeed uncovers patterns in group cohesion in the sense of emerging multimodal group synchrony and related group bonding. The obtained insights offer an understanding of group interaction dynamics and adaptivity. Future research may consider larger groups and more variations of synchrony detection functions to widen the obtained findings.

7 Computational Analysis of Real-World Scenarios for Synchrony-Induced Social Interaction Adaptivity

In this volume [44] part VI, the methods and models developed in previous parts are applied to analyse two real-world cases about events with elderly in more detail. First, this volume Chapter "Joint Action and Memory Recall for Elderly by Detecting Interpersonal Synchrony: A Higher-Order Adaptive Network Model" [101] addresses computationally recreating, simulating and subsequently inspecting a real-life scenario. In order to recreate such scenario computationally, an adaptive agent model involving representation of mental mechanisms inspired by neurological processes.

Next, this volume Chapter "Joint Action and Memory Recall for Elderly by Detecting Interpersonal Synchrony: A Higher-Order Adaptive Network Model" [104] explores the potential of adaptive network modeling for joint action and memory recall among elderly through detecting interpersonal synchrony. With the aging population increasing, there is a crucial need to focus on the health and social interaction of older adults. Based on research of the significance of social interaction and memory use for the elderly, as well as the role of interpersonal synchrony in joint action, this paper aims to analyse computationally how to enhance positive effects of social interactions among older individuals by applying an adaptive network model. The research examines the concept of interpersonal synchrony and its impact on joint action, memory, and emotional well-being in elderly populations. Through simulation experiments and analysis, the study demonstrates the potential benefits for music in memory recall for older adults with cognitive decline, highlighting the importance of social interaction and emotional resonance. This study offers a valuable contribution to understanding and improving social interactions and memory recall among the elderly.

8 Outlook: Limitations and Future Directions

A considerable amount of empirical research has studied social interaction and more specific topics such as interpersonal synchrony, affiliation and bonding. Interpersonal synchrony is a complex process, in which many variables work simultaneously on each other. A central idea within this complexity regards that higher-order (synchronizing) processes such as language are grounded in lower-order (synchronizing) processes such as movement. In other words, the lower-order processes scaffold the higher-order processes. Because of this complexity, it is needed to collect and analyze all data simultaneously that can support the theory, which remains difficult. In this volume, we have considered how this problem may be resolved by using the perspectives offered by AI-based modeling in this study of interpersonal synchrony.

Adaptation processes induced by social interaction such as in short-term affiliation during single interaction episodes or in the long-term bonding over multiple interaction episodes form a cyclic complex adaptive dynamic system in which also synchronization plays a main role. Understanding and describing such a complex adaptive system requires handling timing elements which in empirical research based on statistical analyses often is left unaddressed. For analysis of these main elements within social interaction and their adaptive temporal and cyclic mutual effects, analysis methods based on adaptive dynamical causal modeling from AI has shown itself as a useful tool.

Like any research, the research described in this volume is subject to limitations and leaves much room for further research. In this section, we will outline such limitations and along with these limitation, we provide suggestions for future research directions.

8.1 From Dyads to Groups of Agents

A first important limitation is that, in almost all our simulation studies for the (adaptive) agent models, only two agents interacted simultaneously with each other. Even in the adaptive agent model when relationship-independent adaptivity of interaction behavior was studied, although there was a group of four agents, the interactions occurred according to interaction episodes for randomly picked pairs from these four agents (Chapter "A Multi-timescale Multi-adaptive Dynamical System Approach for Relationship-Specific and Relationship-Independent Adaptivity in Affiliation and Bonding"). Thus, in the latter case, no option was created to let an agent interact with more than one agent simultaneously. One notable exception is Chapter "The Emergence of Group Synchrony and Adaptivity for Group Bonding: A Multi-Adaptive Network Model" where a first step was made to address group synchrony and group bonding.

To study how interpersonal synchrony and social interaction behavior adaptivity evolves within such groups of agents more extensively, more agents can be designed with different trait characteristics specified by their network characteristics. Their emerging collective behavior within simulation experiments when they are simultaneously involved can be analyzed. In this way, the focus shifts more from within-agent modeling to within-dyad modeling to within-group modeling of interpersonal synchrony and its adaptivity in interaction behavior. Then adaptivity in interaction behavior can arise at multiple levels: within-dyad adaptivity and within-group adaptivity.

Such interaction in groups of agents instead of only within dyads goes along with multiple additional research challenges. It can be examined which measurement functions can be used to detect the interpersonal synchrony (transition) scores over the whole group, or how to define and model group bonding. It can also be studied what happens when only one agent gets a stimulus who is able to interact with other agents, and that these agents in turn interact with the remaining agents. Is there then a cascading of emergence of interpersonal synchrony, and how does this affect adaptivity in interaction behavior? Alternatively, some agents can receive the same stimulus and other ones different stimuli, and the possibilities of interaction can also be manipulated. Experiments with groups of agents as started by Chapter "The Emergence of Group Synchrony and Adaptivity for Group Bonding: a Multi-Adaptive Network Model" result in numerous possibilities for future research.

8.2 Experimental Variations

A second important limitation is that the present volume explored only a limited number of experimental simulations in testing the emerging behavior in the considered agent models. Thus, it is still possible to explore more experimental variations for each (adaptive) agent model. Throughout the chapters, several experimental setups already have been included. For example, one agent who got a stimulus over intervals while the other agent was only able to interact with this agent. Or both agents who repeatedly were in equal sized episodes of a common stimulus/no interaction, common stimulus/interaction, different stimuli/no interaction and different stimuli/interaction. Furthermore, the time interval lengths of stimulus exposure could differ between the agents or the stochastic noise added in the environment in other experimental setups. We further experimented with manipulations in which response modalities interaction can occur in a certain time interval (Chapter "The Role of Timing in Synchrony Detection and Social Interaction Adaptivity: Including Time Lags in a Second-Order Adaptive Agent Model").

Nevertheless, there is still room for more extensive simulation experiments. In future research, more variations in different experimental setups of simulation for the same agent model, and more variations of the agent models with regard to the same simulation experimental setup can be explored. Such further comparison of experimental variations can result in noticing more details of emerging agent behavior under specific circumstances.

8.3 Additional Mental States

A third important limitation is that the agent models used in the present volume were rather complex, i.e., they had already a lot of states. This complexity was due to the different action modalities (movement, affective responses and language responses), and on top of this multiple forms of behavioral adaptivity or synchrony detection. Due to the already present complexity and the desire to keep agent models as simple as possible for interpretation, not all possible mental states were added to the agent models. Obviously, in reality persons have more mental states than the ones introduced in the current agent models. Future agent models could capture other characteristics of persons by adding a limited number of extra states in the agent models, for example, belief states for interpretation of representation states. As another instance, for models involving synchrony-induced social interaction adaptivity, there could be separate representation states to interpret an agent's own specific execution state and the corresponding sensed execution state of another agent such as done in Chapters "A Second-Order Adaptive Network Model for Bonding Based on Subjectively Perceived Homophily" to "Human Learning, Attunement and Social Bonding based on Adaptive Homophily: A Multi-adaptive Network Model" for homophily-induced social interaction adaptivity. Another interesting area to be added more is emotion regulation [24, 63]. Here Chapter "Interpersonal Emotion Regulation Through Attunement of Emotions in Human-Bot Interaction: An Agent-Based Network Model" and especially Chapter "Interpersonal Emotion Regulation via Multimodal Interaction with a Bonding Bot: An Adaptive Network Model" form a good starting point, but much more can be done in this direction. Furthermore, interpersonal synchrony has been related to many other concepts, such as anticipatory motivation and person perception [10]. Future adaptive agent models can be inspired by such concepts and theories in which there is an interplay between interpersonal synchrony and additional states.

8.4 Connecting with Empirical Data

A fourth important limitation is that it is difficult to gather real-time data sets over time periods that cover important aspects of the proposed agent models. This is mainly because this research is multidisciplinary incorporating elements from neuroscience, psychology and AI, Even more complicated, multiple response modalities (movement, affect and language) are modeled within the adaptive agent models, and the activation levels of these modalities normally fluctuate over time. This means that—on top of the multidisciplinary aspect—complexity is introduced by different modalities that require time-series data sets that are collected longitudinally with intensive measurements.

The current work focuses on the development of (adaptive) agent models, based on general mechanisms from the literature. As validation, the emergence of specific behavioral adaptivity is then compared to theoretical expectations from the literature. If the emerging patterns in the (adaptive) agent models are in line with the literature, this serves as an initial validation of the (adaptive) agent model. Future work should verify whether such synchrony patterns and behavioral adaptivity in the agents are in accordance with empirical data. It will be challenging to collect such extensive data sets, but it is important as a follow-up step. Through optimization methods used for parameter estimation applied to values of network characteristics in a model, it can be quantified to what extent the activation values of states of the agent model and the scores on variables for these states in the empirical data set are the same and which changes in the values will make the differences smaller.

8.5 *Incorporating More Machine Learning Algorithms*

A fifth important limitation is that the present adaptive agent models use already different (biologically inspired) types of machine learning, but the field of machine learning offers other directions as well. On the go learning methods already have been used in almost all of the chapters, for example, first-order learning such as Hebbian learning, synchrony-induced learning, and transition-induced learning, and second-order learning such as adaptation speed learning, adaptation strength learning, and homophily tipping point learning. Nevertheless, other on-the-go machine learning methods may be tried as well, such as reinforcement learning methods (e.g., Q-learning). In contrast, yet other machine learning methods work on given large data sets. Such types of learning methods can be incorporated as well when within the agent models, part of the history is represented as an internal or external data set. For example, a person may remember how often a certain type of error was made, or an organization may have a data set of errors made in the past by all employees. Based on such data set belief states in the agent, models can be generated based on application of (supervised or unsupervised) machine learning methods.

8.6 *Virtual Agents*

A sixth and last notable limitation is that this volume was centered around the development of adaptive agent models including homophily or synchrony (transition) detection. These agent models can form the basis of virtual agents or robots in the future. In Chapter “Learning to Synchronize: Adaptive Joint Decision-Making by Virtual Agents” of this volume, we have already visualize how the breakthrough of a client during a therapy session was accomplished to demonstrate how the functioning of an adaptive agent model can be shown. Moreover, Chapters “Interpersonal Emotion Regulation through Attunement of Emotions in Human-Bot Interaction: an Agent-Based Network Model” and “Interpersonal Emotion Regulation via Multimodal Interaction with a Bonding Bot: An Adaptive Network Model” focused on agent models that can provide a basis for bots visualised by avatars that interact with humans in an adaptive manner, such that, for example, they can bond with humans.

Online therapeutics would be a setting in which an avatar can interact with a human client, and this avatar is then equipped with such an adaptive agent model. In the therapy session that we have visualized in Chapter “Learning to Synchronize: Adaptive Joint Decision-Making by Virtual Agents”, the functioning of the avatars was relatively basic. Obviously, many more elements can be addressed. In human interactions, virtual agents should be able to behave human-like. Meaning they are equipped to saying words and sentences (for example, by incorporating Large Language Models), displaying more facial expressions than smiling and moving more kinds of body parts. Furthermore, these virtual agents should be able to read the movements of humans through some detection method, or sentiment analysis to get the affect from the verbal expressions from humans, or analysis of the pitch. With the current adaptive agent models, virtual agents can detect their interpersonal synchrony with their human clients and intentionally try to synchronize. Experiments in which humans and virtual agents interact to see whether they can adapt their social interaction behavior from both sides, to what extent a human feels a bond and being in sync with a virtual agent can also be conducted. Future work may also study the degree to which humans experience avatars as displaying human behavior (and thereby indirectly validating the adaptive agent model).

8.7 Methods for Detecting Synchrony

Agent models can be used to simulate different aspects that are considered important for the overall aim: emerging multimodal synchrony, subjective synchrony detection and behavioural adaptivity and disruptions or transitions of synchrony detection. Exploring these aspects has also been done to some extent, for example, in this volume some variations of these functions were made: in this volume Chapter "The Role of Timing in Synchrony Detection and Social Interaction Adaptivity: Including Time Lags in a Second-Order Adaptive Agent Model" [49] for different time lags and in Chapter "The Emergence of Group Synchrony and Adaptivity for Group Bonding: A Multi-Adaptive Network Model" [65] for group synchrony for four agents. Some other detection methods that work across multiple dyads and for more than a single variable simultaneously within a dyad (multidimensional cross-recurrence quantification analysis) are possible as well. In further work, researchers can also experiment with other synchrony detection methods that transcend single variables in dyads. More in general, further studies are required in the field because it is still not very clear when to use which synchrony detection method and for which parameter values, although attempts are increasing to formulate guidelines [11, 15, 82].

9 Concluding Remarks

In the opening of this volume, we observed how the feeling of being on the same wavelength can be nearly universally recognized and yet remains still rather mysterious. This contradiction is caused because it is a complex process in interaction with emerging and adapting behavior. In this volume, complex dynamical agent models have turned out to be a useful tool to study such emerging and adaptive processes in interpersonal dynamics. In silico experiments complement in vivo experiments to unravel the underlying processes of interpersonal synchrony and social interaction adaptation.

Moreover, the subjective agent perspective provides a different perspective than the more traditional objective approach. The subjective agent and objective approaches offer insights that mutually reinforce and complement each other. Our scientific understanding of interpersonal dynamics remains far from complete. Yet, the present volume has shown how this understanding can be deepened by modeling interpersonal dynamics within adaptive dynamical agents. Virtually everyone is familiar with the feeling of being on the same wavelength with someone. With the help of adaptive dynamical systems models, we can discover the scientific meaning of this intuitive feeling, which is of foundational significance to social life.

More background from biological and philosophical perspectives on the use of a dynamical system architecture with multiple adaptation or control levels can be found in Bechtel and Bich [9, 12–14, 96, 97]. Although in this volume [44] only two adaptation or control levels are used, in other research where the role of environment-induced epigenetic changes for the development of mental disorders such as Schizophrenia, ASPD, Binge eating, Insomnia, Psychopathy, SUD is analysed, a fifth-order control architecture turned out useful, e.g., [6, 16, 22, 41, 57, 81].

Acknowledgements The research presented in this volume (Hendrikse, Treur, Koole, 2025) was supported by NWO Open Competition project 'The Rhythm of Relating: How Emotional Sharing Emerges From Interpersonal Synchrony in Movements, Physiology, and Neural Activations' (Project number 406.18.GO.024).

References

1. Abraham, W.C., Bear, M.F.: Metaplasticity: the plasticity of synaptic plasticity. Trends Neurosci. **19**(4), 126–130 (1996)
2. Accetto, M., Treur, J., Villa, V.: An adaptive cognitive-social model for mirroring and social bonding during synchronous joint action. In: Proceedings of the 9th International Conference on Biologically Inspired Cognitive Architectures, BICA'18, vol. 2. Procedia Computer Science, vol. 145, pp. 3–12 (2018)
3. Alejandre-García, T., Kim, S., Pérez-Ortega, J., Yuste, R.: Intrinsic excitability mechanisms of neuronal ensemble formation. eLife **11**, e77470 (2022)
4. Andersen, S.M., Berk, M.S.: The social-cognitive model of transference: experiencing past relationships in the present. Curr. Dir. Psychol. Sci. **7**(4), 109–115 (1998)
5. Andersen, S.M., Przybylinski, E.: Experiments on transference in interpersonal relations: implications for treatment. Psychotherapy **49**(3), 370 (2012)
6. Augustinus, F., Hendrikse, S.C.F., Treur, J.: Psychopathic genes: a multi-level adaptive dynamic modelling approach. In: Proceedings of the 8th Computational Methods in Systems and Software Conference, CoMeSySo'24. Lecture Notes in Networks and Systems, vol 1471, pp 477–498. Springer Nature (2025)
7. Bain, A.: Mind and Body: The Theories of Their Relation. HS King & Company. Farnborough Eng. Gregg International (1873)
8. Barsalou, L.W.: Grounded cognition. Ann. Rev. Psychol. **59**, 617–645 (2008)
9. Bechtel, W., Bich, L.: Organisms need mechanisms; mechanisms need organisms. In: Cordovil, J.L., Santos, G., Vecchi, D. (eds.) New Mechanism: Explanation, Emergence and Reduction. History, Philosophy and Theory of the Life Sciences, vol. 35, pp. 85–108. Springer Nature (2024)
10. Beckes, L., Coan, J.A.: Relationship neuroscience. In: Mikulincer, M.E., Shaver, P.R., Simpson, J.A., Dovidio, J.F. (eds.) APA Handbook of Personality and Social Psychology, vol. 3: Interpersonal relations, Ch 5, pp. 119–150. American Psychological Association (2015)

11. Behrens, F., Moulder, R.G., Boker, S.M., Kret, M.E.: Quantifying physiological synchrony through windowed cross-correlation analysis: statistical and theoretical considerations. BioRxiv, 2020-08 (2020)
12. Bevilacqua, A., Wilkinson, S.J., Dimelow, R., Murabito, E., Rehman, S., Nardelli, M., et al.: Vertical systems biology: from DNA to flux and back. In: Hetherington, A., Grierson, C. (eds.) Practical Systems Biology, pp. 79–106. Taylor & Francis (2008)
13. Bich, L., Bechtel, W.: Control mechanisms: explaining the integration and versatility of biological organisms. Adapt. Behav. **30**(5), 389–407 (2022a)
14. Bich, L., Bechtel, W.: Organization needs organization: understanding integrated control in living organisms. Stud. Hist. Philos. Sci. **93**, 96–106 (2022b)
15. Bizzego, A., Azhari, A., Esposito, G.: Assessing computational methods to quantify mother-child brain synchrony in naturalistic settings based on fNIRS signals. Neuroinformatics 1–10 (2021)
16. Blei, Z.A., Hendrikse, S.C.F., Treur, J.: Modeling epigenetic modification due to childhood abuse and its relation to ASPD: A fifth-order adaptive network model. In: Proceedings of the 15th Biologically Inspired Cognitive Architectures for AI Conference, BICA*AI'24. Studies in Computational Intelligence, vol 477, pp 59–70. Springer Nature (2024)
17. Bössenecker, G., Mreijen, P., Treur, J.: A second-order adaptive network model for connecting social media platforms and users through bonding by faked homophily. Procedia Comput. Sci. **213**, 32–40 (2022). https://doi.org/10.1016/j.procs.2022.11.035
18. Bössenecker, G., Mreijen, P., Treur, J.: Connecting social media platforms and users through bonding by faked homophily: a second-order adaptive network model. In: Hen-drikse, S.C.F., Treur, J., Koole, S.L. (eds.) New Analysis and Modeling Directions in Social Interaction Science: Emergent Multimodal Interpersonal Synchrony, Affiliation, and Bonding, Ch 9. Springer Nature (2025)
19. Boukouvala, A: Imitation of affects and mirror neurons: exploring empathy in Spinoza's theory and contemporary neuroscience. Philosophia **45**(3), 1007–1017 (2017)
20. Bowlby, J.: Attachment. Basic Books (2008)
21. Brass, M., Spengler, S.: The inhibition of imitative behaviour and attribution of mental states. In: Striano, T., Reid, V. (eds.) Social Cognition: Development, Neuroscience, and Autism, pp. 52–66. Wiley-Blackwell (2009)
22. Călin, R.S., Hendrikse, S.C.F., Treur, J.: Environmental and Epigenetic Influences in Binge Eating: An Adaptive Network Modeling Approach. To appear (2025)
23. Chen, S., Andersen, S.M.: Relationships from the past in the present: significant other representations and transference in interpersonal life. In: Advances in experimental social psychology, vol. 31, pp. 123–190. Academic Press (1999)
24. Coan, J.A.: Towards a neuroscience of attachment. In: Cassidy, J., Shaver, P.R. (eds) Handbook of Attachment: Theory, Research, and Clinical Applications, 3rd edn, pp. 242–269. Guilford, New York, NY (2016)
25. Courtney, A.L., Meyer, M.L.: Self-other representation in the social brain reflects social connection. J. Neurosci. **40**(29), 5616–5627 (2020)
26. Damasio, A.R.: Descartes' Error: Emotion, Reason, and the Human Brain. Penguin, London (1994)
27. Damasio, A.R.: The Feeling of What Happens: Body and Emotion in the Making of Consciousness, Harvest Harcourt, San Diego (1999)
28. Dikker, S., Montgomery, S., Tunca, S.: Using synchrony-based neurofeedback in search of human connectedness. In: Nijholt, A. (ed.) Brain Art: Brain-Computer Interfaces for Artistic Expression, pp. 161–206. Springer International Publishing (2019)
29. Dimopoulos, F., Eler, E., Timmers, M., Treur, J., Koole, S.L.: An agent-based network model for interpersonal emotion regulation in a human-bot interaction. In: International Conference on Computational Collective Intelligence, Proceedings of the 15th International Conference on Computational Collective Intelligence, ICCCI'23. Lecture Notes in AI, pp. 68–79. Springer Nature Switzerland, Cham (2023)

30. Eler, E., Dimopoulos, F., Timmers, M., Treur, J., Koole, S.L.: Interpersonal emotion regulation through attunement of emotions in human-bot interaction: an agent-based network model. In: Hendrikse, S.C.F., Treur, J., Koole, S.L. (eds.) New Analysis and Modeling Directions in Social Interaction Science: Emergent Multimodal Interpersonal Synchrony, Affiliation, and Bonding, Ch 4. Springer Nature (2025)
31. Eler, E., Treur, J., Hendrikse, S.C.F., Donker, T., Koole, S.L.: Interpersonal emotion regulation via multimodal interaction with a bonding bot: an adaptive network model. In: Hendrikse, S.C.F., Treur, J., Koole, S.L. (eds.) New Analysis and Modeling Directions in Social Interaction Science: Emergent Multimodal Interpersonal Synchrony, Affiliation, and Bonding, Ch 19. Springer Nature (2025)
32. Fairhurst M.T., Janata P., Keller P.E.: Being and feeling in sync with an adaptive virtual partner: brain mechanisms underlying dynamic cooperativity. Cereb Cortex. **23**(11), 2592–2600 (2013). https://doi.org/10.1093/cercor/bhs243
33. Feldman, R.: Parent–infant synchrony: biological foundations and developmental outcomes. Curr. Dir. Psychol. Sci. **16**, 340–345 (2007)
34. Fried, I., Mukamel, R., Kreiman, G.: Internally generated preactivation of single neurons in human medial frontal cortex predicts volition. Neuron **69**, 548–562 (2011)
35. Gallese, V., Goldman, A.: Mirror neurons and the simulation theory of mindreading. Trends Cogn. Sci. **2**, 493–501 (1998)
36. Garcia, R.: Stress, metaplasticity, and antidepressants. Curr. Mol. Med. **2**(7), 629–638 (2002)
37. García, E., Di Paolo, E.A.: Embodied coordination and psychotherapeutic outcome: beyond direct mappings. Front. Psychol. **1257** (2018)
38. Goldman, A.I.: Simulating Minds: The Philosophy, Psychology, and Neuroscience of Mindreading. Oxford University Press, New York (2006)
39. Grandjean, D., Sander, D., Scherer, K.R.: Conscious emotional experience emerges as a function of multilevel, appraisal-driven response synchronization. Conscious. Cogn. **17**(2), 484–495 (2008)
40. Hansel, C., Yuste, R.: Neural ensembles: role of intrinsic excitability and its plasticity. Front. Cell. Neurosci. **18**, 1440588 (2024)
41. Hassouna, S., Hendrikse, S.C.F., Treur, J.: The Dynamics of Epigenetic Influence in Insomnia: A Higher-Order Adaptive Modeling Perspective. In: Itthipuripat, S., Ascoli, G.A., Li, A., Pat, N., Kuai, H. (eds) Brain Informatics. Proceedings BI 2024. Lecture Notes in Computer Science, vol 15541, pp 312–325. Springer Nature (2024)
42. Hebb, D.O.: The organization of behaviour. **4**(60), 78–60 (1949)
43. Hendrikse, S.C.F., Kluiver, S., Treur, J., Wilderjans, T.F., Dikker, S., Koole, S.L.: Learning to synchronize: adaptive joint decision-making by virtual agents. In: Hendrikse, S.C.F., Treur, J., Koole, S.L. (eds.) New Analysis and Modeling Directions in Social Interaction Science: Emergent Multimodal Interpersonal Synchrony, Affiliation, and Bonding, Ch 5. Springer Nature (2025a)
44. Hendrikse, S.C.F., Treur, J., Koole, S.L. (eds.): New Analysis and Modeling Directions in Social Interaction Science: Emergent Multimodal Interpersonal Synchrony, Affiliation, and Bonding. Springer Nature (2025b)
45. Hendrikse, S.C.F., Treur, J., Koole, S.L.: The adaptive interplay of interpersonal synchrony, short-term affiliation and long-term bonding: a multi-timescale multi-adaptive agent model. In: Hendrikse, S.C.F., Treur, J., Koole, S.L. (eds.) New Analysis and Modeling Directions in Social Interaction Science: Emergent Multimodal Interpersonal Synchrony, Affiliation, and Bonding, Ch 12. Springer Nature (2025c)
46. Hendrikse, S.C.F., Treur, J., Koole, S.L.: A multi-timescale multi-adaptive dynamical system approach for relationship-specific and relationship-independent adaptivity in affiliation and bonding. In: Hendrikse, S.C.F., Treur, J., Koole, S.L. (eds.) New Analysis and Modeling Directions in Social Interaction Science: Emergent Multimodal Interpersonal Synchrony, Affiliation, and Bonding, Ch 13. Springer Nature (2025d)
47. Hendrikse, S.C.F., Treur, J., Wilderjans, T.F., Dikker, S., Koole, S.L.: The emergence of multimodal interpersonal synchrony during social interaction: a network model. In: Hendrikse,

S.C.F., Treur, J., Koole, S.L. (eds.) New Analysis and Modeling Directions in Social Interaction Science: Emergent Multimodal Interpersonal Synchrony, Affiliation, and Bonding, Ch 3. Springer Nature (2025e)

48. Hendrikse, S.C.F., Treur, J., Wilderjans, T.F., Dikker, S., Koole, S.L.: Growing connectedness through synchrony: the role of subjective analysis of intrapersonal synchrony and interpersonal synchrony within an adaptive agent model. In: Hendrikse, S.C.F., Treur, J., Koole, S.L. (eds.) New Analysis and Modeling Directions in Social Interaction Science: Emergent Multimodal Interpersonal Synchrony, Affiliation, and Bonding, Ch 11. Springer Nature (2025f)
49. Hendrikse, S.C.F., Treur, J., Wilderjans, T.F., Dikker, S., Koole, S.L.: The role of timing in synchrony detection and social interaction adaptivity: including time lags in a second-order adaptive agent model. In: Hendrikse, S.C.F., Treur, J., Koole, S.L. (eds.) New Analysis and Modeling Directions in Social Interaction Science: Emergent Multimodal Interpersonal Synchrony, Affiliation, and Bonding, Ch 14. Springer Nature (2025g)
50. Hendrikse, S.C.F., Treur, J., Wilderjans, T.F., Dikker, S., Koole, S.L.: The effect of transition dynamics in synchrony on social interaction adaptivity: a multi-adaptive network model. In: Hendrikse, S.C.F., Treur, J., Koole, S.L. (eds.) New Analysis and Modeling Directions in Social Interaction Science: Emergent Multimodal Interpersonal Synchrony, Affiliation, and Bonding, Ch 15. Springer Nature (2025h)
51. Hesslow, G.: Conscious thought as simulation of behaviour and perception. Trends Cogn. Sci. **6**, 242–247 (2002)
52. Hove, M.J., Risen, J.L.: It's all in the timing: Interpersonal synchrony increases affiliation. Soc. Cogn. **27**(6), 949–960 (2009)
53. Iacoboni, M.: Mirroring People: The New Science of How We Connect with Others. Farrar, Straus & Giroux, New York (2008)
54. Iacoboni, M., Dapretto, M.: The mirror neuron system and the consequences of its dysfunction. Nat. Rev. Neurosci. **7**(12), 942–951 (2006)
55. IJzerman, H., Koole, S.L.: From perceptual rags to metaphoric riches: bodily, social, and cultural constraints on sociocognitive metaphors: comment on Landau, Meier, and Keefer (2010). Psychol. Bull. **137**, 355–361 (2011). https://doi.org/10.1037/a0022373
56. Izawa, J., Rane, T., Donchin, O., Shadmehr, R.: Motor adaptation as a process of reoptimization. J. Neurosci. **28**(11), 2883–2891 (2008). https://doi.org/10.1523/JNEUROSCI.5359-07.2008
57. Jaganjac, I., Hendrikse, S.C.F., Treur, J.: An adaptive dynamical system model for development of schizophrenia: epigenetics and false memories. Cogn. Syst. Res. 88, 101288. (2024)
58. Keller, P.E., Novembre, G., Hove, M.J.: Rhythm in joint action: psychological and neurophysiological mechanisms for real-time interpersonal coordination. Philos. Trans. R. Soc. Lond. Ser. B Biol. Sci. **369**(1658), 20130394 (2014). https://doi.org/10.1098/rstb.2013.0394
59. Kernberg, O.F., Yeomans, F.E., Clarkin, J.F., Levy, K.N.: Transference focused psychotherapy: overview and update. Int. J. Psychoanal. **89**(3), 601–620 (2008)
60. Keysers, C., Gazzola, V.: Social neuroscience: mirror neurons recorded in humans. Curr. Biol. **20**, 253–254 (2010)
61. Keysers, C., Gazzola, V.: Dissociating the ability and propensity for empathy. Trends Cogn. Sci. **18**(4), 163–166 (2014)
62. Kirschner, S., Tomasello, M.: Joint music making promotes prosocial behavior in 4-year-old children. Evol. Hum. Behav. **31**, 354–364 (2010). https://doi.org/10.1016/j.evolhumbehav.2010.04.004
63. Koole, S.L., Tschacher, W.: Synchrony in psychotherapy: a review and an integrative framework for the therapeutic alliance. Front. Psychol. **7**, 862 (2016)
64. Lisman, J., Cooper, K., Sehgal, M., Silva, A.J.: Memory formation depends on both synapse-specific modifications of synaptic strength and cell-specific increases in excitability. Nat. Neurosci. **21**, 309–314 (2018)
65. Mattera, F., Hendrikse, S.C.F., Treur, J.: The emergence of group synchrony and adaptivity for group bonding: a multi-adaptive network model. In: Hendrikse, S.C.F., Treur, J., Koole,

S.L. (eds.) New Analysis and Modeling Directions in Social Interaction Science: Emergent Multimodal Interpersonal Synchrony, Affiliation, and Bonding, Ch 16. Springer Nature (2025)
66. Mukamel, R., Ekstrom, A.D., Kaplan, J., Iacoboni, M., Fried, I.: Single-neuron responses in humans during execution and observation of actions. Curr. Biol. **20**, 750–756 (2010)
67. Mukeriia, Y., Treur, J., Hendrikse, S.C.F.: The Interplay of bonding based on subjective homophily and learning during a varied rhythmic action: an adaptive network model. In: Hendrikse, S.C.F., Treur, J., Koole, S.L. (eds.) New Analysis and Modeling Directions in Social Interaction Science: Emergent Multimodal Interpersonal Synchrony, Affiliation, and Bonding, Ch 8. Springer Nature (2025a)
68. Mukeriia, Y., Treur, J., Hendrikse, S.C.F.: Human learning, attunement and social bonding based on adaptive homophily: a multi-adaptive network model. In: Hendrikse, S.C.F., Treur, J., Koole, S.L. (eds.) New Analysis and Modeling Directions in Social Interaction Science: Emergent Multimodal Interpersonal Synchrony, Affiliation, and Bonding, Ch 10. Springer Nature (2025b)
69. Niedenthal, P.M.: Embodying emotion. Science **316**, 1002–1005 (2007)
70. Pacherie, E.: The phenomenology of joint action: self-agency vs. joint-agency. In: Seeman, A. (ed.) Joint Attention: New Developments, pp. 343–389. MIT Press (2012)
71. Poppa, T., Bechara, A.: The somatic marker hypothesis: revisiting the role of the 'body-loop' in decision-making. Curr. Opin. Behav. Sci. **19**, 61–66 (2018)
72. Prince K., Brown S.: Neural correlates of partnered interaction as revealed by crossdomain ALE meta-analysis. Psychol. Neurosci. **15**(1), 1 (2022). https://doi.org/10.1037/pne0000282
73. Ravreby, I., Shilat, Y., Yeshurun, Y.: Liking as a balance between synchronization, complexity and novelty. Sci. Rep. **12**(1), 3181 (2022)
74. Rayner, K., Smith, T.J., Malcolm, G.L., Henderson, J.M.: Eye movements and visual encoding during scene perception. Psychol. Sci. **20**(1), 6–10 (2009)
75. Rizzolatti, G., Sinigaglia, C.: Mirrors in the Brain: How Our Minds Share Actions and Emotions. Oxford University Press, Oxford (2008)
76. Robinson, B.L., Harper, N.S., McAlpine, D.: Meta-adaptation in the auditory midbrain under cortical influence. Nat. Commun. **7**(1), 13442 (2016)
77. Salter Ainsworth, M.D., Bowlby, J.: Child Care and the Growth of Love. Penguin Books, London (1965)
78. Salter Ainsworth, M.D., Bowlby, J.: An ethological approach to personality development. Am. Psychol. **46**(4), 333–341 (1991)
79. Salter Ainsworth, M.D., Blehar, M., Waters, E., Wall, S.: Patterns of Attachment. Erlbaum, Hillsdale, NJ (1978)
80. Salter Ainsworth, M.D.: Infancy in Uganda. Johns Hopkins, Baltimore (1967)
81. Schlöder, D., Hendrikse, S.C.F., Treur, J.: Higher-order adaptive dynamical system modeling of the role of epigenetics in substance use disorder. In: Proceedings of the 15th Biologically Inspired Cognitive Architectures for AI Conference, BICA*AI'24. Studies in Computational Intelligence, vol 477, pp 358–368. Springer Nature (2024)
82. Schoenherr, D., Paulick, J., Strauss, B.M., Deisenhofer, A.-K., Schwartz, B., Rubel, J.A., Lutz, W., Stangier, U., Altmann, U.: Identification of movement synchrony: validation of windowed cross-lagged correlation and -regression with PeakPicking algorithm. PLoS ONE **14**, e0211494 (2019)
83. Shatz, C.J.: The developing brain. Sci. Am. **267**(3), 60–67 (1992)
84. Shelton, J., Kumar, G.: Comparison between auditory and visual simple reaction times. Neurosci. Med. **1**(1), 30–32 (2010)
85. Sjostrom, P.J., Rancz, E.A., Roth, A., Hausser, M.: Dendritic excitability and synaptic plasticity. Physiol. Rev. **88**(2), 769–840 (2008)
86. Tarr, B., Launay, J., Dunbar, R.I.M.: Silent disco: dancing in synchrony leads to elevated pain thresholds and social closeness. Evol. Hum. Behav. **37**(5), 343–349 (2016)
87. Thompson, J.J., Blair, M.R., Henrey, A.J.: Over the hill at 24: persistent age-related cognitive motor decline in reaction times in an ecologically valid video game task begins in early adulthood. PLoS ONE **9**(4), e94215 (2014)

88. Tichelaar, C., Treur, J.: On the interaction of bonding and mirroring in joint decision making: an adaptive self-modeling network modeling approach. In: Hendrikse, S.C.F., Treur, J., Koole, S.L. (eds.) New Analysis and Modeling Directions in Social Interaction Science: Emergent Multimodal Interpersonal Synchrony, Affiliation, and Bonding, Ch 6. Springer Nature (2025).
89. Treur, J.: Temporal factorisation: a unifying principle for dynamics of the world and of mental states. Cogn. Syst. Res. **8**(2), 57–74 (2007a)
90. Treur, J.: Temporal factorisation: realisation of mediating state properties for dynamics. Cogn. Syst. Res. **8**(2), 75–88 (2007b)
91. Treur, J.: Network reification as a unified approach to represent network adaptation principles within a network. In: Theory and Practice of Natural Computing: Proceedings of the 7th International Conference TPNC'18. Lecture Notes in Computer Science, vol. 11324, pp. 344–358. Springer International Publishing (2018)
92. Treur, J.: Multilevel network reification: representing higher-order adaptivity in a network. In: Aiello, L., Cherifi, C., Cherifi, H., Lambiotte, R., Lió, P., Rocha, L. (eds.) Complex Networks and Their Applications VII: Volume 1 Proceedings The 7th International Conference on Complex Networks and Their Applications, COMPLEX NETWORKS'18. Studies in Computational Intelligence, vol. 812, pp. 635–651. Springer International Publishing (2019)
93. Treur, J.: Modeling higher-order adaptivity of a network by multilevel network reification. Netw. Sci. **8**, S110–S144 (2020a)
94. Treur, J.: Network-Oriented Modeling for Adaptive Networks: Designing Higher-Order Adaptive Biological, Mental and Social Network Models. Springer Nature, Cham (2020b). https://www.researchgate.net/publication/334576216
95. Treur, J.: Modeling multi-order adaptive processes by self-modeling networks. Machine learning and intelligent systems: proceedings of MLIS 2020. In: Tallon-Ballesteros, A.J., Chen, C.-H. (eds.) Frontiers in Artificial Intelligence and Applications, vol. 332, pp. 206–217. IOS Press (2020c)
96. Treur, J.: On structure, dynamics, and adaptivity for biological and mental processes: a higher-order adaptive dynamical system modeling perspective. In: Proceedings of the 46th Annual Meeting of the Cognitive Science Society (2024a). https://escholarship.org/uc/item/9911192q
97. Treur, J.: Using ultilevel temporal factorisation to analyse structure and dynamics for higher-order adaptive and evolutionary processes. In: Nguyen, N.T., et al. (eds.) Computational Collective Intelligence. Proceedings of the 16th International Conference on Computational Collective Intelligence, ICCCI 2024. Lecture Notes in AI, vol. 14811, pp. 378–392. Springer Nature, Cham (2024b) https://doi.org/10.1007/978-3-031-70819-0_29
98. Treur, J.: A second-order adaptive network model for bonding based on subjectively perceived homophily. In: Hendrikse, S.C.F., Treur, J., Koole, S.L. (eds.) New Analysis and Modeling Directions in Social Interaction Science: Emergent Multimodal Interpersonal Synchrony, Affiliation, and Bonding, Ch 7. Springer Nature (2025)
99. Tschacher, W., Ramseyer, F., Koole, S.L.: Sharing the now in the social present: duration of nonverbal synchrony is linked with personality. J. Pers. **86**(2), 129–138 (2018)
100. Valdesolo, P., Ouyang, J., DeSteno, D.: The rhythm of joint action: synchrony promotes cooperative ability. J. Exp. Soc. Psychol. **46**(4), 693695 (2010)
101. Vervoordeldonk, D., Treur, J., Hendrikse, S.C.F., Roelofsma, P.H.M.P.: On the interplay of synchrony and adaptation for reliving memory and religious experience through multimodal interaction: an adaptive network model. In: Hendrikse, S.C.F., Treur, J., Koole, S.L. (eds.) New Analysis and Modeling Directions in Social Interaction Science: Emergent Multimodal Interpersonal Synchrony, Affiliation, and Bonding, Ch 17. Springer Nature (2025)
102. Williams, L.E., Huang, J.Y., Bargh, J.A.: The scaffolded mind: higher mental processes are grounded in early experience of the physical world. Eur. J. Soc. Psychol. **39**(7), 1257–1267 (2009)
103. Wiltermuth, S.S., Heath, C.: Synchrony and cooperation. Psychol. Sci. **20**(1), 1–5 (2009)

104. Xu, Y., Treur, J., Hendrikse, S.C.F., Roelofsma, P.H.M.P.: Modeling joint action and memory recall for elderly by detecting interpersonal synchrony: a higher-order adaptive network model. In: Hendrikse, S.C.F., Treur, J., Koole, S.L. (eds.) New Analysis and Modeling Directions in Social Interaction Science: Emergent Multimodal Interpersonal Synchrony, Affiliation, and Bonding, Ch 18. Springer Nature (2025)

MIX
Papier aus verantwortungsvollen Quellen
Paper from responsible sources
FSC® C105338

If you have any concerns about our products,
you can contact us on
ProductSafety@springernature.com

In case Publisher is established outside the EU,
the EU authorized representative is:
Springer Nature Customer Service Center GmbH
Europaplatz 3, 69115 Heidelberg, Germany

Printed by Libri Plureos GmbH
in Hamburg, Germany